Informatik-Fachberichte

Herausgegeben von W. Brauer
im Auftrag der Gesellschaft für Informatik (GI)

95

Kommunikation in Verteilten Systemen I

Anwendungen, Betrieb, Grundlagen
GI/NTG-Fachtagung
Karlsruhe, 13.–15. März 1985
Proceedings

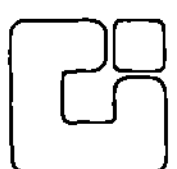

Herausgegeben von
D. Heger, G. Krüger, O. Spaniol und W. Zorn

Springer-Verlag
Berlin Heidelberg New York Tokyo

Herausgeber
Dirk Heger
Fraunhofer-Institut für Informations- und Datenverarbeitung (IITB)
Sebastian-Kneipp-Str. 12–14, 7500 Karlsruhe 1

Gerhard Krüger
Universität Karlsruhe, Institut für Informatik III
Kaiserstr. 12, 7500 Karlsruhe 1

Otto Spaniol
RWTH Aachen, Lehrstuhl Informatik IV
Templergraben 64, 5100 Aachen

Werner Zorn
Universität Karlsruhe, Institut für Informatik III
Kaiserstr. 12, 7500 Karlsruhe 1

CR Subject Classifications (1982): C.2, B.4, H.4, K.6, C.3

ISBN-13: 978-3-540-15197-5   e-ISBN-13: 978-3-642-70285-3
DOI: 10.1007/978-3-642-70285-3

2145/3140 – 5 4 3 2 1 0

VORWORT

Die Veranstaltungen der GI/NTG-Fachgruppe "Rechnernetze" haben mittlerweile schon
Tradition: sie sind die mit Abstand größte Reihe von wissenschaftlichen Konferenzen
zum Bereich "Kommunikation in verteilten Systemen" im deutschsprachigen Raum. Auch
die zunehmende Zahl von kommerziellen Kongressen hat ihre Attraktivität nicht schmä-
lern können.

Die Dokumentation der bisherigen Tagungen (IFB-Bände 22, 40 und 60) enthielten immer
eine Fülle von Grundlagenuntersuchungen, Erfahrungsberichten sowie Informationen
über zukünftige Entwicklungen. Sie initialisierten Diskussionen über alle Themen
des Bereichs Datenkommunikation meist lange bevor die betreffenden Sachfragen einer
größeren Öffentlichkeit bekannt wurden und verschafften damit den Tagungsteilnehmern
und auch den Lesern der Konferenzbände einen unschätzbaren Informationsvorsprung.
Dasselbe wird sicher rückwirkend auch von der jetzigen Konferenz gesagt werden
können, bei der die Aspekte der Anwendung und des Betriebs im Vordergrund stehen.
Schwerpunkte sind die Themenbereiche "Netze und Netzverbunde" sowie "Kommunikationsdienste", denen jeweils mehrere Sitzungen gewidmet sind. Mit dem durch die Ein-
führungen neuer Technologien verursachten Strukturwandel unserer Gesellschaft und
seinen Auswirkungen befassen sich mehrere Referate vor allem in der Eröffnungs- und
Schlußsitzung. Obwohl ein großer Teil der Vortragsmeldungen nicht angenommen wurde,
war es erstmals erforderlich, fast durchgehend Parallelsitzungen zu veranstalten
und die Referate auf zwei Tagungsbände aufzuteilen; auch dies dokumentiert das
breite Spektrum und die Bedeutung der behandelten Thematik.

Die bisher immer in Berlin organisierte Veranstaltung wurde erstmals im west-
deutschen Raum durchgeführt, wobei wegen der Konzentration von Forschungsgruppen
sowie von innovativen Kommunikationsprojekten kein geeigneterer Ort als Karlsruhe
hätte gefunden werden können; im Namen des Programmausschusses möchte ich daher
den Karlsruher Informatikern für die sorgfältige Vorbereitung und die ausgezeichnete
Durchführung der Veranstaltung ganz besonders danken.

Aachen, im Januar 1985   Für die Herausgeber:
                                                Otto Spaniol

<u>**Programmausschuß:**</u>

D. Baum, Uni Trier
B. Butscher, HMI, Berlin
H. Diel, IBM, Böblingen
O. Drobnik, Uni Karlsruhe
N. Gerner, Siemens, München
G. Glas, DFVLR, Oberpfaffenhofen
D. Heger, FhG-IITB, Karlsruhe
H.-G. Hegering, TU München
P. Hohn, DAK, Hamburg
E. Holler, KfK, Karlsruhe
P. Jilek, Siemens, München
G. Krüger, Uni Karlsruhe
P. Kühn, Uni Stuttgart
H.W. Meuer, Uni Mannheim
E. Raubold, GMD, Darmstadt
S. Schindler, TU Berlin
P. Schnupp, Interface, München
J.C.W. Schröder, DANET, Darmstadt
O. Spaniol, TH Aachen
R. Speth, Uni Düsseldorf
W. Staudinger, FTZ, Darmstadt
J. Swoboda, Uni Siegen
J.K. Wild, Augsburg
W. Zorn, Uni Karlsruhe

<u>**Vorsitzender des Programmausschusses:**</u>

Otto Spaniol, TH Aachen

<u>**Tagungsleiter:**</u>

Gerhard Krüger, Uni Karlsruhe

<u>**Organisationskomitee:**</u>

Dirk Heger, FhG-IITB, Karlsruhe
Ulrich Stier, Uni Karlsruhe
Werner Zorn, Uni Karlsruhe

<u>**Tagungsbüro**</u>

Universität Karlsruhe
Fakultät für Informatik
"Kommunikation 85"
Kaiserstr. 12
<u>7500 Karlsruhe 1</u>
Tel.: 0721/ 69 92 84

INHALTSVERZEICHNIS

STAND UND ENTWICKLUNGSTENDENZEN
DER TELEMATIK - PROTOKOLLE

Karlo Németh

Unternehmensbereich Kommunikations- und Datentechnik
Siemens AG, München

Übersicht:

Dieser Beitrag gibt einen Überblick über den neuesten Stand
und allgemeine Anwendungsgebiete der Telematik-Protokolle, die
anläßlich der Plenarsitzung des CCITT im Oktober 1984 nach
vierjähriger Arbeit verabschiedet wurden. Die Telematik-Protokolle basieren auf den Teletex-Protokollen und ermöglichen
den Austausch und die originaltreue Wiedergabe von Teletex-,
Faksimile- und gemischten Text/Faksimile-Dokumenten.

Die Protokolle nach den CCITT-Empfehlungen T.70 und T.62 stellen den Telematik- wie auch allgemeinen OSI-Anwendungen einen
zuverläßigen und vielseitigen Transfer-Dienst über Daten-,
Fernsprech- sowie ISDN-Netze zur Verfügung. Das "Document
Interchange Protocol for the Telematic Services" nach der
CCITT-Empfehlung T.73 bietet umfangreiche und erweiterungs-
fähige Funktionen zur Beschreibung, Strukturierung, Erzeugung und zum Austausch gemischter Text/Bild-Dokumente.

Das Telematik-Gebäude wird vervollständigt durch Empfehlungen
für die Leistungsmerkmale von Teletex-, Faksimile- und Mixed-
Mode-Endgeräten, die für die Kompatibilität bei der Kommunikation unerläßlich sind.

1. Einführung

In der Studienperiode 1977 - 1980 erarbeitete die Studienkommission
VIII des CCITT die Protokolle für den Teletex-Dienst und gab sie als
Empfehlungen S.70 "Network-Independent Basic Transport Service for
Teletex", (Schichten 1 bis 4 des OSI-Referenzmodells), S.62 "Control
Procedures for the Teletex Service" (Schicht 5), S.61 "Character
Repertoire and Coded Character Sets for the International Teletex
Service" (Schicht 6) und S.60 "Terminal Equipment for Use in the
Teletex Service" (Schicht 7) heraus /1,2/. Diese Protokolle stellen
den ersten international standardisierten Satz von Protokollen aller
Schichten des OSI-Basis-Referenzmodells dar. Sie ermöglichen automatischen und sicheren Speicher-zu-Speicher Austausch von zeichenco-
dierten Dokumenten in allen Sprachen, die lateinische Schriftzeichen
benutzen. Inhalt, Format und Layout der ausgetauschten Dokumente sind
garantiert gleich. Der Austausch kann über leitungsvermittelte oder
paketvermittelte Datennetze sowie über analoge Fernsprechnetze erfolgen. Als Übertragungsgeschwindigkeit wurde im internationalen Verkehr
2400 bit/s vereinbart.

Die Teletex-Protokolle S.70 und S.62 wurden wegen ihrer allgemeinen
Gültigkeit als Basis für die Entwicklung der Telematik-Protokolle T.70
und T.62 gewählt. Neue Funktionen wurden dabei grundsätzlich durch
Erweiterungen der bestehenden gewonnen, so daß die Protokolle mit
höherem Leistungsumfang immer abwärtskompatibel zu Protokollen mit
niedrigerem Leistungsumfang sind. Auch die Harmonisierung der Telematik-Protokolle mit ISO-Protokollen wurde vorangetrieben /3/.

Ganz neu entstanden sind die Protokolle nach den CCITT-Empfehlungen

- T.71   "LAPB Extended for Half-Duplex Physikal Level Facility",
- T.72   "Terminal Capabilities for Mixed-Mode of Operation",
- T.73   "Document Interchange Protocol for the Telematic Services",
- T.5   "General Aspects of Group 4 Facsimile Apparatus" und
- T.6   "Facsimile Coding Schemes and Coding Control Functions for
         Group 4 Facsimile Apparatus".

Bild 1 zeigt die Einordnung der Telematik-Protokolle in das OSI-
Basis-Referenzmodell. Die Empfehlungen T.100 und T.101 für Bildschirmtext werden in diesem Beitrag nicht behandelt.

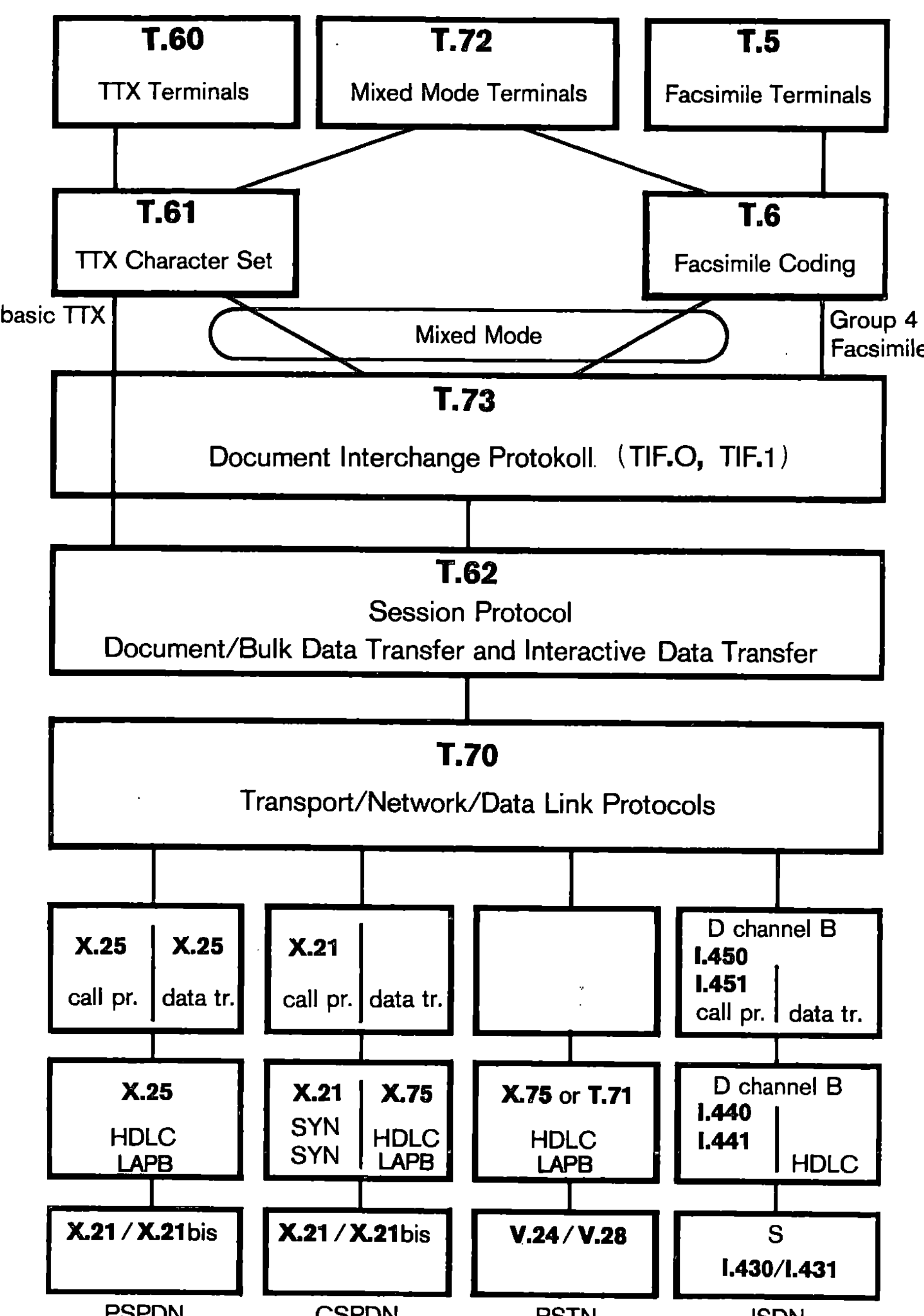

Bild 1. Die Telematik-Protokolle im OSI-Referenzmodell

2. Transportorientierte Protokolle

Die transportorientierten Protokolle der Schichten 1 bis 4 nach der
CCITT-Empfehlung T.70 "Network Independent Basic Transport Service for
the Telematic Services" ermöglichen den Zugang zu leitungs- und paket-
vermittelten Datennetzen sowie zu analogen Fernsprechnetzen und bieten
den anwendungsorientierten Schichten netzunabhängigen, gesicherten und
transparenten End-zu-End Daten-Transfer. Diese Protokolle haben gegenüber ihren Vorgängern, den S.70-Protokollen, insgesamt gesehen keine
wesentlichen Ergänzungen erfahren. Die Parameter der HDLC/LAPB-Proze-
dur sind z.B. die gleichen geblieben. Für paketvermittelte Netze gelten weiterhin die X.25-Protokolle. Die nationalen Postverwaltungen
entscheiden darüber, ob die Version X.25/80 oder X.25/84 verwendet
wird.

Neu hinzugekommen ist die Empfehlung T.71 "LAPB Extended for Half-
Duplex Physical Level Facility", die nun auch Halb-Duplex-Betriebs-
weise mit 2400 bit/s in analogen Fernsprechnetzen erlaubt, z.B. unter
Benutzung eines V.26 bis-Modems. Zur Datensicherung wird die existierende HDLC/LAPB-Prozedur in ihrer ursprünglichen Form beibehalten und ·
mit Hilfe eines zusätzlichen "Half-Duplex Transmission Module" (HDTM)
für den Halb-Duplex-Mode adaptiert. Der HDTM-Modul befindet sich
zwischen der HDLC/LAPB-Prozedur und der physikalischen Schicht und
regelt im wesentlichen das Senderecht der LAPB-Prozedur. Die Kombination LAPB + HDTM wird auch LAPX (Kürzel für LAPB extended) genannt.

Das Basis-Protokoll der Schicht 4 ist kompatibel mit dem ISO-Trans-
port-Protokoll der Klasse 0 /4,5/. Dieses Protokoll verfügt jetzt
optional über erweiterte Adressierung und enthält ausführliche
Zustandsgraphen, die nicht nur die peer-to-peer, sondern auch die
internen Aktionen zwischen den angrenzenden Schichten sowie die benötigten lokalen Beziehungen beschreiben. Eine zusätzliche umfassende
Beschreibung liefern Zustandstabellen, ergänzt durch Service-Primitive
an den Schnittstellen (Service-Access-Points) zu den darüber und darunter liegenden Schichten.

Die T.70-Protokolle können ohne weiteres auch im ISDN verwendet werden, s. Bild 1. Da die ISDN-B-Kanäle leitungsvermittelt sind, können
die bestehenden Protokolle der Schichten 2 und 3 für leitungsvermittelte Datennetze (CSPDN) übernommen werden. Im ISDN-Pilotversuch der
DBP, der 1986 mit kompatiblen Telematik-Anwendungen beginnen wird,
sollen die T.70-Protokolle eingesetzt werden.

3. Das Protokoll der Session-Schicht

Das Telematik-Session-Protokoll T.62 "Control Procedures for Teletex
and Group 4 Facsimile Services" ist selbstverständlich voll kompatibel
mit dem Teletex-Session-Protokoll S.62. Es enthält zusätzlich einen
interactiven Teil, ein neues Parameterfeld "Session User Data" (SUD),
Service-Primitive an der Schnittstelle zu Presentation/Anwendungs-
Schichten und es wird beschrieben durch detaillierte Zustandsgraphen,
u.a. mit Timern, sowie durch Zustandstabellen.

Das Interaktive Session-Protokoll (ISP) bietet einen sehr effektiven
Mechanismus zur Anforderung und Übergabe der Sendeberechtigung oder
des Sende-Tokens. Die Seite, die die Sendeberechtigung im Halbduplex-
Mode nicht hat, kann diese anfordern, indem sie das Protokollelement
"Response Session User Information" (RSUI) mit dem Parameter "Session
Control" auf "1" gesetzt aussendet. Daraufhin übergibt die andere
Seite die Sendeberechtigung durch das Protokollelement "Command
Session User Information" (CSUI) mit dem auf "1" gesetzten Parameter
"Session Control". Die anfordernde Seite bestätigt die Übergabe nicht
explizit, sondern implizit, indem sie einfach Nutzdaten aussendet.

Der im S.62-Protokoll bereits vorhandene Berechtigungsübergabe-Me-
chanismus mit expliziter Bestätigung (über das Protokollelemente-
Paar "Command Session Change Control" (CSCC)/"Response Session
Change Control Positive" (RSCCP)) steht beim ISP weiterhin zur Verfügung.

Welche der Berechtigungsübergabe-Mechanismen verwendet wird, hängt
von der jeweiligen Anwendung ab. Bei automatischem Bulk-Data-Transfer
empfiehlt sich die explizite Bestätigung, bei interaktivem Dialog
unter Beteiligung von Menschen reicht eine implizite Bestätigung aus.

Nun besteht auch die Möglichkeit, Nutzdaten beschränkter Länge losgelöst von der Sendeberechtigung zu senden, um z.B. kurze Kontroll-
Informationen jederzeit mit minimalen Protokollaufwand übermitteln zu
können. Dieser sogenannte Typed-Data-Transfer erfolgt über das Proto-
koll-Element "Command Session Typed Data" (CSTD).

In der Session kommen die kommunizierenden Anwendungen zum ersten
Mal in unmittelbaren Kontakt. Es ist daher naheliegend, bereits bei
der ersten Berührung Informationen auszutauschen, die nicht nur die
Session unmittelbar, sondern auch die Presentation und Anwendung

betreffen. Neben den schon bestehenden Möglichkeiten in der Session-
Eröffnungsphase wie z.B. Dienste-Indikator, Terminal-Identifizierung,
Datum und Uhrzeit, zusätzliche Codes für Steuerzeichen usw., wurde ein
neues Parameterfeld variabler Länge "Session User Data" (SUD) einge-
führt. In diesem Parameterfeld können beliebige Eigenschaften der
Presentation und der Anwendung angezeigt werden. Sie werden von der
Session transparent an die darüberliegenden Schichten übergeben.
Bild 2 zeigt die Einbettung des SUD-Parameterfeldes in das Proto-
kollelement "Command Session Start" (CSS). Das SUD-Feld trägt in
diesem Fall Parameter, die den Mixed-Mode charakterisieren.

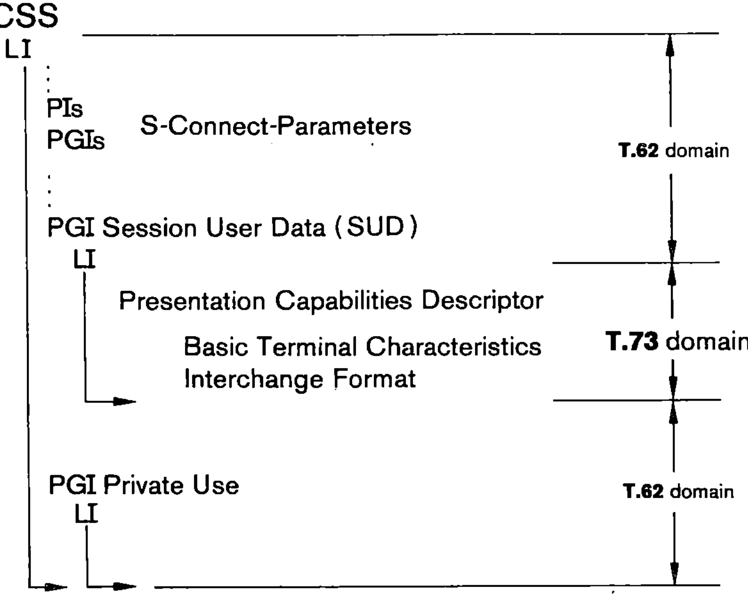

LI: Lenght Indicator
PI: Parameter Identifier;   PGI: Parameter Group Identifier

Bild 2. Das SUD-Parameterfeld im "Command Session Start"

Die Einführung des SUD-Parameterfeldes sowie der Session-Service-
Primitive führte zu einer weiteren Harmonisierung zwischen dem Telematik-Session-Protokoll und dem generalisierten ISO-Session-Protokoll
/6,7/. Im Zuge der Harmonisierung wurde auch vereinbart, daß die entsprechenden Elemente beider Protokolle gleich codiert werden und zwar
nach den Regeln für die Codierung des Teletex-Protokolls S.62.

Bild 3 zeigt den Zusammenhang zwischen den Elementen des Telematik-
und des ISO-Session-Protokolls. Aus diesem Bild ist auch die Auftei-

Functional unit	T.62 name	SPDU name
Kernel	CSS RSSP RSSN CSE RSEP CSA RSAP CSUI-CDUI	CONNECT ACCEPT REFUSE FINISH DISCONNECT ABORT ABORT ACCEPT DATA TRANSFER
Negotiated release	— — —	NOT FINISHED GIVE TOKENS PLEASE TOKENS
Use of half-duplex	CSUI RSUI	GIVE TOKENS PLEASE TOKENS
Use of duplex		No additional associated SPDUs
Expedited data	—	EXPEDITED DATA
Typed data	CSTD	TYPED DATA
Capability data exchange	CSUI-CDCL RSUI-RDCLP	CAPABILITY DATA CAPABILITY DATA ACK
Minor synchronize	CSUI-CDPB RSUI-RDPBP CSUI RSUI	MINOR SYNC POINT MINOR SYNC ACK GIVE TOKENS PLEASE TOKENS
Major synchronize	CSUI-CDE RSUI-RDEP — CSUI RSUI	MAJOR SYNC POINT MAJOR SYNC ACK PREPARE GIVE TOKENS PLEASE TOKENS
Resynchronize	— — —	RESYNCHRONIZE RESYNCHRONIZE ACK PREPARE
Activity management	CSUI-CDS CSUI-CDC CSUI-CDR RSUI-RDRP CSUI-CDD RSUI-RDDP CSUI-CDE RSUI-RDEP — CSUI RSUI CSCC RSCCP	ACTIVITY START ACTIVITY RESUME ACTIVITY INTERRUPT ACTIVITY INTERRUPT ACK ACTIVITY DISCARD ACTIVITY DISCARD ACK ACTIVITY END ACTIVITY END ACT PREPARE GIVE TOKENS PLEASE TOKENS GIVE TOKENS CONFIRM GIVE TOKENS ACK

Bild 3. Elemente der Session-Protokolle

lung des gesamten Funktionsumfangs des ISO-Session-Protokolls in
Funktions-Einheiten (Functional Units, (FU)) ersichtlich. Während
das Telematik-Session-Protokoll im wesentlichen einen Basisteil für
Dokumenten- bzw. Bulk-Data-Transfer und einen interaktiven Teil
(ISP) beinhaltet, setzt sich das ISO-Session-Protokoll aus mehreren
Bausteinen, d.h. Funktions-Einheiten zusammen. Aus diesen Funktions-
Einheiten sind drei Subsets zusammengestellt: Basic Combined Subset
(BCS), Basic Synchronized Subset (BSS) und Basic Activity Subset
(BAS). BCS ist der einfachste Subset und besteht aus der Kernel-FU,
Halbduplex-FU und der Duplex-FU; BSS aus der Kernel-FU, Negotiated-
Release-FU, Halbduplex-FU, Typed-Data-FU, Minor-Synchronize-FU,
Major-Synchronize-FU und der Resynchronize-FU; BAS aus der Kernel-
FU, Halbduplex-FU, Typed-Data-FU, Capability-Data-FU, Minor-Syn-
chronize-FU, Exceptions-FU und der Activity-Management-FU. BAS ist
unter Befolgung bestimmter Regel kompatibel mit dem Basisteil des
Telematik-Session-Protokolls. BCS entspricht weitgehend dem Telema-
tik-ISP.

Das Telematik-Session-Protokoll nach der CCITT-Empfehlung T.62
bietet den von einem allgemeinen Session-Protokoll geforderten
Funktionsumfang mit den Schlüsselfunktionen Token-Management,
Synchronisation sowie Activity-Management und kann nicht nur für
Telematik-Dienste sondern auch für OSI-Zwecke verwendet werden.

4. Das Document Interchange Protocol

Das "Document Interchange Protocol for the Telematic Services" nach
der CCITT-Empfehlung T.73 beschreibt ausgehend von einem abstrakten
Architektur-Modell gemischte Text/Bild-Dokumente in mehreren Stufen
und erzeugt einen Datenstrom zur Übertragung dieser Dokumente. Die
objektorientierte Dokumenten-Architektur (Office Document Architecture, (ODA)) unterscheidet klar die Struktur und den Inhalt eines
Dokumentes /8,9/. Jedes Dokument hat eine logische und eine Layout-
Struktur. Beide Strukturen sind hierarchisch aufgebaut und können
durch baumartige Graphen dargestellt werden, s. Bild 4. Den Knoten
der Graphen sind logische bzw. Layout-Objekte zugeordnet und die
Kanten repräsentieren die Beziehungen zwischen den Objekten. Es wird
zwischen zusammengesetzten und Basisobjekten unterschieden. Inhaltsstücke können nur Basisobjekten zugeordnet werden und sind immer homogen codiert, d.h. nur zeichen- oder nur faksimilecodiert.

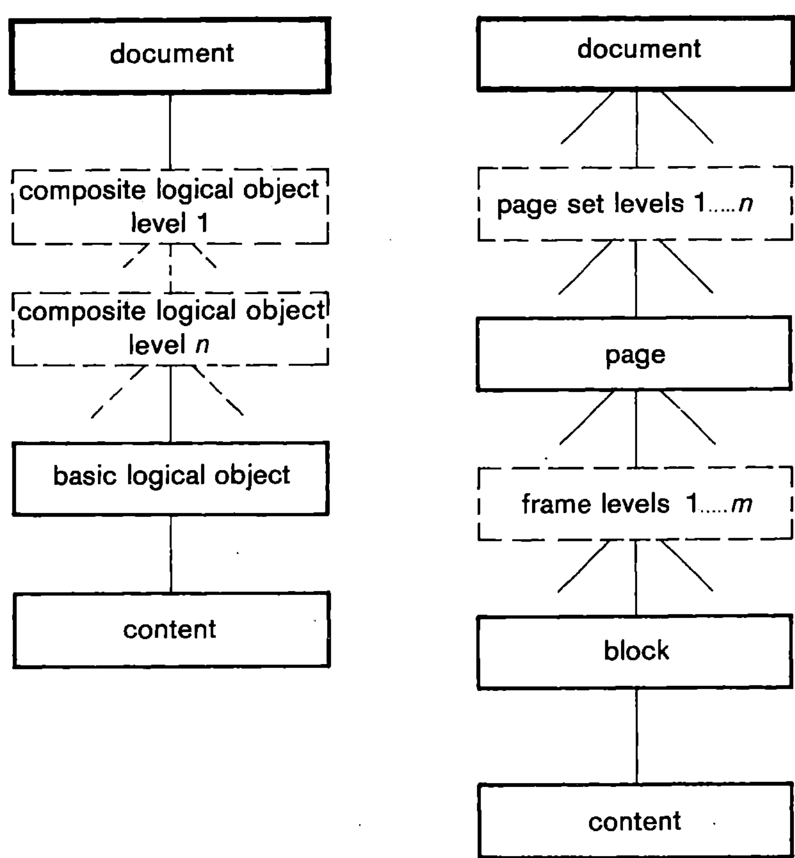

Bild 4. Logische und Layout-Struktur

Die Struktur und der Inhalt von Dokumenten wird übertragen mittels Protokollelementen vom Typ "Descriptor" bzw. "Text Unit". "Descriptoren" repräsentieren spezifische oder generische Objekte und die ihnen zugehörigen Attribute, "Text-Units" Inhaltsstücke mit dazugehörigen Attributen.

Zur formalen Spezifikation der Protokollelemente wird eine erweiterte Backus-Naur-Form-Notation (BNF-Notation) nach der CCITT-Empfehlung X.409 verwendet /10/. Sie erlaubt, zusammengesetzte Datentypen aus grundlegenden Datentypen nach vorgegebenen Regeln zu generieren. Tabelle 1 zeigt z.B. die BNF-Notation für einen spezifischen Layout-Descriptor. Descriptoren können im allgemeinen gedeutet werden als verschachtelte Listen von Attributen.

```
LayoutDescriptor              : : = SEQUENCE {
      layoutObjectType              LayoutObjectType,
      layoutDescriptorBody          LayoutDescriptorBody, OPTIONAL }

Layout ObjectType             : : = INTEGER { document (0), pageSet (1), page (2),
                                         frame (3), block (4) }

LayoutDescriptorBody          : : = SET {
      objectIdentifier              ObjectReferenceName OPTIONAL,
      referencesTo                  CHOICE {
         subordinateObjects         [0] IMPLICIT SEQUENCE OF NumericString,
         contentPortions            [1] IMPLICIT SEQUENCE OF NumericString }
                                                               OPTIONAL,
      referenceToGenericObject      [2] IMPLICIT ObjectReferenceName OPTIONAL,
      position                      [3] IMPLICIT MeasurePair OPTIONAL,
      dimensions                    [4] IMPLICIT MeasurePair OPTIONAL,
      transparent                   [5] IMPLICIT Transparent OPTIONAL,
      presentationAttributes        [6] IMPLICIT PresentationAttributes OPTIONAL,
      defaultValueLists             [7] IMPLICIT SEQUENCE OF DefaultValueList OPTIONAL,
      userReadableComments          [8] IMPLICIT CommentString OPTIONAL }

ObjectReferenceName           : : = [APPLICATION 1] IMPLICIT PrintableString
                                    -- digits 0 to 9 with space as a delimiter --

MeasurePair                   : : = SEQUENCE { Measure, Measure }

Measure                       : : = CHOICE {
      fixedMeasure                  [0] IMPLICIT INTEGER,
      variableMeasure               [1] IMPLICIT INTEGER }

Transparent                   : : = INTEGER { transparent (0) }
                                    -- other values for further study --

CommentString                 : : = IA5String
                                    -- same character set as PrintableString --
                                    -- plus carriage return and line feed --
```

Tabelle 1. BNF-Notation für Layout-Descriptoren

Die Codierung der Protokollelemente erfolgt immer mit Hilfe von
drei Komponenten, die in der Reihenfolge "Identifizierer", "Länge"
und "Inhalt" angeordnet werden müssen. Der "Identifizierer" kennzeichnet den Datentyp und die "Länge" gibt die Anzahl der darauf-
folgenden Daten. Der "Inhalt" enthält entweder Inhaltsstücke des
Dokuments oder bei zusammengesetzten Datentypen weitere, ineinander
verschachtelte Datentypen.

Für die derzeit vorgesehenen kompatiblen Anwendungen im Rahmen der
Telematik-Dienste wurden aus dem vollen Funktionsumfang der ODA zwei
konkrete Austauschformate definiert.

Das "Text Image Format 0" (TIF.0) dient zum Austausch reiner, nicht-
strukturierter Faksimile-Dokumente und besteht dementspechend nur

aus spezifischen Layout-Descriptoren für Dokumente und Seiten. Die
"Text-Units" werden direkt den Seiten-Descriptoren zugeordnet. Jede
Seite beinhaltet immer einen einheitlichen Faksimile-Code, z.B. den
Modified-READ-Code (MRC). TIF.0 ist vorgeschrieben für die Gruppe-4-
Faksimilegeräte der Klasse 1.

Das "Text Image Format 1" (TIF.1) unterstützt den Austausch und originaltreue Wiedergabe von gemischten Text/Faksimile-Dokumenten und
besteht aus spezifischen und generischen Layout-Descriptoren für
Dokumente, Seiten und Blöcke, angeführt durch einen Document-Pro-
file-Descriptor. Die generische Layout-Struktur ist optional. Die
Text-Units tragen entweder den Teletex-Zeichen-Code nach der CCITT-
Empfehlung T.61 oder den Faksimile-Code nach der CCITT-Empfehlung T.6.

Für beide Austauschformate ist jeweils ein Satz von obligatorischen
und optionalen Attributewerten aufgelistet.

5. Endgeräte-Eigenschaften

Die Endgeräte-Eigenschaften für Teletex sind spezifiziert in der
Empfehlung T.60 (früher S.60), für Fernkopierer der Gruppe 4 in T.5
und für Mixed-Mode in T.72. Die Empfehlung T.60 ist so gut wie identisch mit S.60 und wird hier nicht weiter erläutert.

Die Fernkopierer der Gruppe 4 sind nach der Empfehlung T.5 eingeteilt in 3 Leistungsklassen. In Klasse 1 fallen reine Fernkopierer mit der Standard-Übertragungs-Auflösung von 200 Punkten/25,4 mm
(Kompatibilität mit bestehenden Fernkopierern der Gruppe 3!) und
optionalen Übertragungs-Auflösungen von 240, 300 und/oder 400 Punk-
ten/25,4 mm. Die Übertragungs-Auflösung (pel transmission density)
die gesendet oder empfangen wird, muß nicht unbedingt gleich sein
mit der tatsächlichen Auflösung der Abtaster oder der Aufzeichnungs-
einheiten. Klasse-2-Fernkopierer können zusätzlich zu dem reinen
Faksimile-Sende/Empfangs-Betrieb der Klasse 1 auch noch Mixed-Mode-
Dokumente empfangen. Klasse-3-Fernkopierer sind eigentlich voll-
wertige Mixed-Mode-Endgeräte nach der Empfehlung T.72.

Mixed-Mode-Endgeräte nach der Empfehlung T.72 müssen in der Lage
sein, gemischte Dokumente nach dem Austauschformat TIF.1 zu erzeugen, zu senden und zu empfangen. Die Übertragung der Mixed-Mode-
Dokumente erfolgt wie bei Teletex automatisch von Speicher zu

Speicher ohne Beeinträchtigung des Lokalbetriebes. Die Mindestkapazität des Kommunikationsspeichers soll 1 Mbit betragen, oder ausreichen, um eine volle Faksimile-Seite abspeichern zu können. Alle Mixed-Mode-Endgeräte müssen die Faksimile-Übertragungs-Auflösungen von 240 <u>und</u> 300 Punkten/25.4 mm empfangen können.

6. Zusammenfassung

Die Telematik-Protokolle T.70 und T.62 der Schichten 1 - 5 verfügen über sichere und universelle Mittel zum Transfer von Teletex-, Faksimile- und gemischten Text/Faksimile-Dokumenten über verschiedene Netztypen, inclusive ISDN. Sie werden auch zum Informationsaustausch zwischen Message Handling Systemen als "Reliable Transfer Server" (RTS) eingesetzt und können darüber hinaus für allgemeine Bulk-Data-Anwendungen, wie z.B. File-Transfer sowie interaktive Anwendungen, wie z.B. Datenbank-Abfragen, verwendet werden.

Das T.73 "Document Interchange Protocol" bietet umfangreiche Funktionen zur Beschreibung, Strukturierung und zum Austausch von beliebig zusammengesetzten Mixed-Mode-Dokumenten. Diese Funktionen ermöglichen nicht nur die originalgetreue Wiedergabe des Layouts von Mixed-Mode-Dokumenten sondern auch dessen Weiterverarbeitung. Die Erstellung und Weiterverarbeitung von Mixed-Mode-Dokumenten wird in der Zukunft durch Hinzufügen der logischen Struktur noch vielseitiger und effizienter werden. ECMA und ISO arbeiten derzeit intensiv an Austauschformaten mit logischen und Layout-Strukturen (Text Processing Formats, (TPF)). Die Ergebnisse dieser Aktivitäten werden in die Arbeit des CCITT in der neuen Studienperiode einfließen, so daß auch in der Zukunft wohlabgestimmte internationale Standards entstehen werden.

L i t e r a t u r :

/1/ Schenke, K., Rüggeberg, R. und Otto, J.: "Teletex, ein
     neuer internationaler Fernmeldedienst für die Textkom-
     munikation". Jahrbuch der Deutschen Bundespost 1981.

/2/ Moore, D.J.: "Teletex - A worldwide link among office
     systems for electronic document interchange". IBM Sys-
     tems Journal, Vol. 22, Nos 1/2, 1983, S. 30 - 45.

/3/ Németh, K.: "On the usage of Teletex protocols as
     general purpose protocols for open systems intercon-
     nection". Informatik-Fachberichte Nr. 60: Kommunika-
     tion in verteilten Systemen - Anwendungen und Betrieb,
     Berlin, 1983, S. 74 -86.

/4/ ISO International Standard 8072: "Transport service
     definition".

/5/ ISO International Standard 8073: "Connection oriented
     transport protocol specification".

/6/ ISO International Standard 8326: "Basic connection
     oriented session service definition".

/7/ ISO International Standard 8327: "Basic connection
     oriented session protocol specification".

/8/ Horak, W. und Krönert, G.: "An object oriented of-
     fice document architecture model for processing and
     interchange of documents". Proc. of the Second ACM
     Conference on Office Information Systems, June 1984,
     Toronto, Canada.

/9/ Németh, K.: "Advanced document interchange within
     Telematic services". Proc. of the IEEE Int. Confe-
     rence on Communications (ICC '84), Vol. 3, May 1984,
     Amsterdam.

/10/ CCITT Recommendation X.409: "Message Handling Systems:
     presentation transfer syntax and notation". Torremo-
     linos-Malaga, 1984.

SINGLE SYSTEM IMAGE FOR A NETWORK OF PROCESSORS

H. Diel, G. Kreissig, N. Lenz, B. Schoener

IBM, Böblingen

Abstract:

Support of distributed data processing, especially in a local area network,requires major modifications and extensions of existing operating system
structures to provide,for example, automatic data set exchange, remote spooling,
and terminal passthrough. Most of these facilities can be considered under the
common goal of achieving a single system image for a network of processors.

The paper describes a systematic approach for the design of a single system image
for an operating system supporting a network of processors. The network global
objects are organized into two virtual control blocks,called virtual system communication area and virtual system catalog. Operations on these virtual control
blocks are provided via a general abstract data-type interface. In addition,
these operations support network-wide synchronization and consistency requirements.

INTRODUCTION

A network of processors provides a **single system image,** if it appears to the external
user like a single system (e.g. like running on a single processor). Provision of a
single system image is the task of distributed data processing software, primarily
of the operating system.
The 'single system image' idea has some commonalities with the Distributed Operating
System concept. However, whereas with Distributed Operating Systems (see WATSON 1981)
the distributed parts of an operating system are not separately viable (or if they
are they do not form a single system image), with single system image the existence of
multiple complete operating systems is assumed.Thus, the major problem to be solved
is to make a multitude of global resources of the same type look like a single one, or
to merge the multiple resources into a single one without losing the autonomy of the
individual operating systems.
This paper considers only networks controlled by the same kind of operating system,i.e. a homogeneous network. It is believed that without an extreme degree of
standardization (which is at present not available) this is the only case where single system image can be achieved. Aside from requesting a single type of operating
system, however, it is assumed that the different nodes of the network normally differ in many aspects, e.g. processor scale, type of processor architecture, and
attached devices. Except for a few reasonable deviations (to be discussed later) a
user of the operating system should have transparent access to all resources within
the network. He should not see the difference whether his processor is part of a net-

work or not, or whether specific resources (e.g. printer,data sets) are attached to his local processor or to some remote processor.

HARDWARE CONFIGURATIONS

The concepts described in this paper apply to a large variety of network configurations and types of processor coupling such as
>   Local Area Networks,
>   Loosely Coupled Multiprocessing,
>   Tightly Coupled Multiprocessing.

These configurations could be varied in many respects. Provision of a maximum degree of network transparency or even single system image is normally an important goal of any such configuration. It is desirable that modern operating systems support the multitude of different configurations by a general common concept.

PHYSICAL NETWORK

In the context of this paper the above mentioned hardware configurations are called **physical networks.** A physical network consists of a number of processors connected by some physical media. The kind of physical media (coax, I/O bus, memory,etc.) as well as the kind of topology (master/slave processors, central network controller, equally entitled nodes,etc.) and the kind of communication protocol are irrelevant.

FACILITIES TO SUPPORT SINGLE SYSTEM IMAGE

Support of distributed data processing requires major modifications and extensions of existing operating systems. In the following a list of operating systems facilities is given which has to be considered under the goal to achieve **network transparency.** The maximum degree of network transparency is the provision of a single system image.

Common, Shared Disks
> Disks are utilized by multiple processors of the network. One processor manages the shared access to the disks. If this processor is dedicated to managing the disk access, it is called a file server.
> With specific configurations disks may also be shared such that it is not feasible or desirable to have a single processor managing the disk access.

Common Files
> In addition to common, shared disks (as described above) it should be possible to share individual files.The type of sharing (read, write, global, record-level) depends on the kind of file and the sharing options supported by the operating system. Different requirements may exist for user files, system files, libraries, or data bases (see Reference 2).

Message and Transaction Routing

The same facilities for message and transaction routing which are available at a single processor have to be provided across processors as well. This implies that the list of users that have logged on has to be kept network-global.

Inter-Process Communication

The same facilities for inter-process communication (like send, receive, signal etc.) should be provided for processes residing in different processors as well as between processes on a single processor.

Common Job/Logon Scheduling

A global policy and associated functions determine which work is executed on which processor. As an extension of this function network-global workload balancing could be supported.

Transparent Terminal Link

This facility enables log on from any terminal to any of the processors within the network. With specific processor configurations this facility may be provided largely or completely by the I/O attachment hardware.

Common/Remote Spooling

Print output is routed to a processor which does the spooling for a particular printer commonly for the entire network. If this processor is dedicated to support of the printer, it is called a print server.

Common System Administration

Besides requesting that an operator console may be physically attached to any processor, this facility assumes that a single administrator (process) is responsible for the complete network.

The above list of distributed data processing facilities is incomplete in so far as their support would not yet provide a single system image. Many further (mostly smaller) extensions and modifications are necessary to ensure single system image.
One of the major concerns of this paper is that single system image cannot be attempted by a long list of individual enhancements, but only by a general and common concept applied to various areas of the operating system.Such a concept is described in the following text.

SCOPE AND LIFETIME OF INFORMATION

All the functions described above can be related to some global data or resources that these functions represent or control. The totality of information managed by the operating system can be classified according to different scopes and lifetimes. In the following the terms **scope, lifetime** of objects and **logical network** are defined.

EXECUTION CONTEXT

Since it is assumed that in case of errors in a physical network the remaining subnets remain viable, we have to distinguish between the physical network and those viable

subnets. We define a **logical network** to consist of the active nodes of a subnet of a
physical network, where the connections between those nodes are also active. In the
extreme case a logical network may consist only of a single node.

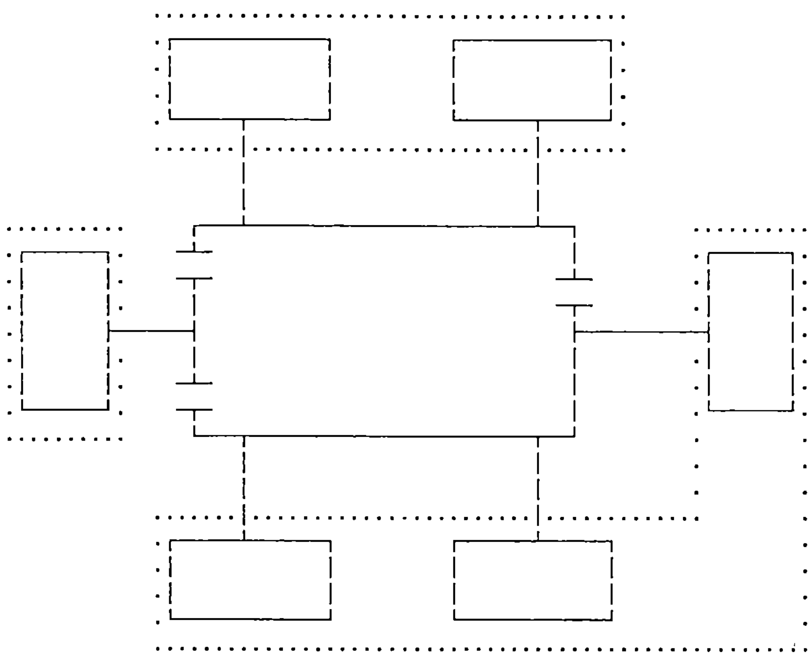

Figure 1. Broken physical network

Figure 1 shows an example of one physical network with 6 nodes and 3 broken connections, and therefore 3 separate logical networks.
The totality of work performed within a logical network of processors is structured
into execution contexts. The smallest entity of work execution, the **Process** is
defined as the basic type of execution context. Further types of execution contexts
(e.g. node) are defined as sets of processes, or more generally as sets of already
defined execution context types. The exact structure of the execution contexts is
dependent on the type of operating system controlling the logical network. However,
the structure given below is a reasonably general example.

Process	-	Basic Unit
User Session	-	Set of Processes
Node	-	Set of User Sessions
Logical Network	-	Set of Nodes

Figure 2. Example of Execution Context Hierarchy

Although this grouping is not just a set relation, but has some meaning associated with it, this is irrelevant in this context. The above structure shows an example which may be in accordance with most operating systems supporting distributed processing. Other operating systems may support different structures, for example additional types of contexts. Also, it should be noticed that the structure of execution contexts does not necessarily have to be limited to hierarchical groupings only.

SCOPE OF INFORMATION

The information managed by an operating system (e.g. datasets, catalogs, control' blocks) can be associated with a certain scope relative to an execution context. If EC_i is an execution context and EC_m is the mother execution context of EC_i, i.e. EC_m = $\{EC_1, \ldots EC_i \ldots EC_n\}$, information I is defined to have scope EC_i.
Scope (I) = EC_i, if

(a) I is accessible from within EC_i, but not necessarily from all the sister execution contexts EC_j (j = 1 to n but j $\neg$= i)

(b) The information I is unique within EC_i. This means that at a given point in time all parts of EC_i see the same value/version of I. ·

Information I is defined to have network global scope,
SCOPE(I) = NETWORK ,if

(a) I is accessible from within all execution contexts of the logical network.

(b) I is unique within the logical network.

Example:
$$EC_1 = \text{User-Session}_1 \text{ at Node}_m$$
$$EC_n = \text{User-Session}_n \text{ at Node}_m$$
$$I = \text{a dataset allocated within User-Session}_i \text{ at Node}_m$$

SCOPE (I) = EC_i

because I (as an allocated dataset) is accessible by EC_i but

not by EC_j $(j = 1$ to n, $j \neg= i)$.

Note, that the above definitions do not preclude certain information from being associated with multiple execution contexts. This has been allowed in order to support objects (e.g. data sets) shared between multiple execution contexts (e.g. processes).

Figure 3 shows some examples of operating system resources and associated scope and lifetime. It shows,under Scope and Lifetime,the **type** of execution context which is associated with the respective information. It should be noted, however,that according to the definitions given above the scope of **particular** information is always defined to be a **particular** instance of an execution context.
The examples shown are reasonably general, but not necessarily true for all types of operating system. There are operating systems in which certain information shown in Figure 3 does not exist (e.g. Allocated Dataset), or specific information might be associated with a different scope.

LIFETIME OF INFORMATION

The information managed by an operating system can be associated with a certain lifetime relative to an execution context.
Information I is defined to have the lifetime EC_i,

LIFETIME (I) = EC_i , if this information

(a)   has scope EC_i, i.e. Scope (I) = EC_i

(b)   cannot exist longer than EC_i

(c)   may exist longer than any subcontext EC_j with $E_i = \{EC_1,....E_n\}$, $j = 1$ to n.

Information I is defined to have the lifetime 'Persistent' if
SCOPE (I) = EC_i and I may exist longer than EC_i.

For the purpose of this paper the execution context 'Logical Network' is assumed to exist from the point in time when the first node belonging to the logical network is started up to the point in time when the last node that joined the logical network is shut down.

Type of Information	Scope	Lifetime
Unallocated Dataset	Logical Network	Persistent
Allocated Dataset	User-Session	User-Session
Opened Dataset	Process	Process
Unconnected Catalog	Logical Network	Persistent
Connected Catalog	User-Session	User-Session

Local-Device-Configuration
    i.e. the information
    describing the set of
    all devices available
    at a specific node.
	Node	Persistent

Global-Device-Configuration
    i.e. the information
    describing the set of
    all devices available
    within the network.
	Logical Network	Logical Network

User list
    i.e. the information
    describing the set of
    all users that have
    logged on.
	Logical Network	Logical Network

Catalog list
    i.e. the information
    describing the set of
    catalogs available in
    the network.
	Logical Network	Logical Network

Global Application list	Logical Network	Logical Network
Local Application list	Node	Persistent

Enroll-table
    i.e. the information
    describing the set of
    all users that may
    log on.
	Logical Network	Persistent

Program Function Key Setting	User-Session	Persistent
Spool Queue	Logical Network or Node	Persistent
Dispatching Information	Node	Node

Figure 3. Scope and Lifetime of System Information

NETWORK - GLOBAL OBJECTS

Each network has objects which contain information describing the resources of the system. This information has a scope relative to an execution context. As a consequence these objects are either local to some execution context smaller than a logical network (e.g. process, user-session, node) or they are global to a logical network.

Three kinds of Network - Global objects are distinguished:

Processor Objects

There are objects which, despite their network - global scope, can be associated with a particular processor. Typical examples of such objects are the control blocks associated with servers (e.g. file servers and print servers). In general, there are special functions in a distributed data processing operating system that support the access to and manipulation of the various resources. These objects are excluded from the following object classification because their treatment is not a unique problem of the single system image concept but a general problem of distributed operating systems.

Network objects

These are objects whose lifetime is identical with the lifetime of the logical network. The objects are created when the first node of a logical network is activated and they disappear when the logical network (i.e. its last node) is shut down.
Examples of this kind of network-global object are:

- the list of active users
- the list of active nodes
- the list of catalogs available in the network
- the device configuration of the network.

Permanent objects

These objects remain in existence when the logical network is shut down.
Examples of this kind of network-global object are network tailoring and configuration parameters, user-enrollment tables, and data sets containing network-global control information (e.g. performance or accounting data).

The difference between network objects and permanent objects can be illustrated as follow:
The network objects are related to the logical network whereas the permanent objects are related to the physical network. If,in a physical network,a connection is interrupted so that one set of nodes cannot communicate with another set of nodes,but each set is viable on its own,we obtain two logical networks. Each of these networks will have its set of network objects but both might have the same permanent objects.

VIRTUAL SYSTEM COMMUNICATION AREA

The **Virtual System Communication Area** V_SYSCOM is defined as the collection of network - global objects with a lifetime equal to the network lifetime, i.e. network objects. This means that the Virtual System Communication Area is created when the logical network (i.e. the first processor) is started,and disappears when this network is shutdown.

In some operating systems the term **system communication area** is the name of a control block that contains system-global control information.

The term **virtual** is used to express the fact that in distributed data processing configurations this kind of control information is not necessarily represented in a single global control block.

There are many implementation alternatives for the actual realization of the Virtual System Communication Area. The optimal implementation depends very much on the type of processor coupling (e.g. LAN connection, or loosely coupled MP, or tightly coupled MP).

It is proposed to provide a well-defined interface to access and manipulate the Virtual System Communication Area in order to be independent of the particular type of processor coupling.

MODEL FOR V_SYSCOM

The network objects of a logical network are described by the V_SYSCOM.

A V_SYSCOM is a set of entities,each containing a set of specific information

$$V\_SYSCOM = \{A_1, \ldots , A_n\} \text{ with } A_i = \{b_{i1}, \ldots , b_{ik}\}$$

The entities and their number are system-built parameters and cannot be changed during the existence of a logical network ;they cannot be added or deleted. The number of elements within each entity is variable, can be zero (empty set),and the entity can grow and shrink during the lifetime of the logical network.

An element b_{ij} of an entity can contain different kinds of information but it is regarded as an atom and operations on it are atomic. It is assumed that any element within an entity can be manipulated independently of any other element.Therefore,only operations on elements are provided; there is no need for atomic operations on groups of elements, nor is there a need for a locking mechanism for an entity.

Typical examples of an entity of the V_SYSCOM are the network objects mentioned earlier (e.g. the list of active nodes, the list of active users,etc.).

IMPLEMENTATION ALTERNATIVES

The V_SYSCOM must be accessible for each process on each node within a reasonable time although changes to the entities must be sequential and consistent.

Centralized single V_SYSCOM

In a network with shared disks or with a file server the V_SYSCOM can be placed there as one control block. This guarantees consistency and fair access by all processes. But it ties the viability of the network to the existence of the file server. This is contradictory to the requirement that a single node should be viable too.

Circulating V_SYSCOM

In this implementation,which is similar to the multiprocessor architecture described in LECOUFFE (Reference 5), the V_SYSCOM is constantly moving between the nodes,either as one control block or as separate entities. A process can manipulate the V_SYSCOM when it reaches its node. The difficulties are:

1. The time dependencies;
   It must be ensured that the V_SYSCOM reaches each node within a reasonable time and if it is delayed,the node must decide how long it will wait before it decides that the object is lost (e.g. because of interrupted connection) and that it therefore has to create a new one.
2. Merging V_SYSCOMs if a connection between two networks is established.

The advantages are that only one version of the V_SYSCOM exists and that it does not rely on the existence of a specific node or process.

Handshaked V_SYSCOM

The V_SYSCOM is kept on one node called the **network master**,which controls the entire logical network. If the node is shut down it selects another node and passes the V_SYSCOM to that node. This node then becomes the network master. The problem with this solution is that in the event of an unexpected shutdown (system crash, connection failure), it may not be possible to pass the V_SYSCOM. In this case each node must have the capability of becoming the network master; the remaining nodes must select one of them and provide the information to it so it can build a new V_SYSCOM.

Replicated V_SYSCOM

The V_SYSCOM exists on each node and manipulation to it must be communicated through the network so that all nodes contain the same information.This implementation has the advantages that each node can access it without a network-dependent delay,and that in the event of network interruptions each resulting network has its own V_SYSCOM.

The problem here is the propagation of changes and the assurance that each version of the V_SYSCOM reflects the actual state of the logical network. This can be achieved in two ways:

1. Each node has a read copy of the V_SYSCOM but only a network master is allowed to update it. A change is sent by the requesting node to the network master,which performs the update and then distributes the updated version across the logical network. This solution simplifies the approach of the handshake implementation because in the event of an unexpected shutdown each node already possesses a version of the V_SYSCOM and it is not necessary to rebuild it.
2. Each node can update its version of the V_SYSCOM and then distribute only the update through the network. Having received an update,each node is responsible for applying it to its own version of the V_SYSCOM. Each node is also responsible for reporting conflicting updates.

ACCESS PRIMITIVES FOR VIRTUAL SYSTEM COMMUNICATION AREA

A V_SYSCOM primitive operation is an atomic operation which transforms the V_SYSCOM from one state into another:

$$V\_SYSCOM \xrightarrow{\quad op \quad} V\_SYSCOM'$$

The primitives are read/write an element of an entity and add/remove an element to or from an entity.

In the following definitions of the primitives, the parameters in upper case are names and the parameters in lower case are values:

add(Ai, z) a new element with the inital value z is added to the set of entity Ai.

$V\_SYSCOM' = \{A_1, \ldots, A'_i, \ldots, A_n\}$ with

$A'_i = A_i \vee \{b_{ik+1}\}$ and $b_{ik+1} = z$

z = read(Ai, j) the value of the j-th element of the entity Ai is returned to z

$V\_SYSCOM' = V\_SYSCOM$

remove(Ai, j) the j-th element of entity Ai is removed.

$V\_SYSCOM' = \{A_1, \ldots, A'_i, \ldots, A_n\}$ with $A'_i = A_i \, / \, \{b_{ij}\}$

update(I , J , x , z) the value of the j-th element of entity Ai is compared to the value x and if they are equal, it is updated by the value z.

if $b_{ij} = x$ then $V\_SYSCOM' = \{A_1, \ldots, A'_i, \ldots, A_n\}$ with

$A'_i = \{b_{i1}, \ldots, b'_{ij}, \ldots, b_{ik}\}$ and $b'_{ij} = z$

merge(S_1, S_2) the two V_SYSCOMs S_1 and S_2 are merged to form a new V_SYSCOM which is the superset of both.

$V\_SYSCOM' = S_1 \vee S_2$

IMPLEMENTATION CONSIDERATIONS

The primitives add, read, and remove do not require a specific state of an element to perform their functions. However, for the primitive update the state of the element has to be checked before the update can take place. This can be implemented in two ways:

1. The entire element must be locked so that no other operation can occur and change the state of that element. In this case it is not only required to protect the update against another update, but also against a read and remove.
2. The combination of the compare operation and the update operation is available as an atomic operation.

VIRTUAL SYSTEM CATALOG

The Virtual System Catalog is defined as a collection of permanent network global objects. For a given operating system there usually exists a set of permanent information (e.g. enrollment table, network tailoring parameters) possessing network global scope. This information would be kept in the Virtual System Catalog. In addition there might be special user datasets or information generated by specific applications which have the same characteristics. The Virtual System Catalog can be used for these as well.

The model given in the last paragraph for the Virtual V_SYSCOM is also applicable for the Virtual System Catalog. It can be seen as a set of sets. Also the access primitives add, read, remove and update remain the same. Because the Virtual System Catalog remains in existence when the logical network is shut down, its implementation alternatives differ from those of the Virtual System Communication Area.

IMPLEMENTATION ALTERNATIVES

It seems reasonable to require from a distributed data processing network that its nodes should be capable of living without the rest of the physical network. Because of this requirement the Virtual System Catalog should be separated from the Virtual System Communication Area and should be accessible also in stand-alone mode.

The following major implementation alternatives are apparent:

Single Virtual System Catalog on file server

If it is acceptable that a File Server must be active in order to work with a node, then the Virtual System Catalog could reside on that File Server. The updates of the Virtual System Catalog are done sequentially and as soon as a node becomes active, the current version of the System Catalog is available. The drawback of this solution is, of course, that the availability of a node is coupled to the availability of the File Server.

Replicated Virtual System Catalog with network global time

The solution preferred in this paper is to have the Virtual System Catalog logically (i.e. virtually) unique, but physically replicated to all network nodes. This allows a node to disconnect from the network and later on, when the network has shut down, it can be reactivated with the local copy of the Virtual System Catalog.

The problem of keeping the multiple copies of the System Catalog consistent has to be solved in that case. A clean solution is possible if a network global time is available. The time could be supplied by a special hardware feature in the physical network which would always be accessible in any active subnet (i.e. in any logical network). In this case the copies of the Virtual System Catalog contain a time stamp. When a node becomes active, it communicates the time stamp of its Virtual System Catalog version to the other active nodes and receives the latest copy (if its own copy is not the latest one). If it has the latest copy or updates the Virtual System Catalog, then it sends its copy to all other active nodes.

Replicated Virtual System Catalog without network global time

Things become more complicated when there is no global time available. Theoretically the problem can be solved by using an ordering mechanism for the events of updating the Virtual System Catalog as proposed by LAMPORT (see Reference 3). LAMPORT has proposed a protocol which ensures sequential processing of updates. The problem with this protocol is that every node has to be active and has to acknowledge the update request before an update can take place. This requirement will generally not be fulfilled. The protocol could only be applied to the set of nodes being active at some point in time (i.e. to the logical network). When a new node becomes active, it might possess a copy of the Virtual System Catalog, which contains updates not known to the other active nodes, and vice versa. These concurrent updates cannot be brought into a sequential order and some kind of decision process (possibly by user involvement) has to be invoked in order to restore the uniqueness of the Virtual System Catalog. In any case the active nodes must determine which copy of the Virtual System Catalog is the currently valid copy.

Eliminate the Virtual System Catalog

Depending on the underlying operating system characteristics it might also be worthwhile thinking about a totally different way of solving these problems. As said before, there are objects with network global scope which can be associated with a particular processor. An example is the information describing the print belts of printers belonging to a Print Server. This information can be kept local to the server since if the server is not operational, this information cannot be used anyhow. This kind of solution could also be sought for the information in the Virtual System Catalog. For example, a user-enrollment table must not necessarily exist as system global information. The information enabling a user to log on can reside in his user catalog. This means that no user can log on without access to his user data. This restriction might make perfect sense. Another example is the accounting information. It could also be kept in the user catalog. The drawback then is that accounts can only be settled when the corresponding user is active.

With newly designed operating systems it might be feasible to handle all permanent network global information in this way and thus eliminate the concept of the Virtual System Catalog altogether.

SUMMARY

The preceding sections describe a general concept for support of single system image for a network of processors. It can be applied to different types of network configurations and different kinds of processor interconnections. The basis of the described concept consists of a clean categorization of the resources within the network according to their scope and lifetime. Single system image is then supported by a generalized common treatment of all the network global information. For this purpose two virtual control blocks, Virtual System Communication Area and Virtual System Catalog are introduced. An abstract data-type kind of interface is defined to access these virtual control blocks.

The virtual control blocks could be realized in many different ways depending on the underlying network configuration. Some examples of possible realizations have been shown.

REFERENCES

1. Abraham, S.M. and Dalal, Y.K., "Techniques for decentralized management of distributed systems", Proc. IEEE COMPCON, Spring 1980, pp.430-437.

2. Diel, H.,Kreissig, G., Lenz, N.,Scheible, M.,Schoener, B., "Data Management Facilities of an Operating System Kernel", Proc. of ACM SIGMOD '84, June 1984.

3. Lamport,L.,"Time, Clocks, and the Ordering of Events in a Distributed System", Communications of the ACM, July 1978

4. Lampson, B.W. and Sproull, R.F., "An open operating system for a single-user machine," Proc. 7th ACM Symp. on Operating Systems Principles, Asilomar, California, December 1979, pp. 98-105.

5. Lecouffe,M.,"A multiprocessor architecture using a circulating memory", Trends in Information Processing Systems, Springer Verlag 1981

6. Popek, G., Walker,B., English,R., Kline, C., and Thiel, G. "The LOCUS Distributed Operating System", Proc. 9th ACM Symp. Operating System Principles, ACM SIGOPS Operating Syst. Rev. 17, 5 (Oct. 1983), 49-89.

7. Spector,A.Z., "Performing remote operations efficiently on a local computer network, "Comm. ACM., Vol. 25, pp. 246-260, April 1982.

8. Svobodova, L. "A reliable object-oriented data repository for a distributed computer system," in Proc. 8th ACS Symp. Operating Systems Principles, ACM SIGOPS Operating Syst. Rev. 15,5 (Dec. 1981), 47-58.

9. Watson,R.W.,"Distributed System Architecture Model",in Distributed Systems - Architecture and Implementation,Springer Verlag 1981.

An Optical Local Area Network

Andreas Paepcke
Richard D.Crawford
Charles A.Freeman
Frank J. Lee
Robert D.Paull

Hewlett-Packard Laboratories
1501 Page Mill Road, Palo Alto CA 94304

An optical link for the interconnection of workstations, terminals, mainframes and other devices is described. It allows for the flexible networking of nodes whose physical locations may be changed without the overhead of cable rerouting. CHIPNET, an experimental local area network based on this optical link is outlined.

1 Introduction

A common practical problem of local area network installation has been the layout of cables. Safety codes require expensive installation procedures. Appropriate routing schemes must be worked out to accommodate all sites within the constraints of the physical building layout.

Once the cable has been installed, these problems will often recur over the life span of a network. One example are large office areas which are popular in many companies. One advantage of large office spaces with office partitions has been the adaptability of this setup to changing needs. Partitions may be moved to change the layout of rooms. For computer networks this very flexibility poses problems. The cable installation difficulties are encountered repeatedly with every change. A similar situation is encountered in manufacturing facilities where floor layouts must be changed to make room for new machinery.

Hewlett-Packard Laboratories has developed the experimental Chipnet network to solve this problem. Originally an attempt was made to use radio frequency band electromagnetic waves in the implementation of a local area network. While technical feasibility was soon established, it became clear that official frequency allocation was a problem which would take years to solve. Efforts were therefore directed to developing an optical link technology which would use light as a transmission carrier. This paper describes the result of this effort. It also outlines the present software capabilities which were developed once the optical link was available.

2 Link Model

Figure 1 shows the operating principal of the optical link.

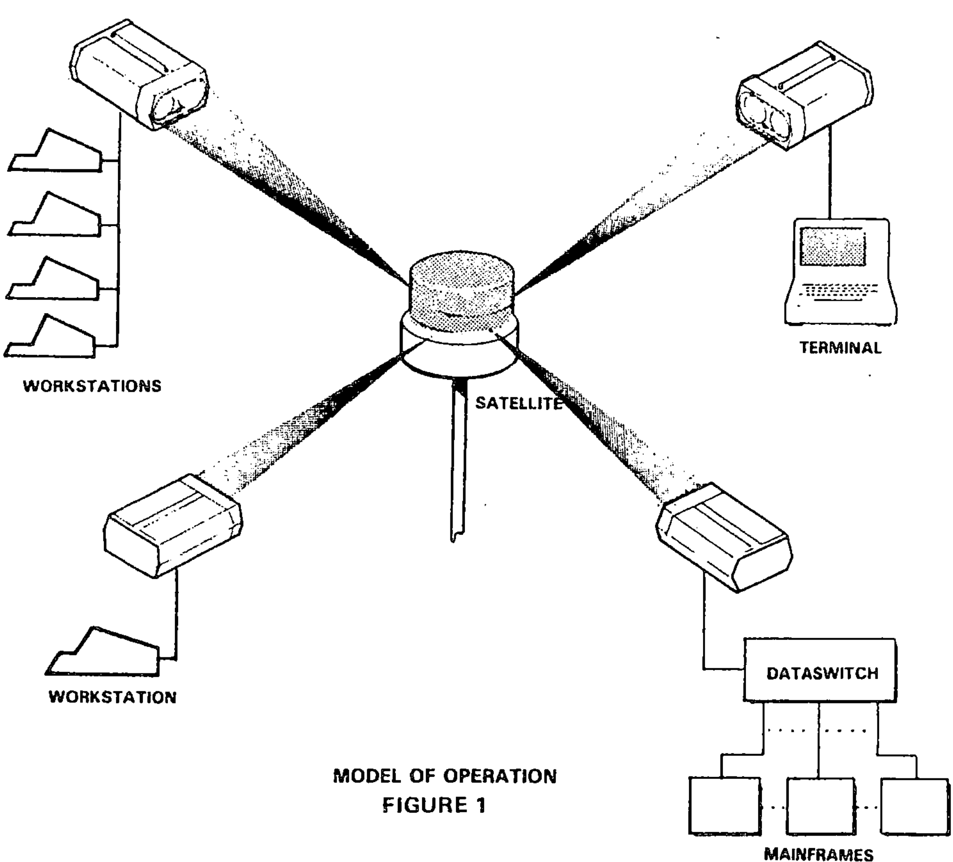

Two optical components are used: datalights and satellites. A datalight is an optical transceiver which is attached to a single station or a cluster of stations. A datalight generates light by means of a light emitting diode which is switched by a binary signal. A lens focuses the light which is then transmitted to the satellite. The satellite is an optical repeater. The received signal at the satellite is amplified, regenerated, and retransmitted in all directions at a different optical wavelength. All datalights in view receive the satellite transmissions. The datalight receivers amplify and convert the optical signal into a binary electrical signal. The maximum distance between a datalight and the satellite is 50 meters.

Datalights are installed at the station sites using any convenient attachment. In rooms with office partitions they are mounted on poles and clamped onto a partition wall. They may also be attached to support pillars or desks. All datalights must, of course, point to the satellite. The aiming procedure is not critical and is made easy by two gun sights installed on the casing of the datalight. The divergence of the light beam as it travels to the satellite ensures that the required aiming accuracy is well within practical limits for free hand adjustment. The authors had the opportunity to observe the optical network during a 30 second earthquake of magnitude 6.3 on the Richter scale. File transfers proceeded throughout the quake without interruption.

Satellite installation is as simple as datalight mounting. The equipment may be placed on a stand and left in a convenient location or it may be hung from the room ceiling.

The optical link as described so far would have a serious limitation. A network based on this link would be confined to one room. A gateway would then be required just to cover adjacent rooms in a building. This problem was solved by designing all hardware drivers in such a way that they would operate either on optical devices or on coaxial cable. No hardware adjustment is necessary in either case. In order to cover several rooms, then, a satellite is placed in every room. All satellites are connected by a coaxial cable. The cable may have a length of up to one kilometer. With this arrangement the various rooms appear logically as one big hall. If the 50 meter datalight - satellite limitation in one very large room is a limitation, two satellites may be installed in one room.

The transfer rate of the optical medium is one megabit/second. At the start of the project, available optical elements were limited to this speed. Now faster devices are available which could allow ten Mbit/sec speeds at some cost in range. After software was developed for the optical link, it became clear that the speed of the medium did not present a bottleneck for individual transfers. The software on Motorola MC68000 microprocessor based workstations was able to use only up to 70% of the theoretical speed with no disc I/O, no additional traffic and a very simple protocol. After a full scale protocol was implemented which allowed for automatic routing and high level host-host communication, only 28% of the theoretical 1Mbit transfer rate was used when disc I/O was involved. Up to 41% of the bandwidth was taken up without disc accesses.

3 Node Interfaces

For nodes to interface to the optical link, access hardware was developed which encodes the digital information in a form appropriate for transmission and which drives the optical and cable transceivers. Such interfaces were built for the Hewlett-Packard HP9000 series 200 workstations and for terminals which use the optical link to communicate with hosts of various vendors.

The media access control used on the network is carrier-sense multiple access broadcast with collision avoidance (CSMA/CA) [ALME]. A station wishing to access the medium will listen for traffic and hold until no carrier is sensed. It will then begin transmission. The difference between this approach and carrier-sense multiple access with collision detect (CSMA/CD) enters during transmission: In contrast to CSMA/CD stations, a CSMA/CA station will not listen for collisions while it is transmitting. This means that collision recovery will not begin until all stations involved have finished their transmission. In [KLEI] the properties of various multiple access methods are examined.

The terminal interfaces are housed in boxes so as not to require any modifications of the terminals themselves. They are connected to the terminals by standard RS-232 lines. Mainframe hosts are similarly connected to the link interfaces through their RS-232 ports. The network between terminals and hosts is transparent to the user. A Z80 microprocessor is used in the interface to handle all networking functions. No routing capabilities are provided in the terminal-host protocols. Most of the software effort was therefore expended on the transport layer. A virtual circuit service is implemented there which guarantees sequenced delivery. Since user keyboard inputs are transmitted to the hosts verbatim, no higher level protocols are required for the terminal interface other than some network management capabilities. Figure 2 shows a block diagram of the terminal interface.

Serial input/output (SIO) chips handle data transfers between the terminal and the network interface. Read-only memory contains enough code to download the interface software from a code server through the network into the random access memory of the interface. After the download process is complete, a custom made real time operating system handles packetization, sequencing, retransmissions and acknowledgments. Outgoing packets are passed through an SIO into a differential Manchester encoder/decoder [SAND]. This coding scheme guarantees a zero crossing on every bit allowing clock recovery from the received signal for synchronization. The encoded bit stream is output to the network by the optical transceiver hardware. The terminal networking was completed and operational before the 802 networking standards [L802] were available. Terminal link level control protocols do therefore not conform to this standard.

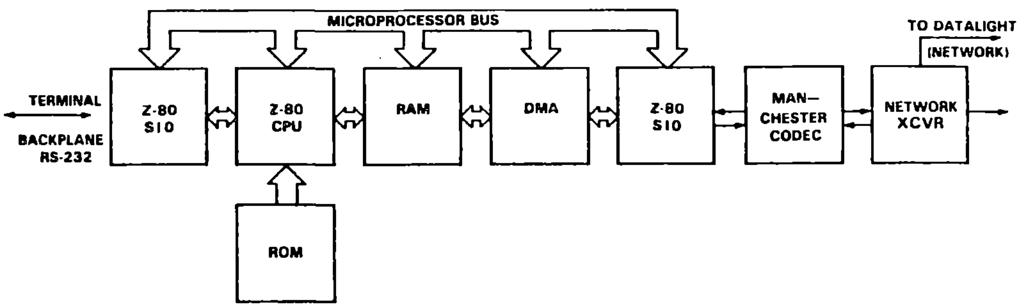

TERMINAL HARDWARE INTERFACE

FIGURE 2

The workstation interface to the optical link also uses a Z80 microprocessor. It is implemented on a set of two printed circuit boards which are inserted in the I/O slots of the workstation backplane. They interface directly to the MC68000 bus. A section of the on board random access memory is accessible from the host and the Z80. Incoming packets are transferred by DMA to that common memory. Outgoing packets are also transferred by DMA to the network access hardware. The host/Z80 interface is interrupt driven. The designers were careful to pipeline activities on the interface card as much as possible. It was thereby possible to avoid bottlenecks in the card. Transfer speed limitations are imposed by higher level host software only. Figure 3 is a block diagram of the workstation interface card.

The software for the workstation link level control was developed after the terminal software. Its development proceeded while the 802 standards were being drafted. Since workstations and terminals were to coexist on one medium, the designers of the workstation link level control were constrained in their decisions. As development deadlines approached, the workstation link level protocol was frozen at the status of Draft D of the 802 standard with some unavoidable deviations.

Figure 1 shows that several stations may share a datalight. Clusters of workstations and terminals may be interconnected by coaxial cable thus spreading the cost of one datalight over several users. Up to 30 meters of cable may be used with no one stub exceeding 10 meters. There are no further constraints on the configuration of a cluster. The system was designed to avoid the requirement for cable terminators. This was done to ensure that non expert users would be able to configure a network without help and intimate knowledge of transmission line theory.

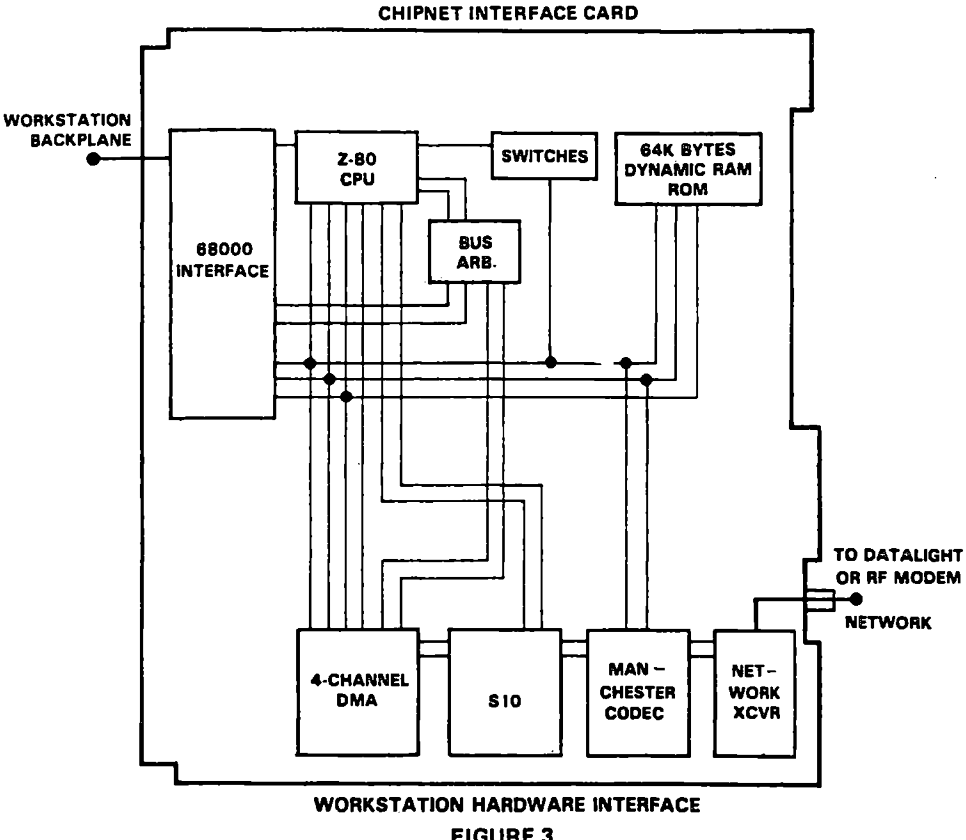

WORKSTATION HARDWARE INTERFACE

FIGURE 3

4 Network Software

The Chipnet network of terminals, mainframes and workstations was developed based on the optical link described above. Using standard RS-232 interfaces, a dataswitch and an autodial modem were attached to the network. The dataswitch allows terminal users to access a large collection of mainframes of several vendors through the network. The modem is operated remotely through Chipnet. Users are normally working on machines local to the plant. Only occasionally do they need to access off site computers. Attaching the modem to the network provided a good way of sharing this resource without the overhead of physically passing it around the building. While all terminal networking software is contained in the interface box, the workstation interface contains only the link layer software. All additional services are implemented on the hosts.

Chipnet uses several link technologies other than the optical link which are not covered in this paper. Gateway capabilities were therefore required for workstations. Automatic routing through switching nodes was also implemented in the network layer. Heartbeat packets allow detection of node crashes and rerouting through alternate paths where possible.

Above the transport layer a high level host-host protocc¹ was implemented which allows remote file access and file transfer capabilities. File transfer utilities were further improved by providing integrated file conversion options to accommodate various mutually incompatible file formats.

Some workstations on Chipnet have multi-user capabilities while others are single user stations. In order to allow the single user stations to take advantage of the more powerful multi-user nodes, remote login capabilities were added. A user may log into any multi-user node through the network without interfering with work in progress at the destination. After login is complete, a user may execute commands and programs at the remote site. Any output is transmitted back through the network and may be displayed on the local screen or directed into a file. Processes may be started in the background at a remote site and other work may be done locally during the remote execution. These basic capabilities have been used successfully to implement network print and computing services. Various resources like printers, mass storage, access to other networks etc. have been distributed and are available to all users.

When new nodes are added or when nodes are physically moved, the network is not disturbed due to the flexibility of the optical link.

Careful consideration was given to the software presentation of the network to users. All utilities have been written and organized in such a way that system programmers may access low level efficient capabilities while ensuring that application programmers have available to them utilities which hide the network. Several applications have been written which use the network as obvious easily integrated extensions of local activities. Figure 4 shows the outline of the workstation software.

Chipnet was implemented on Hewlett-Packard's HP9000 series 200 workstations. These machines are available with the HP-UX operating system which is UNIX*-like and as Pascal

---

* UNIX is a trademark of AT&T Laboratories

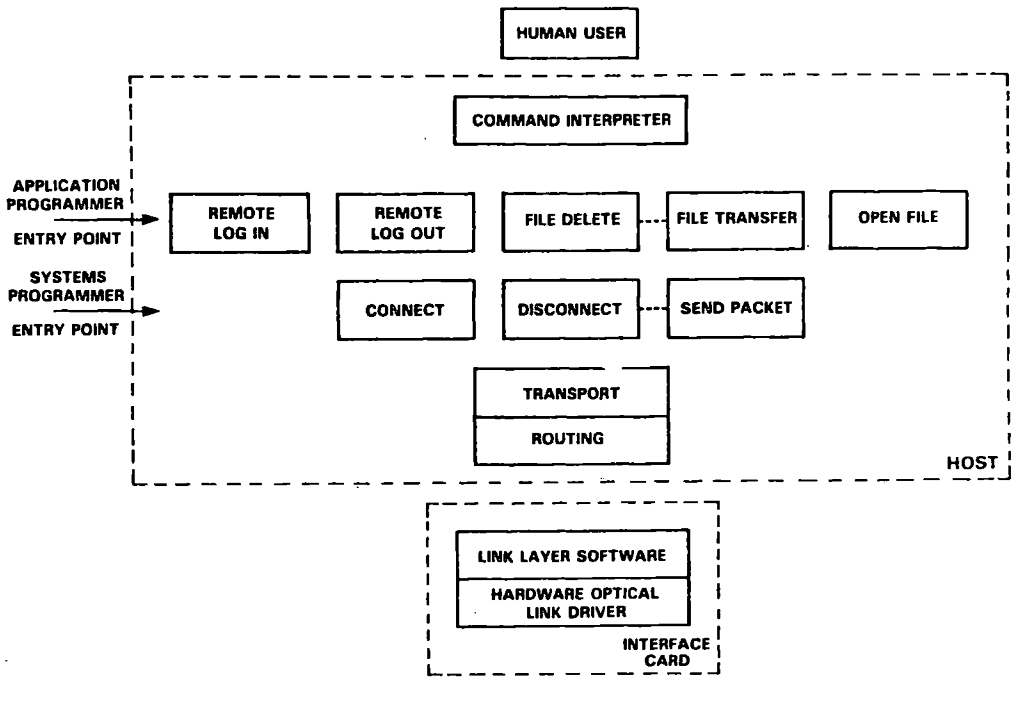

WORKSTATION SOFTWARE STRUCTURE

FIGURE 4

workstations. While Pascal machines are single-user systems, HP-UX based stations allow for background processes and multiple users.

The software is structured in modules which may be mapped onto the OSI reference model for open systems interconnection [OSIM]. The protocols for routing and transport are documented in [JAMP]. These protocols were developed at Hewlett-Packard. They map into the network, transport and session layers of the OSI model. The protocol for remote file operations is also proprietary [HWAN] and covers the presentation layer of the reference model.

The command interpreter of figure 4, finally, maps to the application layer of the model. It allows the user to control network operations by typing commands causing remote login, file transfers, remote process execution etc. Below this module in figure 4, a set of applications level utilities are shown which provide networking functionality for applications programmers. These utilities hide the details of connection establishment, packetization, sequencing etc. They build on a group of systems level functions which interact with the transport layer and cover details like requests for connections with remote hosts. Transport and routing provide sequencing, retransmissions and other related operations.

5 Conclusion

Chipnet, a network of terminals, mainframes, workstations and special hardware devices was outlined in this paper. The network uses an optical link for transmission of information. The major advantage of this link is its flexibility of configuration. Problems of cable layout in changing environments are solved. New nodes are added non intrusively.

The network has improved productivity by making available to everyone resources previously isolated and available only to one node. New users have no problems installing the Chipnet interfaces as most hardware and software were designed to be handled by non-experts.

6 Acknowledgments

More engineers have been involved in this project than can be mentioned here. J. Davis, B. Williams, V. Ostoich, J. Nishimoto, R. Jamp, K. Field, R. Cheng, L. Miller, M. Farrell, W. Yen, G. LaBelle, C. Corsetto, P. Szente, C. Hwang and L. Chen have been among the engineers working on the project.

Capable management was provided by Z. Fazarinc, R. Eschenbach, C. Yen, R. Chiu, T. Horniak and B. Brown.

Their contributions of time, expertise and spirit were indispensable for the development of Chipnet.

7 References

[ALME] G.T. Almes, E.D. Lazowska: "The Behavior of Ethernet-Like Computer Communications Networks". In Proc. 7th Symp.Oper.Syst. Principals, Dec. 1979, pp 66-81.

[HWAN] C. Hwang, C. Lin: "Preliminary Programming Interface to the Distributed File Server - Draft". Internal report Hewlett-Packard Corp.

[JAMP] R. Jamp: "XPN Internet Architecture - Draft". Internal report Hewlett-Packard Corp.

[KLEI] L. Kleinrock, F. Tobagi: "Random Access Techniques for Data Transmission over Packet-Switched Radio Channels". Proc. NCC, pp 187-201, 1975.

[L802] "IEEE Project 802 Local Area Network Standard". Draft D, 1982 and later.

[OSIM] "OSI Reference Model - The ISO Model of Architecture for Open Systems Interchange", IEEE Trans.Comm. Apr. 1980, pp 425-432.

[SAND] L. Sanders: "Encoding Schemes in Serial Data Communications". Electron.Eng., Vol.53, No.654, pp 117-127, July 1981.

Modellierung realzeitspezifischer Verkehrslasten zur Untersuchung
der Eignung von lokalen Netzen in verteilten Realzeitsystemen

----

Jürgen Suppan-Borowka
RWTH Aachen
Lehrstuhl für Informatik IV

0. Übersicht

Ein Verkehrslastmodell mit inhomogenen Stationen und inhomogenen, stations- und
nachrichtentypbezogenen Lasten zur Leistungsanalyse von lokalen Netzen in verteilten Realzeitsystemen wird vorgestellt. Zuvor wird die Struktur und Kontrollhierarchie derartiger Realzeitsysteme analysiert und eine Reihe von realzeittypischen
Verkehrsanforderungen werden abgeleitet. Das Lastmodell geht aus von einer Menge
von Nachrichtentypen, denen als Attribute Service-Request-Klassen (Prioritäten,
Wartezeitanforderungen u.ä.) und Transaktionsklassen zugeordnet werden. Die zu'
bewältigende quantitative Verkehrslast wird modelliert durch die nachrichtentyp-
und stationsbezogene Bildung von Ankunftsprozessen, Nachrichtenlängenverteilungen
über 5 Klassen und gewichteten Informationsflußverteilungen. Das so definierte
Lastmodell ermöglicht die gezielte Gestaltung rein realzeitspezifischer Verkehrslasten zur Beurteilung des Realzeit-Leistungsverhaltens lokaler Netze gegenüber
ausgewählten Leistungsanforderungen.

1. Einführung

Verteilte Realzeitsysteme - insbesondere zur Steuerung und Überwachung von Fließ-
und Prozeßfertigungen - stellen hinsichtlich der erforderlichen Zuverlässig- und
Leistungsfähigkeit hohe Anforderungen an ein Kommunikationsnetz (die u.a. erheblich
von Anforderungen an Bürokommunikationssysteme abweichen). Die PROWAY-Normungs-
gruppe der IEC, die an einem Industrie-Netz-Standard arbeitet, hat derartige realzeitspezifische Anforderungen an lokale Netze spezifiziert (/PRO83/,/GRA80/) und
die erforderliche Spannbreite der Leistungsfähigkeit zum Ausdruck gebracht. In
Deutschland trägt die Entwicklung des PDV-Bus-Standards /PDV83/ den spezifischen
Ansprüchen verteilter Realzeitsysteme Rechnung.

Innerhalb dieses breiten und hochstehenden Anforderungsspektrums befaßt sich die vorliegende Arbeit mit der Untersuchung der realzeitorientierten Leistungsfähigkeit von lokalen Netzen bezogen auf die Protokolle der ISO-OSI-Schichten "data link layer" und "physical layer" /ISO82/, gleichbedeutend mit den IEEE 802 Schichten "logical link control", "media access control" und "physical signaling" /IEE82/. Insbesondere einige der in der letzten Zeit vorgestellten lokalen Netze bzw. Medien-Zugangsprotokolle, die ihre Realzeiteignung gezielt betonen (u.a./KIE83/, /NIE84/, /TOB83/), machen deutlich, daß eine Leistungsbewertung und ein Vergleich dieser Protokolle nicht unproblematisch ist. Eine konkrete Aussage über die realzeitorientierte Leistungsfähigkeit der zur Zeit vorrangig auf diesem Gebiet konkurrierenden Protokolltypen "CSMA/CD mit Nachrichtenprioritäten", "token passing mit Nachrichtenprioritäten auf Bus-Systemen gemäß IEEE 802.4 (IEE82)" und zentralisierte Bus-Protokolle mit Querverkehr /PDV83/, die über die Wertung des Vorhandenseins allgemeiner Protokolleigenschaften wie garantierte Wartezeit, Stabilität usw. hinausgeht (s. dazu /SPA82/), ist nur auf der Basis einer umfangreichen und differenzierten Leistungsanalyse möglich (simulativ oder analytisch). Die in der Literatur zahlreich vorhandenen Analysen sind auf Grund der verschiedensten einschränkenden Annahmen nur schwer miteinander vergleichbar. Erst recht fällt die Beurteilung der Eignung eines konkreten Netzes oder Protokolls für eine gegebene Anwendungssituation schwer.

Voraussetzung für eine Analyse ist die Schaffung eines auf die Realzeiterfordernisse zugeschnittenen Lastmodells. Eine allgemeine Lastmodellierung im Rahmen eines Modells mit homogenen Stationen und homogenen Lasten ohne die Berücksichtigung realzeitspezifischer Erfordernisse erscheint in diesem Zusammenhang wenig geeignet. Die Arbeiten von Heger und Watson /HEG83/, die auch die Aktivitäten der COST11 Arbeitsgruppe "Leistungsanalyse von lokalen Netzen" beeinflußten (COST=Cooperation Scientifique et Technique) und die Veröffentlichungen von Färber /FÄR84/ und Polke /POL84/ unterstützen diese Argumentation. Heger und Watson entwickeln ein Modell zur Gestaltung von Verkehrslasten, das auf vier Transaktionstypen, Nachrichtenprioritäten, poisson- oder periodisch-verteilten Ankünften, der Unterscheidung kurz, mittel und langer Nachrichten und der Angabe der Verkehrsflußverteilung (gleichförmig oder zentral) beruht. Aus diesem Modell leiten sie eine Reihe von universellen Benchmarks ab, die auch im COST11 Projekt ihren Einsatz fanden /IIT84/. Die Ergebnisse von COST11 zeigen, daß zur Unterscheidung der Realzeiteignung bestimmter Zugangsprotokolle ein modifizierter und speziell auf Realzeiterfordernisse zugeschnittener Ansatz notwendig ist.

Die vorliegende Arbeit modifiziert und erweitert den Ansatz von Heger und Watson, deren Intention auch mehr auf der Entwicklung universeller Benchmarks liegt (z.B. auch zur Beurteilung von lokalen Netzen für die Büroautomatisierung). Es wird nachgewiesen, daß es möglich ist, eine Reihe rein realzeitspezifischer Leistungsanforderungen aus der Analyse der Struktur und Kontrollhierarchie verteilter Realzeitsysteme abzuleiten, wobei Bezug genommen wird auf die Arbeiten von Christensen /CHR81/, Prince und Sloman /PRI81/ und Walze /WAL83/. Als Ergebnis wird ein Lastmodell zur gezielten Gestaltung von differenzierten Lasten vorgestellt, das es möglich macht, das Leistungsverhalten der zu untersuchenden Netze gegenüber ausgesuchten realzeitspezifischen Verkehrsanforderungen zu beurteilen. Im Vordergrund stehen dabei die Anforderungen an obere Grenzen für die Leistungsparameter (Wartezeiten auf Netzzugang, Transaktionsdauer, effektive transaktionsorientierte Bitdatenraten, relativer Durchsatz u.ä.) bei der Übertragung aus Realzeitsicht wichtiger Nachrichten (Alarm-Meldungen, Notfall-Aktionen, Echtzeit-Protokolle usw.).

2. Analyse der Struktur und Kontrollhierarchie verteilter Realzeitsysteme

Die Arbeiten von Walze /WAL83/, Christensen /CHR81/, Prince und Sloman /PRI81/, Färber /FÄR84/ und Polke /POL84/ bestätigen die Annahme, daß es typische Strukturen von verteilten Realzeitsystemen gibt. Die Realzeitanforderungen von Fließ- und Prozeßfertigungen bilden den Ausgangspunkt und werden als Grundlage einer Analyse dieser Strukturen herangezogen. Ziel der Analyse ist die Untersuchung der Verteilung von Kontrollaufgaben über die Stationen eines verteilten Realzeitsystems, um daraus konkrete Verkehrsanforderungen abzuleiten.

Als Baustein jeder Fertigung wird die Fertigungsgruppe definiert:

Definition Fertigungsgruppe :

    Eine Fertigungsgruppe besteht aus einer Reihe von Bearbeitungsstationen ST1...STn, die in einer ereignisgebundenen Ordnung zueinander stehen. D.h. das zu fertigende Produkt durchläuft die voneinander abhängigen Bearbeitungsstationen der Fertigungsgruppe in einer produktionstechnologisch vorgegebenen festen Reihenfolge. Es werden besonders hervorgehoben:

    - ST1 : die erste Station ist die Ladestation
    - STn : die letzte Station ist die Entladestation

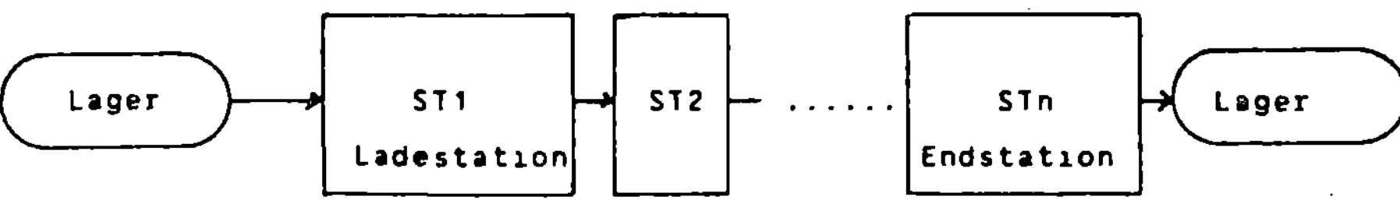

Bild 2.1 : Fertigungsgruppe

Definition Fertigung :

Eine Fertigung besteht aus einer Menge von Fertigungsgruppen. Diese Gruppen hängen ggf. über Zwischenpuffer zusammen. Eine direkte Taktabhängigkeit zwischen den Gruppen besteht nicht.

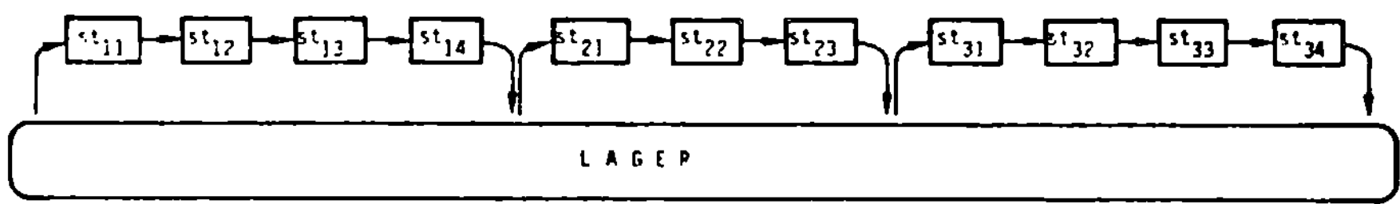

Bild 2.2 : Beispiel-Fertigung mit 3 Fertigungsgruppen

Im Sinne der uns gegebenen Aufgabenstellung ist es notwendig, festzustellen, welche Kontrollaufgaben es in einer derartigen Fertigung gibt, und in welcher Beziehung diese zueinander stehen. Folgende hierarchisch geordnete Kontrollinstanzen sind einzugrenzen:

- Kontrollebene 1: Direkte Kontrolle der zu einer Bearbeitungsstation gehörenden Maschinen. Diese Kontrolle beinhaltet die Überwachung aller aktuellen Maschinenzustände und die Steuerung der Maschinenaktionen (Prozeß-schnittstelle). In der Literatur wird diese Ebene zum Teil noch in eine Prozeßkontroll- und eine Feldebene aufgeteilt (/FÄR84/,/POL84/).

- Kontrollebene 2: Kontrolle einer Fertigungsgruppe.
  Diese Kontrolle umfaßt:
  - die Überwachung der korrekten Arbeitsweise aller Bearbeitungsstationen einer Fertigungsgruppe
  - die Transportüberwachung und Steuerung
  - die Synchronisation der Arbeit der Bearbeitungsstationen
  - die Steuerung der Kontrollfunktionen der Ebene 1 (z.B. durch Vorgabe von Zielparametern)

- Kontrollebene 3: Gesamtkontrolle.
  Diese Kontrollebene umfaßt:
  - die Abstimmung der Fertigungsgruppen aufeinander
  - die Überwachung der Fertigungsgruppen
  - die Sammlung aller fertigungsrelevanten Daten
  - die Auswertung der aktuellen und langfristigen Daten der Fertigung
  - für das Management
  - für die Fertigungsüberwachung
  - die Festlegung der Aufgabe der einzelnen Fertigungsgruppen

Die Kontrollebenen bilden eine Kontrollhierarchie:

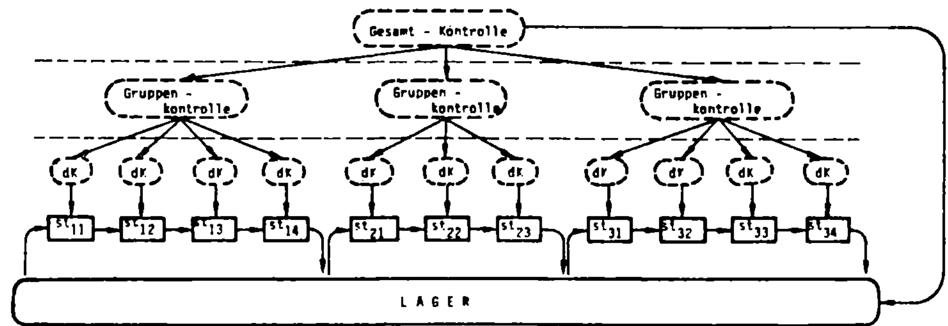

dK = direkte Kontrolle

Bild 2.3 : Kontrollhierarchie

Die einzelnen Kontrollaufgaben werden in einem verteilten Realzeitsystem von Prozeßrechnern wahrgenommen. Dies heißt:

- Die Aufgaben der Kontrollebene 1 werden von Prozeßrechnern, den sogenannten Steuerrechnern wahrgenommen, die mit einer entsprechenden Peripherie zur Erfassung von Meßwerten (A/D-Wandler, digitale Eingänge, ...) und zur Abgabe von Steueranweisungen (D/A-Wandler, digitale Ausgänge,...) versehen sind oder ihre Prozeßschnittstelle über sogenannte Feldmodule realisieren. Die Steuerrechner einer Fertigungsgruppe stehen in einer zeitlichen Abhängigkeit zueinander (Taktzeit); dies macht für jeden einzelnen die Abstimmung seiner Aktivitäten mit dem direkt benachbarten Steuerrechner notwendig (Taktsynchronisation).

- Die Aufgaben der Kontrollebene 2 werden für jede Fertigungsgruppe von dem sogenannten Gruppenrechner wahrgenommen. Er kommuniziert mit den Steuerrechnern zu folgenden Zwecken:
  - Statusabfrage zur Feststellung möglicher Ausfälle oder fehlerhafter Systeme,
  - Abfrage der für die Beurteilung der augenblicklichen Situation der Fertigungsgruppe relevanten Meßwerte und Auswertung dieser Meßwerte,
  - Abfrage von kritischen Prozeßzuständen (Alarmmeldungen) bzw. Reaktion auf die Meldung kritischer Prozeßzustände
  - Synchronisation der durch die Steuerrechner kontrollierten Prozesse
  - Vorgabe von Zielparametern (Soll-Daten)
  - Überwachung und Steuerung des Transports zwischen den Steuerrechnern

Der Prozeßrechner der Kontrollebene 2 benötigt Peripheriegeräte zur
- Abspeicherung der erfaßten Daten
- Darstellung der Situation des Fertigungsprozesses für einen Beobachter (Kontrollwartenfunktion) auf einem Graphikbildschirm oder Drucker.
- Erfassung von manuell eingegebenen Steueranweisungen (Tastatur, Maus, o.ä.)
- Ausgabe von Protokollen

- Für den Rechner der Kontrollebene 3, den sog. Zentralrechner, besteht nicht die Notwendigkeit der ereignisorientierten Reaktion. Zu seinen Aufgaben zählen:
  - die Bereitstellung einer Entwicklungsumgebung für Steuerungsprogramme der Kontrollebenen 1+2
  - die an den Unternehmenszielen orientierte Optimierung der Arbeit der einzelnen Fertigungsgruppen
  - die Erfassung des Gesamtstatus der Fertigung
  - die Sammlung der zur Beurteilung der Fertigungssituation notwendigen Daten (langfristige Speicherung auf Band/Platte z.B.)
  - die Auswertung der aktuellen und vergangenen Fertigungssituation (Historie) für das Management und das Fertigungspersonal unter Benutzung geeigneter Hilfsmittel (Bildschirm oder Plottgraphiken o.ä.)
  - das Laden von Steuerungsprogrammen in die Prozeßrechner der Ebenen 1+2.(down line load)
  - die Initiierung der Tätigkeit der Steuer-, Gruppen- und Transportrechner.

Die den Rechnern auf den jeweiligen Kontrollebenen zugewiesenen Aufgaben werden durch Prozesse realisiert. Um eine begriffliche Verwechslung mit den Fertigungsprozessen zu vermeiden, werden Prozesse im EDV-Sinne im folgenden als Task bezeichnet.
Die Aufgabenstellung eines verteilten Realzeitsystems bleibt in der Regel über einen Zeitraum von Monaten oder Jahren konstant. Änderungen entstehen durch den begrenzten Wechsel der Fertigungstechnologie, den Austausch oder die Hinzufügung von Fertigungsmaschinen usw.. Derartige Änderungen sind entsprechend selten. Dynamische Lasten in Verbindung mit stark dynamischen und ggf. heterogenen Anforderungsprofilen existieren nicht.
Aus diesen Gründen und unter Berücksichtigung der für die einzelnen Kontrollebenen definierten Aufgabenstellung wird deshalb davon ausgegangen, daß die Zuordnung der existierenden Tasks zu den Rechnern der einzelnen Kontrollebenen <u>statisch</u> ist, d.h. diese Tasks migrieren nicht. Diese Ansicht wird von Christensen /CHR81/ und Prince /PRI81/ geteilt.
Das Ergebnis der bisherigen Untersuchungen ermöglicht die Abgrenzung des Begriffs einer Station:

Definition Station:

Eine Station in einem verteilten Realzeitsystem ist die Menge der auf einem Rechner der Kontrollebenen 1,2 oder 3 ablaufenden Tasks zur Erfüllung der Kontrollaufgaben des jeweiligen Rechners. Demzufolge lassen sich unterscheiden:

- Steuerstationen,

- Gruppenstationen,

- Zentralstationen.

Man sagt, eine Station st_i kommuniziert mit einer Station st_j, wenn es eine Task auf Station st_i gibt, die mit einer Task auf Station st_j kommuniziert.

Man sagt, eine Station st_i kontrolliert eine Station st_j, wenn

- es mindestens eine Task auf Station st_i gibt, die einer Task auf Station st_j Anweisungen geben kann, die diese zu befolgen hat,

- es keine Task auf Station st_j gibt, für die dies bzgl. einer Task auf Station st_i gilt.

Die bisherigen Erkenntnisse ermöglichen es, den vagen Begriff der Struktur eines verteilten Realzeitsystems präziser zu fassen:

Definition : Verteiltes Realzeitsystem (zum Systembegriff s. /STE83/):

Ein verteiltes Realzeitsystem S_{PA} ist ein Tripel (ST, KR, CR) mit

1. ST ist die Menge der Stationen im System:

$$ST = \{ st_1, \ldots, st_n \} \quad n \in \mathbb{N}$$

2. KR ist Kontrollrelation, $KR \subseteq (ST \times ST)$, mit

$$(st_i, st_j) \in KR :\Longleftrightarrow st_i \text{ kontrolliert } st_j$$

Die Kontrollrelation wird beschrieben durch die Kontrollmatrix

$$KR_M = (kr_{ij}) \quad 1 \le i,j \le n$$
$$kr_{ij} \in \{0,1\}$$
$$kr_{ij} = 1 \Longleftrightarrow (st_i, st_j) \in KR$$
$$kr_{ij} = 0 \Longleftrightarrow (st_i, st_j) \notin KR$$

Die Menge der Stationen ST läßt sich derart in Teilmengen $ST_o, \ldots, ST_m$ $m \in \mathbb{N}$ zerlegen, daß gilt

(a) $ST_i \cap ST_j = \emptyset$ für $i <> j$ und $0 \le i,j \le m$

(b) $ST_o = \{ st \in ST \,/\, \text{es gibt kein } st' \in ST \text{ mit } (st,st') \in KR \}$

$ST_i = \{ st \in ST \,/\, (\exists\, st' \in ST_{i-1} : (st, st') \in KR) \text{ \underline{und}}$

$(\text{falls } st'' \in ST \text{ mit } (st, st'') \in KR$

$\Longrightarrow st'' \in ST_j, \; j \le i-1) \}$

für $1 \le i \le m$

(c) Jede Menge ST_i , $0<=i<m$, läßt sich in Partitionen $ST_{i_{st_1}}$, ..., $ST_{i_{st_v}}$ zerlegen:

(c1) $ST_{i_{st_k}} \cap ST_{i_{st_l}} = \emptyset$ für $st_k <> st_l$

(c2) $ST_{i_{st_k}} = \left\{ st \in ST_i , st_k \in ST_{i+1} \text{ und } (st_k , st) \in KR \right\}$

(c3) $\bigcup (ST_{i_{st_1}} , ... , ST_{i_{st_v}}) = ST_i$

Das verteilte Realzeitsystem S_{PA} ist bezüglich der Kontrollhierarchie gemäß dieser Definition hierarchisch und zyklenfrei. Die Hierarchie ist aufgebaut aus Hierarchieebenen gemäß (a) und (b) und Hierarchiegruppen innerhalb der Ebenen gemäß (c).

3. CR ist Kommunikationsrelation, $CR \in (ST \times ST)$, mit

$(st_i , st_j) \in CR :<==> st_i$ kommuniziert mit st_j ohne Einbeziehung anderer Stationen. Wir sagen: st_i und st_j sind topologisch benachbart.

Die Kommunikationsrelation eines verteilten Realzeitsystems wird beschrieben durch die Topologiematrix $TM = (tm_{ij})$, $1 <=i,j<=n$.

$$TM = \begin{bmatrix} tm_{11} & \cdots & tm_{1n} \\ \cdot & & \cdot \\ \cdot & & \cdot \\ \cdot & & \cdot \\ tm_{n1} & \cdots & tm_{nn} \end{bmatrix}$$

$$tm_{ij} \in \{0,1\}$$
$$tm_{ij} = 1 <==> (st_i , st_j) \in CR$$
$$tm_{ij} = 0 <==> (st_i , st_j) \in CR$$

3. Charakterisierung von Nachrichten

Aus den vorangegangenen Überlegungen geht hervor, welche Stationen in einem verteilten Realzeitsystem existieren (wer kommuniziert) und in welcher Kontrollbeziehung diese Stationen zueinander stehen (warum wird kommuniziert). Die angegebenen Kontrollaufgaben machen es möglich, den Typ und die Art der zu übertragenden Nachrichten näher zu erfassen. Ein ähnliches Vorgehen ist in den Arbeiten von Christensen, Prince, Sloman und Walze (/CHR81/, /PRI81/, /WAL83/) zu beobachten, die aber hinsichtlich der Zahl und der Charakterisierung der Nachrichten zu leicht abweichenden Ergebnissen kommen.

Allgemein gibt es sicherlich ein breites Spektrum an Möglichkeiten, Nachrichtentypen für ein verteiltes Realzeitsystem aus der hierarchischen Verteilung der Kontrollaufgaben abzuleiten. Es wird als hinreichend angesehen, wenn die Typisierung die Erfassung und Gestaltung einer typischen Verkehrslast ermöglicht, auch wenn damit nicht jeder Einzelfall erfaßt wird:

- Kommando-Nachricht

  Eine Kommando-Nachricht wird von einer Task versandt, um eine andere Task zu einer Aktion zu veranlassen (z.B. Sollwertveränder"ng) oder eine Kurznachricht anzufordern (typisch: Statusabfrage).

- Kommando-Antwort

  Eine Kommando-Antwort beinhaltet die kurzfristige Reaktion auf ein Kommando. Diese besteht aus einer Kurznachricht (Hardwarestatus, Softwarestatus, Quittung). Erfordert ein Kommando eine mittelfristige Reaktion, die gleichzeitig auch einen größeren Nachrichtenumfang hat, so erfolgt erst eine Quittierung und die Reaktion wird als eigenständige Nachricht ausgeführt.

- Alarm

  Ein Alarm ist eine Nachricht, die einen Fehlerzustand meldet. Für die Feststellung von Alarm-Situationen wird zugelassen, daß neben der expliziten Alarm-Meldung auch ein Alarm durch zyklische Statusabfrage durch eine übergeordnete Station festgestellt werden kann.

- Bedien-Anforderung

  Eine Task kann an eine andere Task eine Bedien-Anforderung stellen. Eine Bedien-Anforderung ist insbesondere sinnvoll, wenn in einem verteilten hierarchischen System ein Datentransport an eine hierarchisch übergeordnete Task nur von dieser veranlaßt werden kann. Die untergeordnete Task meldet in diesem Fall nur, daß sie eine Bearbeitung wünscht oder daß bestimmte Daten bereit stehen und wartet auf das entsprechende Kommando.

- Synchronisations-Signale

  Synchronisationssignale dienen zur Synchronisation von Task-Aktionen, z.B. zwischen aufeinanderfolgenden Steuerstationen ("ich bin fertig, Produkt steht zur Weitergabe bereit").

- Protokolle

  Zur Generierungszeit wird für jedes Protokoll die Datenmenge festgelegt, die es zu umfassen hat. Ein Protokoll wird erzeugt, um einer anderen Station den aktuellen Wert dieser Daten mitzuteilen. Nach dem angenommenen zeitlichen Grad der Aktualität werden folgende Protokolltypen definiert:

  - Echtzeit-Protokoll

    Ein Echtzeit-Protokoll umfaßt alle Daten, die notwendig sind, um die augenblickliche Situation bezogen auf die jeweilige Kontrollaufgabe der Station zu beschreiben. Echtzeit-Protokolle werden in relativ kurzen festen Zeitabständen erzeugt. In kritischen Situationen können diese Zeitabstände verringert werden.

- Bericht-Protokoll

  Bericht-Protokolle umfassen Daten, die sich nur in größeren Zeitabständen ändern, die von den erzeugenden Stationen über den Protokollzeitraum aufbereitet werden. (z.B. Daten über die Auslastung/Produktivität der Station, Werkzeugzustand, usw.)

- Schnappschuß-Protokoll

  Ein Schnappschuß-Protokoll wird in Reaktion auf ein Schnappschußkom-mando erzeugt. Der Erfassungszeitraum ist durch das Schnappschuß-Kommando festgelegt.

- Notfall-Aktion

  Eine Notfall-Aktions-Nachricht ist ein Kommando, das bei der empfangenden Task eine sofortige Aktion auslöst. Typischerweise werden solche Nachrichten nach Meldung bzw. Erkennung einer Alarm-Situation versandt und bedürfen bevorzugter Behandlung.

- Schnappschuß

  Mit einem Schnappschuß-Kommando ruft eine hierarchisch höherstehende Station von mehreren untergeordneten Stationen gleichzeitig sogenannte Schnappschuß-Protokolle ab. Auf diese Weise wird die schnelle Lokalisierung von Fehlerursachen unterstützt. Das Schnappschuß-Kommando beinhaltet in der Regel eine Synchronisation der Erfassungszeiträume für die Protokolle auf den angesprochenen Stationen (Schnappschuß-Fenster).

- Graphik-Files

  Die verschiedenen im verteilten Automatisierungssystem vorhandenen Graphik-Bildschirme (Kontrollwarte, Management usw.) werden in der Regel von einer zentralen Graphikstation (Bildspeicher) mit Bild-Information versorgt.

- Daten-Files

  Es wird davon ausgegangen, daß den auf den einzelnen Stationen ablaufenden Tasks Daten-Files zugeordnet sind, die zu Beginn der Programmausführung den Startzustand beschreiben (Prozeßparameter), die während der Ausführung verändert werden können (Solldatenvorgabe durch den Bediener, selbstlernende adaptive Tasks) und die nach Beendigung der Ausführung ggf. in der dann vorliegenden Version (Lernergebnis) abgespeichert werden.

- Programme

  Es besteht die Möglichkeit, Programme von einer hierarchisch übergeordneten Station in eine untergeordnete Station zu laden (down line load). Auf diese Weise wird eine von der Steuerungsaufgabe unabhängige Programmentwicklung unterstützt und die Verwaltung von zeitweise nicht benötigter Hintergrundinformation auf Platten o.ä. zentralisiert.

Diese Nachrichtentypen lassen sich wie in der folgenden Tabelle angegeben

charakterisieren (s. a. /CHR81/, /PRI81/, /WAL83/). Die gegebenen Nachrichtentypen machen eine fünfstufige Wertung jedes Attributes notwendig.

Nachricht	zulässige Warte-zeit auf Über-tragung	Auftreten	Länge	Anforderung an sichere Übertragung
Kommando	kurz	sehr häufig	sehr kurz - kurz	sehr hoch
Kommando-Antwort	sehr kurz	nach Kom-mando	kurz	sehr hoch
Alarm	sehr kurz	selten	kurz	sehr hoch
Bedien-Anforderung	kurz	mittel	sehr kurz	sehr hoch
Synchronisations-signale	kurz	mittel, in Abhängigkeit von Prozeß-Taktzeit	sehr kurz	sehr hoch
Notfall-Aktion	sehr kurz	nach Alarm oder selten	sehr kurz - kurz	sehr hoch
Echtzeit-Protokoll	kurz	häufig	mittel	sehr hoch
Bericht-Protokoll	mittel	selten	mittel - lang	mittel
Schnappschuß-Protokoll	sehr kurz	nach Alarm	mittel	sehr hoch
Graphik-Files	hoch	selten	sehr lang	mittel
Daten-Files	hoch - sehr hoch	sehr selten	lang - sehr lang	sehr hoch
Programme	hoch - sehr hoch	sehr selten	sehr lang	sehr hoch

Tabelle 3.1 : Charakterisierung der Nachrichentypen

Eine weitere Quantifizierung der einzelnen Attribute ist naturgemäß schwer. Einige
Anhaltspunkte lassen sich dennoch ermitteln:

- Die PROWAY-Gruppe (/GRA80/, /PRO83/) kennzeichnet die zulässige Wartezeit auf
  Übertragungen unter Berücksichtigung gewisser Rahmenbedingungen (Bandbreite,
  Nachrichtenlänge etc.) auf 20 ms. Eine derartige Zeitspanne, die sich primär auf
  die für die zuverlässige Arbeitsweise des Systems "wichtigen" Nachrichten beziehen sollte, wird in der Tabelle als kurz gekennzeichnet.
- In Anlehnung an Christensen /CHR81/,Sloman und Prince /PRI81/ könnten die Nachrichtenlängen wie folgt quantifiziert werden:

sehr kurz   8 ... 64   Bit

kurz   64 ... 128   Bit

mittel   128 ... 1024 Bit

lang   1024 ... 10 K Bit

sehr lang   >10 K Bit

- Bezüglich der Häufigkeit des Auftretens der einzelnen Nachrichtentypen ist es
  hinsichtlich einer Modellierung der Verkehrslast weniger notwendig, die absoluten
  Häufigkeiten zu kennen als vielmehr die relativen. Dabei ist zu beachten, daß
  diese stationsbezogen sind, d. h. daß sich die quantitative Aufteilung des
  Nachrichtenflusses zwischen den Stationen je nach der Kontrollfunktion der einzelnen Stationen stark unterscheidet.

Bei der Auswertung der Tabelle ist zu beachten, daß nicht jeder Nachrichtentyp von jeder Station ausgehen kann und daß ebenfalls nicht jede Station alle Nachrichtentypen empfängt. Die reale Verkehrsfluß-Aufteilung ist an der Kontrollaufgabe der einzelnen Stationen orientiert.

4. Resultierende Forderungen an ein Lastmodell

Die Forderungen werden stichpunktartig aufgelistet:
- die in 3 definierten Nachrichtentypen müssen getrennt modellierbar sein, dabei muß jeder Station eine stationsspezifische Menge von Nachrichtentypen zugeordnet werden können.
- Modellierungsparameter sind mindestens
  - Nachrichtenpriorität
  - Anforderung an Wartezeit
  - Häufigkeit des Auftretens
  - Nachrichtenlänge
  - Verkehrsfluß-Aufteilung
- für einzelne Modellierungsparameter gilt:
  - folgende Typen der Häufigkeit des Auftretens sind zu unterscheiden:
    - periodisches Auftreten
      - konstant periodisch oder beschränkt periodisch
    - Ereignisse, die als Folge des Auftretens eines anderen Ereignisses mit einer Verzögerung bzgl. dieses Ereignisses auftreten. Für die Verzögerung gilt, daß die Reaktionswahrscheinlichkeit für die nähere Zukunft hoch ist und für die fernere Zukunft schnell abnimmt.
    - nicht-periodisches Auftreten
      innerhalb eines anstehenden Zukunftsintervalls ist die Wahrscheinlichkeit des Auftretens hoch, danach nahezu null.
      Unterscheide:
      - innerhalb des Intervalls gleich wahrscheinlich (z. T. als Ausdruck fehlender weiterer Information)
      - innerhalb des Intervalls mit schnell abnehmender Wahrscheinlichtkeit
  - die möglichen Nachrichtenlängen müssen in 5 Klassen modellierbar sein. Innerhalb der einzelnen Klassen sind Wahrscheinlichkeitsverteilungen einzusetzen.
  - die Häufigkeit des Auftretens einzelner Nachrichtentypen und ihre Länge müssen für die einzelnen Stationen getrennt modellierbar sein.
  - die Verkehrsfluß-Aufteilung muß für jeden Nachrichtentyp auf jeder Station getrennt gestaltbar sein.

5. Diskussion des Lastmodells

Eine Verkehrslast in einem verteilten Realzeitsystem wird durch folgende Komponenten modelliert:

a) Stationen

Es werden Stationen $ST = \{st_o, \ldots, st_n\}$ definiert.

b) Nachrichtentypen

Es werden Nachrichtentypen $NT = \{nt_1, \ldots, nt_k\}$ definiert.

c) Den Nachrichtentypen werden Transaktionsklassen und Service-Request-Klassen zugeordnet. Als Transaktionsklassen werden in Modifikation und Erweiterung des Ansatzes von Heger und Watson /HEG83/ definiert:

    TA1 = Einzelnachricht

    TA2 = Einzelnachricht mit Sofort-Antwort

    TA3 = Antwort-Nachricht

    TA4 = Antwort-Nachricht mit Sofort-Antwort

    TA5 = Grupen-Nachricht (Nachricht an mehrere oder alle Stationen)

    TA6 = Serien-Nachricht <=> Folge von TA2

Eine Service-Request-Klasse ist ein Dreitupel $SRK = (p, t_W, t_A)$ mit

p = Priorität einer Nachricht; Dynamische Prioritäten werden nicht ausgeschlossen

t_W = maximale Wartezeit auf den Medienzugang

t_A = maximale Wartezeit bis zum Erhalt der Sofort-Antwort

d) Jedem Nachrichtentyp wird für jede Station ein Ankuftsprozeß A_{st}^{nt} zugeordnet, der mit Hilfe der folgenden Ankunftsverteilungen A modellierbar ist:

$$A = \{(w_1, w_2, V) \; / \; w_1, w_2 \in \mathbb{R}, \; V \in \{n, r, c, d, m\}\} \quad \text{mit:}$$

$(w_1, w_2, V) = (\mu, \sigma, n) \quad <==>$

    die Zeitabstände sind normalverteilt mit dem Erwartungswert μ und der Varianz σ^2. σ^2 beschreibt eine beschränkte Periodizität.

$(w_1, w_2, V) = (max, min, r) \quad <==>$

    die Zustände sind gleich wahrscheinlich aus dem diskreten Intervall (min, ..., max).

$(w_1, w_2, V) = (const, 0, d) \quad <==>$

    die Zeitabstände sind konstant, alle const. Zeiteinheiten wird eine Nachricht erzeugt.

$$(w_1, w_2, V) = (a, 0, m) \quad <==>$$

die Zeitabstände sind exponential verteilt mit Parameter a.

$$(w_1, w_2, V) = (delay, a, d) \quad <==>$$

nach einem Ereignis tritt mit einer erwarteten Verzögerung delay ein Folgeereignis auf. Delay ist exponential verteilt mit Parameter a.

e) Nachrichtenlänge

Jedem Nachrichtentyp $nt \in NT$ wird eine Längenfunktion L_{nt} zugewiesen mit

$$L_{nt} \in \left\{ (L_i, w, V) \ / \ i \in \left\{ 1, \ldots, 5 \right., \ w \in \mathbb{N}, \ V \in \left\{ n, r, m, c \right\} \right\} \text{ mit}$$

$$(L_i, w, V) = (L_i, \sigma^2, n) \quad <==>$$

die Länge ist normalverteilt mit dem Erwartungswert L_i und der Varianz σ^2.

$$(L_i, w, V) = (L_i, 0, r) \quad <==>$$

die Länge ist gleich wahrscheinlich aus dem Intervall $(L_{i_{min}}, \ldots, L_{i_{max}})$.

$$(L_i, w, V) = (L_i, \sigma^2, m) \quad <==>$$

die Länge ist exponential verteilt mit dem Erwartungswert L_i und der Varianz σ^2.

$$(L_i, w, V) = (L_i, 0, c) \quad <==>$$

die Länge ist konstant L_i.

Die Werte L_i, $i \in \left\{ 1, \ldots, 5 \right\}$, beziehen sich auf 5 Längenbereiche: sehr kurz, kurz, mittel, lang, sehr lang. Mit jedem Längenbereich verbunden ist eine minimale und eine maximale Länge $L_{i_{min}}$ bzw. $L_{i_{max}}$.

f) Gewichtete Informationsfluß-Verteilung

Jede Nachricht entsteht im Rahmen einer Kommunikation zwischen einem Task auf Station st_i mit einem oder mehreren Tasks auf einer oder mehreren anderen Stationen (point-to-point, multipoint, broadcast-Kommunikation). Die Gesamtmenge der potentiellen Partner-Tasks und Stationen ist statisch, da die Task-Verteilung im System statisch ist. Daraus folgt unter Ausdehnung dieses Gedankens auf die nachrichtentyp-bezogene Kommunikation, daß es möglich sein muß, für jeden Nachrichtentyp nt anzugeben, welche Stationen Nachrichten dieses Typs von Station st_i gesandt bekommen. Aus der alle Nachrichten umfassenden Topologiematrix TM wird eine nachrichtentyp-bezogene Topologiematrix TM_{nt}. In Anlehnung an den in diesem Zusammenhang eingeführten Sprachgebrauch sprechen wir dementsprechend auch von _topologischer Nachbarschaft bezüglich des Nachrichtentyps nt_. Die Matrix TM_{nt} ist stabil, eine Eigenschaft, die auf die statische Task-

Zuordnung zurückzuführen ist(s.a. /PRI81/). Diese Eigenschaft ist für die präzise Modellierung der Verkehrslast von besonderer Bedeutung, da sich dynamisch ändernde Informationsflußverteilungen (dynamic information flow patterns) nur schwer hätten erfassen lassen.

Somit sind die Voraussetzungen geschaffen, um für jeden Nachrichtentyp nt eine Kommunikationsmatrix K_{nt} zu entwickeln, die einer gewichteten Matrix TM_{nt} entspricht:

$$K_{nt} = (K_{ij}^{nt})_{0 \leq i,j \leq n} \quad \text{mit}$$

$$n = \text{Stationszahl}$$

K_{ij}^{nt} = Anteil der Übertragung von Nachrichten des Typs nt von Station st_i nach st_j an der Gesamtheit aller von Station st_i versandten Nachrichten des Typs nt. Es gilt natürlich:

$$0 \leq K_{ij}^{nt} \leq 1$$

$$K_{ii}^{nt} = 0,$$

$$\sum_{j=1}^{n} K_{ij}^{nt} = 1 \quad \text{für jedes } i = 1,\ldots,n$$

g) Kommunikationsrate

Unter Zusammenfassung der durch K_{ij}^{nt} beschriebenen Lastverteilung innerhalb eines Nachrichtentyps $nt \in NT$ und der definierten Erzeugungsfunktion A_{st}^{nt} ergibt sich die Kommunikationsrate von gewünschten Übertragungen vom Nachrichtentyp nt von Station st_i nach Station st_j (Zahl der Übertragungswünsche pro Zeiteinheit, hier ms):

- reale Kommunikationsrate: $KRr_{ij}^{nt} = e_i^{nt} * K_{ij}^{nt}$

- erwartete Kommunikationsrate bei normal oder exponentiell verteilter Erzeugung:

$$KRn_{ij}^{nt} = 1/\mu^{nt} * K_{ij}^{nt}$$

- erwartete Kommunikationsrate bei gleichverteilter Erzeugung:

$$KRd_{ij}^{nt} = 2/(max+1) * K_{ij}^{nt}$$

- konstante Kommunikationsrate bei konstant verteilter Erzeugung:

$$KRc_{ij}^{nt} = 1/const * K_{ij}^{nt}$$

h) Bit-Datenrate

Die normierte quantitative Last pro Zeiteinheit wird für jede Station st_i und jeden Nachrichtentyp nt bezüglich der Kommunikation mit einer Station st_j beschrieben durch die Bit-Datenrate (Bits pro Zeiteinheit, hier pro ms):

Sei L_{nt} wie unter 5 definiert die Längenfunktion, l_{nt} die ermittelte reale Länge.

- Reale Bit-Datenrate

$$BDr_{ij}^{nt} = KRr_{ij}^{nt} * l_{nt} = e_i^{nt} * K_{ij}^{nt} * l_{nt}$$

- Erwartete Bit-Datenrate bei exponentiell oder normalverteilter Erzeugung:

$$BDe_{ij}^{nt} = KRe_{ij}^{nt} * L_i = \left(1/\mu_{nt}\right)K_{ij}^{nt} * L_i$$

- Erwartete Bit-Datenrate bei gleichverteilter Erzeugung:

$$BDd_{ij}^{nt} = KRd_{ij}^{nt} * (L_i - L_{min} + 2)/2 = \frac{L_i - L_{min} + 2}{max + 1} K_{ij}^{nt}$$

- Konstante Bit-Datenrate bei konstant verteilter Erzeugung:

$$BDc_{ij}^{nt} = KRc_{ij}^{nt} * L_i = L_i/const * KR_{ij}^{nt}$$

6. Schlußbemerkungen

Die vorgestellte Arbeit faßt einen Diskussionsstand zusammen. Sicherlich wird man über einige Aspekte noch nachdenken müssen:

- das Lastmodell sollte gegebenenfalls erweitert werden, so daß eine Unterscheidung zwischen Nutzdaten und Protokoll-Rahmendaten zur Bestimmung des effektiven Durchsatzes möglich wird,
- zur Zeit bietet sich die Simulation als die vorrangig einzusetzende Analyse-

methode an. Es wird zu untersuchen sein, inwieweit eine stochastische Analyse möglich ist,

- die Service-Request-Klassen könnten um Attribute zur Beschreibung der Zuverlässigkeitsanforderungen erweitert werden,

- das Lastmodell gewährleistet die gute Gestaltbarkeit einer künstlichen Realzeit-Last (i.S.v. Benchmarks).Es wird nachzuweisen sein, daß damit eine signifikante Verbesserung der Beurteilungsmöglichkeiten lokaler Netze erzielt werden kann, bzw. zu untersuchen sein, wann diese verfeinerten Lastmuster zu signifikant anderen Ergebnissen als bisher führen, (vgl. /IIT84/)

- interessant wird die Frage sein, ob mit dem Lastmodell auch reale Lasten erfaßt werden können und inwieweit der vorgetragene Ansatz von Bedeutung für praktische Probleme sein kann,

- die Angabe der Leistungsmaße steht noch aus. Als geeignet angesehen werden vorläufig die transaktionsbezogene mittlere Datenbitrate und der mittlere effektive Durchsatz relativ zur Kanalkapazität. Weitere Hilfsmasse wären die nachrichtentyp-bezogene Wartezeit auf den Medienzugang bzw. auf Transaktionsbeendung und Masse, die den Overhead eines Protokolls charakterisieren (Kollisionszahl, Nutzdatenrate vs. Overheaddatenrate, Think-Time etc.).

Literatur:

/CHL80/ I.Chlamtac, W.R.Franta : Message Based Priority Access to Local Networks, Comp. Comm., Vol.3, No.2, Apr.1980

/CHR81/ H.Christensen : Requirement Analysis of Industrial Control / Production Management Systems, Ninth Meeting of the International Purdue Workshop on Industrial Computer Systems, 1982

/FÄR84/ G.Färber: Architektur zukünftiger Prozeßrechnersysteme, Prozeßrechner 84, Informatik-Fachberichte, Springer-Verlag 1984

/GRA80/ M.Graube : PROWAY - A Local Network for Process Control, Compcon Spring 1980

/HEG82/ D.Heger : Klassifizierung und Leistungsbewertung von Bus-Systemen, in: Struktur und Betrieb von Rechensystemen, GI/NTG-Fachtagung, NTG-Fachbericht Bd. 80, Ulm 1982

/HEG83/ D.Heger K.Watson : Modelling of load patterns and benchmarks for performance of local area networks, in: Messung, Modellierung und Bewertung von Rechensystemen, 2. GI/NTG-Fachtagung, Informatik Fachberichte, Bd.61, Springer Verlag 1983

/IEE82/ IEEE Project 802 : Local Network Standards, Draft D IEEE Standard 802, Dezember 1982

/ISO82/ ISO/TC97/SC16 : Data Processing Open Systems Interconnection. Basic Reference Model, Draft Proposal, ISO/DP7498, 1982

/IIT84/ IITB (Editor): Final Report for COST11, bis: performance analysis of LANs for real time environments. IITB Karlsruhe, Report No. 9797, 1984

/KIE83/ Kiesel, Kühn, Schroeder, Schwanke : Architecture of the Communication Subsystem for Local Area Networks Operating under a new CSMA-Protocol with Dynamic Priorities, in: Kommunikation in verteilten Systemen, Informatik-Fachberichte, Bd. 67, 1983, Springer Verlag

/NIE84/ I.G.Niemeggers, C.A.Vissers : TWENTENET : a LAN with message based priorities, design and performance considerations, ACM Symposium on Communication Architectures & Protocols, Juni 1984, Montreal, Quebec

/PDV83/ PDV-Berichte : Dezentrale PDV-Systeme, Abschlußdokumentation, Herausgeber : H.Walze, Kernforschungszentrum Karlsruhe, KfK-PDV224 April 1983

/POL84/ M.Polke: Informationshaushalte technischer Prozesse, Prozeßrechner 84, Informatik-Fachberichte Bd. 86, Springer-Verlag 1984

/PRI81/ S.M.Prince, M.S.Sloman : Communication Requirements of a Distributed Computer Control System, Imperial College London 1981

/PRO83/ Proway Document 65a : Part1 : General Description and Functional Requirements, IEC-Group 65, 1983

/SPA82/ O.Spaniol : Konzepte und Bewertungsmethoden für lokale Rechnernetze, Informatik-Spektrum, Heft 5, 1982

/STE83/ F.Stetter: Softwaretechnologie, BI-Wissenschaftsverlag, Reihe Informatik Bd. 33, 2. Auflage 1983

/TAN81/ A.S.Tanenbaum : Network Protocols, Computing Surveys, Vol. 13, No. 4, 1981

/TOB82/ F.A.Tobagi : Carrier Sense Multiple Access with Message Based Priority Functions, IEEE Trans. on. Com., Vol.30, Januar 1982

/WAL83/ H.Walze: Realzeitspezifische Anforderungen und Lösungsmerkmale für Prozeßdatenbusse und lokale Netzwerke, RTP, Jahrgang 25, Heft 6,83

Realisierung
eines ISO/CCITT-orientierten Schicht-5-Dienstes
im BS2000

Eckart Giese
Gesellschaft für Mathematik und Datenverarbeitung
Rheinstr.75,6100 Darmstadt
Dezember 1984

Inhalt

Einleitung

In 1980 hat die technische Beratungskommission der Postgesellschaften (CCITT) mit der Festlegung des TELETEX-Dienstes ein Datentransportprotokoll (T.70) und ein 'session/document'-Protokoll (T.62) empfohlen.

Das Transportprotokoll T.70 ist 1983 in die Spezifikation des ISO-Transportprotokolls als einfachste Klasse (class 0) zur Erbringung des ISO-Standard-Transportdienstes übernommen worden.

Bei den ISO-Aktivitäten zur Standardisierung der Kommunikationsteuerungsschicht (auch Sitzungsschicht bzw. Session-Schicht genannt) wurde das Subset BAS (Basic Activity Subset) definiert, in dem das Protokoll T.62 als kompatible Teilmenge enthalten ist.

Der Stand dieser Normungsaktivitäten und die innerhalb von DFN gewählte Festlegung auf die Protokolle T.70 bzw. T.62/BAS hat bei der GMD dazu geführt, Implementation der Protokolle T.70 und T.62 auf drei verschiedenen Rechnern (Siemens 7541 mit BS2000,IBM 4341 mit VM,PDP11/44 mit UNIX) mit dem Ziel in Angriff zu nehmen, die Dienste der Transportschicht und der Kommunikationssteuerungsschicht an einer Schnittstelle zum Anwendungsprogrammierer bereitzustellen.

Damit soll zum einen **DFN-Kompatibilität** zur Bereitstellung von bereits definierten und vielleicht künftigen DFN-Diensten und zum anderen **TTX-Kompatibilität** zur Bereitstellung von TTX-Dienstleistungen auf Großrechnern und zum Aufbau von auf TTX-basierenden 'mail'-Dienstleistungen (CCITT-Empfehlung X.430) erreicht werden.. Die beiden Schichten ordnen sich wie folgt in das ISO-Referenzmodell ein:

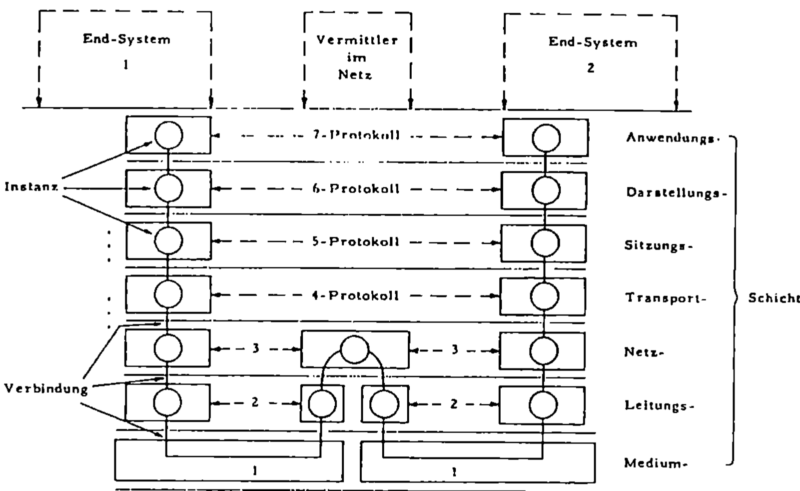

Abb. 1 : Die Schichten des ISO-Referenzmodells

Wie aus Abbildung 1 ersichtlich, benötigt die Kommunikationssteuerungsschicht die Dienste der Transportschicht, um ihre Leistungen erbringen zu können.
Die Transportschicht wiederum benutzt die Dienste der unterlagerten Netzschicht.
Im folgenden wird beschrieben, welche Leistungen auf welche Weise dem Anwendungsprogrammierer an den Programmiererschnittstellen zur Transportschicht und zur Kommunikationssteuerungsschicht angeboten werden.
Skizziert wird dafür eine Implementation der Protokolle T.70 und T.62 auf einem Siemens-Rechner 7541 mit BS2000.
Ein wichtiges Ziel bei der Implementation war dabei, die Kommunikationsdienste an einer Programmiererschnittstelle anzubieten, die bei neuen Betriebssystemversionen und bei Änderungen an der herstellerspezifischen Kommunikationssoftware weitgehend stabil gehalten werden kann.
Die hier vorgestellte Lösung ist funktionell äquivalent auf einem IBM-Rechner 4341 mit dem Betriebssystem VM und -hinsichtlich der Transportschnittstelle- auch im Betriebsystem UNIX (PDP 11/44,PCS) implementiert. In den folgenden Abschnitten wird zunächst kurz auf die prinzipiellen Möglichkeiten bei der Wahl der Dienstschnittstelle und dann auf die tatsächlich implementierten Schnittstellen für die Transportdienste und die Kommunikationssteuerungsdienste eingegangen.

1. Wahl des Schnittstellentyps

Ein wesentlicher Schritt auf dem Wege zu Protokollimplementationen besteht in der Festlegung der Schnittstellen, an denen lie Dienste der Protokollschichten angeboten werden sollen. Bevor eine solche Festlegung erfolgen kann, müssen die Randbedingungen und die Umgebung, in die die Implementation einzubetten ist, sowie die Anwendungsziele auf das Genaueste bekannt sein, um eine den Möglichkeiten und Erfordernissen angepaßte Lösung zu finden.

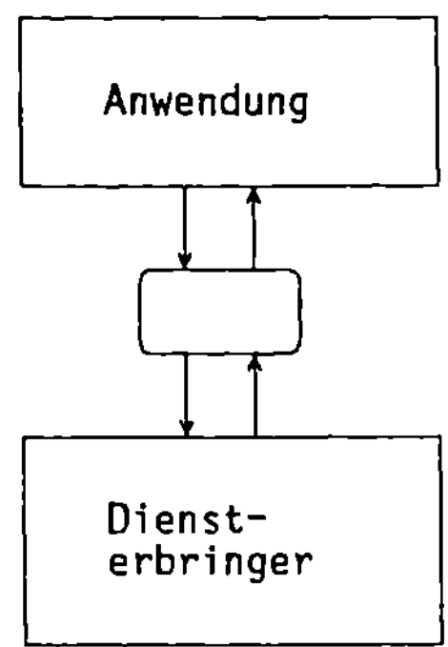

Abb. 2: Schnittstelle Anwendung - Diensterbringer

Hinsichtlich der Gestalt der Schnittstelle sind zwei Grundtypen unterscheidbar:
- Prozeß-Schnittstelle
- Prozedur-Schnittstelle

Bei der Prozeß-Schnittstelle bilden Anwendung und Diensterbringer zwei verschiedene Prozesse, die über eine geeignete Interprozeß-Kommunikationsschnittstelle miteinander kommunizieren.

Unter Prozedur-Schnittstelle wird hier verstanden, daß ein Dienst durch Aufruf einer Prozedur angefordert wird. Die Kontrolle geht dabei an den Diensterbringer und wird von diesem explizit nach Bearbeitung der Dienstanforderung zurückgegeben. Die Arbeitsweise ist also streng sequentialisiert; Anwendung und Diensterbringer sind eng aneinander gekoppelt.

Die Entscheidung für eine der beiden Schnittstellentypen sollte nach sorgfältiger Abwägung der jeweiligen Vor- und Nachteile, insbesondere unter Betrachtung der Problemkreise
- Programmiersprache
- Adressierung
- Flußregelung
- Handhabung
- Performance

erfolgen.

Natürlich müssen auch die Möglichkeiten des Betriebssystems, auf dem ein Diensterbringer implementiert werden soll, und die Nähe des Diensterbringers zur Anwendung (siehe Abb.1) berücksichtigt werden.
So sollte beispielsweise auf Systemen,die nur eine beschränkte Anzahl von Prozessen optimal verwalten können oder deren Interprozeß-Kommunikationsschnittstelle nur schwerfällig zu handhaben ist,die Anzahl der Prozesse minimiert werden.

Generell könnte zur Zweckmäßigkeit des einen oder anderen Schnittstellentyps gesagt werden:

- je 'näher an der Anwendung', desto eher eine Prozedurschnittstelle, da sie einen wesentlich geringeren Komplexitätsgrad hat und damit recht einfach für Anwendungsprogrammierer zu verwenden ist
- je 'tiefer im System/näher an der Leitung', desto eher eine Prozeßschnittstelle, um die Möglichkeiten des Betriebssystems voll ausnutzen zu können.

Für die Einbettung der Transportschicht und der Kommunikationssteuerungsschicht bieten sich folgende Möglichkeiten an:

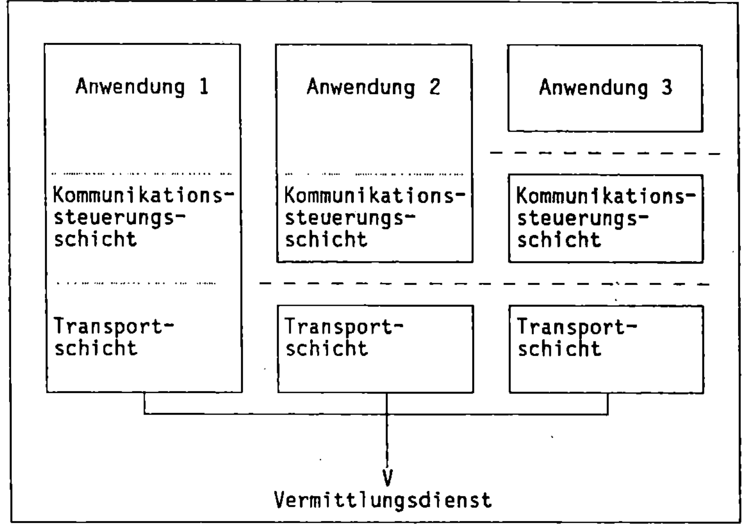

· · · ·   Prozedurschnittstelle
— — — — — Interprozeß-Kommunikationsschnittstelle

Abb. 3 : Schnittstellenvarianten

2. Einbettung der Transportschicht

Die Aufgabe der Transportschicht ist die transparente Übertragung von Daten zwischen zwei Benutzern.
Zur Erfüllung dieser Aufgabe sind in der Protokollbeschreibung des ISO-Transportprotokolls der Klasse 0 die drei Dienste

- CONNECT (Verbindungsaufbau)
- DISCONNECT (Verbindungsabbau)
- DATA (Transport von Daten)

definiert.
Die Transportschicht erbringt dabei ihre Dienstleistungen über 'Dienstprimitive' an der Schnittstelle zur Schicht 5, wobei die zulässigen Reihenfolgen dieser Dienstprimitive, nicht aber deren Implementation genau festgelegt sind.
Für die Implementation ist zwingend, daß sie sich 'nach außen' gemäß den Regeln der Protokollzustandsmaschine verhält.

Folgende **Lösung** wurde gewählt:

Die Dienste der Transportschicht werden dem Anwendungsprogrammierer an einer PASCAL-Prozedurschnittstelle durch den sog. Transport Service Handler (TSH) anwendungsunabhängig angeboten.
Wie in Abb. 4 dargestellt, werden Kommunikationsbeziehungen innerhalb des Rechners durch die beteiligten TSH abgehandelt.
Für die externe Kommunikation übernimmt ein als 'Gateway' bezeichneter eigenständiger Prozeß die Protokollabhandlung gemäß T.70 und die Codierung und Decodierung der Transportprotokollelemente.
Wichtig : Für den Anwendungsprogrammierer ist der unterschiedliche Ablauf nicht sichtbar.
Als Kommunikationsvehikel zwischen TSH-TSH und TSH-Gateway wird die Siemens-Kommunikationssoftware DCAM benutzt.

Bei der Realisierung dieser Schnittstelle wurden folgende **Randbedingungen** eingehalten :

- keinerlei Änderungen am Betriebssystem für die Implementation
- Verwendung des von Siemens bereitgestellten Zugangs zu X.25-Netzen als Netzdienst (Schicht 3)
- Beachtung geplanter Weiterentwicklungen der herstellerspezifischen Kommunikationssoftware durch intensive Kontakte mit Siemens

Die Einbettung kann wie folgt dargestellt werden :

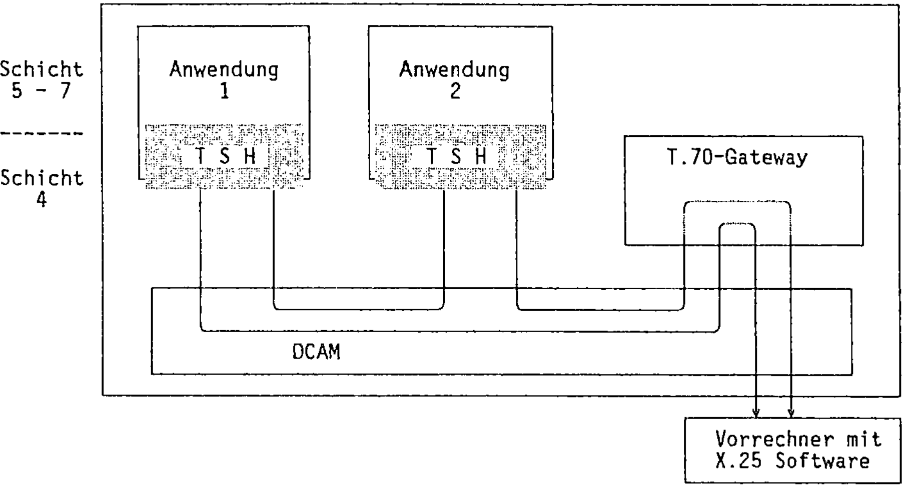

Abb. 4 : Die Einbettung der Transportschicht

Dieser Schnittstellentyp wurde gewählt,

- da eine Prozedurschnittstelle durch den Anwendungsprogrammierer recht einfach benutzbar ist

- da eine derartige Prozedurschnittstelle auch bei Veränderungen der Kommunikationssoftware des Herstellers mit großer Wahrscheinlichkeit stabil gehalten werden kann

- da für die externe Kommunikation vermutlich in absehbarer Zeit - nach Abschluß der laufenden Standardisierungsarbeiten - integrierte Lösungen der T.70-Protokollabhandlung im Front-End-Rechner bereitgestellt werden und ein Übergang auf diese Lösung bei der gewählten Implementation ohne Änderungen an der Schnittstelle zum Anwendungsprogrammierer machbar scheint.

Bemerkungen:

- gemäß den Begriffen des ISO-Architekturmodells beinhaltet das Anwendungsprogramm damit die Instanzen der ISO-Schichten 5-7 und ist also Benutzerinstanz der Transportschicht

- ein auf dem Anwendungsprogramm gestarteter Prozeß ist eine Inkarnation der jeweiligen Benutzerinstanz

- ein Prozeß ist über genau soviele unterschiedliche Transportdienstzugangspunkte erreichbar, wie er unterschiedliche 'Kommunikationsverhalten' zeigt (die wiederum durch die höheren Protokolle beschrieben sind)

Funktionalität der Implementation

Die eben skizzierte Lösung bietet dem Anwendungsprogrammierer folgende Leistungen
an der Dienstschnittstelle :

- Bereitstellung der Transportdienste CONNECT, DATA, DISCONNECT zum Auf- und Abbau von Verbindungen und dem Versand von Daten über mit Parametern versehenen
  Prozeduren
- Adressierung von lokalen und externen Partnern nach den gleichen Konventionen
  an der Prozedurschnittstelle
- Vereinbarung eines oder mehrerer Transportdienstzugangspunkte (TSAP = Transport Service Access Point) durch einen Anwendungsprozeß
- parallele Verbindungen innerhalb eines TSAP
- lokale, verbindungsspezifische Flußkontrolle
- Zeitüberwachung von Verbindungen/Diensten
- 'eventing'-Mechanismen zur Verarbeitung asynchron eintreffender Ereignisse
- Übertragung 'beliebig' großer Datenblöcke

Im weiteren Verlauf dieses Kapitels wird nun beschrieben, wie diese Funktionalität
durch die Implementation realisiert wird.

Realisierung der **Dienst-Prozeduren**

Der Aufbau der Dienstprozeduren wird am Beispiel des CONNECT-Dienstes aufgezeigt :

Der CONNECT-Dienst wird mit Hilfe der Prozeduren TSCONREQ/TSCONACC realisiert:

(1)   TSCONREQ (conid,partner) : Aufbauwunsch für eine Verbindung
(2)   TSCONACC (conid)   : Akzeptieren eines beliebigen
                                   Verbindungsaufbauwunsches

Erläuterungen:
(1)   Die Adresse des Partners (partnername,prozessorname) wird als Parameter
       übergeben; 'conid' entspricht einer eindeutigen Verbindungskennung.
(2)   Die Bereitschaft zur Annahme eines Verbindungsaufbauwunsches eines beliebi-
       gen Partners wird erklärt.

Realisierung der **Definition von TSAP's :**

Vor dem ersten Verbindungsaufbau zu einem Partner muß die Anwendung einen TSAP
bei der Transportschicht anmelden (Prozedur TSOPEN). Analog dazu ist TSOPEN
auch aufzurufen, um für einen externen Partner adressierbar zu sein.

Realisierung der **Flußregelung**

Zur Flußregelung wurde ein Flußregelungsmechanismus – ein Quittungsspiel mit bestimmter Fenstergröße – eingerichtet.

Wenn das Fenster ausgeschöpft ist, werden Sendewünsche der Anwendung abgelehnt (Meldung : Stau aufgetreten).

Erst nach Eintreffen und Verarbeitung des Stauauflösungsereignisses darf die Anwendung auf dieser Verbindung wieder Daten senden.

Dieser Mechanismus wirkt also verbindungsspezifisch, so daß der Stau auf einer Verbindung nicht auch andere Verbindungen beeinflußt.

Die empfangsseitige Flußkontrolle ist analog dazu geregelt.

Realisierung der Behandlung von **parallelen Verbindungen** :

Um mehrere parallele Verbindungen zu einer Zeit weitgehend unabhängig voneinander bedienen zu können, ist ein Mechanismus zur Verarbeitung asynchron eintreffender Ereignisse installiert, der wie folgt arbeitet.

- der TSH startet bei seiner Initialisierung Systemdienstleistungen, mit denen asynchron eintreffende Ereignisse über Programm-Interrupts empfangen werden können (sog. 'contingencies')
- das Eintreffen eines asynchronen Ereignisses startet eine derartige 'contingency', die ihrerseits wiederum einen Eintrag in eine prozeßspezifische Warteschlange macht und dann die Kontrolle wieder abgibt; solche asynchronen Ereignisse können sein
    - Eintreffen von Verbindungsaufbauwunsch oder von Daten
    - Abbruch der Verbindung, Timeout
- mit dem Aufruf einer WAIT-Routine wartet die Anwendung auf eingehende Ereignisse und erfährt, ob und wenn auf welcher Verbindung ein Ereignis eingetroffen ist
- die Anwendung gibt den Auftrag zur Verarbeitung an den TSH

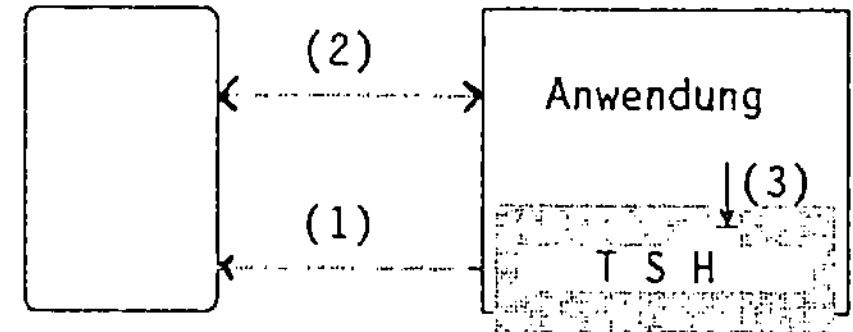

Abb. 5 : Verarbeitung asynchroner Ereignisse

(1) Empfangen eines asynchronen Ereignisses und Eintragen in eine Warteschlange
(2) Warten auf eingehende Ereignisse mit WAIT-Routine
(3) Wenn Ereignis angezeigt,dann Auftrag an TSH zur Verarbeitung dieses Ereignisses

3. Einbettung der Kommunikationssteuerungsschicht

Die wesentlichen Aufgaben der Kommunikationssteuerungsschicht sind
- Etablierung von Kommunikationsbeziehungen
- Steuerung,Sicherung und Wiederanlauf der Anwendungskommunikation

Zur Erfüllung dieser Aufgaben enthält die Protokollbeschreibung der Kommunikationsteuerungsschicht zahlreiche Dienste mit ihren Dienstprimitiven und deren Parametern, die hier nicht alle dargestellt werden können. Ein Teil der innerhalb von T.62 möglichen Dienste und deren Zusammenhang sind im folgende Bild unter Verwendung der ISO-Terminologie dargestellt.

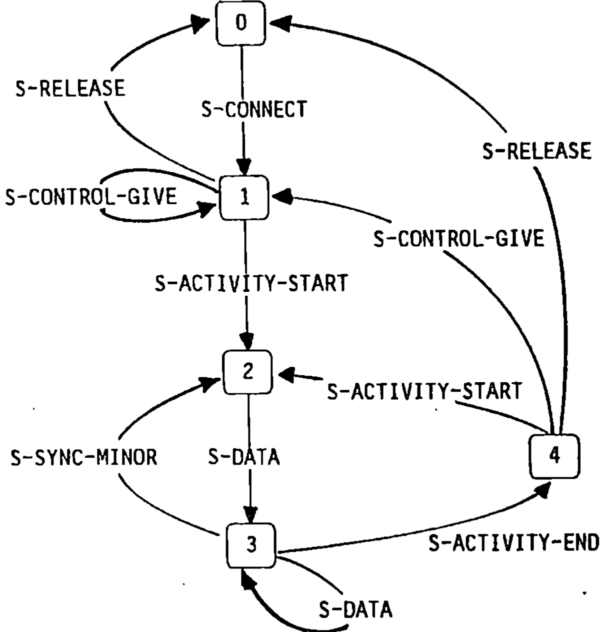

Abb. 6 : Dienste der Kommunikationsteuerungsschicht (Auszug)

Daneben sind natürlich u.a. noch Fehlerdienste wie S-USERABORT und S-EXCEPTION-REPORT sowie Synchronisationsdienste wie S-ACTIVITY-INTERRUPT/DISCARD und S-ACTIVITY-RESUME definiert.

Für die Implementation der Kommunikationssteuerungsschicht wurde folgende Lösung gewählt:

Die Dienste der Kommunikationssteuerungsschicht werden an einer **PASCAL-Prozedur-Schnittstelle** angeboten.

Die Implementation beinhaltet im einzelnen
- die eigentliche Prozedur-Dienstschnittstelle, die für jeden Dienst eine Prozedur mit den zugehörigen Parametern anbietet

- einen Protokollteil, der von allen Dienstprozeduren aufgerufen wird
    - in dem die Zulässigkeit des angeforderten Dienstes geprüft wird und
    - in dem die Protokollelemente erzeugt bzw. entschlüsselt werden
- einen Teil zur Bedienung der Transportschnittstelle (siehe Kap.2)

Dieser Schnittstellentyp wurde gewählt, da
- der Informationsaustausch zwischen Anwendung und Session direkt erfolgen kann
- eine Unterstützung der Anwendung bei Dienstabwicklungen und Datensicherungs-
  maßnahmen einfach realisiert werden kann und
- die Schnittstelle für den Anwendungsprogrammierer leicht bedienbar ist

Funktionalität der Implementation

Die gewählte Lösung bietet dem Anwendungsprogrammierer folgende Leistungen an der
Schnittstelle :

- Bereitstellung aller 'session'-Dienste gemäß vorheriger Beschreibung
- parallele Verbindungen
- Vereinbarung unterschiedlicher Transportdienstzugangspunkte für unterschiedliches Kommunikationsverhalten
- Zeitüberwachung von Verbindungen/Diensten
- lokale, verbindungspezifische Flußkontrolle
- Bereitstellung von unterschiedlich komplexen Verarbeitungsmodi für Dienste,
  die erst nach Bestätigung durch den Partner abgeschlossen sind
- Unterstützung der Anwendung bei nach Fehlern für den Wiederanlauf einzuleiten-
  den Synchronisations- und Sicherungsmaßnahmen

Im weiteren Verlauf dieses Kapitels wird nun beschrieben, wie diese Funktionalität
durch die Implementation realisiert wird.

Realisierung der **Dienst-Prozeduren**

Der Aufbau der Dienstprozeduren wird am Beispiel der CONNECT-Prozedur aufgezeigt :


```
CONNECT (partadresse, capabilities, conid, rc)
partadresse   : Adresse des Partners, entsprechend den BS2000-Konventionen als
                symbolischen Namen
capabilities  : Eigenschaften der Sitzung, die beim Verbindungsaufbau verhandelt
                werden müssen (z.B. Fenstergröße)
conid         : eindeutige logische Kennung dieser Verbindung
rc            : Returncode
```

Realisierung der **Verarbeitungsmodi**

1. SYNCHRONER MODUS :

Der synchrone Modus bietet dem Anwendungsprogrammierer eine vereinfachte Handhabung der Dienste der Kommunikationssteuerungsschicht.

Dieser Modus wird in der Regel immer dann gewählt, wenn nur eine Verbindung zu einer Zeit unterhalten werden soll.

Fordert die Anwendung einen bestätigten Dienst an, so erhält sie die Kontrolle erst dann zurück, wenn der Dienst vollständig ausgeführt wurde oder eine Ausnahmesituation eingetreten ist (z.B. Resynchronisation,Abbruch).

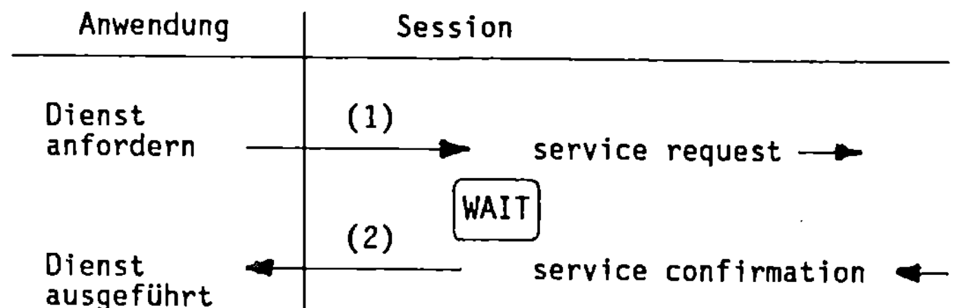

Abb. 7 : synchroner Verarbeitungsmodus

(1) Dienstanforderung prüfen,Protokollelement erzeugen und verschicken
(2) Protokollelement decodieren, prüfen ggf. ordnungsgemäße Ausführung des Dienstes anzeigen

2. ASYNCHRONER MODUS

Im asynchronen Verarbeitungsmodus werden die durch den TSH gebotenen Leistungen zur Verarbeitung asynchroner Ereignisse an die Anwendung hochgereicht, wodurch allerdings die Programmierung komlexer wird.

Der asynchrone Modus ist immer dann zu verwenden, wenn mehrere parallele Verbindungen weitgehend unabhängig voneinander bedient werden sollen.

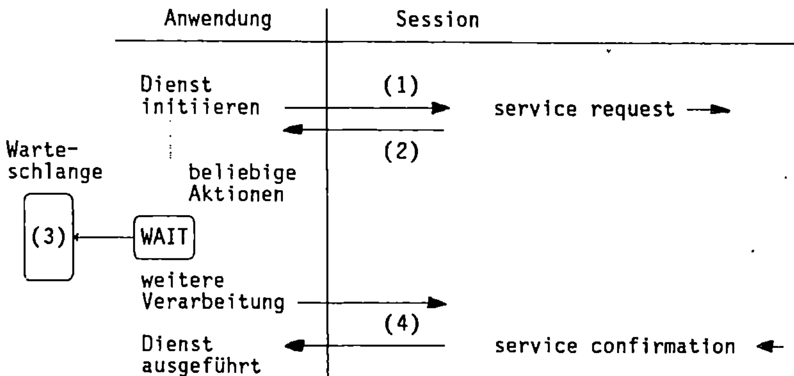

Abb. 8 : asynchroner Verarbeitungsmodus

(1) wie bei Abb. 7
(2) nach dem Versenden sofortige Rückgabe der Kontrolle
(3) Anzeige von asynchronen Ereignissen durch den TSH
(4) wie (2) bei Abb. 8

Realisierung der **Wiederanlaufunterstützung** von Anwendungen

Die zur Wiederherstellung der Anwendung nach Fehlern erforderlichen Synchronisations- und Sicherungsmaßnahmen sind in die Systemleistung unterhalb der Anwendungsprogrammierer-Schnittstelle gezogen.
Bei der Entwicklung der Anwendung legt der Anwendungsprogrammierer die entsprechenden Maßnahmen fest und stellt sie in sog. EXIT-Routinen bereit.
Die Kommunikationssteuerungsschicht ruft dann zu fest definierten, für vorbereitende Sicherungsmaßnahmen relevanten Zeitpunkten während des Ablaufs einer Sitzung, z.B. bei

- Eröffnen einer 'activity'
- Schließen einer 'activity'
- Bestätigung eines Synchronisationspunktes
- Abbruch der Sitzung

diese EXIT-Routinen auf, übergibt ihnen die erforderliche Zustandsinformation über die aktuelle Sitzung und ermöglicht damit die vorher definierten zustandsabhängigen Aktionen der Routinen.

4. Stand/Verfügbarkeit

TRANSPORTSCHICHT :

T.70-Gateway und TSH sind in 1983 implementiert und 1984 getestet worden.
Der T.70-Gateway ist als permanenter TP-Prozeß mit Restart-Funktion installiert und wird bei Systemstart automatisch 'hochgefahren'.
Installationsvoraussetzungen sind die Bereitstellung der Siemens X.25-Software und der DCAM-Versionen 6 oder 7.
Bemerkung : Bei Bereitstellung des Siemens-X.21-Anschlusses nach T.70-Konventionen ist die Adressierung von Partnern im DATEX-L-Netz (z.B. TTX-Gerät an DATEX-L) direkt möglich.

KOMMUNIKATIONSSTEUERUNGSSCHICHT :

Die Implementation wurde im ersten Quartal 84 abgeschlossen.
Umfangreiche Tests mit verschiedenen externen Partnern wurden anschließend erfolgreich durchgeführt, sodaß die Software seit dem 3.Quartal 84 in einer Bibliothek zur Verfügung steht.
Installationsvoraussetzung ist die Implementation der Transportschicht.

Literatur

/1/ CCITT - Telegraph and Telematic Services Terminal Equipment
Recommendations of the S and T Series
Geneva 1981

/2/ GILT-Session-Description
COST-11-bis,March 82

/3/ Recommendation for the GILT-Session-Service-Interface
Arbeitspapiere der GMD Nr.31

/4/ Benutzungsanleitung der Ebene 4 Prozedurschnittstelle in PASCAL-Programmen
Arbeitspapiere der GMD Nr.90

/5/ DCAM Usage Conventions to achieve a Transport Service
GMD-DA-F2G3

/6/ ISO-Schriften: ISO/TC 97/SC 16
Information Processing Systems
Open System Interconnection
ISO 8072 : Transport Service Definition
ISO 8073 : Connection Oriented Transport Protocol Specification
ISO 8326 : Basic Connection Oriented Session Service Definition
ISO 8327 : Basic Connection Oriented Session Protocol Specification

"Einführung eines Portkonzeptes im BS2000 als Basis zur Konstruktion eines heterogenen verteilten Systems"

H. Grund, F.A. Jirka

GMD — Gesellschaft für Mathematik und Datenverarbeitung mbH Bonn

0. Abstract

Unter der Zielsetzung, so verschiedene Betriebssysteme wie etwa
UNIX* und BS2000 über ein LAN so zu verbinden, daß eine evolutive
Weiterentwicklung der Einzelsysteme zu einem verteilten System möglich erscheint, werden Portkonzepte als Architekturhilfe vorgestellt und kurz diskutiert. Ausgehend von der gegenwärtigen Situation im BS2000 wird die schrittweise Einführung einer solchen Konzeption in dieses Betriebssystem dargestellt und aufgezeigt, wie
eine kompatible Integration erreicht wird, so daß bereits im
tasklokalen Bereich erhebliche Systemverbesserungen zu verzeichnen
sind. Durch Einführung von speziellen Kommunikationsobjekten in das
BS2000 innerhalb dieser Konzeption wird auf der Basis eines
Portkonzeptes eine effiziente, auch asynchrone, Interprozesskommunikation erzielt, die als Grundlage einer LAN-weiten Kommunikation dienen kann, die auf dem Prinzip des "message passing" aufgebaut ist.

Stichworte:

Local Area Net, Portkonzept, verteilte Systeme, Message passing,
Interprozesskommunikation

1. Überblick

In Abschnitt 2 dieses Berichtes wird gezeigt, daß es sinnvoll ist, ein verteiltes System aufzubauen, das aus unterschiedlichen Einzelsystemen zusammengesetzt ist. Im nächsten Abschnitt 3 wird das Portkonzept eingeführt und dargestellt, wie mit Hilfe einer solchen Konzeption in bestehende Systeme Schichten eingezogen werden können, die eine Entwicklung in Richtung verteiltes System eröffnen.
In den Abschnitten 4 und 5 wird die Realisierung dieser Konzeption im BS2000 besprochen, wobei zunächst nur die tasklokale Situation erläutert wird. Im nächsten Abschnitt wird auf die Nachrichtenbehandlung durch einen Makrointerpreter eingegangen. In Abschnitt 7 wird dargestellt, wie die Aufwärtskompatibilität gesichert wird. Der Ausbau zur Interprozesskommunikation wird in Abschnitt 8 erläutert, während der Übergang in die netzweite Kommunikation in Abschnitt 9 behandelt wird.

2. Zielsetzung: Verteilte Systeme

In der GMD sind für Forschung und Entwicklung eine Vielfalt von Betriebssystemen – MVS, BS2000, VMS , UNIX* und dessen Derivate – sowie verschiedene Spezialsysteme, wie z.B. LISP-Maschinen oder PERQ Rechner, im Einsatz. Es gibt daher ein natürliches Interesse, diese Systeme über ein Netzwerk miteinander zu verbinden, um sowohl ein Maximum an Diensten netzweit zur Verfügung zu stellen, als auch die Flexibilität und Dynamik der kleineren Systeme mit den gestandenen Anwendungen und Serviceleistungen der Großsysteme zu koppeln. Zur Illustration seien hier zwei Beispiele genannt:

- Nutzung von "Rasterdisplays", wie sie etwa im Bereich der "Personal Computer" über "Window Manager" zugänglich sind, auch von den Großsystemen wie z.B. BS2000, in welchem sowohl der Anschluß als auch die Unterstützung von Fremdhardware große Schwierigkeiten bereitet.

- Nutzung der (Laser-) Druckerkapazität, die z.Zt. über die BS2000 Systeme der GMD erreichbar ist, auch durch die kleineren Systeme

Bei der Konstruktion solcher Schnittstellen in vorhandenen Betriebssystemumgebungen sollten folgende Designziele beachtet werden:

- Schnittstellen bestehender Anwendungen (Programme) sollen ohne Änderung der Programme in Konzepte des Verbundsystems überführbar sein, wobei Performance-Verluste tragbar sind.

- Auf den neuen Schnittstellen des Verbundkonzepts beruhende Anwendungen sollen bezüglich der Performance verglichen mit den alten Schnittstellen nicht wesentlich schwächer sein, aber im Anwendungsspektrum deutliche Vorteile bieten.

Diese Zielsetzung ist von der Hoffnung geprägt, unterschiedliche Betriebsysteme evolutiv in eine Richtung bringen zu können, die den homogenen Fall bevorzugt und es im inhomogen Fall gestattet, die Vorzüge des anderen Partners zu nutzen.

3. Das Portkonzept

Im Sinne dieser Systemarchitektur sind vor allem die Beziehungen eines Programms zu dem Betriebssystem, auf dem es abläuft, von Interesse. Unter diesen wiederum sind diejenigen Schnittstellen von größter Wichtigkeit, über welche die Ein- und Ausgaben gesteuert werden.
In modernen Systemkonzepten (→ [RAS83]) gerade für verteilte Systeme wird in diesem Zusammenhang häufig auf den PORT-Gedanken zurückgegriffen, der ja schon seit längerem (→ [BAL71]) diskutiert wird und in aktuellen Systemen in gewissen Ausprägungen, also nicht als allgemeines Kommunikationskonzept, enthalten ist.
Als Beispiel seien hier Standard-IN und Standard-OUT im UNIX* erwähnt, die als Realisierung einfacher Ports aufgefaßt werden können, wobei über das Konstrukt der "PIPE" der Standard-OUT eines Programms mit dem Standard-IN eines weiteren nachbearbeitenden Programms verbunden werden kann.

Aus den Anwendungsbeispielen ist ersichtlich, daß innerhalb dieser Zielsetzung sehr hohe Übertragungsraten erforderlich werden, wie sie heute nur im Bereich der "Local Area Networks" (LAN) mit üblicherweise ca. 10 Millionen Bit/sek. (1,25 MB/sec) erreicht werden, der Weg über vorhandene "Wide Area Networks" (WAN) mit üblichen 9600 Bit/sec. sicher unbefriedigend bleiben muß. Da für alle genannten kleineren Systeme, vor allem im wichtigen UNIX* Bereich, mit ETHERNET* ein geeignetes LAN-Produkt (ISO-Level 1 bis 2) verfügbar ist, gilt es innerhalb eines Verbundkonzepts auch für die Großsysteme ETHERNET-Anschlüsse zu realisieren.
Hierbei ist zu beachten, daß wegen der erforderlichen hohen Übertragungsraten i. d. R. nur direkte Kanalanschlüsse sinnvoll sind, da die Durchsatzraten der z. Zt. verwendeten Front-end Prozessoren (z.B. TRANSDATA 9687 maximal 20 KB/sec.) um Größenordnungen zu gering sind.

Es ist selbstverständlich, daß die auf ETHERNET* aufsetzenden Schichtungen (ISO-Level 2, 3 bis 4) sicher sein müssen und Informationen nicht verloren gehen können. Die hier üblichen Protokolle (z.B. TCP/IP, XNS, CHAOS) sind nach dem "Server/Client" Prinzip konstruiert, und damit speziell den LAN Erfordernissen angepaßt und stehen im Vorfeld der Normungsanstrengungen, die zunächst mehr auf den WAN-Bereich ausgerichtet sind. In diesem Sinne konstruierte LANs verhalten sich zum WAN hin wie ein System, sind also unter der hier diskutierten Zielsetzung als eine Einheit aufzufassen.
Die an einem LAN angeschlossenen Systeme bilden insgesamt ein verteiltes System, in welchem die Dienstleistungen nicht mehr zentral von einem Rechner erbracht werden, sondern dezentral von einer Reihe von "Servern" (z.B.: Druck-Server, Datenbank-Server, File-Server, ...). Vorhandene Einzelsysteme müssen evolutiv in die neue Architektur eingepaßt werden. Hierbei sind Schnittstellen zu schaffen, die im Falle gleichartiger Systeme eine stark betriebssystemkonforme Anpassung bis hin zur völligen Transparenz gewährleisten, als Beispiele seien hier für UNIX* – UNIX* die "Newcastle Connection" und für BS2000 – BS2000 der MRS-Verbund genannt. Im inhomogenen Fall aber müssen sie mindestens die Einrichtung von Servern mit interessanten Dienstleistungen erlauben, wobei Server für "Remote Login" und "Filetransfer" selbstverständlich sind.

Der wesentliche Vorteil eines PORT-Konzeptes liegt in der Verwendung eines sehr allgemeinen, also unspezifischen, abstrakten Betriebsmittels, "PORT" genannt, das wegen seiner Unabhängigkeit von physikalischen Randbedingungen mit einer einfachen Zugriffsmethode von Programmen aus zur Kommunikation mit der Programm-Außenwelt verwendet werden kann. Hierbei bleiben die spezifischen Ausprägungen dieser Außenwelt dem Programm verborgen.

Es ist klar, daß dann zur Zeit der Programmausführung eine **Bindung** der verwendeten PORTS an konkrete Objekte der Betriebssystemumgebung erfolgen muß. Es ist wichtig zu sehen, daß diese Bindung dem Prinzip der "latest Binding time" folgend, i.d.R. außerhalb des Programmes vorgenommen wird, wodurch in Abhängigkeit vom Bindungsvorgang und der bindungsfähigen Objekte des Betriebssystems beliebiges "Routing" der Daten auch netzweit möglich wird, sofern entsprechende Objekte zur Verfügung stehen. Das nach der Bindung entstande Tupel (PORT,Objekt) wird KANAL genannt und stellt ein neues Objekt des Systems dar, das durch den Bindevorgang entstanden ist. Erweitert man diese Konzeption auf Mehrfachbindungen von PORTs, indem man Tupel der Form

$$(PORT, Objekt_1, Objekt_2, \ldots, Objekt_n)$$

zuläßt, so verliert man nicht an Allgemeinheit, wenn man sich darauf beschränkt, daß bereits gebundene PORTs nicht noch einmal gebunden werden können. Unter dieser Beschränkung gibt es eine 1-1 Zuordnung von gebundenen PORTs zu KANÄLEN, so daß die KANÄLE über die jeweiligen PORTnamen identifiziert werden können.

Eine Reihe von interessanten Fragestellungen ergibt sich zum einen aus der Klassifikation der bindefähigen Objekte und zum anderen aus der Vollständigkeit der Bindung, d.h. welche Eigenschaften unter der Bindung manipulierbar bleiben können oder festgeschrieben werden müssen. Diese Fragen können in dieser Arbeit nicht umfassend dargestellt, sondern nur in einzelnen Aspekten kurz angedeutet, bzw. für das Betriebssystem BS2000 etwas näher erläutert werden.

Bindefähige Objekte sind z.B. häufig Files, Standard Ein- und Ausgabegeräte und von ganz besonderem Interesse PORTs und KANÄLE, wobei sich unterschiedliche Konsequenzen für die Zugriffsmethode ergeben, ob PORTs oder KANÄLE zugelassen sind. Im Falle eines KANALs können die Zugriffsfunktionen durch Abbildungen auf die Zugriffsfunktionen und Eigenschaften des gebundenen Objekts rekursiv definiert werden, während im anderen Fall (PORT,PORT) einerseits eine

Vielzahl von Fragen zu klären ist, andererseits aber auch eine gewisse Gestaltungsfreiheit gegeben ist (→ [SIL81]).

Werden **alle** Außenbeziehungen eines Programms über Ports realisiert, so ergibt sich zwischen Betriebssystem und Anwenderprogramm eine abstrakte Zwischenschicht, in welcher einerseits Abbildungen auf konkrete Objekte des Betriebssystems, wie z.B. Files, vollzogen werden können, andererseits aber auch beliebiges Steuern und Umlenken der aktuellen Nachrichtenströme möglich ist, wodurch im Prinzip eine gewisse Betriebssystemunabhängigkeit erzielt wird.
Hierbei werden die betroffenen Objekte des Betriebsystems durch Ports eingeschalt, wobei von ihrer Funktionalität nur soviel nach "außen" dringt, wie an der Portschnittstelle tatsächlich benötigt wird. Im Rahmen einer solchen Modellbildung wird die Komplexität von Programmen reduziert, da die verschiedenen Zugriffsmethoden zugunsten **einer** einheitlichen und einfachen Zugriffsmethode zu Ports ersetzt werden können.

4. Realisierung eines Portkonzeptes im BS2000

Im folgenden soll am Beispiel des BS2000 aufgezeigt werden, wie diese Vorstellungen in ein aktuelles Betriebssystem aufwärtskompatibel integriert werden können. Gerade im BS2000 findet man im Bereich der sequentiellen Ein- und Ausgaben, die zunächst einmal im Vordergrund stehen sollen, eine sehr große Vielfalt von Quellen und Senken, wobei sich die Zugriffsmethoden (Aktionsmakros) stark unterscheiden (z.B: im "Data Management" PUT/GET bei ISAM,SAM; PAM bei PAM; EAM bei EAM — im "Sysfile Management" WRLST bei SYSLST; $EOUT/WROUT bei SYSOUT, $READ bei SYSCMD, RDATA bei SYSDTA — innerhalb "Data Communication" SEND/RECEIVE bei DCAM ; SEVNT/REVNT bei IPK). Auf dem Umweg über Files sind zwar im Bereich des "Sysfile Management" bescheidene Routing-Möglichkeiten gegeben, die jedoch unsystematisch und schwerfällig sind.

Auf Grund der besonderen Situation im BS2000 — in einer TASK kann nur **ein** Anwenderprogramm ablaufen — und aus Implementationsgründen haben wir die beiden Funktionen der PORT-Definition und -Bindung in einer Funktion zusammengefaßt, die dann im Sinne obiger Ausführungen einen KANAL definiert. Diese Funktion wird sowohl in der Kommandosprache über das Kommando /CHANNEL als auch an der System-schnittstelle (SVC) über einen entsprechenden Aktionsmakro zur Verfügung gestellt. Kanäle werden über ihren Namen, der innerhalb einer Task eindeutig ist, in allen Funktionen identifiziert. Jeder Kanal enthält Verweise auf die Objekte des BS2000, die von ihm umhüllt werden. Zugelassene Objekte sind z.Zt. Umhüllungen von SYSOUT, SYSIN, SYSLST (alle Sysfile Management), SAM-Files (demnächst auch ISAM- und PAM-Files), Synchronem DCAM und einem Nachrichten-vernichter, der in dieser Form auch ein neues Objekt im BS2000 darstellt (vergl. Filename *DUMMY). Von besonderer Bedeutung ist (s.o.), daß ein KANAL auch als bindefähiges Objekt des BS2000 aufgefaßt wird. Dieses ermöglicht weitergehende Anwendungen, die über die existierenden Funktionen hinausgehen.

Diese Umhüllungen sind z.Zt. nur insoweit ausgeprägt, als sie für die Bewältigung sequentieller Nachrichtenströme notwendig sind. An einen PORT kann aus technischen Gründen (einfachere Implementation) jeweils nur ein Objekt gebunden werden. Eine Ausnahme bilden hier die Kanäle selbst, da hierbei zur flexiblen Steuerung von Nachrichtenströmen mehrere Kanäle an einen PORT gebunden werden können, wobei Rekursionen in der Bindungsstruktur verboten sind ("multicast" Struktur). Die Grundidee ist hierbei die rekursive Verwendung des Objekts "KANAL", bei der ein Objekt "KANAL" durch einen anderen KANAL vollständig eingeschalt werden kann. Auf diese Weise erhält man ein Netz von Kanälen, das einen rekursionsfreien Graphen realisiert, wobei die Knoten durch PORTs, die weitere Kanäle binden, und die Blätter durch PORTS, die andere Objekte des BS2000 binden, repräsentiert werden.

Zur Manipulation solcher Netze stehen die kanal- und netzbezogenen Operationen CREATE, CONNECT, LIST, MODIFY, DISCONNECT und DELETE sowie sinnvolle Kombinationen derselben zur Verfügung. CONNECT aktiviert ein Subnetz vollständig, es wird also nur dann erfolgreich ausgeführt, wenn CONNECT für alle Kanäle des involvierten Subnetzes durchgeführt werden kann. Auf dieser rekursiven Eigenschaft der Kanalbindungen beruht die Möglichkeit, Nachrichtenströme in beliebiger Weise aufteilen oder zusammenführen zu können.

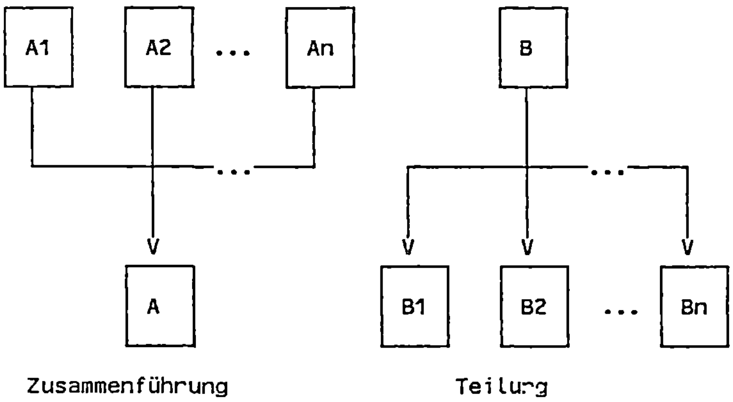

Fig. 1 Manipulation von Nachrichtenströmen

5. Transportfunktionen

Zum Transport der Nachrichten stehen die Operationen WRITE, READ
und TRANSPORT (Kommandos wie auch Aktionsmakros) zur Verfügung, wobei jeweils eine Nachricht beliebiger Länge behandelt wird. WRITE
in ein Subnetz wird so interpretiert, daß die Nachricht ausgehend
von einem Kanal durch alle Kanäle transportiert wird und in Kopien
an den Blättern ausgegeben wird. Analog hierzu gilt für READ, daß
an allen Blättern Nachrichten eingelesen werden und im Sinne der
Definition des Subnetzes konkateniert werden; es wird also an dem
Kanal, von dem gelesen wird, nur eine Nachricht ausgegeben. Zur Lösung des oft angesprochenen Problems (s.a. die Konstrukte BUNCH und
WHICH in (→ [BOR83])), eine Nachricht auf einem von vielen Kanälen
zu selektieren, sobald sie eintrifft ("multiple wait problem") sei
auf die SHELFs verwiesen, die in Abschnitt 8 vorgestellt werden.
Als Beispiel sei hier folgende Kanalreferenzstruktur gegeben: Der
Kanal A verweist auf die Kanäle B und C. Alle Kanäle haben einen
entsprechenden Pre- und Postfix (s.a. Nachrichtenbehandlung) und
können sowohl gelesen als auch beschrieben werden.

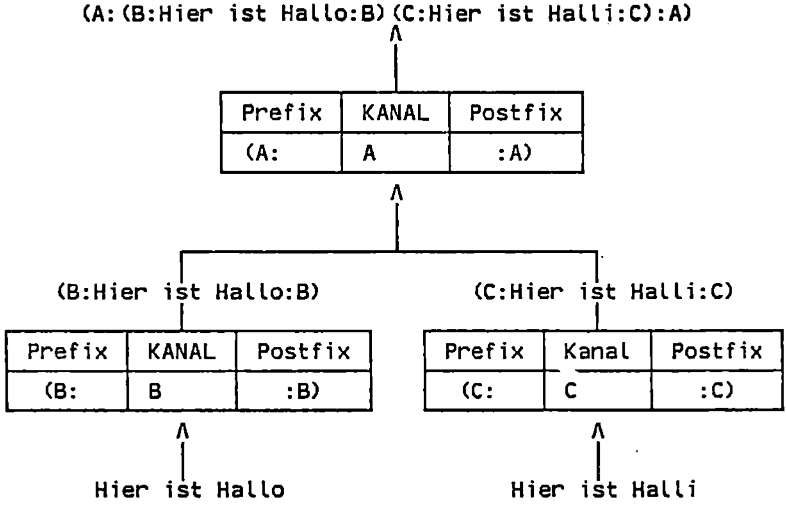

<u>Fig. 2</u> Wirkung von READ

Wird von A gelesen, so erhält man die Nachricht "(A:(B:Hier ist Hallo:B)(C:Hier ist Halli:C):A" wenn an Kanal B die Nachricht "Hier ist Hallo" und an Kanal C die Nachricht "Hier ist Halli" eingegeben wurden.

Die Verträglichkeit von READ und WRITE über das Subnetz hinweg wird durch die Operation CONNECT garantiert, weshalb diese Transportfunktionen nur dann ausgeführt werden können, wenn die Funktionsfähigkeit des Subnetzes gesichert ist. Die Operation TRANSPORT wird durch ggf. wiederholte Hintereinanderschaltung der Operationen "READ von einem PORT" und "WRITE auf einen anderen PORT" realisiert, wobei die explizite Angabe von Datenbehältern vermieden werden kann, also Nachrichten von einem Subnetz auf einfache Weise in ein anderes transportiert werden können.

Zur Erzielung einer höheren Flexibilität kann ein Subnetz dynamisch konfigurierbar werden, indem man den Referenzen eines Kanals auf andere Kanäle jeweils einen beliebigen Selektor zuordnet und den Selektstatus des Kanals entsprechend setzt. Während einer Transportoperation werden dann genau die Verbindungen ausgewählt, deren Selektor mit dem Selektstatus übereinstimmt. Der Selektstatus kann explizit durch die MODIFY Funktion oder implizit durch die Transportfunktionen selbst oder durch einen Nachrichteninterpreter (s. Nachrichtenbehandlung) geändert werden. Sowohl der Selektstatus als auch die jeweiligen Selektoren sind stets lokal in Bezug auf **einen einzigen** Kanal definiert und bilden daher zusammen den Selektionsmechanismus eines Kanals. Ist eine Selektion während einer Transportoperation an einem Kanal nach seinem Selektstatus nicht erfüllbar, so wird der Selektstatus dieses Kanals auf den Selektstatus des übergeordneten Kanals übertragen. Gleichzeitig wird die Transportoperation am untergeordneten Kanal abgebrochen und am übergeordneten Kanal wieder aufgenommen. Existiert ein übergeordneter Kanal nicht, so wird die Operation insgesamt mit Fehler abgebrochen.

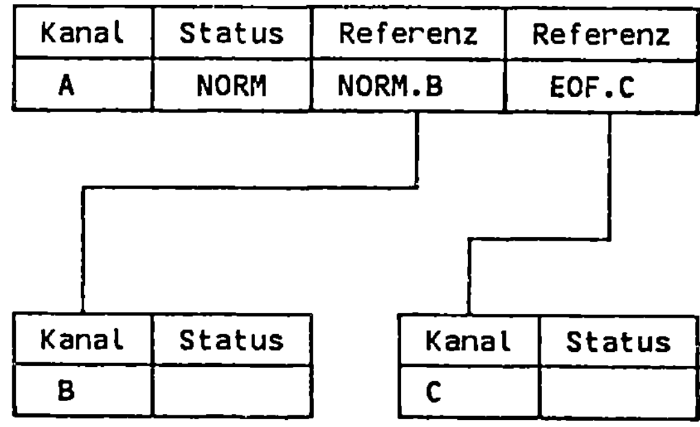

<u>Fig. 3</u> Es wird nur Kanal B ausgewählt

In diesem Beispiel wird der Selektstatus von Kanal B auf 'EOF' gesetzt, wenn bei einer READ Operation eine Bedingung auftritt, die mit 'End-of-File' vergleichbar ist. Da an Kanal B keine Referenzen mit Selektor 'EOF' definiert sind, ist die READ Operation an diesem Kanal nicht mehr erfüllbar. Deshalb wird der Selektstatus in den übergeordneten Kanal übertragen und der Selektstatus von Kanal A auf 'EOF' geändert. Weil an Kanal A jedoch eine Referenz mit dem Selektor 'EOF' definiert ist, wird die READ Operation am Kanal A wiederaufgenommen und damit über den Kanal C weiter durchgeführt. Durch diese flexiblen Bindungsstrukturen kann eine Konkatenation von endlichen Nachrichtenströmen erreicht werden, im Sonderfall der umhüllten Files also die bisher im BS2000 nicht vorhandene Konkatenation von Files durchgeführt werden.

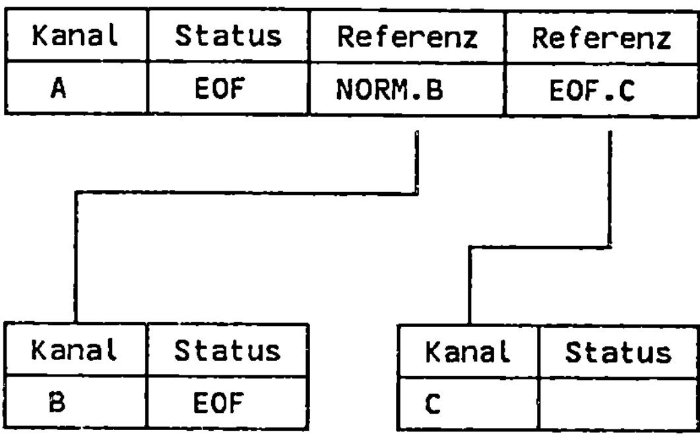

<u>Fig. 4</u> Durch EOF bei Kanal B wird der Selektstatus von B und A auf
EOF gesetzt. Daher wird jetzt nur noch Kanal C ausgewählt.

6. Nachrichtenbehandlung an Kanälen

An jedem Kanal ist eine einfache Nachrichtenbehandlung möglich, die
es gestattet, jede durchlaufende Nachricht mit einem beliebigen
Pre- und Postfix (auch sedezimal) zu versehen, was sowohl zur Kenn-
zeichnung von Quelle oder Senke als auch zur Adaption an bestimmte
Objekteigenschaften verwendet werden kann (z.B.: Schreiben in die
Systemzeile einer Datensichtstation TD 9750).

Für die Zukunft ist zusätzlich eine allgemeine Nachrichtenbehandlung auf der Basis eines "template Macro" Interpreters vorgesehen,
der beliebige Nachrichtentransformationen kanalspezifisch durchzuführen gestattet (→ [JIR84a]). Besondere Merkmale dieses Verfahrens
sind, daß die jeweils spezifische Syntax der Makrodefinitionen
dynamisch änderbar ist und keine syntaktischen Einschränkungen vorgegeben sind.
Es können für jeden Kanal verschiedene Makrosätze dynamisch aktiviert und deaktiviert werden. Bei der Makrodefinition müssen sowohl
die Parameter als auch die Erscheinungsform, in der der Makro im
Eingabestrom erkannt werden soll, definiert werden. Jede Nachricht,
die einen Kanal erreicht, wird vor der Weitergabe an den Interpreter übergeben, der, dem zugehörigen Macrosatz entsprechend, Trans-
formationen wie z.B. Textsubstitutionen oder n:m Abbildungen aus-

führt. Die vom Interpreter dann an den Kanal wieder zurückgegebenen
Nachrichten werden der Kanaldefinition folgend weitertransportiert.

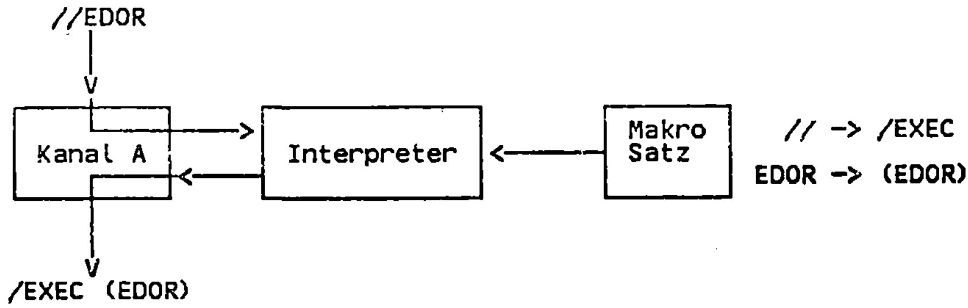

Fig. 5 Arbeitsweise des Interpreters

Im obigen Beispiel sind für den Interpreter zwei Makros (Produk-
tionsregeln) definiert, die, auf die Nachricht "//EDOR" angewendet,
die Nachricht "/EXEC (EDOR)" ergeben.
Der bereits konzipierte Interpreter erlaubt den Anschluß externer
Funktionen, wobei für den hier besprochenen Einsatzfall Funktionen
bereitgestellt werden sollen, die vom Interpreter her die Manipul-
tion des Kanals gestatten, an dem der Interpreter gerade aktiv ist.
Anwendungsbeispiel ist hier die von **Nachrichteninhalten abhängige**
Aufteilung eines Nachrichtenstroms in mehrere Teilströme durch entsprechende Änderung des Selektstatus in Abhängigkeit von dem Inhalt
der gerade durch den Kanal zu transportierenden Nachricht, während
ja die einfache Aufteilung von Nachrichtenströmen bereits durch
Kanalreferenzen und den Selektstatus möglich ist.
So läßt sich z.B ein Bildschirm in mehrere Abschnitte aufteilen,
indem man jeder Zeile einen Kanal zuordnet und einen Abschnitt
durch einen Kanal realisiert, der Referenzen auf die entsprechenden
"Zeilen-Kanäle" definiert hat. Durch den Einsatz des Interpreters
an einem "Abschnitts-Kanal" kann nun sowohl ein Umbruch als auch
die Auswahl der zu diesem Abschnitt gehörenden nächsten Zeile erfolgen.

7. Integration in bestehende BS2000 Zugriffsfunktionen

Eine wesentliche Schwierigkeit der Implementierung ist die evolutive Einführung dieses Portkonzeptes in das existierende BS2000, wobei bestehende Schnittstellen auf die neue Kanalschnittstelle abzubilden sind. Dazu müssen in die benötigten Schnittstellen Filter
eingebaut werden, die sich dynamisch für die jeweilige Zugriffsmethode an- und abschalten lassen. Innerhalb dieser Filter werden ankommende Nachrichten, in Abhängigkeit von der Zugriffsmethode, in
die der Filter eingesetzt ist, in READ/WRITE Operationen umgelenkt,
wobei der jeweils benutzte PORT sowohl von der Zugriffsmethode als
auch von der Umgebung abhängt, in der der ursprüngliche Funktionsaufruf erfolgt ist. So wird z.B. eine übliche Ausgabe im BS2000 mit
dem Aktionsmakro WROUT ausgegeben.

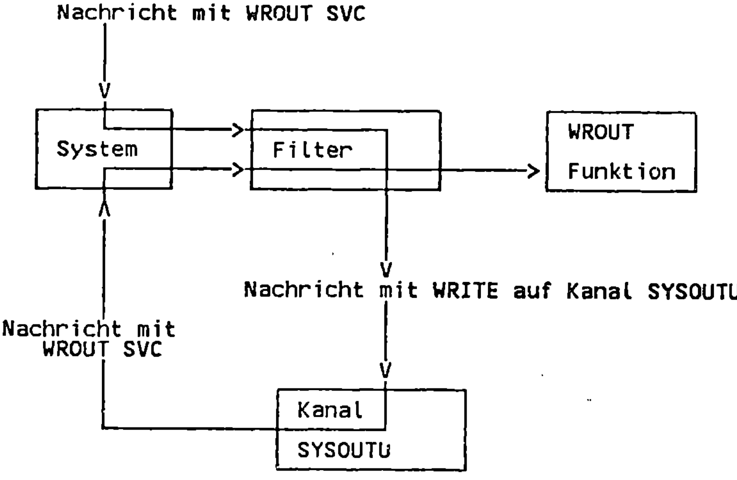

Fig. 6 Arbeitsweise des WROUT Filters
(Der Kanal SYSOUTU umhüllt hierbei den BS2000 SYSOUT)

Bei eingeschaltetem Filter für die Funktion WROUT wird jede ankommende Nachricht mit einer WRITE Operation auf den Kanal SYSOUTU
weiter transportiert. Durch entsprechende Definition des Kanals SY-
SOUTU können nun die Nachrichten in beliebiger Weise umgelenkt werden, also z.B. gleichzeitig auf den Bildschirm, eine Protokolldatei
und einen Hardcopy Drucker. In der dargestellten Standardvorein-
stellung wird der Kanal SYSOUTU nicht weiter in einen anderen Kanal

geleitet, sondern auf den BS2000 Standard SYSOUT mit der Funktion WROUT abgebildet. Durch diesen Trick wird eine Zwischenschicht eingeführt, die das Systemverhalten an der Benutzeroberfläche gegenüber dem üblichen BS2000 unverändert erscheinen läßt, aber die Möglichkeiten zur beliebigen Änderung in sich birgt, wie das folgende Beispiel zeigt.

Eine Nachricht des Operateurs (Kommando /BCST) wird an den Anwender über die Funktion $EOUT weitergeleitet. Bei eingeschaltetem Filter wird durch eine Umgebungsanalyse für Aufruf des $EOUT SVC's festgestellt, daß die "Broadcast" Funktion ausgeführt werden sollte. Daher erfolgt die WRITE Operation in diesem Fall auf den Kanal SYSBCST, und nicht auf den Kanal SYSOUTS, der üblicherweise für diese Funktion gewählt wird. Referiert der Kanal SYSBCST jedoch einen anderen Kanal, der analog zu Fig. 6 definiert ist, so ergibt sich wieder das ursprüngliche Systemverhalten. Es ist aber durchaus möglich, den PORT SYSBCST an einen beliebigen anderen Kanal anzubinden, um die Meldungen des Operateurs z.B. in ein File oder in die Systemzeile einer Datensichtstation umzurouten.

Der Filter wirkt nicht, ist also durchlässig, wenn der benötigte Kanal nicht definiert ist, oder wenn die ankommende Nachricht durch eine vorhergehende Aktion des Filters selbst erzeugt wurde. Die Filter sind hierbei so konstruiert, daß sie mit Standard Systemschnittstellen (RZ-Exits) eingebaut werden können. Wo diese Schnittstellen noch nicht zur Verfügung standen, wurde im Vorgriff auf diese ein Exitmechanismus implementiert. Über ein spezielles Kommando (/SETOPT) können alle Filter gemeinsam oder jeder einzeln an- und abgeschaltet werden.

8. Ausbau zur Interprozesskommunikation

Die innerhalb der Konzeption bisher dargestellten Objekte des BS2000, die an PORTs gebunden werden können, gestatten leider erst tasklokale Anwendungen des Portkonzeptes, wobei allerdings bestehende Anwendungen kompatibel integriert sind. Die im BS2000 vorhanden Konzepte zur Intertaskkommunikation, wie DCAM UND IPC, sind in dieser Konzeption nicht verwendbar, da sie an die Existenz von P1-Anwenderprogrammen gebunden sind. Da auch Files oder Jobvariable für diesen Zweck nur bedingt einsetzbar sind, da entweder das

Fassungsvermögen oder die asynchronen Eigenschaften unzureichend sind, werden zum Zweck der Interprozesskommunikation neue Objekte, die SHELFs heißen, in das BS2000 eingeführt. Die Konstruktion dieser Objekte wurde so vorgenommen, daß ein großer Teil der Eigenschaften auf bereits BS2000 intern vorhandene Komponenten, wie "name manager", Börsenkonzept und "P2-Contingencies" abbgebildet werden konnte.

Ein SHELF realisiert einen benannten Puffer mit den möglichen Geltungsbereichen tasklokal, global für die erzeugende "Userid" und global für alle "Userids". Er kann eine beschränkte (n<255) aber variable Anzahl von beliebig langen (V-Record) Nachrichten aufnehmen, die sowohl in LIFO- als auch FIFO-Technik eingeordnet werden können. Die erzeugende Task ist gleichzeitig der Eigentümer eines SHELF, zu dem in Analogie zu Files und auch weitergehende Zugriffrechte vergeben werden können. SHELFS werden vernichtet, sobald sie nicht mehr in Benutzung sind und die erzeugende Task entweder beendet wird oder einen entsprechenden expliziten Auftrag erteilt. SHELF Eigenschaften und Verwaltungsfunktionen wie CREATE, USE, DELETE, MODIFY und LIST werden über das BS2000 Kommando /SHELF definierbar und zugänglich. Nach der Definition (CREATE Funktion) können und müssen diejenigen Prozesse, die den SHELF benutzen wollen, diesen importieren (USE Funktion), wobei die tasklokalen Benutzungsaspekte definiert werden. Dies gilt auch für die erzeugende Task.

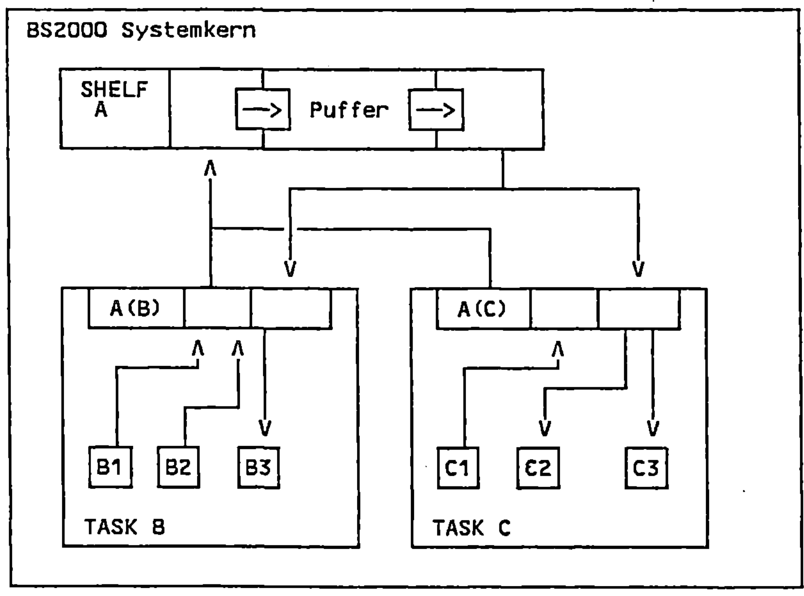

<u>Fig. 7</u> SHELF A, der von den Tasks B und C über die Kanäle Bi und Ci
erreichbar ist

Auf ein SHELF kann von beliebig vielen Tasks zugegriffen werden,
wenn die Zugriffsrechte, die der Eigentümer vergeben hat, dies erlauben. Daher werden Zugriffsrechte SHELF-bezogen definiert und
aufbewahrt, während die Zugriffsbedingungen, also die Art und Weise
der SHELF Nutzung, jeweils taskspezifisch unterschiedlich ausgeprägt sein können und daher tasklokal definiert und aufgehoben werden. SHELFS können an PORTs gebunden werden, so daß der Zugriff zu
SHELFs über Kanäle möglich ist, eine eigenständige Zugriffsmethode
also nicht entwickelt zu werden braucht.
An einem SHELF können die Zustände "Puffer voll" und "Puffer leer"
erreicht werden. Das Eintreten dieser Zustände und deren Aufhebung
stellen Ereignisse dar, auf die mit synchronen (WAIT, ABORT) und
asynchronen Maßnahmen reagiert werden kann, wobei letztere die verschiedenen Möglichkeiten nutzen, die im BS2000 gegeben sind, um ein
Software-Signal auszusenden. Es kann ein beliebiges BS2000 Kommando
als nächstes Kommandos ausgeführt werden (vergl. z.B. "BREAK-Event"
in Prozeduren in Verbindung mit Kommando /SETOPT (→ [GRU83])). Es
ist möglich, sowohl eine der in diesem Konzept enthaltenen Trans-

portoperationen, als auch das Setzen einer Jobvariablen bei Eintreten des Ereignisses unmittelbar über MCLP (MCLP erlaubt die Ausführung von Kommandos durch Programme) ausführen zu lassen, wobei bei allen Operationen weitere Ereignisse ausgelöst werden können. Speziell für die Nutzung durch Anwenderprogramme kann man alternativ einen "post signal" für das P1-Eventing veranlassen.

Es ist klar, daß mit diesem hier vorgestelltem SHELF Konstrukt "deadlocks" leicht hergestellt werden können. Es soll jedoch darauf hingewiesen sein, daß die vorgestellten Konzepte und Objekte eine Basis für höhere Anwendungsschichten im Sinne eines verteilten Systems bilden sollen. Wie also "deadlock"- und auch anderen Gefahren begegnet werden soll, wird kurz im Abschnitt 10 angegangen.

9. Einbindung in ein Netzkonzept

Mit den vorgestellten Mitteln lassen sich innerhalb des BS2000 Nachrichtenströme systemweit lenken, aufteilen und zusammenführen, wobei gleichzeitig über die Synchronie der SHELFs Möglichkeiten zur Flußkontrolle gegeben sind. Durch die Möglichkeit PORTs an SHELFs binden zu könnnen, erhält man implizit über die SHELF-Eigenschaften auch die Möglichkeit BS2000-Tasks an PORTS zu binden, wobei allerdings i.d.R. weitergehende Protokollstrukturen notwendig werden können. Durch Einbindung eines (oder auch mehrerer) Netzwerkkontrollprogramme (NCPs), die als eigenständige Tasks laufen können, in dieses Konzept können Nachrichtenströme ("message passing") über ein NCP netzweit verteilt werden, in dem man Netzaddressen oder Server Namen, je nach Fähigkeit des involvierten NCP, mit SHELFs assoziiert. Diese Aufgabe kann wiederum von einem bestimmten ("well-known")m Server selbst wahrgenommen werden.

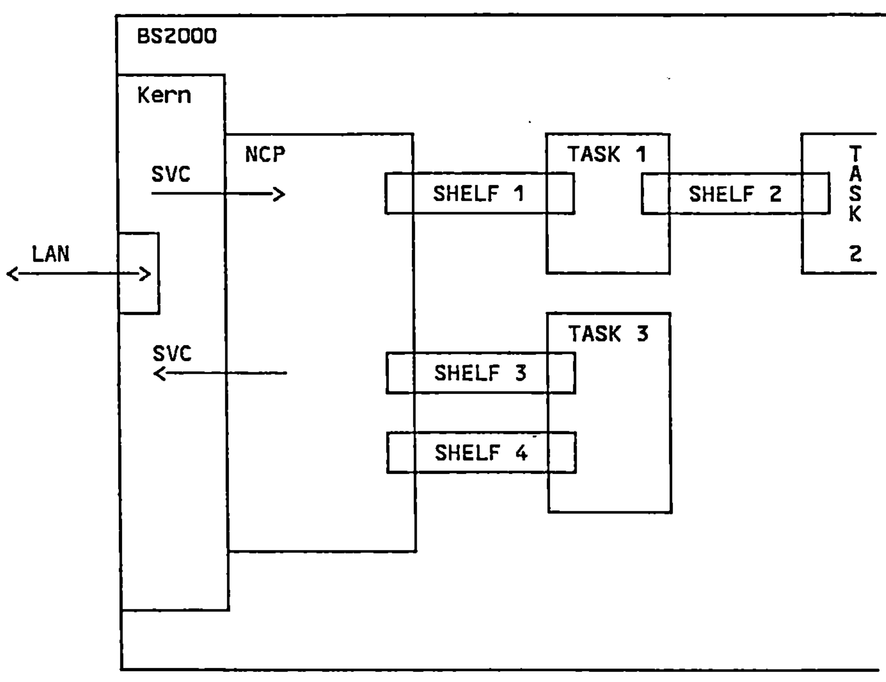

Fig. 11 Anbindung von TASKs über SHELFs an ein NCP

Da die Basis der "Newcastle-Connection" im UNIX-Bereich, das Basic Block Port Protocoll (BBP) (→ [SHR82]), in alle UNIX-Systeme leicht portierbar erscheint, wird dieses Protokoll wegen seiner Einfachheit zunächst einmal für den Anschluß eines ETHERNET an das BS2000 z.Zt (Juli 84) für den bereits funktionierenden Kanalanschluß (der Universität Saarbrücken einen herzlichen Dank) als ein erstes NCP implementiert. Im ISO-Sinne werden die Ebenen 1 und 2 durch den Hardwareanschluß und die Ebene 3 durch das BBP realisiert. Die nächsten beiden Ebenen (4 und 5) werden durch ein "Remote Procedure Call" Konzept (→ [SPE84a]) abgedeckt, wobei strukturierte Nachrichten netzweit versendet werden. Für die beiden oberen Ebenen (6 und 7) wurde für den UNIX Bereich ein Konzept ("ESPERANTO") entwickelt (→ [SPE84b]), das mit solchen "Remote Procedure Calls" aufgebaut ist. Für diese ESPERANTO Schicht bleibt allerdings noch zu prüfen, wieweit sie auch für den inhomogenen Verbund mit BS2000 zu nutzen ist.

10. Ausblick

Für die weitere Zukunft muß noch untersucht werden, ob zusätzlich zu den sequentiellen Transportfunktionen noch Funktionen für den strengen Dialog eingebracht werden müssen (WRITE-READ).

Die Einbettung der vorgestellten Konstrukte in die Kommandosprache des BS2000 verursacht durch die aktuellen Randbedingungen eine beträchtliche Unhandlichkeit der jeweiligen Kommandos. Darüberhinaus wird durch die Mächtigkeit der einzelnen Komponenten und deren Zusammenwirken nicht <u>die</u> Anwendungssicherheit (z.B. "deadlock"-Gefahr bei SHELFs) erzwungen, die man i.d.R. erwarten könnte. Da der in Abschnitt 6 eingeführte Nachrichteninterpreter natürlich auch in den Eingabestrom für BS2000 Kommandos eingehängt werden kann, werden z.Zt. höhere Sprachkonstrukte untersucht (→ [JIR84b]), die die notwendige Anwendungssicherheit bieten und eine komfortable Basis für die Konstruktion von "Client-Server" Modellen darstellen.

Eine exakte Beschreibung der dargestellten BS2000 Kommandos ist in (→ [GRU83]) und (→ [SPE84]) zu finden.

11. Literatur

BAL71   R.M. Balzer:
          <u>Ports - a Method for Dynamic Interprogram Communicatrion and Job Control</u>
          AFIPS Conference, (1971)

GRU83   H. Grund, F. Jirka, A. Klier, H. Knobloch, E. Schenk:
          <u>Erweiterungen und Änderungen am BS2000 in der GMD</u>
          Arbeitspapiere der GMD, Nr. 29, (1983)

MUE83   H. Mühlenbein:
          <u>Evolution im Bereich arbeitsteiliger Systemarchitekturen</u>
          GMD Spiegel, Sonderheft, (1983)

RAS81   R.F. Rashid:
          <u>ACCENT</u>: A communication oriented network operating system kernel
          ACM 0-89791-062-1-12/81-0064, (1981)

SPE84   E. Spenke, F.A. Jirka, H. Grund, A. Klier:
        CHANNEL/WRITE/READ/SHELF - Ein/Ausgabesteuerung im BS2000
        Arbeitspapiere der GMD, to be published, (1984)

SIL81   A. Silberschatz:
        Port directed communication
        THE COMPUTER JOURNAL, VOL.24, No.1,, (1981)

GRA83   JK.P. Gray, P.J. Hansen, P. Homann, M.A. Lerner, M.
        Pozefsky:
        Advanced program-to-program Communication in SNA
        IBM Systems Journal 22 pp. 298-318, (1983)

BOR83   A. Borg, J. Baumbach, S. Glazer:
        A Message System Supporting Fault Tolerance
        ACM SIGOPS 17 No.5 pp. 90-99, (1983)

JIR84a F.A. Jirka:
        SUBTIL: A Message Stream Interpreter
        Arbeitspapiere der GMD, to be published, (1984)

JIR84b F.A. Jirka:
        A Substitution System in a Message Passing Environment
        to be published, (1984)

SHR82   S.K. Shrivastava, F. Panzieri:
        The Design of a Reliable Remote Procedure Call Mechanism
        IEEE Transaction on Computers, Vol. C-31, No. 7, (1982)

SPE84a M. Spenke:
        Programmer's Interface and Communication Protocol for Remo-
        te Procedure Calls
        to bepublished, (1984)

SPE84b M. Spenke:
        The ESPERANTO Net: A Networking Concept for inhomogenous
        Operating Systems
        to be published, (1984)

BEZIEHUNG ZWISCHEN EINEM ABSTRAKTEN KOMMUNIKATIONSMODELL

UND SEINER KONKRETEN IMPLEMENTIERUNG

E.Hinsch,E.Giessler,A.Jägemann,E.Mäser

Institut für Systemtechnik der Gesellschaft für
Mathematik Datenverarbeitung mbH Bonn
Rheinstrasse 75
6100 Darmstadt

Inhaltsverzeichnis

Kurzfassung:

Im Rahmen des Projektes "Technische Kommunikationsinfrastruktur (TEKIN)" des Institutes F2 der GMD haben wir die Transportinstanz, wie sie durch DIN ISO 7498, DIN ISO 8072, die CCITT- Empfehlung T.70 und DIN ISO 8073 gegeben ist, im Betriebssystem UNIX implementiert. In der Norm ist nur der Transportdienst für die Kommunikation zwischen Teilnelmern in verschiedenen Endsystemen beschrieben. In unserer Implementation unterstützt der Transportdienst auch die Kommunikation zwischen Teilnehmern im gleichen Endsystem.

Im folgenden soll am Beispiel dieser Implementation die Problematik der Abbildung eines abstrakten Kommunikationsmodells auf eine konkrete Betriebssystemumgebung gezeigt werden.

Dazu werden die für die Transportinstanz relevanten Konstrukte der Norm (Schichten, Instanzen, Dienstschnittstellen, Dienstzugangspunkte, Adressierung, Verbindungszugangspunkte und der Ablauf der Dienstelemente) dargestellt. Dann werden die Konstrukte des UNIX Betriebsystems, soweit sie in unserer Implementation verwendet wurden, beschrieben. Darauf aufbauend wird unsere Implementation beschrieben und die Beziehung zwischen den Konstrukten der Norm und unserer Implementation hergestellt.

Abschließend sind einige Leistungsparameter unserer Implementation aufgeführt.

Abstract:

Within the framework of the 'TEKIN'project (GMD institute F2), we have implemented the transport entity according to the specifications in DIN ISO 7498, in DIN ISO 8072, in DIN ISO 8073, and in the CCITT recommendation T.70. This has been done in the UNIX operating system.

In the standard the transport service is described only with respect to transport service users in different computer systems. The transport servic.. in our implementation additionally supports communication between transport service users in the same computer system.

In the following we wish to demonstrate some of the problems, which arise by mapping an abstract communication model into a concrete operating system environment.

For this purpose we describe the constituent elements of the standard (layers, entities, service interfaces, service access points, addressing, connection endpoints, and the sequence of service primitives). We then list the elements of the UNIX system, which were available for our implementation. Subsequently we describe the technical details of our implementation and show the relationship between the elements of the standard and those utilized in our implementation.

Finally we present some performance parameters of the transport entity which we have implemented.

1. Einführung

Im Rahmen des Projektes "Technische Kommunikationsinfrastruktur (TEKIN)" des Institutes F2 der GMD haben wir die Transportschicht, wie sie durch DIN ISO 7498 [1], DIN ISO 8072 [2], die CCITT- Empfehlung S.70 [3] und DIN ISO 8073 [4] gegeben ist, im Betriebssystem UNIX implementiert. In der Norm ist nur der Transportdienst für die Kommunikation zwischen Teilnehmern in verschiedenen Endsystemen beschrieben. In unserer Implementation unterstützt der Transportdienst auch die Kommunikation zwischen Teilnehmern im gleichen Endsystem.

Im folgenden soll am Beispiel dieser Implementation die Problematik der Abbildung eines abstrakten Kommunikationsmodells auf eine konkrete Betriebssystemumgebung gezeigt werden.

Dazu werden die für die Transportinstanz relevanten Konstrukte der Norm (Schichten, Instanzen, Dienstschnittstellen, Dienstzugangspunkte, Adressierung, Verbindungszugangspunkte und der Ablauf der Dienstelemente) dargestellt. Dann werden die Konstrukte des UNIX Betriebsystems, soweit sie in unserer Implementation verwendet wurden, beschrieben. Darauf aufbauend wird unsere Implementation beschrieben und die Beziehung zwischen den Konstrukten der Norm und unserer Implementation hergestellt.

Abschließend sind einige Leistungsparameter unserer Implementation aufgeführt.

2. Beschreibung der Transportschicht nach DIN ISO 7498 und 8072

2.1 Einordnung in das Architekturmodell

In Bild 1 sind die Schichten und Protokolle des Basis Referenz Modells [1] dargestellt.

Instanzen sind die aktiven Elemente einer Schicht, die mit der über- und unterlagerten Instanz interagieren.

Die Transportschicht (Schicht 4) erweitert Endsystemverbindungen, die von Endsystem zu Endsystem führen, zu Teilnehmerverbindungen, die von Teilnehmer zu Teilnehmer führen. Ein Teilnehmer wird durch die drei überlagerten Instanzen (Verarbeitung, Darstellung, Kommunikationssteuerung) repräsentiert, d.h. die Transportschicht adressiert und verbindet Teilnehmerinstanzen.
Die Dienste der Transportschicht werden erbracht durch (siehe Bild 2)
- die Interaktion der Transportinstanzen miteinander, die über den Austausch von Transportprotokollelementen erfolgt,
  und
- die Interaktion jeder der beiden beteiligten Transportinstanzen mit dem

überlagerten Dienstbenutzer (Kommunikationssteuerungsinstanz) und dem unterlagerten Diensterbringer (Vermittlungsinstanz), die über den Austausch von Dienstelementen erfolgt.

2.2 Dienstzugang, Adressierung und Verbindungsendpunktkennung

Am Dienstzugangspunkt werden die Dienste der Schicht N durch die Instanz der Schicht N einer Instanz der Schicht N+1 angeboten. Ein Dienstzugangspunkt ist zu jedem Zeitpunkt nur einer Instanz der Schicht N+1 zugeordnet (siehe Bild 3). Jeder Dienstzugangspunkt trägt eine Adresse, die ihn auffindbar macht und eine Instanz der Schicht N+1 identifiziert. Ordnet eine Instanz der Schicht N Dienstzugangspunkte der Schicht N-1 bestimmten Dienstzugangspunkten der Schicht N zu, so ist ein Dienstzugangspunkt der Schicht N über die Adresse des Dienstzugangspunktes der Schicht N-1 und einer Kennung des Dienstzugangspunktes der Schicht N erreichbar. Auf den zu implementierenden Transportdienstes angewandt, heißt das:
- die Schicht N+1 entspricht der Kommunikationssteuerungsschicht,
- die Schicht N entspricht der Transportschicht,
- die Schicht N-1 entspricht der Vermittlungsschicht,
- ein Teilnehmer (Kommunikationssteuerungsinstanz), identifiziert durch einen Transportdienstzugangspunkt, kann über die Adresse des Vermittlungsdienstzugangspunktes und eine zusätzliche Kennung erreicht werden.

Jede Verbindung wird durch eine Verbindungsendpunktkennung in einem Dienstzugangspunkt identifiziert. Diese Verbindungsendpunktkennung (endpointidentifier) ist nur am jeweiligen Dienstzugangspunkt während einer Verbindung bekannt und dient dazu, gleichzeitig bestehende Verbindungen zu unterscheiden.

2.3 Dienste der Transportschicht

Dienste und Dienstelemente werden beschrieben gemäß ISO DP 8509 [5]. Es gibt folgende Typen von Dienstelementen (siehe Bild 4) :

- REQUEST   Anforderung eines Dienstes der Schicht N an die Schicht N-1

- INDICATION   Anzeigen der Schicht N-1 an die Schicht N, daß von der Partnerinstanz ein Dienst angefordert wurde

- RESPONSE   Antwort der Schicht N auf eine INDICATION der Schicht N-1

- CONFIRMATION Anzeigen der Schicht N-1 an die Schicht N, daß eine Antwort der Partnerinstanz auf ein REQUEST angekommen ist.

Es werden bestätigte ('confirmed') und unbestätigte ('unconfirmed') Dienste unterschieden. Ein bestätigter Dienst setzt sich zusammen aus den Dienstelementen:
REQUEST, INDICATION, RESPONSE, CONFIRMATION.
Ein unbestätigter Dienst setzt sich zusammen aus:
REQUEST und INDICATION.

Die Transportschicht bietet die folgenden Dienste an:
- Verbindungsaufbau,
- Verbindungsabbau und
- Transport von Daten.

Diesen Diensten entsprechen die folgenden Bezeichnungen:
- CONNECT,
- DISCONNECT und
- DATA.

Der in den Diagrammen von S.70 verwendete Dienst EXCEPTION wird auf den Dienst DISCONNECT abgebildet.
Der Dienst CONNECT ist ein bestätigter Dienst; die Dienste DISCONNECT und DATA sind unbestätigt.

2.4 Dienste der Vermittlungsschicht

Die Vermittlungsschicht verknüpft gesicherte Systemverbindungen (von Endsystem zu Transitsystem und zwischen Transitsystemen) zu Endsystemverbindungen.

Die Transportinstanz interagiert mit der Vermittlungsinstanz, um das Transportprotokoll abzuwickeln, über folgende Dienste :
- CONNECT,
- DISCONNECT,
- RESET und
- DATA.

2.5 Das Transportprotokoll

Das hier implementierte Transportprotokoll ist in CCITT Empfehlung S.70 beschrieben. Für dieses Transportprotokoll ist charakteristisch, daß eine Teilnehmerverbindung genau auf eine Endsystemverbindung abgebildet wird und es keine Protokollelemente für einen Verbindungsabbau gibt. Die Teilnehmerverbindung und die Endsystemverbindung werden immer gleichzeitig beendet.

3. Randbedingungen für unsere Implementation und verwendete Elemente des UNIX-Systems

Folgende Voraussetzungen waren für unsere Implementation gegeben:
- das UNIX-System III war verfügbar,
- die Programme sollten in der Programmiersprache C geschrieben werden,
- die Vermittlungsdienste waren uns über eine X.25 Treiberschnittstelle der Firma Interactive Systems zugänglich.

Wir haben die folgenden Möglichkeiten des UNIX Systems III ausgenutzt:

- das Management von Prozessen:
  Ein Prozeß kann andere Prozesse (Söhne) kreieren. Bei der Kreation eines Prozesses können Parameter übergeben werden z.B. 'Filepointer', die den gemeinsamen Zugriff auf ein 'File' gestatten. Dies wird bei der Bedienung der X.25 Treiberschnittstelle ausgenutzt. Prozesse können sich selbst beenden oder von anderen Prozessen beendet werden. Dies wird benötigt, da die Lebensdauer der Prozesse in unserem System entprechend ihrer Aufgabenstellung unterschiedlich ist.

- das File Management:
  Speziell wurde die Kommunikation von Prozessen über FIFO-files ausgenutzt. Prozeßschnittstellen sind realisiert mit FIFO-files, über die Schnittstellenelemente ausgetauscht werden. Es gibt bei uns innerhalb von Prozessen Programmschnittstellen, die durch in C geschriebene Unterprogrammpakete realisiert sind und die die Bedienung der Prozeßschnittstellen erleichtern sollen.

- die Möglichkeit ein Timersignal und damit Unterbrechungen zu erzeugen. Dies dient der Überwachung von Verbindungen und Prozessen.

4. Zuordnung vom abstrakten Modell der Norm zur Implementation

Im folgenden wird unsere Implementation zunächst im Hinblick auf das Basisreferenzmodell nach DIN ISO 7498 beschrieben. Danach wird skizziert, wie die Dienste der Transportschicht (DIN ISO 8072) erbracht werden.

4.1 Implementationsmodell im Hinblick auf DIN ISO 7498

Unsere Implementation erlaubt sowohl die Kommunikation zwischen Teilnehmern in verschiedenen Endsystemen (nicht lokal) als auch die Kommunikation zwischen Teilnehmern im selben Endsystem (lokal). Die Dienstschnittstelle zur Transportschicht ist für lokale und nicht lokale Kommunikation identisch.

Die Komponenten des Modells sind FIFO-files (Named Pipes) und Prozesse. In Bild 5 ist der Aufbau der Komponenten für den Fall dargestellt, daß ein

Anwendungsprozeß (AP) mit einem Anwendungsprozeß in einem anderen Endsystem kommuniziert. Bild 5 zeigt außerdem die Einordnung der Komponenten und Schnittstellen in die ISO Kommunikationsarchitektur.

Der Zusammenhang zwischen lokaler und nicht lokaler Kommunikation wird aus Bild 6 ersichtlich. Bei der lokalen Kommunikation befinden sich die Anwendungsprozesse in demselben Endsystem. Daher entfällt die Vermittlungsschicht und der Transportsender (TS) eines APes fällt zusammen mit dem Transportempfänger (TE) des korrespondierenden APes. Diese TS und TE, die in einem Prozeß realisiert sind, haben nur noch eine Eingangswarteschlange.

4.1.1 Zuordnung von Prozessen und Schnittstellen zu Instanzen und Schichten

Den dargestellten Prozessen werden die folgenden Aufgaben zugewiesen:

- in den Anwendungsprozessen werden alle Aufgaben erledigt, die den ISO-Schichten 5 - 7 entsprechen,
  d.h ein Anwendungsprozeß repräsentiert eine Kollektion der Instanzen der Schichten 5, 6 und 7.

- der Communication Manager (CM) und die Transportprozesse (Transportsender (TS) und Transportempfänger (TE)) erbringen zusammen die Dienstleistungen der Schicht 4. Für jeden AP gibt es pro Verbindung genau einen TS und einen TE. Dieses TS/TE-Paar handelt die Protokollelemente ab.

  Der CM hat Gedächtnis über die kommunikationsfähigen Anwendungen bzw. Prozesse im System und muß deswegen über die Dauer einzelner Verbindungen hinaus verfügbar sein (ein ewig lebender Systemprozeß). Weiterhin verwaltet er alle Teilnehmerverbindungen. Zu diesem Zweck wertet er alle Verbindungsaufbau- und Verbindungsabbaupakete aus. Er kreiert ggf. Anwendungs- und Transportprozesse und zerstört sie.

  Die Transportinstanz wird also durch den CM und die Transportprozesse realisiert.

Die Transportinstanz fordert die Dienste der Vermittlungsinstanz über Unterprogramme an, die den X.25 Treiber bedienen.

Da die Transportinstanz unabhängig von den Teilnehmerinstanzen ist, liegt die Schichtgrenze zwischen der Schicht 4 und 5 an der Schnittstelle zwischen Prozessen und nicht in einem Prozeß.

In Bild 7 ist die Zuordnung von Instanzen und Schichten zu Prozessen dargestellt.

4.1.2 Adressierung, Dienstzugangspunkte, Verbindungsendpunkte

Eine Anwendung wird durch ein lauffähiges Programm repräsentiert. Ein lauffähiges Programm kann jedoch mehrere Anwendungen repräsentieren. Eine Verbindung kann nur zu einem Prozeß aufgebaut werden, der auf dem entsprechenden Anwendungsprogramm gestartet wurde. Aus der Sicht der ISO-OSA wird eine Anwendung eindeutig über einen Transport-Dienstzugangspunkt ('transport service access point' (TSAP)) erreicht, der seinerseits durch eine Teilnehmer-Adresse bezeichnet wird. In unserer Implementation wird eine Teilnehmer-Adresse gebildet aus
- X.25-DTE-Adresse und X.25-Protokollkennung und
- dem Adreß-Erweiterungs-Parameter des Transport-Protokolls.
Dieser Adressierungs-Parameter wird kurz als TSAP bezeichnet, da er den TSAP identifiziert. Bei Verbindungsaufbau wird einer Verbindung in einem System ein eindeutiger Endpunkt zugeordnet:
-   zur Transportschicht - die Endpunktkennung der Endsystemverbindung -
    bei uns realisiert durch ein X.25 file, das genau einer Verbindung
    zugeordnet ist.
-   zur Anwendung - die Endpunktkennung der Teilnehmerverbindung -
    bei uns realisiert durch eine eindeutige Nummer.

4.2 Erbringen der Dienste in unserer Implementation

Die Dienste der Transportschicht sind Verbindungsaufbau/abbau und der Transport von Daten.

Im folgenden soll beschrieben werden
-   die Kommunikation zwischen Prozessen zur Abwicklung einer Verbindung
    einschließlich der Überwachung von Prozessen und Verbindungen
-   die Zuordnung von Dienstelementen zu Schnittstellenelementen in unserer
    Implementation wie sie an der Schichtgrenze 4/5 (siehe Bild 5)
    sichtbar werden und
-   die Nutzung der Transportdienste mit Hilfe angebotener Unterprogramme.

Die für uns relevanten Schnittstellen sind in Bild 5 gegeben. Es gibt Prozeß- und Programmschnittstellen.

Programmschnittstellen liegen in einem Prozeß und werden durch Unterprogrammaufrufe realisiert.

An der Schnittstelle zur Transportschicht ist die Programmschnittstelle durch Unterprogramme realisiert, die FIFO-files bedienen und die Schnittstellenelemente bearbeiten. An der Schnittstelle zum Vermittlungsdienst ist die Programmschnittstelle durch Unterprogramme gegeben, die den X.25 Treiber bedienen.

Die Prozeß-Schnittstelle ist beschrieben durch Schnittstellenelemente, die die Realisierung von Dienstelementen darstellen, und ihre Ablauffolgen. Die Schnittstellenelemente werden vom Erzeuger-Prozeß in einer empfängerprozeß-spezifischen Warteschlange abgelegt und vom Empfängerprozeß in der Reihenfolge, wie sie abgelegt wurden, abgeholt und bearbeitet. Jede Warteschlange wird durch einen UNIX-FIFO-file realisiert.

4.2.1 Abwicklung und Überwachung von Verbindungen

Die Kommunikation zwischen den Prozessen zur Abwicklung einer Verbindung soll im folgenden übersichtsweise beschrieben werden. Die Abläufe zwischen den Prozessen werden in drei Abschnitte eingeteilt:
- Kreieren und Zerstören von Prozessen
- Ablauf der Schnittstellenelemente beim Verbindungsaufbau und
- Ablauf der Schnittstellenelemente in der Datenphase.

Kreieren und Zerstören von Prozessen

Die Lebensdauer von Prozessen in unserem Kommunikationssystem ist verschieden. Der CM, der die Kommunikation steuert, ist ein "ewig" lebender Systemprozeß ("ewig" ist definiert als Zeitraum zwischen zwei Systemzusammenbrüchen). Prozesse, die Anwendungen repräsentieren, können eine Lebensdauer haben, die - abhängig von der jeweiligen Anwendung - Verbindungen überdauern. APe können entweder vom Benutzer - bei aktiven Anwendungen ist dies immer der Fall - oder durch den CM kreiert werden - das kann bei passiven Anwendungen als Folge eines ankommenden Verbindungsaufbauwunsches auftreten.

Der Transportprozeß TS ist nur für die Dauer einer Verbindung lebendig. Der Transportprozeß TE wartet auf den Aufbau einer Verbindung und lebt bis zum Abbau der Verbindung.

Im Falle eines ausgehenden Verbindungsaufbauwunsches kreiert der CM den Transportprozeß TS, der die Verbindung aufbaut. Bei einem ankommenden Verbindungsaufbauwunsch kreiert der CM einen weiteren Transportportprozeß TE, sodaß immer zwei TE auf ankommende Aufbauwünsche warten. In diesem Fall handelt der TE den Verbindungsaufbau ab. Der jeweilige verbindungsaufbauende Transportprozeß kreiert den zugehörigen Transportprozeß für die andere Datenfluß-Richtung (im aktiven Fall den TE, im passiven Fall den TS). Beim normalen Verbindungsabbau werden die Transportprozesse vom CM inaktiviert.

Ablauf der Schnittstellenelemente in der Aufbauphase

In den Bildern 8 und 9 sind die Abläufe der Schnittstellenelemente durch die Komponenten des Modells in der Verbindungsaufbauphase für die lokale und die nicht lokale Kommunikation dargestellt. Die Abfolge der Schnittstellenelemente zwischen einem AP und den Transportprozessen ist im lokalen und nicht lokalen

Fall gleich. Durch die Zusammenlegung der Prozesse der Transportschicht im Lokalfall ändert sich der Ablauf der Schnittstellenelemente zwischen den Prozessen der Transportschicht (wie aus den Bildern ersichtlich).

Ablauf der Schnittstellenelemente in der Datenphase

In den Bildern 10 und 11 sind die Folgen der Schnittstellenelemente in der Datenphase für die lokale und die nicht lokale Kommunikation dargestellt.

Die Flußregelung

Die in der Norm für jede Übertragungsrichtung unabhängig definierte Flußregelung, die die Geschwindigkeit, mit der der sendende Dienstbenutzer Dateneinheiten verlustfrei in eine Verbindung einspeisen kann, an die Abnahmegeschwindigkeit des empfangenden Dienstbenutzers anpassen kann, wurde in unserer Implementation - wie in Bild 10 gezeigt - realisiert. Aus dem Bild 10 ist der Mechanismus der Flußregelung anhand der ausgetauschten Schnittstellenelemente für den nicht lokalen Fall ersichtlich.

Der Anwendungsprozeß schickt eine Dateneinheit (Transportinterfacedataunit) an den TS, der sie quittiert, nachdem er sie erfolgreich an den X.25 Treiber abgesetzt hat. Erst nach Erhalt der Quittung darf der AP die nächste Dateneinheit absetzen. Dies ist eine explizite Flußregelung mittels eines separaten Schnittstellenelements. Zwischen TS und X.25 Treiber wird ein impliziter Flußregelungsmechanismus benutzt. Die Übergabe einer Dateneinheit an den X.25 Treiber erfolgt über ein 'geblocktes write'; d.h. der TS verharrt solange im 'write', bis das Schreiben entweder erfolgreich war oder Fehler aufgetreten sind.

Der AP benutzt den m-bit Mechanismus, um Transportschnittstellendateneinheiten zu einer Transportdienstdateneinheit zu verketten. Diese Notwendigkeit tritt allerdings nur dann auf, wenn eine Dateneinheit die Länge von 4kbyte überschreitet. Die vom AP gesendete Dateneinheit wird vom TS in Transportprotokolldateneinheiten zerlegt. Der TE liest eine Dateneinheit aus dem X.25 file, wertet das m-bit der Transportschicht aus und liest Dateneinheiten von X.25 bis entweder die Transportdienstdateneinheit vollständig ist oder eine Länge von 4kbyte überschritten ist. Die so erhaltene Dateneinheit schreibt er dann in die Warteschlange des APes. Da dieses Schreiben nur eine Verbindung berührt, kann der TE an der Warteschlange (zeitüberwacht) warten. Hat der AP die Dateneinheit erhalten und kann eine neue entgegennehmen, dann sendet er eine Quittung an den TE, der nach Erhalt dieser Quittung versucht, im X.25 file die nächste Dateneinheit zu lesen. Dieser Ablauf gestattet die volle Ausnutzung der Kapazität des FIFO-files (5kbyte) in Senderichtung wie in Empfangsrichtung für die Schnittstellendateneinheit abzüglich benötigter Kapazität für Steuerinformationen. Die maximale Länge einer Transport-schnittstellendateneinheit wurde auf 4kbyte festgesetzt.

Im lokalen Fall (siehe Bild 11) vollzieht sich im Prinzip der gleiche, aber wie in Bild 6 gezeigt, symmetrisch ergänzte Mechanismus. Der TS kann 'seinem' Anwendungsprozeß die Freigabe-Quittung geben, sobald der zugehörige TE die Daten in den FIFO-file 'seines' Anwendungsprozesses schreiben konnte. Ein nächstes Schreiben des TE in den FIFO-file 'seines' Anwendungsprozesses ist aber erst nach Erhalt der Freigabe-Quittung von diesem erlaubt.

Der Vollständigkeit halber sei hier nochmals darauf hingewiesen, daß einem Anwendungsprozeß alle Schnittstellenelemente, also auch die zu unterschiedlichen Verbindungen gehörenden, über eine Empfangs-FIFO-file zugestellt werden. Die Zugehörigkeit zu einer Verbindung wird aus der Verbindungsendpunktkennung erkannt. Dieser Mechanismus funktioniert verklemmungsfrei, da der AP verbindungsunspezifisch aus seinem FIFO-file liest und über den Quittungsmechanismus sichergestellt ist, daß ein Empfangspuffer zur Verfügung steht. Der AP kann seine Dateneinheiten - gesteuert durch die Flußregelung - immer an die entsprechenden Sendeprozesse absetzen, da auch hier keine Blockierungen auftreten.

Die Zeitüberwachung

Die Zeitüberwachung dient dazu, Verbindungen zu überwachen. Diese Überwachung beinhaltet die Kontrolle darüber, ob der Ablauf der Schnittstellenelemente innerhalb eines vorher definierten Zeitrasters erfolgt. Dies erfordert eine Überwachung
- der an der Verbindung beteiligten Prozesse im eigenen System und
- der über die Vermittlungsschicht eintreffenden Schnittstellenelemente.
Die Verbindungen werden im TS/TEpaar zeitüberwacht. Eine Zeitüberwachung, die während des Verbindungsaufbaus eintritt, führt grundsätzlich zum Abbau der Verbindung. Das Ansprechen einer Zeitüberwachung auf Grund des Nichtfunktionierens eines APes im eigenen System führt zum Abbau sämtlicher Verbindungen dieses APes und zu seiner Inaktivierung. Da in der Datenphase Senden und Empfangen durch zwei parallele Prozesse abgewickelt werden, wird die Aktivität in Sende- und Empfangsrichtung getrennt überwacht. Das Ansprechen der Zeitüberwachung nur in Senderichtung bzw. nur in Empfangsrichtung wird nicht als Fehlverhalten gewertet. Erst das gemeinsame Auftreten beider Ereignisse führt zum Abbau der Verbindung.

4.2.2 Zuordnung von Dienstelementen zu Schnittstellenelementen an der Schichtgrenze 4/5

Die Schnittstelle 4/5 ist identisch für lokale und nicht lokale Kommunikation. In Bild 5 ist die Einbettung dieser Schnittstelle in das System (mit den Komponenten Prozesse und Kommunikationspfade) dargestellt.

Die Schnittstellenelemente und ihre Parameter werden im folgenden angegeben. Der lokale Parameter Prozeßidentifikation wird für die Namensgebung der Eingabewarteschlangen und für das Zerstören von Prozessen benötigt. Der lokale Parameter "endpoint identifier" dient der Zuordnung von Schnittstellenelementen zu Verbindungen.

Die Elemente werden nach folgenden Kriterien klassifiziert und entsprechend gekennzeichnet:
 '+' : normgerechte Schnittstellenelemente
 '$' : Schnittstellenelemente, die keine Entsprechung in der Norm finden
 '%' : Elemente, die zwischen den Prozessen der Transportschicht ausgetauscht
       werden

**Übersicht über die Schnittstellenelemente und Elemente, die
zwischen den Prozessen der Transportschicht ausgetauscht werden**

      Verbindungsaufbau und Abbau
+%   CONNRQ   connect request
+%   CONNEI   connect indication
+%   CONRSP   connect response
+%   CONNCP   connect confirmation
+%   DISCRQ   disconnect request
+%   DISCIN   disconnect indication
$ DIQUIT   lokale Quittung auf ein DISCIN
%   OKOKOK   aktivieren des zugehörigen TS oder TE

      Verwalten einer Verbindung
%   MELDT1   T1 abgelaufen
%   T1RUEC   Verbindung wieder aktiv
%   FEHLVR   Verbindungsspez. Fehler aufgetreten

      Bekanntmachung einer Anwendung bei dem CM
$ FALLRQ   file allocate request
$ FALLCP   file allocate confirmation positive
$ FALLCN   file allocate confirmation negative

      Bekanntmachung eines Prozesses (Ebene 5-7) bei dem CM
$ PALLRQ   process allocate request
$ PALLCP   process allocate confirmation positive
$ PALLCN   process allocate confirmation negative

      Abmelden einer Anwendung bei dem CM
$ FDALRQ   file deallocate request
$ FDALCP   file deallocate confirmation positive
$ FDALCN   file deallocate confirmation negative

```
        Abmelden eines Prozesses (Ebene 5-7) bei dem CM
$       PDALRQ    process deallocate request
$       PDALCP    process deallocate confirmation positive


        Verwalten der Prozesse, die Verbindungen unterhalten
%       FEHLPR    Prozeß fehlerhaft


        Datenphase
+       DATARQ    Daten senden
+       DATAIN    Daten empfangen
$       QUITTU    lokale Quittung auf Daten
```

4.3 Nutzung der Transportdienste mit Hilfe angebotener Unterprogramme

Die Nutzung des Transportdienstes läuft im allgemeinen wie folgt ab:
Eine bzw. mehrere Anwendungen, repräsentiert durch ein lauffähiges Programm, werden der Transportschicht (s. Schnittstellenelemente) durch Angabe des bzw. der zugehörigen TSAP-Adressen und durch Beschreibung ihres Kommunikationsverhaltens bekannt gemacht. Das lauffähige Programm wird durch Starten eines oder mehrerer Prozesse aktiviert, die sich ebenfalls bei der Transportschicht anmelden müssen. Danach sind die Dienstleistungen der Transportschicht, wie Verbindungsauf/abbau und Datentransport, für die Prozesse verfügbar. Anwendungsprozesse können auch durch die Transportinstanz auf Grund eines ankommenden Rufes aktiviert werden.

Im allgemeinen Fall können einer Anwendung mehere TSAPs zugeordnet werden. Dies ist in der gegenwärtigen Implementation noch nicht berücksichtigt. Sie wird aber diesbzgl. erweitert werden.

Das Kommunikationsverhalten einer Anwendung, identifiziert durch den zugehörigen TSAP, wird beschrieben durch :

- Anzahl der gleichzeitig bedienbaren Verbindungen/TSAP,
- durch die Klassifizierung eines TSAPs als aktiv (immer rufender TSAP) oder als passiv (immer gerufener TSAP) oder als aktiv und passiv (rufender oder gerufener TSAP) (Kollisionen!),
- sein Zeitverhalten (es wird eine Zeitspanne angegeben, die es gestattet zu überwachen, ob eine Verbindung noch aktiv ist).

Im folgenden werden die Unterprogramme, die in C geschrieben sind und die Dienste der Transportschicht verfügbar machen, dargestellt. Zunächst wird die Bedeutung der Argumente der Unterprogramme erklärt.

Bedeutung der Namen der Argumente :

aus	Auslösungsgrund (aufbauend auf S.70)
	(=0 - Grund nicht angegeben,
	=1 - Endeinrichtung besetzt,
	=2 - Endeinrichtung nicht in Ordnung,
	=3 - Adresse unbekannt,
	=4 - lokaler Fehler,
	=5 - Netzverbindung abgebaut,
	=6 - Reset,
	=8 - falsche Daten oder TBR,
	=9 - Lese/Schreibfehler X.25,
	=10- Timerablauf in der Datenphase).
buf	Datenpuffer
fromtsap	Dienstzugangspunkt des rufenden Prozesses
info	Auslöseinformation
lang	Länge des Datenpuffers
linf	Länge der Auslöseinformation
locid	lokale Verbindungsnr. des Anwendungsprozesses
mbit	Endekennzeichnung einer Transportschnittstellendateneinheit
	(=0 - nicht letzte Schnittstellendateneinheit einer
	Dienstdateneinheit
	=1 - letzte Schnittstellendateneinheit einer
	Dienstdateneinheit)
mdup	max. Anzahl gleichzeitig existierender Prozesse
	eines lauffähigen Anwendungsprogramms
mvproz	max. Anzahl der Verbindungen pro Prozeß
name	Name des lauffähigen Anwendungsprogramms
part	Prozeßart (=0 - aktiv,
	=1 - passiv,
	=2 - beides)
pos	Kennzeichnung des empfangenen Dienstelements
timer	max.Zeitdauer,während der die Verbindung unbenutzt
	sein kann
	(=-1 - Voreinstellung wird genommen,
	= 0 - Timer ausgeschaltet)
totsap	Transportdienstzugangspunkt des zu rufenden Prozesses
tsap	Transportdienstzugangspunkt
	(X.25-Adresse + erweiterte Adressierung) Die erweiterte Adressierung ist zu Beginn mit einem '*' zu kennzeichnen.Folgt hinter '*' nichts, so wird die Anwendung TELETEX adressiert.

Unterprogramme für Management- und Informationsdienste :

- Anmelden eines lauffähigen Anwendungsprogramms:
  ```
  int e4falloc (tsap,file,mdup,mvproz,part,timer)
      char *tsap,*file;
      int mdup,mvproz,part,timer;
      return (-1)      nicht erfolgreich
      return (0)       erfolgreich
  ```

- Abmelden eines lauffähigen Anwendungsprogramms:
  ```
  int e4fdalloc (tsap)
      char *tsap;
      return (-1)      nicht erfolgreich
      return (0)       erfolgreich
  ```

- Anmelden eines Prozesses (Ebene 5-7):
  ```
  int e4palloc (tsap)
      char *tsap;
      return (-1)      nicht erfolgreich
      return (0)       erfolgreich
  ```

- Abmelden eines Prozesses (Ebene 5-7):
  ```
  int e4pdalloc ()
      return (-1)      nicht erfolgreich
      return (0)       erfolgreich
  ```

- Ausgabe der verbindungsspezifischen Daten der Anwendung:
  ```
  int auskomtab ()
  ```


Unterprogramme für Kommunikationsdienste :

- Verbindungsaufbauwunsch:
  ```
  int conrq (totsap,locid,timer1)
      char *totsap;
      int *locid,timer1;
      return (-1)      nicht erfolgreich
      return (0)       erfolgreich
  ```

- Erzeugen einer positiven Quittung auf ein connect indication:
  ```
  int conrsp (locid,timer1)
      int locid,timer1;
      return (-1)      nicht erfolgreich
      return (0)       erfolgreich
  ```

- Erzeugen einer negativen Quittung auf ein connect indication:
  ```
  int discrq (locid,aus,info)
      char aus,*info;
      int locid;
      return (-1)     nicht erfolgreich
      return (0)      erfolgreich
  ```

- Verbindungsabbauwunsch:
  ```
  int discrq (locid,aus,info)
      char aus,*info;
      int locid;
      return (-1)     nicht erfolgreich
      return (0)      erfolgreich
  ```

- Anforderung Daten senden:
  ```
  int datarq (buf,lang,mbit,locid)
      char *buf;
      int lang,mbit,locid;
      return (-1)     Fehler ( nicht erfolgreich)
      return (0)      erfolgreich
  ```

- Warten auf Lesen aus der Eingangsfifo (geblockt) und auswerten
  ```
  int warten (pos,locid,buf,lang,mbit,aus,fromtsap,linf,info)
      char *buf,*aus,*fromtsap,*info;
      int *pos,*locid,*lang,*mbit,*linf;
      return (-1)     nicht erfolgreich
      return (0)      erfolgreich
  ```

Wert von pos	empfangenes Schnitt-stellen-Element	mitgelieferte Parameter
0	connei	locid,fromtsäp
1	conncp	locid
2	discin	locid,aus,linf,info
3	datain	locid,buf,lang,mbit
4	quittu	locid

- Lesen aus der Eingangsfifo (ungeblockt) und auswerten
  ```
  int lesen (pos,locid,buf,lang,mbit,aus,fromtsap,linf,info)
      char *buf,*aus,*fromtsap,*info;
      int *pos,*locid,*lang,*mbit,*linf;
      return (-1)     nicht erfolgreich
      return (0)      fifos leer
      return (1)      erfolgreich
      Werte von pos : siehe Unterprogramm warten;
  ```

```
- Senden einer Quittung auf eine empfangene Dateneinheit
  int e4dataq (locid)
      int  locid
      return (-1)        nicht erfolgreich
      return (1)         erfolgreich
```

5. Zustand und Eigenschaften der Implementation

5.1 Zustand der Implementation

Die Transportinstanz ist implementiert. Der dafür benötigte Zeitaufwand war
2.5 MJ. Die Implementation hat bisher folgende Tests durchlaufen:

- manuelle Tests
  Dienstelemente wurden manuell erzeugt sowohl bei der zu testenden
  Transportinstanz als auch beim Testpartner. Dabei hatten wir als Partner bei
  unseren Tests folgende Transportinstanzen :
  . die eigene Transportinstanz
    Dies ist die einzige Möglichkeit auch den lokalen Transportdienst zu prüfen.
    Der nicht lokale Transportdienst wurde ebenfalls geprüft.
  . TESDI [6] Test- und Diagnoseprógramme (GMD-Entwicklung)
  . Transportinstanz in Siemens-BS2000 [7] (GMD-Entwicklung)
- automatische Tests
  Dienstelemente wurden durch Programme erzeugt. Dabei hatten wir als Partner
  bei unseren Tests folgende Transportinstanzen :
  . die eigene Transportinstanz
  . TESDI-Test- und Diagnoseprogramme (nur für nicht lokale Kommunikation)
  Geprüft wurden lokale und nicht lokale Kommunikation. Diese Tests dienten
  dazu, um Synchronisationsfehler aufzudecken, die zwischen Prozessen auftreten
  können und die bei manuellen Tests nicht auftreten, weil genügend Zeit
  verfügbar ist.

Unsere Transportinstanz ist bisher nur im Testbetrieb eingesetzt worden.

5.2 Eigenschaften der Implementation

Wir haben die Eigenschaften unserer Implementation nach folgenden
Gesichtspunkten gegliedert:
- Speicherbedarf,
- Leistungsparameter und
- Zeitverhalten.

Speicherbedarf

Zur Abwicklung der Transportdienste müssen in unserem System immmer folgende
Prozesse vorhanden sein: der CM und zwei TE.

    Speicherbedarf für CM: 33000 byte
    Speicherbedarf für ein TE: 15800 byte
    Speicherbedarf für ein TE: 15800 byte
                                   ____

    Gesamt   **64600 byte**

Ein Anwendungsprozeß, der die Transportdienste über unsere angebotenen
Unterprogramme nutzen möchte, muß das entsprechende Programmpaket laden. Die
Größe des Programmpakets beträgt **29400 byte.**
Pro Verbindung wird ein TS und ein TE benötigt. Das ergibt für den
Speicherbedarf pro Verbindung:
für eine lokale Verbindung: 18500 byte
für eine nicht lokale Verbindung: 26000 byte

Leistungsparameter

max. Anzahl der Verbindungen / Prozeß   : 9
max. Anzahl von nicht lokalen Verbindungen   : 31
max. Anzahl von lokalen Verbindungen   : abgeschätzt 80
max. Größe der Schnittstellendateneinheit   : 4kbyte
max. Größe der Protokolldateneinheit   : 2kbyte

Die Leistungsparameter der Transportinstanz werden mitbestimmt durch die
folgenden Systemparameter:
- maximale Anzahl der Prozesse   : 200
  Die maximale Anzahl der Prozesse ist ein einstellbarer Parameter,
  der in unserem System auf 200 eingestellt ist.
- maximale Anzahl geöffneter Filepointer / Prozeß : 20
- maximale Anzahl der X.25 Verbindungen   : 31
- Kapazität der FIFO-files   : 5kbyte

Zeitverhalten

Zeitmessungen wurden auf der PDP11/44 für die lokale und nicht lokale
Kommunikation in der folgenden Testkonfiguration durchgeführt:
- beide Anwendungsprozesse lagen im gleichen Endsystem,
- im Fall der lokalen Kommunikation verließen die ausgetauschten Dateneinheiten
  das Endsystem nicht,
- bei der nicht lokalen Kommunikation flossen die Daten über eine X.25 VC zum
  X.25-'inhouse'-Vermittler und wurden von dort weitervermittelt. Die
  Anschlußleitung vom Endsystem zum 'inhouse' - Vermittler hatte eine Kapazität
  von 9,6 kbit/s.

Um reproduzierbare Ergebnisse zu erhalten, wurden die Messungen zu einem
Zeitpunkt durchgeführt, in dem unsere Anwendungen die einzigen Benutzer des
Systems waren.

Die Zeitdauer für den Verbindungsaufbau wurden in der Anwendung gemessen, die
die Verbindung aufgebaut hat - und zwar der Zeitraum vom Absetzen des
Verbindungsaufbauwunsches an die Transportschicht bis zum Eintreffen der
Quittung.
Die Systemuhr zeigt nur Sekunden an.
Für den Verbindungsaufbau wurden bei der lokalen Kommunikation 2s und bei der
nicht lokalen 8s benötigt.

Bei der Messung der Zeitdauer für die Übertragung von Dateneinheiten war - der
eine der beiden kommunizierenden Anwendungsprozesse nur Datensender - der andere
nur Datenempfänger.

Es wurden gemessen:

- die Sendedauer pro Anzahl Dateneinheiten,
    d. h. die im Datensender benötigte Zeitdauer, um eine bestimmte Anzahl
    Dateneinheiten abzusetzen.
- die Empfangsdauer pro Anzahl Dateneinheiten,
    d. h. die im Datenempfänger benötigte Zeitdauer, um eine bestimmte Anzahl
    Dateneinheiten anzunehmen.
- die Übertragungsdauer pro Anzahl Dateneinheiten,
    d.h. die Zeitdauer vom Beginn des Sendens der 1. Dateneinheit im Datensender
    bis zum Empfang der letzten Dateneinheit (einschließlich lokaler Quittierung)
    im Datenempfänger.

Bei der lokalen Kommunikation wurden je Messung jeweils 1000
Transportdienstdateneinheiten der Längen 1byte bzw. 4kbyte übertragen.

Im lokalen Fall hatten die zugehörigen Sende-, Empfangs- und Übertragungsdauer
die gleichen Werte.

Die Übertragungsdauer für 1000 x 1byte betrug 77s .
Die Übertragungsdauer für 1000 x 4kbyte betrug 221s .

Bei der nicht lokalen Kommunikation wurden jeweils 100 Transportdienstdateneinheiten gesendet. Für Transportdienstdateneinheiten von 1byte
stimmten Sende-, Empfangs- und Übertragungsdauer überein.

Die Übertragungsdauer für 100 x 1byte betrug 11s .
Die Übertragungsdauer für 100 x 4kbyte betrug 733s.
(die Sende- bzw. Empfangsdauer betrug 731s)

Zusammengefaßt ergibt sich daher für den Durchsatz:
- Bei lokaler Kommunikation ist der erreichbare Durchsatz bei maximaler Länge einer Interfacedataunit gleich 18,5kbyte/s.
- Bei nicht lokaler Kommunikation kann bei maximaler Länge der Interfacedataunit und zwei jeweils in Gegenrichtung mit Daten belasteten virtuellen Kanälen über einer 9,6kbit/s Leitung ein Durchsatz von 4,5kbit/s je Kanal erzielt werden.

6. Bilder

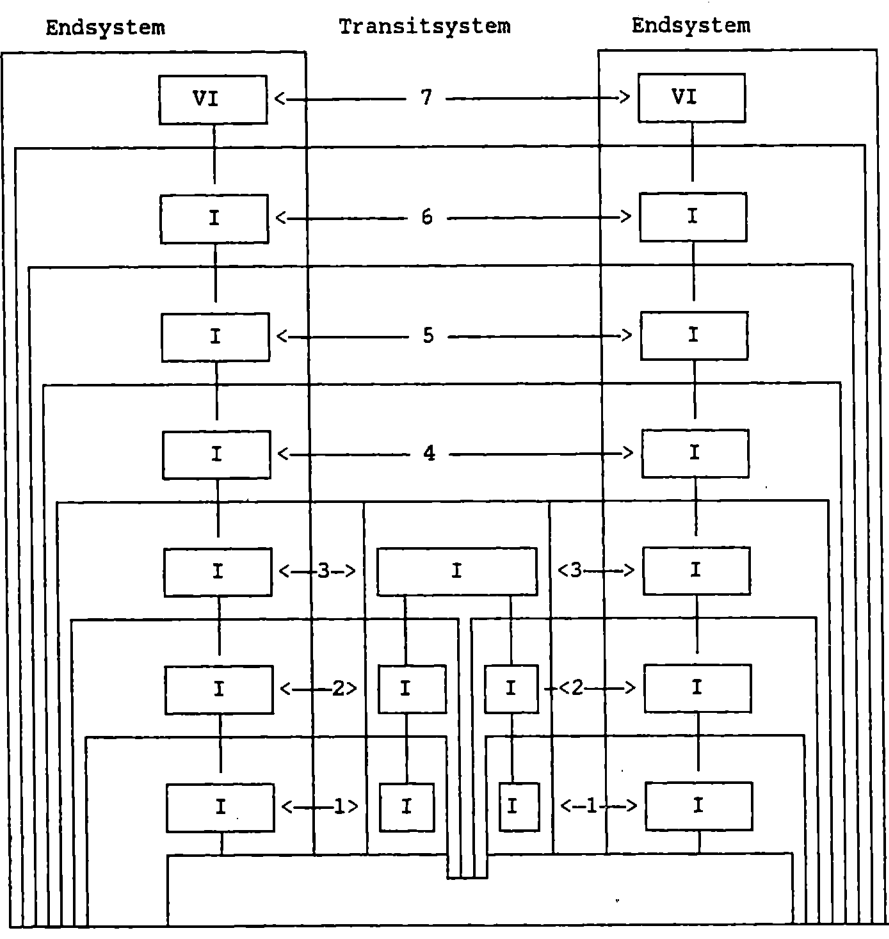

1 - Bitübertragungsprotokoll/-schicht
2 - Sicherungsprotokoll/-schicht
3 - Vermittlungsprotokoll/-schicht
4 - Transportprotokoll/-schicht
5 - Kommunikationssteuerungsprotokoll/-schicht
6 - Darstellungsprotokoll/-schicht
7 - Anwendungsprotokoll/-schicht
VI - Anwendungsinstanz
I - Instanz

Bild 1: Die Schichten und Protokolle des Referenzmodells

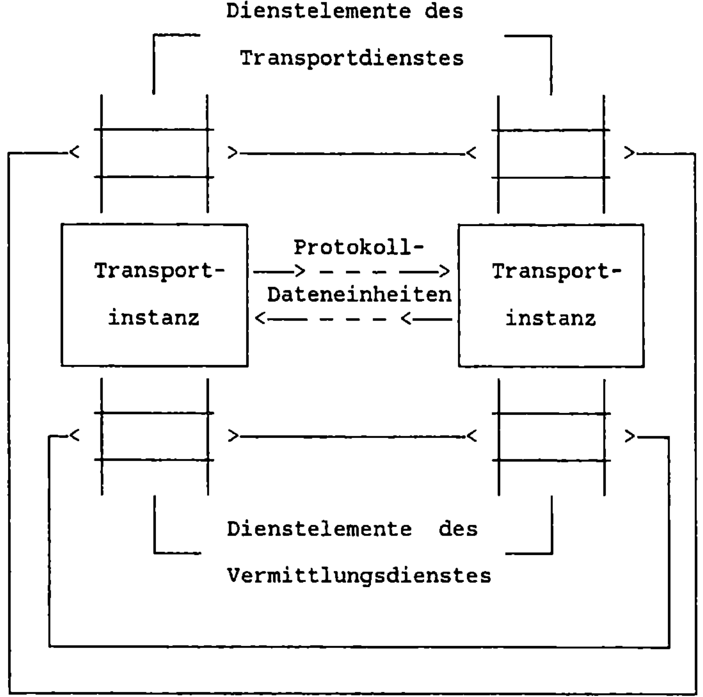

Bild 2: Einbettung der Transportinstanz

Kommunikations-
steuerungsschicht

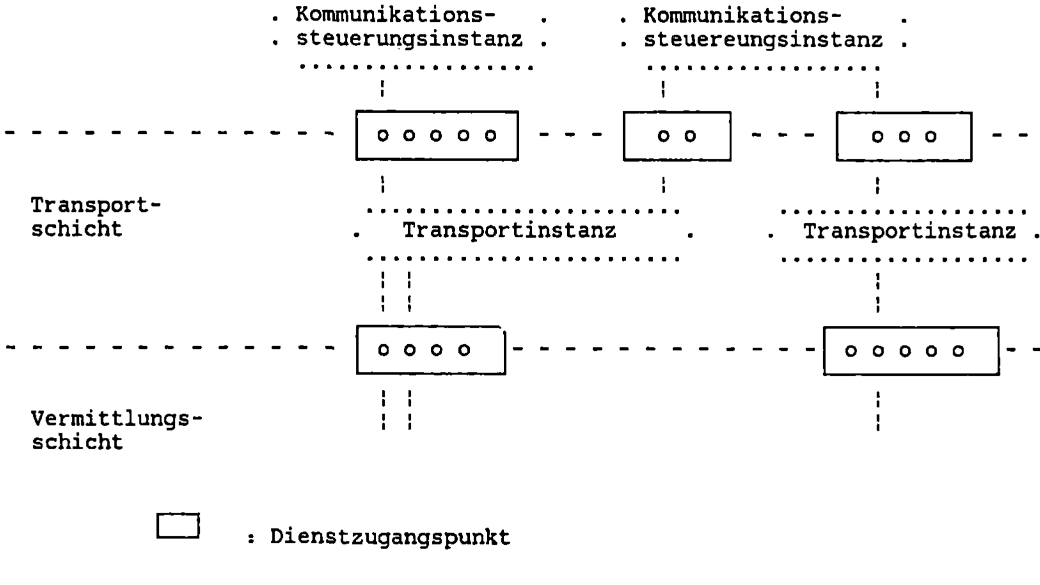

☐ : Dienstzugangspunkt

. . : Instanz

o : Verbindungsendpunktkennung

| : Verbindung

- - : Schnittstelle

Bild 3 : Dienstzugang und Verbindungsendpunkt

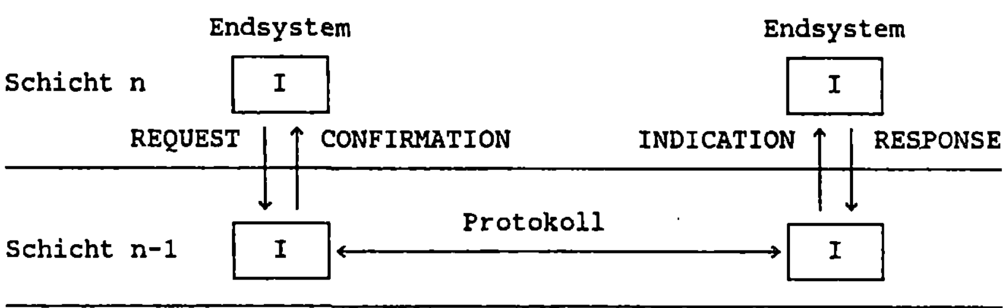

Bild 4: Dienstelemente

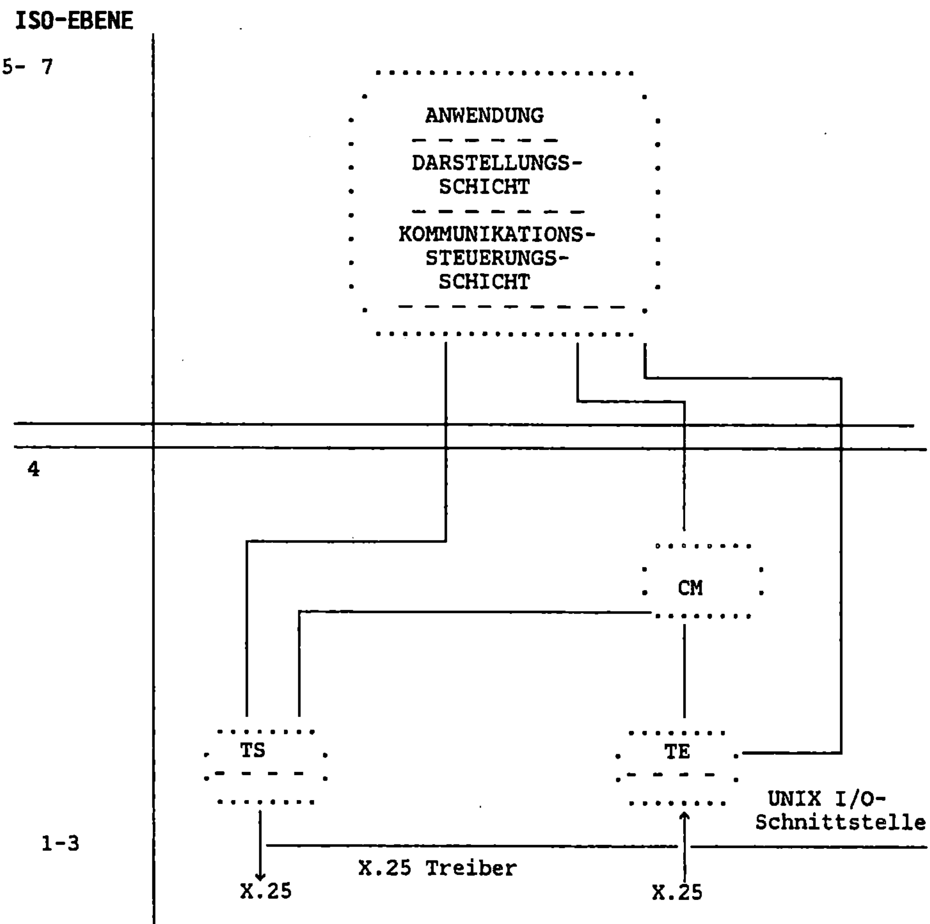

Bild 5: Kommunikationssystem, Prozesse und Schnittstellen

—————	Prozeßschnittstelle
– – – –	Programmschnittstelle
····· · · ·····	Prozeß
—————	Kommunikationspfad
TS	Transportsender
TE	Transportempfänger
CM	Communication Manager
AP	Anwendungsprozeß

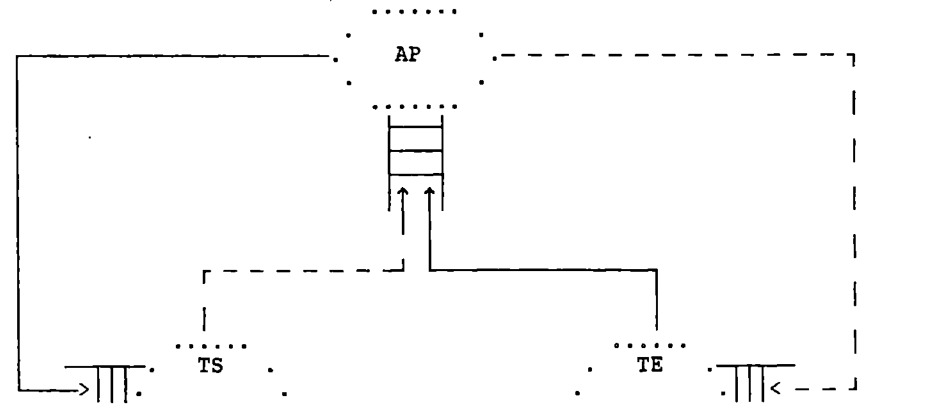

Vermittlungsschicht : entfällt bei lokaler Kommunikation;
bei nicht lokaler Kommunikation wird auf
den flußkontrollierten Vermittlungsdienst über
den X.25 Treiber zugegriffen.

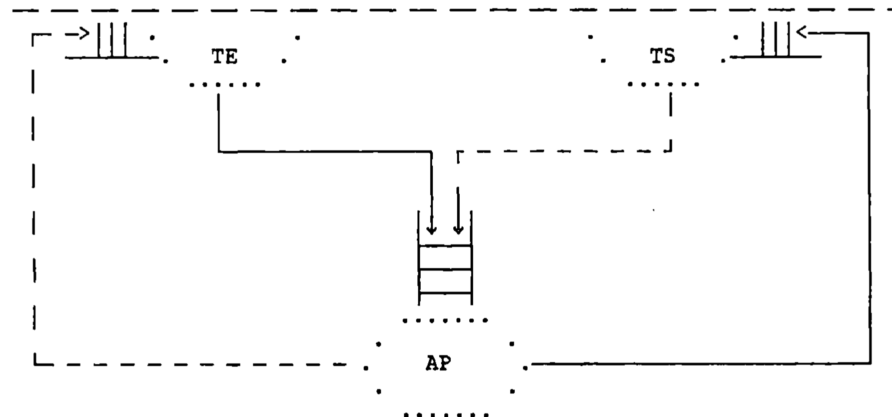

Bild 6 : Prozeßkonstellation in der Datenphase

————— : Datenpfad
------- : Quittungspfad

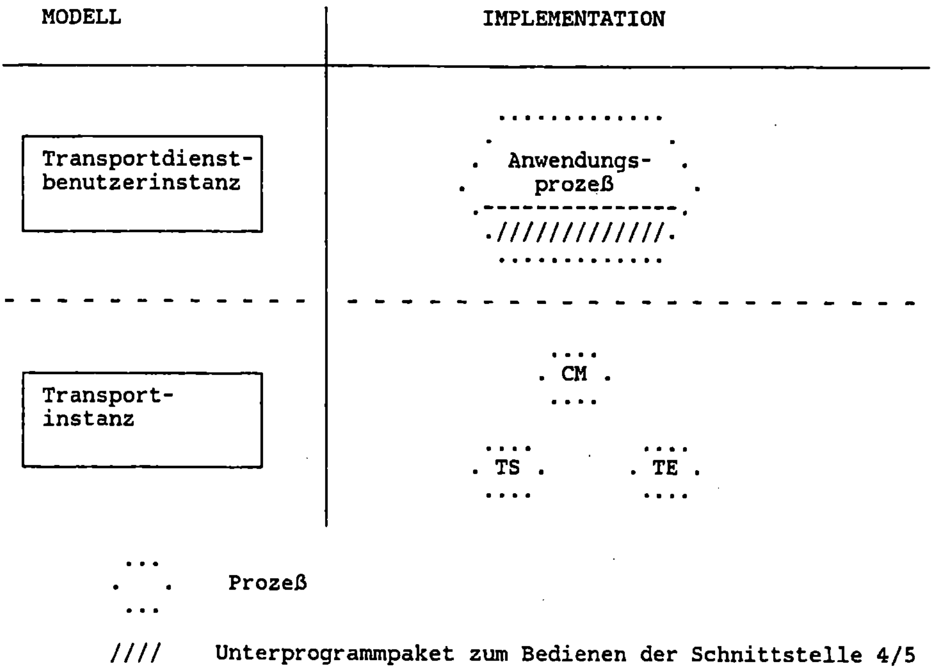

MODELL | IMPLEMENTATION

Transportdienst-
benutzerinstanz

Anwendungs-
prozeß

////////////

Transport-
instanz

CM

TS   TE

...
.   .   Prozeß
...

////   Unterprogrammpaket zum Bedienen der Schnittstelle 4/5

Bild 7 : Zuordnung von Instanzen und Schichten zu Prozessen

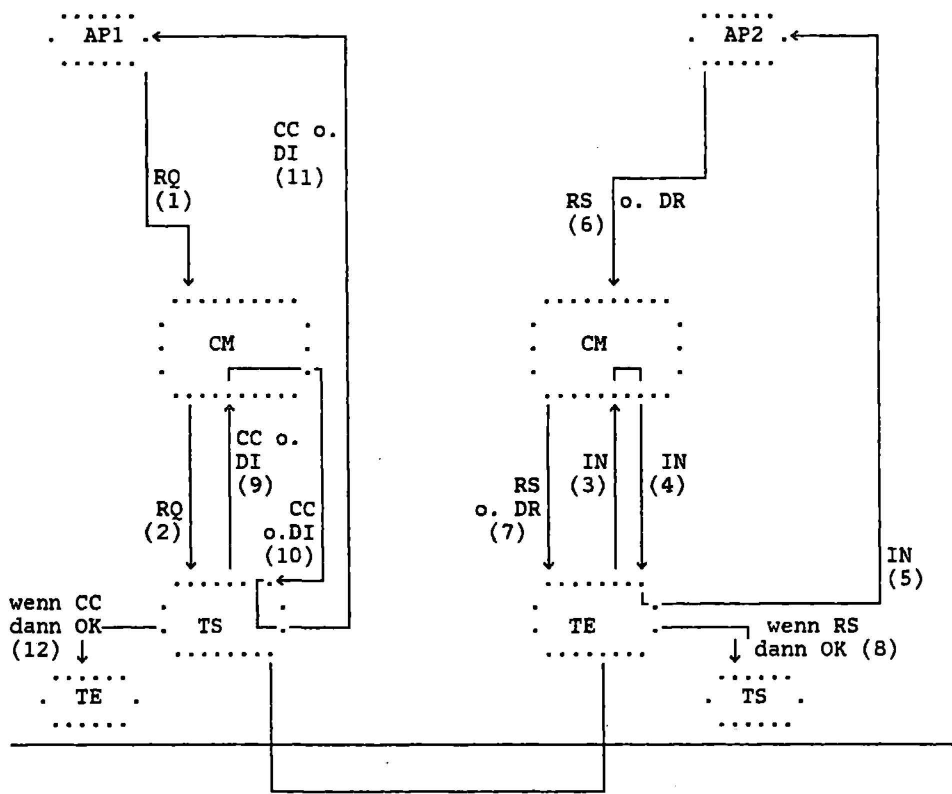

Vermittlungsschicht

Bild 8: Folge der Schnittstellenelemente für den Verbindungsaufbau im nicht lokalen Fall

verwendete Abkürzungen: RQ - CONNECT REQUEST
CC - CONNECT CONFIRMATION
DI - DISCONNECT INDICATION
IN - CONNECT INDICATION
RS - CONNECT RESPONSE
DR - DISCONNECT REQEST
OK - Meldung Aufbau erfolgreich

Die Zahlen, die in Klammern angegeben sind, repräsentieren die zeitliche Reihenfolge.

Der Austausch der Transportprotokollelemente zwischen TS und TE ist nicht dargestellt.

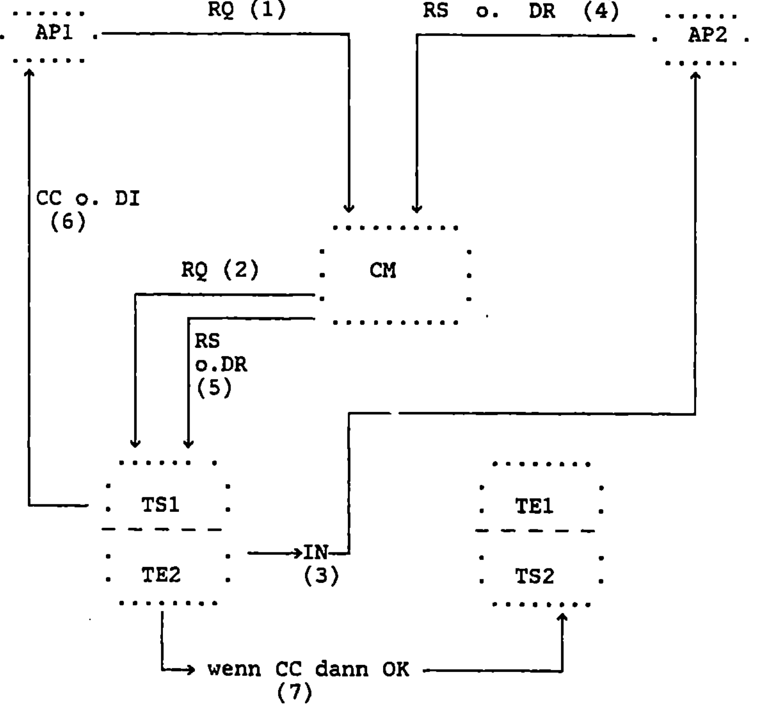

Bild 9: Folge der Schnittstellenelemente für den Verbindungsaufbau im lokalen Fall

verwendete Abkürzungen:

RQ	-	CONNECT REQUEST
CC	-	CONNECT CONFIRMATION
DI	-	DISCONNECT INDICATION
IN	-	CONNECT INDICATION
RS	-	CONNECT RESPONSE
DR	-	DISCONNECT REQUEST
OK	-	Meldung : Aufbau erfolgreich

Die Zahlen, die in Klammern angegeben sind, repräsentieren die zeitliche Reihenfolge.

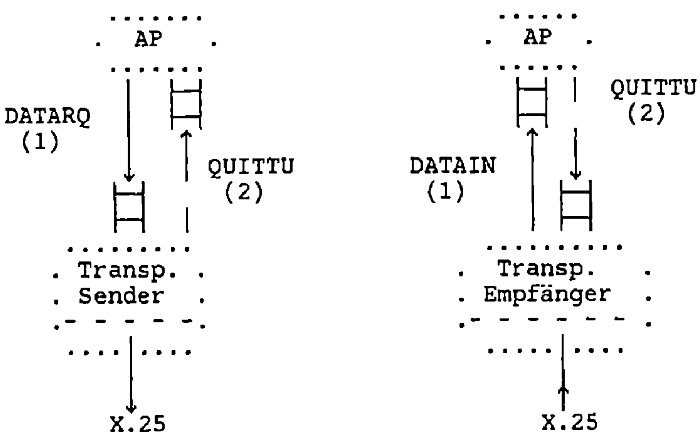

**Bild 10 : Prozesse und Warteschlangen für eine Verbindung
in der Datenphase bei nicht lokaler Kommunikation**

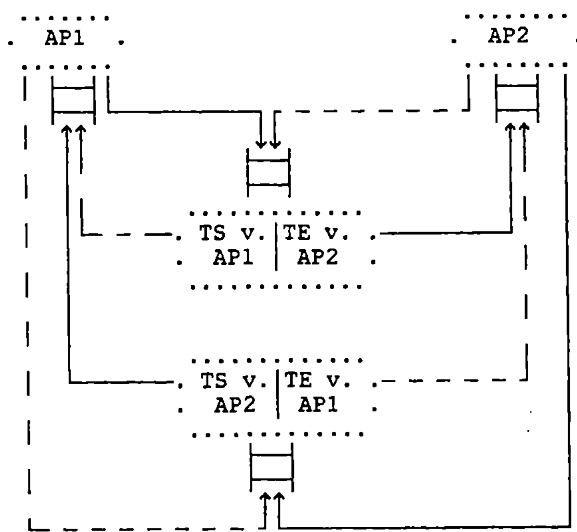

**Bild 11 : Prozesse und Warteschlangen für eine Verbindung
in der Datenphase bei lokaler Kommunikation**

————— : Datenpfad

— — — — : Quittungspfad

7. Literaturverzeichnis

[1] Basic Reference Model DIN ISO 7498

[2] DIN ISO 8072: Information processing systems - Open systems inter-
     connection - Transport service definition, ISO/TC 97 1983

[3] CCITT Yellow Book: Recommendation S.70
     "Network-Independent Basic Transport Service for TELETEX", Genova 1981

[4] DIN ISO 8073: Information processing systems - Open systems inter-
     connection - Connection oriented transport protocol specification,
     ISO/TC 97 1983

[5] DIN ISO 8509: Information processing systems - Open systems inter-
     connection - Service conventions

[6] U.Faltin,E.Faul,A.Giessler,E.Giessler,I.Günther,E.Hinsch,W.Orth,
     H.Parslow: TESDI -Test and Diagnosis services for higher level
     protocols, GMD-F2

[7] R.Vonthien: DCAM Usage Conventions to Achieve a Transport Service,
     GMD-F2 Juni 84

[8] Dr.E.Raubold: Programmschnittstelle zum ISO-"OPEN SYSTEMS"-
     Transport-Dienst, GMD-F2 April 84

[9] Burkhardt,H.J.;Eckert,H.;Prinoth,R.: Modelling of OSI-Communication
     Services and Protocols using Predicate/Transition Nets
     Proceedings of the IFIP WG 6.1 Fourth International Workshop on
     Protocol Specification, Testing and Verification,
     Mt. Pocono, USA, 1984,   North-Holland, 1984

[10] Faul-Luers,E.;Prinoth,R.: Ableitung von Implementationsvorgaben
     aus modularisierten Produktnetzen, GMD-F2 November 1984

[11] Eckert,H.,Prinoth,R.: Produktnetze-Definition eines PROSIT-
     Beschreibungsmittels, GMD-F2 April 1984

SNA ENHANCEMENTS

FOR

DISTRIBUTED DATA PROCESSING

K. M. Roehr

IBM Deutschland GmbH

Pascalstr. 100

7000 Stuttgart 80

ABSTRACT

Systems Network Architecture (SNA) defines the behavior of networks of
heterogeneous, loosely coupled processors. After a brief introduction
of SNA, this paper describes three recent enhancements to SNA which
will further increase the capabilities of SNA products and users to
interact in distributed data processing system environments.

The first enhancement allows applications in different, independent
SNA networks to communicate with one another without having to be
modified. In fact, after installation of this Network Interconnection
function, a user need not be aware that the session partner is in a
separate network.

The second enhancement is the definition of a general-purpose program-
to-program protocol that serves as a base for future application programs cooperating in a distributed system environment. This Advanced
Program-to-Program Communication (APPC) interface also serves as a
foundation upon which additional distribution services can be provided

by IBM, other suppliers of hardware and software, and owners of individual networks.

The third enhancement provides a general asynchronous (i.e., queued) data distribution facility for all applications needing such a service, e.g., office systems, file transfer, job networking, and network management. SNA Distribution Services (SNADS) are provided by a set of architected service transaction programs using APPC.

INTRODUCTION

Rapid improvements in computer and terminal function and the steadily declining cost of data processing equipment have resulted in a large growth in applications and computer hardware. This trend is expected to continue in the years to come. Parallel to this growth Systems Network Architecture (SNA) has been steadily enhanced since its first announcement in 1974. Starting from a set of functions supporting distribution of data processing between applications in a single central processor and multiple distributed cluster controllers, SNA today supports a fully intermeshed network of data processing equipment. After an introduction into the basic principles of SNA, a short overview of this development will be given.

The latest three major enhancements of SNA are discussed in more detail in the main part of this paper:

The first enhancement was triggered by the requirement for application programs in one SNA network to be accessible from terminals or application programs in another network. This requirement was satisfied by interconnecting the networks via gateways implementing the new NETWORK INTERCONNECT architecture. This architecture allows separate management and control of individual networks, as established in the past, but facilitates the dynamic creation of communication paths across several networks if the need arises.

The second enhancement was caused by the requirement that distributed application programs should be able to communicate via an SNA network without regard to the particular processor they are presently running on. This requirement was satisfied by providing a standard, common set of primitives to be used by all newly developed distributed application programs. The gerenal use of this new ADVANCE PROGRAM-to-PROGRAM COMMUNICATION (APPC) interface is expected to enhance the usage of distributed data processing applications greatly in the future.

The third enhancement was caused by the common need of many distributed applications (such as office systems, file transfer, and job networking) to distribute data asynchronously. Existing applications have generated their own distribution mechanism which satisfies certain specific needs but precludes data interchange with users of other distributed application products. The new SNA DISTRIBUTION SERVICES (SNADS) architecture fills this void by defining a common set of architected transaction programs that run in an Logical Unit Type 6.2 environment using the Advanced Program-to-Program Communication services also described in this paper. Common use of these distribution services make general data interchange between applications feasible, without having to duplicate design and implementation efforts in every new product.

The following descriptions will be limited to discuss the major characteristics of these latest SNA enhancements. More complete discussions can be found in References 4, 9, and 12.

SNA OVERVIEW

To appreciate the latest advances of SNA fully, a good unterstanding of the existing architecture is important. In the following the network elements and their internal structure will be outlined as a base for the discussions in the remainder of this paper. More detailed introductions into SNA can be found in References 1 and 2.

ELEMENTS OF THE NETWORK

An SNA network in this context is considered to consist of a number of nodes, such as terminals, controllers, and processors and the links that connect them. Depending on their capabilities, SNA differentiates between two main kinds of nodes: SUBAREA nodes and PERIPHERAL nodes.

Subarea nodes are the major building blocks of a network and are capable of routing information from one subarea node to another. Subarea nodes are typically implemented in host processors or communication controllers. Most existing peripheral nodes are only able to receive and transmit data from a subarea node and always have to depend on an associated subarea node for routing messages to other parts of the network. Some recently announced peripheral nodes (Scanmaster, Displaywriter and S/38) have the capability to send data directly to themselves. Peripheral modes consist mostly of terminals or cluster controllers.

Together, a subarea node and its associated peripheral nodes represent a SUBAREA. Routing between subareas is accomplished by using a 16-bit NETWORK ADDRESS. N=1...8 bits of this network address are used to identify the subarea. The remaining 16-N subarea element address bits are used to address local resources within the subarea, like terminals or transaction programs. The address split between subarea address and element address is selected by the network administrator and has to be the same throughout the entire network. This split determines the possible number of subareas within a network and the maximum number of addressable elements within each of the subareas.

Peripheral nodes use an 8-bit LOCAL ADDRESS to communicate with other nodes in the network. Their associated subarea node receives this local address and transforms it into a full network address before forwarding the message through the network. Similary, when the sub-area node receives a message for one of its associated peripheral nodes, it translates the received element address part of the net-work address into a corresponding local address. The component within the subarea node performing this function is called BOUNDARY FUNCTION. The effect of this address translation service in the boundary function is that changes to the network address space to not affect addresses within peripheral nodes and have only to be coordinated between the subarea nodes of the network.

As a further means to shield the user from details of the underlying physical configuration, each network element is identified externally by a NETWORK NAME which is associated with an internal network address. The mapping between network names and network addresses is provided by a separate, centralized directory within a SYSTEM SERVICES CONTROL POINT (SSCP) residing in one of a group of subarea nodes, usually within a host node. All nodes for which an SSCP provides direct name-to-address mapping services are considered to belong to the same DOMAIN. For target nodes outside of an SSCP's domain, the directory points to another SSCP in the network which can provide the mapping service directly or will be able to pass the requests to an SSCP that can.

Besides providing directory services, each SSCP is a focal point for handling configuration management and problem determination within its domain of control. To accomplish these tasks, the SSCP cooperates with PHYSICAL UNITS (PU), which are local agents located in each network node contained in its domain of control. Each node may also contain one or more LOGICAL UNITS (LU), which provide end users, e.g., terminal operators, transaction programs, or various device media, access to the SNA network. Each node element, SSCP, PU, and LU is associated with a unique network address and is therefore called a NETWORK ADDRESABLE UNIT (NAU). NAUs communicate with one another using logical connections called sessions, i.e., LU-LU, LU-SSCP, SSCP-PU, and SSCP-SSCP sessions. NAUs are the origin or the destination of all information transmitted via distributed network elements collectively called the PATH CONTROL NETWORK.

Summarizing, an overview of an SNA network is given in Figure 1:

o   Subarea nodes and peripheral nodes each containing one PU, controlling node resources, and one or more LUs, providing ports for network users.

o   Subareas consist of one subarea node and its associated peripheral nodes.

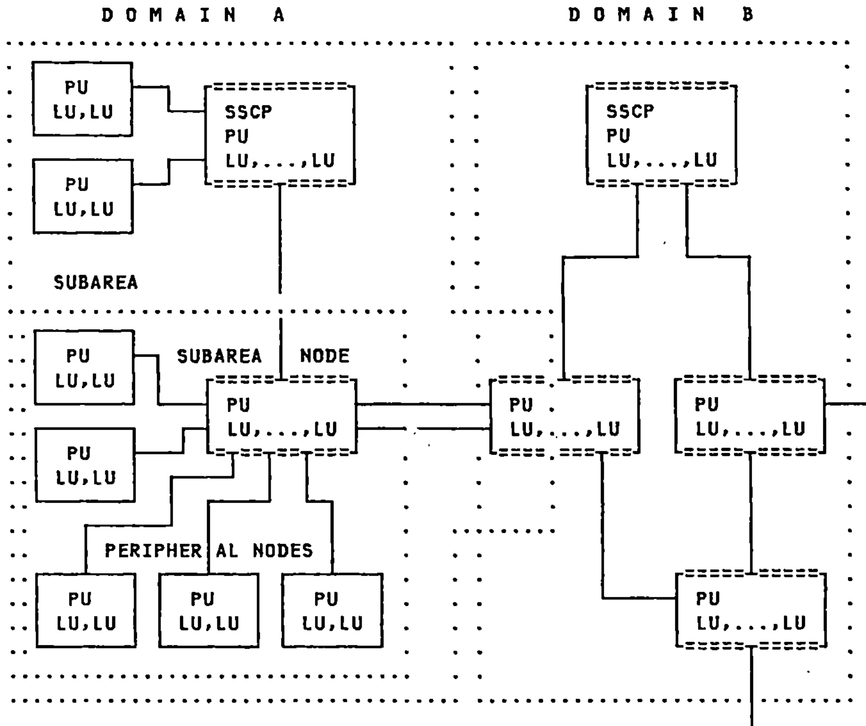

Figure 1 :   SNA Network Overview

o   Domains consisting of one or more subareas, where one of the
     subarea nodes contains the SSCP, which controls all the re-
     sources in its domain.

INTERNAL STRUCTURE OF NETWORK NODES

Functions within SNA nodes are organized into layers, as given in

Fig. 2. The following are examples of services provided in the various
layers of SNA:

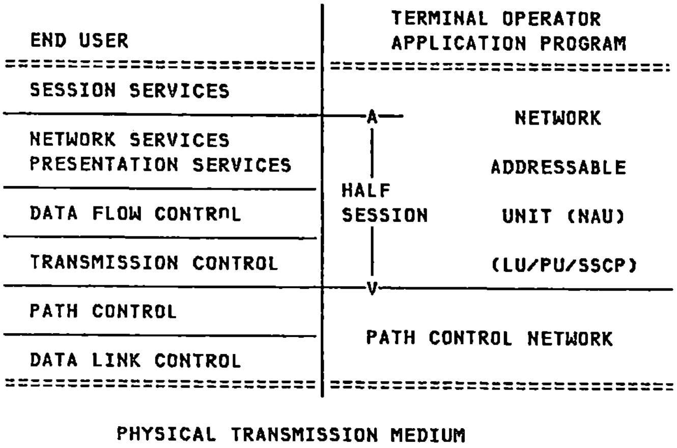

```
                                         TERMINAL OPERATOR
             END USER                    APPLICATION PROGRAM
      ===========================  ============================:
      SESSION SERVICES
      ------------------------------A---         NETWORK
      NETWORK SERVICES             |
      PRESENTATION SERVICES        |         ADDRESSABLE
      ---------------------------  HALF
      DATA FLOW CONTROL            SESSION    UNIT (NAU)
      -------------------------    |
      TRANSMISSION CONTROL         |          (LU/PU/SSCP)
      -------------------------  --V--
      PATH CONTROL
      -------------------------    PATH CONTROL NETWORK
      DATA LINK CONTROL
      ===========================  ============================

             PHYSICAL TRANSMISSION MEDIUM
```

Figure 2: Layers Within an SNA Node

o   SESSION SERVICES in LUs coordinate session initiation and
     session termination requests from end users with the help of
     the SSCP owning the domain. There exists one session services
     component per NAU.

o   NETWORK SERVICES comprise configuration and maintenance services
     (SSCP/PU) as well as management services (SSCP/LU).

o   PRESENTATION SERVICES in LU-LU sessions provide data stream
     mapping, as well as device- or program-oriented data formating.

o   DATA FLOW CONTROL provides send/receive concurrency control,
     message sequence numbering, request/response correlation,
     chaining of related user data, and enforcement of the trans-
     action oriented bracket protocol.

o   TRANSMISSION CONTROL provides session level sycnronization,
     pacing, message size control, sequence number checking, and
     enciphers and deciphers end-user data.

Together, network services, presentation services, data flow control, and transmission control constitute a HALF-SESSION component,
which depends on the specific interactions between a particular end-
user pair. There may be more than one half-session active in each NAU.

o   PATH CONTROL provides network routing and flow control for basic
     information units between NAUs by use of an additional trans-
     mission header, containing adresses, priority classes, window
     size indicators, and sequence numbers. There exists one on path
     control component per link in each subarea node (in SNA gateway
     nodes, discussed below, one path control component per network).

o   DATA LINK CONTROL schedules data transfer over a link between
     two nodes and performs error control for the link. There will
     be one link specific data link control component for each link,
     which controls the traffic through additional link headers and
     link trailers.

Path control and data link control components throughout the network
comprise the PATH CONTROL NETWORK.

At session set-up time, the type of session to be used is established. This session type depends on the nature of the parties that are trying to communicate. Depending on the selected LU-LU session type, a certain subset of services, formats and protocols will be used for the session. SNA defines presently a number of different LU-types which accomodate three major end-user groups: display terminals, printers, and programs. LU type 6, for example, provides a general-purpose inter-program protocol. Its latest release, LU 6.2, also called Advanced Program-to-Program Communication (APPC), will be discussed in this paper.

SNA HISTORY OVERVIEW

Systems Network Architecture was first announced by IBM in 1974. Since then, the original set of functions, which supported distribution of data processing between applications in a single central processor (S/360 under DOS) and multiple distributed cluster controllers (3600s), has been enhanced by the addition of many new functions. The major announcement milestones of this development are:

o   Single host-to-remote terminal connection, via system /370 channel, communication controller, SDLC link, and terminal cluster controller. Definition of first LU. (September 1974)

o   Attachment of remote communication controllers via SDLC links. (March 1975)

o   Attachment of remote cluster controllers via switched, dial-up links. Definition of LUs 1, 2, 3. (September 75)

o   Attachment of SNA networks to public packet-switched X.25 networks. (Februar 77)

o   Support of multiple-hosts connected via links between communication controllers. This allowed terminals controlled by one
host to access applications in any host in the network as well
as direct communication between host processors. Multiple SSCPs
establish multiple domains of control and cooperate in establishing and terminating sessions. Definition of LU types 4, 6,
and 7. (December 77)

o   Improved network management support for error recovery and
dynamic reconfiguration of peripheral nodes. Support of parallel
sessions allows multiple concurrent conversations between users
in hosts. (November 78)

o   Extended path control capabilities, like parallel links between
subareas, multiple active routes between two subareas and transmission priorities, can be used to increase network capacity
and reliability. (June 79)

o   Support of the CCITT Recommendation X.21 interface, which allows
users to choose any data link control and higher level protocol.
Support for Centrale Modem, and major Communication Network
Management functions. (December 79)

o   A standard, general-purpose LU-LU session protocol referred to as
ADVANCED PROGRAM-to-PROGRAM COMMUNICATION (APPC) to be used for
future communication program products. (July 82)

o   SNA DISTRIBUTION SERVICE (SNADS) provides a general asynchronous
     data distribution facility for SNA applications using the APPC
     interface. (November 83)

o   Support for interconnecting several SNA networks via gateways.
     This NETWORK INTERCONNECT support affects only the path control
     network and is accomplished without impact to any application
     program. (December 83)

A more detailed discussion of these enhancements up to 1980 can be
found in Reference 3. The characteristics of the latest three SNA
enhancements Network Interconnect, APPC, and SNADS will be discussed
in the following sections in the sequence of their occurrence in the
layered SNA structure.

SNA NETWORK INTERCONNECTION

During the time since its first introduction in 1974, SNA networks
have been established by many users. The rapid improvements in computer and terminal functions together with declining costs resulted
in large growth of these networks. To support this trend, IBM has
announced its intent to extend the range of possible subarea node
adresses. In addition, there exists a definite requirement for
application programs in one SNA network to be accessible from terminals or application programs in another network. This is generally
achieved by interconnecting the networks via gateways (see Ref. 4 and
5). A gateway in this context must have the capability to accept SNA
messages from one network, to translate these messages into a form
that can be unterstood in the other network, and to transmit them to
the appropriate destination.

REQUIREMENTS ON INTERCONNECTION

The prime requirements for interconnecting networks is to avoid major
disruption of existing operating procedures and to maintain network
independence and network management autonomy within each network,
while facilitating full communication between the networks. That is,
an application programmer or a terminal operator should be able to
use the same methods for requesting and controlling SNA sessions,
independent of whether the session partner is located in the same
or in an other network. This implies that existing application programs and terminals are able to participate in inter-network sessions
using the same conventions, formats, and protocols that were used before the networks were interconnected.

Similarly, network management should not be significantly more difficult in interconnected networks. Specifically, the following should
be true between interconnected networks:

o   Individual address spaces of networks remain independent; that
     is, duplicate network addresses are allowed to exist.

o   Network name spaces of networks remain independent; that is,
     duplicate network names are allowed to exist.

o   Mutual knowledge of SSCP control domains is not required between
     networks.

o   Physical and logical configuration changes should remain in-
     visible outside of the network. For example, changing the
     address split between subarea and subarea element addresses,

modifying route or domain definitions, or adding new links or nodes, should not necessitate corresponding changes in other networks.

o   Malfunctions in one network should not affect other networks; that is, an ailing network should not be able to paralyze control or flood buffers in neighboring networks.

While it is important to maintain independence between networks, it should also be possible to collect error and performance statistics across networks, to perform fault isolation in inter-network sessions, and to run tests in other networks.

In addition it is desirable that individual networks can be used as independent building blocks to create larger, more global configurations. Flexibility should exist to connect networks in series or in parallel, to have more than one gateway between networks or to use one gateway to interconnect more than two networks.

GATEWAY SOLUTION

The solution to the requirement stated above is to insert a gateway function between the networks, which translates names and addresses for all messages sent from one network to the other. Figure 3 shows the distribution of addresses and names after a gateway was installed and a session was initiated between LUA and LUB.

In network A, for instance, LUB is known by its alias name NAMELB' instead of by its original name NAMELB. Similarly, LUA is known in network B by its alias name NAMELA' instead of by its original name

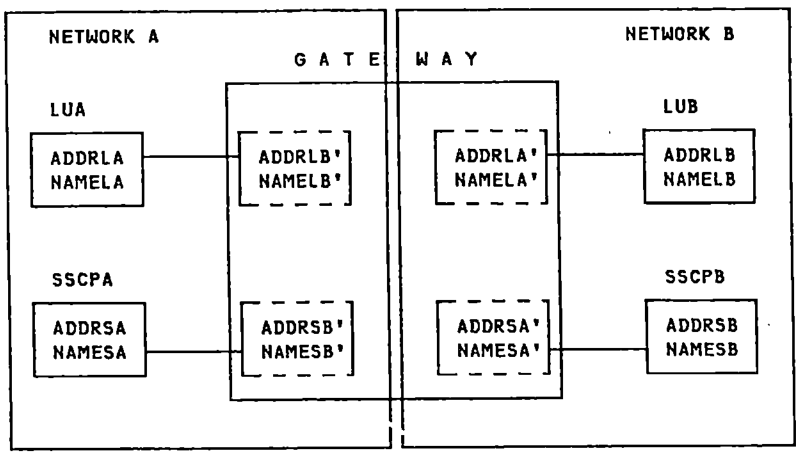

Figure 3: Names and Addresses Used in Internetwork Sessions.

NAMELA. Session traffic from LUA to LUB will have different adresses in the transmission header fields, depending upon through which network it is currently moving. In network A, the transmission header contains the real address of LUA, ADDRLA, and the alias address for LUB, ADDRLB'. After reception by the gateway, the transmission header is translated to addresses unterstood in network B, that is, the alias address of LUA, ADDRLA', and the real address for LUB, ADDRLB. The fact that LUB and SSCPB are really located in another network is not apparent to LUA.

From the viewpoint of network A, the gateway represents a subarea of this network. The LUs in network B, with which sessions have been established, appear to be terminals attached to the gateway subarea.

Each alias address in the gateway that is assigned to LUs and SSCPs of network B or any other additional network has to be assigned from the space of element addresses available in the gateway subarea belonging to network A. This limits the number of alias addresses that can be used at a gateway at any one time. However, by dynamically reusing the element addresses in a gateway subarea, this limitation is alleviated for all practical purposes. At session initiation time, the gateway assigns a free element address from the pool of addresses at the subarea to represent a NAU of the other network. At session termination time, when there is no further need for the address, the address is returned to the free pool of element addresses and can subsequentily be used to represent another NAU.

To avoid duplication of network names in interconnected networks, a user is free to choose alias network names for NAUs of the other network. For example, at network installation time a user of network A may decide to refer to a CICS application in the neighboring network B as CICSB to differentiate it from a CICS application that exists already in his own network. At the gateway SSCP user-defined tables are created that map alias logical unit names to actual logical unit names. At session initiation time the gateway SSCP uses these tables in conjunction with associated routing tables for forward the request from a NAU in one network to the proper SSCP in the other network. A similar translation process occurs when a cross-network session is terminated.

As was discussed above, the gateway has basically two functions (compare Fig. 4):

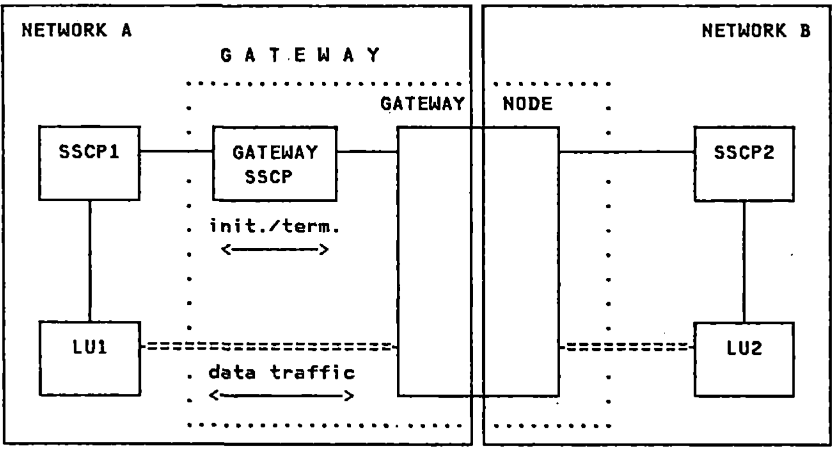

Figure 4: Operation of Gateway Components

o   To handle SSCP-SSCP request traffic between networks, especially
     during session setup and takedown. For this traffic the gateway
     handles name translations, address translations and rerouting of
     requests to the proper SSCP in the other network. The gateway
     component performing these functions is called the gateway SSCP.
     It is physically located within the associated host computer
     (VTAM, Ref. 6).

o   To handle LU-LU session traffic between networks, once sessions
     have been set up by the gateway SSCP. For this traffic the gate-
     way performs the address translations in the transmission
     headers. The gateway component performing this function is
     called the gateway node. This gateway node component allows the
     gateway to look like a subarea node in each network to which it

attaches. The gateway node is physically located within a communications controller. The connection between gateway and gateway SSCP can be directly via a channel or indirectly via a telecommunications link through an SNA network.

Figure 4 shows only one gateway node and one gateway SSCP in network A used for interconnecting the two networks A and B. In principle, one gateway can be used to interconnect more than one network. In this case only one network will ~ontain the gateway SSCP, while the gateway node can be shared among all interconnected networks. Sometimes it may be advantageous to connect two networks by more than one gateway, for example, to improve performance of a connection or to connect two disjoint parts of a network via a conveniently located link in an adjacent network. The amount of sharing and parallelism that is practical will be mainly determined by cost/performance trade-off considerations.

In general, operator commands processed by a control point in one network cannot control resources in another network. This isolation of one network from another is a fundamental, basic objective. For operators that have to control resources in interconnected networks outside of their own networks, the Network Communications Control Facility (NCCF) program product can be provided by IBM (Ref. 7). NCCF allows an operator at a terminal in one domain of an SNA network to issue operator control commands to access resources in other domains of the same or other networks.

Problem determination for sessions through a gateway is inherently more complex because the messages leave the jurisdiction of one network and enter the jurisdiction of another. In addition, problem determination is being complicated by the change of addresses in the transmission header when a message moves from one network to another. To deal with these additional complexities, the existing problem determination tool Network Logical Data Manager (NLDM) has been updated accordingly (Ref. 8).

ADVANCED PROGRAM-TO-PROGRAM COMMUNICATION

LU type 6.2, also known as Advanced Program-to-Program Communication (APPC), represents the latest release of LU type 6. Access to this function is provided via the APPC protocol boundary which is intended for use by general transaction service programs like distribution services (Ref. 9), network services (Ref. 7), file transfer (Ref. 10), job entry (Ref. 11), or any arbitrary user program. The entire system below the APPC boundary can be understood as a distributed operating system which provides services to programs wishing to communicate via an SNA network (see Ref. 12 and Fig. 5).

Requirements for a general program-to-program service are outlined below and the APPC interface and its associated transaction verbs are described.

PROGRAM-TO-PROGRAM COMMUNICATION REQUIREMENTS

The most important requirements for a general inter-program communication service are as follows:

o   A high level of abstraction which shields the application programmer from the details in the implementation of the service.

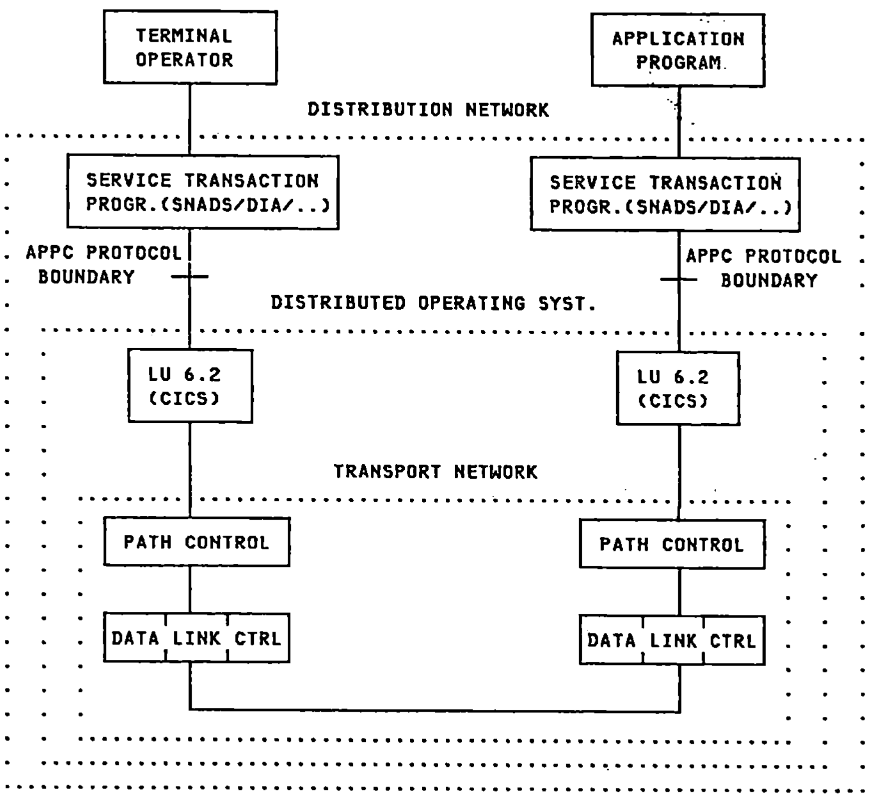

Figure 5 : Communication Accross SNA Network

o   Efficient execution of protocols by using a minimum number of data transfers between nodes.

o   Provision for a complete set of functions like:

-   Sending and receiving of data,

-   Notifying partners if conversation fails,

- Selection of transport characteristics such as delay,
  packet size, cost, and security,

- Mapping and formatting support for data to be send or
  received,

- Allowing more than one parallel session between nodes,

- Commitment control via checkpoint and resynchronization
  support,

- Transaction identification for use in tracing, activity
  logging, and accounting

o   Adequate control of session traffic, and receipt of information
    on status of transaction by both partners.

o   Provision of optional subsets, allowing for ease of implementa-
    tion over a wide range of product sizes.

o   Support of application programs using earlier program-to-program
    interface.

DEFINITION OF ADVANCED PROGRAM-TO-PROGRAM COMMUNICATION

The requirements stated above are satisfied by the set of pro-
gramming-language-like statements, called verbs. In the following a
brief description of these verbs and their usage will be given. More
complete description can be found in Reference 13 and 14.

ALLOCATE activates a conversation between the issuing program and a
named partner program. The named partner program is started and given
addressability to the program that started the conversation. The
issuer of ALLOCATE has the initial permission to send. Besides

carrying the LU-NAME and the transaction program name associated with the partner program, the ALLOCATE verb carries a MODE_NAME parameter identifying the type of transport service the conversation should use: if encryption is to be used, if low delay of messages is required, or if bulk data is to be handled.

SEND_DATA transfers data into a local buffer and returns control to the transaction program. Data consists of logical records, each record consisting of a two byte length field followed by a data field. The data are actually sent if either the allocated buffer space at the sender side is exhausted or a verb is issued which causes the buffers to be flushed (e.g., CONFIRM, DEALLOCATE, SYNCPT).

RECEIVE_AND_WAIT transfers the issuing program into a state where it is waiting for data or control information to arrive. Incoming data will be collected and transferred to the program.

REQUEST_TO_SEND asks the partner program for the right to send. This request does not cause an immediate interrupt and the partner may or may not grant this request. If the request is granted, the partner program informs the requesting program of this fact by including a parameter in a subsequent verb (e.g., CONFIRM, SEND_DATA, ...) sent over the session.

CONFIRM ends a message and asks the partner to assure that no errors have been detected in the message. The receiving program will answer with CONFIRMED if no error was detected, or it may signal an error by issuing a SEND_ERROR verb in return.

SEND-ERROR informs the remote transaction program that the local

transaction program detected an error. When the program is in receive
state, the rest of the incoming message is discarded and the remote
transaction program is informed that an error has been detected. In
send state, the send buffers are sent to the other side and the
logical record is truncated. SEND_ERROR always leaves the local program in send state, which now may decide to send further information
to allow its partner program to recover from the error situation.

DEALLOCATE ends the conversation. The allocation can be either completed with this verb, or it can be deferred until the program
issues a CONFIRM, FLUSH, or SYNCPT verb.

FLUSH causes all buffered data and control information to be sent.

SYNCPT (syncpoint) makes all accumulated changes to protected resources permanent, unless data errors or resource failures require
a rollback to a boundary defined by the previous SYNCPT execution.

A simple one-way conversation between two transaction programs A and
B may demonstrate the use of the APPC verbs defined above:

TP A issues the following verbs:

(1) ALLOCATE(TPN=B)   Request to start conversation with
                                    TP B is placed in send buffer.
                                    Nothing is sent.

(2) SEND_DATA   Data to be sent is placed in send
                                    buffer behind allocation request.

	Nothing is sent (buffer large enough).
(3) DEALLOCATE(TYPE=FLUSH)	Ends conversation. Buffer contents are put on link. TP A remains active.

```
        MESSAGE((3)(2)(1))
TP A ──────────────▶ TP B
```

Transfer of buffer contents from TP A to TP B. TP B is started.

TP B issues the following verbs:

(4) RECEIVE_AND_WAIT (WHAT_RECVD=DATA_COMPLETE)	TP B waits for data to arrive. Return Parameter indicates that a complete logical record was received.
(5) RECEIVE_AND_WAIT (WHAT_RECVD=DEALLOCATE_FLUSH)	Waits for more data to arrive Return parameter indicates that conversation was ended by A.
(6) DEALLOCATE(TYPE=LOCAL)	TP B also ends the conversation.

This example shows how large-enough buffers can result in only one data transmission across the network. Alternately, if buffers are a critical resource, several packets may have to be transferred for each transaction. This way APPC performance can be tuned by adjusting packet sizes and the number of available buffers without affecting the sequence of APPC verbs.

Besides the basic conversation verbs discussed above, APPC supports also a set of mapped conversation verbs, which hide certain options and details from the program. Mapped conversations are designed to be used by application programs written in high-level languages. If the data mapping option is used in a mapped conversation, a map name is sent with the transmitted data so that the receiving map support can understand the format of the data. Product support of mapped conversation verbs is optional.

APPC also supports a set of control operator verbs intended for use by programs that assist an operator in performing functions related to the control of LU, including:

o   Changing the number of permissible sessions between two LUs,

o   Examining and changing LU operating parameters,

o   Activating and deactivating LU-LU sessions.

Product support of control operator verbs is optional.

All products implementing APPC are required to support a minimum basic set of functions. Use of this basic set of functions by a transaction program assures its ability to communicate with all programs using an APPC interface, including those that may be added at a later time.

Functions that are not part of the base set are optional. Optional functions have been grouped into a limited number of option sets. A product that wishes to implement any particular function within an option set must implement all functions in this set. This way, connectivity between all programs using the same option set is assured.

The Advanced Program-to-Program Communication interface described here provides a foundation upon which additional distributed services can be provided by IBM or other suppliers. Examples of products providing the APPC interface already today are S/38, CICS, DISPLAY-WRITER, and SCANMASTER. One specific example of a general product architecture using this interface is the SNA Distribution Service discussed next.

SNA DISTRIBUTION SERVICES (SNADS)

The wider use of office systems and other distributed applications has generated the need for a common architecture for interchanging data asynchronously among diverse systems and products. SNADS is IBM's answer to these needs.

Asynchronous communication implies that the sender may submit a request to the distribution service without participation or knowledge of the receiver. Requests are queued within the distribution service and are forwarded along the route toward the receiver as communication paths become available. Depending on the priority of the data and the availability of communication resources, data may be delayed at intermediate points along the route. At the receiving end, the distribution service queues the data and remains responsible for a safe delivery until the destination application program has accepted the data.

SNA Distribution Services are provided by a set of architected transaction programs which use the APPC services described above. The set of distribution transaction programs and the underlying SNA distributed operating system together form an SNA Distribution Network (compare Fig. 5). The requirements leading to the SNADS design and the major concepts used in implementing it are discussed below.

REQUIREMENTS FOR DISTRIBUTION SERVICES

The most important requirements for a general distribution service architecture are:

o   Accommodation of any type of data, such as documents, files, and digitized audio.

o   Efficient handling of a wide range of distribution sizes, starting from short messages to large size bulk data.

o   Explicit naming of specific receiving program must be possible.

o   Provision for sending specific handling instructions along with each type of request sent across the network.

o   Ability to distribute information to multiple destinations using one single distribution package.

o   Facility to return a standard notification message identifying the status of distribution requests.

o   Ease of managing the network, usage by application programs, terminal operators, and system administrators should only be minimally affected by changes in the network configruation, movement of users, addition or deletion of resources should be localized to the affected entities of the network and should not have any affect on application code.

o   Efficient use of network resources, such as storage space and access paths, line capacity, and processor cycles.

o   Capability to be easily extended to satisfy future requirements. Provisions have to be made to allow graceful evolution of the architecture without having to modify its basic structure.

The SNA Distribution Services architecture satisfying the above requirements and will be described below.

DESCRIPTION OF SNA DISTRIBUTION SERVICES ARCHITECTURE (SNADS)

In the following a brief description of SNA Distribution Services and
the underlying architectural concepts will be given. More complete
descriptions can be found in References 9 and 15.

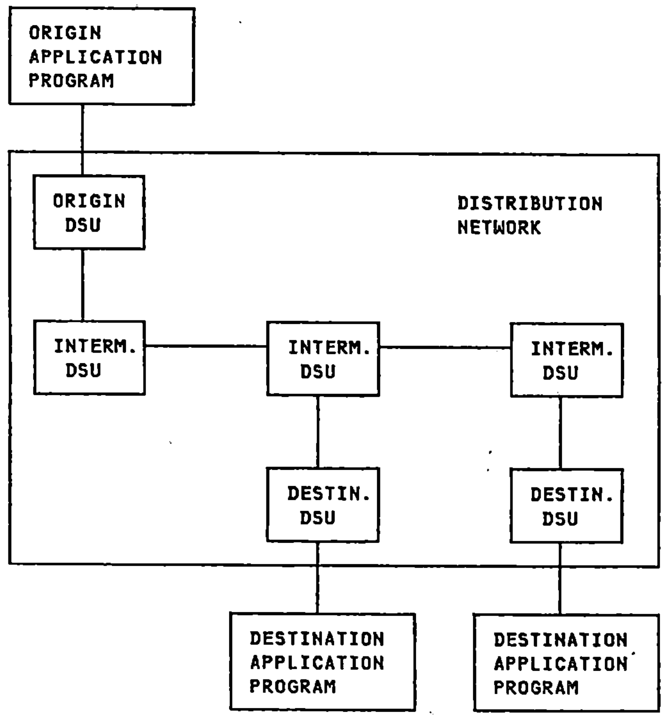

Figure 6 : Data Transfer over Distribution Network

Distribution of data across the distribution network is accomplished by transmitting the data from one DISTRIBUTION SERVICE UNIT (DSU) to the next (see Figure 6). A DSU consists of distribution transaction programs running in an LU 6.2 environment. The interface between application transaction programs and distribution services consists of three verbs:

DISTRIBUTE_DATA initiates an asynchronous distribution containing application data. This verb contains a number of parameters that are supplied by the sending application:

o   Name of user where distribution originates

o   Name of user and the application that are to receive the distribution

o   User supplied correlation ID, to be returned in feedback (optional)

o   Service level, defining priority, protection, and DSU storage capacity needed for routing the request

o   Name of user and application to receive feedback (optional)

o   Service level required for feedback (optional)

o   Parameters needed by the distribution units at both ends to access and store information (optional)

After execution the DISTRIBUTE_DATA verb is used for returning the following parameters from the DSU to the origin application:

o   Return code specifying the result of the verbs execution, e.g.: OK, Invalid Origin Name, System Error, ...

o   Request identifier allowing the origin application to identify the distribution request by date, time, and sequence number

DISTRIBUTE_STATUS operates similarly as DISTRIBUTE_DATA but initiates status distribution within a SNADS network. Typical status types are: Syntax Error, Function Not Supported, Routing Error, and Invalid Destination Unit Name.

RECEIVE_DISTRIBUTION receives a distribution queued for one or more users. The applicable queue is identified by a parameter of the verb. After execution, the verb is used for returning the distribution parameters, originally sent by the origin application (see DISTRIBUTE_DATA verb above), from the destination DSU to the destination application(s).

Each of the above distribution interface verbs is accepted by the DSU and placed in the appropriate queue. Each DISTRIBUTE verb will be resolved by the DSU into a series of APPC interface verbs sent to the underlying LU 6.2 implementation, as soon as an LU 6.2 session becomes available. A DISTRIBUTE_DATA verb, for sending a data object over the distribution net, may be resolved, for example, into the following sequence of APPC verbs:

```
ALLOCATE
SEND_DATA (PREFIX)
SEND_DATA (DISTRIBUTE COMMAND)
SEND_DATA (OBJECT)
SEND_DATA (SUFFIX)
CONFIRM
```

where the terms in parentheses represent one interchange unit. Its encoding structure is defined by the Document Interface Architecture (DIA) described in Ref. 16.

In addition to the verbs used by applications for distributing data, another set of operator verbs is intended for managing SNADS resources like tables and queues within the DSUs (see Ref. 15).

NAMING AND ADDRESSING

In SNADS the distribution user name is defined as a two-part, hierarchical name consisting of Distribution Group Name (DGN) and subordinate Distribution Element Name (DEN). The Distribution Group Name can be used to define user groups, such as departments or divisions. The distribution user name DGN.DEN identifies the user and does not change if a user moves from one location to another or if the configuration of the distribution network is changed.

The user address, which identifies the user's location, is defined as a two-part, hierarchical Distribution Service Unit Name (DSUN) consisting of a Routing Group Name (RGN) and a subordinate Routing Element Name (REN). Presently the Routing Group Name is optional, but it can be used to group Distribution Service Units according to geographical location, which leads to smaller routing tables. DSU names are used for routing the data between the various DSUs in a network.

The independence between user names and addresses assure insulation of application transaction programs from changes in the distribution network, but it requires the use of directory tables at each DSU which correlate user names with associated DSU names. These tables have to be updated whenever a user moves and connects to another DSU.

ROUTING AND DIRECTING

The distribution service is responsible for routing a distribution
from the input queue at the origin DSU to the output queue at the
destination DSU. To accomplish this task, a routing component in each
DSU has the function to determine the next DSU names for all incoming distributions, to route them to the appropriate distribution
queues, and to start the appropriate program for further processing
of these queues. During this process, one entry from an input queue
may cause several entries in different distribution queues if multiple receivers were specified.

If a distribution request is received directly from an application
transaction program, directing services first have to determine the
destination addresses (i.e., the destination DSUNs) from the specified user names. If a distribution request is received from another
DSU, routing services may have to replace the destination DSUN if
a user was moved to another DSU and the distribution has to be re-
directed.

Once the proper DSUN is determined, routing services uses this name
together with the distribution service level (DSL) to find the name
of the next-DSU queue (QNAME), the LU_NAME of the next DSU, and the
MODE_NAME required for transport services (see Fig. 7). LU_NAME and
MODE_NAME are parameters of the ALLOCATE verb sent across the APPC
interface to start a synchronous communication with the next DSU.

When a distribution arrives finally at a destination DSU, directing
services enqueue the distribution on the proper output queue(s) and
invoke the destination transaction program defined by a parameter in

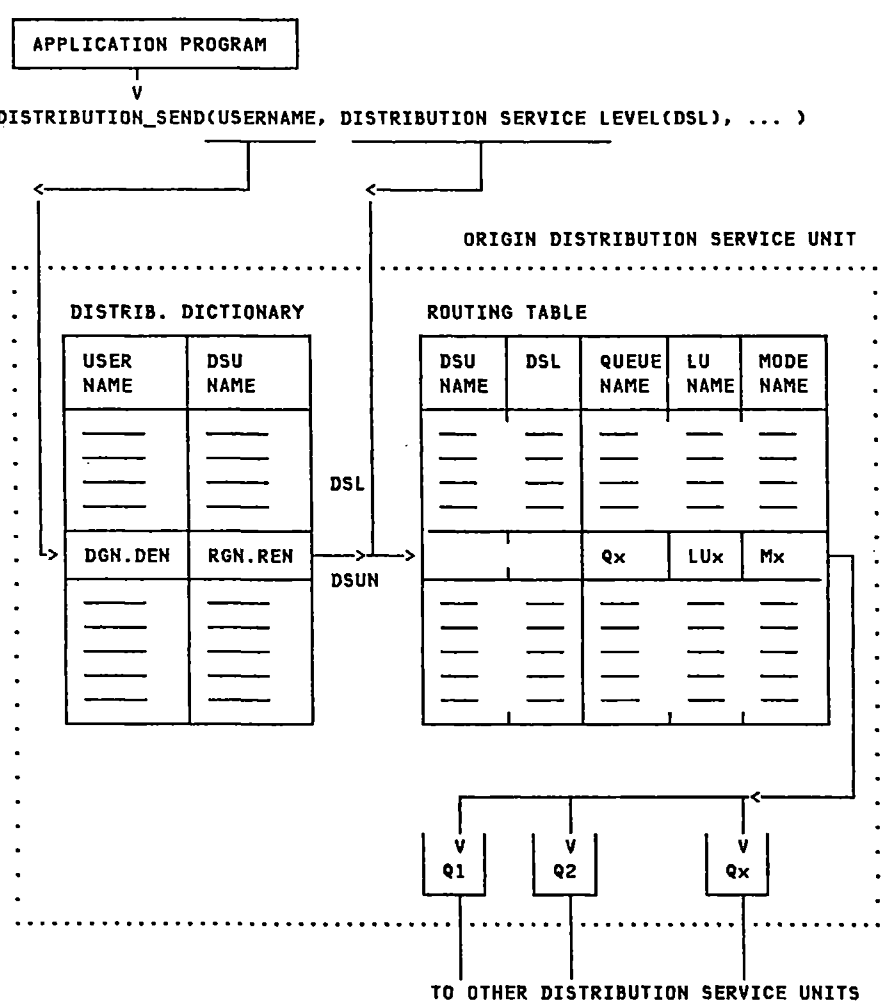

Figure 7 : Processing at Origin Distribution Service Unit

the distribution command. This Transaction Program Name (TPN) gives
the originator the ability to select different processing for diffe-

rent distributions addressed to the same recipients.

<u>SERVERS</u>

In the past, most asynchronous distribution systems have required
that the distribution object be moved from the user's space to a
temporary buffer, such as a spool file. Such a scheme allows the design of a relatively simple control program but requires at least
twice the object's size for data storage. For very large objects
these additional storage requirements may become intolerable. Therefore, SNADS uses a set of servers to access the user's data space
directly.

Servers are application-specific programs that access the distribution objects. Servers are invoked from a DSU by using a set of verbs,
such as INCREMENT_OBJ_LOCK, READ_INITATE, READ, or WRITE. From the
viewpoint of SNADS, servers act as source or sink for the object data
stream. SNADS uses the server name to determine which server to call,
but otherwise has no knowledge of the contents of the object data
stream or any server specific processing of the data. Typical functions performed by servers at origin and destination are:

o   Encoding and decoding of data contents

o   Data stream transformations such as compression or de-
     compression

o   Processing of server specific profiles (such as profiles defined
     in the Document Interchange Architecture, Ref. 16).

o   Interfacing with local data management components, like data base
     access methods or library services

The most simple and general type of a server is used to read or write byte-perfect copies of data streams without regard to the semantics of distribution objects. Such a general server is required at every intermediate DSU and is considered to be part of the distribution service.

First applications of SNADS are in the office systems environment, where data is interchanged using formats described in the Document Interchange Architecture. The first products using SNADS are DISOSS Version 3 Release 2 and the IBM 5520 Release 5. The intent to support SNADS was announced for IBM's 8100 DPPX/SP, S/36, S/38, and Series/1.

SUMMARY

After an outline of existing Systems Network Architecture (SNA), a short historic overview of SNA was given. Before discussing each of the three latest enhancements of SNA, Network Interconnect, Advanced Program-to-Program Communication (APPC), and SNA Distribution Services (SNADS), the requirements leading to the selected solutions are outlined in each case. One intent of the paper is to show how these latest enhancements, which make it easier to design and implement new distributed applications, represent logical extensions of existing SNA services.

The first enhancement, Network Interconnect, is accomplished by changes that affect only network components contained in the path control component of transport network services. Similarly, the second enhancement, APPC, only affects the LU services above the transport network, and the third enhancement, SNADS, only affects components above the APPC interface (see Fig. 5). These latest developments are another example of how the layered structure of SNA allows the step-by-step, planned and orderly expansion of networking capabilities to satisfy the ever expanding requirements of advanced distributed data processing applications.

ACKNOWLEDGEMENTS

The author feels expecially indepted to the authors of previously published material on discussed topics: R.J. Sundstrom and G.D. Schultz (Ref. 3), J.H. Bemjamin, M.L. Hess, R.A. Weingarten, und W.R. Wheeler (Ref. 4), J.P. Gray, P.J. Hansen, P. Homan, M.A. Lerner, and M. Pozefsky (Ref. 9), B.C. Housel and C.J. Scopinich (Ref. 12). In addition many persons have contributed to the architecture and implementation of SNA. Each of the above references contains a detailed list of the main contributors to the latest SNA enhancements. Finally, the author has to thank E.L. Miller for technical guidance, G.D. Schultz for editorial help, and M.A. Lerner for management encouragement received during this work.

REFERENCES

1. System Network Architecture: Concepts and Products, GC30-3072, IBM Corporation; available through IBM branch offices.

2. System Network Architecture: Technical Overview, GC30-3073, IBM Corporation; available through IBM branch offices.

3. R.J. Sundstrom and G.D. Schultz, "SNA's First Six Years: 1974 - 1980", Predeedings of the Fifth International Conference on Computer Communications, Atlanta, GA, October 27 - 30, 1980, pp. 578 - 585.

4. J.H. Benjamin et al., "Interconnecting SNA Networks", IBM

Systems Journal, vol. 22, No. 4, 1983, pp. 344 - 366.

5. Network Program Products General Information, GC27-0657, IBM
    Corporation; available through IBM branch offices.

6. ACF/VTAM Version 2 General Information: Introduction, GC27-0608,
    IBM Corporation; available through IBM branch offices.

7. NCCF General Information Manual, GC27-0429, IBM Corporation;
    available through IBM branch offices.

8. NLDM General Information Manual, GC30-3081, IBM Corporation;
    available through IBM branch offices.

9. B.C. Housel and C.J. Scopinich, "SNA Distribution Service", IBM
    Systems Journal, vol. 22, No. 4, 1983, pp. 319 - 343.

10. File Transfer Program, GH12-5129, IBM Corporation; available
    through IBM branch offices.

11. Network Job Entry for JES2, GC23-0100, IBM Corporation; available
    through IBM branch offices.

12. J.P. Gray et al., "Advanced Program-to-Program Communication in
    SNA", IBM Systems Journal, vol. 22, No. 4, 1983, pp. 298 - 318.

13. Introduction to Advanced Program-to-Program Communication,
    GC24-1584, IBM Corporation; available through IBM branch offices.

14. SNA: Transaction Programmer's Reference Manual for LU Type 6.2,
    GC30-3084, IBM Corporation; available through IBM branch offices.

15. SNA Format and Protocol Reference Manual: Distribution Services,
    SC30-3098, IBM Corporation; available through IBM branch offices.

16. T. Schick and R. Brockish, "The Document Interchange Architec-
    ture: A Member of a Family of Architectures in the SNA Environ-
    ment", IBM Systems Journal, vol. 21, No. 2, 1982, pp. 220 - 244.

<u>COMMUNICATION SUPPORT FOR DISTRIBUTED DATABASE SYSTEMS</u>[*]

K. Rothermel
Universität Stuttgart
Azenbergstr. 12, D-7000 Stuttgart-1
Fed. Rep. of Germany

<u>ABSTRACT</u>

A communication kernel for distributed database systems is presented. This kernel provides communication primitives that support the creation, execution and termination of nested transactions. Moreover, this kernel maintains so-called transaction state tables, which are used to recall uncompleted transactions during restart recovery. Besides a detailed description of these primitives, the paper presents an addressing mechanism based on the new concept of functional port classes.

1. INTRODUCTION

Up to now a lot of different communication concepts have been developed. Unfortunately, the majority of these concepts are convenient for session oriented systems rather than for transaction oriented systems, such as distributed database systems, reservation systems or banking systems. However, as more attention is paid to transaction oriented applications, new techniques for transaction oriented communication continue to be developed. These developments are taking place in operating systems (e.g. see /Panz82/, /Rash81/, /Tane81/, /Walk83/), programming languages (e.g see /Lisk83/, /ADA83/, /Andr81/) and communication models (e.g. see /Nels81/, /Roth84a/, /Spec82/).

In this paper a communication kernel for distributed database systems (DDBSs) is presented. Besides low-level communication primitives, the presented kernel provides high-level primitives for supporting the creation, execution and termination of nested transactions. Moreover, the kernel maintains so-called transaction state tables, which are used to recall uncompleted transactions during restart recovery. On the one hand, these primitives are gerneral and flexible enough to support a great variety of communication patterns, and therefore are a good foundation on which to implement protocols for DDBSs efficiently. On the other hand, they are not so simple that implementations using them are difficult and time-consuming.

---

[*] This work was supported in part by the United States Army through its European Research Office under Grant DAJA 37-82-C-0419.

In order to support various kinds of transaction oriented systems the proposed kernel includes only 'mechanisms' and no 'policies'. Thus, it might be a first step towards a kernel that provides general services for transaction processing systems. In the context of ISO's Open Systems Interconnection Architecture the functions of the kernel would be located in the Application Layer, in which they would form a sublayer providing services for transaction processing systems.

Up to now the kernel supports two types of transactions, nested transactions and a very simple type of flat transactions. Due to space limitations, in this paper we will concentrate only on nested transactions. A detailed description of the primitives supporting the other transaction type can be found in /Roth84b/. For a discussion of the motivation for the proposed kernel, we must also refer to a companion paper /Roth84c/.

The remainder of the paper is organized as follows. The next section introduces the objects provided by the kernel, and presents an addressing mechanism based on the new concept of functional port classes. Section 3 describes the communication primitives provided by the kernel, and finally, Section 4 gives a brief summary.

2. OBJECTS

The basic objects of the proposed kernel are transactions, processes, basic modules, ports and functional port classes. The first part of this section introduces transactions, processes, basic modules and ports, and developes a conceptual framework for structuring DDBSs. In the second part the new concept of a functional port class is presented.

2.1 TRANSACTIONS, PROCESSES, BASIC MODULES AND PORTS

The users interact with the DDBS by executing atomic transactions. An atomic transaction is an execution consisting of a set of actions which take place indivisibly in the presence of both failures and concurrent executions. That is, either all of the actions are executed or the transaction has no effect, and other transactions executing concurrently cannot modify or observe intermediate states of the execution. The transaction model used by the kernel will be described in detail in Sec. 3.2.1.

Basic module objects are the basis for structuring DDBSs. A DDBS consits of a collection of basic modules cooperating with each other for the purpose of processing transactions. From an implementation point of view a basic module is logically associated with a sequential program and consists of a collection of processes (schedulable tasks), which are executions of the program associated with the basic module. That is, all processes residing in the same basic module are executions of the same sequential program. Each basic module exists entirely at a single node.

From a behavioral point of view a basic module defines and exports a set of functions for the use by transactions. When a transaction requires a certain function, it issues a request to a basic module exporting the required function. During its execution a transaction may issue several requests to the same basic module. With regard to transaction processing basic modules can be organized in different ways (for a detailed discussion see /Roth84d/). In the following we will describe two basic types of module:

- The process structure of the basic module is static. The requests of a transaction can be executed by different processes of the basic module. Conversely, each process may execute on behalf of several transactions, simultanously. We will refer to this as type1-module. For example, the System D software for transaction processing consists of three distinct types of modules, application modules, data manager modules, and storage modules /Andl81/, each of which is implemented as a type1-module.

- The process structure of the basic module is dynamic: For each transaction that requires a service provided by the module a process is created that executes all requests of this transaction; the process lives at most as long as this transaction is active. Consequently, all requests of a transaction are executed by the same process of the basic module, and conversely, each process executes on behalf of one transaction. We will refer to this as type2-module. The Relational Base Machine of POREL /Walt84/ is organized as type2-module, for instance. Most parts of R* /Lind84/ and Distributed INGRES /Ston77/ are also implemented as type2-modules.

Processes can communicate with one another by exchanging messages via their ports. A port is an object into which messages can be placed and from which messages can be removed. Associated with each port is a queue on which reside messages sent to that port but not yet removed from it by a process. Ports are of two types, private or common. A private port is owned by the creating process, and only the owner may remove messages from a private port. Common ports can be opened and closed by processes, and only a process that has opened a common port may remove messages from it. In contrast to private ports, common ports may be shared by multiple processes.

2.2 FUNCTIONAL PORT CLASSES

In this section, first the addressing concept used by the kernel is presented, and then the membership of ports in fp-classes is discussed. Finally, this section describes two basic types of fp-classes.

2.2.1 Addressing

The addressing concept used by the kernel is a form of global addressing /Andr83/, which is the underlying communication concept applied in many current message-based systems (e.g. see /Rash81/, /Panz82/, /Tane81/, /Mao80/). The majority of these systems use ports (or similar objects) to identify functions: a process requesting a certain function must send a request to a port logically associated with this function. For the requestor the identity of the process providing this function is insignificant. What is significant is that this process provides the requested function.

In DDBSs frequently the same function is identified by a collection of ports. For example, consider a type2-module consisting of k processes each of which owns a separate port for each function provided by the module. In this example each function provided by the module is identified by k ports. As will be seen later, arranging ports identifing the same function into groups allows a very general kind of addressing.

The communication concept used by the kernel provides mechanisms for arranging ports into so-called functional port classes (fp-classes). Instead of ports this concept uses fp-classes to identify functions. An fp-class consists of a collection of ports, each of which corresponds to one or more transactions. A port residing in an fp-class c that corresponds to a transaction t is called the <u>correspondent</u> of t in c. At any point of time each transaction has at most one correspondent in each fp-class. When a transaction requires a certain function, it sends a request to its correspondent in the fp-class associated with the required function. For identifing ports two types of identifers are provided:

- <u>Port Numbers</u>: The port number of a port is generated by the kernel at port creation time. Most systems using global addressing only provide this type of identifer.
- <u>Functional Addresses</u>: A functional address consists of an (fp-class identifier, transaction identifier)-pair, which identifies the correspondent of a transaction in an fp-class. Of course, only ports residing in fp-classes can be identified by fuctional addresses.

Assume that a basic module provides the functions f1,...,fn, each of which is identified by a separate fp-class. The fp-class identifing function fi is denoted by ci. A transaction t that requires a function fi must send a request to its corresponent in ci, which is identified by a (ci,t)-pair. For the following discussion we will define the interface between a transaction and a module to be the set of identifers the transaction must know to be able to request the functions provided by the module. The interface of the basic module fulfills the following properties if functional addressing is employed for identifing ports:

(P1) The interface of the basic module is identical for each transaction. To be able to request functions f1,...,fn, each transaction must know the same set of identifiers, namely {c1,...,cn}. This holds for each type of basic module.

(P2) The interface of the basic module is independent of its internal structure. That is, an existing basic module can be reimplemented as a basic module of any other type unless this has any effect on its interface.

Since basic modules can be used to compose more complex modules, the proposed addressing concept allows the implementation of an arbitrary complex functional unit that provides the same interface for each transaction. Moreover, the interface of a functional unit is absolutely independent of the functional unit's internal structure, i.e. the internal stucture of the functional unit can be modified arbitrarily unless this has any effect on its interface.

2.2.2 Types of Correspondents

Ports can be inserted into and removed from fp-classes. At each point of time a port may be member of at most one fp-class. Two types of members are possible:

- <u>Private Correspondents</u>: A private correspondent corresponds to exactly one transaction. Each fp-class contains at most one private correspondent per transaction.
- <u>Common Correspondents</u>: Each fp-class contains at most one common correspondent. A common correspondent corresponds to all transactions for which no private correspondent exists in the fp-class.

For discussing the memberships of ports we will use the following terminology. A port may become a member of an fp-class by nomination. As will be seen later, a port can be nominated in two ways, explicitly by the application and implicitly by the kernel. The membership of an port in a fp-class can by terminated by removing the port from the fp-class. Note, removing a port from an fp-class does not mean that the port is destroyed.

Fig. 1 shows the mapping of the correspondent of transaction t in fp-class c to the members of c. If there exists a private correspondent of t in c (p-correspondent$_t$), this member represents the corresponent of t in c, independent whether or not c contains a common correspondent (c-correspondent). If no private correspondent of t exists in c, two cases must be considered: if c contains a common correspondent, the correspondent of t is represented by this member; otherwise there exists no correspondent of t in c.

correspondent of t	c-correspondent existent	c-correspondent non-existent
p-correspondent$_t$ existent	p-correspondent$_t$	p-correspondent$_t$
p-correspondent$_t$ non-existent	c-correspondent	---------

Figure 1. Representation Matrix

2.2.3 Types of FP-Classes

For the following discussion we will introduce the notion of a 'destination-non-existent' message. A functional addressed message is called 'destination-non-existent' if the correspondent the message is addressed to does not exist. Note, a message addressed to the correspondent of transaction t in fp-class c is only 'destination-non-existent' if neither a private correspondent of t nor a common correspondent exists in c. We do not use the term 'misaddressed' here, since, as will be seen later, a 'destination-non-existent' message can be well-addressed, too.

The kernel provides two basic types of fp-classes, conservative fp-classes and creative fp-classes. With regard to the way they nominate correspondents, conservative and creative fp-classes can again be subdivided into nominating and non-nominating fp-classes. Due to space limitations, in this paper we will only describe two types of fp-classes, conservative non-nominating fp-classes and creative nominating fp-classes. A description of conservative nominating fp-classes and creative non-nominating fp-classes can be found in /Roth84d/.

Conservative fp-classes conserve the process structure of the system when a 'destination-non-existent' message arrives. A 'destination-non-existent' message addressed to this type of fp-class is misaddressed and hence discarded. The correspondents in a <u>conservative non-nominating fp-class</u> cannot be nominated by the kernel but only can be nominated explicitly by the application. Conservative non-nominating fp-classes are used to model type1-modules. Assume, for example, that the functions provided by a type1-module are identified by one conservative non-nominating fp-class which only contains a common correspondent. All 'not busy' processes in the module listen to the common correspondent, which corresponds to all transactions. When a request arrives, one of these processes receives and executes the request. Therefore, a process can receive requests of arbitrary transactions, and conversely, the requests of a transaction can be executed by arbitrary processes.

The creative fp-class objects are the basis for the dynamic creation of processes. Each creative fp-class is logically associated with a basic module. Whenever a 'destination-non-existent' message arrives, a creative fp-class creates a new process in the associated basic module. The correspondents in a <u>creative nominating fp-class</u> can be nominated explicitly by the application as well as implicitly by the kernel. Whenever a 'destination-non-existent' message arrives, the kernel implicitly nominates a correspondent. In particular, the following actions are performed by the kernel when a 'destination-non-existent' message m addressed to the correspondent of a transaction t in a creative nominating fp-class c arrives:

(1) A private port is created and m is placed into this port.
(2) The new port is nominated to the private correspondent of t in c.
(3) In the basic module associated with c a new process is created. This process becomes the owner of the new port and therefore has exclusive access to m <u>and</u> the following messages of t directed to c.

Creative nominating fp-classes are suitable to model type2-modules, which create one process for each requesting transaction. Assume, for instance, that the functions provided by a type2-module are identified by a creative nominating fp-class which is logically associated with the module. When the first request of a transaction arrives, the kernel automatically creates a port representing the transaction's correspondent. Furthermore, in the basic module it establishes a process, which has exclusive access to the first and all following requests of the transaction. Note, since the transaction's correspondent already exists when the second request arrives, only the first request of a transaction causes a new port and a new process to be created.

Of course, type2-modules can also be implemented without creative nominating fp-classes. However, then an additional monitor process is needed, which receives the first request of a transaction, establishes a new process in the basic module, creates a new port and forwards the message to this port. It should be clear that such a solution needs more context switches and extra message passing and hence is less efficient than a solution with creative fp-classes.

3. COMMUNICATION PRIMITIVES

The kernel provides two types of communication primitives, low-level primitives and high-level primitives. The low-level primitives are described in the first part of this section. The transaction model used by the kernel and the high-level primitives based on this transaction model are described in the second part of this section.

3.1 LOW-LEVEL PRIMITIVES

In order to control transaction processing in DDBSs a lot of protocols, such as protocols for concurrency control, recovery control or commit control, are needed. Since these protocols follow a multitude of different communication patterns, it is necessary to have general and flexible communication primitives that are a good foundation for implementing protocols efficiently. In /Roth84a/ we examined several communication concepts with regard to their suitability in the context of transaction oriented systems and showed that a simple asynchronous send primitive is the only operation that supports all indicated patterns provided extra message passing is to be avoided. This send primitive does not guarantee reliable message transfer, i.e. messages can be lost and duplicated and can arrive out of order. Further arguments for low-level primitives can be found in /Salt81/ and /Lisk79/.

1. Send (SetOfPortId, TId, MsgType, Message)
2. Listen (SetOfPortNo, Timeout) -> (PortNo)
3. RemoveMessage (PortNo) -> (TId, MsgType, Message, FuncAdr)
4. Listen&Remove (SetOfPortNo, Timeout) -> (PortNo, TId, MsgType, Message, FuncAdr)

Figure 2. Low-Level Communication Primitives

The low-level communication primitives provided by the kernel are shown in Fig. 2. Since multidestination patterns occur in a large number of protocols for DDBSs, the kernel provides a send primitive that supports multicasting: _Send_ multicasts the specified message to the message ports designated by SetOfPortId. The destination ports' may be identified by port numbers or functional addresses. Send is non-blocking and does not guarantee reliable message delivery. Parameters TId and MsgType are optional - they can be used to associate the message with a transaction identifier and a (user-defined) message type, respectively.

The _Listen_ primitive listens to the set of ports designated by SetOfPortNo. A process issuing Listen is blocked if no message is available from one of the specified ports. The issuing process waits for a message to arrive or times out after Timeout units of time, whichever happens first. The port number of the port from which a message is available is returned in parameter PortNo. The _RemoveMessage_ primitive removes a message from the port specified by PortNo and returns the message in parameter Message. The transaction identifier and the type of the message is returned in TId and MsgType. If the destination port of the message is identified by a functional address, this address is returned in FuncAdr. The _Listen&Remove_ primitive is a combination of Listen and RemoveMessage. When a message is available from one of the specified ports, Listen&Remove immediately removes the message.

3.2 HIGH-LEVEL COMMUNICATION PRIMITIVES

The communication primitives described in the previous section are unreliable, i.e. an application using them must worry about lost, duplicated and out-of-sequence messages. On the one hand, this property is necessary the provide the flexibility required to support a great variety of communication patterns, efficiently. On the other hand, if only these low-level primitives are provided, the development of complex applications may become difficult and time-consuming. To avoid this, the kernel additionally provides a set of high-level communication primitives which efficiently support transaction processing in DDBSs. In this section, first we will introduce the concept of nested transactions, and then present the primitives supporting the execution of nested transactions in a distributed environment.

3.2.1 Nested Transactions

In the context of database systems a transaction represents an atomic unit of work, which transfroms the database from a consistent state to another consistent state. Atomic transactions form the basic units of both recovery and concurrency control and can be characterized by two semantic properties:

- All-or-nothing property: A transaction either completes entirely or has no effect, i.e. either all changes of a transaction are applied to the database or none of them, independent of communication failures and node crashes.
- Indivisibility property: The effect of executing transactions concurrently must be the same as if the transactions are executed one after the other in some serial order.

In a nested transaction model, each transaction can invoke subtransactions, which can invoke subtransactions, and so on. Subtransactions provide a mechanism for introducing concurrency within a transaction, as well as an added measure of robustness. A transaction whose initiator is not a transaction is called top-level transaction. In the following we will use the term 'transaction' to denote both top-level transactions and subtransactions. Tree terminology will be used in discussing relationships between transactions. Transactions having subtransactions are called parents, and their subtransactions are their children. We will also speak of ancestors and descendants.

Transactions can terminate either normally by committing or abnormally by aborting. Subtransactions appear atomic to the surrounding transaction and may commit and abort independently. A subtransaction may abort without affecting the outcome of the surrounding transaction. However, the commitment of a subtransaction is relative; even if the subtransaction commits, aborting one of the transactions containing it will undo its effects. Updates become permanent only when the enclosing top-level transaction commits. Therefore, top-level transactions are special, they are the only irrevocable transactions.

3.2.2 Primitives

For describing the high-level primitives we will introduce the notion of an agent. An agent is a functional unit, which is made up of one or more basic modules residing at the same node. Each transaction is executed entirely at a single agent, which is called the transaction's home agent. In order to execute a transaction, an agent may create any number of subtransactions to be executed at other agents. The agent that creates a subtransaction (and runs the subtransaction's parent) is called the parent agent of the subtransaction. The kernel, which exists at each node of the computer network, provides services supporting the interaction between the agents running related transactions. In order to interact with each other, the agents use the services provided by their local kernel. In particular, the kernel provides the following services:

<u>High-level communication primitives:</u>
The kernel provides primitives supporting the creation, execution and termination of nested transactions. These primitives guarantee a reliable message delivery, i.e. an application using these primitives need not worry about damaged, duplicated, lost and out-of-order messages.

- <u>Maintainance of state tables:</u>
Since transaction atomicity must be preserved despite of node crashes, nodes must remember that they are involved in transaction processing despite the loss of volatile storage during node restart. Therefore, we define the local stable state of a transaction at a node to be the state to which the transaction will return following restart recovery. Since stable states must survive node crashes, they have to be recorded in stable storage /Lamp81/. The initial stable state of a transaction is 'UNKNOWN'. This corresponds to a transaction that has no state information recorded in stable storage. Also the state is 'UNKNOWN' when the transaction is completed. At each node there exists a <u>transaction state table (TST)</u>, which is maintained by the local kernel. The information stored in the local TST allows the application to recall transactions during restart recovery. The TST is included in the kernel, because it is also used by the kernel to recover from node failures.

- <u>Protocol recovery from node failures:</u>
In order to preserve the atomic property of transactions, recovery facilities are used to be able to bring the data stored in the system to a state that reflects only the effects of committed transactions following node failures. These facilities can be subdivided into protocol recovery facilities and data recovery facilities. The former recovery techniques guarantee that, despite of node failures, the agents interact with each other in a defined and correct manner, while the latter ones are used to be able to undo or redo the effects of

transactions. A widely applied data recovery technique is based on undo/redo-logging /Gray78/. Another technique using intention lists is described in /Lamp81/. The protocol recovery facilities are provided by the kernel, whereas the data recovery facilities must be implemented by the application itself. The strict separation of data recovery and protocol recovery allows the application to choose its own data recovery techniques, which hence can be tailored to the specific needs of the application.

In the remainder of this section we will sketch the kernel's high-level communication interface. Fig. 3 depicts the high-level primitives provided by the kernel, and Fig. 4 illustrates interactions between agents performing related transactions.

A top-level transaction can be created by calling the CreateTop primitive. When CreateTop terminates, parameter TId contains the globally unique identifier of the new transaction. In order to perform a transaction, an agent may create any number of subtransactions to be executed at other agents. The CreateSub primitive creates a subtransaction of the transaction identifed by parameter ParentTId; parameter HomeId designates the node of the subtransaction's home agent. After termination of CreateSub, TId contains the globally unique identifier of the new subtransaction.

After an agent has created a subtransaction, it may request the subtransaction's home agent to perform actions under the subtransaction. An agent may request the execution of an action by calling the Work primitive, which issues a WORK request message describing the action to be performed to the corresponding home agent (see Fig. 4a). Parameter ToPortId identifies the destination port, and parameter Action specifies the action to be performed within the subtransaction identifed by parameter TId. The kernel guarantees that an agent having sent a WORK request will receive a DONE, COMMITTED, ABORTED or UNAVAILABLE response from the port identifed by ResPortId.

The parent agent of a subtransaction may send a sequence of WORK requests to the subtransaction's home agent. When the parent agent has sent all WORK requests it desires to sent, it can call the Finish primitive, which issues a FINISH request message to the subtransaction's home agent (see Fig. 4b). Parameter ToPortId iden-tifes the destination port, and parameter TId designates the subtransaction to be finished. The Work&Finish primitive is a combination between Perform and Finish, and delivers a WORK&FINISH request message to the destination port specified in the pri-mitive (see Fig. 4b). An agent having sent a FINISH or WORK&FINISH request message will receive a COMMITTED, ABORTED or UNAVAILABLE response from the port designated by parameter ResPortId. The Work, Finish and Work&Finish primitives are non-blocking, i.e. an agent may issue several request in parallel. However, at most one response message may be outstanding per subtransaction at any instant.

1. CreateTop () -> (TId)
2. CreateSub (ParentTId, HomeId) -> (TId)
3. Work (TId, ToPortId, ResPortId, Action)
4. Work&Finisch (TId, ToPortId, ResPortId, Action)
5. Finish (TId, ToPortId, ResPortId)
6. Done (TId, ReturnData)
7. CommitSub (TId, ReturnData)
8. AbortAtHome (TId)
9. AbortAtParent (TId)
10. CommitTop (TId, ResPortId)
11. Ready (TId)
12. DropOut (TId)
13. Forget (TId)
14. DefineControlPort (TId, PortId)
15. Recall (State) -> (SetOfTIds)

Figure 3. High-Level Communication Primitives

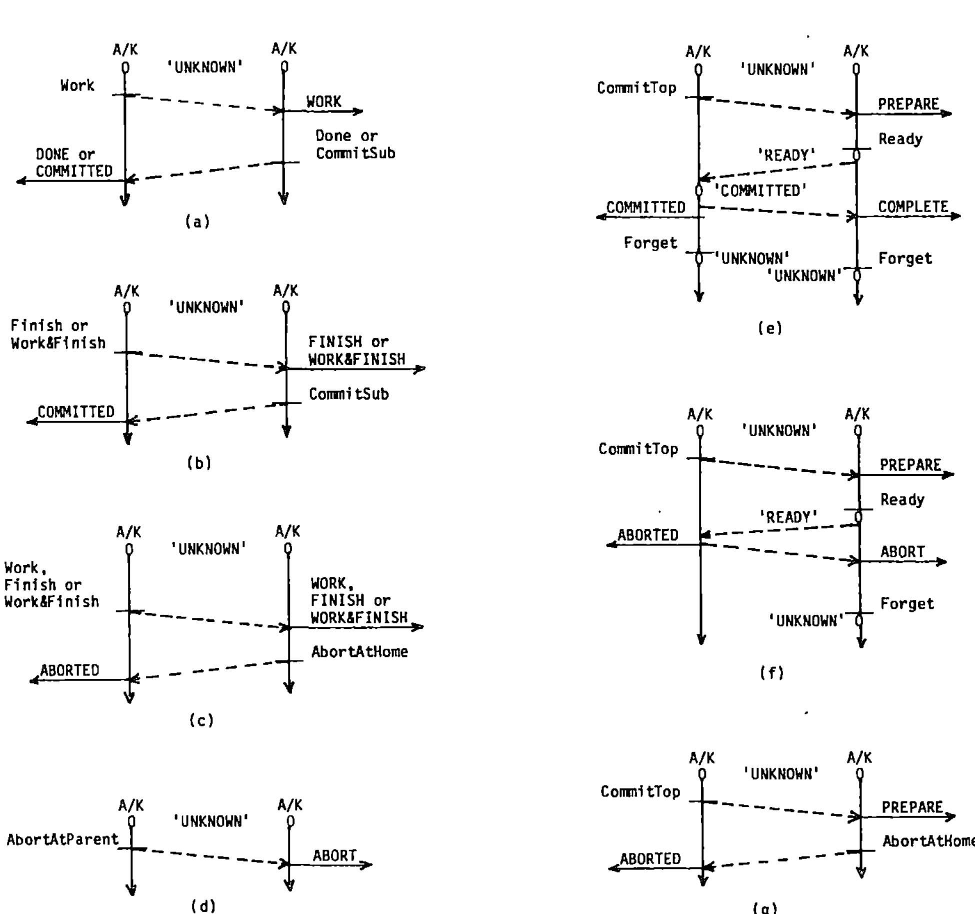

A/K...Agent/Kernel Interface

Figure 4. Agent/Agent Interactions

When an agent receives a WORK request message, it performs the required action under the subtransaction specified in the message. After having satisfied the WORK request, the agent may proceed in two different ways. It can call the Done primitive which returns a DONE message informing the subtransaction's parent agent that the required action has been executed; DONE contains the return data specified by parameter ReturnData. Alternatively, if the agent can determine that no further actions will be performed within the subtransaction, it can independently finish the subtransaction. When an agent receives a FINISH request message, it finishes the subtransaction specified in the message. If the agent receives WORK&FINISH instead of FINISH, it performs the action described in the message before finishing the subtransaction.

Subtransactions can be finished by committing or aborting. A subtransaction can be committed unless its children are all committed. This allows the application to decide which children are essential and which are not. On the other hand, a subtransaction cannot be committed unless all of its children are locally resolved (aborted or committed). This restriction essentially simplifies the concurrency control and recovery control algorithms for nested transactions (for a detailed discussion see /Moss81/ and /Lisk84/). When the children of a subtransaction are all locally resolved, the home agent can commit the subtransaction by calling the <u>CommitSub</u> primitive, which sends a COMMITTED message back to the subtransaction's parent agent (see Fig. 4a and 4b). The COMMITTED message contains the return data specified by parameter ReturnData, e.g. COMMITTED may include the results of the last WORK or WORK&FINISH request. Note, the effects of committed subtransactions are not permanent - committed subtransactions can be aborted due to crashes or explicit aborts. Changes become permanent only when the enclosing top-level transaction commits. Moreover, the commitment of a subtransaction does not cause the subtransaction's effects to become visible to all other transactions. In the synchronization algorithm described in /Moss81/, for example, the locks of a subtransaction are not released but are only moved to the parent when the subtransaction commits.

If the home agent of a subtransaction is unable to commit the subtransaction, it finishes the subtransaction by aborting. When a transaction is aborted, all of its descendants will be aborted too. The home agent of a transaction is allowed to unilaterally abort the transaction by calling the <u>AbortAtHome</u> primitive up to the moment when it has agreed to participate in the commit procedure of the top-level transaction (for details see below). The AbortAtHome primitive aborts the transaction identified by parameter TId at the transaction's home agent. The kernel guarantees that all agents running a descendant of the aborted transaction will eventually receive an ABORT message. Moreover, if the parent agent of the aborted transaction is waiting for a response to a WORK, FINISH or WORK&FINISH request, an ABORTED message is sent back to the parent agent (see Fig. 4c).

When the parent agent of a transaction has lost the interest in the execution of the transaction, it can unilaterally abort the transaction by calling the <u>AbortAtParent</u> primitive. The parent agent may abort the transaction up to the moment when it initiates commitment for the transaction's parent transaction (by calling Commit or CommitTop). AbortAtParent aborts the transaction identified by parameter TId at the transaction's parent agent. The kernel guarantees that the transaction's home agent and each agent running a descendant of the transaction will eventually receive an ABORT message (see Fig. 4d).

If a WORK, FINISH, or WORK&FINISH request cannot be satisfied because the transaction's home agent is not available (e.g. due to a node crash or a network partitioning), an UNAVAILABLE message is returned to the transaction's parent agent. If one of these request cannot be satisfied since the transaction has aborted at its home agent (due to crash or explicit abort), an ABORTED message is returned to the parent agent.

As subtransactions top-level transaction cannot be committed unless their children are all locally resolved. A subtransaction becomes locally committed at its parent agent when a COMMITTED message is received from the subtransaction's home agent. Note, since the effects of committed subtransactions are not permanent, it can happen that a subtransaction which is locally committed at its parent agent has aborted at its home agent. A subtransaction becomes locally aborted at its parent agent when an ABORTED or UNAVAILABLE message arrives, or the parent agent aborts the (locally uncommitted or committed) subtransaction by calling the AbortAtParent primitive. A subtransaction is said to have 'committed to the parent', if it is locally committed at its parent agent. A subtransaction is called to have 'committed to the top' if it has 'committed to the parent', and so have all its ancestors up to, but not including the top-level transaction. The transactions that have 'committed to the top' are called the needed descendants of the top-level transaction. A top-level transaction commits only if all of its needed descendants can be completed. If at least one of the needed descendants cannot be completed, the top-level transaction aborts. When a subtransaction completes, its effects become permanent.

When the home agent of the top-level transaction, called the top agent, has performed all work it desires to do, it recoverably prepares the top-level transaction for commit. A recoverably prepared transaction can be completed or aborted, regardless of crashes. Of course, the activities necessary to prepare a transaction for commit depend on the applied data recovery technique. After having recoverably prepared the top-level transaction, the top agent waits until the children of the top-level transaction are all locally resolved, and then initiates the commit process by calling the <u>CommitTop</u> primitive. A message indicating whether or not the top-level transaction has committed (COMMITTED or ABORTED) is retruned to the port identifed by parameter ResPortId.

In order to commit a top-level transaction, the kernel residing at the top-level transaction's home node communicates with the agents of the needed descendants, called the participants, according to a two-phase commit protocol /Gray78/. The kernel acts as commit coordinator, which makes the final commit/abort decision. During the first phase of the two-phase commit protocol, the coordinator asks all participants whether they are able to complete their subtransaction. Each participant becomes recoverably prepared to complete or abort and awaits the coordinator's decision. Once all participants are prepared to complete, the coordinator makes its decision and the participants are notified to complete during phase two of the commit protocol.

When CommitTop is called, the coordinator sends PREPARE request messages to all participants - each participant is asked whether it is able to complete its transaction (see Fig. 4e - 4g). A participant receiving a PREPARE request message recoverably prepares the transaction specified in the message. After becoming prepared, the participant calls the <u>Ready</u> primitive, which atomically changes the stable state of the transaction from 'UNKNOWN' to 'READY', and then returns an answer-yes message back to the coordinator (see Fig. 4e and 4f). Residing in the 'READY' state implies, that the agent will not abort or complete the transaction until it receives the coordinator's decision. Furthermore, no matter what happens, the agent is able to complete or abort the transaction. Since complete and abort are the same thing for read-only transactions, this type of transaction need not pass through the 'READY' state. Therefore, if the transaction is read-only, the agent can drop out of the two-phase commit protocol early by calling <u>DropOut</u> instead of Ready. The DropOut primitive returns a special answer-yes message informing the coordinator that the agent does not need to participate in the second phase of the commit protocol. If the agent aborts the transaction after receiving the PREPARE message, an answer-no message is returned back to the coordinator (see Fig. 4g). Also a answer-no is returned, if the PREPARE message cannot be deliverd because the transaction is unknown; this situation can occur only if the transaction has aborted at its home agent.

The coordinator collects the answers from the participants. If any answer-no arrives, then the top-level transaction will be aborted. If, however, all the PREPARE requests are acknowledged with answer-yes, the top-level transaction can be committed. As soon as the coordinator kernel knows that all needed descendants are either 'READY' or read-only, it commits the top-level transaction by atomically changing the stable state of the transaction from 'UNKNOWN' to 'COMMITTED' (see Fig. 4e). After that, the coordinator returns a COMMITTED message to the return port specified in the CommitTop call, and sends a COMPLETE message to all agents participating in the second phase of the commit protocol. If, however, at least one answer-no arrives, the coordinator kernel aborts the top-level transaction. In this case, the coordinator returns an ABORTED message to the return port specified in CommitTop and issues an ABORT message to the agents of 'READY' transactions (see Fig. 4f).

When a participant receives a COMPLETE message, it completes the indicated transaction, and then invokes the <u>Forget</u> primitive, which changes the stable state of the transaction from 'READY' to 'UNKNOWN'. If a participant receives ABORT instead of COMPLETE, it undoes the effects of the transaction before calling Forget. When the top agent receives a COMMITTED message, it completes the top-level transaction, and then invokes the Forget primitive, which changes the transaction's stable state from 'COMMITTED' to 'UNKNOWN'. If it receives an ABORTED message, it undoes the effects of the top-level transaction.

To be able to receive PREPARE, COMPLETE and ABORT messages, agents must define so-called control ports. An agent can define the control port of a transaction by calling the <u>DefineControlPort</u> primitive. When DefineControlPort terminates, the port identfied by parameter PortNo represents the control port of the transaction identfied by parameter TId, i.e. each PREPARE, COMPLETE or ABORT message concerning the transaction identified by TId will be placed into the port designated by PortNo.

As mentioned before, the kernel performs protocol recovery from node crashes. Whether a trasaction survives a crash or aborts depends on the transactions's stable state recorded in the local TST at the time of the crash. An uncompleted transaction residing in the 'UNKNOWN' state is aborted when its home node crashes. The kernel guarantees that each agent running a descendant of a crashed transaction will eventually receive an ABORT message from the descendant's control port. In contrast to transactions residing in the 'UNKNOWN' state, 'READY' and 'COMMITTED' transactions survive crashes. 'COMMITTED' transactions will be successfully completed after restart. The kernel guarantees that the agents running a needed descendant of of a 'COMMITTED' transaction will all receive a COMPLETE message, regardless of crashes. A transaction residing in the 'READY' state after restart cannot be terminated before the decision of the coordinator arrives. The kernel guarantees, that each 'READY' transaction will receive the coordinator's decision, ABORT or COMPLETE, from the transaction's (new defined) control port after restart.

Like the kernel the application needs the information stored in the local TST for performing data recovery after restart. In order to allow the application to read the local TST, the kernel provides the <u>Recall</u> primitive, which returns the identifiers of the transactions residing in the stable state specified by parameter State ('COMMITTED' or 'READY').

For receiving messages agents use the primitives Listen, RemoveMessage and

Listen&Remove (see Sec. 3.1). When a message is received, the message type, which is chosen from the set WORK, FINISH, WORK&FINISCH, PREPARE, COMPLETE, ABORT DONE, COMMITTED, ABORTED and UNAVAILABLE is returned in parameter MsgType; the transaction identifier, which is included in each message, is returned in parameter TId. If the received message contains a data field (WORK, COMMITTED, DONE), the content of this field is returned in Message.

4. SUMMARY

The objects and the communication primitives of a kernel supporting the communication in DDBSs have been presented. The proposed kernel provides unreliable low-level communication primitives as well as reliable high-level primitives that support the creation, execution and termination of nested transactions. The former primitives are needed to efficiently support the great variety of communication patterns that occur in DDBSs, while the latter ones are necessary to make the implementation of DDBSs easier.

Primitives for nested transactions are already offered by some distributed operating systems (e.g LOCUS /Muel83/, ARGUS /Lisk84/, EDEN /Jess82/). However, these primitives are rather complex and mix data processing and communication processing. For example, an 'abort-transaction' primitive not only notifies the corresponding agents, but also undoes the effects of the aborted transaction in the database. Therefore, these primitives are not convenient to be part of a kernel that only includes 'mechanisms' and no 'policies'.

REFERENCES

/ADA83/   "Reference Manual for the Ada Programming Language" (ANSI/MIL-STD-1815A), U.S. Department of Defense, Washington, D.C., 1983.
/Andl81/   Andler, S., et al., "System D: A Distributed System for Availability", IBM Research Report RJ 3313, San Jose, 1981.
/Andr81/   Andrews, G.R., "Synchronizing Resources", ACM Transactions on Programming Languages and Systems, 3:4, 1981.
/Andr83/   Andrews, G.R., Schneider, F.B., "Concepts and Notations for Concurrent Programming", ACM Computing Surveyes, 15:1,1983.
/Gray78/   Gray, J., "Notes on Database Operating Systems", in: "Operating Systems: An Advanced Course", Lecture Notes in Computer Science 60, 1978.
/Jess82/   Jessop, W.H., et al., "The EDEN Transaction Based File System", Proc. 2nd Symp. on Reliability in Distributed Software and Database Systems, 1982.
/Lind84/   Lindsay, B.G., et al., "Computation and Communication in R *:A Distributed Database Manager, ACM Trans. on Computer Systems, 2:1, 1984.
/Lamp81/   Lampson, B., "Atomic Transactions", in: "Distributed Systems - Architecture and Implementation", Lecture Notes in Computer Science 105, 1981.
/Lisk79/   Liskov, B., "Primitives for Distributed Computing", Proc. of the 7th Symposium on Operating Systems Principles, 1979.

/Lisk83/ Liskov, B., Schleifer, R., "Guardians and Actions: Linguistic Support for Robust, Distributed Programs", ACM Transactions on Programming Languages and Systems, 5:3, 1983.

/Lisk84/ Liskov, B., "The ARGUS Language and System", Programming Methodology Group Memo 40, M.I.T., Department of Computer Science, 1984.

/Mao80/ Mao, T.W., Yeh R.T., "Communication Port: A Language Concept for Concurrent Programming", IEEE Trans. Softw. Eng., 7:4, 1981.

/Moss81/ Moss, J.E.B., "Nested Transactions: An Approach to Reliable Computing", MIT/LCS/TR-260, 1981.

/Muel83/ Mueller, E.T., et al., "A Nested Transaction Mechanism in LOCUS", Proc. 9th ACM Symp. on Operating Systems Principles, 1983.

/Nels81/ Nelson, B. J., "Remote Procedure Call", Xerox Cooperation Technical Report CSL-81-9, 1981.

/Panz82/ Panzieri, F., Shrivastava, S.K., "Reliable Calls for Distributed UNIX: An Implementation Study", 2nd Symp. on Reliability in Distributed Software and Database, 1982.

/Rash81/ Rashid, R., Robertson, G., "Accent: A Communication Oriented Network Operating System Kernel", Proc. of the 8th Symposium on Operating Systems Principles, 1981.

/Roth84a/ Rothermel, K., "A Communication Model for Transaction Oriented Applications in Distributed Systems", Proc. 17th Hawaii International Conference on System Sciences, 1984.

/Roth84b/ Rothermel, K., "Communication Primitives Supporting the Execution of Atomic Actions at Remote Sites, Proc. ACM SIGCOMM 84 - Symp. on Communications Architectures and Protocols, 1984.

/Roth84c/ Rothermel, K., Walter, B., "A Kernel for Transaction Oriented Communication in Distributed Database Systems", Proc. Distributed Computing Systems, 1984.

/Roth84d/ Rothermel, K., "Functional Port Classes: A Communication Concept for Distributed Transaction Processing Systems", submitted for publication, 1984.

/Salt81/ Saltzer, J.H., Reed, D.P., Clark, D.D.,"End-to-End Arguments in Systems Design", Proc. 2nd Int. Conf. on Distributed Computing Systems, 1981.

/Spec82/ Spector, A.Z., "Performing Remote Operations Efficiently on a Local Computer Network", Communications of the ACM, 25:4, 1982.

/Ston77/ Stonebraker, M., Neuhold, E.J., "A Distributed Database Version of INGRES", Proc. 2nd Berkeley Workshop on Distributed Databases and Computer Networks, 1977.

/Tane81/ Tanenbaum, A. S., Mullender, S., "An Overview of the AMOEBA Distributed Operating System", ACM Operating Systems Review, 15:3, 1981.

/Walk83/ Walker, B., et al., "The LOCUS Distributed Operating System", Proc. 9th Symposium on Operating Systems Principles, 1983.

/Walt84/ Walter, B., Neuhold, E.J., "POREL: A Distributed Database System", C. Mohan (ed.), Recent Advances in Distributed Database Management, IEEE Press, 1984.

PROGRAMM-PROGRAMM-KOMMUNIKATION IN VERTEILTEN

SINIX -SYSTEMEN

K. Danassy und G. Henning

Geschäftsbereich Datenverarbeitung
Systemtechnik Datenfernverarbeitung

Siemens AG München

1. Einführung

Bedingt durch die schnelle Weiterentwicklung leistungsfähiger 16-
und 32-Bit Mikroprozessoren gewinnt das Betriebssystem UNIX[*] auch
im Bereich kommerzieller Anwendungen, insbesondere bei Arbeitsplatz-
systemen, zunehmend an Bedeutung. Von diesem Benutzerkreis werden
jedoch Anforderungen an UNIX gestellt, die im zur Zeit häufigsten
Einsatzfall als Entwicklungsrechner nicht gefordert werden. Kommerzielle Datenverarbeitungssysteme benötigen überwiegend eine Verknüpfung räumlich entfernter Teilsysteme im Sinne einer "verteilten Verarbeitung" (1).

Dieses Konzept nutzt sowohl die Möglichkeiten leistungsfähiger
Großrechner im Rechenzentrum als auch die der Dezentralisierung durch
unabhänige EDV-Systeme, z.B. Arbeitsplatzcomputer mit UNIX und geht
dabei von folgenden Grundgedanken aus:

> Funktionen, die hohen Koordinierungsaufwand oder hohe
> Rechner- oder Speicherleistung erfordern, werden zentral abgewik-
> kelt. Beispiele dafür sind komplexe Verarbeitungsprozesse, unter-
> nehmensweite Datenbanken und das systemweite Management der Be-
> triebsmittel.

---

[*] UNIX ist das Warenzeichen der Bell Laboratories

Funktionen, die endbenutzerorientiert sind oder die autonom abgewickelt werden können, werden so nahe wie möglich zum Benutzer hin verlagert. Beispiele dafür sind Datenerfassung, Textbearbeitungsdienste und Bürodienste.

Voraussetzung für das effiziente Zusammenwirken zentraler und dezentraler Anwendungen ist die kommunikative und administrative Verbindung der dezentralen Systeme untereinander und mit den zentralen Rechenzentren. Dies ist die Aufgabe des in den einzelnen Teilsystemen implementierten Kommunikationssystems, das auf den Diensten öffentlicher und privater Netze zur Datenübertragung aufbaut. Daher wurde bei der Entwicklung des Betriebssystems SINIX für Arbeitsplatzcomputer und Personalcomputer einer der Schwerpunkte auf die Verbesserung der Kommunikationsfähigkeit gelegt. Der Beitrag beschreibt das Kommunikationskonzept des Arbeitsplatzcomputers 9786 mit dem Betriebssystem SINIX auf Basis des Datenfernverarbeitungssystems TRANSDATA® unter dem Vorbehalt einer endgültigen Festlegung im Rahmen der vertrieblichen Freigabe.

2. Übersicht Datenfernverarbeitungssystem TRANSDATA

Das System TRANSDATA von Siemens dient als Beispiel für die Synthese von zentraler, dezentraler und verteilter Verarbeitung (Bild 1).

Die dazu erforderliche Kommunikationsleistung erbringt ein Kommunikationssystem, das auf den Diensten öffentlicher und privater Netze zur Datenübertragung aufbaut. Mit Hilfe dieses Systems kommunizieren Programme in zentralen und dezentralen Rechnersystemen miteinander und mit Menschen als Benutzern von Datenendgeräten.

Als zentrale Systeme dienen Rechner der Siemens-Systeme 7.500/ 7.700/ 7.800. Als dezentrale Rechner werden vor allem die Siemens-Systeme TRANSDATA 9.600 und Personalcomputer PC-X und PC-MX verwendet. Was bereits in den Datenstationsrechnern des Siemens-Systems TRANSDATA 9.600 als Möglichkeit sichtbar und nutzbar ist, findet sich in deutlicher Ausprägung bei den neueren Mitgliedern der TRANSDATA Familie, den Personalcomputer Siemens PC-X und PC-MX. Sie sind dezentrale Rechner, die in erster Linie für die Datenverarbeitung vor Ort konzipiert wurden.

Als Betriebssystem dient SINIX, entwickelt auf der Basis des inzwischen zum Weltstandard avancierten Betriebssystems UNIX. SINIX wurde gegenüber UNIX um einige wichtige Funktionen erweitert und für den Betrieb als kommerzieller Arbeitsplatzcomputer angepaßt.

Ein weiterer Schritt in diese Richtung bis zur vollständigen Integration in das Datenfernverarbeitungssystem gelingt mit dem jüngsten und leistungsfähigsten Arbeitsplatzcomputer, dem APC TRANSDATA 9786. Die folgenden Kapitel beschreiben das Funktionskonzept und die Lösungswege dieses Produktes insbesondere für die Kommunikation in verteilten Systemen.

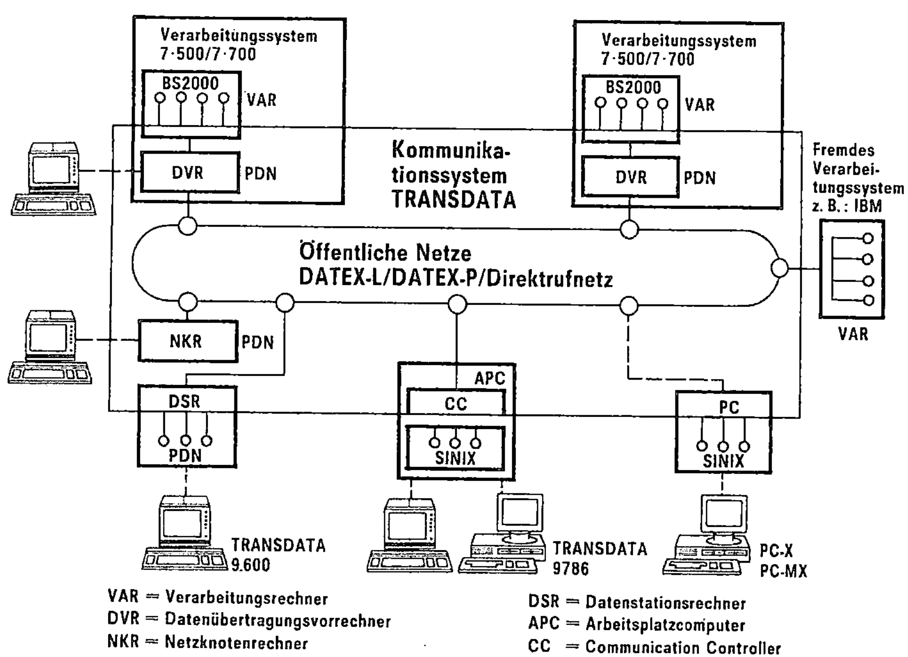

Bild 1: Übersicht TRANSDATA

3. <u>Funktionskonzept und Lösungswege für die Programm-Programm Kommunikation</u>

Ein SINIX-Rechner, der als Arbeitsplatzcomputer in einem Verbund mit "verteilter Verarbeitung" eingesetzt wird, muß die traditionell starken UNIX-Funktionen des lokalen Betriebs (Programmentwicklung und Verarbeitung) mit der Funktion der Netzkopplung in sich vereinen. Basisleistung für die Zusammenarbeit mit anderen dezentralen und zentralen Rechnern ist eine Programm-Programm-Kommmunikationsmethode. Diese Basisleistung kann in Stufen unterschiedlicher Mächtigkeit realisiert werden.
In einfachster Form verhalten sich Programme in den dezentralen SINIX-Rechnern gegenüber ihren Kommunikationspartnern wie Datenstationen. Dabei bilden diese Programme die Eingaben eines Datenstationsbenutzers (das "Terminal User Interface", TUI) nach und verarbeiten ("emulieren") das Geräteprotokoll. Dieser Aufwand ist vergleichsweise klein gegenüber der Implementierung von Transport- und Netzprotokollen (Bild 2).

PC-MX und PC-X bieten bereits auf dieser Grundlage für einige ausgewählte Systemanwendungen die Kopplung mit dem TRANSDATA-Netz:

. Mit Hilfe der 9750- und 8122-Emulation können die Arbeitsplätze über SINIX mit Anwendungen in Datenfernverarbeitungssystemen, z.B. Transaktionsanwendungen in BS2000-Rechnern, kommunizieren.

. Mit Hilfe des Filetransfers werden Dateien zwischen diesen Rechnern ausgetauscht.

Die Anwendung dieses Konzeptes setzt die entsprechende Konfigurierung und Generierung des Gesamtsystems voraus. Der steuernde Datenübertragungsvorrechner bedient eine begrenzte Anzahl namentlich festgelegter Datenstationen. Eine allgemein verwendbare Programmschnittstelle für verteilte Anwendungen bietet dieses Konzept jedoch noch nicht.

Ein dezentrales System mit "verteilter Verarbeitung" sollte über den einfachen Lösungsweg der Emulation hinausgehen. Einer beliebigen Anzahl von Programmen unter beliebigen Namen sollte eine freizügige Programm-Programm Kommunikation in Form einer Programmschnittstelle angeboten werden.

Dies ist nur dadurch möglich, daß man das SINIX-System zu einem
Transportendsystem ausbaut.

Über das Transportsystem werden zusätzliche Leistungen ermöglicht:

Beliebige anwendungsspezifische Programm-Programm-Kommunikation

Steuerung entfernter Datenstationen (in Planung)

Administration der SINIX-Rechner von einer Zentrale aus

Die Realisierung des Transportsystems folgt der Konzeption von
ISO/OSI, die durch Realisierung der ISO-level 4 und 3 eine beliebige Anzahl TSAP's (Transport Service Access Point) ermöglicht.

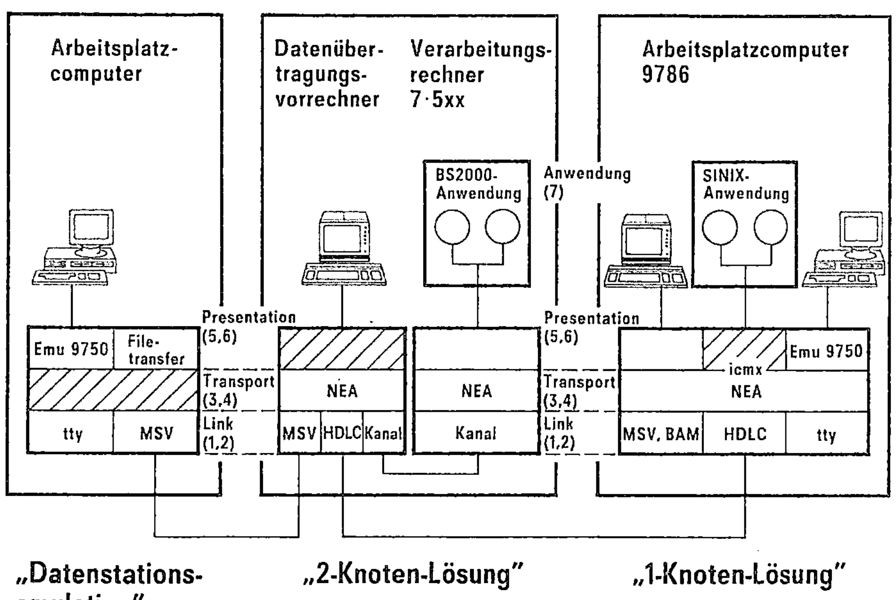

Bild 2: Funktionskonzept Programm-Programm-Kommunikation

4. Das Kommunikationskonzept des APC TRANSDATA 9786

Hardware-Konzept

Die Verarbeitungsfunktionen erbringt ein leistungsfähiger 32-Bit-
Mikroprozessor unter Steuerung des Betriebssystems SINIX. Der Ver-
arbeitungsprozessor steuert die Lokalperipherie und die Arbeits-
platzstationen. Die Kommunikationsfunktionen sind in einen Kommunikationsprozessor (CC) ausgelagert (Bild 3). Der Kommunikationsprozessor bildet ein Transportsystem nach ISO und übernimmt die
Anbindung an öffentliche und private Netze und unterschiedliche
Kommunikationssysteme. Er wird durch das CCP (Communication Control
Program) gesteuert. Die Programme zur Steuerung des Verarbeitungs-
prozessors und des Kommunikationsprozessors liegen in einem gemeinsamen Speicher, über den die Anforderungen der beiden Prozessoren
untereinander ausgetauscht werden. Ein gemeinsamer Datenbereich
speichert die für beide Prozessoren erforderlichen Daten.

An den zentralen Bus sind Verarbeitungs- und Kommunikationsprozessor, Speicher und die peripheren, intelligenten Steuerungen angeschlossen.

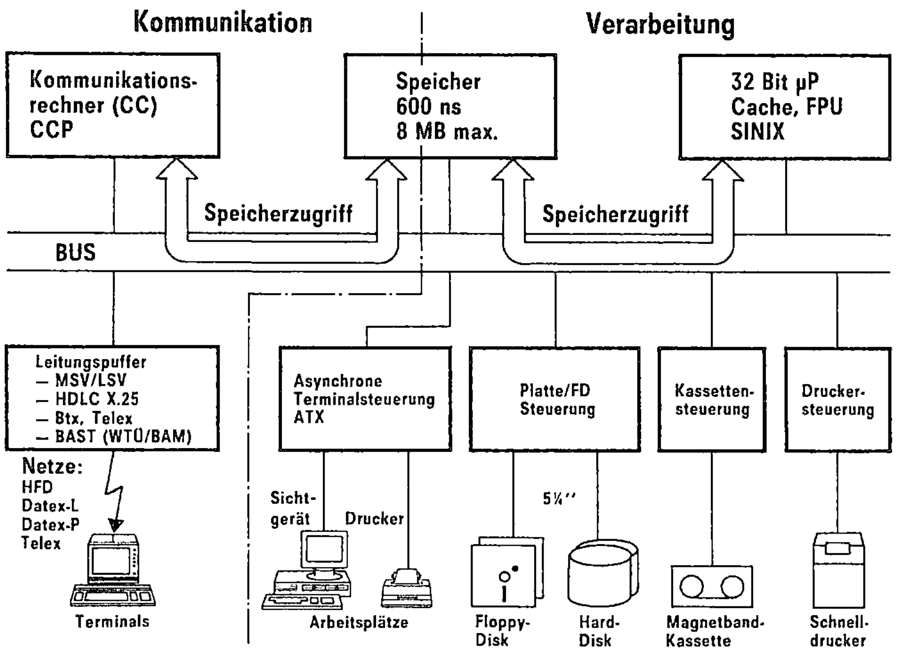

Bild 3: Hardware-Konzept APC 9786

<u>Software-Konzept</u>

Der Baustein CMX (Communication Method SINIX) läuft im Verarbei-
tungsprozessor unter SINIX und stellt SINIX-Anwendungen eine Programmschnittstelle (icmx) zu den Diensten der Transportschicht
zur Verfügung. Er besteht implementierungstechnisch aus einem
"Treiber" (SINIX-Terminologie) für Pseudodateien zum Netzanschluß
(network character special files) und einem Systemprozess ICM
(Incoming Call Manager), der ankommende Verbindungsaufbauwünsche
behandelt (Bild 4).

Die in diesem Beitrag vorgestellte Programmschnittstelle (Ebene
4/5) wird mit Hilfe von System-Calls, die in SINIX zur Bedienung
von Geräten zur Verfügung stehen, realisiert. Mit Hilfe von icmx
können Teilnehmer in SINIX sowohl mit Teilnehmern in anderen Endsystemen als auch mit Teilnehmern im selben Endsystem kommunizieren.
Zunächst vorgesehene Teilnehmer sind Programme (Anwendungen) oder
Datenstationen an entfernten Rechnern und Anwendungen in SINIX-
Rechnern.

Die Schnittstelle icmx ist so beschaffen, daß genormte ISO-Transport-
protokolle für Wide Area Network und Lokal Area Network unterlagert
werden können, so daß die Kommunikation im Sinne eines offenen
Systems mit allen Partnern, die diese Protokolle beherrschen,
möglich wird. Als erste Stufe wurde die Variante icmx(sys) entwickelt.

Diese privilegierte Schnittstelle wird ausschließlich von SINIX-
Anwendungen genutzt, die Bestandteile der Systemsoftware sind,
z.B. Filetransfer, Terminalemulation, Administrationsprogramme.
Zur allgemeinen Benutzung wird eine auf icmx(sys) aufsetzende,
durch Bibliotheksroutinen realisierte Schnittstelle icmx(lib)
zur Verfügung stehen. Es ist geplant, diese Schnittstelle für
alle Mitglieder der SINIX-Familie zu implementieren.

Die Transportfunktionen, Netzfunktionen und die Portfunktionen
(also alle Funktionen der Schichten 4, 3 und 2 entsprechend dem
Referenzmodell nach ISO 7498) werden durch die Software-Bausteine
im Kommunikationsprozessor erbracht.

Die Kommunikation läuft verbindungsorientiert in drei Phasen ab:
Verbindungsaufbau, Datenaustausch und Verbindungsabbau.

In SINIX und im CCP sorgen Software-Bausteine für den Austausch
von Anforderungen und Informationen zwischen den beiden Prozessoren. Dieser Mechanismus wird unter anderem auch benutzt, um eine
Auftragsschnittstelle zwischen den Anwendungen in SINIX und dem
CCP zu realisieren.

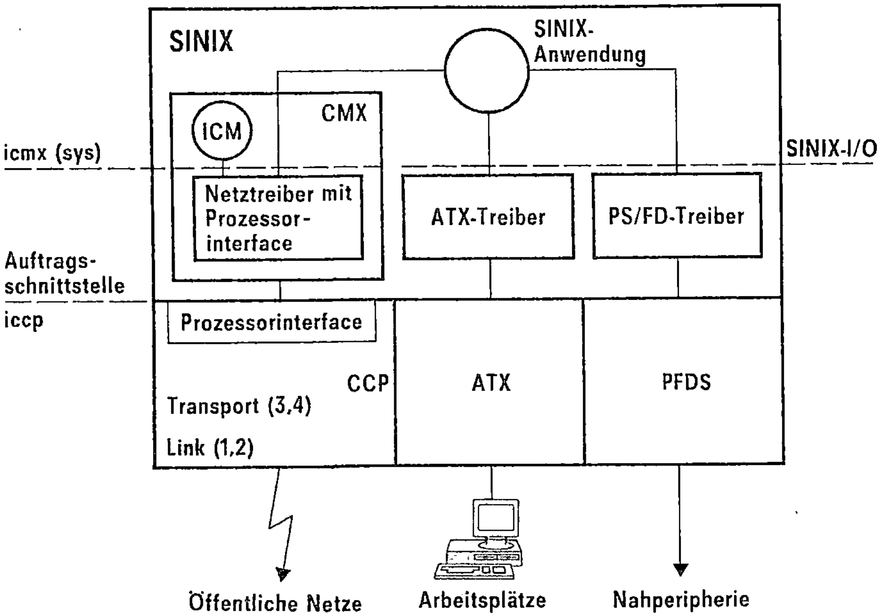

Bild 4: Software-Konzept APC 9786

Programmschnittstelle icmx

Kommunikationspartner in SINIX ist die Anwendung, im allgemeinen
repräsentiert durch eine Gruppe von verwandten Prozessen, im Spezialfall durch genau einen Prozeß. Der Name einer Anwendung wird
durch das Programm, das die Anwendung steuert, frei wählbar definiert.

Die Programmschnittstelle icmx(sys) besteht aus System-Calls der
SINIX Standard I/O, die auf die Pseudodateien für Netzanschluß
(/dev/net??) angewendet werden.

Diese Pseudodateien für Netzanschluß bieten den Anwendungen die
Möglichkeit, über "Transportkanäle" verbindungsorientiert mit Partnern zu kommunizieren. Die Zahl der zur Verfügung stehenden Transportkanäle (minor device number der Pseudodatei) ist eine Konfigu-
rationskonstante. Die Liste der Transportkanäle ist in der Datei
/etc/netfile enthalten gemäß dem Konzept des Terminal I/O mit
/etc/ttys.

Die Anwendungen bewerben sich konkurrierend um das Betriebsmittel
Transportkanal, in dem sie den nächsten verfügbaren Transportkanal
durch Eröffnung mit open(2) für sich reservieren. Der Transportkanal ist Eigentum der Anwendung. Diese Zuordnung trifft die Anwendung mit dem Sprachmittel ioctl(2)/S_OWNER.

Eine Anwendung kann mehrere Verbindungen besitzen, für die jeweils
ein Transportkanal benötigt wird. Auch eine Anwendung, die aus
nur einem Prozess besteht, kann mehrere Verbindungen haben. Über
die icmx(sys)-Schnittstelle wird lediglich die Zuordnung der Transportkanäle zu den Anwendungen geregelt. Die Zuordnung der Transportkanäle zu den Prozessen obliegt der Anwendung selbst (Bild 5).

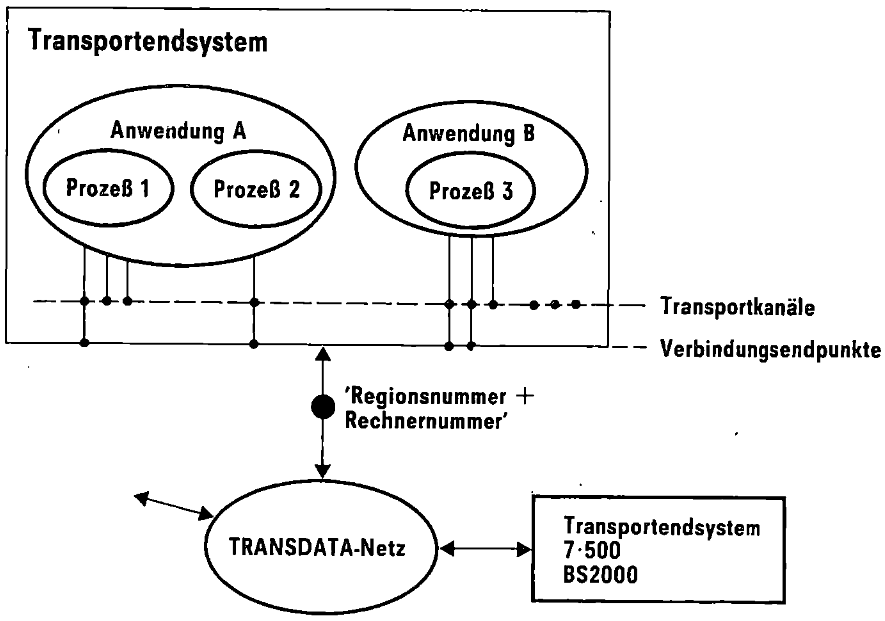

Bild 5: Anwendungen - Prozesse - Verbindungen

Die Adressierung von Anwendungen erfolgt nach TRANSDATA-Konventio-
nen durch den Namen des Endsystems (Regionsnummer und Rechnernum-
mer) und durch den lokal eindeutigen Namen der Anwendung.

Der Ablauf einer Kommunikation wird aus dem folgenden Beispiel
deutlich. In der Datenphase dienen die unveränderten SINIX-Aufrufe
read(2) und write(2) zum Empfangen und Senden von Nachrichten.
Unabhängig vom Zähler im read(2) wird maximal eine Nachricht gelesen, so daß die Anwendung das Ende einer Nachricht erkennen kann,
und keine Zeichenkette "ohne Ende" zugestellt bekommt.

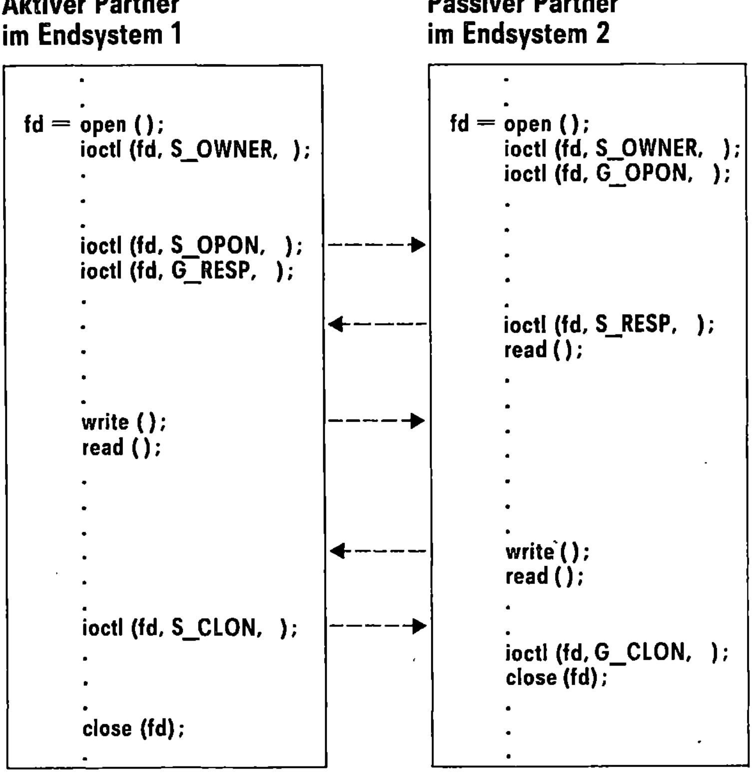

Bild 6: Beispiel: Programmierung mit icmx(sys)

Dies ist ein einfaches Beispiel der Kommunikation zweier Anwendungen über genau eine Verbindung, wobei in den beiden Anwendungen
jeweils _ein_ Prozeß an der Kommunikation beteiligt ist. Die Empfangsaufrufe G_OPON, G_RESP und read(2) sind "blockierend", d.h. die
Anwendung wird in einen Wartezustand versetzt, falls das Ereignis
noch nicht eingetroffen ist.

Die Programmschnittstelle icmx(sys) bietet jedoch in Kombination
mit Standard-SINIX-Funktionen der Prozeßsteuerung und Signalisierung
noch weitere Möglichkeiten:

Es gibt Ereignisse, deren Eintreten vom zeitlichen Verhalten des
Partners abhängt und von der Anwendung nicht beeinflußt werden kann.
Die Anwendung hat zwei Möglichkeiten. Sie kann entweder Empfangsaufrufe für Ereignisse tätigen mit der Folge, daß sie bis zum Eintreten des Ereignisses in den Wartezustand versetzt wird ("synchro-
nes Warten"), oder sie kann sich bei Eintreten bestimmter Ereignisse über Signale benachrichtigen lassen.

Danach getätigte Empfangsaufrufe werden sofort bedient. Für die
asynchrone Mitteilung dieser Ereignisse wird der Signalmechanismus von SINIX verwendet. Die Anwendung muß durch geeignete Wahl der
Signalnummer dafür sorgen, daß die Behandlung der Ereignisse der
Kommunikation über icmx(sys) mit der Behandlung anderer Signale
nicht kollidiert.

Es bleibt der Anwendung überlassen, wie sie sich intern organisiert.
Sie kann das in _einem_ Prozeß oder in _mehreren_ Prozessen tun. Überall dort, wo zeitlich parallele Arbeit gewünscht wird, müssen entweder die blockierenden Aufrufe (die Wartezustände beinhalten wie
ioctl(2)/G_RESP, G_OPON, ...) in einem separaten Prozeß getätigt
werden, oder die Signalisierung der Ereignisse angefordert werden.
Alle Funktionen von icmx(sys) können in _einem_ Prozeß genutzt werden. Während der Wartezustände ist die Anwendung jedoch auch für
alle anderen Verbindungen blockiert. Wartezustände kann man dadurch
vermeiden, daß man die blockierenden Aufrufe erst tätigt, wenn das
Eintreffen des Ereignisses über den Signalmechanismus von icmx(sys)
gemeldet wird. Eine Anwendung mit nur einem Prozeß kann über verschiedene Transportkanäle Mehrfachverbindungen unterhalten.

Mittels der Prozeßverwaltung von SINIX kann sich eine Anwendung
in mehreren Prozessen organisieren, um Aufgaben zeitlich parallel
abzuwickeln. Während ein Prozeß auf das Eintreffen eines Ereignisses
wartet, kann in anderen Prozessen für andere Verbindungen gearbeitet
werden. Die Aufgaben des Verbindungsaufbaus und -Abbaus kann zweck-
mäßigerweise ein "Primär-Prozeß" erledigen.

Maximal 6 Ereignisse können signalisiert werden:
- Antwort auf einen abgehenden Ruf
- Ankommender Ruf
- Normale Daten
- Vorrangdaten
- Transportquittung
- Verbindungsabbau

Die Anwendung kann für eine beliebige Auswahl aus diesen Ereignissen die Benachrichtigung über Signale vereinbaren. Die Signalisierung von Ereignissen ist verbindungsspezifisch und erfolgt für eine Prozeßgruppe. Falls mehrere Anwendungen zu einer Prozeßgruppe ·
gehören, müssen sie sich auf eine gemeinsame Signalauswahl einigen.
Der Anwendung steht es frei, die Signalisierung der Ereignisse
über dasselbe Signal oder über unterschiedliche Signale festzulegen.
Diese Festlegungen trifft die Anwendung mit ioctl(2)/S_SIGTYPE.
Werden unterschiedliche Ereignisse über dasselbe Signal gemeldet,
kann eine Fallunterscheidung erfolgen. Die Signalursache wird über
ioctl(2)/G_SIGC gemeldet.

5.   Anwendung dieses Konzeptes

ACX (Administration Center SINIX)

Die administrative Einbettung des APC TRANSDATA 9786 ist ein
Beispiel für die Implementierung bestehender Anwendungskonzepte
in SINIX auf Basis der Programm-Programm-Kommunikation (2).
Für die Administration (im Sinne des TRANSDATA-Netzmanagements)
des APC 9786 wurde die Administrationskomponente ACX geschaffen,
die als "Filiale" mit einem TRANSDATA Administrationszentrum
zusammenarbeitet und SINIX-spezifische Aufträge ausführt.

Implementierungstechnisch ist ACX eine SINIX-Anwendung, die eine
Verbindung zum TRANSDATA Administrationszentrum im lokalen CCP
aufbaut, Aufträge des Netzverwalters, des Systemverwalters und des
Wartungstechnikers entgegennimmt und die Ergebnisse dem Administrator zurückmeldet. ACX löst diese Funktionen mit Hilfe der Aufrufe ioctl, read und write über die Schnittstelle icmx(sys).

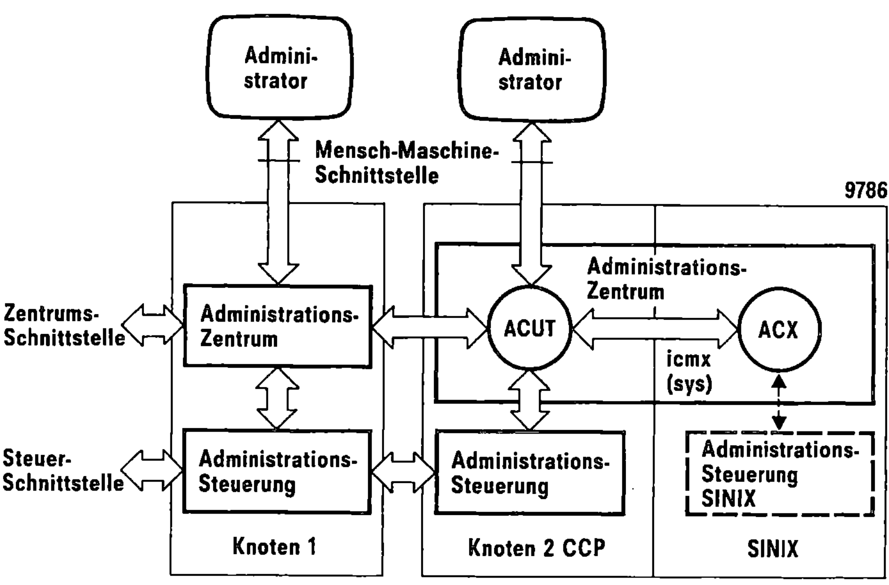

Bild 7: Anwendungsfall: Administrative Kopplung mit TRANSDATA

<u>Remote login server</u>

Die Implementierung eines "remote login servers" ist ein Beispiel
für den passiven Verbindungsaufbau einer SINIX-Anwendung.

"Server" ist ein Anwendungstyp, der eine Dienstleistung für mehrere
"clients" gleichzeitig anbietet. Dazu muß die Anwendung ständig
bereit sein, ankommende Rufe für eine bestimmte Maximalzahl von
Klienten zu empfangen. Daher muß sie einen Prozeß mit nur dieser Aufgabe betrauen.

Ist ein Ruf angekommen, dupliziert sich der Prozeß mit fork(2)
und überläßt die Verbindung dem Kindprozeß, der noch für die richtige
Prozeßumgebung sorgt. Der Vaterprozeß schließt den Transportkanal
des Kindprozesses und geht auf den anderen, noch geöffneten Trans-
portkanälen auf Empfang.

Dieses Verfahren ist vergleichbar mit der Rolle des init-Prozesses
bei der Erzeugung der lokalen Login-Prozesse für die TTY-Terminals,
erlaubt jedoch die dynamische Zuschaltung entfernter Datenstationen.

open (2)	**Transportkanal eroeffnen**
ioctl (2) request = S_OWNER	**Definition der Anwendung und 7uordnung eines Transportkanals zur Anwendung**
S_OPON	**Aktiver Verbindungsaufbau (connection request)**
G_OPON	**Passiver Verbindungsaufbau (connection indication)**
S_RESP	**Antwort auf ankommenden Ruf (connection response)**
G_RESP	**Warten auf Antwort (connection confirmation)**
G_EXP	**Lesen Vorrangdaten**
S_EXP	**Senden Vorrangdaten**
S_SIGTYPE **G_SIGC** **S_SIGREL**	**Ereignisbehandlung durch Signale**
S_CLON	**Aktiver Verbindungsabbau**
G_CLON	**Passiver Verbindungsabbau**
read (2)	**Nachrichten lesen**
write (2)	**Nachrichten senden**
close (2)	**Transportkanal schliessen**

Bild 8: Übersicht über die ICMX(sys)-Schnittstelle

Literatur

(1) Henning, G.: Ist das Rechenzentrum überholt?
    Ein Plädoyer für verteilte Verarbeitung.
    Das Rechenzentrum 7. Jahrgang (1984) S. 99-110, Heft 2/84

(2) Andiel, J.: Benutzerkreise des Netzmanagements von
    Rechnernetzen.
    telcom report 7 (1984), Heft 6/84.

<u>CONCURRENCY CONTROL MECHANISM FOR AN</u>

<u>AVAILABLE DISTRIBUTED DATA BASE SYSTEM</u>

Jean-Michel FEUVRE
NADIR Project
INRIA
BP 105
78153-LE CHESNAY CEDEX
France

<u>Abstract</u>

That paper presents algorithms used in an available database system to process and control operations. The structure of the system and the algorithms are designed to allow transaction processing to proceed in case of one site failure. We shall focus on the required network functionalities, the concurrency control and transaction atomicity problems. The protocols used on each site will be formally described using evaluation nets.

I - <u>INTRODUCTION</u>

Many database applications need continuously accessible data : reservation systems, banking systems, inventory control systems... In those systems, the inaccessibility of some part of the data (due to a machine crash or a disk destruction) may result in the impossibility of satisfying customers' needs. So we need to develop, or adapt data base systems to offer a good availability.

Some projects are currently working on that subject : HAS [AGHI82], and System D. [ANDL82]. HAS (Highly Available System) is a data base system with hardware and software redundancy. It uses an Auditor on each processor to detect failure and coordinate their recovery. System D is a distributed processing system where data and some software modules are duplicated. This experimental system is implemented on a network of loosely coupled minicomputers.

Our objective is to achieve high availability in a distributed data base system, composed of geographically distant computers interconnected with a satellite network [FEU84]. We plan to take advantage of 2 new services offered by satellite networks like TELECOM1 [LOMB83] : broadcasting and high transmission speed (up to 2 Mbits/s).

In this paper, we shall particularly focus on concurrency control mechanism specially designed for such systems and on mechanism used to control atomicity of transactions.

We can find many concurrency control algorithms in the literature [ALSB76], [LELA78], [STON79], [BERN80], [BERN81]. Their differences rely on :

- The kind of data distribution the system manages : partitionned, fully duplicated or generally distributed.

- The use of locks to control concurrency or of time-stamps to order the transactions.

- The methods used to control deadlocks (if using locks) : prevention or detection.

- The decision taken when a site fails : abortion of transactions, blocking of transaction waiting for the site to restart or queuing of work (messages) for that site.

That last point is very important as long as availability is concerned. For, if transactions are aborted or blocked, the service is stopped from the user point of view. In the last case, the down site will spend a very long time to process all queued messages before it is anew available for user's needs.

As long as atomicity is concerned, distributed data base systems use variants of the 2 phase commit protocol [LAMP76]. In case of site failures, these algorithms generally abort transactions, hence destroying some work yet performed.

The main idea of these algorithms (concurrency control and 2 phase commit) is to prevent inconsistency by either aborting all transactions that may be dangerous for the data base integrity or storing enough information to allow down sites to recover a correct state (in a generally long recovery phase). These methods decreases the system availability. Hence, we shall use a totally different idea. Every down site, which is cause of the trouble will immediately lost ownership of its data, which will be regenerated on another site which works correctly. So down sites having no data, cannot damaged data integrity. Data regeneration will be performed quickly by transferring a backup copy on a high speed satellite link.

Section 2 of this paper will present the structure of our system, and explain briefly its processing principles. In section 3 we shall detail the processing of one operation (read, write, commit and abort) when no breakdown happens. Finally, section 4 will present the extra work to be done in case of one site failure.

2 - Structure of an available distributed data base system

As in TANDEM computers, where the availability is provided by hardware duplication [KATZ77], [KATZ78], the availability of our system will rely on data duplication. The system will control this duplication as it will need, to keep a minimum number of fragment copies.

2.1. Data structure

In the database, data are partitionned in a set of logical fragments, which are the logical units of data distribution. The system permanently manages two copies of each fragment : a primary copy and a backup copy. Each of these copies is stored on a different site. On each site is stored a global schema that describes all fragments together with the localisation of their two copies.

When a fragment copy is lost, due to a site failure, it will be regenerated by the site that contains the remaining copy of the lost fragment. This site will apply the following algorithm :

- updates in its schema the localisation of this lost fragment.

- modifies the status of its remaining copy from backup to primary if necessary.

- chooses a new site to receive the new backup copy.

- transfers a copy of the fragment to that site.

- when the transfer is over, broadcasts the new localisation of the fragment on the network.

When every lost fragment copy is regenerated, the system is anew in its normal state. Two copies of each fragment are available.

2.2. Locking mechanism

The concurrency control mechanism used by the system rely on the primary copy locking mechanism presented in [ALSB], at least when no failure occurs. Each operation on a fragment is transmitted to the site containing the primary copy where a lock is requested.

When the lock is granted, the operation can be performed, and for write operation, is transmitted to the backup site to update the backup copy. On this backup copy, operations are processed in the same order than on the primary copy, by using a numbering scheme given by the locking mechanism.

If the lock is refused, the system can use any of the mechanism found in the literature, provided that it guarantees no transaction waits indefinitely. It may indifferently decide to abort the transaction or to wait some time before submitting again the lock request. In the last case that can result in a deadlock situation, a deadlock avoidance a detection algorithm must be used. These algorithms generally cause a transaction to be aborted to avoid or to suppress the deadlock.

2.3. Network functionalities

The network must transmit messages and data from one site to another site without introducing errors. It must inform the sender of either the correct delivery of messages or the impossibility to deliver them. That can be achieved by using acknowledgments on every messages and timeouts. Messages cannot be delivered to the destination site for three reasons : failure of the destination site, saturation of that site (which may be assimilated to the previous case), and network failure. That last case may leads to the partition problem where one or more running sites are isolated from the rest of the network. A more complex problem may arise if the site is detected as failed by some sites and not by others. That leads to some inconsistency between the sites states stored on each site. In that paper, we shall suppose we can trust the network. When a site failure is detected, the site is really failed for everybody. The consequences of a lying network will be studied in a further work.

Failure detection through message transmission is sometimes inefficient, especially when a site waits for a reply from another site, and has no message to send to that site.

To get rid of that problem, we shall use time outs to watch over distant site. But, as we can predict neither the time needed to process the operation nor the waiting time on the distant site, we have to modify usual time out mechanisms. When the time out fires, instead of considering the distant site has failed (it may be overloaded), we shall send it an empty message and wait for the network acknowledgment. If the network cannot deliver the message, then the distant site has really failed, otherwise, it has not yet processed our operation and we reactivate the time out.

As far as satellite communication characteristics are concerned, broadcasting facility and high transmission rate are used. Broadcasting facility is used to transmit information (failure of one site, new localisation of a fragment...) to every site on the network. High transmission rate speeds up fragment transfers. As a matter of fact, when a site fails, generation of all lost copies will compel to transfer a great volume of data, equivalent to the amount of data that site stored. So big transfers could only be performed in a reasonable time by high speed networks.

2.4. Notations

A transaction is a sequence of read and write operations, each being performed on one fragment, and ended by a either commit or an abort command. That last command can be issued either by the user or by the system as a consequence of a deadlock between transactions.

In this section, and in the next one, we shall detail the processing of four operations : Read (Fi, Tj), Write (Fi, Tj), Commit (Tj) and Abort (Tj). Fi stands for the fragment being read or written, and Tj for the transaction to which the operation belongs.

Operations on a fragment (like read or write) involve 3 sites that will be denoted by : Sinit, Sref and Scop. Sinit is the site on which the transaction had been initiated. Sref is the site containing the primary (or reference) copy of the fragment and Scop contains its backup copy.

3 - Operation processing without breakdown

3.1. Processing a write operation

In this paragraph, we shall detail the processing of a write operation on the 3 involved sites (Sinit, Sref and Scop). On each site, the protocol to be used will be described using evaluation nets [NUTT73]. We think this formalism greatly improves the understanding of complex asynchronous protocols. In this paragraph, the protocols are still easy to understand, but in the next section, where we shall add failure recovery, their complexity will increase. Moreover, this formalism can be helpful for proving correctness of protocols.

- Sinit

The protocol to be used on that site is presented on figure 1. Operation Write (Fi) is created by the user. Then it is given and Identifying Number (denoted by IDN) unique throughout the network, in transition t1. This IDN is composed with the concatenation of the site number and a local counter incremented every time an operation is created on the site. After fragment localisation has been fetched in the schema (Sref and Scop) the operation is transmitted to site Sref. Transition t2 depends on the network acknowledgment. If message has been delivered then we enter a wait state, otherwise we consider a failure has occured on site Sref. We give up this hypothesis which will be studied in next section. On arrival of an Execution Ack from site Sref, transition t3 is fired, and we reach the end of the protocol. We can inform the user of the correct processing of its order.

If site Sref fails, the execution Ack will never arrives. That failure will be detected by the modified Time Out mechanism presented above and represented by the state T.0 in figure 1. Transition t4, fired by this time out state leads to a failure state which will be studied in next section.

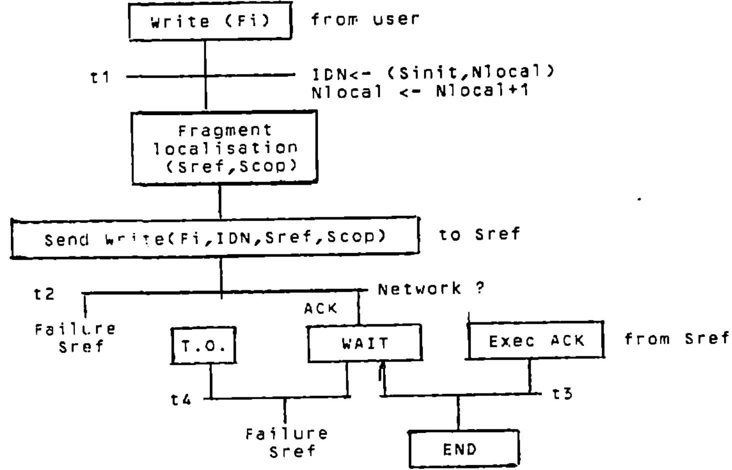

Figure 1 - Write processing on Sinit

- Sref

Figure 2 describes the protocol used on site Sref. The process is activated by the arrival from Sinit of an order : Write (Fi) with its IDN. This order is immediately submitted to the lock mechanism.

If the lock is granted, the operation is given a number, which is an operation counter on the fragment. This number will be useful on site Scop to process operation in the same order than on site Sref, hence avoiding inconsistency that may arise from the two copies updating. Transition t1 can now fire, incrementing the fragment operation counter.

If the lock is refused, the concurrency control mechanism activated through transition t2 enters a reject processing mode in which it can take any decision it wants : delay the operation before submitting it again, abort the transaction, activate a deadlock detection mechanism...

Write processing is divided into two parallel actions. The first one is the local processing of the write order, while in the second one, we transmit that order to site Scop. If the network could deliver the message then we consider the operation will be correctly processed on Scop and we do not wait for any execution ack. Hence copies updating introduces no extra delay (we may consider that one message sending is shorter than any write processing on the local site). Transition t4 represents synchronisation between local processing and good reception of the write order on site Scop. Then we can send an Exec Ack to Sinit, and the protocol ends when this message is delivered.

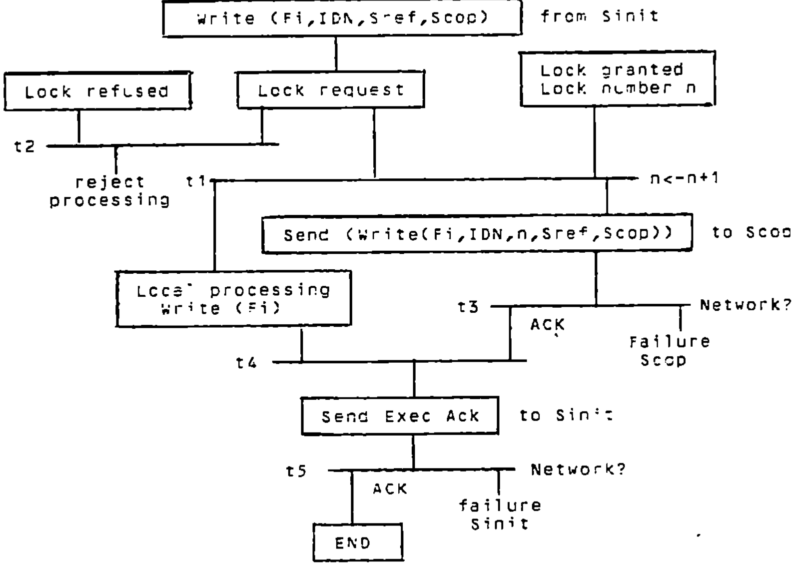

Figure 2 - Write processing on Sref

- Scop

The protocol on site Scop (cf fig. 3) is very simple. Operations are processed according to their locks number. For each fragment, we keep the number of the last operation processed on it. We shall denote it by $n(Fi)$ for fragment Fi. As operations are consecutively numbered by the lock mechanism, we can only process operation numbered $n(Fi) + 1$. This test is made in transition t1. If an operation arrives in advance, it will enter a wait state. Each time an operation is processed, we leave the wait state (transition t2) to test if it is our turn to be processed. If it is, we are processed, otherwise we return to wait state.

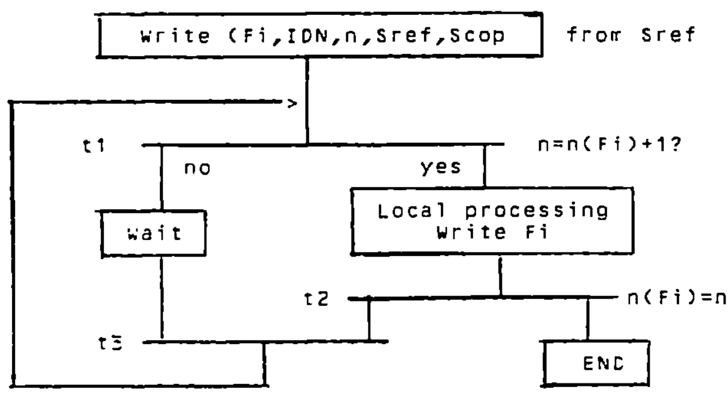

Figure 3 – Write processing on Scop

3.2. Processing other operations

– Read operation

A read operation is performed on site Sref, and site Scop is not concerned with it. Hence, on Sinit, the operation is processed exactly as a write operation. On Sref, we just keep the local part of the write operation processing (suppressing the message exchange with Scop).

– Commit or Abort operations

These operations end transactions that have read or updated many fragments. They must be transmitted to every site containing one of these fragments. We shall study the commit operation, the abort being processed in exactly the same way.

Let Si be the set of all fragments having been read or written by transaction Ti. Commit (Ti) is decomposed on Sinit in a set of elementary commit orders [Commit (Ti,Fj) : Fj is in Si]. These elementary commit orders are only concerned with one fragment, and are processed in exactly the same way as a write operation on that fragment. The global commit order finished when all elementary commits are acknowledged on Sinit.

4 – Operation processing in case of one site failure

In this section, we shall present the actions performed on every site when a failure is detected. Then we shall show what has to be added to the protocol described in the previous section for the operation to be correctly processed when one failure happens. To end this section, we shall say a few words of the critical phase of this algorithm, when a second failure may damage the data base.

4.1. Detection of a failure

In the previous section, we have shown how failure of a site was detected, through time outs on message exchange. When a site detects a failure, it broadcasts this information over the network, and enters a failure recovery phase. When receiving the broadcasted message, each site of the network enters that recovery phase which is described in Figure 4.

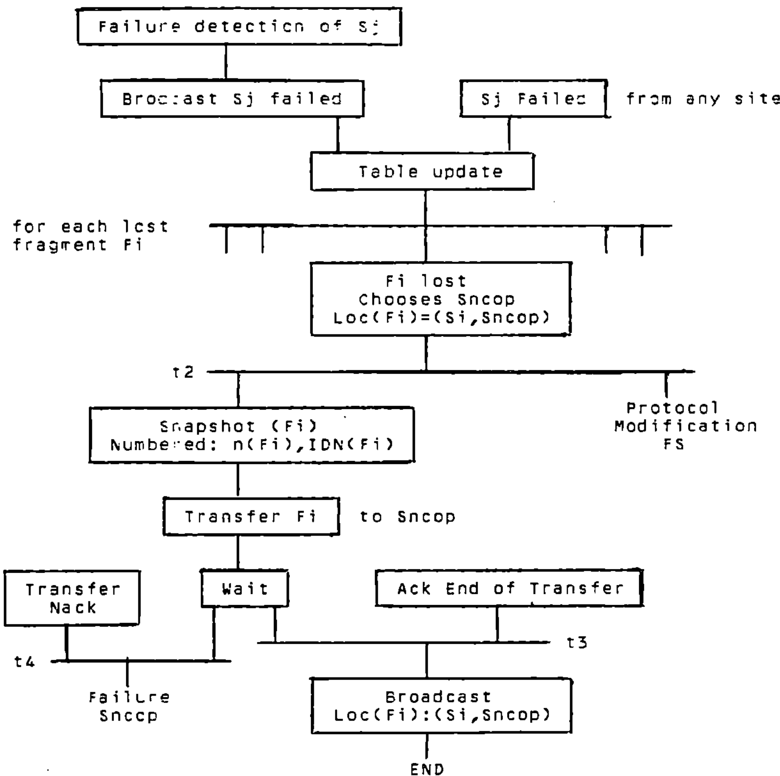

Figure 4 - Failure recovery on site Si

In table updating state, the schema is fetched for all fragments being stored (either as a primary copy or as a backup copy) on the failed site, we shall denote by Sj. Their localisation couple which is either (Sx,Sj) or (Sj,Sx) is replaced by (Sx,0) indicating that Sx now has the primary copy (if it had before the backup one). Then the site selects among these lost fragments those it owns, and begins to regenerate them (transition t1). It first chooses a new site (denoted by Sncop) to receive the new copy of the fragment. The way we choose that site has no consequences on our algorithms. But it may influence the performance of the system : if we overload a site with many fragments for exemple. This performance problem is out of the scope of that paper.

We then create a snapshot [ADIB80] of the fragment, storing it together with two numbers denoted by $n(Fi)$ and $IDN(Fi)$. $n(Fi)$ has been defined in last section as the number of the last operation performed on the fragment. $IDN(Fi)$ is the IDN of that last operation. We shall see in the next paragraph the use of these numbers to detect duplicated orders. We had chosen to use snapshot not to block operation on the fragment while we transfer it. Morevoer, the locking mechanism may allow operations (read or write) to be processed together with the snapshot creation, hence not delaying transaction processing. When the transfer is over (transition t3), the site broadcasts the fragment new localisation (Si, Sncop) on the network. The system is again in normal state, possessing two copies of the fragment. The period while there only exist one copy of the fragment is the critical phase of our system. We shall study it briefly in the last paragraph of that section.

In parallel with the snapshot creation, transition t2 activates the protocol modifications. This state will appear as an entry point (denoted by FS : Failure State) in the modified evaluation nets describing protocols to apply in case of one site failure (see figure 5, 6 and 7).

4.2. Read processing

Two sites are involved in read order on fragment Fi : Sinit and Sref where the read takes place. We shall study consequences of Sinit failure in paragraph 4.4 on Commit and Abort processing. If Sref fails, then Sinit sends again the order to Scop which will soon be the reference site for that fragment.

4.3. Write processing

- Sinit

Sref failure is the only one that causes protocol modifications on Sinit. The write order sent to Sref may have not been transmitted to Scop and hence been in lost after Sref failure. As Sinit cannot detect that loss, it must send again the order to Scop. That appears on the protocol description in transitions t5 and t6. These transitions, fired by the failure state (FS in figure 4) put again the order in the localisation state of the protocol. As localisation has been updated by the recovery protocol (before FS) to (Scop,0), the order will be sent again to Scop using the normal protocol.

It may happen that the order had correctly been transmitted to Scop. In that case it will be duplicated by our protocol. We shall see in paragraph on Scop, how to detect duplication with the IDN-value.

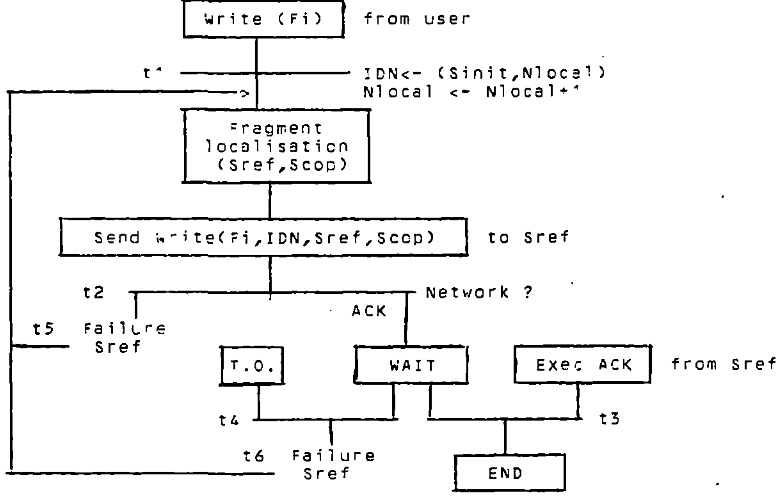

Figure 5 - Write procesing on Sinit

- Sref

Scop failure causes Sref to apply the failure protocol (cf figure 4) to choose a new copy site and to regenerate a copy of the lost fragment. As fragment generation and operation processing on the fragment are asynchronous operations, we must be sure that every operation processed on Sref will also be processed on Sncop. As soon as a failure is detected, all operations present on the site (either waiting or being processed) are reentered in the protocol after their Scop field has been updated to

Sncop. This appears in transitions t6, t7 and t8 fired by failure state FS. The snapshoted fragment having been numbered with the last processed operation, it will be easy to detect duplicated orders on site Sncop and not to process them. Using these operation numbers, site Sref could also detect the duplicated operations by transition t7. This test on operation number is added in the macro state representing local processing.

Sinit failure detected by transition t5 causes the current transaction to be aborted. We can take that decision because, the order having not been acknowledged, the transaction cannot be in the commit phase.

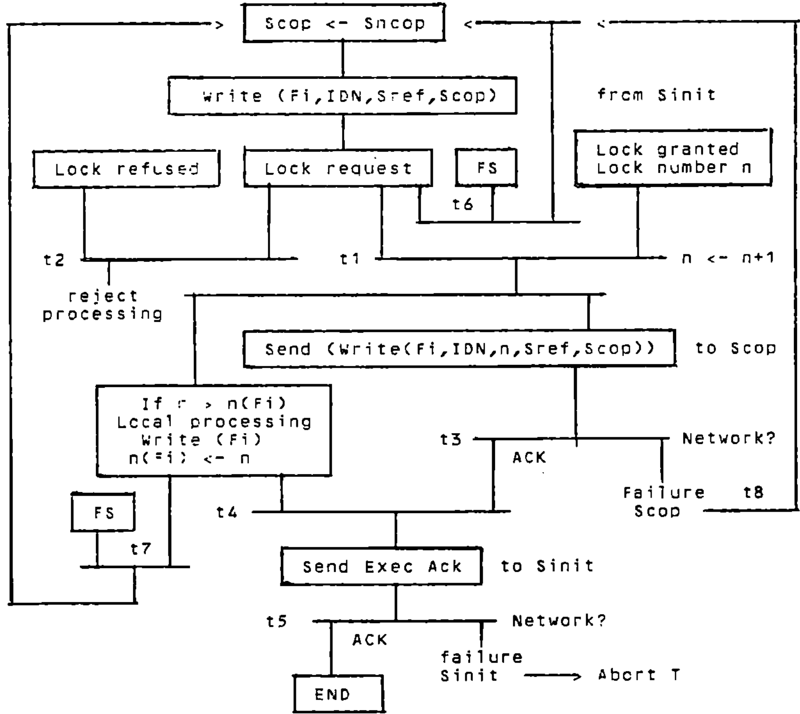

Figure 6 - Write processing on Sref

- Scop

When Scop is informed of Sref failure, it becomes responsible of the lost fragment and of access on that fragment. It has hence to initialise the locking mechanism. To ensure that Scop will be able to recover easily all locks that were put on fragments by site Sref, the locking mechanism is extended. Each time Sref grants a lock on a fragment, it must inform Scop. For write processing, this locking information can be piggybacked in the write order sent to Scop while for read operation an extra message must be sent to Scop. When the site state changes from Scop to Sref, locks table are yet initialised and the mechanism proceeds without any delay.

Orders not yet processed on Scop at the failure detection time, will have to be sent to the site Sncop (to update the snapshoted copy) and acknowledged to Sinit (because these acknowledgments may have never been sent by the failed site Sref). Hence, these operations will follow the usual protocol of site Sref being processed in the order of their number. Transition t4, fired by the failure state FS enters the first order in FIFO queue. Then transition t5 causes all waiting orders to enter that queue according to their numbering. Orders leaving the queue are submitted to the lock mechanism (locks are always granted, because the orders have yet be numbered on Sref by the lock mechanism), and processed using an Sref protocol. When queue is empty, the site can process new orders arriving from any Sinit site. This appears in our figure, in transition t6 fired by the empty queue condition.

Sinit can send two kinds of orders : orders created by a user, or orders duplicated by Sinit after detection of Sref failure. We must detect orders of that last kind that have yet be processed. In fact, such orders can only belong to one transaction (because of the lock mechanism). If a transaction is only allowed to send an order on a fragment when the previous one has been acknowledged, then we can have only one duplicated order. To detect it, operation numbers are useless because orders are not numbered on Sinit, we must use IDN. For that reason we shall store with each fragment the IDN of the last performed operation. As we tested the operation number in the local processing state, we shall also test their IDN. If a site can have more than one unacknowledged order at a time, we can use a window mechanism limiting their number (to Nmax) and we shall store with each fragment the IDN of the Nmax last processed orders.

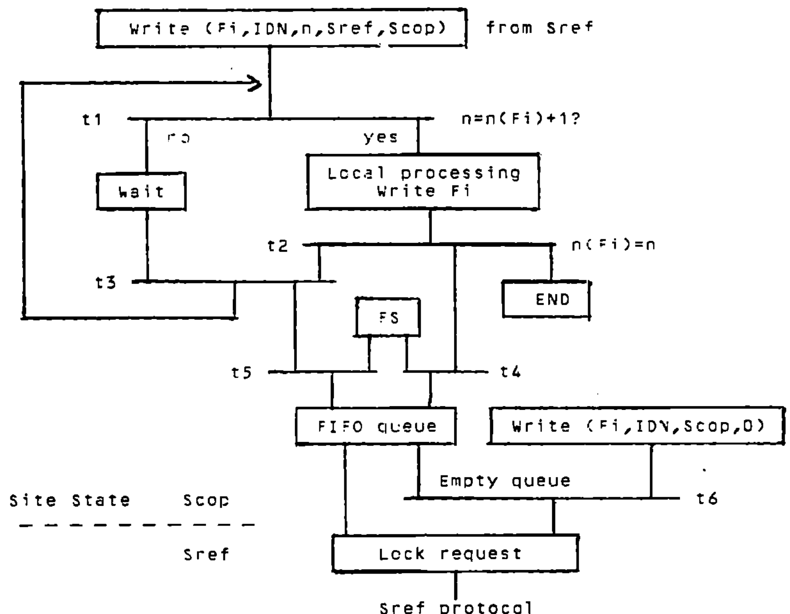

Figure 7 - Write processing on Scop

- Sncop

Sncop will receive, in any order, one copy of fragment Fi and orders to be processed on that fragment. While the fragment transfer is not finished (detected by localisation couple of Fi = (Scop,0)), write orders are delayed. Transition tl, fired by that localisation update, initialises processing of those orders. The site is now the Scop site for Fi. As orders arriving from Scop, may or may not have been processed on the transmitted fragment, we shall use the fragment number and the orders numbers to detect duplicated orders (see Local processing state in protocol Scop).

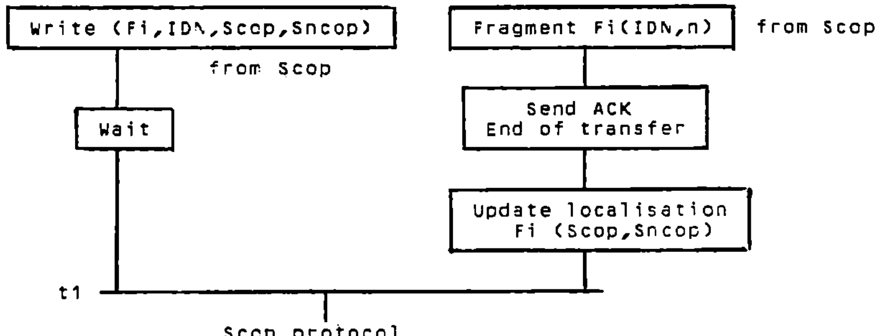

Figure 8 - Write processing on Sncop

4.4. Commit and abort processing

Each elementary commit (or abort order) concerning only one fragment can be processed as any write order.

A particular problem arises when the Sinit site, which coordinates the transaction, fails.

It is useless to process a transaction for a failed site, so transaction can be aborted by all involved sites. But, if the transaction was in the commit phase (commit initialised by Sinit) some site may have received the commit order and some may not. Due to our algorithm, site having received a commit order will do process it. We must ensure that every other site will also commit. Such decision cannot be taken individually by any site without querying everybody, a very costly mechanism. Hence we shall use a backup structure for commit coordination.

When the transaction is initialised, Sinit chooses a backup coordinator site. The Commit order will be sent to that site before being broadcasted on the network. When all elementary commit orders have been acknowledged, Sinit will inform the backup coordinator.

When Sinit failure is detected, if the backup coordinator has received the commit order it will broadcast it on the network, otherwise it will broadcast an abort order. This mechanism guarantees that every site will take the same decision.

4.5. Critical phase and simultaneous failures

Our system is supposed to keep permanently 2 copies of every fragment for failure survival. Of course, when a failure happens, this property becomes false while the lost copies have not been regenerated. This period, between the site failure and the end of transfer of one fragment to site Sncop is called the critical phase for that fragment. As a matter of fact, if Sref fails during it, anything becomes wrong : we loose data.

We must study mechanisms to get rid of that critical phase. We can either increase our system resiliency by using more copies or by reducing the duration of that phase (hence decreasing the probability of data loss), or detect data loss and use conventionnal recovery mechanism with backups and logs to restore lost data.

5 - Conclusion

We have presented a mechanism (concurrency control and transaction commit) allowing transaction never to be stopped by one site failure. The aim of such a mechanism is to improve availability of a database system from the user point of view.

No formal proof of the algorithm have been given, but we think its formal description with evaluation nets will greatly help such a network.

Many studies are still to be done. They concern evaluation of the availability improvement resulting from the structure we presented, evaluation of the critical phase duration and of the probability of a second crash during it. To choose a new backup site, many algorithms can be tested for their performance consequences. We can also study recovery mechanism to offer full security of the system. We can also extend our algorithm to accept more than one backup copy and offer resiliency to simultaneous failures.

References

[ADIB80] ADIBA M.E. and LINDSAY B.G. : "Database snapshot". IBM research report. R.J 2772. March 80.

[AGHI82] AGHILI H. and al. : "A highly available database system". IBM research report. RJ 3275. Dec. 82.

[ALSB76] ALSBERG P.A. and DAY J.D. :"A principle for resilient sharing of distributed resources". Proceedings of 2nd Int. Conf. on Software Engineering. October 1976.

[NADL82] ANDLER S. and al. : "System D : a distributed system for availability". Proc. 8th VLDB Mexico. Sept. 82.

[BERS80] BERNSTEIN P.A., SHIPMAN D.W. and ROTHNIE J.B. : "Concurrency control in a system for distributed database (SDD-1)". ACM Transaction on Database Systems. Vol. 5, N°1, march 1980, 18-51.

[BERS81] BERNSTEIN P.A. and GOODMAN N. : "Concurrency control in distributed database systems". ACM Computing Surveys, vol. 13, N°2, june 1981, 185-221.

[KATZ77] KATZMAN J.A. : "System Architecture for non stop Computing". Proc. COMPCON. Feb. 1977.

[KATZ78] KATZMAN J.A. : " A fault tolerant computer system". Proc. Int. Conf. on system sciences. Hawai. June 1978.

[FEUV84] FEUVRE J.M. : "Availability in a distributed database system". 3rd seminar on Distributed Data Sharing Systems. Parma (Italy) March 1984.

[LAMP76] LAMPSON B.W., STURGIS H.E. : "Crash recovery in a Distributed Data Storage System" Xerox research report. Palo Alto 1976.

[LELA78] LE LANN G. : "Algorithms for distributed data-sharing systems which uses tickets". Proc. 3rd Berkeley Workshop on Distributed Data Management and Computer Networks. San Francisco. August 1978. 259-272.

[LOMB83] LOMBART D. and al. : "TELECOM1 - Multi-services Network". Proc. Int. Symposium on Satellite and Computer Communication. Versailles. April 83.

[NUTT73] NUTT G.J. and NOE J.D. : "Macro E-nets for representation of parallel systems". IEEE Trans on Computers. Vol. C22 N° 8. August 1973.

[STON79] STONEBRAKER M. : "Concurrency control and consistency of multiple copies of data in distributed INGRES". IEEE Trans. on Soft. Eng. Vol SE-3 n°3. May 1979. 188-194.

ISDN - DIE FOLGERICHTIGE WEITERENTWICKLUNG DES DIGITALISIERTEN FERN-
      SPRECHNETZES FÜR DAS KÜNFTIGE DIENSTLEISTUNGSANGEBOT DER
      DEUTSCHEN BUNDESPOST

K. H. Rosenbrock
Deutsche Bundespost
Bundesministerium für das Post-
und Fernmeldewesen in Bonn

Zusammenfassung

Nach einer ersten einleitenden Begriffsbestimmung wird das digitalisierte Fernsprechnetz mit seinen Netzbestandteilen vorgestellt, weil
es die wesentliche Voraussetzung für das ISDN bildet.
Es schließt sich eine weitergehende allgemeine Beschreibung des ISDN
an. Im folgenden wird dann auf das künftige Dienstleistungsangebot
der Deutschen Bundespost im einzelnen eingegangen: auf die möglichen
Dienste und Dienstmerkmale im ISDN. Einige Schlußfolgerungen und ein
kurzer Ausblick auf die Weiterentwicklung des ISDN runden den Beitrag
ab.

Summary

ISDN - THE LOGICAL EVOLUTION OF THE DIGITAL TELEPHONE NETWORK FOR
      THE FUTURE SERVICES OF THE DEUTSCHE BUNDESPOST

After an introductory definition of the basic characteristics of
an ISDN the digital telephone network with its components (digital
transmission, digital switching, and common channel signalling) is
dealt with. This is done because the ISDN will evolve from the
digital telephone network. In the following the ISDN is described
in more detail. With these statements a reply is given to the
question "What is the ISDN?".

In addition, the future services to be offered by the Deutsche
Bundespost in the ISDN are presented in some detail. In this respect
the emphasis is put on new services operating with the typical ISDN
bit rate of 64 kbit/s using the internationally standardized
subscriber interface S_0. This section will be complemented by a
description of possible service attributes within the ISDN.

The paper ends with some conclusions and an outlook at the future
evolution of the ISDN towards broadband communication.

ISDN - DIE FOLGERICHTIGE WEITERENTWICKLUNG DES DIGITALISIERTEN FERNSPRECHNETZES FÜR DAS KÜNFTIGE DIENSTLEISTUNGSANGEBOT DER DEUTSCHEN BUNDESPOST

K.H. Rosenbrock
Deutsche Bundespost
Bundesministerium für das Post- und Fernmeldewesen, Bonn

Inhaltsverzeichnis

1 Einleitung

Die Abkürzung ISDN steht für den weltweit eingeführten Begriff
"Integrated Services Digital Network", der mit "Dienste-integrieren-
des digitales (Fernmelde-)Netz" übersetzt werden kann.

Aufgrund unterschiedlicher Voraussetzungen wird die Entwicklung zum
ISDN bei den vielen Fernmeldeverwaltungen und Betriebsgesellschaften
wahrscheinlich unterschiedlich verlaufen, dennoch ist man sich
international darüber einig, daß das ISDN durch folgende wesentliche
Merkmale gekennzeichnet ist:

o Es entwickelt sich aus dem digitalisierten Fernsprechnetz.
o Es wird durch möglichst wenige Schnittstellen beschrieben, die
  aber international genormt sind.
o Es erlaubt den Anschluß vielfältiger Endgeräte über eine genormte
  Schnittstelle (Prinzip der einheitlichen Kommunikationssteckdose).
o Es bietet neben dem Fernsprechen eine Reihe neuer sowie bestehender
  Dienste und Dienstmerkmale an.
o Etwas abstrakt kann das ISDN auch als ein universelles digitales
  (Fernmelde-)Netz angesehen werden, welches 64-kbit/s-Verbindungen
  (für die verschiedenen Dienste) zwischen beliebigen Endpunkten
  bereitstellt.
o Das ISDN ist medienunabhängig, d. h. als Übertragungsmittel
  kommen im Grundsatz alle heute und künftig eingesetzten Übertragungsmedien in Frage.
o Die Erweiterung auf Breitbanddienste ist vorgesehen.
o Der vollständige Übergang zum ISDN wird Jahrzehnte dauern.

Da sich das ISDN aus dem digitalisierten Fernsprechnetz weiterentwickelt, soll im folgenden zunächst das digitalisierte Fernsprechnetz der Deutschen Bundespost vorgestellt werden. Anschließend
wird erläutert, welche Ergänzungsmaßnahmen für den Übergang zum
ISDN erforderlich sind. Es folgt die Darstellung des künftigen
Dienstleistungsangebots der Deutschen Bundespost im ISDN. Abgeschlossen wird der Beitrag mit Schlußfolgerungen und einem Ausblick
auf die Weiterentwicklung des ISDN.

2 Das digitalisierte Fernsprechnetz als Voraussetzung für das ISDN

Obwohl das herkömmliche Fernsprechnetz der Deutschen Bundespost
eine derart hohe Dienstgüte gewährleistet, daß es auch heute weltweit
noch keinen Vergleich zu scheuen braucht, handelt es sich hier
um ein Systemkonzept, welches den künftigen erweiterten Anforderungen
der Kunden nicht mehr genügen wird. Die Deutsche Bundespost hat

deshalb im Jahre 1979 entschieden, das Fernsprechnetz mit Vorrang
zu digitalisieren. Sie folgt damit dem weltweiten Trend zur Digital-
technik.

Das digitalisierte Fernsprechnetz wird im wesentlichen durch folgende
Netzbestandteile geprägt: Digitale Vermittlungstechnik, digitale
Übertragungstechnik und Zentralkanalzeichengabe. Die Anschlußlei-
tungen sind dabei jedoch noch analog. Wie sieht nun die Situation
bei der Deutschen Bundespost bezüglich der drei Netzkomponenten
aus?

2.1 Digitale Übertragungstechnik

Die Deutsche Bundespost sieht den Einsatz folgender digitaler
Übertragungssysteme vor:

- PCM 30 (2 Mbit/s),
- PCM 480 (34 Mbit/s),
- PCM 1920 (140 Mbit/s) und
- PCM 7680 (565 Mbit/s).

Nach der Grundsatzentscheidung zugunsten eines digitalisierten
Fernsprechnetzes werden seit 1982 im <u>regionalen</u> Fernsprechnetz
der Deutschen Bundespost grundsätzlich nur noch digitale Übertragungssysteme beschafft. Ein Grundausbau der gesamten Fernebene
wird zunächst mit den Übertragungssystemen PCM 30 und PCM 480
bis zum Jahre 1985 erreicht und soll anschließend ständig erweitert
werden.

Noch in diesem Jahr sollen auch im <u>überregionalen</u> Fernsprechnetz
die Übertragungssysteme PCM 1920 und 7680 eingesetzt werden. Als
Übertragungsmedien dienen hierfür: symmetrische Kabel (2 Mbit/s),
Koaxialkabel (34, 140 und 565 Mbit/s), Glasfaserkabel (34, 140
und 565 Mbit/s), digitaler Richtfunk (2, 34 und 140 Mbit/s) und
Satelliten.

2.2 Digitale Vermittlungstechnik

Die digitale Vermittlungstechnik ist gekennzeichnet durch das
vierdrähtige Durchschalten von 64-kbit/s-Kanälen; d. h. dieselbe
Bitrate wie bei der digitalen Übertragungstechnik. Digitale Vermittlungssysteme bieten sowohl dem Kunden als auch dem Netzbetreiber
eine Reihe neuer Dienst- und Leistungsmerkmale.

Um möglichst schnell zu einer soliden Systementscheidung unter
Wettbewerbsbedingungen zu kommen, hat die Deutsche Bundespost
ein jeweils einjähriges "Präsentationsverfahren" für die Fern-
und Ortsvermittlungstechnik durchgeführt. Das Präsentationsverfahren
hat schließlich zu der Entscheidung der Deutschen Bundespost zu
Gunsten des Systems EWSD der Firma Siemens und des SYSTEMs 12
der Firma Standard Elektrik Lorenz geführt.

Das weitere Vorgehen der Deutschen Bundespost ist durch folgende
Zielvorgaben geprägt:

o Beschaffungsbeginn für
   - digitale Fernvermittlungstechnik in der zweiten Hälfte 1984,
   - digitale Ortsvermittlungstechnik in der ersten Hälfte 1985.

o Beschaffungsübergang auf digitale Vermittlungstechnik innerhalb
   von 5 Jahren; d. h. ab 1990 soll nur noch digitale Vermittlungstechnik beschafft werden.

o Wegen der hohen Zuwachsraten im Fernverkehr wird der Einsatz
   digitaler Vermittlungstechnik in der Fernebene mit Vorrang vorgenommen.

o Vollständiges Digitalisieren des Fernsprechnetzes bis spätestens
   zum Jahre 2020.

Bis zum Jahre 1990 werden bereits mehr als je 100 digitale Fern-
und Ortsvermittlungsstellen im Netz der Deutschen Bundespost installiert sein. Digitale Ortsvermittlungstechnik wird grundsätzlich an digitale Fernvermittlungstechnik angebunden. Ihr Einsatz
beginnt mit dem Auswechseln ganzer Vermittlungsstellen in den
großen Ortsnetzen und wird anschließend auf mittlere und kleine
Ortsnetze ausgedehnt.

2.3 Zentralkanalzeichengabe

Im Gegensatz zur sprechkreisgebundenen Zeichengabe wird bei der
Zentralkanalzeichengabe die Signalisierung für viele Sprechkreise
über einen gemeinsam benutzten Zeichengabekanal abgewickelt. Sie
eignet sich besonders für Vermittlungssysteme mit Rechnersteuerung,
weil der große Zeichenvorrat und die wesentlich erhöhte Leistungsfähigkeit der Zentralkanalzeichengabe es ermöglichen, daß die
Vermittlungsrechner unmittelbar miteinander Nachrichten austauschen
können. Für digitale Netze ist das Zeichengabesystem Nr. 7 vom
CCITT standardisiert worden. Mit seinem Einsatz ist im Netz der
Deutschen Bundespost im Jahre 1986/87 zu rechnen.

2.4 Das integrierte digitalisierte Fernsprechnetz

Die digitale Übertragungs- und Vermittlungstechnik sind einander
sehr ähnlich geworden. Entsprechendes gilt für die Zentralkanalzeichengabe, die auch als eine Untermenge der Vermittlungstechnik
betrachtet werden kann. Etwas vereinfachend ist in allen Fällen
davon auszugehen, daß Bitströme mit einer Übertragungsrate von
64 kbit/s im Zeitmultiplex übertragen, vermittelt oder verarbeitet
werden. Damit der Vorgang geordnet ablaufen kann, bedient man
sich hierbei der Synchronisation.

Aufgrund des Zusammenwachsens, insbesondere der Übertragungs-
und der Vermittlungstechnik, spricht man hier von einer Integration
beider Techniken oder auch vom integrierten digitalen Fernsprechnetz.
Der englische Begriff hierfür lautet: "Integrated Digital Network
(IDN)". Hierbei handelt es sich um eine erste technikbezogene
Integrationsstufe. Die zweite Stufe der Integration ergibt sich,
wenn das (integrierte) digitalisierte Fernsprechnetz dazu herangezogen wird, das Zusammenfassen (die Integration) verschiedener
Fernmeldedienste - neben dem Fernsprechen - in einem Netz zu ermöglichen.

Wie umfangreiche Studien gegen Ende des letzten Jahrzehnts nachgewiesen haben, können im digitalisierten Fernsprechnetz die technischen Vorzüge der gemeinsam für den Fernsprechdienst eingesetzten
digitalen Übertragungs- und Vermittlungstechnik wirtschaftlich

ausgenutzt werden. Der Beginn des digitalisierten Fernsprechnetzes
wird durch den Einsatz der digitalen Vermittlungstechnik bestimmt
und kann mit dem Jahreswechsel 1985/86 angesetzt werden. Zu diesem
Zeitpunkt ist jedoch noch nicht der Serieneinsatz des Zentral-
kanalzeichengabesystems Nr. 7 vorgesehen.

Mit dem digitalisierten Fernsprechnetz kann die Deutsche Bundespost
eine fernmeldetechnische Infrastruktur bereitstellen, die durch
digitale Verbindungen von der Ursprungs- zur Zielvermittlungsstelle
mit einer Standard-Übermittlungsgeschwindigkeit von 64 kbit/s
gekennzeichnet ist. Die Teilnehmeranschlußleitungen werden hierbei
in der Regel noch analog betrieben.

Die früher durchgeführten Untersuchungen, daß das Digitalisieren
des Fernsprechnetzes wirtschaftlich sinnvoll ist, sind inzwischen
von der Praxis bestätigt worden. Der Einsatz digitaler Vermittlungstechnik ist bereits unter ungünstigsten Bedingungen - in analoger
Umgebung - wirtschaftlicher als die herkömmliche elektromechanische
Vermittlungstechnik. Die Wirtschaftlichkeit wird jedoch noch erheblich verbessert, wenn die digitale Vermittlungstechnik gemeinsam
mit der digitalen Übertragungstechnik eingesetzt wird. Das Digitalisieren des Fernsprechnetzes ist für sich allein gesehen eine wirtschaftliche und zweckmäßige Maßnahme, die zunächst völlig losgelöst
von den Möglichkeiten der Dienstintegration im ISDN betrachtet
werden muß.

Verglichen mit dem herkömmlichen analogen Fernsprechnetz bietet
das digitalisierte Fernsprechnetz neben den erheblichen wirtschaftlichen und betrieblichen Vorteilen eine Reihe weiterer Vorzüge
für den Anwender:

o verbesserte Übertragungsgüte und Verständlichkeit auch bei größten
  Entfernungen.
o kürze Verbindungsaufbauzeiten.
o weniger Blockaden im Netz.

Die letzten beiden Verbesserungen werden ermöglicht durch ein
hohes Maß an Intelligenz im Netz aufgrund rechnergesteuerter Vermittlungsstellen sowie durch eine leistungsfähige Zeichengabe.

3 Was ist das ISDN?

Einleitend ist bereits eine erste grobe Bestimmung des Begriffs
"ISDN" gegeben worden.

Der gleichzeitige Einsatz mehrerer verschiedener Fernmeldenetze
bedeutet in der Regel einen zum Teil beachtlichen Mehraufwand
z. B. für die Entwicklung, die Technik, die Planung und den Betrieb
der verschiedenen Netze und ihrer Bestandteile, insbesondere bei
der Vermittlungstechnik und bei der Teilnehmeranschlußleitung
einschließlich der Zeichengabe. Es ergeben sich verhältnismäßig
hohe Kosten für die Hauptanschlüsse, insbesondere wenn die Anzahl
der Teilnehmer eines eigenständigen Netzes sehr gering ist. Deshalb
ist es verständlich, daß die Fernmeldeverwaltungen überall nach
wirtschaftlichen Wegen und Möglichkeiten suchen, möglichst viele
Fernmeldedienste in einem einheitlichen Netz anzubieten; d. h.
Diensteintegration.

Nach der Grundsatzentscheidung der Deutschen Bundespost im Jahre
1979, das Fernsprechnetz aus wirtschaftlichen Gründen zu digitalisieren, ergibt sich folgerichtig der Schluß, auch die vorhandenen
Teilnehmeranschlußleitungen des digitalisierten Fernsprechnetzes
zu digitalisieren und damit den entscheidenden Schritt zum ISDN
zu gehen. Da die Kupferdoppeladern der Teilnehmeranschlußleitungen
den mit Abstand größten Anteil der bisher erbrachten Investitionen
des Fernmeldenetzes darstellen, bietet es sich an, die Kupferdoppeladern mehrfach auszunutzen.

Den Schlüssel zum ISDN stellt der sogenannte ISDN-Basisanschluß
dar. Über die gewöhnliche Kupferdoppelader der Teilnehmeranschlußleitung können in digitaler Form 2 Nutzkreise, sogenannte Basiskreise
B_1 und B_2 mit einer Bitrate von je 64 kbit/s sowie ein zusätzlicher
Steuerkreis D_o mit einer Bitrate von 16 kbit/s angeboten werden.
Der Steuerkreis D_o ist für ein außerordentlich leistungsfähiges
Zeichengabeverfahren auf der ISDN-Anschlußleitung vorgesehen,
für das sogenannte D-Kanal-Protokoll. Kennzeichnendes Merkmal
für den ISDN-Basisanschluß ist darüber hinaus die genormte Teilnehmerschnittstelle S_o, die als "Universalsteckdose" für verschiedene
ISDN-Endeinrichtungen angesehen werden kann.

An einen ISDN-Basisanschluß können bis zu 8 Endgeräte angeschlossen
werden, dabei sind neben Einzeldienstendgeräten auch Mehrdiensteend-
geräte vorstellbar.

Neben dem ISDN-Basisanschluß gibt es auch noch Primärmultiplexan-
schlüsse (Digitalsignalverbindungen mit 2 Mbit/s) mit 30 Basiskreisen
für den Anschluß mittlerer und großer Nebenstellenanlagen. Hierauf
soll jedoch nicht näher eingegangen werden.

Ausgehend von den - hier nicht dargestellten - erheblichen Vorteilen
die sich bereits durch das Digitalisieren des Fernsprechnetzes
ergeben, wird nun erläutert, welche allgemeinen Gründe außerdem
noch für die Integration von Fernmeldediensten im ISDN sprechen:

o Vermeiden eigener Netze für verschiedene Dienste.

o Wirtschaftliches Ausnutzen der Einrichtungen des digitalisierten
  Fernsprechnetzes, insbesondere der Kupferdoppelader beim Teilnehmeranschluß.

o Das ISDN bietet neue Dienste und Dienstmerkmale verhältnismäßig
  schnell und zu geringen zusätzlichen Kosten an.

o Abwickeln mehrerer Dienste auf einem einzelnen Anschluß.

o Wahlweises Benutzen zweier Basiskreise.

o Gleichzeitige Kommunikation zweier Endgeräte.

o Erweiterte und verbesserte Dienstmerkmale.

4 Das künftige Dienstleistungsangebot der Deutschen Bundespost
   im ISDN

4.1 Mögliche Dienste im ISDN

Das ISDN wird wesentlich gekennzeichnet durch eine transparente
übermittelte Bitrate von 64 kbit/s. Es kann damit als Träger für

alle jene Fernmeldedienste herangezogen werden, deren Informationen
in digitaler Form übertragbar sind und bei denen die 64 kbit/s
als Bitrate hierfür ausreichen.

Es wird in diesem Beitrag bewußt darauf verzichtet, den Begriff
"Dienst" besonders zu bestimmen. Der interessierte Leser wird
auf die neuen CCITT-Empfehlungen I.200 ff. verwiesen, die sich
mit den Diensten im ISDN und ihren Definitionen befassen. Hier
soll aus Gründen der Vereinfachung kein Gruppieren der Dienste
wie im CCITT - anhand des ISO-Schichtenmodells - vorgenommen werden,
sondern es wird lediglich zwischen folgenden Gruppen unterschieden
werden:

o neue 64-kbit/s-Dienste, die für die S_o-Schnittstelle optimiert
  sind,

o Dienste, die sich z. B. durch das Anschalten bestehender Endeinrichtungen an das ISDN ergeben, bei denen sich jedoch die Dienstmerkmale von denen des bisherigen Dienstes unterscheiden, und

o bisherige Dienste.

Es wird in jedem Fall notwendig sein, die einzelnen Dienste genau
zu kennzeichnen und eindeutig zu definieren.

4.1.1 Neue 64-kbit/s-Dienste im ISDN

Bei den neuen Diensten kann es sich wiederum um 2 verschiedene
Kategorien handeln:

o Weiterentwicklung bestehender Dienste, die insbesondere von
  der erhöhten Übermittlungsgeschwindigkeit profitieren, und

o völlig neue Dienste.

Den 64-kbit/s-Diensten kann im ISDN eine Fülle von Dienstmerkmalen
angeboten werden (s. Abschnitt 4.2).

o <u>ISDN-Fernsprechen</u>

Vergleicht man den Fernsprechdienst im ISDN mit dem des analogen
oder dem des gemischten analogen/digitalen Fernsprechnetzes,
so profitiert der Fernsprechdienst im ISDN insbesondere von
der weitspannenden (Ende-zu-Ende) digitalen Übermittlung, die
ihre Vorzüge in folgenden Merkmalen hat:

- die höhere Sprachqualität führt zu einer besseren Verständlichkeit,
- ein verbessertes Signal-Geräusch-Verhältnis sorgt für ein
  störungsfreies Gespräch,
- da die Dämpfung nahezu entfernungsunabhängig ist, kann ein
  Beeinträchtigen der Verbindung durch mangelhafte Lautstärke
  vermieden werden.

Den bisherigen Betrachtungen liegt eine unveränderte Fernsprech-
bandbreite von 300 bis 3 400 Hz zugrunde. Mit Hilfe anderer
oder erweiterter Kodierungsverfahren wird es künftig - auch
bei einer Bitrate von 64 kbit/s - möglich sein, eine höhere
Sprachbandbreite, z. B. 7 kHz, zu übertragen.

o <u>ISDN-Datenübermittlung</u>

Die Bitrate von 64 kbit/s eröffnet für das Übertragen von Daten
eine Vielfalt neuer Möglichkeiten im geschäftlichen wie im
privaten Bereich. Mit 64 kbit/s steht ein außerordentlich leistungsfähiger Übertragungsweg zur Verfügung, der ein Vielfaches
dessen übertragen kann, was heute üblich ist. Es darf angenommen
und gehofft werden, daß eine ISDN-Standard-Datenübermittlung
mit 64 kbit/s dazu beiträgt, die z. Z. vorhandene Vielzahl unterschiedlicher Übertragungsgeschwindigkeiten mittelfristig
entscheidend zu verringern und damit zu einer spürbaren Dienste-
und Produktbereinigung (mit den wirtschaftlichen Vorteilen insbesondere im Endgerätebereich) beizutragen.

Im geschäftlichen Sektor wird die 64-kbit/s-Datenübermittlung
die Bürokommunikation mitgestalten. In den privaten Haushalten
eröffnen sich außerordentlich vielseitige neue Anwendungsmöglich-
keiten, z. B. durch den Anschluß von Heimcomputern für Fortbil-
dungsprogramme, Haushaltsführung, Steuererklärungen usw. sowie
durch den Zugang zu elektronischen Bibliotheken und privaten

Datenbanken und schließlich durch die stark im Vormarsch befindlichen Telespiele.

o <u>ISDN-Teletex</u>

Z. Z. findet die schnellste Textübertragung im Teletex-Dienst
des integrierten Text- und Datennetzes statt. Mit einer Bitrate
von 2,4 kbit/s werden dort etwa 10 Sekunden für das Übermitteln
einer DIN-A4-Seite benötigt. Bei 64 kbit/s wird hierfür die
Größenordnung "kleiner als 1 Sekunde" erreicht. Auch ein 64-kbit/s-
Teletexdienst wird sich besonders für die künftige Bürokommunikation eignen, bei der es entscheidend darauf ankommt, eingehende
Informationen in der jeweils zweckmäßigsten Art darzustellen,
zu speichern, zu verändern, zu ergänzen, weiterzusenden usw.

o <u>ISDN-Telefax</u>

Die heutigen Kopier- oder Telefaxdienste leiden darunter, daß
- abhängig von der verfügbaren Geräteklasse - für das Übermitteln
einer DIN-A4-Seite noch 1 bis 3 Minute(n) benötigt werden. Im
ISDN sieht es dagegen wesentlich günstiger aus. Mit der ISDN-
Bitrate werden etwa nur noch 10 bis 1 Sekunde(n) für das Fernkopieren einer DIN-A4-Seite gebraucht. Mit dieser Fernkopier-
geschwindigkeit, die schon fast die Größenordnung heutiger ortsgebundener Kopiergeräte erreicht, eröffnen sich weitere Anwen-
dungsmöglichkeiten, insbesondere im Bereich der Bürokommunikation.

o <u>ISDN-Textfax</u>

Mit dem Textfax-Dienst werden sowohl Texte als auch Bildpunkte
übermittelt. Er stellt damit eine Kombination aus Teletex und
Telefax dar und eigent sich besonders zum Übermitteln von Dokumenten, z. B. mit Briefkopf, Texten, Skizzen, Bildern und Unter-
schriften.

o <u>ISDN-Bildschirmtext</u>

Der Bildschirmtext-Dienst wird erheblich von der höheren Übertragungsgeschwindigkeit profitieren können, z. B. durch schnelleren
Bildwechsel und bessere bildliche Darstellung als bisher.

o <u>ISDN-Bilddienste</u>

Die Ausführungen zum Telefax- und zum Bildschirmtextdienst sind
zum Teil auch hier anwendbar. Es ist eine Reihe unterschiedlicher
Anwendungen vorstellbar:

- Festbildübermittlung
- langsames Bewegtbild (Bildwechsel z. B. alle 8 Sekunden),
- Fernzeichnen und
- Fernskizzieren.

o <u>ISDN-Fernwirken</u>

Hinter dem Begriff "Fernwirken" verbirgt sich eine Reihe interessanter Anwendungsgebiete. Die Deutsche Bundespost untersucht
z. Z. intensiv geeignete Einsatzmöglichkeiten im herkömmlichen
Fernsprechnetz (TEMEX) und wird Fernwirkdienste später auch
im ISDN anbieten.

o <u>Zusammenfassung</u>

Es ist durchaus vorstellbar, einige der hier beschriebenen Dienst-
arten zu kombinieren. Als Endeinrichtungen werden dann Mehr-
diensteendgeräte oder auch Multifunktionsterminals benötigt.
Hierbei bieten sich z. B. die Kombinationen Fernsprechen und
Bildschirmtext zum Bildfernsprecher sowie Teletex und Telefax
zum Textfaxendgerät an.

Die Deutsche Bundespost wird sich bemühen, die neuen 64-kbit/s-
Dienste mit Vorrang einzuführen, weil bei ihnen sämtliche ISDN-
spezifischen Vorteile genutzt werden können. Dabei bietet sich
ein schrittweises Vorgehen an; d. h. die hier aufgeführten möglichen Dienste werden nicht schlagartig, sondern nacheinander
eingeführt, um die damit verbundenen Schwierigkeiten beherrschen
zu können.

Der ISDN-Basisanschluß stellt dem Fernmeldekunden 2 Basiskreise
B_1 und B_2 zur Verfügung. Damit ist es möglich, auf beiden Nutz-
kreisen 2 wechselseitige Kommunikationen (d. h. in Vor- und
Rückwärtsrichtung) mit einer Bitrate von jeweils 64 kbit/s gleich-

zeitig und unabhängig voneinander - auch zu verschiedenen Zielen -
auf der vorhandenen Kupferdoppelader durchzuführen. Bei beiden
Kommunikationsformen kann es sich um einen einheitlichen Dienst
oder um 2 verschiedene Dienste handeln. Auch hier ist es möglich,
daß sich die Dienstart von Verbindung zu Verbindung oder gar
während einer Verbindung ändert.

Mit der gleichzeitigen Kommunikation unterschiedlicher Dienste
auf dem ISDN-Basisanschluß wird z. B. möglich, daß ein Fernge-
spräch durch das gleichzeitige Übertragen einer Zeichnung oder
Grafik auf dem zweiten Basiskreis ergänzt oder unterstützt wird.
Ebenfalls ist vorstellbar, daß auf nur einem 64-kbit/s-Kreis
die Dienstart kurzzeitig gewechselt wird, um eine andersartige
Information zwischenzeitlich zu übermitteln. An einen ISDN-Basisanschluß können mehrere - auch verschiedene - ISDN-Endgeräte
über einheitliche Steckdosen ohne besondere Anpassungseinrichtung
angeschlossen werden (theoretisch bis zu 8).

4.1.2 Der mögliche Einsatz herkömmlicher Endgeräte

Beim Umwandeln einer analogen Anschlußleitung in einen digitalen
Anschluß im ISDN wird es sich in der Regel empfehlen, entsprechende
ISDN-Endgeräte, z. B. digitale Fernsprechapparate, einzusetzen.

Verfügt jedoch der Kunde über verhältnismäßig teure herkömmliche
Endgeräte, z. B. Fernkopierer, kann ihr Ersetzen durch geeignete
neue ISDN-Endgeräte zur Investition führen, die mancher Teilnehmer
scheuen wird. Deshalb ist es naheliegend, Möglichkeiten für den
Einsatz herkömmlicher Endgeräte im ISDN zu untersuchen.

Herkömmliche Endgeräte können mit Hilfe einer Endgeräteanpassung
TA an die Teilnehmerschnittstelle S_0 des ISDN angeschlossen werden.
Die Deutsche Bundespost hat 2 verschiedene Endgeräteanpassungen
vorgesehen: die TA (a/b) und TA X.21.

o Die __Endgerätenapassung TA (a/b)__ ermöglicht das Anpassen herkömmlicher Endeinrichtungen, die für den Anschluß an das analoge
  Fernsprechnetz vorgesehen sind, an die S_0-Schnittstelle des

ISDN-Basisanschlusses. Über die TA (a/b) können folgende herkömmliche Endgeräte eingesetzt werden:

- Faxgeräte der Gruppen 2 und 3
- Serien-/Parallel-Modems zum Anschalten von Datenendeinrichtungen
- Anschlußboxen für Bildschirmtext-Endgeräte.

o Die <u>Endgeräteanpassung TA X.21</u> ermöglicht das Anpassen herkömmlicher Endeinrichtungen, die mit einer Schnittstelle gemäß der CCITT-Empfehlung X.21 ersehen sind, an die S_o-Schnittstelle des ISDN-Basisanschlusses. Bei den Endgeräten handelt es sich um Datenendeinrichtungen mit einer Datenübertragungsgeschwindigkeit von 2,4 und 64 kbit/s (Benutzerklassen 4 und 12 gemäß CCITT-Empfehlung X.1).

4.1.3 Bisherige Dienste

Grundsätzlich kommen alle bestehenden Fernmeldedienste auch für das ISDN in Betracht, solange sie mit einer Bitrate von 64 kbit/s übertragbar sind. Es bietet sich an, daß insbesondere jene Dienste, die bisher schon im analogen Fernsprechnetz geführt werden, auch auf das ISDN übernommen werden. Hierzu gehören: Fernsprechen, Datenübermittlung, Fernkopieren, sowie der Zugang zu Bildschirmtext.

Z. Z. wird untersucht, inwieweit Nichtfernsprechdienste, z. B. Datendienste gemäß Benutzerklasse X.21, Telex, Teletex, DATEX-P, aus dem bestehenden integrierten Daten- und Fernschreibnetz der Deutschen Bundespost wirtschaftlich in das ISDN überführt werden können. Hierbei zeichnet sich ab, daß der Telexdienst nicht in das ISDN übernommen werden wird. Paketvermittlung ist noch nicht im ISDN vorgesehen.

Es ist jedoch aus Gründen der Planungssicherheit für den Anwender davon auszugehen, daß alle bestehenden Dienste - unabhängig davon, in welchem Netz sie heute realisiert sind - auch noch während einer angemessenen Übergangszeit in unveränderter Weise dem Kunden angeboten werden. Die Übergangszeit wird dabei etwa dieselbe Größenordnung haben wie die Abschreibungszeiten für die Endgeräte.

4.2 Mögliche Dienstmerkmale im ISDN

Jeder Dienst kann durch eine Reihe von Dienstmerkmalen gekennzeichnet
und ergänzt werden. Dienstmerkmale sind als Unter- oder Teilmengen
eines Dienstes zu verstehen. Dienstmerkmale können in folgende
Gruppen aufgeteilt werden:

o Anschlußdienstmerkmale
o Verbindungsdienstmerkmale und
o Informationsdienstmerkmale.

Im folgenden sollen sie etwas veranschaulicht werden. Hierbei
ist anzumerken, daß eine Menge der vorgestellten Dienstmerkmale
bereits mit herkömmlicher Technik realisiert werden kann.

4.2.1 Mögliche Anschlußdienstmerkmale

Mit den Anschlußdienstmerkmalen wird der Anschluß für einen bestimmten
Dienst näher gekennzeichnet. Neben den dienstspezifischen Endgeräten
werden z. B. Aussagen über die Anschlußart gemacht:

o Verbindungsart: Wähl- oder Festverbindung
o Vermittlungsart: leitungs- oder paketvermittelt.

4.2.2 Mögliche Verbindungsdienstmerkmale

Hierunter wird eine Reihe im ISDN möglicher Dienstmerkmalegruppen
verstanden, welche die Verbindung betreffen:

o schneller und bequemer Verbindungsaufbau, z. B.
  - Kurzwahl
  - Direktruf
  - Wahlwiederholung
  - Durchwahl

o Möglichkeiten der besonderen Verbindungsvollendung, z. B.
  - Warten auf Freiwerden des B-Teilnehmers mit Hinweis (Anklopfen)
  - Umlenken im Besetztfall
  - Umleiten von Anrufen zu einem anderen Anschluß
  - Anrufweiterschaltung

o Möglichkeiten der Nachrichtenübermittlung mit Zwischenspeicherung,
  z. B.
  - Voice-mail
  - Telebox
  - Registrieren ankommender Verbindungswünsche

o Möglichkeiten zum Einschränken bestimmter Verbindungen, z. B.
  - Ruhe vor dem Telefon

o Mögliche Sonderverbindungen, z. B.
  - individuelle Gebührenübernahme durch den B-Teilnehmer
  - Erinnerungen oder automatischer Terminkalender
  - automatisches Wecken
  - Konferenzverbindungen
  - geschlossene Benutzergruppen oder Teilnehmerbetriebsklassen

o Dienst- und Netzübergänge
  Damit der Teilnehmer im ISDN möglichst viele Kommunikationspartner
  erreichen kann, bietet es sich im wirtschaftlich vertretbaren
  Rahmen an, Netzübergänge vorzusehen, z. B. zu und vom:

  - analogen Fernsprechnetz
  - DATEX-L-Netz (leitungsvermittelt).

4.2.3 Mögliche Informationsdienstmerkmale

Auch hier wird eine Reihe möglicher Dienstmerkmalegruppen unterschieden, welche Informationen übermitteln und zum Teil bereits
im herkömmlichen Netz realisiert sind.

o mögliche Gebühreninformationen, z. B.
  - automatisches Unterrichten über die aufgelaufenen Gebührenein-
    heiten oder DM-Beträge für den A-Teilnehmer während oder nach
    der Verbindung

o Netzinformation, z. B.
  - Ansage geänderter Rufnummer

o Auskünfte, z. B.
  - Fernsprechansagen

o Identifizieren, z. B.
  - Identifizieren der Rufnummer des A-Teilnehmers zur Anzeige
    beim B-Teilnehmer vor, während oder nach der Verbindung
  - Dienstekennung.

5 Schlußfolgerungen und Ausblick

Mit den bisherigen Ausführungen soll verdeutlicht werden,
o daß eine umfangreiche Palette neuer, z. T. erheblich verbesserter
  Dienste und Dienstmerkmale dem Kunden und Anwender mit dem ISDN
  angeboten werden kann,
o daß die bestehenden Fernmeldenetze mittel- bis langfristig im
  ISDN aufgehen werden und
o daß außerdem im ISDN die Techniken, insbesondere die Übertragungs-
  und die Vermittlungstechnik immer mehr zusammenwachsen werden.
  Man kann deshalb auch von Übermittlungstechnik sprechen.

Die Deutsche Bundespost beabsichtigt, ab 1988 mit dem Übergang
vom digitalisierten Fernsprechnetz zum ISDN zu beginnen. Aus wirtschaftlichen und betrieblichen Gründen wird es sich hierbei um
einen kontinuierlichen Übergang handeln; d. h. es kann nicht erwartet werden, daß der Gesamtumfang des im ISDN möglichen Dienstleistungsangebots sofort bundesweit flächendeckend bereitgestellt
wird. Wegen der benötigten Planungssicherheit für den Anwender
wird darauf hingewiesen, daß jene Dienste, die durch die neuen
und verbesserten Dienste im ISDN mittelfristig substituiert werden,
auch noch für eine angemessene Übergangszeit angeboten werden.
Damit kann die Abschreibungszeit vorhandener Endeinrichtungen
voll ausgeschöpft werden.

Um das einwandfreie Funktionieren aller neuen ISDN-Netzbestandteile
überprüfen zu können, plant die Deutsche Bundespost ein ISDN-Pilot-
projekt, das in den Ortsnetzen Mannheim und Stuttgart mit jeweils

etwa 400 Teilnehmern durchgeführt werden wird. Der Aufbau der
technischen Einrichtungen ist für die 2. Hälfte des Jahres 1986
vorgesehen. Der Probebetrieb wird ab 1987 aufgenommen werden.

Zu Beginn des nächsten Jahrzehnts, wenn Glasfaserkabel und optische
Nachrichtenübermittlungssysteme wirtschaftlich konkurrenzfähig
zur Verfügung stehen, kann das ISDN derart erweitert werden, daß
eine Integration aller 64-kbit/s- und breitbandigen Nutzungsformen
(Fernsprechen, Daten-, Text- und Festbildkommunikation, Bildfern-
sprechen und Videokonferenz) im Breitband-ISDN möglich ist.

Das ISDN stellt eine wesentliche Voraussetzung für das Einführen
vermittelter Breitbanddienste dar, denn wichtige Bestandteile
des ISDN sind auch für künftige Breitbandkommunikationen geeignet.
Außerdem benötigen Breitbanddienste im Grundsatz die gleiche Struktur
wie das ISDN. Somit empfiehlt sich das folgerichtige Weiterentwickeln
des ISDN zu einem diensteintegrierenden breitbandigen Fernmeldenetz,
bei dem im wesentlichen die heutigen Kupferkabel von Glasfaser-
kabeln abgelöst und die Vermittlungsstellen um entsprechende breitbandige Koppelfelder ergänzt werden.

Als letzter Schritt können schließlich in einem integrierten Breitbandfernmeldenetz ab 1992 auch Fernseh- und Hörfunkprogramme verteilt
werden, d. h. die bis dahin aus wirtschaftlichen Gründen getrennt
von der Individualkommunikation verlaufende Weiterentwicklung
der Breitbandverteilnetze könnte dann in ein gemeinsames universelles Fernmeldenetz, das integrierte Breitbandfernmeldenetz,
einmünden.

Literaturhinweise

Der Bundesminister für das Post- und Fernmeldewesen, Referat 247:
"ISDN - Die Antwort der Deutschen Bundespost auf die Anforderungen
der Telekommunikation von morgen"

H. Schön, Zeitschrift für das Post- und Fernmeldewesen, Heft 6/84:
"Die Deutsche Bundespost auf ihrem Weg zum ISDN"

K. H. Rosenbrock, Jahrbuch der Deutschen Bundespost 1984: "ISDN
- eine folgerichtige Weiterentwicklung des digitalen Fernsprechnetzes"

Sammlung der Vortragsmanuskripte der gemeinsamen Tagung des Münchner
Kreises und der Nachrichtentechnischen Gesellschaft (NTG) 11/84
in München, Springer-Verlag, K. H. Rosenbrock: "Die Entwicklung
der Dienste im ISDN"

TELEBOX als Beispiel eines öffentlichen Mitteilungs-Übermittlungs-Systems

Walter Tietz, Fernmeldetechnisches Zentralamt Darmstadt

1 Gesamtzusammenhang

Dienste im Bereich der Telekommunikation waren seit jeher sehr
vielfältig. Im Laufe der letzten Jahre haben sie besonders viele
Anreicherungen erhalten. Neue Dienste sind hinzugekommen. Die
Idee der Telematik-Dienste hat Raum gewonnen. Teletex ist der
anschaulichste Vertreter dafür. Datenkommunikationsdienste, die
seit 20 Jahren Dateldienste heißen, tragen zur Infrastruktur der
Text- und Datenkommunikation bei.

Systeme, Datenverarbeitungsanlagen haben Aufgaben von Menschen
übernommen, haben in größerem Maßstab Vorgänge zu automatisieren
erlaubt, haben auch zur besseren Kommunikation der Menschen beigetragen. Für die wesentlich verbesserte Fern-Massen-Kommuni-
kation zwischen Menschen über Fernmeldewege ist im wesentlichen
seit der Erfindung des Telefons im großen Maßstab nichts mehr
beigetragen worden. In diesem Feld der Mitteilungen zwischen
Person und Person spricht man seit einigen Jahren von Messaging,
Message Handling, Mitteilungs-Übermittlung und meint damit vorwiegend den Austausch von textlichen (aber auch sprachlichen)
Information über zwischenspeichernde Systeme, wobei die Zuführung und Entnahme der Informationen ganz dicht an den Menschen
herangeführt wird. Im Idealfall begleitet das Ein-/Ausgabegerät
den Menschen als ein Gebrauchsgegenstand, so wie es heute beispielsweise eine Uhr tut.

Die Normung hat sich der Sache angenommen, nachdem schon eine
ganze Reihe von Entwicklungen geschehen sind /1,2/. Dienstleistungen mit Hilfe von Message Handling entstehen.

Die Deutsche Bundespost bietet Mitteilungsübermittlung im Rahmen
der öffentlichen Telekommunikations-Dienstleistungen an. Beispiele sind der Mitteilungsdienst im Bildschirmtext und TELEBOX.

2 Weltweite Standardisierung

2.1 CCITT-Empfehlungen

Im Internationalen Fernmeldevertrag, Nairobi 1982 ist im Artikel
11 (Nr. 84) festgelegt:

"Der Internationale Beratende Ausschuß für den Telegrafen- und
Fernsprechdienst (Comité Consultatif International Téléphonique
et Télégraphique - CCITT) ist beauftragt, über technische, betriebliche und tarifliche Fragen der Fernmeldedienste Studien
durchzuführen und Empfehlungen herauszugeben".

Die Empfehlungen des CCITT werden in einer Vollversammlung der
Studienkommissionen beschlossen, die in der Regel alle drei (in
der Praxis vier) Jahre stattfinden. Die III. Vollversammlung des
CCITT ist vom 25.5. bis 26.6.1964 in Genf abgehalten worden. Die
IV. Vollversammlung fand vom 14.10. bis 25.10.1968 in Mar del
Plata statt. In Genf trat die V. Vollversammlung zusammen und
tagte vom 04.12. bis 15.12.1972. Auch die VI. Vollversammlung
trat in Genf zusammen und tagte vom 26.9. bis zum 08.10.1976.
Die VII. Vollversammlung fand wiederum in Genf vom 10.11. bis
21.11.1980 statt. Die VIII. und bisher letzte Vollversammlung
wurde in Malaga-Torremolinos vom 08.10. bis 19.Oktober 1984 abgehalten.

Vom CCITT werden Bücher herausgegeben, welche die gültigen Empfehlungen, die Studienfragen, Beiträge aus der Arbeit der Studienkommissionen, denen mit Zustimmung des Direktors des CCITT
ein allgemeines Interesse zukommt, enthalten. Die Bücher werden
in einer besonderen Farbe aufgelegt. Die Ausgabe von New Delhi

1960 bestand aus einer Reihe von "Rotbüchern", die Ausgabe Genf
1974 aus einer Reihe von "Blaubüchern". Die Vollversammlung in
Mar del Plata wurde in "Weißbüchern" dokumentiert. Im Jahre 1972
wurden "Grünbücher" aufgelegt, im Jahre 1976 "Orangebücher". Die
im Jahre 1980 beschlossenen Empfehlungen waren in "Gelbbüchern"
enthalten. Die Empfehlungen von 1984 sind wieder in "Rotbüchern"
zusammengefaßt:

Band	I	Protokoll und Berichte der Vollversammlung, Liste der Studienkommissionen, Liste der Studienfragen, Allgemeine Empfehlungen
Band	II.1	Allgemeine Tarifgrundsätze, Tarife und Abrechnung, Vermietung von Fernmeldeleitungen
Band	II.2	Internationaler Fernsprechdienst, Betrieb
Band	II.3	Internationaler Fernsprechdienst, Netzmanagement, Verkehr
Band	II.4	Telegrafendienste, Betrieb und Dienstgüte
Band	II.5	Telematikdienste, Betrieb und Dienstgüte
Band	III.1	Allgemeine Eigenschaften von internationalen Fernsprechverbindungen und -leitungen
Band	III.2	Internationale analoge Trägerfrequenzsysteme, Eigenschaften der Leitungen
Band	III.3	Digitale Netze, Übertragungssysteme, Multiplexeinrichtungen
Band	III.4	Übertragung von Signalen zu anderen Zwecken als zum Fernsprechen, Übertragung von Rundfunkprogrammen und Fernsehsignalen
Band	III.5	Diensteintegriertes digitales Fernmeldenetz (ISDN)
Band	IV.1	Unterhaltung, Allgemeine Grundsätze, internationale Übertragungssysteme, internationale Telefonleitungen
Band	IV.2	Unterhaltung, Internationale Telegrafie- und Faksimilesysteme, internationale Mietleitungen
Band	IV.3	Unterhaltung, Internationale Ton- und Fernsehleitungen
Band	IV.4	Spezifikationen von Meßeinrichtungen
Band	V	Fernsprechübertragungsgüte
Band	VI.1	Allgemeine Empfehlungen zu Fernsprechvermittlung und -zeichengabe
Band	VI.2	Kennzeichensysteme Nr. 4 und Nr. 5
Band	VI.3	Kennzeichensystem Nr. 6
Band	VI.4	Kennzeichensysteme R 1 und R 2
Band	VI.5	Digitale Transitvermittlungen für nationale und internationale Nutzungen
Band	VI.6	Zusammenarbeit zwischen Kennzeichensystemen
Band	VI.7	Schaltkennzeichensystem Nr. 7
Band	VI.8	Spezifikation des Schaltkennzeichensystems Nr. 7
Band	VI.9	Digitales Kennzeichensystem für den Zugang
Band	VI.10	Programmsprache SDL
Band	VI.11	Funktionelle Spezifikation der Programmsprache SDL

Band VI.12 Höhere Programmsprache des CCITT (CHILL)
Band VI.13 Mensch-Maschine-Sprache (MML)
Band VII.1 Telegrafieübertragung
Band VII.2 Telegrafievermittlung
Band VII.3 Endgeräte und Protokolle für Telematikdienste
Band VIII.1 Datenübertragung über das Fernsprechnetz
Band VIII.2 Datenübermittlungsnetze, Dienste und Leistungsmerk-
            male
Band VIII.3 Datenübermittlungsnetze, Schnittstellen
Band VIII.4 Datenübermittlungsnetze, Übertragung, Zeichengabe,
            Vermittlung, Unterhaltung, Verwaltung
Band VIII.5 Datenübermittlur.gsnetze, offene Kommunikationssy-
            steme
Band VIII.6 Datenübermittlungsnetze, Zusammenarbeit zwischen
            Netzen, mobile Übertragungs-Systeme
Band VIII.7 Datenübermittlungsnetze, Mitteilungs-Übermittlungs-
            Systeme
Band IX   Schutz gegen Beeinflussungen
Band X.1   Begriffe und Definitionen
Band X.2   Index der Rotbücher

Jeder Band enthält
die Empfehlungen und weitere Ergänzungen.

Die Empfehlungen des CCITT sind in Serien eingeteilt, die mit
einem Großbuchstaben gekennzeichnet sind. Dem Buchstaben folgt
zur eindeutigen Unterscheidung eine Zahl.

Innerhalb der Serien können Abschnitte gebildet sein. Die Serien
behandeln die folgenden Themen:

Serie		CCITT-Band
A	Organisation der Arbeit des CCITT	I
B	Ausdrucksmittel (Definitionen, Voka- bular, Symbole, Klassifizierung)	I
C	Statistiken	I
D	Vermietung (Überlassung) internatio- naler Fernmeldewege	II.1
E	Fernsprechbetrieb, Tarife	II.2 - II.3

F	Telegrafenbetrieb, Tarife	II.4
G	Fernsprechübertragung über drahtge- bundene Verbindungen, Satelliten- und Funkverbindungen	III.1 - III.3
H	Einsatz von Leitungen für Telegrafie (einschl. Bildtelegrafie)	III.4
I	Diensteintegrierende digitale Netze (ISDN)	III.5
J	Ton- und Fernsehübertragung	III.4
K	Schutz gegen Störungen	IX
L	Schutz gegen Korrosion	IX
M	Unterhaltung von Fernsprechleitungen und Trägerfrequenzsystemen	IV.1 - IV.2
N	Unterhaltung von Ton- und Fernsehüber- tragungswegen	IV.3
O	Eigenschaften von Meßgeräten	IV.4
P	Fernsprechübertragungsgüte, Teil- nehmereinrichtungen und Fernsprech- ortsnetze	V
Q	Fernsprech-Zeichengabe, Fernsprech- vermittlung	VI.1 - VI.9
R	Telegrafenkanäle	VII.1
S	Apparate der alphabetischen Telegrafie	VII.1
T	Faksimileapparate	VII.2
U	Telegrafievermittlung	VII.2
V	Datenübertragung über das Fernsprech- netz	VIII.1
X	Datenübertragung über öffentliche Datenübermittlungsnetze	VIII.2 - VIII.7
Z	Programmsprachen für rechnergesteuerte Vermittlungen	VI.10 - VI.13

Die Empfehlungen gelten zunächst nur für den internationalen Bereich. Sie haben nicht den Rang von Dienstvorschriften. Es können im Einzelfall Gründe vorliegen, die eine Fernmeldeverwaltung
zu Abweichungen von diesen Empfehlungen veranlassen. Inwieweit
CCITT-Empfehlungen im nationalen Bereich angewendet werden,
bleibt jeder Fernmeldeverwaltung überlassen. Vor allem bei der
Projektion dieser Regelungen auf nationale Probleme ist alle

Vorsicht anzuwenden, um Mißdeutungen zu vermeiden. Keinesfalls
dürfen sie dort als Arbeitsunterlage betrachtet werden, wo der
Umfang ihrer Gültigkeit im Bereich der DBP nicht bekannt ist. In
der Bundesrepublik sind letztlich verbindlich immer nur die jeweils gültigen Bestimmungen der DBP.

An der Arbeit des CCITT nehmen die Verwaltungen von 159 Ländern,
56 anerkannte Betriebsgesellschaften, 147 wissenschaftliche oder
industrielle Gesellschaften und 20 internationale Organisationen
teil.

Die Aufgabengebiete werden von eine großen Anzahl von Studienkommissionen wahrgenommen. Die folgende Aufzählung wird verbunden mit einer stichwortartigen Aufzählung einzelner (nicht aller)
Aufgaben.

Studienkommission I
   Definition und betriebliche Aspekte von Telegrafendiensten,
   Datenübertragungsdiensten und Diensten der Telematik
Studienkommission II
   Fernsprechbetrieb und Dienstgüte
Studienkommission III
   Grundsätzliche Fragen der Tarife, einschließlich der Abrechnung
Studienkommission IV
   Unterhaltung von internationalen Fernmeldelinien; automatische
   und halbautomatische Netze
Studienkommission V
   Schutzmaßnahmen gegen Gefährdung und elektromagnetische Stö-
   rungen
Studienkommission VI
   Außenanlagen
Studienkommission VII
   Datenübermittlungsnetze
Studienkommission VIII
   Endeinrichtungen für Dienste der Telematik (Faksimile, Tele-
   tex, Videotext usw.)
Studienkommission IX
   Telegrafennetze und Endeinrichtungen
Studienkommission X
   Sprachen und Methoden für Anwendungen der Telekommunikation

Studienkommission XI
  ISDN sowie Fernsprechvermittlung und Zeichengabe
 Studienkommission XII
   Fernsprech-Übertragungsgüte und Endgeräte
 Studienkommission XVI
   Fernsprechleitungen
 Studienkommission XVII
   Datenübertragung über das Fernsprechnetz
 Studienkommission XVIII
   Digitale Netze, einschließlich ISDN
 Studienkommission CMTT (gemeinsam mit CCIR)
   Fernsehübertragung und Tonübertragung
 Studienkommission CMV (gemeinsam mit CCIR)
   Definitionen und Symbole
 Plankommissionen
   für verschiedene Bereiche der Welt; in Verbindung mit CCIR;
   vom CCITT verwaltet.
 Spezialkomitee S
   Struktur der Studienkommissionen
 Spezialkomitee PC/WATTC-88
   Vorbereitung der Konferenz WATTC-88

Die folgenden gemischten Arbeitsgruppen bestehen:

GR TAF   Tariffragen (Afrika)
GR TAL   Tariffragen (Lateinamerika)
GR TAS   Tariffragen (Asien und Ozeanien)
GR TEUREM Tariffragen (Europa und mediterane Gebiete)
GAS (3, 7, 9, 10, 11) Autonomome spezialisierte Arbeitsgruppen
   für das Studium vorwiegend wirtschaftlicher Zusammenhänge.

Die einzelnen Studienkommissionen arbeiten nach einer Liste der
Studienfragen, die jeweils von der vorangegangenen Vollversammlung verabschiedet worden sind. Bei genügender Unterstützung
können auch andere Probleme in das Studienprogramm eingefügt
werden.

Für die Studienperiode 1984 - 1988 wird für die Studienkommission VII (Datenübermittlungsnetze), für die Studienkommission
XVII (Datenübertragung über das Fernsprechnetz), für die Studienkommission VIII (Endeinrichtungen für Dienste der Telematik)

und für die Studienkommission I (Definition und betriebliche Aspekte von Telegrafendiensten, Datenübertragungsdiensten und Diensten der Telematik) insgesamt 120 Studienfragen formuliert worden, von denen erwartet werden kann, daß die Arbeit an ihnen zur Weiterentwicklung des weltweiten Fernmeldewesens beitragen werden.

2.2 Standards für Mitteilungs-Übermittlungs-Systeme

In der vergangenen Studienperiode des CCITT sind große Anstrengungen unternommen worden, neue Standards für Message Handling zu entwickeln. Die Bemühungen wurden durch umfangreiche Zuarbeit von Verwaltungen, anerkannten privaten Betriebsgesellschaften, Herstellerfirmen und anderen internationalen Organisationen unterstützt. Als Resultat liegen jetzt 8 Empfehlungen vor, die von der Vollversammlung des CCITT im Oktober 1984 angenommen und bestätigt worden sind:

X.400   Mitteilungs-Übermittlungs-Systeme:
Systemmodell - Dienstelemente der Schichten

X.401   Mitteilungs-Übermittlungs-Systeme:
Basis-Dienstelemente und wahlfreie Leistungsmerkmale

X.408   Mitteilungs-Übermittlungs-Systeme:
Codierte Informationen - Regeln der Umsetzung

X.409   Mitteilungs-Übermittlungs-Systeme:
Syntax der Darstellung und der Notation

X.410   Mitteilungs-Übermittlungs-Systeme:
Zusammenwirken zwischen den Schichten

X.411   Mitteilungs-Übermittlungs-Systeme:
Schicht für den Transfer von Mitteilungen

X.420   Mitteilungs-Übermittlungs-Systeme:
Schicht für die Interpersonelle Mitteilungs-Übermittlung

X.430   Mitteilungs-Übermittelungs-Systeme:
Zugangsprotokoll für Teletex-Terminals

Eine Reihe neuer Studienfragen behandeln auch Sachverhalte der
Mitteilungs-Übermittlung, sodaß mit einer zielstrebigen Fortentwicklung der bisher vorliegenden Standards gerechnet werden kann.
Vor allem müssen ergänzende Standards zu "Verzeichnissen" erarbeitet werden, ohne die eine weltweite Verbreitung von Mittei-
lungs-Übermittlungs-Systemen nicht vorstellbar ist /1, 2/.

3   TELEBOX

3.1 Konzeption

Die Deutsche Bundespost hält ein großes Spektrum an Dienstleistungen für die Text- und Datenkommunikation bereit.

Marktbeobachtungen im Geschäftsbereich lassen in letzter Zeit
ein immer deutlicheres Interesse an Möglichkeiten eines persönlichen Mitteilungsaustausches erkennen. Es wird die ständige
Erreichbarkeit von Kommunikationspartnern im In- und Ausland gefordert, ohne deren Mobilität einzuschränken.

Die Deutsche Bundespost hat sich daher entschlossen, über ihre
öffentlichen Wählnetze unter dem Namen TELEBOX Leistungsmerkmale
eines personenbezogenen "elektronischen Briefkastens"
anzubieten.

In verschiedenen Ländern werden derartige Dienstleistungen, die
dort überwiegend dem Sammelbegriff "Electronic Mailbox Systems"
zugerechnet werden, bereits seit einiger Zeit von den Fernmeldeverwaltungen oder den öffentlichen anerkannten Betriebsgesellschaften angeboten.

Die Deutsche Bundespost hat schon begonnen, möglichst frühzeitig
das nationale TELEBOX-System mit anderen im Ausland betriebenen
gleichartigen Systemen zu verbinden.

Gute Voraussetzungen für eine weltweite Kommunikation unter
Kopplung der nationalen Systeme bieten die in den internationalen Standardisierungsgremien des CCITT erarbeiteten und verab-
schiedeten Empfehlungen über "Message Handling Systems".

Die Deutsche Bundespost hat sich zur sofortigen Übernahme dieser
Standards für das nationale System entschieden.

Die erste Ausbaustufe eines TELEBOX-Systems stellt ein öffentliches End-Systemteil (UA) bereit, mit dem eine Reihe von Dienstelementen verfügbar werden. Da für Funktionen, die dem Benutzer
örtlich zu Verfügung stehen und die für einen Transfer in die
Ferne den Ausgangspunkt darstellen, bisher keine internationalen
Normen vorliegen, ist der Vorrat an Merkmalen zunächst von dem
am Markt durch Ausschreibung erlangten System abhängig. Von Anfang an wird aber die Erweiterung des Systems in die internationale MHS-Umgebung angestrebt.

Weitere Ausbaustufen (etwa ab 1985 verfügbar) sehen den Übergang
in gleichartige Systeme anderer Länder vor.

Die spätere Heranführung privater Systemteile auf der Basis der
internationalen Standards ist Bestandteil des Konzeptes. Die angegebenen Protokolle P1 und P2 sind in CCITT-Empfehlungen X.400
ff festgelegt.

Ein Mitteilungs-Übermittlungs-System auf der Basis der internationalen Empfehlungen kann die Drehscheibe zwischen verschiedenen Netzen und Diensten bilden.

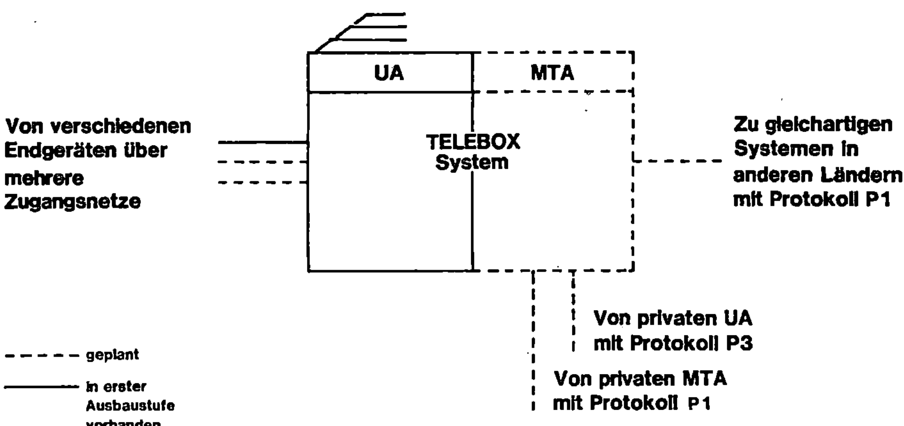

Bild 1   TELEBOX-Konzept

3.2 Zugangsformen für TELEBOX

In der ersten Ausbauphase kann über die drei vorhandenen Fernmeldenetze mit Hilfe von asynchronen Datenendgeräten (300 und
1200 bit/s) Zugang zum TELEBOX-System erreicht werden.

Von DATEX-P-Hauptanschlüssen aus kann das TELEBOX-System außerdem erreicht werden, wenn die Anschaltung mit Hilfe der Kommunikationsprotokolle P10/P20B geschieht und sich die Datenendein-
richtung konform mit dem TELEBOX-System verhält.

Alle verwendeten Einrichtungen müssen für die jeweiligen Netze
zugelassen sein.

Für jede der Zugangsgruppen gibt es eigene Rufnummern. In der
ersten Phase sind die TELEBOX-Dienstleistungen zusätzliche Dienste zu den betreffenden Dateldiensten.

Andere Geschwindigkeitsstufen sollen später einbezogen werden.

Ein "Teilnehmer" in einem der öffentlichen Zugangsnetze kann
"BOX-Inhaber" werden. Die Person, die mit Zustimmung des Teil-
nehmers/BOX-Inhabers die BOX benutzt und das Paßwort verwaltet,
wird Benutzer genannt.

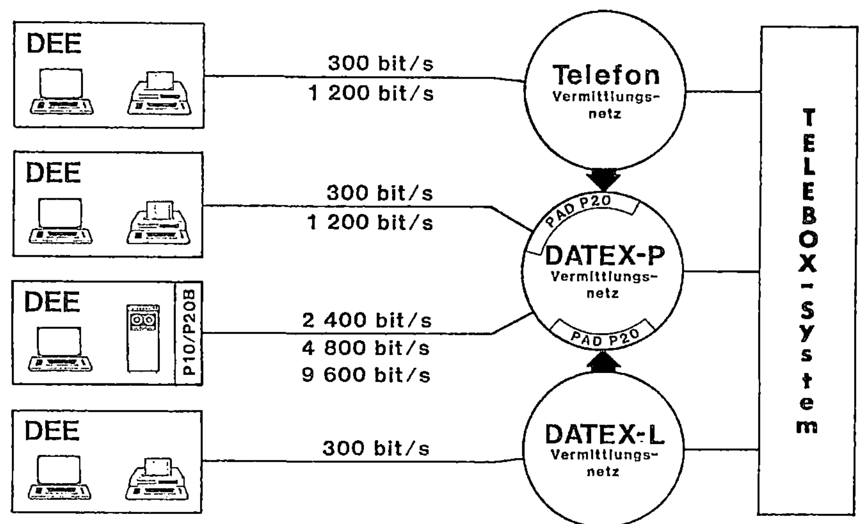

Bild 2   Zugangsformen für TELEBOX

Wenn das TELEBOX-System die Aufgabe einer Drehscheibe zwischen
verschiedenen Diensten wirklich erfüllen soll, müßten noch weitere Zugangsformen hinzukommen.

Anwärter dafür sind:
- Bildschirmtext
- Teletex
- Telex

3.3 Teilleistungen von TELEBOX

Die Fähigkeiten des TELEBOX-Systems, die einem Benutzer über
seine BOX zugänglich sind, stellen sich als Teilleistungen
dar. Eine kurze (nicht erschöpfende) Aufzählung mach deutlich,
was TELEBOX dem Benutzer im wesentlichen ermöglicht.

- **Abfrage von Kopfzeilen vorliegender Mitteilungen**
- **Lesen der Mitteilungen**
- **Beantworten der Mitteilungen**
- **Weiterleiten der Mitteilungen unter Beifügen von Zusätzen**
- **Eingeben von abzusendenden Mitteilungen**
- **Versenden von Mitteilungen**
  an eine oder mehrere TELEBOX-Adressen
- **Editieren und Formatieren von Texten**
- **Speichern von Mitteilungen und Texten**
- **Schwarzes Brett**
- **Unterstützungsfunktionen**
- **Verzeichnisse**

Bild 3   Teilleistungen von TELEBOX

Weitere Teilleistungen lassen sich dieser Liste anfügen. Einige
werden erst bei späterem Ausbau des Systems zur Verfügung stehen.

Wenn einzelne Funktionen im Detail angesprochen werden sollen,
spricht man z.B. von Dienstelementen, Merkmalen, zusätzlichen
Diensten, Funktionszweigen.

Durch zunächst ein, an zentraler Stelle in Mannheim installier-
tes System werden die TELEBOX-Leistungsmerkmale realisiert.

Jeder Nutzer des TELEBOX-Systems erhält eine eigene Adresse, die
ihn zusammen mit dem persönlichen Paßwort als berechtigten Nutzer des Systems ausweist und ihm dadurch das Eingeben und Aus-
lesen von Mitteilungen im System ermöglicht.

Eine Mitteilung erreicht ihren Empfänger, indem sie in das System eingegeben, mit einer oder mehreren Ziel-Adressen versehen
und "abgesendet" wird. Sie kann vom Empfänger gelesen oder ausgedruckt werden, wenn dieser sich unter eigener Adresse und seinem persönlichen Paßwort mit dem System in Verbindung setzt.

Bei diesem Kommunikationsvorgang sind unmittelbar nur Absender
und Empfänger beteiligt, also nur die Person, welche die Mitteilung erzeugt und die Person(en), für welche die Mitteilung bestimmt ist. Deshalb spricht man auch von einem personenbezogenen
Mitteilungsdienst.

Die Übersicht über die wesentlichen Merkmale lassen sich verein-
facht aus dem folgenden Bild erkennen.

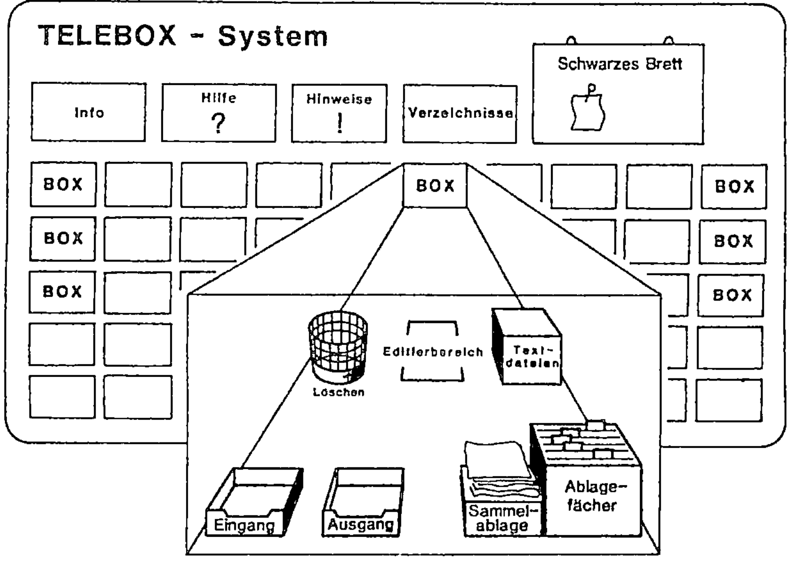

Bild 4   Übersicht

3.4 Gebühren für die öffentliche Dienstleistung

Die Leistungen des TELEBOX-Systems müssen dem BOX-Inhaber in Rechnung gestellt werden.

Ein Teilnehmer eines bestehenden Fernmeldedienstes kann in Ablehnung an sein Teilnehmerverhältnis Inhaber einer oder mehrerer BOXEN werden.

Besitzt ein Teilnehmer oder eine Teilnehmergruppe mehr als eine BOX, so wird nur eine einzige Fernmelderechnung ausgefertigt.

Die verbindlichen Gebührenangaben sind in der Fernmeldeordnung und der Verordnung für den Fernschreib- und Datexdienst enthalten. Man spricht dort von "Einrichtungen zur Zwischenspeicherung".

Einmalige Gebühren
- Bereitstellungs- und Änderungsgebühr

Monatliche Gebühren
- Grundgebühr je Box
- Mindest-Benutzungsgebühr
- Anschaltegebühr (je Minute)
- Adressiergebühr (je Adresse)
- Speichergebühr (je Tag)
- (Zuschläge)

Zugangsgebühren
- je Verbindung (abhängig vom Zugangsnetz)

Bild 5 Gebührenelemente

Für das Bereitstellen oder Ändern einer (oder gleichzeitig mehrerer) Adresse(n) berechnet die Post einmalig 65 DM. In der Testphase kamen keine weiteren TELEBOX-Gebühren auf, es waren nur die Verbindungsgebühren in den Zugangsnetzen zu entrichten.

Ab 01.10.84 kostet jede Adresse monatlich 40 DM Grundgebühr, zuzüglich 40 DM pauschale Nutzungsgebühr, insgesamt also 80 DM.

Erst mit Beginn des Wirkbetriebes (01.10.1986) werden die Nut-
zungsgebühren voll in Rechnung gestellt. Folgende Nutzungsge-
bühren sind vorgesehen:

- Belegungsgebühr (Anschaltgebühr)
  je Minute   0,30 DM
- Speichergebühr je Einheit und Tag   0,03 DM
  Eine Einheit umfaßt 2 K (= 2048 Zeichen)
- Adressiergebühr je Adresse   0,10 DM
Mindesnutzungsgebühr je Abrechnungszeitraum
(ca. 30 Tage) und Adresse   40,00 DM

Die reguläre Aufnahme des Übermittlungsdienstes in öffentliche
Systeme anderer Länder ist vorgesehen. Die Übermittlungsgebühr
je Mitteilung wird dann betragen:

- in Europa   0,70/0,10 DM
- in den USA   1,20/0,25 DM
- in Kanada   1,25/0,25 DM .
- in der übrigen Welt   1,45/0,35 DM

Die erste Zahl gibt die Mindestgebühr für eine Mitteilung mit
max. 2048 Zeichen an, die zweite Zahl die Gebühr für jede weitere angebrochene oder volle Einheit von 1024 Zeichen.

4   Literaturverzeichnis

/1/ Empfehlungen des CCITT, Serie X.400 ff (demnächst in deut-
    scher Übersetzung im R.v. Decker's Verlag, G. Schenck, Hei-
    delberg verfügbar)

/2/ Tietz, W.: Stand der Normung im Bereich des "Message Hand-
    ling"; NTZ-Nachrichtentechnische Zeitschrift, Heft 1/1984

/3/ TELEBOX, Fernmeldeprospekte der Deutschen Bundespost, Nr.
    193 und 194

/4/ Einführung TELEBOX, Fernmeldetechnisches Zentralamt, T 21-2,
    Serie F1
Dezember 1984

DAS DEUTSCHE FORSCHUNGSNETZ (DFN)

E. Jessen

Institut für Informatik der
Technischen Universität

München

Kurzfassung:

Das Projekt DFN dient dem Aufbau eines heterogenen öffenen Rechnernetzes im Wissenschaftsbereich der Bundesrepublik nach ISO bzw. CCITT, sowie der Förderung neuartiger Nutzungsformen eines solchen Netzes. Die grundlegende Philosophie des Projektes
wird beschrieben. Strukturierung und zeitliche Staffelung der Grunddienste des DFN
werden vorgestellt. Das stufenweise Vorgehen zur Errichtung eines Message Handling
Systems nach X. 400 wird erklärt. Die Ansätze innovativer Nutzungsformen und die ihnen unterliegenden Arbeiten zur verteilten graphischen Datenverarbeitung werden beschrieben. Über den Betriebsanlauf der sogenannten Nullgeneration wird berichtet,,
Vorkehrungen für den späteren Betrieb werden genannt, und es werden erste Erfahrungen mit der DFN-Entwicklung dargestellt.

Summary:

The German Science Net (DFN)

The DFN project is to construct a heterogeneous, open computer network in the science
area of the FRG, in accordance with ISO and CCITT, and is to promote innovative usage
of such net. The philosophy underlying the project is described. The layering and the
temporal sequence of the basic network services of the DFN are presented. The stepwise procedure towards the construction of a X. 400-Message Handling System is explained. Examples of projected innovative usage are given with reference to the underlying work in distributed graphical data processing. Some zero generation operational
experience is reported, provisions for full operation are described, and some development experience is presented.

1. Deutsches Forschungsnetz: Ziele, Randbedingungen, Folgerungen.

Das Deutsche Forschungsnetz (DFN), das wesentlich auf die Initiative von Prof.Dr.Karl
Zander (Hahn-Meitner-Insitut Berlin) zurückgeht, soll eine rechnergestützte Kommunikationsinfrastruktur für die Wissenschaft in der Bundesrepublik bieten und die innovative Nutzung einer solchen Infrastruktur fördern.

Dabei soll vorzugsweise die überregionale Kommunikation unterstützt werden, wobei die
heutigen und künftigen Gegebenheiten lokaler Kommunikationsstrukturen berücksichtigt
werden müssen. Als Endsysteme des Deutschen Forschungsnetzes treten Rechner verschie-
denster Größenordnung vom Personal Computer bis zum Großrechner auf; funktionell können sie als Rechner oder nur als Terminal eingesetzt werden; lokale Rechnernetze können ähnlich als Endsysteme, mit möglicherweise verteilten Netzdiensten, angeschlossen
sein oder auch als Subnetze, die den Weg zu Endsystemen eröffnen. Das DFN muß also
die Kommunikation sehr heterogener Endsysteme ermöglichen. Im Grundsatz kann diese
Aufgabe durch Abbildung der lokalen Kommunikationsformen auf beliebige, auch herstel-
lerbegrenzte Kommunikationsprotokolle gelöst werden, insbesondere wenn man Gateway-
rechner heranzieht. Andere Projekte haben gezeigt, daß solche Lösungen - insbesondere
bei eingeschränkter Funktionalität und Teilnehmerart - schnell und kostengünstig eingerichtet werden können, vgl. etwa EARN [Hultzsch H 85]. Dagegen entschied sich das
DFN von vornherein für das Konzept der offenen Systeme nach dem ISO-OSI-Referenzmo-
dell und den dazu konsistenten CCITT-Empfehlungen und ISO-Standards. Dieses Vorgehen
hat - 1983 begonnen - zwar den Nachteil langsameren Betriebsanlaufs, da noch wichtige
Standards anstehen und es an Erfahrungen fehlt, aber als Vorteil:

- einmal eingeführte Protokolle sind stabil, da sie keinen herstellerstrategischen
  Manipulationen unterliegen

- es ergeben sich einfache Übergänge zu anderen nationalen Netzen in Europa.

- zum Vorteil des Wissenschaftsbereichs werden unnötige Abhängigkeiten von Herstellern vermieden bzw. abgebaut

- von den Fernmeldeverwaltungen (CCITT)vorgegebene Empfehlungen und Techniken werden
  nutzbar (X.25, T. 70, T.62..). So kann das DFN die Nachrichtenvermittlung - nach
  heutigem Stande - vom DATEX-P-Netz besorgen lassen.

Der wichtigste Vorteil für den Wissenschaftsbereich ist aber, daß sich die besondere
Bereitstellung und Pflege von Kommunikationsdiensten langfristig erübrigt, indem
ISO-konsistente Dienste von den Herstellern in ihr Produktspektrum aufgenommen werden; im Jahr 1983 ist dies durch die Kooperation zwölf europäischer Hersteller im
Rahmen von ESPRIT zur frühzeitigen Festlegung von ISO-Protokoll-Optionen ebenso belegt worden wie durch die Ankündigung von IBM, die ISO-Protokolle der Ebenen 4 und 5
für die MVS-Systeme bereitzustellen.

Technisch gesehen, besteht daher die Entwicklung des DFN im wesentlichen aus der Entwicklung von Software-Komponenten, die innerhalb der Betriebssoftware von Rechnern und lokalen Netzen agieren und im überregionalen Verkehr die standardisierten Protokolle ausführen. Das erforderliche, systemspezifische Wissen für diese Entwicklung ist nicht zentral bereitstellbar. Das DFN ist daher ein verteiltes Entwicklungsvorhaben, in dem dort konzipiert, spezifiziert, entwickelt und gewartet wird, wo das Systemwissen und die Systemumgebungen vorhanden sind. Die entstehenden Softwaremoduln müssen industrietypischen Qualitäts- und Wartungsnormen gehorchen und müssen ggf. an neue Versionen ihrer Systemumgebung angepaßt werden. Dies spricht zugunsten von Entwicklungen beim Systemhersteller oder in Softwarehäusern. Neben die Bereitstellung von Kommunikations-Grunddiensten tritt die Förderung ihrer innovativen Nutzung. Rechnernetze sind in Deutschland ein neues Arbeitsmittel in der Wissenschaft, mit dem Aktualität, Qualität und Produktivität der wissenschaftlichen Arbeit deutlich verbessert werden können. Für das Arbeitsmittel müssen die Fachdisziplinen oftmals erst ihre spezifischen Arbeitstechniken entwickeln. Je wirksamer,innovativer und beispielhafter für andere diese sind, umso eher verdienen sie besondere Förderung. Ein wichtiges Kriterium für das Gewicht solcher Ansätze sind Qualität und Breite der Gruppen, die die neuartige Nutzung vorbereiten. Sie sollen sich selbst formen und organisieren. Es gibt zahlreiche erfolgversprechende Beispiele dafür.

Hier liegt übrigens auch das Potential für neue wissenschaftliche Ergebnisse in DFN, die die Entwicklung der eigentlichen Kommunikationsinfrastruktur kaum bietet.

Selbstorganisation ist auch der Nutzerschaft des DFN im ganzen auferlegt. Das Bundesministerium für Forschung und Technologie, das die Entwicklung des DFN 1983 bis 1988 bezuschußt, hat von Anfang an darauf gedrungen, daß die potentiellen Nutzer - wissenschaftliche Einrichtungen wie z.B. Universitäten, andere Forschungseinrichtungen - eine Organisation schaffen, in der sie verantwortlich Konzipierung, Entwicklung, Betrieb und Förderung neuartiger Nutzung durchführen. Eine solche Organisation wurde mit der Gründung des "Vereins zur Förderung des Deutschen Forschungsnetzes (DFN-Verein)" Anfang 1984 geschaffen. Der Verein hat nach dem Stand von Oktober 1984 ca 50 Mitglieder.

2. Das Basis-DFN

Das Basis-DFN umfaßt die Dialogdienste (zeilenweiser Dialog und Virtuelles Terminal), den Dateitransferdienst (File Transfer) und den Stapelfernverarbeitungsdienst (Remote Job Entry) sowie die ihnen unterliegenden Dienste (z.B. X.25, X.28, X.29, T.70, ISO Level 4, Level 5).

Während die erstgenannten, im Sinne der OSI-Hierarchie höheren Dienste überwiegend

noch nicht standardisiert sind, liegen zu den Diensten der Schichten 3, 4, 5 der OSI-
Hierarchie CCITT-Empfehlungen bzw. ISO Standards vor, so daß das DFN sich hier teilweise fertiger Produkte und Dienstleistungen bedienen kann. Wo die erstgenannten
Dienste dringend gebraucht werden, hat sich das DFN für die vorübergehende Einführung
von Kommunikationsprotokollen entschieden, die im Rahmen der PIX-Entwicklung entstanden sind. Das trifft für den Remote-Job-Entry-Dienst zu, der zunächst durch ein
PIX-Protokoll realisiert wird, bis er - voraussichtlich erst gegen Ende des Jahrzehnts
- durch ISO-JTM (Job Transfer and Manipulation) abgelöst werden kann, und ebenso für
den File Transfer, der ein DFN-eigenes, vom Arbeitskreis Basis-DFN erarbeitetes Protokoll darstellt, das wahrscheinlich beginnend mit der 2. Hälfte 1987 durch Herstel-
lerimplementationen des ISO-File Transfer FTAM (File Transfer, Access, and Manipulation) ersetzt werden kann. Beide vorübergehend eingesetzte Protokolle werden in der
sogenannten "Nullgeneration" des DFN auf ca. 50 Rechnern heute schon verwendet,
allerdings zunächst noch in einer vorläufigen, ebenfalls aus dem PIX-Vorhaben über-
nommenen Softwareumgebung, die auf sechs Systemfamilien verfügbar ist (vgl. Abb. 1).

Abb. 2 zeigt die Strukturierung der Software für File Transfer und Remote Job Entry
in der nullten, ersten und zweiten Protokollgeneration. Die zugehörigen Softwaremoduln
müssen innerhalb aller in Abb. 1 genannten Systemfamilien und in wichtigen lokalen
Netzen realisiert werden.

Neben den bereits genannten Diensten sind Dialogdienste wesentlich; sie erlauben einem DFN-Teilnehmer, von einem Terminal mit einem entfernten Teilnehmersystem oder einem Mail-System zu kommunizieren. Hierzu bietet sich der zeilenweise Dialog nach
X.3/X.28/X.29 an, der von fast allen in Abb. 1 gezeigten Rechensystemen möglich ist,
bzw. von direkt an das DATEX-P-Netz angeschlossenen Terminals. Für gewisse Anforderungen, z.B. für den Zugang zu Fachinformationssystemen, wird ein Virtual-Terminal-
Dienst gebraucht, der im DFN für verschiedene Terminalklassen entwickelt wird. So
wird unter anderen die Umsetzung von IBM 3270-bezogenen Darstellungen in Siemens 8161
- bezogene Darstellungen und umgekehrt möglich.

Eine wachsende Anzahl von DFN-Teilnehmern möchte DFN-Dienste über lokale Netze in Anspruch nehmen. Das DFN muß ihnen die Basis-Dienste ebenso zugänglich machen wie den
Benutzern der direkt an das DFN angeschlossenen Rechner. Lokale Netze können Subnetze
des DFN darstellen, in welchem Falle sich für die am lokalen Netz operierenden Rechner ähnliche Entwicklungsaufgaben stellen wie für die autonomen angeschlossenen Rechner; sie können auch verteilte Endsysteme des DFN sein, in denen dann einzelne Rechner die DFN-orientierten Dienste für das gesamte lokale Netz aufführen (z.B. Server
für File Transfer, Remote Job Entry). Das Vorgehen des DFN ist hier noch in der Konzeptionsphase.

DFN-Protokolle sind für alle DFN-Entwicklungen verbindlich in einem Protokollhandbuch

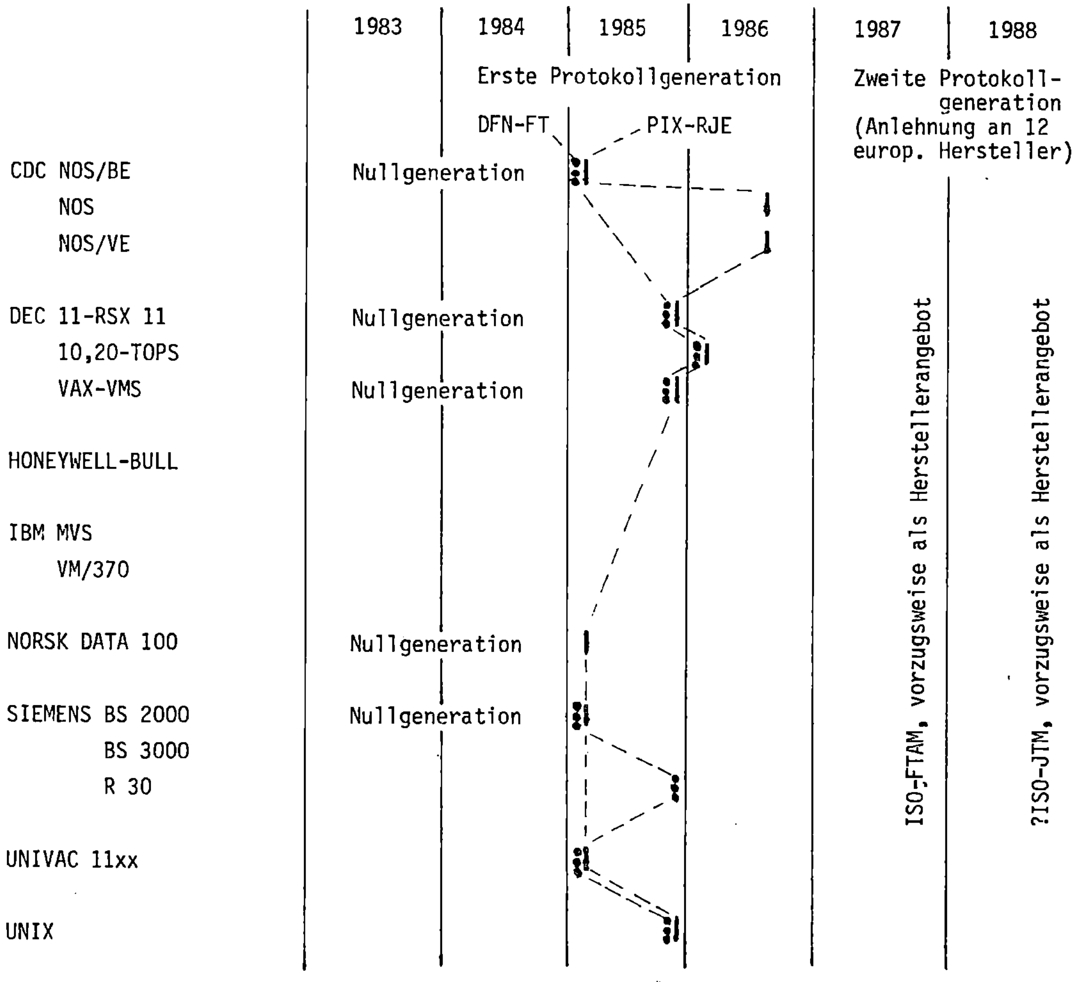

Abb. 1: Entwicklungsabschlußtermine für File Transfer und Remote Job Entry, erste und zweite Protokollgeneration, DFN-Hostanschlüsse, Stand November 84. Terminplanung für IBM/Siemens BS 3000-Systeme ausgesetzt wegen fehlender IBM-Unterlagen.

		Nullgeneration ab 1983	1. Generation ab 1985	2. Generation ab. 1987
ISO	7	DFN-FT PIX-RJE	DFN-FT PIX-RJE	FTAM ab 1987 JTM ab 1988?
	6			
	5			ISO 5 BSS, BCS
	4	Message Link Protokoll (PIX)	T. 70 = ISO 4 Class 0	ISO 4 Class 2
	1..3	X.25 (DATEX - P)		(ISDN?)

Abb. 2: Protokollhierarchie für File Transfer und Remote Job Entry; nullte,
erste und zweite Protokollgeneration

niedergelegt; mit dem DFN wird eine "Protokolltestmaschine" entwickelt, an der verbindliche Abnahmen von Kommunkationsbausteinen durchgeführt werden, soweit nicht auf
Abnahmeverfahren autorisierter Einrichtungen (FTZ) zurückgegriffen werden kann.

3. Mail-Dienste im DFN

Im Wissenschaftsbereich gibt es verschiedene, meistens lokal eingeführte Systeme zum
Nachrichtenaustausch (Mail-Dienste). Viele sind überregional über den zeilenweisen
Dialog (X.3/X.28/X.29) zugänglich. Das DFN wird diesen Systemen nicht ein weiteres
ähnlich begrenzter Zielsetzung hinzufügen. Es nimmt sich statt dessen vor, ein Message Handling System MHS auf der Basis der X.400-Empfehlung der CCITT einzuführen und
dieses zur Integration bestehender Mail-Dienste einzusetzen.

Dazu hat das DFN zunächst untersuchen lassen, welche Funktionen die bestehenden Mail-
Systeme haben, MHS-X.400 eingeschlossen, und wie sich diese zu den Anforderungen des
Wissenschaftsbereichs verhalten. Ergebnis ist, daß MHS-X.400 um einen Adreß-("Direc-
tory-")Dienst und um Mechanismen zur Gruppenkommunikation erweitert werden muß. Das
bereits vorliegende MHS-X.400 der University of British Columbia wird auf Eignung für
das DFN untersucht.

Da nach jetzigem Stand anzunehmen ist, daß MHS-X.400 für DFN erst im Herbst 1986 zur Verfügung steht, empfiehlt das DFN seinen Teilnehmern Zwischenschritte in Richtung auf den endgültigen Leistungstand. Seit Mitte 1984 ist DFN Pilotbenutzer des TELEBOX -Dienstes der Bundespost; Ende 1984 gab es ca. 100 TELEBOX-Teilnehmer aus dem DFN, das seine interne Projektkommunikation seit Januar 1985 über TELEBOX abwickelt.

Die Bundespost wird TELEBOX voraussichtlich an das MHS-X.400 anschließen; die GMD entwickelt einen Gateway zwischen ihrem KOMEX-MHS und EARN unter Benutzung von MHS-X.400-gemäßen Protokollen;damit werden auch diese beiden Systeme an X.400 gekoppelt. Der Anschluß weiterer Systeme ist in Planung. Damit ergibt sich in Stufen das in Abb. 3 gezeigte Gesamtsystem.

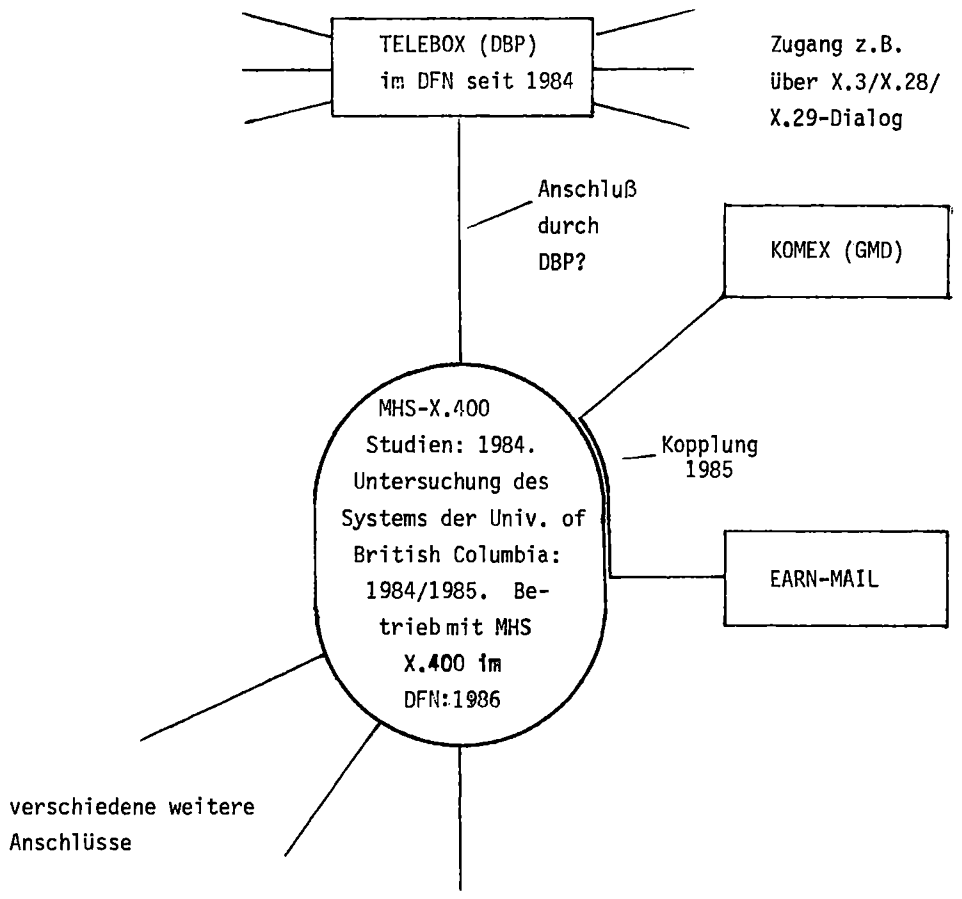

Abb. 3: Schrittweiser Aufbau von Maildiensten im DFN und Integration durch Message Handling System MHS-X.400.

4. Neue Nutzungsformen

Die Förderung neuer Nutzungsformen von rechnergestützter Kommunikation und überregional verteilter Datenverarbeitung ist ein wichtiges langfristiges Ziel. Im DFN sind es nach dem jetzigen Stande die folgenden Gruppen, die sich, in der Regel selbst organisiert, durch eigene Planung und Entwicklung auf die Benutzung eines überregionalen Rechnernetzes in neuartiger Form vorbereiten:

-Rechnergestützer Entwurf im Maschinenbau: hier werden Verfahren für Darstellung und Austausch von Beschreibungen von Entwurfsobjekten in einem offenen Netz unter Rückgriff auf DFN-Protokolle vorbereitet; hiermit hängt inhaltlich eng eine Kooperation von Schiffbautechnikern zusammen.

--Hochenergiephysik: Hier besteht - unabhängig vom DFN - seit längerem eine Arbeitsgemeinschaft HEPNET, die sich jetzt auf die Nutzung des DFN vorbereitet.

-Künstliche Intelligenz

--Entwurf von VLSI-Schaltungen: Als "Pasinger Kreis" betreibt eine Arbeitsgemeinschaft von Firmen und Instituten verteilten Entwurf von VLSI-Schaltungen, in dem auf Datenhaltungen und Entwurfsoperatoren an verschiedenen Orten zurückgegriffen wird; der Betrieb unter Benutzung der Nullgeneration der DFN-Dienste ist aufgenommen worden. Auch das EIS-Projekt wird das DFN einsetzen .

--Chemie: Eine Kooperation zur Errichtung eines verteilten Unterstützungssystems für die chemische Forschung ist begonnen worden.

An den Beispielen zeigt sich übrigens, daß Betrieb und Wartung großer Anwendungssoftwaresysteme so aufwendig sein und so kritisch von ortsgebundener Erfahrung abhängen kann, daß die Inanspruchnahme nicht am Ort betreibbarer Anwendungssoftware ein wesentliches Motiv für verteilte Verarbeitung darstellen wird.

Den genannten Beispielen liegt eine wichtige Klasse verteilter Dienste zugrunde, die Graphik in Netzen betrifft. Hier werden im Rahmen des Deutschen Forschungsnetzes drei Projekte verfolgt:

- Anpassung des graphischen Kernsystems (GKS) für zweidimensionale Raster- und Strichgraphik an graphische Datenverarbeitung in einem verteilten System. Dazu werden Verfahren zur Übermittlung von Graphik-Dateien unter Rückgriff auf im DFN bereitgestellte Dienste entwickelt, und das graphische Kernsystem wird durch Einführung neuer Schnittstellen so zerlegt, daß Verarbeitung und Ein/Ausgabe auf verschiedenen

Systemen ablaufen können. Damit können große Verarbeitungsleistungen, spezielle
Programme und spezielle Ein/Ausgabegeräte netzweit einsetzbar gemacht werden.
GKS-Systeme werden über das Netz in einem neuartigen graphischen Benutzerdialog
(graphisches Virtuelles Terminal) zugänglich gemacht.

- Entwicklung von Verfahren zur Übermittlung und Archivierung von dreidimensionalen
  Modellierdaten, wie sie für die Produktbeschreibung im Meschinenbau gebraucht werden, unter Rückgriff auf den ANSI-IGES-Standard (Initial Graphics Exchange Specification).

- Entwicklung von Verfahren zur Übertragung von Dokumenten, d.h. gemischten Text-
  Bild-Unterlagen.

5. Betrieb des DFN

Im Herbst 1984 gab es zwei größere Komplexe von Rechenzentren, die Software der
DFN-Nullgeneration für Netzbetrieb einsetzten:

- der Nordrhein-Westfalen-Verbund unter der Federführung der Universität Düsseldorf,
  der über RJE die Leistung des Vektorrechners der Universität Bochum Benutzern im
  Lande zugänglich macht (ca. 1500 Jobs je Monat, Herbst 84)

- Anschluß niedersächsischer Rechenzentren   über RJE an den Berliner Vektorrechner
  (ca. 1000 Jobs je Monat, Herbst 84).

Der DFN-Verein konnte in 1983 und 1984 ca. 3 Mio. DM an Investitionen und DATEX-P-
Gebühren an Rechenzentren des Wissenschaftsbereichs vergeben;hiervon wurden insbesondere X.25-Anschlüsse beschafft.

Für den breiten DFN-Betrieb werden Vorbereitungen getroffen, die noch nicht vollständig und noch nicht abgeschlossen sind. Aus den bereits genannten Gründen werden auch
die Betriebsfunktionen dezentral organisiert. Genauere Konzepte sind entstanden für
Software-Abnahme und Wartung, für Abrechnung und für ein Netzinformationssystem.

Die Software-Abnahme und Wartung wird an sogenannten "Referenzmaschinen" angesiedelt,
die die aktuelle Betriebssoftwareumgebung nachbilden können und die bereits entwickelte DFN-Software etwaigen Veränderungen der rechnerspezifischen Grundsoftware anpassem.
Bei diesen liegt auch die Zuständigkeit für systembezogene Fehleranalyse und -behe-
bung und für Schulung.

Die Kosten der Benutzung des DFN sollen nach dem Verursacherprinzip und möglichst
nach zweiseitigen Regelungen verrechnet und erstattet werden. Das ist zwin -

gend bei Berechnung der DATEX-P-Kosten durch die Bundespost. Im Grundsatz sind dieselben Prinzipien auch anwendbar auf die Inanspruchnahme von Betriebsmitteln bei
anderen Rechenzentren; allerdings sind verteilte Dienstleistungen absehbar, so zunächst im MHS-Bereich, bei denen zweiseitige Verrechnungen nicht sachgemäß sind. Die
Verwendung von Rechnernetzen wird ganz wesentlich erleichtert werden, wenn mindestens
in einer Anfangsphase im Rahmen von Gegenseitigkeit bis zu einer bestimmten Grenze
fremde Kapazitäten frei in Anspruch genommen werden können. Der DFN-Verein bemüht
sich um Einrichtung einer Pilotphase bis 1988, in der die Universitätsrechenzentren
gegenseitig vereinfachte Verfahren anwenden können; in dieser Pilotphase sollen recht-
lich, technisch und wirtschaftlich taugliche Abrechnungsverfahren aufgrund prakti-
scher Erfahrungen erarbeitet werden. Neben den genannten Kosten fallen noch solche
für die Wartung der DFN-Software an und globale Betriebsleistungen. Sie müssen durch
Umlage bei den Benutzern pauschal verrechnet werden.

In 1984 sind auch erste Planungen für ein Netzinformationssystem entstanden. Ein
solches Zentrum soll zunächst nur statische Daten über DFN-zugängliche Rechner, Programme und Datenbasen enthalten, über Personen, Betriebseinschränkungen usw. Es wird
voraussichtlich auf mehreren am DFN arbeitenden Rechnern eingerichtet. Der Zugriff
zum Netzinformationssystem geschieht über den zeilenorientierten Dialog.

6. Erfahrungen mit der Entwicklung des DFN

Das DFN ist notwendigerweise ein Verbundprojekt, da der für die Entwicklung erforderliche Sachverstand nicht in einer Institution konzentriebar ist. Die Partner des
Projekts haben weit auseinanderliegende wirtschaftliche, betriebliche und wissenschaftliche Interessen am DFN, sie haben sehr verschiedene Infrastrukturen und Füh-
rungsmechanismen. Ihnen steht die Zentrale Projektleitung gegenüber, die verbindliche
Leistungen fordern muß, termingerechte, normkonsistente, langfristig zu wartende Produkte.

Zugleich treten im DFN rein wissenschaftliche Ziele - obwohl der Titel "Forschungs"-
netz eine andere Auffassung suggerieren kann - gegenüber der Entwicklung einer be-
triebstauglichen Infrastruktur zurück. Das wird nur wenig dadurch eingeschränkt, daß
das Vorhaben an verschiedenen Stellen durch analytische Teilvorhaben abgesichert
wird. Grundsätzliche Fragen der Gestaltung der Dienste stellen sich - angesichts der
fortschreitenden Standardisierung- kaum noch. So stellt die Errichtung des DFN keine
wissenschaftliche Herausforderung dar, wohl aber die innovative Nutzung; auch ist
die Kooperation so vieler Partner im Wissenschaftsbereich eine Herausforderung.

Diese Kooperation hat sich im DFN bisher sehr bewährt. Insbesondere die Arbeitskreise

des DFN, die sehr viel zur Gestaltung des Projektes in seiner ersten Phase beigetragen haben, sind hier zu nennen, und die wirksame Selbstorganisation einiger
Nutzergruppen des DFN. Die Industrie hat sich früh und mit erheblichen Beiträgen
am DFN-Verein beteiligt.

Besonders wichtige, glückliche Startbedingungen für das Projekt waren die frühzeitige
Verfügbarkeit einer aktionsfähigen, in Netzprojekten erfahrenen Projektleitung in
Berlin und die frühzeitige Erklärung des Bundesministeriums für Forschung und Tech-
nologie, das DFN mit erheblichen Mitteln zu fördern.

Literatur:

Hultzsch, H.: EARN-Status und Perspektiven, GI-NTG-Fachtagung "Kommunikation in verteilten Systemen", Karlsruhe, 11. bis 15.03.85.

Zum Deutschen Forschungsnetz berichten die folgenden Beiträge der GI-NTG-Fachtagung
"Kommunikation in verteilten Systemen", Karlsruhe, 11. bis 15.03.85 über Einzelpro-
bleme:

Bauerfeld, W.L.: Einbettung von lokalen Netzwerken im DFN

Conrads, D.: Funktionalität und Bewertung von Message-Systemen

Egloff, P.: Graphische Dienste im DFN

Kaufmann, P., Tschichholz, M.: Der DFN-Message-Dienst im MHS-Kontext

Knop, J., und Mitarbeiter der Rechenzentren der Universitäten Aachen, Bielefeld,
Bochum, Düsseldorf, Köln: NRW-Jobverbund im Deutschen Forschungsnetz-Erfahrungen und
Probleme beim Betrieb

Truöl, K.: Zum Betrieb des Deutschen Forschungsnetzes

Warnking, A., Santo, H.: Konzept für den Directory Service im Message-Handling-System
des Deutschen Forschungsnetzes

Wosnitzy, L.: Gruppenkommunikation im MHS-Kontext

Allgemeine Information über das Deutsche Forschungsnetz findet sich vor allem in

Deutsches Forschungsnetz-DFN-, Kurzbeschreibung, Zentrale Projektleitung DFN, Berlin
1984.

Ullmann, K.: Deutsches Forschungsnetz (DFN) – eine anwendungsorientierte Entwicklung
von Kommunikationsdiensten, 14. Jahrestagung der Gesellschaft für Informatik, Braunschweig, 1984.

Technische Detailinformation zum Deutschen Forschungsnetz ist zugänglich über:
DFN-Verein, Zentrale Projektleitung, Pariser Straße 44, 1000 Berlin 15

EARN, Status und Perspektiven
Franz Busch, Hagen Hultzsch, Roland Wolf
GSI Darmstadt

Abstract

EARN ist ein Rechnernetz, das 1984 in Europa aufgebaut wurde und der Kommunikation zwischen Wissenschaftlern dient. Es basiert auf den gleichen Techniken (Store and Foreward) wie sie im Amerikanischen BITNET verwendet werden. BITNET und EARN bilden ein einheitliches weltweites Netz, an das momentan ca. 430 Knotenrechner angeschlossen sind. Daneben existieren Gateways in andere Netze. In EARN/BITNET können Nachrichten, Dateien und Stapelaufträge übertragen werden.

EARN is a European computer network that has been built in 1984 for better communications between scientists. EARN is based on the same store and foreward techniques as the American BITNET. EARN and BITNET actually form a unique worldwide network with about 430 nodes presently connected. There are Gateways to other networks too. EARN/BITNET is capable for transmission of messages, files and batch jobs.

Einleitung

Das Rechnernetz EARN (European Academic and Research Network), ist ein weiträumiges Kommunikationsmedium und dient dem weltweiten wissenschaftlichen Erfahrungs- und Informationsaustausch zwischen Universitäten und anderen öffentlichen Forschungseinrichtungen. EARN soll insbesondere die Durchführung und Planung von Kooperationsforschungsvorhaben, die Analyse und Diskussion wissenschaftlicher Ergebnisse und das Abfassen von Publikationen größerer Gruppen erleichtern.

Der Wunsch nach einem schnell verfügbaren, grenzübergreifenden Rechnernetz wurde aus Kreisen der Europäischen Forschungsinstitute wiederholt geäussert.

Die Firma IBM hat diesen Vorschlag aufgegriffen und im Rahmen ihres Forschungsförderungsprogramms erhebliche finanzielle und technische Mittel bereitgestellt, so daß EARN Anfang 1984 ins Leben gerufen werden konnte. Bei der Realisierung wurde ausschliesslich auf erprobte Standard-Techniken zurückgegriffen, wie sie auch im IBM-internen, weltweiten Computernetz VNET, oder im Amerikanischen Forschungsnetz BITNET (1) verwendet werden. Der Aufbau von EARN wurde 1984 begonnen und weitgehend vollzogen. EARN erfreut sich seither, ebenso wie BITNET, eines kräftigen, stetigen Zustroms neuer Teilnehmer.

Der Deutsche Teil von EARN soll später in das Deutsche Forschungs Netz DFN (2), das noch in der Diskussionsphase ist, soweit als möglich integriert werden. Mit EARN sollte in der Übergangszeit bis zum DFN Produktionsbetrieb ein dringend benötigtes Kommunikationsmedium bereitgestellt werden. Die wesentlichsten Merkmale von EARN sind in Anhang A.1 zusammengestellt.

Technische Daten und Funktionscharakteristika

EARN besteht aus einer Reihe unabhängiger Knotenrechner, die über postalische Festverbindungen in offener Baum-Topologie miteinander verbunden sind. EARN ist in nationale Teilnetze untergliedert, die jedes einen nationalen Zentralknoten beinhalten. Diese dienen der Abwicklung des grenzüberschreitenden Verkehrs und zur Bereitstellung zentraler Dienste. Die Topologie des Europäischen Basisverbunds und des Deutschen Teilnetzes sind im gegenwärtigen Ausbaustand in Abb.1 dargestellt.

EARN basiert auf den "Store and Forward" Techniken des RSCS (Remote Spooling Communication Subsystem) des Betriebssystems VM (Virtual Machine), und dem NJE/NJI (Network Job Entry/Network Job Interface) der Job Entry Subsysteme JES2 oder JES3 unter MVS (Multiple Virtual Storage). Alle Rechenanlagen, die eines dieser Protokolle unterstützen, können in das Netz integriert werden. Dazu gehören neben anderen Systemen insbesondere die Rechner von Siemens unter BS3000 und BS2000, CDC CYBER, DEC VAX und UNIVAC.

In der Sicht des Sieben-Schichtenmodells des OSI (Open Systems Interconnect) wird auf der Datalink-Ebene das Leitungsprotokoll BSC1 verwendet. Die RSCS-

EARN
TRONDHEIM
STOCKHOLM
KOPENHAGEN
DUBLIN
OXFORD
NIJMEEGEN
WASHINGTON
PARIS
DARMSTADT
LINZ
BITNET
GENF
NEW YORK
HAIFA
MADRID
ROM
KOPENHAGEN
KIEL
DESY UNI
HAMBURG
OLDENBURG
OSNABRÜCK
BERLIN
TU
GRZ
HMI
HANNOVER
TU BRAUNSCHWEIG
MÜNSTER
NIJMEGEN UNI
UNI
GÖTTINGEN
ESSEN UNI UNI DORTMUND
DÜSSELDORF UNI
UNI HAGEN
JÜLICH KFA
UNI KÖLN
UNI
GMD
AACHEN RWTH
BONN
WASHINGTON
GSI
DARMSTADT GMD
TH
IBM
ERLANGEN
MANNHEIM DKFZ HEIDELBERG
UNI UNI FHS HEILBRONN
LINZ
KFK
UNI
KARLSRUHE UNI
FHS AALEN
UNI
STUTTGART
UNI TÜBINGEN
IPP
DFVLR MÜNCHEN
GENF
OBERPFAFFENHOFEN
MPI-Physik
GNOM/F.B. 84

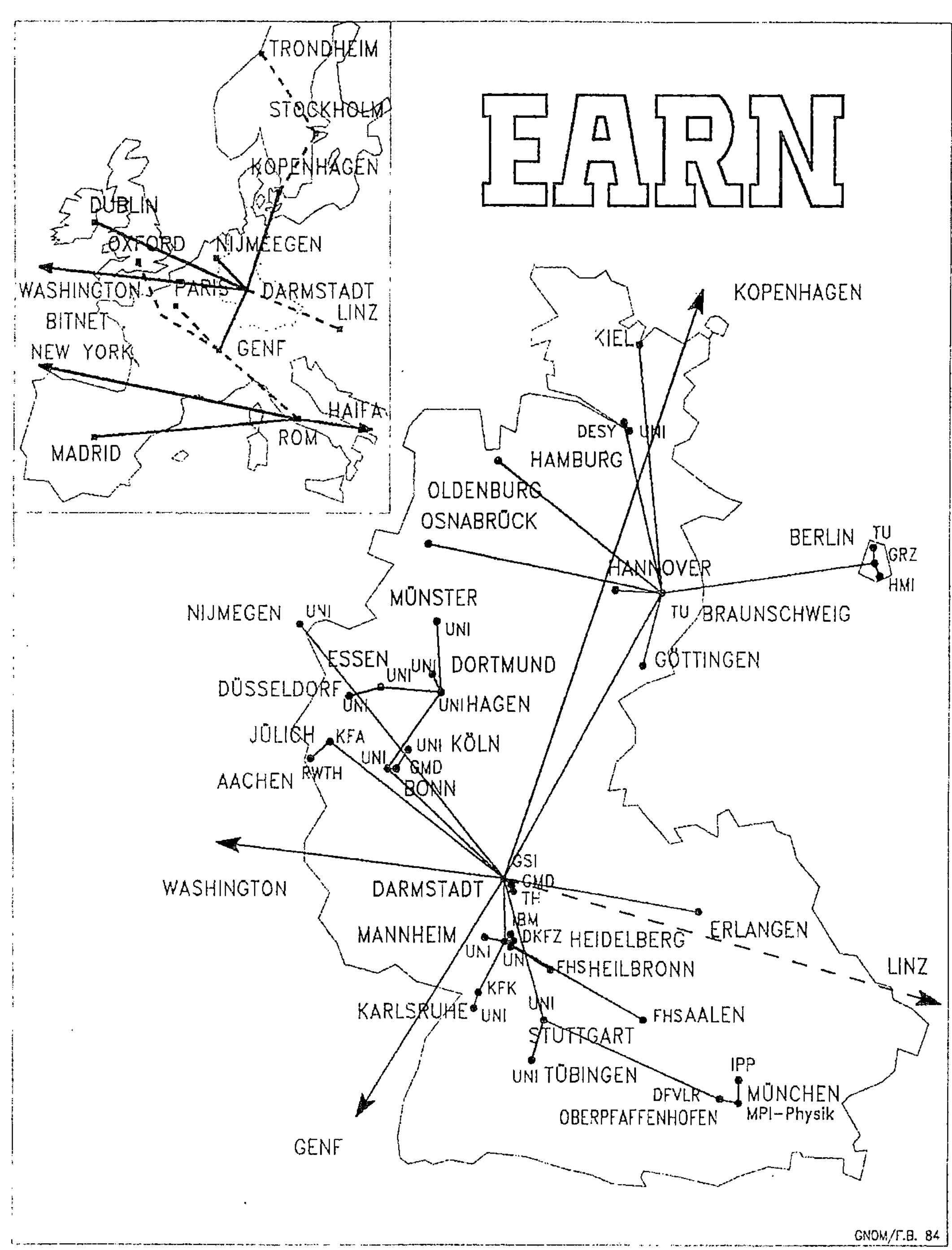

Abb.1

und NJE/NJI-Protokolle kommen in den darüberliegenden Schichten, insbesonde
der Netzwerks- und Transportschicht, zum Tragen. Die Transportwege zu jedem
anderen Rechner sind in den Netztabellen (Routingtables) der beteiligten Rechner
festgehalten. Die übermittelten Daten werden an jedem Rechner auf Magnetplatte
zwischengespeichert (Spooling), bis die Übermittlung zum nächsten Rechner erfolgreich abgeschlossen ist. Sollte die Verbindung zum nächsten Rechner unter-
brochen sein, werden die Daten so lange zwischengespeichert, bis die Verbindung
wiederhergestellt ist. In der Regel wird eine Übertragungsgeschwindigkeit von
9600 bit pro Sekunde benutzt; einige weniger frequentierte Übertragungsstrecken
begnügen sich mit einer niedrigeren Übertragungsgeschwindigkeit.

EARN, als weltweites Kommunikationsmedium zwischen Wissenschaftlern konzipiert,
dient in der Hauptsache zur Übertragung von paketierter Information (Files).
Darunter verstehen sich Daten aller Art, insbesondere Texte, Programme und
Forschungsergebnisse. Es können Stapelaufträge an entfernte Rechner übermittelt
und Rechenergebnisse verschickt werden (Remote/Network Job Entry, RJE/NJE).
Der direkte, interaktive Zugriff eines Endbenutzers auf entfernte Rechner ist jedoch aufgrund der technischen Gegebenheiten (Protokollinkompatibilitäten) nur
begrenzt möglich (VM Pass-Through) und daher nicht vorgesehen. Ausserdem
würde das Netz, das auf Standleitungen mit maximal 9.6 KBaud Übertragungsrate
aufbaut, durch interaktive Sitzungen zu stark belastet.

Die Adressierung eines entfernten Partners erfordert die Kenntnis des Knotenna-
mens des Zielrechners und die Benutzerkennung des Adressaten an diesem Rechner. Beides kann in eleganter Weise durch das Anlegen einer "Nickname"-Datei
erleichtert werden. In Anhang A.2 ist als Beispiel die Benutzerführung am MVS
Rechner der GSI Darmstadt unter TSO/E mittels angebotener ISPF Menuetafeln
dargestellt. Ein fullscreenfähiges Datensichtgerät ist zwar praktisch aber nicht
Voraussetzung zur Benutzung von EARN, da alle EARN-Kommandos auch zeilenorientiert eingegeben werden können.

Wie die Erfahrung im Umgang mit EARN zeigt, ist die Möglichkeit Dateien zu
transferrieren die weitaus wichtigere Funktion in einem Netz. Wer unbedingt
interaktiv an entfernten Rechnern arbeiten will, kann mit geeigneter Terminalein-
richtung den DATEX-P Dienst der DBP in Anspruch nehmen. Fullscreenanwen-
dungen sind in einem weltweiten Netz wegen der Gerätevielfalt ohnehin nicht ohne
erheblichen Aufwand realisierbar. Zudem wäre der Anwender in erheblichem
Maße mit den Eigenheiten anderer Rechner, beispielsweise den Funktionstastenbe-
legungen und der Kommandostruktur belastet. Wesentlich benutzerfreundlicher ist

es, wenn der Anwender die erworbenen Kenntnisse im Umgang mit der Multifunktionalität seines Rechners nutzt, mit der zusätzlichen Möglichkeit Dateien und Stapelaufträge über das Netz verschicken zu können.

Am Deutschen Zentalknotenrechner, einem kleineren Rechner vom Typ IBM 4331 der bei GSI in Darmstadt aufgestellt ist, sind eine Reihe von zentralen Diensten verfügbar bzw. in der Entwicklung. Dazu gehören ein Informationszentrum für Netzkonfigurationen (virtuelle Maschine "NETSERV"), ein Knotennamen- und Benutzerverzeichnis, Benutzeranleitungen, Standards, sowie ein "Schwarzes Brett" für allgemeine Neuigkeiten. Daneben wird ein Computer-Konferenzsystem für weltweiten Gebrauch entwickelt und installiert. Standardformate für elektronische Briefe und für graphische Daten werden noch diskutiert. Daten über Netznutzung, Auslastung, transportiertes Datenvolumen sowie über Verfügbarkeit von Knoten und Leitungen, werden hier zentral erfasst, ausgewertet und der Allgemeinheit zugänglich gemacht. Als Beispiel eines zentralen Dienstes sind einige Funktionen der Maschine NETSERV in Anhang A.3 zusammengestellt.

Organisation und Management

Die Teilnahme am EARN steht allen öffentlichen wissenschaftlich/technischen Einrichtungen offen, ist für alle nicht kommerziellen Aufgabenstellungen verfügbar und kann von allen Mitgliedern der akademischen Einrichtungen benutzt werden. Studenten soll die Benutzung von EARN allerdings nur im Rahmen von Projekten gestattet sein, zu denen neben anderem auch Diplom- und Doktorarbeiten gehören. Es gibt keine Sicherheitsgarantien für die übertragenen Daten. Datenschutzaspekte stehen voll in der Verantwortung des Übermittlers der gegebenenfalls selber für die Verschlüsselung und Dekodierung sensibler Daten Sorge tragen muß.

Für den Betrieb von EARN sind die teilnehmenden Institutionen gemeinsam verantwortlich. Zugunsten des gemeinsamen Nutzens haben sich diese bereit erklärt, EARN-relevante Resourcen wie Fernanschlüsse, Modems, Speicherplatz auf Magnetplatte, auch für den Datenverkehr dritter, möglichst rund um die Uhr verfügbar zu halten. Die Bereitstellung weiterer Kapazitäten im Netz, wie die Ausführung vom Netz übermittelter Stapelaufträge, steht auf der Basis projektorientierter, bilateraler Abmachungen im Belieben der jeweiligen

Institution. Die nicht projektgebundene, missbräuchliche Nutzung, kann etwa durch Passwort- und Benutzerkennungsschutz, wirksam unterbunden werden.

Repräsentiert wird EARN durch das EARN-Direktorium (Board of Directors), das sich aus jeweils einem Vertreter jeden beteiligten Landes zusammensetzt. Das EARN-BOD legt auch die Nutzungs- und Zugangskriterien fest. Die Kosten für die internationalen Leitungsverbindungen trägt IBM für einen Zeitraum von vier Jahren. In der Bundesrepublik werden die Kosten für die meisten Leitungsverbindungen zwischen den Rechenzentren und dem Koordinierungsrechner für die internationalen Leitungsverbindungen ebenfalls für einen Zeitraum von vier Jahren von IBM Deutschland getragen. Die durch den Datentransport entstehenden Lastanteile an den beteiligten Rechnern liegen in der Regel im Bereich weniger Prozent, bei größeren Anlagen unterhalb von einem Prozent.

Das technische Management erfolgt durch sogenannte Knotenadministratoren, die am heimischen Rechner die EARN-relevanten Belange abdecken, wie Pflege der Knotentabellen der Netzsoftware und Beratung der Benutzer. In der Regel wird dies von den für DFÜ zuständigen Personen nebenbei bewältigt. In der Bundesrepublik laufen Informationen über Änderungen im Netz, wie das Hinzukommen neuer Knoten, beim Administrator des Zentralen Knotenrechners zusammen, werden aufbereitet und etwa in Form aktueller Knotentabellen an jeden Knotenrechner verschickt, oder auf dem Zentralrechner abrufbar bereitgehalten. Dies geschieht in Deutschland mit Hilfe der virtuellen Maschine "NETSERV" am Deutschen Zentralknoten in Darmstadt.

Status

EARN ist weitgehend realisiert. Von besonderem Interesse ist die Anbindung an das Amerikanische Forschungsnetz BITNET über Darmstadt und ROM, die die Kommunikation mit derzeit bereits über 300 Knotenrechnern an Amerikanischen ·Universitäten ermöglicht. BITNET und EARN bilden technisch gesehen ein einziges gemeinsames Netz. Momentan sind in Spanien 5, in Italien 11, in der Schweiz 11, in Israel 10 und in der Bundesrepublik 64 Knoten miteinander verbunden. Der Anschluss von Dänemark über Darmstadt und in der Folge Schweden und Norwegen stehen unmittelbar bevor. Dasselbe gilt für die Niederlande, Österreich und

Irland. Die Anbindung der nationalen Netze in Belgien, Frankreich und England ist ebenfalls vorgesehen.

Über Gateway-Verbindungen sind weitere Netze erreichbar, oder werden es in naher Zukunft sein. So kann über BITNET-Gateways mit Teilnehmern im ARPANET (3) und CSNET kommuniziert werden. In EARN sind Gateways zum Britischen Netz JANET (4), zum Deutschen Forschungsnetz DFN (2) und anderen nationalen oder regionalen Netzen in Arbeit.

Verzögerungen bei der Inbetriebnahme der internationalen Leitungsverbindungen sind durch die Tarifierungsüberlegungen der CEPT entstanden, nach denen ursprünglich neben den Standleitungsgebühren volumenabhängige Kosten in der Höhe von DM 0.12 pro 1000 Zeichen in Europa berechnet werden sollten. Diese Tarifierungsüberlegungen hätten allerdings für die Strecke Darmstadt - Genf beispielsweise zu Gesamtkosten in der Grössenordnung von einer Million DM pro Jahr geführt, ein Betrag, der weder aus den geplanten Ansätzen noch mittel- und längerfristig aus den Haushalten der beteiligten Institutionen finanzierbar gewesen wäre. Die Gespräche mit der Deutschen Bundespost und der CEPT haben zu einem Überdenken dieser Planungen geführt und haben Tarifierungen entstehen lassen die zu vertretbaren Kostenbelastungen geführt haben. Die Deutsche Bundespost hat die internationalen Leitungsverbindungen kooperativ und zügig bereitgestellt, so daß die Leitung nach CERN in Genf, nach USA sowie Skandinavien, den Niederlanden, Österreich und Irland verfügbar geworden sind bzw. werden.

Die mangelnde Flexibilität oder Kooperationbereitschaft anderer nationaler Postbehörden, so der Norditalienischen, Französoischen und Britischen, resultiert in einer Netzkonfiguration, die etwas von der ursprünglichen Planung abweicht. So sind momentan Spanien, Italien und Israel von Deutschland aus gegenwärtig nur durch einen Umweg von 150.000 Km Länge via Satellit über die USA erreichbar, Belgien, Frankreich und Großbritannien vorläufig noch nicht.

Die Verfügbarkeit des Netzes ist insgesamt befriedigend. Sie ist besonders hoch dort wo Rechner mit VM Betriebsystemen miteinander verbunden sind. Etwas problematisch stellen sich MVS Systeme dar, da zum Beispiel Betriebssystemtests eine vorübergehende Abschaltung der Kommunikationsleitungen erfordern. In Anhang A.4 ist die Verfügbarkeit der wichtigsten Strecken in der Bundesrepublick aus der Sicht des Darmstädter Zentralknotens für den Zeitraum von 16.11. bis 1.12.84 dargestellt.

Die Nutzungsintensität während dreier Monate spiegelt sich im Transportvolumen am Knotenrechner DEARN in Darmstadt wieder, die in Anhang A.5 dargestellt ist. Im November 1984 wurden 320 Millionen Zeichen durch den Deutschen Zentralknoten übertragen, was einem Tagesmittel von 2.5 Bibeln entspricht. Die Bibel enthält etwa vier Millionen Zeichen. Im Dezember 1984 war das Tranportvolumen bereits nahezu auf das zweifache angestiegen. In Anhang A.6 ist eine Liste aller Knotenrechner im EARN/BITNET Netzwerk mit der jeweils verwendeten Betriebssoftware gegeben.

Ausblick

Ein wichtiges Ziel ist, durch Gespräche zwischen CEPT und EARN die Entwicklung von volumenabhängigen Tarifen voranzubringen, die einerseits den zukünftigen Entwicklungen zum ISDN gerecht werden, andererseits jedoch den datenintensiven Kommunikationsbedarf von Wissenschaft und Technik finanzierbar machen. Vergleichsweise günstige Kommunikationskosten führen ohne Zweifel zu einer Entfaltung avantgardistischer Forschungs- und Entwicklungskonzeptionen in Europa. Dies zu erreichen muß Ziel der Überlegungen auch im Zusammenhang mit der Integration von EARN und DFN sein.

Das in der Zukunft zu erwartende starke Wachstum der beteiligten Knotenrechner am EARN/BITNET macht Überlegungen zur eleganteren Behandlung der Netzwerktabellen im EARN unumgänglich, denn bereits heute sind mit derzeit ca. 430 beteiligten Knotenrechnern fast täglich Änderungen der Netzwerktabellen jedes beteiligten Rechners erforderlich. Glücklicherweise führen Neuzugänge in der Regel nicht zu einer Umstrukturierung der bestehenden Verbindungen, sodaß ein Überarbeiten der Knotentabellen in monatlichem Abstand zumeist ausreichend ist.

Ein großes Problem ist die Beseitigung von Inkompatibilitäten und Einschränkungen zwischen RSCS, JES2 und JES3. So können in JES2 lediglich insgeamt 1000 Knoten über Nummern definiert werden, was angesichts der neben EARN existierenden separaten Netze beteiligter Institutionen und anderer über Gateways angeschlossener Netze bereits jetzt problematisch ist und neben anderem auch eine EARN-weite einheitliche Numerierungskonvention verhindert. Im Zusammenspiel zwischen Knoten mit unterschiedlicher Netzsoftware wie RSCS, JES2 oder JES3 ergeben sich relativ häufig eigentlich vermeidbare Problemsituationen, die unter anderem durch mangelde Verifizierung der Kompatibilität der drei unterschiedlichen Produkte durch die Entwicklungsgruppen entstehen. IBM ist in der Folge gefordert die unterschiedlichen Komponenten der Netzwerksoftware besser aneinander anzugleichen.

Mittel- und längerfristig sind Techniken, wie sie bereits heute im DFN vorgesehen sind, nämlich die Nutzung des DATEX-P Dienstes der DBP, der die unteren drei OSI-Schichten abdeckt, unumgänglich. In der Tat ist damit zu rechnen, daß IBM seine Netzsoftware in Richtung ISO/OSI ausbaut, sodaß neben einem Funktionsgewinn, wie interaktiver Verkehr, auch die Nutzung des DATEX-P Netzes der DBP möglich wird. Das EARN-BOD hat sich im Rahmen der Tarifierungsgespräche mit der CEPT verpflichtet, diese Umstellung voranzutreiben und baldmöglichst zu realisieren.

Anhang

A.1   Zusammenfassung der wesentlichsten Merkmale von EARN
A.2   Benutzerführung mittels Menuetafeln am Beispiel GSI
A.3   Beispiel eines zentralen Dienstes: die virtuelle Maschine NETSERV
A.4   Verfügbarkeit der wichtigsten Strecken im Deutschen EARN aus der Sicht des Darmstädter Zentralknotens.
A.5   Nutzungszuwachs im Zeitraum 9.- 11. 1984
A.6   Knotenverzeichnis EARN/BITNET

Literatur

(1)   BITNET-Because it's Time, Perspectives in Computing, Vol.3, No.1, March 1983
(2)   Deutsches Forschungsnetz DFN, Projektvorschlag 1982
(3)   J. McQuillan, D.C. Walden, The ARPA Network Design Decisions, Computer Networks, Vol.1, 1977
(4)   The ICF Network, Electronics and Power, Nov-Dec 1981

A.1

Das **European Academic and Research Network** ist ein Computerverbund zwischen wissenschaftlichen und akademischen Institutionen, und dient der weltweiten Kommunikation.

Verfügbare Funktionen:

* Transfer von Dateien, Texten, Programmen, Stapelaufträgen und Rechenergebnissen
* Elektronische Übermittlung von Nachrichten
* Direkter Zugang in das Amerikanische Forschungsnetz **BITNET** Gateway—Verbindungen zu ARPANET, CSNET, JANET, DFN u.a.
* Am zentralen Deutschen Knotenrechner bei GSI werden u. a. folgende zentrale Dienste bereitgehalten:
  - Benutzeranleitung
  - zentrales Namens— und Knotenverzeichnis
  - elektronisches schwarzes Brett
  - Computer—Konferenzen

Technische Eigenschaften:

* Sichere Übertragung durch Zwischenspeicherung der Daten in den Knotenrechnern (Store and Forward Technik).
* Übertragungsraten: 300 — 1000 Zeichen pro sec und Leitung
* Leitungsprotokoll: BSC1
* Erforderliche Software: NJI/NJE unter MVS, RSCS unter VM, oder entsprechende Emulationen; u.a. verfügbar für SIEMENS BS200/BS300,

A.2 Benutzerführung durch Menütafeln bei GSI

```
-------------------- E A R N  COMMUNICATION MENUE --------------------

                EUROPEAN ACADEMIC AND RESEARCH NETWORK  E A R N

    SELECT APPLICATION ===> n_
                                                    USERID    - RZ02
        H  HELP      - WHAT IS EARN ?               PREFIX    - RZ02
                     - HOW CAN I COMMUNICATE IN EARN ?  TIME   - 15:45
        N  NAMES     - CREATE AND MODIFY A NAMES FILE TERMINAL - 3278
        T  TRANSMIT  - SEND A MESSAGE OR DATASET     FF KEYS   - 24
        R  RECEIVE   - RECEIVE A MESSAGE OR A FILE   DATE      - 85/01/04
        S  NETSERVE  - NETWORK SERVICES MENUE
        L  LOGLIST   - LIST THE LOG FILE  ( ENTER LOG NAME ==> NCTE     )
        D  DIRECTORY - USER DIRECTORY SERVICE PANEL (NOT YET AVAILABLE)


        WELCOME IN THE   E A R N   NETWORK !!!

        NEWS: NOW THE USA, CANADA, ITALY, SPAIN, ISRAEL AND SWITZERLAND
              CAN BE REACHED !!!! FOR THE NODENAMES SEE E.S (THE COUNTRY
              INITIAL FOR ISRAEL IS L)
    PRESS END KEY TO TERMINATE
```

```
-------------- CREATE, CHANGE OR VIEW A NICKNAME ENTRY --------------------

     N  - CREATE A NEW ENTRY       L      - LIST THE WHOLE NAMES FILE
     S  - SHOW THE NEXT ENTRY      C      - CLEAR THE FIELDS
     M  - MODIFY THE CURRENT ENTRY 'NAME' - FIND THE NICKNAME ENTRY: 'NAME'
     D  - DELETE THE CURRENT ENTRY P      - CHANGE YOUR PROLOG ENTRY

     SELECT OPTION ===> N         (OR ENTER A NICKNAME)

     NICKNAME ==> BOSS          USERID ==> RZ01     AT  NODEID ==> DDAGSI3

     NAME    ==> Dr. Hagen Hultzsch
     PHONE   ==> 06151-359762
     ADDRESS ==> Gesellschaft fuer Schwerionenforschung
     ADDRESS ==> 6100 Darmstadt
     ADDRESS ==> Planckstr. 1_
     ADDRESS ==>
     OR
     LIST    ==>

     NOTEBOOK ==>              (LOG NOTEBOOK)
```

```
------- TRANSMISSION OF MESSAGES AND FILES IN   E A R N --------------------

     FILL IN OR CHANGE THE FOLLOWING FIELDS:

     SPECIFY THE ADDRESSEE
     ENTER A NICKNAME           ==> BOSS
     OR
     ENTER THE USERID           ==>
     AND NODEID                 ==>           OF THE ADDRESSEE
     OR
     ENTER A LIST  ==>

    IF YOU WANT TO SEND A FILE  ============================================

     ENTER THE DATA SET NAME OF
     THE FILE YOU WANT TO SEND  ==> BRIEF.TEXT(WASOWAS)_
    =========================================================================

     ENTER FURTHER OPTIONS     ==>
     LIKE NOEPILOG , MSG OR MEMBERS(MEMBERLIST)

    PRESS ENTER TO WRITE THE MESSAGE TEXT OR SEND THE FILE
    OR PRESS THE END KEY TO TERMINATE
```

A.3

<u>NETSERV</u>

NETSERV ist eine virtuelle Servicemaschine die am Deutschen
Zentralknoten läuft und als Informationscenter und
Mangementhilfe dient. Eine Vielzahl von Dateien koennen per
Kommando abgerufen werden. Kommandobeispiel fuer einen
VM−Benutzer: TELL NETSERV GET PROGRAMS FILELIST
Im Folgenden ist eine Auswahl von Kommandos aufgelistet:

Kommandos für allgemeine Informationen:

HELP   Information über Netserv
GET EARNJES2 HINTS   Sammlung bekannter JES2 Probleme
GET EARN NEWS   Neuigkeiten
GET PROGRAMS FILELIST   Liste verfügbarer Programme
GET DEARN NETMAP   EARN Topologie in Deutschland
GET IEARN NODELIST   Liste der EARN−Knoten in Italien
GET EARNCORD NAMES   Liste der EARN Koordinatoren
GET CEARNAD NAMES   Schweizer Knotenadministratoren

Kommandos für Knotenadministratoren:

GET NODENTRY nodeid   Knoteneintragung holen zum
                              Zwecke der Modifizierung
PUT NODENTRY   stellt eine neue oder
                              modifizierte Knotenbeschreibung
                              in eine Äderungserfassungsdatei
GET NODESW REGFORMG   Registrierungsformular zwecks
                              Erfassung eines neuen Knotens
GET NETINIT FILELIST   Liste der abrufbaren Files für
                              Knotentabellengenerierung.

A.4

EARN Verfügbarkeitsstatistik

aus der Sicht von DEARN

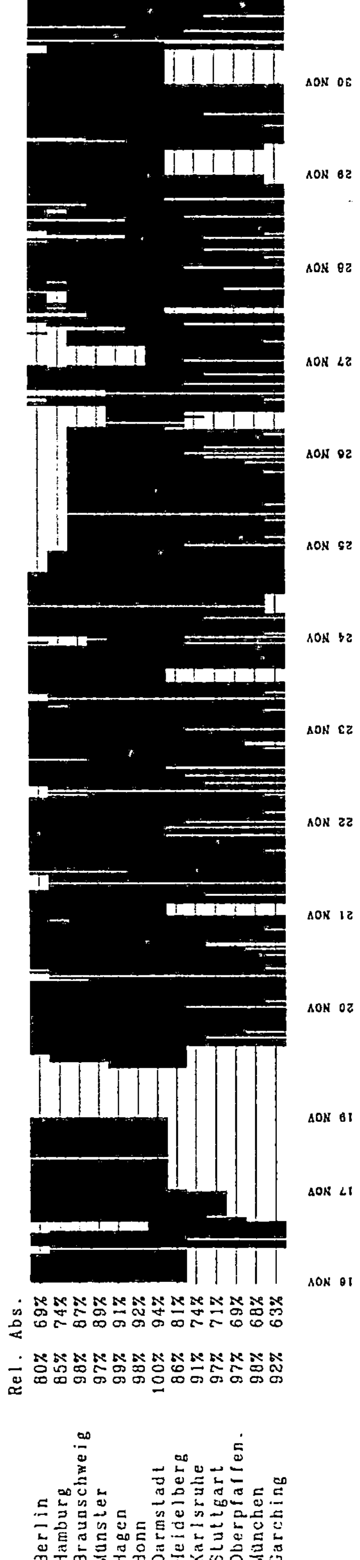

A.5

```
••••••• E A R N  ACCOUNTING INFORMATION •••••••   from 01.09.84  to 30.11.84   Local node: DEARN

All (80 byte) records which passed the local node or were sent from/received there are counted.

Date        Records  |• = 3000 records
-----------------------------------------------------------------------------------------------------
SA 01.09.84     2212 |•
SU 02.09.84     2334 |•
MO 03.09.84    47738 |••••••••••••••••
TU 04.09.84    56108 |•••••••••••••••••••
WD 05.09.84    17355 |••••••
TH 06.09.84    45707 |•••••••••••••••
FR 07.09.84    28138 |•••••••••
SA 08.09.84     9620 |•••
SU 09.09.84       13 |-
MO 10.09.84    33279 |•••••••••••
TU 11.09.84    21799 |•••••••
WD 12.09.84    36152 |••••••••••••
TH 13.09.84    31736 |•••••••••••
FR 14.09.84    41005 |••••••••••••••
SA 15.09.84      348 |-
SU 16.09.84       32 |-
MO 17.09.84    42109 |••••••••••••••
TU 18.09.84    51347 |•••••••••••••••••
WD 19.09.84    79448 |••••••••••••••••••••••••••
TH 20.09.84    31534 |•••••••••••
FR 21.09.84    56224 |•••••••••••••••••••
SA 22.09.84        5 |-
SU 23.09.84        4 |-
MO 24.09.84    62547 |•••••••••••••••••••••
TU 25.09.84   112730 |••••••••••••••••••••••••••••••••••••••
WD 26.09.84    66752 |••••••••••••••••••••••
TH 27.09.84    30025 |••••••••••
FR 28.09.84    38306 |•••••••••••••
SA 29.09.84     1221 |-
SU 30.09.84      974 |-
MO 01.10.84    89386 |•••••••••••••••••••••••••••••
TU 02.10.84    27146 |•••••••••
WD 03.10.84    20321 |•••••••
TH 04.10.84    61276 |••••••••••••••••••••
missing days       0 |
SA 06.10.84    24536 |••••••••
SU 07.10.84      149 |-
MO 08.10.84    77848 |•••••••••••••••••••••••••
TU 09.10.84    57576 |•••••••••••••••••••
WD 10.10.84   104530 |••••••••••••••••••••••••••••••••••
TH 11.10.84    86995 |•••••••••••••••••••••••••••••
FR 12.10.84    72629 |••••••••••••••••••••••••
SA 13.10.84     1723 |•
SU 14.10.84     7670 |•••
MO 15.10.84    74040 |••••••••••••••••••••••••
TU 16.10.84    60698 |••••••••••••••••••••
WD 17.10.84    53857 |••••••••••••••••••
TH 18.10.84    66014 |••••••••••••••••••••••
FR 19.10.84   101399 |•••••••••••••••••••••••••••••••••
SA 20.10.84     2354 |•
SU 21.10.84      981 |-
MO 22.10.84    99779 |••••••••••••••••••••••••••••••••
TU 23.10.84    89514 |•••••••••••••••••••••••••••••
WD 24.10.84    59528 |•••••••••••••••••••
TH 25.10.84   104101 |••••••••••••••••••••••••••••••••••
FR 26.10.84   176480 |••••••••••••••••••••••••••••••••••••••••••••••••••••••••••
SA 27.10.84    14341 |•••••
SU 28.10.84     1016 |-
MO 29.10.84    77174 |•••••••••••••••••••••••••
TU 30.10.84    67756 |••••••••••••••••••••••
WD 31.10.84   180927 |•••••••••••••••••••••••••••••••••••••••••••••••••••••••••••
TH 01.11.84    34743 |•••••••••••
FR 02.11.84    79235 |••••••••••••••••••••••••••
SA 03.11.84    10281 |•••
SU 04.11.84    42013 |••••••••••••••
MO 05.11.84   116045 |••••••••••••••••••••••••••••••••••••••
TU 06.11.84    66929 |••••••••••••••••••••••
WD 07.11.84   143026 |•••••••••••••••••••••••••••••••••••••••••••••••
TH 08.11.84   165429 |••••••••••••••••••••••••••••••••••••••••••••••••••••••
FR 09.11.84    68271 |••••••••••••••••••••••
SA 10.11.84     6615 |••
missing days       0 |
MO 12.11.84   322015 |•••••••••••••••••••••••••••••••••••••••••••••••••••••••••••••••••••••••••••••••••••••••••••••••••••••••••••• =322015
TU 13.11.84   281864 |••••••••••••••••••••••••••••••••••••••••••••••••••••••••••••••••••••••••••••••••••••••••••••••••
WD 14.11.84   260958 |•••••••••••••••••••••••••••••••••••••••••••••••••••••••••••••••••••••••••••••••••••••••••
TH 15.11.84   259761 |•••••••••••••••••••••••••••••••••••••••••••••••••••••••••••••••••••••••••••••••••••••••••
FR 16.11.84   234186 |•••••••••••••••••••••••••••••••••••••••••••••••••••••••••••••••••••••••••••••••
SA 17.11.84     5747 |••
missing days       0 |
MO 19.11.84   243472 |••••••••••••••••••••••••••••••••••••••••••••••••••••••••••••••••••••••••••••••••••
TU 20.11.84   120887 |••••••••••••••••••••••••••••••••••••••••
WD 21.11.84    76390 |•••••••••••••••••••••••••
TH 22.11.84   137505 |•••••••••••••••••••••••••••••••••••••••••••••••
FR 23.11.84   117201 |•••••••••••••••••••••••••••••••••••••••
SA 24.11.84     6750 |••
SU 25.11.84    59108 |•••••••••••••••••••
MO 26.11.84   136933 |••••••••••••••••••••••••••••••••••••••••••••••
TU 27.11.84   179337 |•••••••••••••••••••••••••••••••••••••••••••••••••••••••••••
WD 28.11.84   299135 |•••••••••••••••••••••••••••••••••••••••••••••••••••••••••••••••••••••••••••••••••••••••••••••••••••••
TH 29.11.84   278426 |•••••••••••••••••••••••••••••••••••••••••••••••••••••••••••••••••••••••••••••••••••••••••••••
FR 30.11.84   242469 |•••••••••••••••••••••••••••••••••••••••••••••••••••••••••••••••••••••••••••••••••••
---------------------|----|---------|---------|---------|---------|---------|---------|---------|---------|---------|---------|
                     0       30000     60000     90000    120000    150000    180000    210000    240000    270000    300000
```

A.6

EARN/BITNET nodes as of 29 November 84
(Total nodes: 429 of which 17 are unconnected)

	Nodename	University name	Operating System	Type
1	BBADMIN	CUNY - Baruch College Admin.	VM/SP Release 3	Full node
2	BBADMIN2	CUNY - Baruch College Admin.	VM/SP Release 3	Full node
3	BB003	CUNY - Baruch College	VM/SP Release 3	Full node
4	BKLYN	CUNY - Brooklyn College	VM/SP Release 3	Full node
5	BMACADM	CUNY - Manhattan C.C. Academic	VM/SP Release 3	Full node
6	BM002	CUNY - Manhattan C.C. Admin.	VM/SP Release 3	Full node
7	BROWNVM	Brown University Comp. Center	VM/SP Release 3	Full node
8	BX001	CUNY - Bronx Community College	VM/SP Release 3	Full node
9	CCNY	CUNY - City College of NY	VM/SP Release 1	Full node
10	CUNYJES3	City University of New York	MVS/JES3 SP 1.3.3	Full node
11	CUNYVM	City University of New York	VM/SP Release 3 HPO 3.2	Full node
12	CUNYVMS1	CUNY - Graduate Center	VAX/VMS	Full node
13	HUNTER	CUNY - Hunter College	VM/SP Release 3	Full node
14	KB001	CUNY - Kingsborough C.C.	VM/SP Release 3	Full node
15	LEHMAN	CUNY - Lehman College	VM/SP Release 3	Full node
16	NJECNVM	NJ Educ. Comp. Net (NJECN)	VM/SP Release 3	Full node
17	NJECNVS	NJ Educ. Comp. Net (NJECN)	MVS/JES3 SP 1.3.1	Full node
18	NY001	CUNY - NYC Technical College	VM/SP Release 3	Full node
19	QB001	CUNY - Queensborough C.C.	VM/SP Release 3	Full node
20	QUEENS	CUNY - Queens College	VM/SP Release 3	Full node
21	SI001	CUNY - College of Staten Isl.	VM/SP Release 3	Full node
22	YALEADS	Yale University Administrative	VM/SP Release 2	Full node
23	YALECS	Yale University CS Department	Unix	Full node
24	YALEMVS	Yale Univ. Computer Center	MVS/JES2	Full node
25	YALEVM	Yale Univ. Computer Center	VM/SP Release 3	Full node
26	YORK	CUNY - York College	VM/SP Release 3	Full node
27	PSUMVS	Penn. State University	MVS/JES2	Full node
28	PSUPDP1	Penn. State University	Unix-R6	Full node
29	PSUVAX1	Penn. State University	Unix	Full node
30	PSUVM	Penn State/Computer Center	VM/SP Release 3.1	Full node
31	CORNELLA	Cornell U. Computer Services	VM/SP Release 3	Full node
32	CORNELLC	Cornell U. Computer Services	VM/SP Release 3 HPO 3.2	Full node
33	PUCC	Princeton University/Comp. Ctr	VM/SP Release 3 HPO 3.2	Full node
34	CUVMA	Columbia University (CUCCA)	VM/SP Release 2	Full node
35	CUVMB	Columbia University (CUCCA)	VM/SP Release 2	Full node
36	CUVMC	Columbia University	VM/SP Release 3	Full node
37	CUVMD	Columbia University	VM/SP Release 3	Full node
38	ROCKVAX	Rockefeller University	Berkeley Unix 4.1	Full node
39	BOSTONU	Boston Univ./ACC	VM/SP Release 3 HPO 3.2	Full node
40	UCBCMSA	U.C. Berkeley Computer Center	VM/SP Release 2	Full node
41	UCBCMSB	U.C. Berkeley Computer Center	VM/SP Release 3	Full node
42	UCBUNIXA	U.C. Berkeley Computer Center	Berkeley Unix 2.8	Full node
43	UCBUNIXB	U.C. Berkeley Computer Center	Berkeley Unix 2.8	Full node
44	UCBUNIXC	U.C. Berkeley Computer Center	Berkeley Unix 2.8	Full node
45	UCBUNIXD	U.C. Berkeley Computer Center	Berkeley Unix 2.8	Full node
46	UCBUNIXE	U.C. Berkeley Computer Center	Berkeley Unix 2.8	Full node
47	UCBUNIXG	U.C. Berkeley Computer Center	Berkeley Unix 4.2	Full node
48	SFASYS	U. of California at SF	VM/SP Release 3	Full node
49	SFBSYS	U. of California at SF	VM/SP Release 3	Full node
50	MITVMA	MIT - Info. Systems	VM/SP Release 3	Full node
51	CRNLCS	Cornell University CS Dept.	Berkeley Unix 4.2	Full node
52	CUMC	Cornell Univ. Medical College	VM/SP Release 2.1	Full node
53	UORMVS	University of Rochester	MVS/JES2	Full node
54	UORVM	University of Rochester	VM/SP Release 2	Full node
55	SUNYBING	SUNY Binghamton	VM/SP	Full node
56	HARVARDA	Harvard/HCC	VM/SP Release 3	Full node
57	OHSTVMA	Ohio State University / IRCC	VM/SP Release 3	Full node

58	PSUVMS1	Penn State Engin. Computer Lab	VAX/VMS	Full node
59	MAINE	U. of Maine	VM/SP Release 3	Full node
60	UICMVS	U. of Illinois at Chicago	MVS/JES2	Full node
61	UICVM	U. of Ill. at Chicago	VM/SP	Full node
62	UCONNMVS	University of Connecticut	MVS/JES2	Full node
63	UCONNVM	University of Connecticut	VM/SP Release 2	Full node
64	PSUDEC10	Penn State Engin. Computer Lab	DECsystem10 (Tops 10)	Full node
65	YKTVMV	IBM TJ Watson Research Ctr	VM/SP HPO 2.0	Full node
66	YKTVMX	IBM TJ Watson Research Ctr	VM/SP HPO 2.0	Full node
67	YKTVMZ	IBM TJ Watson Research Ctr	VM/SP	Full node
68	GWUVM	George Washington U.	VM/SP Release 2	Full node
69	CUCHEM	Columbia Univ. Chem. Dept.	VAX/VMS	Full node
70	PORTLAND	U. of Southern Maine Portland	VM/SP Release 3	Full node
71	UMDA	Univ. of MD, CSC	VM/SP Release 3	Full node
72	UMDB	Univ. of MD, CSC	VM/SP Release 3	Full node
73	UMDC	Univ. of MD, CSC	VM/SP Release 3	Full node
74	UMDT	Univ. of MD, CSC	VM/SP Release 3	Full node
75	UMD2	Univ. of MD, CSC	Sperry OS 1100 Release 39	Full node
76	UMD7	Univ. of MD, CSC	Sperry OS 1100 Release 39	Full node
77	HARVUNXW	Harvard University	Berkeley Unix 2.9	Full node
78	UCBJADE	U.C. Berkeley Computer Center	Berkeley Unix 4.2	Full node
79	SLACVM	SLAC	VM/SP Release 3 HPO 3.2	Full node
80	UIUCUXC	U of Ill at Urbana/CSO	Berkeley Unix 4.1a	Full node
81	UIUCVMD	U of Ill at Urbana/CSO	VM/SP Release 3	Full node
82	UIUCVME	U of Ill at Urbana/CSO	VM/SP Release 3	Full node
83	BROWNCS	Brown University CS Department	Unix	Full node
84	YKTVMT	IBM TJ Watson Research Ctr	VM/SP HPO/2	Full node
85	UCSFCCA	U. of California at SF	Berkeley Unix 4.2	Full node
86	UOFT01	U. of Toledo	VM/SP Release 2	Full node,
87	STANFORD	Stanford University	MVS/JES2	Full node
88	UMCVMA	University of Missouri (UMC)	VM/SP Release 2	Full node
89	UMCVMB	University of Missouri (UMC)	VM/SP Release 2	Full node
90	UMMVSA	University of Missouri (UM)	MVS/JES2 SP 1.3.2	Full node
91	UMRVMA	University of Missouri (UMR)	VM/SP Release 2.1	Full node
92	UMRVMB	University of Missouri (UMR)	VM/SP Release 2.1	Full node
93	UMSLVMA	University of Missouri (UMSL)	VM/SP Release 2.1	Full node
94	UMVMA	University of Missouri (UM)	VM/SP Release 3.1	Full node
95	YALEVMX	Yale Univ. Computer Center	VM/SP Release 3	Full node
96	VPIVAX3	Virginia Poly Tech (VPI)	VAX/VMS	Full node
97	VPIVAX5	Virginia Poly Tech (VPI)	VAX/VMS	Full node
98	VPIVM1	Virginia Poly Tech (VPI)	VM/SP HPO 3.2	Full node
99	VPIVM2	Virginia Poly Tech (VPI)	VM/SP Release 3	Full node
100	VPIVM3	Virginia Poly Tech (VPI)	VM/SP Release 3	Full node
101	CUCHMB	Columbia Univ. Chem. Dept.	VAX/VMS	Full node
102	SLACCB	SLAC Crystal Ball Exp.	VAX/VMS	Full node
103	DUKE	Duke University	MVS/JES2	Full node
104	NCSUVM	North Carolina State U.	VM/SP Release 3	Full node
105	SLACMAC	SLAC Magnetic Calorimeter	VAX/VMS	Full node
106	TUCC	TUCC	MVS/JES2 SP 1.3.3	Full node
107	UDCVM	Univ. of the Dist. of Columbia	VM/SP Release 3	Full node
108	UNC	Univ. of North Carolina	MVS/JES2 SP 1.1.1	Full node
109	AKRON	University of Akron	MVS/VSPC	Full node
110	VNET	Gateway to VNET (IBM)	VM/SP Release 3	Full node
111	WATSON	Gateway to IBM Research Div.	VM/SP HPO	Full node
112	WUVMA	Washington University	VM/SP Release 2	Full node
113	WUVMD	Washington University	VM/SP Release 2	Full node
114	ANLOS	Argonne National Lab	MVS/JES3	Full node
115	ANLVM	Argonne National Lab	VM/SP Release 2	Full node
116	WVNVM	WV Computer Network (WVNET)	VM/SP Release 3	Full node
117	NCSUVAX	North Carolina State U.	VAX/VMS	Full node
118	ANLCHM	Argonne National Lab	VAX/VMS,ANL NJE	Full node
119	ANLCMT	Argonne National Lab	VAX/VMS,ANL NJE	Full node
120	ANLHEP	Argonne National Lab	VAX/VMS,ANL NJE	Full node
121	ANLIPNS	Argonne National Lab	VAX/VMS,ANL NJE	Full node
122	ANLPHY	Argonne National Lab	VAX/VMS,ANL NJE	Full node

123 HARVUNXU	Harvard University	Berkeley Unix 4.2	Full node
124 HARVUNXA	Harvard University	Berkeley Unix 2.9 BSD	Full node
125 UMDNJPW1	NJ Univ. of Med. & Dentistry	DOS/VSE	Full node
126 UMDNJVM1	New Jersey U. of Medicine	VM/SP	Full node
127 UMDNJVM2	NJ Univ. of Med. & Dentistry	VM/SP	Full node
128 SUVM	Syracuse University	VM/SP Release 3	Full node
129 VPIVAX4	Virginia Poly Tech (VPI)	VAX/VMS	Full node
130 PENNDRLS	U of Penn/DRL Comp. Facility	VM/SP Release 3.1	Full node
131 PENNDRLN	U of Penn/DRL Comp. Facility	VM/SP Release 3.1	Full node
132 CSUOHIO	Cleveland State U./Comp Serv	VM/SP Release 2	Full node
133 FNALVM	Fermilab	VM/SP	Full node
134 UCHIMVS1	UofC Computation Center	MVS/JES2 SP 1.3.1	Full node
135 UCHIVM1	U. of Chicago	VM/SP Release 2	Full node
136 NCSUMAE	NCSU Mech. & Aero. Engr.	VM/SP Release 2	Full node
137 UCCCMVS	University of Cincinnati	MVS/JES2	Full node
138 CRNLNS	Cornell U./Lab of Nuc. Studies	VAX/VMS 3.6	Full node
139 CUCCVX	Columbia University Adm. Dept.	VAX/VMS	Full node
140 OACVAX	UCLA-OAC	Unix	Full node
141 UCLAMVS	UCLA-OAC	MVS/JES2 SP 1.3.3	Full node
142 UCLAVM	UCLA-OAC	VM/SP Release 3	Full node
143 UTKVM1	University of Tennessee	VM/SP Release 2	Full node
144 YALEVAX5	Yale University Computer Ctr.	VAX/VMS 3.6	Full node
145 RICE	Rice University/ICSA	VM/SP Release 2.1	Full node
146 EARNET	IBM SC - Roma	VM/SP Release 3	Full node
147 FARMNTON	U. of Maine Farmington	VM/SP Release 3	Full node
148 SJRLVM1	IBM San Jose Research Ctr	VM/SP HPO 2.0	Full node
149 SJRLVM4	IBM San Jose Research Ctr	VM/SP HPO 2.0	Full node
150 SJRLVS1	IBM San Jose Research Ctr	MVS/SP	Full node
151 SJRVM3	IBM San Jose Research Ctr	VM/SP Release 3 HPO 3.4	Full node
152 SLACMK3	SLAC Mark-III Detector Exp.	VAX/VMS	Full node
153 SUCASE	Syracuse University (CASE)	VM/SP Release 3	Full node
154 UMUC	U. of Maryland U. College	VM/SP Release 3	Full node
155 TAMMVS1	DP Center/Texas A&M University	MVS/JES3 SP 1.3.2	Full node
156 TAMVM1	CSC/Texas A&M University	VM/SP Release 2	Full node
157 TAMVM2	CSC/Texas A&M University	VM/SP Release 3	Full node
158 TAMVXCGF	Texas A&M Engineering Graphics	VAX/VMS	Full node
159 TAMVXEE	Texas A&M Univ. EE Dept.	VAX/VMS	Full node
160 TAMVXME	Texas A&M Univ. ME Dept.	VAX/VMS	Full node
161 TAMVXPHY	Texas A&M Univ. Physics Dept.	VAX/VMS	Full node
162 TAMVXRSC	Texas A&M Remote Sensing Cent.	VAX/VMS	Full node
163 UHUPVM1	U. of Houston	VM/SP Release 2	Full node
164 UMAB	U. of Maryland Med. School	VM/SP	Full node
165 UTA3081	U. of Texas, Austin	VM/SP	Full node
166 UTA4341	U. of Texas, Austin	VM/SP	Full node
167 UTSA158	U. of Texas, San Antonio	VM/SP	Full node
168 UTSA4341	U. of Texas, San Antonio	VM/SP	Full node
169 AUVM	American University	VM/SP Release 2	Full node
170 NER	Florida NE Regional Data Ctr	MVS/XA	Full node
171 NERVM	Florida NE Regional Data Ctr	VM/SP Release 3	Full node
172 FNALVX13	Fermilab	VAX/VMS	Full node
173 ANLNESC	Argonne National Lab	VM/SP Release 2	Full node
174 UCSFHC	U. of Cal.-SF, Hosp. & Clinics	VM/SP Release 2	Full node
175 UNFVM	University of North Florida	VM/SP Release 3	Full node
176 EDUCOM	EDUCOM, Princeton, N.J.	VAX/VMS 3.5	Full node
177 RICECSVM	Rice U. CS Dept.	VM/SP Release 2	Full node
178 YALASTRO	Yale University Astro. Dept.	VM/SP Release 3	Full node
179 MECAN1	U. of Maine Appl. Net	VAX/VMS 3.4	Full node
180 VANDVMS1	Vanderbilt University	VAX/VMS	Full node
181 SLACASP	SLAC ASP Experiment	VAX/VMS	Full node
182 SLACMKII	SLAC Mark-II Detector	VAX/VMS	Full node
183 SLACSLC	SLAC Linear Collider Project	VAX/VMS	Full node
184 SLACTBF	SLAC Test Beam Facility	VAX/VMS	Full node
185 SLACTWGM	SLAC Two-Gamma Experiment	VAX/VMS	Full node
186 SUHEP	Syracuse University (HEP)	VAX/VMS	Full node
187 USCVM	U. Of Southern California	VM/SP Release 3	Full node

188	ASUACAD	Arizona State University	VM/SP	Full node
189	ASUEJS	ASU/ECC	VM/SP Release 2	Full node
190	ASUCADAM	ASU/ECC	VM/SP Release 2	Full node
191	CRNLTHRY	Cornell Univ./Theory Center	Berkeley Unix 4.2	Full node
192	NCSUADM	North Carolina State Univ.	MVS/JES2	Full node
193	SLACNIKH	SLAC 2-Gamma Experiment	VAX/VMS	Full node
194	SLACUCD	SLAC 2-Gamma Experiment	VAX/VMS	Full node
195	SLACUCSD	SLAC 2-Gamma Experiment	VAX/VMS	Full node
196	UCBRUBY	U.C. Berkeley Computer Center	Berkeley Unix 4.2	Full node
197	UCHISTEM	UofC Crewe Laboratory	VM/SP Release 3	Full node
198	UDACSVM	University of Delaware	VM/SP Release 3 HPO 3.2	Full node
199	UIUCVMC	U of Ill at Urbana/Engineering	VM/SP Release 3	Full node
200	CRNLGSM	Cornell Grad. School of Mgmt.	VAX/VMS	Full node
201	WESLYN	Wesleyan University	VAX/VMS	Full node
202	WISCPSLB	Univ. of Wisc., Phys. Sci. Lab	VAX/VMS	Full node
203	WISCPSLA	Univ. of Wisc., Phys. Sci. Lab	VAX/VMS	Full node
204	WISCVM	CS Dept. - Univ. Wisc. at Mad.	VM/SP Release 3	Full node
205	BROWNCOG	Brown Univ. Ctr - Cogn. Sci.	VAX/VMS 3.4	Full node
206	CDC205	Colorado State University	CDC	Full node
207	CSU	Colorado State University	NJEF	Full node
208	VPICS1	Virginia Polytechnic Institute	VAX/VMS	Full node
209	IBACSATA	IBACSATA - Bari	VM/SP Release 1.3	Full node
210	ICINECA	ICINECA - Bologna	VM/SP	Full node
211	NEUVMS	NE Univ. Dept. of Physics	VAX/VMS	Full node
212	OHSTVMB	Ohio State University, CAD/CAM	VM/SP Release 3	Full node
213	UMASS	University of Massachusetts	NOS 2.2 Level 602	Full node
214	EBOUB012	Universidad de Barcelona	VM/SP Release 3	Full node
215	EEARN	IBM Madrid Scientific Center	VM/SP Release 3	Full node
216	EMDUAM11	Universidad Autonoma - Madrid	VM/SP Release 3	Full node,
217	EMDUPM11	Universidad Politec. - Madrid	VM/SP Release 3	Full node
218	FNAL	Fermilab	VAX/VMS	Full node
219	FNALA	Fermilab	VAX/VMS	Full node
220	FSU	Florida State University	MVS/JES2	Full node
221	UTKVX1	University of Tennessee	VAX/VMS	Full node
222	HAIFAUVM	Haifa University	VM/SP	Full node
223	ISRAEARN	IBM Israel SC - Haifa	VM/SP Release 2	Full node
224	JHUVM	Johns Hopkins University	VM/SP Release 3	Full node
225	TAUNIVM	Tel Aviv University	VM/SP Release 1	Full node
226	TAUNOS	Tel Aviv University	NOS 2.1	Full node
227	TAURUS	Tel Aviv University	Berkeley Unix 4.2	Full node
228	TECHMVS	Technion - Haifa	MVS/JES2 (guest)	Full node
229	TECHNION	Technion - Haifa	VM/SP Release 2	Full node
230	TECHUNIX	Technion - Haifa	Unix	Full node
231	WEIZMANN	Weizmann Institute of Science	VM/SP Release 2	Full node
232	WISDOM	Weizmann Inst. Dept. of Math.	Unix	Full node
233	UMRVMC	University of Missouri (UMR)	VM/SP Release 2.1	Full node
234	UCLASSCF	UCLA Social Sciences Facility	VM/SP Release 3	Full node
235	USCVAXQ	U. Of Southern California	VAX/VMS	Full node
236	MUVMS1	Marshall University	VAX/VMS 3.5	Full node
237	UMKCVAX1	University of Missouri (UMKC)	VAX/VMS 3.6	Full node
238	HARVHEP	Harvard/High Energy Physics	VAX/VMS	Full node
239	MITECCF1	MIT - East Campus Comp. Facili	VM/SP Release 3	Full node
240	CITHEX	Caltech High Energy Physics	VAX/VMS	Full node
241	HAMLET	Caltech	VAX/VMS	Full node
242	UMEE	Univ. of MD	Unix	Full node
243	CANADA01	University of Guelph	VM/SP Release 3	Full node
244	CARLETON	Carleton University		Full node
245	CATE	Center for Adv. Tech. Educ.	VM/SP	Full node
246	CENCOL	Centennial College	VM/SP	Full node
247	HUMBER	Humber College	VM/SP Release 3	Full node
248	MCGILL1	McGill University	VM/SP Release 2	Full node
249	MCMASTER	McMaster University	VAX/VMS 3.6	Full node
250	NSNCC	Louisiana St. U., Baton Rouge	MVS/JES2 SP 1.3.3	Full node
251	NSNCCVM	Louisiana St. U., Baton Rouge	VM/SP Release 3	Full node
252	QUCDN	Queens University	VM/SP Release 3 HPO 3.2	Full node

253 QUCIS	Queens University	Unix	Full node
254 RYERSON	Ryerson Polytechnic	VM/SP	Full node
255 TSSNRC00	NRC, Ottawa	TSS/370	Full node
256 UOGUELPH	University of Guelph	VM/SP Release 3	Full node
257 UOGVAX2	University of Guelph	Berkeley Unix	Full node
258 UOTTAWA	University of Ottawa	VM/SP Release 3	Full node
259 UTORONTO	University of Toronto	VM/SP	Full node
260 WATACS	Univ. of Waterloo, ACS	VM/SP Release 3	Full node
261 WATARTS	Univ. of Waterloo, Arts VAX	Berkeley Unix 4.2	Full node
262 WATCSG	Univ. of Waterloo, CSG	VM/SP Release 3	Full node
263 WATDCS	Univ. of Waterloo, DCS	VM/SP Release 3 (SSI)	Full node
264 WATDCSU	Univ. of Waterloo, DCS	Berkeley Unix 4.2	Full node
265 WATMNET	Univ. of Waterloo, MICRONET	VM/SP Release 3	Full node
266 WATROSE	Univ. of Waterloo, Math	Berkeley Unix 4.2	Full node
267 YORKVM1	York University	VM/SP Release 3	Full node
268 UFFSC	UF Faculty Support Center	VM/SP Release 3	Full node
269 MITLNS	MIT - Lab. for Nuc. Sci.	VAX/VMS 3.6	Full node
270 UCF1VM	Univ. of Central Florida	VM/SP Release 3	Full node
271 UCF2VM	Univ. of Central Florida	VM/SP Release 3	Full node
272 UFENG	UF College of Engineering	VM/SP Release 3	Full node
273 EBOUB011	Universidad de Barcelona	VM/SP Release 3	Full node
274 ISUMVS	Iowa State University	MVS/JES2	Full node
275 UIAMVS	University of Iowa	MVS/JES2	Full node
276 BOSTCIML	Boston Univ./CIML	VM/SP Release 3 HPO 3.2	Full node
277 CUCEVX	Columbia U. - Civil Eng.	VAX/VMS	Full node
278 NYSPI	NY Psychiatric Institute	VM/SP Release 3	Full node
279 TUCCVM	TUCC	VM/SP Release 3	Full node
280 UIUCHEPG	U of Ill at Urbana/HEP	VAX/VMS	Full node
281 VANDVM1	Vanderbilt University	VM/SP Release 3	Full node
282 BITNIC	BITNET Network Support Center	VM/SP Release 3	Full node
283 IRISHMVS	Univ. of Notre Dame Comp. Ctr	MVS/JES2 SP 1.3.3	Full node
284 IRISHVM	Notre Dame PC Lab	VM/SP Release 2	Full node
285 SLACSLD	SLAC SLD Detector	VAX/VMS	Full node
286 WISCMSE	Univ. of Wisc., Eng. Dept.	VM/SP Release 3	Full node
287 MCGILL2	McGill University	VM/SP Release 2	Full node
288 MCGILLA	McGill University	MUSIC	Full node
289 SITVXA	Stevens Institute of Tech.	VAX/VMS	Full node
290 JHUP	Johns Hopkins Univ. HEP	VAX/VMS 3.6	Full node
291 QUCDNCMC	Queens University CMEC	VAX/VMS	Full node
292 PENNHEP1	U of Penn/HEP	VAX/VMS 3.5	Full node
293 PENNLRSM	U of Penn/LRSM	VAX/VMS 3.6	Full node
294 CITCSSTV	Caltech Comp. Support Services	VAX/VMS	Full node
295 SITVXB	Stevens Institute of Tech.	VAX/VMS	Full node
296 UMKCVAX2	University of Missouri (UMKC)	VAX/VMS 3.6	Full node
297 UMKCVAX3	University of Missouri (UMKC)	VAX/VMS 3.7	Full node
298 YORKVM2	York University	VM/SP Release 3	Full node
299 YULEO	York University	VAX/VMS 3.6	Full node
300 YUORION	York University	VAX/VMS 3.6	Full node
301 VASSAR	Vassar College	VAX/VMS	Full node
302 FNALCDF	Fermilab	VAX/VMS	Full node
303 BNL	Brookhaven National Laboratory	Berkeley Unix	Full node
304 FNALBSN	Fermilab	VAX/VMS	Full node
305 HBUNOS	Hebrew University	NOS 2.2	Full node
306 ICNUCEVM	Nat'l U. Comp. Ctr - Pisa	VM/SP	Full node
307 ICNUCEVS	Nat'l U. Comp. Ctr - Pisa	MVS/JES2	Full node
308 IFIIDG	Inst. Doc. Guiridica - Firenze	VM/SP	Full node
309 IMIBOCCO	Universita' Bocconi - Milano	VM/SP	Full node
310 IMISIAM	Instituto di Fisica Cosmica	VM/SP	Full node
311 IPACUC	Universita' di Palermo	VM/SP	Full node
312 IPIINFN	INFN - S. Piero a Grado - Pisa	VM/SP	Full node
313 IPIVAXIN	INFN - S. Piero a Grado - Pisa	VAX/VMS	Full node
314 IRMCRA	Ist. Ricerche Aerospaz. - Roma	VM/SP	Full node
315 IRMIAS	Ist. Astro. Spaz. (CNR) - Roma	VM/SP	Full node
316 ITOIMGC	Ist. Meteo. Colonnetti- Torino	VM/SP	Full node
317 NNOMED	Louisiana St. U., Med. Ctr.	MVS/XA	Full node

318	NNOMEDV	Louisiana St. U., Med. Ctr.	VM/SP Release 2	Full node
319	RLG	Stanford University/RLG	MVS/JES2	Full node
320	UORDBV	University of Rochester	VAX/VMS 3.5	Full node
321	WATMTA	Univ. of Waterloo, MTA	VAX/VMS 3.6	Full node
322	YUGEMINI	York University	VAX/VMS 3.6	Full node
323	YUURSA	York University	VAX/VMS 3.6	Full node
324	YUVENUS	York University	VAX/VMS 3.6	Full node
325	BIBLIO31	Centennial College	VM/SP	Full node
326	CALTECH	Caltech	VAX/VMS	Full node
327	SUNYABVA	SUNY Buffalo	VAX/VMS	Full node
328	SUNYABVB	SUNY Buffalo	VAX/VMS	Full node
329	ANLEES	ANL EES	VAX/VMS	Full node
330	ANLEL	ANL Electronics Division	VAX/VMS	Full node
331	ANLMCS	ANL MCS	Berkeley Unix 4.1a	Full node
332	IBAUNIV	Bari University	VM/SP	Full node
333	ICSATAXA	CSATA - Bari	MVS/XA	Full node
334	CBEBDA3T	Berne University	MVS/JES3	Full node
335	CBEBDA3C	Berne University	MVS/JES3	Full node
336	CEARN	CERN	VM/SP	Full node
337	CERNVM	CERN	VM/SP	Full node
338	CGEUGE51	University de Geneve, Suisse	VAX/VMS	Full node
339	CLSEPF51	ETH Lausanne	VAX/VMS	Full node
340	CRVXALFB	CERN	VAX/VMS	Full node
341	CRVXALTP	CERN ALEPH TPC	VAX/VMS	Full node
342	CRVXDEV	CERN OC Developement	VAX/VMS	Full node
343	CZHRZU1A	Zurich University	VM/SP	Full node
344	CZHRZU2B	Zurich University	MVS/JES2	Full node
345	DAAFHT1	FHS Aalen	VM/BSEPP	Full node
346	DBNGMD21	GMD Bonn	MVS/SP 2.1.1	Full node'
347	DBNRHRZ1	RHRZ Universitaet Bonn	VM/SP Release 3	Full node
348	DBNRHRZ2	RHRZ Universitaet Bonn	MVS/SP 1.3.2	Full node
349	DBNUAMA1	Inst. Angewandte Mathemathik	VM/SP Release 3	Full node
350	DBNUOR1	Oekonom. und Operatns Research	VM/SP Release 2	Full node
351	DBSNRVO	TU Braunschweig	DIETZ XOS 5.8	Full node
352	DBSTU1	TU Braunschweig	VM/SP Release 3	Full node
353	DBOHMI41	HMI - Berlin	Siemens BS3000 E40	Full node
354	DBOTUI11	TU Informatik - Berlin	VM	Full node
355	DBOTUM11	TU Maschinen - Berlin	VM	Full node
356	DBOTUS11	TU Schiff u. Meer Berlin	VM	Full node
357	DBOZIB21	ZIB Berlin	MVS/SP	Full node
358	DCZRZTUO	TU Clausthal	CGK BS 3 19.36	Full node
359	DDADVS1	TH Darmstadt FB Informatik	VM/SP Release 3	Full node
360	DDAGMD11	GMD Darmstadt	VM/SP Release 2.1	Full node
361	DDAGSI3	GSI Darmstadt	MVS/SP 2.1.1	Full node
362	DDATHD21	TH Darmstadt	MVS/SP 1.3.3	Full node
363	DDOHRZ11	Universitaet Dortmund	VM/SP	Full node
364	DEARN	EARN Central Node Germany	VM/SP	Full node
365	DERRZE1	Univ. Erlangen - Nuernberg	VM/SP Release 3	Full node
366	DEOHRZ1A	Universitaet Essen	VM/SP Release 3	Full node
367	DEOWTZ1A	Uni Klinik Essen, Tumorzentrum	VM/SP Release 3	Full node
368	DFVLROP1	DFVLR	VM/SP Release 1.1	Full node
369	DGAIPP1S	MPI fuer Plasmaphysik	VM/SP Release 2	Full node
370	DGOGWDO1	Uni. Max-Planck-Gesellschaft	Sperry OS 1100	Full node
371	DHAFEU11	Fernuniversitaet Hagen	VM/SP Release 2	Full node
372	DHDDKFZ1	Krebsforschungzentrum HD	VM/SP Release 3	Full node
373	DHDIBM1	IBM WZH Heidelberg	VM/SP Release 3	Full node
374	DHDIHEP1	Hochenergiephysik Heidelberg	VM/SP Release 3	Full node
375	DHDURZ2	Universitaet Heidelberg	MVS/SP 1.3.3	Full node
376	DHHDESY3	DESY	MVS/SP	Full node
377	DHNFHS1	Fachhochschule Heilbronn	VM/SP Release 3	Full node
378	DHVRRZ01	Universitaet Hannover	CDC NOS/BE 564	Full node
379	DJUKFA11	KFA Juelich - ZAM	VM/SP	Full node
380	DJUKFA21	KFA Juelich - ZAM	MVS/SP	Full node
381	DJUKFA51	KFA Juelich - SNQ	VAX/VMS 3.4	Full node
382	DKAFHS1	Fachhochschule Karlsruhe	VM/SP Release 3	Full node

383	DKAKFK3	Kernforschungszent. Karlsruhe	MVS/SP 1.3.1	Full node
384	DKAUNI11	Universitaet Karlsruhe	VM/SP Release 3	Full node
385	DKAUNI12	Univ. Karlsruhe, Info 3	VM/SP Release 3	Full node
386	DKAUNI13	Univ. Karlsruhe, Info 3	VM/SP Release 3	Full node
387	DKAUNI46	Universitaet Karlsruhe	Siemens BS3000 E40 JES2	Full node
388	DKAUNI48	Universitaet Karlsruhe	Siemens BS3000 E40 JES2	Full node
389	DKOZA1	ZA - Koeln	VM/SP	Full node
390	DMARUM8	Universitaet Mannheim	Siemens BS2000 NJE	Full node
391	DMSWWU1A	Uni Muenster	VM/SP Release 3	Full node
392	DMSWWU2B	Uni Muenster	MVS/SP 1.3.1	Full node
393	DMOMPF11	MPI psychologische Forschung	VM/SP Release 3	Full node
394	DMOMPI11	MPI Physik u. Astrophysik	VM/SP Release 2.1	Full node
395	DOLUNIO	Uni Oldenburg	CGK BS 3 19.36	Full node
396	DOSUNI	Uni Osnabrueck	CGK BS 3 19.36	Full node
397	DSOIKE51	Inst. f. Kernenergetik	VAX/VMS	Full node
398	DSOMPA51		VAX/VMS	Full node
399	DSORUS51	Rechenzentrum Univ., Stuttgart	VAX/VMS	Full node
400	DSORUS1I	Uni Stuttgart	VM/SP Release 2.1	Full node
401	DSORUS1P	Uni Stuttgart	VM/SP Release 2.1	Full node
402	DTUZDV1	Uni Tubingen	VM/SP Release 3	Full node
403	GEN	CERN	MVS/JES2	Full node
404	HNYURC11	Uni Nijmegen	VM/SP Release 2	Full node
405	ASUIC	Arizona State University	VM/SP Release 3	Full node
406	CLUTX	Clarkson University	Unix	Full node
407	CLVM	Clarkson University	VM/SP Release 3	Full node
408	CLVMS	Clarkson University	VAX/VMS	Full node
409	HARVLAW1	Harvard University	VAX/VMS	Full node
410	HARVSC3	Harvard University	VAX/VMS	Full node
411	HARVSC5	Harvard University	VAX/VMS	Full node
412	UOFT02	University of Toledo	VAX/VMS	Full node
413	FEARN	HEC, Paris		Unconnect
414	FREMP11	Ecole des Mines, Paris		Unconnect
415	FRENS11	Ecole Normale Sup, Paris		Unconnect
416	FRMOP11	Cnusc, Montpellier		Unconnect
417	FRONI51	Observatoire, Nice		Unconnect
418	FRORS31	Circe, Orsay		Unconnect
419	GEARN	Great Britain EARN		Unconnect
420	NTSU	North Texas State University		Unconnect
421	REARN	Ireland EARN		Unconnect
422	SEARN	Sweden EARN		Unconnect
423	SMU	Southern Methodist Univ.		Unconnect
424	TTU	Texas Tech University		Unconnect
425	TTUHSC	Texas Tech U., Health Sci. Ctr		Unconnect
426	UCCCVM1	University of Cincinnati	VM/SP Release 3	Unconnect
427	UTCVM	U. of Tennessee, Chattanooga	VM/SP Release 3	Unconnect
428	UTDALVM1	U. of Texas at Dallas/ACC	VM/SP	Unconnect
429	UTELP	U. of Texas at El Paso		Unconnect

Gruppenkommunikation im MHS-Kontext

Lothar Wosnitza
Universität Düsseldorf
Rechenzentrum

Übersicht

Im Rahmen des Deutschen Forschungsnetzes soll ein Message-System auf
der Basis der entsprechenden CCITT-MHS-Empfehlungen X.400 ff errichtet
werden. Im derzeitigen Stand der CCITT-Definitionen wird jedoch ein
für den Forschungsbereich sehr wichtiger Aspekt, nämlich die Unterstützung der Kommunikation in Gruppen, nicht abgedeckt. Das vorliegende Papier stellt die für das DFN entwickelten Modelle zur Unterstützung der Gruppenkommunikation und Möglichkeiten zur Einbettung in das
zugrundeliegende MHS-Modell vor. Darüberhinaus sollen einige spezifische Probleme bei der Realisierung aufgezeigt und Lösungsmöglichkeiten
diskutiert werden.

1 Einleitung

Ein wichtiges Teilvorhaben bei der Errichtung des Deutschen
Forschungsnetzes (DFN) ist die Bereitstellung eines Message-Systems
auf der Basis der CCITT-Empfehlungen X.400 ff [1] für "Message Handling Systems (MHS)".

In der vorliegenden Form berücksichtigen die MHS-Empfehlungen ausschließlich die Erfordernisse der individuellen Kommunikation, die
primär dadurch gekennzeichnet ist, daß ein einzelner Benutzer einen
Brief an einen oder mehrere speziell aufgeführte andere Benutzer des
Kommunikationsdienstes schickt. Eine im Vorfeld des DFN durchgeführte
Studie zur Funktionalität und Bewertung von Message-Systemen (vgl.
[2, 3]) wie auch praktische Erfahrungen im Umgang mit existierenden
Message-Systemen (vgl. [8, 9]) zeigen jedoch deutlich die Bedeutung
der Gruppenkommunikation für Message-Systeme: Einerseits steht im
Forschungsbereich die Arbeit in mehr oder weniger fest zusammengesetz-
ten Gruppen im Vordergrund, andererseits hat das Message-System als
Medium gerade im Bereich der Gruppenkommunikation seine besonderen
Stärken.

Gruppenkommunikation wird in verschiedenen existierenden Message-
Systemen in unterschiedlicher Ausprägung unterstützt. In den meisten
Systemen sind zumindest einfache Verteiler-Funktionen ("distribution
list") vorhanden. Komplexere Gruppenkonstellationen werden in einzel-
nen, meist zentral organisierten Systemen unterstützt. Ein umfassendes
Modell zur Unterstützung der Gruppenkommunikation in einer verteilten
Umgebung (Verbund von Message-Systemen) hat das GILT-Projekt der
Europäischen Gemeinschaft erarbeitet (vgl. [6, 7]). Diskussionen
über erforderliche Ergänzungen zu den MHS-Empfehlungen werden zur Zeit
auch in IFIP WG 6.5 geführt (vgl. z.B. [10, 11]).

Für das DFN wurde in der Konzeptionsphase ein konkretes Modell zur Beschreibung der für die Gruppenkommunikation erforderlichen Mechanis= men erarbeitet (vgl. [4, 5]), das als Ergänzung zur MHS= Funktionalität gleich in der ersten Realisierungsstufe des DFN- Message=Verbunds mit einbezogen werden soll. Dabei wird der MHS=Dienst dahingehend erweitert, daß eine Benutzergruppe geschlossen über einen Sender=/Empfängernamen (O/R=Name) angesprochen werden kann. Die Eigen= schaften der Mitglieder einer Gruppe wie auch Außenstehender sollen so einstellbar sein, daß gebräuchliche Gruppenkonstellationen wie

* einfache Verteiler,

* offene oder eingeschränkte Diskussionsrunden oder

* geschlossene Konferenzen und Teams

nachgebildet werden können.

Wünschenswert ist darüberhinaus ein möglichst hohes Maß an Freiheit in der Modellierung von Gruppenkonstellationen, so daß die bereitste= henden Konzepte auch für speziellere Anwendungen genutzt werden kön= nen.

Ein interessanter Aspekt in zentralen Konferenz=Systemen ist die ge= meinsame Archivierung der Gruppenbeiträge. Archivierung ist insbeson= dere im Zusammenhang mit Anwendungen wie "Schwarzes Brett" ("bulletin board") erforderlich. Außerdem kann z.B. neuen Mitgliedern in Kon= ferenzen die Möglichkeit geboten werden, sich nachträglich in den Dis= kussionsstand einzuarbeiten.
Leider läßt sich die Art der Realisierung in zentralen Systemen nicht problemlos auf eine verteilte Umgebung übertragen. Bereits existieren= de MHS=Realisierungen (wie EAN) zeigen jedoch, daß der Benutzerkomfort mit Hilfe einer lokalen Archivierungskomponente im UA bereits deutlich verbessert werden kann.

Noch im Bereich der Forschung befinden sich Ideen zur Unterstützung komplexerer Kommunikationsabläufe innerhalb von Gruppen, die unter dem Schlagwort "Anwendungsprozeduren" zusammengefaßt werden können. Ein Beispiel dazu ist die "programmierbare Konferenz". Hier sind Hilfsmit= tel zur automatischen Auswertung von Abstimmungen oder Umfragen oder zur Koordinierung oder Steuerung von Abläufen wie etwa "wenn zu diesem Beitrag 5 Kommentare eingegangen sind, dann veranlasse Person XY zu einer Stellungnahme" gemeint. Ein weiterer äußerst interessanter Dienst könnte in der Unterstützung der gemeinsamen Erarbeitung von Dokumenten innerhalb von Gruppen ("joint editing") geboten werden. In beiden Bereichen liegen derzeit jedoch noch keine ausreichend detail= lierten Lösungsvorschläge vor.

2 Aspekte der Gruppenkommunikation

Aus dem Blickwinkel einer möglichen Realisierung von Funktionskompo‌
nenten zur Unterstützung der Gruppenkommunikation werden im folgenden
einige relevante Aspekte und Probleme näher betrachtet.

Grundfunktionen an der Benutzerschnittstelle

Funktionen zur Unterstützung der Gruppenkommunikation müssen natür‌
lich an der Benutzerschnittstelle des Message-Systems verfügbar sein,
wobei sich teilweise eine Entsprechung zu den bereits für die indivi‌
duelle Kommunikation bereitstehenden Funktionen findet.

Generell sind Operationen erforderlich, mit deren Hilfe sich die Be‌
nutzer über die existierenden Gruppen und deren Eigenschaften wie Na‌
me, Beschreibung, Peitrittsmöglichkeiten und momentane Zusammensetzung
informieren können. Dazu muß der Directory-Dienst neben den Informa‌
tionen über einzelne Benutzer auch entsprechende Informationen über
Gruppen bereitstellen.

Je nach Gruppentyp liegt die Verantwortung für die Zuordnung eines
Benutzers zur Gruppe beim Benutzer selbst ("offene" Gruppen) oder bei
einem anderen, für die Organisation der Gruppe zuständigen Benutzer
("geschlossene" Gruppen). Dementsprechend werden für offene Gruppen
Funktionen zur Pegelung der eigenen Mitgliedschaft (Beitritt, Modifi‌
kation, Austritt) benötigt, während die Mitgliedschaft bei geschlosse‌
nen Gruppen über Operationen zum allgemeinen Management von Gruppen
kontrolliert werden kann (s.u.).

Das Erzeugen bzw. die Verarbeitung von Gruppenbeiträgen schließlich
kann analog zum Senden und Empfangen von individuellen Nachrichten
über die bereits im MHS vorhandenen Funktionen realisiert werden.

Verteilung

Jedem O/R-Namen einer Benutzergruppe ist die Liste der (momentanen)
Mitglieder zugeordnet. Alle Beiträge, die an die Gruppe gerichtet
sind, werden den als Empfänger gekennzeichneten Mitgliedern dieser Li‌
ste automatisch zugesandt. Der Einsender des Beitrags ist dabei ge‌
gebenenfalls auszuklammern, damit er seinen Beitrag nicht zurück‌
erhält.

Grundsätzlich besteht die Möglichkeit, daß eine Gruppe als Mitglied
einer anderen Gruppe auftritt, damit Untergruppen bzw. parallele Grup‌
pen mit gegenseitiger Mitgliedschaft gebildet werden können. Die letz‌
tere Konstellation ist insbesondere zur Optimierung des Dienstes in
Domain-übergreifenden Gruppen interessant.
In der Realisierung muß dafür Sorge getragen werden, daß endlose Ver‌
teilungen von Beiträgen bei einer zyklischen Gruppenverkettung ausge‌
schlossen sind. Dazu sollte der Gruppenname nach der Expansion der
Mitgliederliste in Kopf der Nachricht erhalten bleiben, so daß Zyklen

erkannt und gestoppt werden können.

Bei der Zustellung der Beiträge ist zu berücksichtigen, daß der Bezug des Beitrags zur Gruppe beim Empfänger erkenntlich ist, damit der umfassende Kontext des Beitrags erhalten bleibt. Auch zu diesem Zweck sollte der Gruppenname in der Nachricht erhalten bleiben und dem End-Empfänger geeignet präsentiert werden.

Ein weiteres Problem im Zusammenhang mit der Verteilung ist die Handhabung von Bestätigungen. Verschiedene Bestätigungsebenen sind für einen Beitrag denkbar:

- angekommen bei der Gruppe (zur Verteilung),

- angekommen bei den direkten Mitgliedern der Gruppe,

- angekommen auch bei den indirekten "Mitgliedern" (rekursiv).

Lediglich die Realisierung der ersten Ebene ist unproblematisch; Möglichkeiten zur Realisierung der anderen Typen erfordern weitere Untersuchungen.

Lokale Archivierung

Besonderes Merkmal der Gruppenkommunikation ist, daß eine einzelne Nachricht in der Regel nicht für sich allein verständlich ist, sondern, daß sie in einen höheren Kontext eingebettet ist, der durch eine Menge von aufeinander aufbauenden oder zueinander in Bezug stehenden Nachrichten gebildet wird. Der übergeordnete Kontext wird dabei durch den Kommunikationsgegenstand der Gruppe selbst gebildet. Darüberhinaus kann es Unterthemen oder Randgespräche innerhalb der Gruppe geben.

Das Message-System sollte dem Benutzer bei der Erkennung und Aufarbeitung dieser umfassenden Zusammenhänge weitgehend behilflich sein. Grundvoraussetzung dafür ist, daß alle ein- und ausgehenden Nachrichten gruppenbezogen archiviert und später geeignet zugänglich gemacht werden. Dazu muß der Benutzer seine Arbeit auf die Beiträge einer Gruppe konzentrieren können ("select <Gruppe>") und komfortable Funktionen zur Bezugsverfolgung und zum allgemeinen Suchen durch Vorgabe von Nachrichtenattributen oder Stichworten im Text zur Hand haben.

In X.400 wird die Archivierung als lokale Funktion des UA angesehen und liegt deshalb außerhalb der Empfehlungen. Wenn allerdings höhere Funktionen in Richtung Gruppenkommunikation angestrebt werden, muß dieser Komponente mehr Aufmerksamkeit geschenkt werden. Daher wird die lokale Archivierung der Nachrichten in den UAs als besonders wichtige Implementierungsvorgabe angesehen.

Gruppenmanagement

Das Einrichten einer Gruppe besteht aus der Bereitstellung eines O/R-Namens, einer Beschreibung des Gegenstands der Diskussion innerhalb der Gruppe, dem Eintrag der initialen Mitglieder (zusammen mit ihren Rechten innerhalb der Gruppe) sowie die Einstellung der initialen Berechtigungen der anderen MHS-Benutzer in Bezug auf diese Gruppe.

Zum weiteren Management einer Gruppe werden Operationen benötigt, mit deren Hilfe sich die Eigenschaften der Gruppe an aktuelle Erfordernisse anpassen lassen. Dazu gehören

- Ändern des Namens

- Ändern der Beschreibung

- Neuaufnehmen, Ändern und Löschen von Mitgliedereinträgen

- Ändern der Zugriffsberechtigungen generell oder speziell für einzelne Benutzer

- Auflösen der Gruppe

Dienstleistungen zur Einrichtung von Gruppen und zum Gruppenmanagement sollten an der Benutzeroberfläche des MHS verfügbar sein.

Konsistenzerhaltung

Immer wenn ein Name im Message-System gelöscht wird, muß er auch in allen Gruppen gelöscht werden, in denen er als Mitglied verzeichnet ist. Ohne diese Vorkehrung würden die Mitgliederlisten allmählich mit ungültigen Namen "verschmutzt", was zu einer unnötigen Anhäufung von Fehlersituationen führt.
Die "Bereinigung" der Mitgliederlisten wird am besten im Zuge der Verteilung der Beiträge durchgeführt: Treten bei einzelnen Mitgliedern Zustellungsfehler auf, wird der Gruppenorganisator informiert, damit dieser entsprechende Korrekturen vornehmen kann.

Autorisierung

Die Rechte der einzelnen Benutzer in Bezug auf die Gruppe müssen sowohl für Mitglieder als auch für Nicht-Mitglieder der Gruppe geregelt sein.

Spezifisch für jedes einzelne Mitglied sollte einstellbar sein, wer Beiträge in die Gruppe einbringen darf ("Schreiben"), wer Beiträge automatisch zugestellt bekommt ("Empfangen"), wer sich über die Mitglieder der Gruppe informieren darf ("Listen") und wer die Gruppe organisiert, d.h. wer Gruppenmanagement-Operationen ausführen darf ("Organisieren").

Für Nicht-Mitglieder sollte die Möglichkeit, der Gruppe aus eigener Initiative beizutreten, einstellbar sein. Grob kann man Gruppen in "offen" und "geschlossen" einteilen, d.h. jeder bzw. niemand kann sich selbst als Mitglied eintragen. Wünschenswert ist aber auch die Möglichkeit, eine Menge potentieller Mitglieder zu klassifizieren, etwa die Mitglieder einer Liste gesondert aufgeführter Gruppen oder Benutzer, deren O/R-Namen eine gewisse Bedingung erfüllen (z.B. die Zugehörigkeit zu einer bestimmten Organisation). Bei offenen oder beschränkten Gruppen sind zusätzlich die möglichen Rechte im Falle einer Mitgliedschaft anzugeben, insbesondere für "Schreiben", "Empfangen" und "Listen".
Darüberhinaus sollten die Informationsmöglichkeiten über die Gruppe

für Außenstehende abgestuft einstellbar sein, so daß geregelt werden
kann, ob eine Gruppe überhaupt von außen sichtbar ist ("verborgen"),
ob man sich über den Kommunikationsgegenstand der Gruppe informieren
kann (Name und Beschreibung der Gruppe sind "sichtbar"), ob man darüberhinaus Einblick in die Mitgliederliste erhält ("transparent") oder
ob man gar Beiträge einbringen kann, ohne Mitglied zu sein ("schreibbar").

   Zuletzt ist noch der Fall zu betrachten, daß gewisse Mitglieder einer Gruppe ihrerseits Gruppen sind. Dabei ist festzulegen, ob sich die
Rechte der Gruppe auf alle ihre Mitglieder (und gegebenenfalls rekursiv auch auf alle indirekten Mitglieder) beziehen oder lediglich auf
die Gruppe als Einheit. Wegen der unübersehbaren Probleme bei der Realisierung der ersten Möglichkeit ist der zweite Ansatz zu bevorzugen,
zumal diese Lösung auch vom praktischen Standpunkt aus durchaus
akzeptabel ist.
Sinnvolle Konstellationen solcher "Gruppen in Gruppen" werden in der
Regel auf gegenseitiger Mitgliedschaft beruhen. Denkbar ist beispielsweise eine Konstellation, in der eine Gruppe A immer über die Vorgänge
in einer anderen Gruppe B informiert, an der dortigen Diskussion
jedoch nicht aktiv beteiligt werden soll. A müßte dann in B als emp-
fangendes Mitglied, B in A als Mitglied mit Schreibberechtigung verzeichnet sein.

Abrechnung

   Je nach Anwendung der Konzepte für die Gruppenkommunikation sind unterschiedliche Verfahren und Strategien zur Abrechnung der Grup-
pendienste denkbar, wobei prinzipiell die Sender von Beiträgen, die
Empfänger vor Beiträgen und die Gruppe als Einheit über ihren Organisator belastet werden können.

   Anzustreben ist, jedes sinnvolle Abrechnungsverfahren zu ermöglichen. Es ist jedoch zu bedenken, daß die Belastung der Sender oder
Empfänger einen hohen administrativen Aufwand erfordert, solange auf
Protokollebene keine ausreichende Unterstützung gegeben wird. Dabei
ist weniger die Erfassung der Abrechnungsdaten als vielmehr die In-
rechnungstellung problematisch, da die Mitglieder der Gruppe im allgemeinen auf eine Menge unterschiedlicher PRMDs verteilt sind.

   Technisch unproblematisch und für den Forschungsbereich ausreichend
ist, wenn immer die Gruppe als Ganzes für die Kosten der Verteilung
und der Verwaltung der Gruppe aufkommt. Für das Gruppenkonto kann dann
der Organisator der Gruppe verantwortlich gemacht werden.

3 Modellierung von Gruppenkonstellationen

Die Spezifikation einer Gruppe, also die Festlegung aller ihrer
Eigenschaften, läßt sich am besten anhand einer abstrakten Datenstruk=
tur vornehmen. Zur Erfüllung der im vorigen Abschnitt aufgeführten An=
forderungen könnte diese Datenstruktur wie in Bild 1 aussehen:

```
+========================================+================================================+
| Gruppenname:                           | <O/R-Name>                                     |
+========================================+================================================+
| Beschreibung:                          | <freier Text>                                  |
+========================================+================================================+
| Externe Erscheinung:                   |                                                |
|   Potentielle Mitgliedschaft:          | (geschlossen,                                  |
|                                        |  beschränkt auf <Bedingung>                    |
|                                        |               für <Zugriffsrechte>,            |
|                                        |  offen für <Zugriffsrechte>)                   |
|   Sichtbarkeit:                        | (verborgen,                                    |
|                                        |  sichtbar,                                     |
|                                        |  transparent,                                  |
|                                        |  schreibbar)                                   |
+========================================+================================================+
| Mitgliederliste mit                    |                                                |
|   Einträgen:                           | <O/R-Name>, <Zugriffsrechte>                   |
+========================================+================================================+


        <Zugriffsrechte> ::= set of   (Schreiben,
                                       Empfangen,
                                       Listen,
                                       Organisieren)
```

Bild 1 Abstrakte Gruppenspezifikation

Dabei sollen die Werte für die einzelnen Attribute folgende Bedeu=
tung haben:

Potentielle Mitgliedschaft:

geschlossen = der Organisator allein bestimmt über die Mitglied=
               schaft

beschränkt   = jeder, dessen O/R-Name die angegebene Bedingung er=
               füllt, darf sich selbst als Mitglied dieser Gruppe
               eintragen

offen   = jeder darf sich selbst als Mitglied eintragen

Sichtbarkeit

verborgen	– Nicht-Mitglieder erhalten keine Information über die Gruppe (nicht einmal den Namen)
sichtbar	– Nicht-Mitglieder können sich über Name und Beschrei- bung der Gruppe informieren
transparent	– Nicht-Mitglieder können sich zusätzlich über die Mit- glieder der Gruppe informieren
schreibbar	– Nicht-Mitglieder können sich über die Gruppe infor- mieren und Beiträge an die Gruppe richten

Zugriffsrechte

Schreiben	– das Mitglied darf Beiträge in die Gruppe einbringen
Empfangen	– das Mitglied erhält automatisch alle Beiträge der Gruppe
Listen	– das Mitglied kann sich die Mitglieder der Gruppe li- sten lassen
Organisieren	– das Mitglied kann Eigenschaften der Gruppe verändern und über die Mitgliedschaft anderer Benutzer ent- scheiden

Beispiele

Die folgenden Beispiele sollen zeigen, wie sich typische Formen von Gruppenkommunikation durch gezielte Einstellung der Attribute einer Gruppe realisieren lassen.

"Verteiler"

Gruppenname:	geeigneter Name für den Verteiler
Beschreibung:	Zweck des Verteilers
Mitgliedschaft:	geschlossen
Sichtbarkeit:	schreibbar (falls öffentlich), verborgen (falls privat)
Mitglieder: Eigner: Empfänger:	[Schreiben, Empfangen, Listen, Organisieren] [Empfangen]

"Offene Diskussionsrunde"

    Gruppenname: Diskussionsgegenstand als Schlagwort

    Beschreibung: eine detailliertere Ausführung des Dis-
                    kussionsgegenstandes

    Mitgliedschaft: offen für [Schreiben, Empfangen, Listen]

    Sichtbarkeit: sichtbar

    Mitglieder:
      Leiter: [Schreiben, Empfangen, Listen, Organisieren]
      Teilnehmer: [Schreiben, Empfangen, Listen]

"Geschlossene Konferenz"

    Gruppenname: Konferenzgegenstand als Schlagwort

    Beschreibung: eine detailliertere Ausführung des Kon-
                    ferenzgegenstandes, eventuell Benennung einer
                    Kontaktperson für den Eintritt in die Konferenz

    Mitgliedschaft: geschlossen

    Sichtbarkeit: je nach Bedarf

    Mitglieder:
      Leiter: [Schreiben, Empfangen, Listen, Organisieren]
      Teilnehmer: [Schreiben, Empfangen, Listen]

"Elektronisches Journal"

    Gruppenname: der Name des Journals

    Beschreibung: eine detailliertere Ausführung über den gene-
                    rellen Inhalt, wie man abonniert, welche Kosten
                    entstehen, ...

    Mitgliedschaft: offen für [Empfangen] oder geschlossen

    Sichtbarkeit: sichtbar

    Mitglieder:
      Leiter: [Schreiben, Empfangen, Listen, Organisieren]
      Editoren: [Schreiben, Empfangen]
      Abonnenten: [Empfangen]

4 Gruppenkommunikation im MHS-Modell

Das einzige Element in MHS mit Bezug zur Gruppenkommunikation ist das Element "distribution list", das jedoch in der derzeitigen X.400-Empfehlung nicht näher spezifiziert ist, sondern lediglich als funktionale Anforderung an einen zukünftigen Directory-Dienst formuliert ist. Soweit die CCITT-Vorstellungen bekannt sind, genügt die zur Zeit beabsichtigte Funktionalität dieses Verteiler-Dienstes jedoch nicht den Anforderungen der Gruppenkommunikation.
Grundsätzlich sind daher zwei unterschiedliche Wege gangbar:

- Lösung der Gruppenkommunikation außerhalb von MHS, also als spezielle Anwendung oberhalb von MHS

- Lösung der Gruppenkommunikation innerhalb von MHS durch Ausfüllung der Verteiler-Funktion des Directory-Dienstes im Sinne des entwickelten Modells

Wegen der Tragweite dieser Entscheidung werden die Lösungsmöglichkeiten im folgenden diskutiert.

4.1 Diskussion der Lösungsmöglichkeiten

Die Grundidee aller Lösungsvarianten ohne Eingriff in die MHS-Definitionen ist, eine Gruppe als einen speziellen User Agent (im folgenden kurz G-UA genannt) aufzufassen. Ein G-UA arbeitet also nicht für einen menschlichen Benutzer, sondern für einen speziellen Typ von Anwendungsprogramm, das die Funktionen zur Gruppenkommunikation, also Verteilung aller eingehenden Nachrichten, Management und Zugriffskontrolle durchführt.
Lösungen außerhalb von MHS stellen sicher, daß auch nach zukünftigen CCITT-Erweiterungen des MHS-Dienstes um Funktionen zur Unterstützung der Gruppenkommunikation keine Inkompatibilitäten entstehen. Es wird dann allerdings zwei Verfahren für ein und denselben Zweck geben. Darüberhinaus besteht die Gefahr, daß in der Zwischenzeit eine Vielfalt unterschiedlicher Lösungen von verschiedenen Institutionen (DFN, ESPRIT, ...) entwickelt werden, die dann Probleme bei institutionsübergreifender Gruppenkommunikation aufwerfen. Zudem ist für den Benutzer nicht mehr kontrollierbar, in welcher Form ein einzelner G-UA die gruppenbezogenen Operationen ausführt, da der UA gemäß MHS-Konventionen unter privater Verfügungsgewalt steht.

Wählt man dagegen den Directory-Dienst als Basis für die Unterstützung der Gruppenkommunikation, so ist der Bezug zu den Entwicklungen bei CCITT gewahrt.
Der Directory-Eintrag einer Gruppe muß, um den gestellten Anforderungen gerecht zu werden, die abstrakte Datenstruktur-Definition aus Kapitel 3 in einer konkreten Form widerspiegeln. Der Directory-Dienst beinhaltet dann die erforderlichen Mechanismen zur Zugriffskontrolle und stellt Operationen zur Verfügung, die eine Realisierung der Gruppenfunktionalität über die vorhandenen MHS-Komponenten ermöglichen.

4.2 Einbettung der Gruppenkommunikation ins MHS-Modell

Die erforderlichen Teilfunktionen zur Unterstützung der Gruppenkommunikation können den verschiedenen Komponenten des MHS-Modells wie folgt zugeordnet werden:

DSA

System-intern wird eine Gruppe durch einen Eintrag im Directory repräsentiert. Dieser Eintrag ist - wie der Eintrag eines UA - einer bestimmten Domain (bzw. einem MTA) zugeordnet.

Der Directory-Dienst stellt die für das Management einer Gruppe erforderlichen Funktionen bereit und übernimmt die Zugriffskontrolle für Gruppenmanagement- und allgemeine Information-Retrieval-Funktionen.

MTA

Das Modell des MTA wird um eine neue Komponente, den "Gruppendistributor (GD)", erweitert, der für die Verteilung der an die Gruppe gerichteten Beiträge verantwortlich ist.

UA

Der UA trägt mit lokalen Funktionen zur Archivierung und zur selektiven Bearbeitung von Nachrichten bei. Damit ermöglicht er die Konzentration der Arbeit auf die Beiträge einer Gruppe und die Verfolgung von Bezügen zu früheren Beiträgen.

5 Protokolltechnische Realisierung der Verteilung

Bei der gewählten Funktionszuordnung ist die Verteilung der Beiträge eine Funktion des MTA, die er in Kooperation mit dem Directory-Dienst auszuführen hat. Dazu stellt der Directory-Dienst dem MTA auf Vorlage des Gruppennamens die für die Weiterverteilung erforderlichen Informationen (Schreibberechtigung, Empfängerliste usw.) bereit. Das Hauptproblem ist nun, die Weiterverteilung so zu definieren, daß der Tatbestand der Weiterverteilung als Transferinformation innerhalb der Nachricht erhalten bleibt, damit der MTA Zyklen in der Verteilung erkennen kann und der UA den Beitrag im richtigen Zusammenhang präsentieren kann.

Naheliegend ist, die Verteilung protokolltechnisch über die MTL-Protokolle P1 und P3 zu realisieren. Dies wird aller Voraussicht nach auch die zukünftige Lösung bei CCITT sein. Prinzipiell besteht aber auch die Möglichkeit, die Verteilung über das UAL-Protokoll P2 zu realisieren.

5.1 Verwendung der MTL-Protokolle P1 und P3

Im folgenden sind verschiedene Möglichkeiten zur Darstellung der weiterverteilten Nachrichten in den MTL-Protokollen P1 und P3 skizziert.

(1) Einfache Expansion der Empfänger-Liste

Der bearbeitende MTA löscht den Namen der Gruppe in der Empfängerliste oder kennzeichnet ihn als bearbeitet und fügt die vom Directory-Dienst gelieferte Liste von Namen als unbearbeitet hinzu.

Bei diesem Ansatz sind zwar keine Änderungen in den Protokollen P1 und P3 erforderlich, die Information bezüglich der Weiterverteilung über die Gruppe geht jedoch verloren. Diese Lösung scheidet daher von vornherein aus.

(2) Informations-erhaltende Expansion des Gruppennamens

Die Expansion des Gruppennamens wird so vorgenommen, daß der Bezug der expandierten Namen zum Gruppennamen im Umschlag der Nachricht sichtbar wird. Dazu sind Ergänzungen zu den Protokollen P1 und P3 erforderlich.

(3) Weiterverteilung über "Message-Encapsulation"

Einen ähnlichen Effekt wie in Lösung (2) könnte man erreichen, indem man den Typ "UMPDU Content" des P1-Protokolls erweitert, so daß neben den P2-Protokoll-Elementen auch Elemente des Protokolls P1 transportiert werden können. Der MTA könnte dann zur Weiterverteilung einen neuen Umschlag mit den Informationen

- MPDUIdentifier (vom MTA generiert)

- Originator ORName (der Name der Gruppe)

- ContentType (zeigt P1 an)

- SEQUENCE OF RecipientInfo (die Namen des Gruppen-Verteilers)

erzeugen und zusammen mit der alten UserMPDU als Inhalt verschicken.

Keine der Lösungen (2) oder (3) kann über einen zukünftigen öffentlichen MHS-Dienst betrieben werden, der nicht auf die vorgeschlagenen Erweiterungen in irgendeiner Form eingeht. Selbst wenn die eigentliche Verteilung auch dann noch innerhalb der PRMDs vorgenommen wird - wegen der erforderlichen Protokolländerungen werden die öffentlichen MTAs nicht in der Lage sein, Nachrichten mit der erweiterten Information zu bearbeiten.

5.2 Verwendung des UAL-Protokolls P2

Als Ausweg bietet sich an, die Verteilung protokolltechnisch nicht im MTL, sondern im UAL zu realisieren (obwohl der für die Verteilung verantwortliche Gruppendistributor funktional weiterhin im MTA und damit im MTS angesiedelt ist). Die grundsätzliche Idee dabei ist, das Konzept "forwarded IP-message" als Grundlage für die Verteilung heranzuziehen, das bereits in MHS vorhanden ist und der manuellen oder automatischen Weiterleitung von Nachrichten auf UA-Ebene dient. Der Vorteil ist, daß keine Protokollerweiterungen erforderlich werden und somit auch solche Systeme, die nicht speziell für diese Art der Gruppenkommunikation entwickelt wurden (wie EAN), einsetzbar sind. Außerdem läßt sich diese Lösung voraussichtlich leicht auf eine zukünftige CCITT-Lösung im MTL umstellen.

Der Gruppendistributor stellt somit eine neue Komponente auf UA-Ebene dar, die allerdings nicht wie ein UA über einen O/R-Namen angesprochen wird, sondern generisch bei allen O/R-Namen mit der Eigenschaft, Name einer lokal registrierten Gruppe zu sein, stellvertretend für die Gruppe als Empfänger und Sender auftritt.
Die Realisierung der Verteilung wird im folgenden anhand des Weges eines an eine Gruppe gerichteten Beitrags durch das System skizziert (vgl. Bild 2)

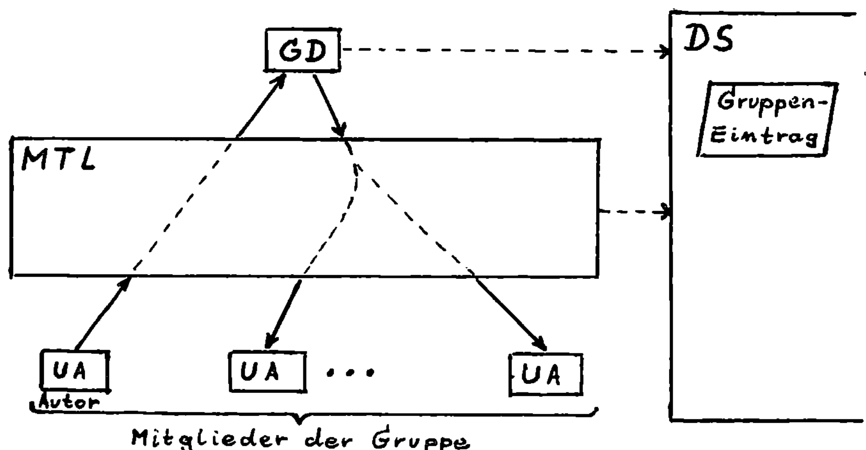

Bild 2 Verteilung von Gruppenbeiträgen

Der Autor des Gruppenbeitrags gibt als Empfänger den O/R-Namen der Gruppe an. Damit wird wie üblich eine entsprechende Nachricht erzeugt und über den MTL in die Domain übermittelt, in der die Gruppe registriert ist. Erst hier ist der bearbeitende MTA nach Konsultation seines DSA in der Lage zu erkennen, daß es sich bei dem benannten Empfänger um eine Gruppe handelt. Daraufhin ist zu prüfen, ob die Gruppe "schreibbar" ist oder ob der Absender als Mitglied mit der Berechtigung "Schreiben" verzeichnet ist. Im positiven Fall wird die Nachricht dann innerhalb des MTA an den Gruppendistributor weitergeleitet. Zu diesem Zeitpunkt wird auch eine eventuell beantragte "Delivery Notification" bzw. eine "Non-Delivery Notification" generiert.

Der Gruppendistributor verhindert endlose Verteilungen (bei zyklischer Mitgliedschaft von Gruppen), indem er die Historie der Nachricht prüft und bei erkanntem Zyklus die Verteilung stoppt. Die Historie kann anhand der ineinandergeschachtelten Nachrichten vom Typ "Forwarded-IP-Message" hergeleitet werden.

Zur Weiterleitung des Beitrags erzeugt der Gruppendistributor eine neue "Forwarded-IP-Message", wobei die ursprüngliche IPM-Nachricht zum IPM-Body wird. Dabei setzt der GD die folgenen Parameter in den IPM-Header ein:

- IPMessageID (neu erzeugt)

- Originator (der O/R-Name der Gruppe)

- autoforwarded (TRUE)

Der IPM-Body und der IPM-Header bilden den Inhalt einer neuen UMPDU, die an alle als Empfänger verzeichneten Gruppenmitglieder (mit Ausnahme des Absenders) adressiert wird. Die Parameter "encoded-information-types", "priority", und "conversion-prohibited" werden dabei aus der ursprünglichen Nachricht abgeleitet. "Disclose-of-other-recipients" wird zur Wahrung der Zugriffskontrolle standardmäßig auf FALSE gesetzt.

Über die normalen Funktionen des MTS wird die Nachricht dann zu den End-Empfängern transferiert. Eventuell resultierende "Non-Delivery-Notifications" können vom MTL wie normale Nachrichten behandelt und dem Gruppendistributor zugestellt werden. Der Gruppendistributor wandelt diese Non-Delivery-Notifications um in eine entsprechende Meldung an den Organisator der Gruppe.

6 Ausblick

Die erforderlichen oder wünschenswerten Funktionen zur Unterstützung der Gruppenkommunikation lassen sich zu verschiedenen "Bausteinen" zusammenfassen.
Ein wichtiger Grundbaustein ist die **Verteilung** von Nachrichten an die Mitglieder der Gruppe. Eingeschlossen sind dabei Funktionen zur Verwaltung der Mitglieder, zur Überwachung der Zugriffsrechte und zur Unterstützung der Benutzer bei der Bearbeitung von Gruppenbeiträgen. Dieser Bereich ist bereits relativ gut verstanden und sollte in jeder MHS-Realisierung berücksichtigt werden.
Weitere wichtige Bausteine wie **gemeinsame Archive** für Gruppen oder **Anwendungsprozeduren** zur Beschreibung komplexer Kommunikationsabläufe in Gruppen sind zukünftig noch zu erarbeiten und zueinander in Beziehung zu bringen.

Literatur

[1]   CCITT, Study Group VII
         Message Handling Systems
         Draft Recommendations X.400, X.401, X.408, X.409,
                               X.410, X.411, X.420, X.430

[2]   DFN, Arbeitsbereich CBMS
         Studie I: Funktionalität und Bewertung von Message-Systemen
         Universität Düsseldorf, April 1984

[3]   Conrads
         Funktionalität und Bewertung von Message-Systemen
         Beitrag zu dieser Tagung

[4]   DFN, Arbeitsbereich CBMS
         Studie II: Konzept für einen DFN-Message-Verbund auf der
         Basis von CCITT-Message-Handling-Systems
         Universität Düsseldorf, Juni 1984

[5]   Kaufmann, P. et al
         Der DFN-Message-Dienst im MHS-Kontext
         Beitrag zu dieser Tagung

[6]   COST-11-bis, Projekt GILT
         Interconnection of Computer based Message Systems
         Green Version
         Universität Düsseldorf, Juni 1983

[7]   Wallerath, P.
         The GILT Abstract Model of a Computer Based Message System
         Proceedings of the European Teleinformatics Conference
         (EUTECO)
         North Holland Pub., 1983

[8]   Palme, J.
         Computer Conferencing is more than Electronic Mail
         Proceedings of the European Teleinformatics Conference
         (EUTECO)
         North Holland Pub., 1983

[9]   Palme, J.
         Individual versus Group Communication
         Text 59358 in COM, QZ, Stockholm, Juni 1984

[10]   Deutsch, D.P.
         Implementing Distribution Lists in Computer-based Message Sy-
         stems
         Proceedings of the IFIP 6.5 Working Conference
         Nottingham, Mai 1984

[11]   IFIP, WG 6.5
         Interpersonal and Group Communication Activity
         Working Document - Version 1
         IFIP/WG 6.5/EUR/84/010, London, September 1984

Funktionalität und Bewertung von Message-Systemen

D. Conrads

Zentralinstitut für Angewandte Mathematik
der Kernforschungsanlage Jülich GmbH
Postfach 1913 - 5170 Jülich 1

Zusammenfassung

Message-Systeme unterstützen ihre Benutzer beim Austausch von Informationen
(Dokumenten) einschließlich der Vorbereitung der Dokumente und ihrer Archivierung.
Zunächst werden für verschiedene Nutzergruppen in einer Forschungsumgebung (Wissenschaftler, Manager, Unterstützungskräfte) die Anforderungen an ein Message-System
formuliert. Weiterhin werden die wichtigsten Funktionen und Eigenschaften von
Message-Systemen erläutert und daraus Kriterien für eine Bewertung solcher Systeme
abgeleitet; schließlich werden einige wichtige existierende Message-Systeme untersucht und bewertet.

Abstract

Message systems support their users in exchanging information (documents) including
the preparation and archival of these documents. First the requirements concerning
message services in R&D environments are identified for different user groups
(scientists, managers, clerical staff). Furthermore functions and properties of
message systems are described and criteria for evaluating such systems are derived.
Finally some of the more important existing message systems are reviewed.

<u>Vorwort</u>

Diese Arbeit enthält die Zusammenfassung eines Berichtes, der gemeinsam von der GMD-Birlinghoven (Institut für Angewandte Informationstechnik), dem HMI-Berlin (Bereich Datenverarbeitung), der KFA-Jülich (Zentralinstitut für Angewandte Mathematik) und der Universität Düsseldorf (Rechenzentrum) im Rahmen des Deutschen Forschungsnetzes (DFN) erstellt wurde. Sie ist somit im Forschungsbereich entstanden und berücksichtigt - soweit das von Bedeutung ist - die Situation in einer Forschungsumgebung. Während die Bewertungskriterien unabhängig von der Einsatzumgebung eines Systems erörtert werden können, hängen die Benutzeranforderungen und die Wertung bestimmter Systemeigenschaften und -funktionen charakteristisch davon ab. Es kann davon ausgegangen werden, daß die Aussagen teilweise auf den kommerziellen Bereich übertragbar sind; es ist aber so, daß bezüglich der Entwicklung, Einführung und Nutzung von Message-Systemen der Forschungsbereich eine gewisse Vorreiterrolle gespielt hat und die damit verbundenen Erfahrungen sowie die sich aus dem Praxisbetrieb ergebenden Korrekturen in Erfahrungsberichten und Begleituntersuchungen recht gut dokumentiert sind.

<u>Einführung</u>

Message-Systeme sind Systeme zur Unterstützung des Austausches von Informationen (Dokumenten), wobei diese Unterstützung nicht auf den eigentlichen Nachrichtentransport beschränkt ist, sondern auch das Erstellen, Formatieren und Absenden (senderseitig), das Empfangen, Darstellen und Auswerten (empfängerseitig) sowie das Archivieren und gezielte Wiederauffinden (sender- und empfängerseitig) umfaßt. Wünschenswert ist, daß neben einfachen Kommunikationsformen wie 'personal mail' oder Vertei-

lung auch Gruppenkommunikation (gemeinhin unpräzise als 'conferencing' bezeichnet)
unterstützt wird. Das Spektrum der Funktionalität von Message-Systemen reicht von
sehr einfachen Systemen (Teletex kann bereits als eine Primitiv-Version eines Message-Systems angesehen werden) bis zu sehr komplexen und umfassenden Werkzeugen zur
Unterstützung der Büroarbeit.

Bei der Konzipierung und in der Anfangsphase der Nutzung von Message-Systemen stehen
durchweg die Belange des aktiven Kommunikationspartners (d.h. des Senders) im Vordergrund. Bei den allmählich zahlreicher werdenden akzeptierten und mit steigenden
Benutzerzahlen arbeitenden Systemen zeigt es sich jedoch, daß die Interessen der
Empfänger in mancher Hinsicht von denen der Sender abweichen, in ihrer Bedeutung
aber gleichrangig zu sehen sind. Es sind hier vor allem zwei Bereiche zu nennen:

- Schutz vor 'Information Overload'
- Schutz der Persönlichkeit.

Information Overload, hervorgerufen durch unerwünschte Informationen (junk mail) ist
in einigen vielbenutzten Systemen bereits ein ernstzunehmendes Problem. Hier müssen
dem Empfänger Werkzeuge an die Hand gegeben werden, die es gestatten, aus dem großen
Strom einlaufender Informationen diejenigen herauszufiltern, die von Wichtigkeit
sind.

Der Aspekt des Persönlichkeitsschutzes wird in zwei Bereichen in besonderem Maße
berührt:

- im Bereich der Directory-Dienste (Teilnehmerverzeichnisse, evtl. mit weitergehen-
  den Attributen versehen) und

- im Bereich der von den Systemen angelegten Logbücher und Statistiken, die - entsprechend ausgewertet - einer weitgehenden Kontrolle des Benutzerverhaltens dienen können.

In beiden Fällen bestehen ernstzunehmende Zielkonflikte zwischen dem Persönlichkeits-
schutz und der Funktionalität und Praxistauglichkeit von Message-Systemen.

Eine letzte Vorbemerkung betrifft die Inhalte der Dokumente. Die nachfolgenden Ausführungen beschränken sich auf Textdokumente. Es kann aber kein Zweifel darüber bestehen, daß die Integration von Bildinformation (Graphik) und Sprache (zumindest in
Form von Sprachannotationen) eine wichtige Bereicherung für Message-Systeme bedeutet; an entsprechenden Erweiterungen wird gearbeitet, es gibt jedoch noch keine
Systeme, die in integrierter Form solche Erweiterungen bieten.

<u>Benutzeranforderungen</u>

Die technischen Eigenschaften von Message-Systemen können nur anhand von Benutzeranforderungen gewichtet werden. Der Wert bestimmter Systemeigenschaften ist nut-
zungs- und damit arbeitsplatzabhängig.
Es ist sinnvoll, die Benutzer in Gruppen mit ähnlichem Anforderungsprofil einzu-
teilen. Im Forschungsbereich können die Benutzer den Kategorien

- Wissenschaftler (Fachkräfte)

- Management

- Unterstützungskräfte

zugeordnet werden, wobei die zu erledigenden Aufgaben forschungsbezogen oder administrativ sein können.

Wissenschaftler arbeiten vorwiegend forschungsbezogen und üben nur in geringerem
Umfang administrative Tätigkeiten aus. Beschaffung von Information und Produktion
von teilweise auch umfangreicheren Papieren (Berichte, Zeitschriftenartikel, Vor-
tragsentwürfe) bilden einen wesentlichen Teil der Arbeit; auch Besprechungen und
Dienstreisen beanspruchen einen erheblichen Anteil der Arbeitszeit. Insgesamt beträgt der Anteil solcher durch die Dienste eines Message-Systems potentiell tangier-
ten Arbeiten über 60 % der Arbeitszeit /5/.

Typisch für Wissenschaftler ist, daß sie für signifikante Zeitabschnitte (etwa im
Rahmen gemeinsamer Vorhaben und Projekte) mit einer begrenzten Anzahl von Kollegen
intensiv zusammenarbeiten müssen. Wenn solche Vorhaben lokal abgewickelt werden,
findet die Kommunikation i.a. auf der Basis direkter Gespräche statt und Hilfsmittel
sind nicht erforderlich. Immer häufiger werden aber insbesondere größere Projekte in
Zusammenarbeit oder zumindest in Abstimmung mit anderen Einrichtungen durchgeführt.
Hier entsteht ein Kommunikationsbedarf, der durch Message-Systeme abgedeckt werden
kann. Voraussetzung für die Akzeptanz eines Message-Systems ist allerdings, daß alle
bevorzugten Kommunikationspartner über das System erreichbar sind; dies hat gegebenenfalls Vorrang vor den funktionalen Eigenschaften eines Message-Systems. Die bisherige Praxis der Nutzung von Message-Systemen durch Naturwissenschaftler (diese
sind bereits seit längerem Nutzer solcher Systeme) zeigt, daß diese Gruppe bzgl.
Unzulänglichkeiten in der Handhabung der Systeme (noch!) ziemlich unkritisch ist.

Der Kommunikationsbedarf eines Managers kann dreifach unterteilt werden:

- Vorwiegend lokale Kommunikation mit administrativem Hintergrund.
  Für die verwaltungsbezogene Kommunikation des Managers ist vor allem wichtig,
  Vorgänge - auch im Zusammenspiel mit Unterstützungskräften - bearbeiten zu können.

- Breit angelegte externe Kommunikation mit vielen Kollegen und Institutionen, die
  dem Zweck dient, Kontakte im wissenschaftlichen und forschungspolitischen Umfeld
  zu pflegen.
  Diese Kommunikation zeichnet sich durch eine große Zahl auch häufiger wechselnder
  Kommunikationspartner aus. Daraus läßt sich als Forderung an ein Message-System
  zum einen eine möglichst große Zahl erreichbarer Kommunikationspartner, zum anderen die Verfügbarkeit von Hilfsmitteln zur Steuerung und Organisation der viel-
  fältigen Kommunikationsaktivitäten ableiten.

- Intensive Kontakte mit wenigen Kollegen im Rahmen eigener wissenschaftlicher
  Aktivitäten.
  Die Anforderungen hier unterscheiden sich nicht wesentlich von denen eines Wis-
  senschaftlers.

Unabhängig vom jeweiligen Kommunikationszweck stellen Manager hohe Anforderungen an
die Qualität der Benutzerschnittstelle. Unzulänglichkeiten in der Handhabung oder
auch nur die Notwendigkeit einer längeren Einarbeitungszeit sind für Führungskräfte
nicht akzeptabel.
Der Anteil der Arbeitsinhalte, die durch Kommunikation geprägt sind, liegt beim
Manager noch deutlich höher als beim Wissenschaftler (50 - 80 %). Da Manager auch
häufig auf Dienstreisen sind, ist eine Zugangsmöglichkeit zum heimischen
Message-System von außerhalb von Vorteil.

Die Forderungen von Unterstützungskräften an ein Message-System korrespondieren mit
den verwaltungsbezogenen Anforderungen aus dem Managementbereich. Der Anteil der
Kommunikation an der Arbeitszeit liegt deutlich niedriger (ca. 30 %), dafür sind
Erstellung, Bearbeitung und Archivierung von Dokumenten Arbeitsschwerpunkte.

In Tab. 1 werden Eigenschaften von Message-Systemen gemäß den Anforderungen der untersuchten Benutzergruppen gewichtet. Eine solche Zuordnung kann nur Tendenzen beschreiben, da sich hinter jeder der Gruppen ein ganzes Spektrum unterschiedlicher
Tätigkeiten verbirgt und überdies auch Temperament und persönlicher Arbeitsstil im
Einzelfall zu einer abweichenden Gewichtung führen können.

	Führungs- kraft	Hilfskraft	Wissen- schaftler
Leichte Benutzbarkeit (Benutzerfreundlichkeit)	●	○	○
Zugriffssicherung	●	○	○
Versandfunktionen	●	●	◐
Archivfunktionen	◐	◐	◐
Bezugskettenbildung zwischen Informationen	◐	○	◐
Formularerstellung und -bearbeitung	○	◐	○
Textbearbeitung	○	●	◐
Organistion von Teamarbeit	◐	○	●
Organisation geregelter Kommunikationsvorgänge	◐	◐	○
Integration von Graphik	◐	○	◐
Integration von Sprache	◐	○	○
● sehr wichtig	◐ wichtig		○ weniger wichtig

Tab. 1.: Wichtung von Leistungsmerkmalen von Message-Systemen
nach Anwendergruppen

Trotz aller Unterschiede in den Anforderungsprofilen dürfen die identifizierten
Gruppierungen innerhalb einer Organisation nicht unabhängig gesehen werden. Für Führungskräfte und Unterstützungskräfte ist dies offensichtlich; aber Unterstützungskräfte leisten oftmals auch Zuarbeit zu den Aufgaben der Wissenschaftler, und
Wissenschaftler sind auch administrativ in ihre Organisation eingebunden und somit
nicht frei von verwaltungsorientierten Tätigkeiten. Es ist deshalb unbedingt zu fordern, daß die verschiedenen Anforderungen durch <u>ein</u> System erfüllt werden, das sich
allerdings den jeweiligen Benutzern in einer dem Nutzungszweck angepaßten Weise präsentieren sollte.

<u>Technische Merkmale (Bewertungskriterien)</u>

Hauptkriterien zur Beurteilung von Message-Systemen sind aus Benutzersicht ihre Funktionalität, ihre Benutzerfreundlichkeit und ihre Sicherheit. Diese werden beschrieben durch

- Benutzer und Datenobjekte in Message-Systemen
- Kommunikationsunterstützung
- Lokale Funktionen und Benutzerschnittstelle
- Statistik und Abrechnung
- Datenschutz und Datensicherheit

wobei jeder dieser Punkte einer eingehenden Diskussion und bei der Darstellung einer systematischen Begriffsbildung bedarf.

<u>Benutzer und Datenobjekte in Message-Systemen</u>

Ein Message-System unterstützt die Benutzer des Systems bei ihrer Kommunikation miteinander und eventuell mit Benutzern anderer Systeme. Diese Kommunikation findet statt durch den Austausch von Datenobjekten zwischen Benutzern.

Bei den Datenobjekten können verschiedene Typen unterschieden werden. Der in allen derzeit angebotenen Systemen realisierte Datenobjekt-Typ "Text-Information" ist Stand der Technik und auf diesen wird - wo eine Unterscheidung wichtig ist - im folgenden Bezug genommen. Andere wichtige Datenobjekt-Typen sind "Graphik", "Bild" und "Sprache". An der Integration dieser Datenobjekt-Typen wird gearbeitet, entsprechende Systeme sind aber noch nicht kommerziell verfügbar.

Der Status eines Datenobjektes ist nicht invariant. Ein Datenobjekt des Typs "Text-Information" ist während der Erstellung bzw. Bearbeitung durch den Benutzer von der Datenobjekt-Art "Text". "Texte" sind somit Datenobjekte im Verantwortungsbereich des Benutzers. Bei der Übergabe eines "Textes" an das Message-System zum Versand wird aus dem "Text" ein unveränderliches "Dokument", das eine systemweit eindeutige Identifikation erhält.
Wird ein Dokument später noch öfter oder weiterversandt, so bleiben alle Bestandteile des Dokumentes einschließlich der Identifikation erhalten. Verändert ein Benutzer jedoch ein Dokument bzw. den Textinhalt eines Dokumentes, so wird bei einem erneuten Versand ein neues Dokument mit einer neuen Identifikation erzeugt. Dokumente sind somit permanente Datenobjekte innerhalb des Verantwortungsbereiches des Message-Systems.

Beim Versand werden dem eigentlichen Dokument Versandinformationen (z.B. Adressat, Versandpriorität u.ä.) hinzugefügt; Dokument plus Versandinformation bilden eine Nachricht. Eine Nachricht ist also ein mit Versandinformationen versehenes versandfertiges Objekt, für welches das Transportsystem innerhalb des Message-Systems die Verantwortung hat.

Den Benutzern von Message-Systemen, die außerhalb der Systeme stehen, werden in umkehrbar eindeutiger Weise systeminterne Darstellungen (im folgenden in Anlehnung an die im CCITT-Umfeld gebräuchliche Bezeichnung 'originator/recipient' Sender/Empfänger genannt) mit systemweit eindeutigen Namen zugeordnet. Diese Sender/Empfänger sind die Quellen und Senken für Datenobjekte. Die Sender/Empfänger-Namen können in eindeutiger Weise auf systeminterne (technische Zustell-)Adressen abgebildet werden. Durch die Entkopplung von Namen (die der Benutzer verwendet) und Adressen (die vom System ansprechbar sind), ist es möglich, die Namenskonventionen benutzergerecht festzulegen, während die Adressen nach systemtechnischen Gesichtspunkten aufgebaut und vergeben werden können.

Die Namensstruktur eines Message-Systems sollte es unterstützen, daß Sender/Empfänger zu Gruppen zusammengefaßt und mit einem eigenen eindeutigen Sender/Empfänger-Namen versehen werden können.

Kommunikationsunterstützung

Die Summe derjenigen Dienstleistungen, mit denen ein Message-System die Kommunikation zwischen Benutzern unterstützt, bestimmt im wesentlichen die Funktionalität eines solchen Systems.

Namensverzeichnisse und Directory-Dienste

Innerhalb eines Message-Systems müssen Informationen über die möglichen Kommunikationspartner vorhanden sein. Unerläßlich ist ein Verzeichnis aller gültigen Sender/Empfänger-Namen und der zugehörigen Systemadressen. Weitergehende Informationen über die durch Sender/Empfänger-Namen repräsentierten Benutzer können sinnvoll sein; solche zusätzlichen Einträge sollten jedoch nur mit Zustimmung der Benutzer erfolgen, und möglicherweise sollten solche Informationen nicht allgemein zugänglich sein.

Zu solchen Verzeichnissen gehören Informationsdienste, die dem Benutzer gezielt die benötigten Informationen zur Verfügung stellen. Weiterhin sind Steuer- und Kontrolldienste für die Anlage und Pflege der Verzeichnisse unerläßlich: Es müssen neue Be-

nutzer eingetragen und alte gelöscht werden können, es muß dafür gesorgt werden, daß
Sender/Empfänger-Namen tatsächlich systemweit eindeutig sind und alle mandatorischen
Einträge vollständig und korrekt sind. Zur Wahrnehmung dieser Aufgaben gibt es in
Message-Systemen die Position eines System-Managers. Die Pflege der Verzeichnisse
ist eine wichtige zentrale Aufgabe, da falsche oder auch nur nicht aktuelle oder
unvollständige Informationen die allgemeine Brauchbarkeit eines Systems stark beeinträchtigen können.

Kommunikationskontexte

Der Austausch von Informationen (Dokumenten/Nachrichten) zwischen Benutzern von
Message-Systemen erfolgt meist nicht kontextfrei, d.h. Nachrichten stehen häufig in
einem logischen oder organisatorischen Zusammenhang mit anderen Nachrichten.

Zu unterscheiden sind hier statische Beziehungen zwischen Dokumenten (die ihren Ausdruck finden in Dokument-Attributen wie 'Antwort auf' oder 'Revision von') und Beziehungen, die durch den Versand von Nachrichten entstehen. Hierzu gehören Beziehungen, die aufgrund von Versandinformationen hergestellt werden können (etwa 'alle
Dokumente, die in einem bestimmten Zeitraum von einem bestimmten Absender eingegangen sind'), aber auch Beziehungen, die sich durch Versandumstände ergeben wie z.B.'
die Nachrichtenhistorie, die - das gleiche Dokument betreffend - sich für verschiedene Sender/Empfänger unterschiedlich darstellen kann (z.B. wenn Nachrichten von
einem Empfänger an Dritte weitergereicht werden).

Die Erhaltung von Kontexten in Message-Systemen bietet den Benutzern die Möglichkeit, aus großen Informationsmengen kontextspezifisch eine Auswahl für die Bearbeitung zu treffen und ist damit ein Hilfsmittel gegen die Überflutung mit Informationen.

Kommunikationsformen

Kommunikation kann sender- oder empfängergesteuert sein. In den meisten Fällen ist
sie sendergesteuert, d.h. Versand und Zustellung einer Nachricht erfolgen auf Initiative des Absenders; dem Empfänger bleibt dann nur (zwingend) die Reaktion, im
Normalfall das Lesen, zumindest aber das Löschen. In manchen Situationen ist jedoch
eine empfängergesteuerte Kommunikation angemessen. Dabei wird eine Nachricht nicht
automatisch zugestellt, sondern verbleibt beim Absender oder im Message-System und
der oder die Empfänger werden allenfalls über das Vorliegen einer Nachricht unter-
richtet. Die Empfänger können dann entscheiden, ob und wann sie die Nachricht übernehmen wollen. Auf diese Weise lassen sich Zeitungen, Schwarze Bretter und ähnliche
Dienste in benutzerfreundlicher Weise unterstützen.

In einem Message-System sollte ein Benutzer sich stets über den Versandzustand von ihm abgesendeter Nachrichten unterrichten können. Manche Systeme liefern automatisch eine Zustellbestätigung, wenn die Nachricht dem Empfänger zugestellt worden ist. Aus der Sicht des Absenders können weitergehende Bestätigungen (wie z.B., daß der Empfänger die Nachricht gelesen hat, evtl. mit Uhrzeit versehen) wünschenswert sein. Im allgemeinen liegt es jedoch im Empfängerinteresse, daß solche Auskünfte wegen der damit verbundenen Kontrollmöglichkeiten nicht gegeben werden.

Message-Systeme unterstützen unterschiedlich komplexe Kommunikationsformen. Basis eines jeden Systems ist der 1:1-Versand, d.h. ein Dokument wird von einem Sender/ Empfänger an einen anderen Sender/Empfänger geschickt (personal mail). Ebenfalls noch zu den einfachen Kommunikationsformen zählt der 1:n-Versand, der über Namenslisten oder über adressierbare Sender/Empfänger-Gruppen (Verteilerlisten) realisiert werden kann; Verteilerlisten können an einen Benutzer gebunden sein (private Verteiler) oder allgemein zugänglich sein (öffentliche Verteiler).

Über die vorgenannten Kommunikationsformen hinaus besteht ein Bedarf an Kommunikationsunterstützung innerhalb bestimmter, zeitlich fixierter oder auch variabler Gruppen von Benutzern. Dazu müssen unter verschiedenen Gesichtspunkten Gruppen definiert werden können, und es muß die Möglichkeit bestehen, Gruppenmitglieder mit Sonder- oder Minderrechten auszustatten. Bei geschlossenen Gruppen muß beispielsweise ein Organisator existieren, der die Berechtigung hat, Mitglieder aufzunehmen oder auszuschließen und der i.a. - da Message-Systeme mit endlichen Betriebsmitteln auskommen müssen - auch darüber zu wachen hat, daß der Informationsbestand nicht unkontrolliert anwächst; ferner sollte es auch möglich sein, korrespondierende Mitglieder mit ausschließlichem Leserecht einzuführen.

Wichtig für die Unterstützung der Gruppenkommunikation ist ein gemeinsames Archiv für alle Beiträge (das gemeinsame Kommunikationsgedächtnis), das es gestattet, bei Bedarf (z.B. wenn neue Mitglieder hinzukommen) die gewesene Diskussion noch einmal aufzurollen.

Gruppenkommunikation kann viele Erscheinungsformen aufweisen. Es können thematisch orientiert offene oder geschlossene Konferenzen nachgebildet werden, es können zeitungsähnliche Nachrichtendienste für feste Benutzergruppen oder frei abonnierbar eingerichtet werden sowie Schwarze-Brett-Funktionen und rundfunkähnliche Verteilerdienste realisiert werden; ebenso können Kommunikationsstrukturen innerhalb von Organisationen modelliert werden, was z.B. eine wirkungsvolle Unterstützung von Büroabläufen gestattet; schließlich fällt auch das gemeinsame Erarbeiten von Texten in diesen Bereich.

Insgesamt muß festgestellt werden, daß die Unterstützung der Gruppenkommunikation ein wichtiges Anliegen und daß eine Weiterentwicklung der Message-Systeme in Richtung auf eine ausgefeiltere und flexiblere Unterstützung der Gruppenkommunikation sehr wünschenswert ist.

In einem Message-System muß jeder Benutzer die Möglichkeit haben, die Kommunikation nach seinen Wünschen organisieren und die die Kommunikation unterstützenden Funktionen und Hilfsmittel (von denen eine Reihe bereits erläutert wurden) bedarfsgerecht einsetzen und steuern zu können. Insbesondere will der Benutzer den Informationsfluß sachgerecht organisieren können (ein hervorragendes Mittel zur Organisation des Informationsflusses ist die Gruppenkommunikation!). Es sollten empfängergesteuerte Eingangsfilter definiert werden können, die den Benutzer vor unerwünschten Informationen schützen; ferner sollen Zustellattribute (Normalversand, Expreßversand, 'Zustellung nicht früher/später als' usw.) angegeben werden können. Nützlich ist auch die Möglichkeit der temporären Delegation von Empfangsrechten (etwa bei vorübergehender Abwesenheit).

Wichtig ist jedoch, daß die an sich wünschenswerte Vielfalt der Möglichkeiten nicht für viele Benutzer zu unüberwindlicher Komplexität in der Handhabung der Systeme führt.

Lokale Funktionen und Benutzerschnittstelle

Die lokalen Funktionen müssen die Erzeugung und Verwaltung von Datenobjekten sowie die Steuerung der Kommunikationsfunktionen in angemessener Weise unterstützen.

Alle Dokumente - und für die Herstellung von Kommunikationskontexten auch die Versandinformationen - müssen permanent archiviert und jederzeit zugreifbar sein. Ferner sollten Retrieval-Funktionen zur Verfügung stehen, die einen komfortablen und sachgerechten Zugriff auf die gespeicherten Informationen ermöglichen. Die Archive können zentral (d.h. gemeinsame Archive für alle Benutzer/Gruppen von Benutzern) oder benutzerspezifisch angelegt sein.

Eine wichtige lokale Aufgabe ist auch die Unterstützung bei der Texterstellung und -manipulation durch das Message-System, die möglichst alle Hilfsmittel der modernen Textverarbeitung bieten sollte; ebenso sollten Möglichkeiten der Formularerstellung und -bearbeitung bestehen.

Ein wichtiger Punkt ist auch die Einbettung des Message-Systems in die lokale Systemumgebung. Sie sollte den Austausch von Informationen mit anderen Subsystemen (z.B. auch mit dem lokalen Dateisystem) zulassen.

Die lokalen Funktionen und die Benutzerschnittstelle sind entscheidend dafür, wie sich ein Message-System dem Benutzer darstellt. Insbesondere für die nicht DV-kundigen Benutzer - und das ist die überwiegende Mehrheit der potentiellen Benutzer von Message-Systemen - ist eine sorgfältige Gestaltung der Benutzerschnittstelle von größter Wichtigkeit. Eine komfortable, dem jeweiligen Kenntnisstand und den Anwendungserfordernissen anpaßbare Benutzerführung ist hier entscheidend.

Statistik und Abrechnung

Sowohl aus technischen wie auch aus Gründen der Rechtssicherheit ist es erforderlich, daß ein Message-System Informationen über Benutzer, Datenobjekte und Kommunikationsvorgänge hält und protokolliert. Dem entgegen steht das Anliegen der Benutzer nach Schutz persönlicher Daten und Bereiche. Es ist aus Datenschutzgründen in jedem Falle erforderlich, daß für die protokollierten Daten sorgfältig und restriktiv festgelegt ist, wer zu welchem Zweck auf diese Informationen zugreifen darf und wann diese Informationen wieder gelöscht werden müssen.

Die Gebührenerfassung eines Message-Systems sollte nach verschiedenen Dienstleistungen differenziert und benutzerspezifisch erfolgen. Die Kosten für den eigentlichen Transport von Nachrichten sollten zu Lasten des Senders gehen; sie dürfen vom Datenumfang, von der Entfernung und evtl. vom Zeitpunkt des Transfers, nicht jedoch vom Transportweg oder der Transportdauer abhängig sein, soweit Variationen dieser Größen im Transportnetz selbst begründet sind. Aus Gründen der Praktikabilität muß jedoch eine Abrechnung in wenigen Klassen - etwa lokal, national, international - gefordert werden. Grundsätzlich sollten die empfängerseitig anfallenden Kosten bis zum ersten Ansehen einer empfangenen Nachricht so niedrig wie möglich gehalten werden.

Datenschutz und Datensicherheit

Datenschutzaspekte wurden bereits mehrfach angesprochen. Insgesamt geht es darum, daß weder Daten eines Benutzers (in Archiven oder beim Transport), noch Informationen über einen Benutzer selbst (in Verzeichnissen und Listen) oder sein Kommunikationsverhalten (in Logbüchern und Statistiken) einem unberechtigten Zugriff ausgesetzt sind; insbesondere muß darauf geachtet werden, daß auch privilegierte Personen (z.B. System-Manager) nicht private Archive der Benutzer einsehen können. Zugangssicherungen zum Message-System und die Möglichkeit, Daten beim Transport und in den Archiven verschlüsseln zu können, sind wichtige technische Hilfsmittel.

Eine selbstverständliche Forderung an jedes Message-System ist, daß Daten, die dem System anvertraut werden, gegen Verlust und Verfälschung gesichert sind und – wenn sie transportiert werden – mit sehr hoher Wahrscheinlichkeit die Adressaten (und nur diese!) unverändert erreichen.

Es muß zum Schluß aber nochmals darauf hingewiesen werden, daß Datenschutz und Datensicherheit nicht nur ein technisches Problem sind und deshalb auch mit technischen Mitteln allein nicht gewährleistet werden können; zur Erzielung eines akzeptablen Standards müssen organisatorische und regulatorische Maßnahmen hinzukommen.

Bewertung existierender Message-Systeme

Die in Tab. 2 aufgelisteten Message-Systeme wurden einer vergleichenden Bewertung unterzogen. Durchgeführt wurde diese Bewertung anhand eines aus den vorher erläuterten Kriterien entwickelten Fragebogens, ergänzt durch eine textliche Beschreibung, in der gezielt die individuellen Charakteristika der Systeme herausgearbeitet wurden, die in der formalisierten Beurteilung keine ausreichende Würdigung finden konnten.

Systembezeichnung	Hersteller
COM	FOA (Schweden)
DIN-A-MIT	Hahn-Meitner-Institut Berlin
Intercomm (TELEBOX)	ITT Dialcom
KOMEX	GMD (Birlinghoven)
PortaCOM	ENEA DATA AB
PROFS	IBM
UNIX-Mail	Bell Laboratories
VAX-Mail	DEC

Tab. 2.: Zusammenstellung der bewerteten Message-Systeme

Die untersuchten Systeme können verschiedenen Gruppen zugeordnet werden:

VAX-Mail und UNIX-Mail gehören zur Gruppe der <u>betriebssystemnahen</u> Message-Systeme. Bei diesen wird der Message-Dienst als zusätzliches Leistungsmerkmal im Rahmen der interaktiven Fähigkeiten des Betriebssystems aufgefaßt. Für eine begrenzte, lokale Benutzerschar gedacht, sind ausgefeilte Directory-Dienste ebenso wie eine spezifische Abrechnung der Message-Dienste entbehrlich. Die Grundfunktionen sind einfach in der Handhabung; höhere Anforderungen an die Funktionalität erfordern eine Kombination von Betriebssystemfunktionen mit den Möglichkeiten des Message-Systems. Anwenderzielgruppen sind somit in erster Linie Benutzer von DV-Anlagen, denen – gute Betriebssystemkenntnisse vorausgesetzt – eine zufriedenstellende Funktionalität geboten wird.

Eine zweite Gruppe innerhalb der untersuchten Systeme bilden die europäischen Message-System-Entwicklungen COM, KOMEX und PortaCOM. Diese Systeme sind eigenständige Message-Systeme, die im <u>Forschungsumfeld</u> entstanden sind. Bei allen drei Systemen bilden die Unterstützung der Gruppenkommunikation und der Kommunikationskontexte Schwerpunkte. Alle drei Systeme sind grundsätzlich für einen breiten Einsatz geeignet. Zusammenfassend kann festgestellt werden, daß diese neueren europäischen Entwicklungen bezüglich der Funktionalität von Message-Systemen die Maßstäbe setzen; die Zahl der Installationen ist aber noch vergleichsweise gering.

Zwischen den beiden vorher erwähnten Gruppen von Message-Systemen steht PROFS. PROFS ist ein eingeständiges Message-System, das sich ziemlich stark auf das darunterliegende Betriebssystem (VM/CMS) abstützt, dies aber nicht notwendig dem einzelnen Benutzer auferlegt, sondern in diesem Zusammenhang dem System-Manager Aufgaben zuweist.

Eine Sonderstellung nimmt DIN-A-MIT ein, das keiner Gruppe zugeordnet werden kann. Es ist als verteiltes System konzipiert und realisiert das Konzept des 'wandernden' Benutzers; in diesem Zusammenhang ist es als Pilotprojekt mit Demonstrationscharakter einzustufen.

Das Intercomm-System gehört zur Gruppe der Message-Systeme, über die öffentliche (PTTs) oder kommerzielle Anbieter Message-Dienste i.a. flächendeckend anbieten. Der Zugang zu diesen Systemen erfolgt grundsätzlich über öffentliche Netze, was für die Realisierung einer komfortablen Benutzerschnittstelle nicht die beste Voraussetzung bietet. Der Vorteil dieser Systeme liegt im flächendeckenden Angebot mit unbeschränkten Zugriffsmöglichkeiten. Diese Zielsetzung legt es nahe, nur wenige, leicht verständliche Funktionen anzubieten, was überdies die in diesem Umfeld wichtige einfache Bedienbarkeit erleichtert.

<u>Zusammenfassung</u>

Es muß festgestellt werden, daß von den derzeit angebotenen Message-Systemen keines in allen wichtigen der dargelegten Kriterien gleichermaßen gut abschneidet, so daß bei einer Auswahl stets der geplante Einsatzbereich zu berücksichtigen ist. Funktionalität und technische Qualitäten sind jedoch für Auswahl und erfolgreiche Einführung eines Message-Systems nicht allein entscheidend. Sehr wichtig sind auch die äußeren Randbedingungen, insbesondere

- die technischen Voraussetzungen (Hardware-Ausstattung, vorhandene Software, aber auch das durch die Historie und die Umgebung vorgeprägte Benutzerverhalten) und

- die organisatorischen Vorbereitungen (Auswahl von Teilnehmern mit entsprechendem Kommunikationsbedarf, Schulung und Betreuung der Teilnehmer).

Benutzeranforderungen und Funktionalität können unabhängig vom Entwurf und vom Implementierungskonzept von Message-Systemen erörtert werden; umgekehrt hat das Implementierungskonzept sehr wohl Auswirkungen auf die Systemeigenschaften und zwar deshalb, weil sich davon abhängig manche Merkmale und Funktionen leicht realisieren lassen, andere dagegen nur sehr schwer oder auch gar nicht.

Stand der Technik sind heute zentrale Message-Systeme, die über zeilenorientierte Benutzerschnittstellen und öffentliche Netze von nahezu jedem Punkt der Erde aus zugänglich sind. Die Überwindung der geographischen Entfernungen findet hier auf Ebene des Zugriffs durch die Benutzer statt. Innerhalb des Message-Systems werden Informationen (Nachrichten) nur lokal zwischen den Archiven der verschiedenen Benutzer bewegt bzw. - wenn ein gemeinsames Archiv vorhanden ist - real überhaupt nicht, sondern nur logisch durch das Herstellen entsprechender Bezüge.

Diese Lösung, d.h. die gemeinsame Nutzung eines bestimmten zentralen Message-Systems durch eine geographisch verteilte Benutzergemeinde, ist verfügbar und praktikabel, aber sie ist nicht zukunftsweisend. In Zukunft werden flexiblere (und für den Benutzer komfortablere) Strukturen benötigt, wie sie im Rahmen der MHS-Modellvorstellungen (CCITT-Empfehlungen X.400 ff) entwickelt wurden.

Literatur

/1/ Babatz; Pankoke-Babatz; Santo; Theidig;
   Experimental Use of KOMEX in the GMD
   Computer Compacts, April 1983, pp. 83

/2/ Conrads, D.; Kaufmann, P.; Pankoke-Babatz, U.;
   Speth, R.; Tschicholz, M.; Warnking, A.; Wallerath, P.
   Funktionalität und Bewertung von Message-Systemen
   DFN-Studie, April 1984

/3/ Hiltz, St. R.; Turoff, M.
   The Evolution of User Behavior in a Computerized Conferencing System
   CACM 24, November 1981, pp. 279

/4/ Palme, J.
   Experience with the Use of the COM Computerized Conferencing System
   FOA Rapport (10166E-M6(h)), December 1981

/5/ Schwetz, R.
   Büroökonomie
   in Bürokommunikation, Springer Verlag, Berlin, 1983

Der DFN-Message-Dienst

im MHS-Kontext

Peter Kaufmann

Michael Tschichholz

Hahn-Meitner-Institut Berlin

Übersicht

Mit der Zielsetzung, in der Bundesrepublik ein Message-System für den
Forschungsbereich nach den CCITT MHS-Empfehlungen X.400 ff [CCIT-'84] zu
errichten, wurden im Rahmen des Deutschen Forschungsnetzes, mit teilweiser
Förderung des BMFT, Untersuchungen durchgeführt, aus denen ein Konzept
für den 'DFN-Message-Dienst' entstanden ist. Diese Untersuchungen und das
daraus erarbeitete Konzept sind in den Studien I und II sowie im Pflichtenheft (vgl. [DFN1-'84], [Conr-'84], [DFN2-'84], [DFNP-'84]) dargestellt.

Dieser Beitrag stellt das CCITT MHS-Modell vor, beschreibt den Funktionsumfang der MHS-Services und DFN-spezifischer Erweiterungen sowie ein
Konzept für das Message-Verbundnetz der ersten Realisierungsstufe. Zwei
wichtige Zielbereiche - der Directory-Dienst und die Gruppenkommunikation - werden an anderer Stelle ausführlich diskutiert (vgl. [Wosn-'84],
[Sant-'84]) und werden deshalb hier nur kurz behandelt.

1 Ausgangspunkte und Randbedingungen

1.1 Aufgabenstellung

In der CCITT Studienperiode 1980-1984 wurde "Question 5, Message Handling Systems (MHS)", von der CCITT Study Group VII zügig vorangetrieben, so daß bereits im Dezember 1983 die Draft Recommendations der X.400-Serie für das Abstimmungsverfahren im Herbst 1984 technisch festlagen [CCIT-'84]. Diese Papiere stellen ein System zusammenhängender Spezifikationen dar, die zum einen Message-spezifische, zum anderen aber auch allgemeinere, jedoch ebenfalls benötigte Festlegungen (wie z.B. Presentation Transfer Syntax and Notation (X.409)) beinhalten. Einige grundsätzliche Vorarbeiten zu diesem Themenkomplex (z.B. Naming) wurden hierbei aus der IFIP WG 6.5 (Message-Systeme) übernommen. Die Papiere beschreiben nicht nur Dienste und Protokolle für den Transfer von Messages zwischen Endbenutzern, sondern auch solche für die Weitergabe von Information in einem "store- and-forward"-Netz auf Ebene 7 des ISO/OSI Referenz-Modells.

Für den Message-Dienst im DFN ergeben sich hieraus insbesondere zwei wichtige Fragestellungen, deren Diskussion mit Blick auf die Ergebnisse der Studie I [DFN1-'84] geführt werden muß.

1. Bietet CCITT MHS die Funktionalität, die für die Benutzung als DFN-Message-Dienst angemessen erscheint, oder sind funktionale Erweiterungen aus DFN-Sicht erforderlich?

2. Ist es mit Hilfe der CCITT MHS-Protokolle möglich, den Grundgedanken zu realisieren, existierende Message-Systeme mit ihren Benutzern in einen DFN-weiten (und auch internationalen) Message-Dienst zu integrieren?

Bei der Verwendung von CCITT MHS sollte bewußt sein, daß in Teilbereichen (teilweise vermutlich aus Zeitmangel) Festlegungen bewußt aufgespart und auf die nächste Studienperiode verschoben werden ("for further study"). Ein wichtiges Beispiel hierfür stellt der sogenannte "Directory Service" dar, der zwar als architekturelles Konzept erwähnt, im Detail jedoch derzeit nicht ausgeführt ist. Da für einen DFN-Message-Betrieb ein - anfangs vielleicht rudimentärer, jedoch längerfristig komfortabler - Directory Dienst nicht nur wünschenswert sondern auch erforderlich ist, müssen derartige Lücken im jetzigen MHS-Standard ausgefüllt werden.

Weiterhin soll im Hinblick auf CCITT MHS diskutiert werden, in welcher Form, mit welchen Einschränkungen und eventuell mit welchen speziellen Interpretationen die vorliegende MHS-Service- und Protokoll-Struktur mit der Vielzahl ihrer (zum Teil optionalen) Parameter auf einen realen DFN-Message-Verbund abgebildet werden kann.

Außerdem ist auf eine aus dem allgemeinen MHS-Architekturrahmen herausfallende Spezifikation (X.430) speziell einzugehen, da diese aus verschiedenen Gründen (siehe unter 3.2.2) für das DFN von Interesse ist. Sie beinhaltet Verfahren zum Übergang zwischen Teletex- und MHS-Kommunikation.

Das Grundkonzept für den DFN-Message-Dienst muß schließlich folgenden
Forderungen Rechnung tragen:

1. Existierende Message-Systeme (mit ihrer zum Teil großen Zahl von
   Benutzern) müssen an die MTA/UA-Protokollautomaten adaptiert werden;
   aus anderer Sicht bedeutet dies: da diese Systeme als bereits existierende lokale Realisierungen der Funktionalität von MTA/UA-Komponenten
   aufgefaßt werden können, ist die Hinzufügung entsprechender Abbildungen
   zwischen lokaler Realisierung und Kommunikations-Standard erforderlich;

2. Message-Austausch zwischen DFN-Systemen (bzw. zwischen den Benutzern)
   muß auf Basis der CCITT MHS-Protokolle ermöglicht werden; neue Realisierungen von MTA-und UA-Komponenten sind dazu erforderlich;

3. Kommunikation mit Benutzern eines (späteren) öffentlichen Message-Dienstes muß über den DFN-Bereich hinaus möglich sein;

4. Teletex-Zugangs-Mechanismen sind einzubeziehen.

1.2 Überblick über das CCITT Message Handling System

CCITT hat inzwischen eine Reihe von Empfehlungsentwürfen definiert, die
ein "Message Handling System" (MHS) beschreiben. Diese sind in der Serie
X.400 ff [CCIT-'84] enthalten.

Die wesentlichen funktionalen Komponenten des Modells sind User Agents
(UAs) und Message Transfer Agents (MTAs); letztere bilden zusammen das
Message Transfer System (MTS). Benutzer des MHS - menschliche Benutzer
oder Anwendungsprogramme - nehmen dessen Dienste über einen User Agent in
Anspruch, der wiederum mit dem Message Transfer System interagiert. Innerhalb des MTS können mehrere MTAs interagieren, um eine Nachricht
gegebenenfalls per "store-and-forward" von dem MTA, der den sendenden UA
bedient, zu einem oder mehreren MTA(s) zu transferieren, von dem oder von
denen die Empfänger-UAs bedient werden. Die zwischen UAs ausgetauschten
Benutzerdaten werden während des Transfervorgangs innerhalb des MHS mit
verschiedenen "Umschlägen" versehen (submission envelope, relaying envelope, delivery envelope), die die Geschichte des Transfers widerspiegeln.

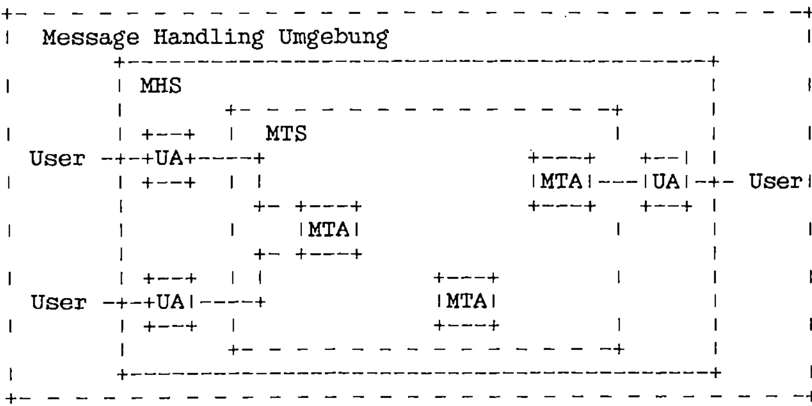

Abb. 1.1: Funktionale Sicht des MHS-Modells

Das MHS-Modell (siehe Abb. 1.1) kann auf mannigfaltige Weise auf die wirkliche Welt umgesetzt werden. Bezogen auf physische Systeme sind "stand-alone-UAs" möglich - Realisierung eines oder mehrerer UAs in einem Rechner oder einem intelligenten Terminal -, ebenso stand-alone-MTAs, aber auch UAs und ein MTA in einem System.

Die Abbildung der Komponenten des MHS-Modells auf physische Systeme kann anhand der Darstellung des Modells in den Schichten des ISO-Referenzmodells veranschaulicht werden. Alle Instanzen und Protokolle des MHS sind im OSI-Application-Layer angesiedelt, aus Beschreibungsgründen werden sie in zwei "Sublayer" eingeteilt:

- dem User Agent Layer (UAL), in dem die auf den Inhalt von Messages bezogene UA-Funktionalität eingeordnet wird, und
- dem Message Transfer Layer (MTL) mit der MTA-Funktionalität des Message-Transfers.

In Abb. 1.2 werden die drei möglichen Typen von Systemen sowie die zwischen ihnen benötigten Protokolle gezeigt: Systeme mit UA- bzw. MTA-Funktionalität allein (S1 bzw. S2) und Systeme mit UA- und MTA-Funktionalität (S3). Systeme mit UA-Funktionen besitzen eine UA-Entity (UAE) im UA-Layer, die die MTL-Dienste benutzt, um mit kooperierenden UAEs über ein UA-UA-Protokoll Pc zu kommunizieren. Als einziges solches Protokoll ist bisher das Interpersonal Messaging-Protokoll P2 definiert. Zu Systemen mit MTA-Funktionen gehört eine MTA-Entity (MTAE) im MT-Layer. Das Protokoll zwischen MTAEs, P1, definiert die Interaktionen zum Relaying von Messages. In Systemen ohne MTA-Funktion stellt die Submission and Delivery Entity (SDE) im MT-Layer der UAE die MT-Dienste durch Kooperation mit einer MTAE über das Submission and Delivery-Protokoll P3 zur Verfügung.

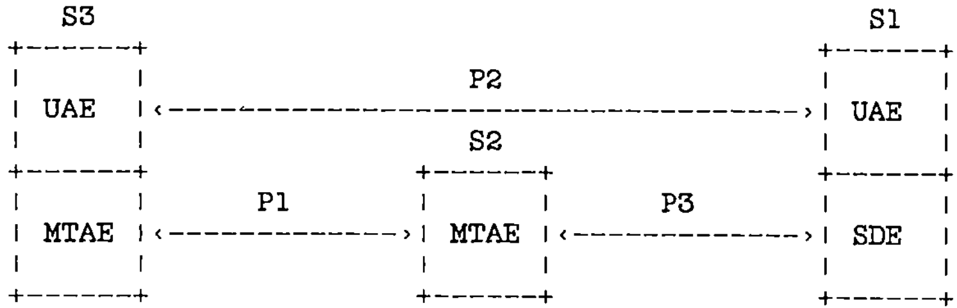

Abb. 1.2: Beschreibung des MHS im OSI-Referenzmodell

Da der Message Handling Service zukünftig von öffentlichen Fernmeldeverwaltungen angeboten werden soll, wird der physischen Gliederung eine Strukturierung in öffentliche, d.h. der Fernmelde-Aufsicht unterstehende, und private Verantwortungsbereiche ("Management Domain" (MD)) überlagert.

Eine Management Domain wird gebildet von mindestens einem MTA und eventuell einigen UAs, die von einer Fernmeldeverwaltung oder einer "Organisation" verwaltet werden. Abhängig davon handelt es sich um eine Administration Management Domain (ADMD) oder eine Private Management Domain (PRMD).

2 Der DFN-Message Dienst

Das DFN-Message-System - als ganzes betrachtet - erbringt den DFN-Message-Dienst für seine Benutzer. Dieser umfassende Dienst wird dem Benutzer über seinen User-Agent angeboten und gliedert sich in drei wesentliche Dienst-Komponenten:

- den MHS-Dienst (im engeren Sinn des Message-Transfers)
- den Directory-Dienst und
- die lokalen Dienste.

Das CCITT MHS-Modell definiert nur die Transfer-orientierten Elemente eines Message-Systems, während die lokalen Dienste und der Directory-Dienst von diesem Modell nicht betrachtet werden. In der CCITT Empfehlung X.401 werden die Service-Elemente des 'Interpersonal Message (IPM) Service' zusammenfassend dargestellt. Sie charakterisieren den Leistungsumfang, den ein MHS nach den bisher entwickelten Auffassungen von CCITT dem potentiellen Benutzer zur Verfügung stellt. Der derzeit dort beschriebene Leistungsumfang umfaßt nicht den aus DFN-Sicht bzw. aus der Sicht des Forschungsbereiches für notwendig oder sinnvoll erachteten Leistungsumfang (siehe Studie 1 [DFN1-'84]). Es gibt sowohl Dienstelemente, die in X.401 vorhanden sind, aber zumindest für eine anfängliche Realisierung als weniger wichtig erachtet werden, als auch wichtige Dienstelemente wie die Gruppenkommunikation, die in X.401 fehlen.

Als wesentliche Erweiterung des Message-Dienstes wurde deshalb die Gruppenkommunikation aufgenommen, die eine einfache Kommunikation zwischen Gruppen im DFN ermöglicht. Sie mußte dabei so in das CCITT MHS-Modell, eingebunden werden, daß die Kompatibilität zum MHS-Standard erhalten bleibt.

Unumstritten ist auch die Forderung nach einem Directory-Dienst mit den erforderlichen Namens-Konventionen. Hierbei können und sollten Arbeiten, wie sie bei IFIP WG 6.5 ([IFIP-1] und [IFIP-2]) vorliegen, berücksichtigt werden. Diese Vorgehensweise ist insbesondere deshalb ratsam, da wegen der Wechselwirkung zwischen den Arbeiten bei CCITT MHS und IFIP WG 6.5, eine Ausrichtung zukünftiger CCITT Arbeiten im Directory-Bereich an IFIP-Vorgaben - zumindest konzeptionell - zu erwarten ist. Von ECMA [ECMA-DIR] sind bereits erste Dienst- und Protokoll-Spezifikationen entwickelt worden.

Unter den 'Lokalen Diensten' sind Funktionen wie Texterstellung, Archivierung, Retrieval u.ä. zusammengefaßt, die einen Großteil des vom Benutzer empfundenen Leistungsumfangs ausmachen. Als wesentliches Konzept wurde hier ein 'Dokument-Konzept' aufgenommen, daß eine global eindeutige Identifikation von Texten ermöglicht. Dadurch können Bezüge (Kontexte, Historie, Referenzen, ...) zwischen den Dokumenten hergestellt werden.

In den folgenden Teilen werden die einzelnen Dienst-Komponenten näher beschrieben, wobei auf die Gruppenkommunikation und den Directory-Dienst hier nur kurz eingegangen werden soll, da sie Gegenstand anderer Vorträge sind (vgl. [Wosn-'84], [Sant-'84]).

2.1 Der DFN-MHS-Dienst

Der DFN-MHS-Dienst bietet eine Reihe von Dienstelementen, die für die Kommunikation relevant sind, sei es für die eigentliche Kommunikation der Benutzer untereinander oder aber für die Kommunikation der sie unterstützenden Applikationsprozesse. Aus der Menge der von CCITT angebotenen Dienstelemente soll im DFN-Bereich eine Teilmenge (Basis A) realisiert werden; zusätzlich werden noch Grundzüge der Gruppenkommunikation angeboten. Dem Benutzer stehen dann die folgenden kommunikationsrelevanten Funktionen zur Verfügung:

- Erzeugen von Nachrichten-Attributen
- Absenden neuer Nachrichten an Personen und Gruppen
- Empfangen von Nachrichten und Bestätigungen
- Anzeigen von Nachrichten-Attributen
- Beantworten bzw. Bestätigen empfangener Nachrichten
- Weiterleiten empfangener Nachrichten

2.1.1 Auswahl der MHS-Dienste für den DFN-MHS-Dienst der Stufe 1

Die Service-Elemente aus X.401, die in einer Grundimplementierung des DFN realisiert werden sollen, werden zur Gruppe Basis A [DFN2-'84] zusammengefaßt. Hierin enthalten sind die Elemente des IPM-Basic-Service sowie weitere sogenannte 'Optional User Facilities', und zwar alle in X.401 mit Essential (E) sowie einige der mit Additional (A) bezeichneten Elemente.

IPM-Basic-Service

Der 'Basic Service' umfaßt unverzichtbare Elemente wie Senden und Empfangen von Nachrichten, die Anzeige der Kodierung und eventueller Konversionen beim Empfänger, das Versehen der Nachricht mit Datum und Uhrzeit für Versand und Empfang, die Benachrichtigung des Absenders bei Unzustellbarkeit usw.

Neben den Service-Elementen sieht der IPM-Service ein relativ breites Spektrum unterschiedlicher 'Body'-Typen zur Darstellung der Benutzer-Information vor, wobei die lokale Unterstützung von den vorhandenen Gerätetypen und entsprechender Software abhängt. Aufgrund der weiten Verbreitung von Endgeräten auf IA5- und Teletex-Basis werden zunächst nur die Typen 'IA5-Text' und 'TTX' in Basis A aufgenommen. Wegen der Bedeutung in Bezug auf Informationen über die Historie von Nachrichten wird 'Forwarded IP-Message' und zum Versand verschlüsselter Nachrichten 'encrypded' als 'Body'-Typen aufgenommen.

Weitere 'Body'-Typen sind in einer nächsten Realisierungsstufe zur Unterstützung von Anwendungen wie Grafik und Textverarbeitung durchaus wünschenswert.

IPM-optional-User-Facilities

Die optionalen Elemente (siehe Tabelle 1 und 2 in X.401 [CCIT-'84]) machen Aussagen über die möglichen Fähigkeiten der UAs, die einzelnen Elemente beim Senden zu erzeugen bzw. beim Empfang zu interpretieren. Die Elemente werden danach unterschieden, ob das Belegen bzw. Interpretieren bei jeder Realisierung unbedingt ermöglicht werden muß (essential) oder ob es im Ermessensspielraum der Realisierung liegt (additional). Volle Kompatibilität mit CCITT erfordert demnach mindestens die Realisierung des

'Basic Service' und aller als notwendig (essential) bezeichneten Elemente. Darüberhinaus werden auch einige der zusätzlichen (additional) Elemente für eine Grundimplementierung im Wissenschaftsbereich als wünschenswert erachtet und zusammen mit den "essential"-Elementen unter Basis A subsummiert.

In die Auswahl Basis A [DFN2-'84] wurden vor allem die (zusätzlichen) Elemente aufgenommen, mit deren Hilfe die Herstellung der Kommunikations- kontexte ermöglicht wird, die Informationen über die Historie einer Nach- richt oder über Bezüge zwischen Nachrichten und ihren Inhalten betreffen. Die nicht aufgenommenen Elemente bieten zusätzlichen Komfort, der für ein annehmbares Message-System im DFN-Rahmen zunächst nicht notwendig erscheint.

2.1.2 Gruppenkommunikation

Die in Studie I im Zusammenhang mit Gruppenkommunikation erwähnten Dienste sind im derzeitigen MHS-Modell nicht enthalten. Für das DFN wird der MHS-Dienst deshalb dahingehend erweitert, daß Benutzergruppen geschlossen über einen "originator/recipient-name" ((O/R)-Name) an- gesprochen werden können. Die Berechtigungen der Mitglieder einer Gruppe wie auch Außenstehender sollten so einstellbar sein, daß gebräuchliche Gruppenkonstellationen wie

- einfache Verteiler,
- offene oder eingeschränkte Diskussionsrunden oder
- geschlossene Konferenzen und Teams

nachgebildet werden können.

Um die Gruppenkommunikation im DFN durch eine kompatible Interpretation des CCITT MHS-Systems realisieren zu können, wurden die verschiedenen Funktionen den Message-Dienst-Komponenten wie folgt zugeordnet:

Directory-Dienst

Eine Gruppe wird durch einen Eintrag im Directory repräsentiert, der die abstrakte Datenstruktur-Definition des Gruppenmodells widerspiegelt. Jede Gruppe erhält einen O/R-Name und ist damit fest einer bestimmten Domain zugeordnet. Der Directory-Dienst stellt die für das Management einer Gruppe erforderlichen Funktionen bereit und übernimmt die Zugriffskontrolle für das Gruppenmanagement. Der jeder Gruppe zugeordnete Organisator ist für die Gruppe verantwortlich und übernimmt auch die Kosten für die Verteilung der Gruppenbeiträge.

Message-Transfer-Agent

Der MTA wird um eine Komponente, den Gruppendistributor, erweitert, der eingehende Gruppenbeiträge (auf UA-Level) als 'Forwarded-IP-messages' an die empfangsberechtigten Gruppenmitglieder weitersendet.

Mit der DFN-Lösung für eine verteilte Umgebung wird es nicht möglich sein, eine gemeinsame Archivierung der Gruppenbeiträge durchzuführen, wie dies in zentralen Konferenz-Systemen [Palm-'83] der Fall ist. Ebenfalls noch im Bereich der Forschung befinden sich Ideen in Hinblick auf die Un- terstützung komplexerer Kommunikationsabläufe innerhalb von Gruppen, die unter dem Schlagwort "Anwendungsprozeduren" zusammengefaßt werden. Hierzu gehören u.a. die "programmierbare Konferenz" (Auswertung von Abstimmungen,

Steuerung von Abläufen) oder auch die gemeinsame Erarbeitung von Dokumenten innerhalb von Gruppen (joint editing).

2.2 Der MHS-Directory-Dienst

Für Message Handling Systeme wird ein Directory-Dienst benötigt, der z.B. Benutzernamen (O/R-Namen) in Adressen (O/R-Adressen) transformiert und den Benutzer darin unterstützt, korrekte O/R-Namen zu finden. Da bisher in den Normungsgremien (CCITT [CCIT-DIR], ISO [ISO-DIR]) noch kein Standard für einen globalen verteilten Directory-Dienst existiert, wird als Übergangslösung im DFN nur ein einfacher MHS-spezifischer Directory-Dienst realisiert, der sich leicht an den zukünftigen verteilten Directory-Standard anpassen läßt.

2.2.1 Adressierung

Um die Benutzer eines Message Systems zu benennen, sollten 'benutzerfreundliche Namen' gewählt werden, die den Benutzer als Person aus menschlicher Sicht und nicht als 'Adresse' aus Computer-Sicht beschreiben. In existierenden Netzen, wie z.B. in ARPANET, werden Benutzer heute jedoch nicht als Person, sondern als Computeradresse, unter Umständen noch mit Routing-Informationen versehen, benannt (z.B. KERMIT.FRGPARC-MAXC). Eine benutzerfreundliche Namensgebung ist jedoch nur möglich, wenn ein Name eine Person in ihrer natürlichen Umgebung (Name, Adresse, Arbeitsplatz, Rolle, usw.) beschreibt.

In Anlehnung an CCITT X.400 werden in [IFIP-2] Konventionen für eine benutzerfreundliche Namensgebung vorgeschlagen, die auch für die Namensgebung im DFN-MHS verwendet werden. Als zunächst im MHS unterstützte Formen von (O/R)-Namen werden im Abschnitt "Initial Selection of O/R-Names" in X.400 zwei Formen von Namen aufgeführt, die architekturelle Form und die terminalorientierte Form (für Telematik-Dienste wie z.B. Teletex).

Eine PRMD im DFN, die als 'naming authority' für ihre Benutzer verantwortlich ist, wird die architekturelle Form für die Benennung der lokalen Benutzer verwenden. Die terminalorientierte Form muß zusätzlich unterstützt werden, um Benutzer in fremden Domains mit terminalorientierter Adressierung und Benutzer von Telematik-Diensten benennen zu können.

Das folgende Beispiel zeigt, wie die (architekturelle) DFN-Namensgebung aussehen wird:

Namens-Attribut:	Beispiel:
Landesbezeichnung	* Deutschland
Administration Domain-Name	* 'DBP' (zur Zeit leer)
Privat Domain-Name	* HMINET
Organisations-Name	HMI
Organisations-Einheit	Bereich D/M
Personen-/Gruppenname	* Berthold Butscher

Die mit '*' gekennzeichneten Attribute sind obligatorisch (PRMD-Name nur im Bereich des DFN).

Neben diesen Attributen sind zusätzliche Attribute wie Organisations-Adresse, -Telefonnummer, persönliche Telefonnummer, oder Titel wünschenswert.

2.2.2 Das MHS-Directory-Modell der Stufe 1 im DFN

Da außer den Konzepten heute seitens CCITT bzw. ISO noch keine Dienst- und Protokollfestlegungen existieren, sollten für diese erste Stufe nur die Festlegungen getroffen werden, die

- für die Benutzer unverzichtbar
- für den IP-Message-Dienst notwendig sind.

Darüberhinaus sollte dieses Konzept zu einem globalen Directory-Dienst aufwärtskompatibel sein.

In der ersten Stufe wird der DSA seine Dienste nur lokalen Benutzern über lokale Schnittstellen zur Verfügung stellen. Da die Dienste noch nicht in Protokolle umgesetzt werden, müssen DSA, MTA und UA auf dem gleichen Host realisiert sein.

Remote-Benutzer, wie z.B. UA's auf anderen Hosts in der PRMD oder Benutzer aus anderen PRMD's, können über einen PAD-Zugang (X.3/X.28/X.29) einfache Informationsabfragen durchführen, jedoch keine Änderungen an Directory-Einträgen durchführen. Zu diesem Zweck wird Benutzern auf dem Host des DSA ein lokaler Dialog-Dienst angeboten, der Anfragen an den Directory-Dienst über eine DIALOG-Komponente ermoeglicht. Der "DIALOG-DUA" wird dabei ähnliche Funktionen anbieten, wie die DUA-Komponente, die von den lokalen UA's verwendet wird.

Abb. 2.1 zeigt die Einbettung des DS auf einem Host. Sind in einer Domain mehrere MTAs vorhanden, so muß jedem MTA ein eigener DSA zugeordnet sein.

```
+--------------------------------------------------------------+
|  HOST (z.B. VAX)                                             |
|                                                              |
|   +------------+     +------------+     +--------------+     |
|   | DFN-MHS-UA |     |    MTA     |     |   DIALOG-    |     |
|   |            |     |            |     |  Komponente  |     |
|   +------------+     +------------+     +--------------+     |
|   |   UA-DUA   |     |  MTA-DUA   |     |   DIALOG-    |     |
|   |            |     |            |     |     DUA      |     |
|   +-----------+----+------------+----+--------------+     |
|   |         DSA (Datenbank,Filesystem)              |     |
|   +-------------------------------------------------+     |
+--------------------------------------------------------------+
```

Abb. 2.1: Komponenten des Directory-Dienstes für Stufe 1

In den Komponenten MTA, UA und der Dialog-Komponente sind jeweils Moduln integriert - Directory-User-Agents (DUAs) -, die den applikationsspezifischen Directory-Dienst auf allgemeine Directory-Operationen abbilden. Den Benutzern und dem MTA werden dabei die folgenden Dienste zur Verfügung gestellt:

Directory-Dienste für den Message-Dienst-Benutzer

1. Funktionen zum Lesen auf dem Namensgraph ('get O/R-Name')
2. Funktionen zum Lesen eines Directory-Eintrages oder von Eigenschaften eines Objektes
3. Hilfsdienste für eine einfache Erzeugung von Objekten (UA, MTA, Gruppen, ...) und Directory-Einträgen
4. Funktionen zum Überprüfen der Zugriffsrechte bei Operationen auf Directory-Eintragen

Directory-Dienste für den Message-Transfer-Agent

1. Funktionen zur lokalen Verwaltung der einem MTA zugeordneten UA's
2. Funktionen für die Behandlung von Nachrichten an Gruppen
3. Funktionen zum Lesen von Routinginformationen

Im Hinblick auf die zukünftige Erweiterbarkeit sollten unbedingt separate DS-Software-Komponenten mit definierten DUA-DSA-Schnittstellen geschaffen werden. Weiterhin sollte der DUA zu seinem jeweiligen Benutzer hin (MTA oder UA) ebenfalls klar definierte Schnittstellen erhalten.

Die Erweiterung des Directory-Dienstes durch Anbindung anderer Applikationen (Basis-Netzdienste, File-Transfer, RJE usw.), DUA-DSA- und DSA-DSA-Protokolle, die Einbindung von Registrier- und Kopien-Diensten, die Erarbeitung einer Lösung der sich daraus ergebenden Datenkonsistenzprobleme sowie andere noch offener Probleme (wie Datenschutz) werden in einer weiteren Entwicklungsstufe behandelt werden müssen.

2.3 Lokale Dienste

==========================

Einen wesentlichen Teil des DFN-Message-Dienstes wird dem Benutzer durch die lokalen Dienste geboten. Die einzelnen Funktionen sind im folgenden aufgelistet und werden ausführlich im Pflichtenheft [DFNP-'84] beschrieben:

1. Identifikation, Authentifizierung, Autorisierung
2. Anzeigen eingegangener Nachrichten und Bestätigungen
3. Archivieren von Dokumenten, Nachrichten und Texten
4. Retrieval auf den Archiven nach Dokumenten, Nachrichten und Texten
5. Löschen von Dokumenten, Nachrichten und Texten
6. Texterstellung, Textbearbeitung (Formatierung, Fehlersuche, etc.), Formulargenerierung, Formularbearbeitung
7. lokales Namensverzeichnis (lokale Verteilerlisten, Nicknames)
8. Einstellen von Benutzerparametern
9. umfangreiche Help-Funktionen

Zur Unterstützung von Archivierung und Retrieval, die ein leichtes Arbeiten mit den zu bearbeitenden Informationen (Texten) ermöglichen sollen, wird im DFN ein (hilfsweises) Dokumenten-Konzept realisiert (siehe dazu Kap. 2.3.2 in [DFN2-'84]). Es soll die Herstellung von Bezügen zwischen den Dokumenten ermöglichen und ist vor allem im Rahmen der Gruppenkommunikation notwendig.

Beispiele für die Verwendung einer Dokumenten-Id sind:

    - in reply to ‹Doc-Id›   - Fortsetzung von ‹Doc-Id›
    - obsoletes ‹Doc-Id›   - Auszug aus ‹Doc-Id›
    - cross ref ‹Doc-Id›

Der Benutzer muß seine Informationen einerseits in angemessener Form ablegen können (z.B. in "Ordnern", "Ablagen", etc.), andererseits müssen den Informationen die Nachrichten- bzw. Dokument-Attribute zugeordnet werden, so daß die Bezüge leicht hergestellt werden können. Dokument-Archive sollten deshalb als spezielle Informationssysteme angesehen werden und wie folgt strukturiert sein (siehe dazu [DFNP-'84]):

    Dokument   = (Doc_ID, Originator, ..., Subject, ...)
    Bezug   = (von_ID, auf_ID, Bezugsart)
    Zustellung = (Doc-ID, Einsender, Zeitpunkt, ...)

3   Der DFN-Message-Verbund - Anwendung des MHS-Modells

Nachdem in Kapitel 2 das MHS-Modell mit den DFN-relevanten Erweiterungen und Dienstleistungen beschrieben wurde, geht es nun darum, eine geeignete Anwendung des DFN-MHS-Modells auf die konkrete DFN-Umgebung zu finden.

3.1   Topologie des DFN-Message-Verbunds

Die potentiellen Anwender des DFN-Message-Systems arbeiten mit einem breitem Spektrum unterschiedlichster Systeme. Dazu gehören Großrechner unterschiedlicher Hersteller unter verschiedenen Betriebssystemen, Local Area Networks mit integrierten Workstations und hochspezialisierten Grafik-Geräten sowie kleine Arbeitsplatz-Rechner und PCs.

Darüberhinaus befindet sich bereits eine Vielzahl existierender Kommunikationssysteme im Einsatz, die wegen der dort schon etablierten Menge von Benutzern und der angebotenen lokalen Funktionalität unbedingt einbezogen werden sollten. Die Vielfalt der vorhandenen Systeme läßt sich wie folgt klassifizieren:

    - internationale Message-Netze, wobei insbesondere die Forschungs-ori-
      entierten Netze wie EARN und CSNET interessant sind,
    - lokal begrenzte CBMS wie KOMEX, PortaCOM und PROFS und ihre Installa-
      tionen im deutschen Raum,
    - öffentliche Telematik-Dienste mit Teletex, als dem zur Zeit wich-
      tigsten Vertreter.

In diese Umgebung hinein ist nun das MHS-Modell zu projizieren. In [DFN2-'84] wird ausführlich diskutiert, welche Möglichkeiten zur Bildung von Management-Domains vorhanden sind.

Aus "administrativer" Sicht ist naheliegend, daß die Wahl von PRMDs den Organisationsstrukturen im DFN entspricht; die Zuständigkeit für die Namensvergabe ist so in "natürlicher" Weise geregelt, und der gewöhnlich umfangreichere innerorganisatorische Nachrichtenaustausch bleibt auf die zugehörige Domain beschränkt.

Da es in der Startphase des DFN-Message-Verbunds keine ADMD geben wird, läßt sich die Kommunikation zwischen den PRMDs über eine direkte Kommunikation der enthaltenen MTAs unter Umgehung einer vermittelnden MD realisieren. Technisch besteht kein Unterschied zwischen PRMD-PRMD und PRMD-ADMD Kommunikation. Es muß nur sichergestellt werden, daß jede PRMD die anderen PRMDs des DFN kennt und den entsprechenden Kontakt-MTA adressieren kann. Dazu müssen entsprechende Directory-Einträge in jeder PRMD vorhanden sein. Eine spätere Einbeziehung der ADMD der Fernmeldeverwaltung ist dann völlig problemlos möglich.

Für die Anbindung von Teletex-Geräten (über das P5-Protokoll) sollte eine vorübergehende Hilfskonstruktion bereitgestellt werden, die in einer ausgewählten PRMD des DFN anzusiedeln ist und die entsprechenden Aufgaben der ADMD übernimmt. Außerdem könnten so auch einfache Übergangslösungen auf der Basis von Teletex-ähnlichen Implementierungen, wie sie auf verschiedenen Rechnern vorhanden sind, betrieben werden (siehe 3.2.2).

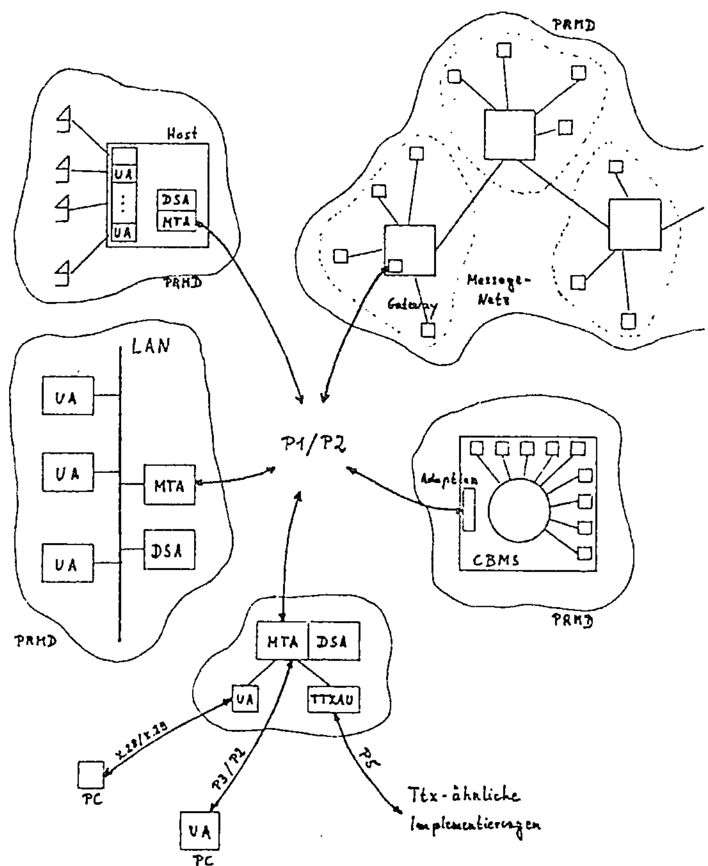

Abb. 3.1: Topologie des DFN-Message-Verbunds

Die Situation für das DFN wird in Abb. 3.1 zusammenfassend dargestellt, in dem die typischen Funktionseinheiten und ihre Einbettung in das MHS-Modell illustriert werden (die Topologie des Verbundes darf nicht mit den tatsächlich vom DFN geplanten Realisierungen verwechselt werden).

3.2 Die Ankopplung vorhandener Kommunikationssysteme

Will man mit Hilfe des DFN-Message-Dienstes möglichst schnell viele Benutzer erreichen, so sind kurzfristig Übergänge zu bereits vorhandenen Kommunikationssystemen zu schaffen, um die dort etablierte Benutzerschar einzubeziehen. Für den deutschen Raum ist daher die Anbindung der hier bereits betriebenen CBMSs wichtig, während im internationalen Bereich der Übergang zu den großen Message-Netzen im Vordergrund steht. Bedeutung wird auch der sofortigen Einbeziehung von Teletex-kompatiblen Implementierungen in den allgemeinen DFN-Message-Dienst beigemessen. In diesem Zusammenhang wäre auch eine Anbindung von DFN-Benutzern, die über Zugang zum Teletex-Dienst verfügen, erstrebenswert.

3.2.1 Übergänge zu vorhandenen CBMS

Wie in Studie I dargelegt, werden lokale CBMSs (neben der über MHS laufenden externen Kommunikation) für organisationsinterne Kommunikation große Bedeutung erlangen, so daß ein auf MHS basierendes Realisierungskonzept der Einbeziehung derartiger Systeme Rechnung tragen muß.

Die Ankopplung vorhandener CBMSs an den DFN-Message-Dienst erfolgt ueber die Protokolle P1/P2. Existierende CBMSs verwalten den Namensraum für ihre Benutzer in eigener Verantwortung, so daß zusammen mit dem PRMD-Namen ohne prinzipielle Probleme eindeutige O/R-Namen für die Benutzer des CBMS gebildet werden können.

Die Probleme bei der Schaffung solcher Übergänge konzentrieren sich in allen Fällen auf zwei große Komplexe:

- Namen und Adressen: keines der relevanten Message Systeme besitzt eine ähnlich umfassende und detaillierte Namens-Struktur,
- Abbildung der Message Attribute: die unterschiedliche Strukturierung der Nachrichten sowie Differenzen in der Semantik der Attribute sind (so gut wie möglich) aufeinander abzubilden.

Darüberhinaus muß aber bemerkt werden, daß weitere Probleme etwa bei der Abbildung von Bestätigungen oder des 'Probe'-Service-Elements auftreten werden.

Bisher ist die Einbeziehung von KOMEX vorgesehen. Das System ist für Siemens BS 2000-Anlagen verfügbar. Ein Verbund von drei KOMEX-Systemen wird seit einigen Jahren in der GMD im Routine-Betrieb eingesetzt. Weitere Installationen befinden sich im Test-Betrieb bzw. in der Vorbereitung. Die Integration weiterer Host-bezogener CBMSs wird in [DFN2-'84] diskutiert.

3.2.2 Teletex

Teletex stellt einen speziellen (allerdings technisch beschränkten) Telekommunikationsdienst zur Verfügung. In der BRD wurde von der DBP ein Teletex-Service mit stetig wachsender Teilnehmerzahl eingerichtet. Verbindungen internationaler Teletex-Netze untereinander sind in Vorbereitung. Die CCITT-Empfehlung X.430 beschreibt ein technisches Verfahren,

Literatur

[CCIT-'84]   CCITT, Study Group VII, Message Handling Systems,
               Draft Recommendations X.400, X.401, X.408, X.409,
                                     X.410, X.411, X.420, X.430

[CCIT-DIR]   CCITT, Study Group VII,
               Interim Rapporteur on Directory Systems
               Directory Systems Working Document, June '84

[Conr-'84]   Conrads, D. et al.
               Bewertung von Message-Systemen
               Beitrag zu dieser Tagung

[DFN1-'84]   DFN, Arbeitsbereich CBMS,
               Studie I: Bewertung von Message-Systemen,
               Deutsches Forschungsnetz, April 1984

[DFN2-'84]   DFN, Arbeitsbereich CBMS,
               Studie II: Konzept für einen DFN-Message-Verbund
               auf der Basis von CCITT Message-Handling-Systems,
               Deutsches Forschungsnetz, Juni 1984

[DFNP-'84]   DFN, Arbeitsbereich CBMS,
               Pflichtenheft für einen DFN-Message-Verbund
               auf der Basis von CCITT Message-Handling-Systems,
               Deutsches Forschungsnetz, Oktober 1984

[ECMA-DIR]   ECMA/TC23/84/110, Directory Service Standard,
               First Draft, September 1984

[IFIP-1]   IFIP WG 6.5, Naming and Directory Services for
               Message Handling Systems,
               Working Paper, Version 4, Juli 1983

[IFIP-2]   IFIP WG 6.5, A User-friendly Naming Convention for
               Public Data Networks,
               Working Paper, Version 2, November 1983

[ISO-DIR]   ISO TC97 SC16, Directory Service for OSI Systems,
               Draft Service Specification
               ISO/TC97/SC16/N1956

[Palm-'83]   Palme, J.
               Computer Conferencing is more than Electronic Mail
               Proceedings of the European Teleinformatics
               Conference (EUTECO)
               North Holland Pub., 1983

[Sant-'84]   Santo, H. et al.
               Der Directory-Dienst im MHS,
               Beitrag zu dieser Tagung

[Wosn-'84]   Wosnitza, L. et al.
               Gruppenkommunikation im MHS-Kontext,
               Beitrag zu dieser Tagung

4 Zusammenfassung

Für eine erste Realisierungsstufe im DFN läßt sich folgendes Vorgehen ableiten:

1) Anbindung existierender CBMSs mit MHS-P1/P2

2) Schaffung von neuen Message-Komponenten auf der Basis MHS-P1/P2

3) Anbindung von TTX-ähnlichen Implementierungen und eventuell auch TTX-Geräten über eine TTXAU

4) Gateway-Projekte zu anderen Forschungsnetzen

Zusätzlich zu den bisher vorliegenden CCITT MHS-Recommodation sind im DFN-Modell Grundzüge der Gruppenkommunikation und ein einfacher Directory-Dienst enthalten. Weiterführende konzeptionelle Arbeiten (umsetzbar in späteren Realisierungen) müssen insbesondere im Hinblick auf einen komfortablen allgemeinen Directory-Dienst und auf Integration fortgeschrittener Methoden der Gruppenkommunikation erfolgen.

Desweiteren sollten komplexere anwendungsorientierte Aktivitäten wie "Electronic Publishing", "Voting", "Joint Editing" oder Vorgängen der Bürokommunikation (z.B. "Umlauf"), die mit dem Schlagwort 'Kommunikations-strukturen' zusammengefaßt werden, in der nächsten Zeit vom Message-Dienst unterstützt werden.

Im Directory-Bereich ist das Fernziel die Errichtung des globalen verteilten Directory-Dienstes über die Kooperation der Domain-lokalen DSA's.

1. BACKGROUND

The Communications research group at University College London has been involved in communications research since the early 1970s; as part of that research, it has had links to many computer networks. The first links were to the US Department of Defence's ARPANET network and to the UK Science and Engineering Research Council's Rutherford and Appleton Laboratory facilities. Since 1975 the group has extended the ARPANET link from a single network into a wider set of concatenated networks (the CATENET). At the same time, networks in the UK have expanded from a few interconnected computers to the public "Packet Switch Stream" (PSS) run by British Telecom which has many international links (IPSS), and a separate universities network (JANET); both of these use the X.25 protocol for access.

It was clear when the initial connection to ARPANET was proposed that other groups in the UK would wish to use the connection. These groups were mainly concerned to use the link to improve the communications between themselves and researchers working in the same field in the US. Therefore we have always allowed our facilities to be shared by other users wishing to access UK facilities from the US and vice-versa. These groups, on both sides of the Atlantic, have provided continuous feedback and further requirements on the experimental services we have provided. We have certainly learnt many valuable things from the effort to provide communications systems to a service standard rather than for one-off experiments.

Initially we provided only a terminal access service, but we have found that bulk transfers are essential between cooperating groups using their local computers. Some of the most serious users have been exchange visitors, who have used our facilities to continue work previously done locally. As networks have developed in the UK, users have wished to have more reliable or higher speed access by using the networks rather than the Public Switched Telephone Network (PSTN), or in some cases just more convenient or cheaper access. This applies to interactive use, but more importantly to bulk transfers where transferring files to users' own machines is vital. The development of mail systems in the UK has extended these requirements and mail forwarding forms a major component of current traffic.

2. THE INTERCONNECTION ARCHITECTURE

The Department has external connections to two X.25 networks and the DARPA CATENET, see Fig. 1. In addition we have been participants in the UNIVERSE Satellite project [1] and are starting a higher speed terrestrial successor (ADMIRAL); these involve further sets of networks and protocols within the Department, but are ignored in this paper.

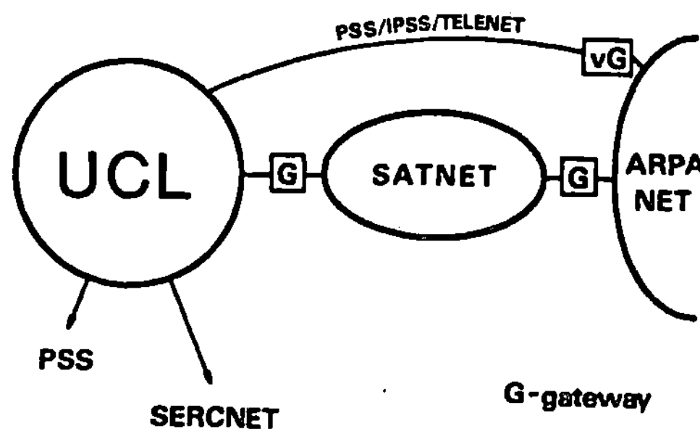

Fig. 1: Schematic of the UK-US International Connection

The two X.25 networks are PSS and JANET. PSS is a national public network from British Telecom (BT), which has an X.75 gateway to IPSS; from this last the national networks of other PTTs can be reached. JANET (Joint Academic Network) is a private X.25 network of the Science and Engineering Research Council (SERC) and the Computer Board. JANET links most of the UK universities and scientific research centres. These two networks use compatible X.25 access protocols. PSS uses the X.121 address scheme, JANET has its own numbering scheme. Within the UK we use a set of higher level protocols, agreed amongst the UK academic community and adopted by the Department of Trade for use pending suitable International Standards Organisation (ISO) standards [2].

The DARPA CATENET [3] is a research system, sponsored by the US Defense Advanced Research Projects Agency (DARPA). The CATENET consists of a number of distinct networks using various types of technology and network level protocols. The networks are interconnected by gateways. Hosts participating in the CATENET use a common set of internet protocols based on datagrams (IP) [4] and a common transport protocol called TCP. Common application protocols are widely used for remote login, file transfer and computer mail.

Finally the Public Switched Telephone Network (PSTN) is also available for several purposes. It is used for Terminal access to UCL facilities, which then allows further access to facilities via the other packet networks mentioned. It is also used directly for relaying messages by a store and forward technique to sites bases not accessible via international packet nets, but reachable by the PSTN; the procedures used (USENET, UUCP) [8], are used on a world-wide basis on UNIX machines.

When we first considered using our hosts on external networks for mail, file transfer and remote login we wished to use the same host on all the external networks. At the same time we considered how we could arrange for a number of hosts to be available on all the networks. We soon realised that placing network connections and real-time protocol software in a multi-access host considerably degrades its performance to all the users. Running a number of different protocols in the same host would not be practical. The problems of multiple connections, multiple hosts and multiple protocols required a new approach.

To solve the multi-network/performance problem we started a research effort to use a local area network (LAN) with **access machines**. Each network connection is controlled by an access computer running the relevant network protocols - X.25 for PSS and JANET, TCP/IP for the CATENET. Each network protocol implementation has a standard interface to higher level protocol implementations, we call this interface "Clean and Simple" [5]. The applications protocols (mail, file transfer, and remote login) are run on the hosts. The access machines and hosts are all connected to a Cambridge Ring LAN. A special protocol, Interprocess · Clean and Simple (IPCS), allows applications processes to control network connections in the access machines across the LAN. Figure 2 shows the internal networks, with the hosts and access machines. Communication between the hosts and the access machines uses specific UK Cambridge Ring network protocols. In this way each host is effectively 'on' PSS, JANET, and the CATENET.

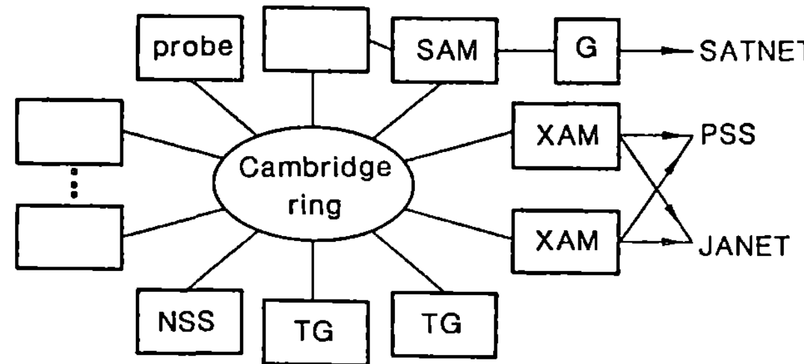

XAM = X25 Access Machines
SAM = SATNET Access Machines
 TG = Terminal Gateway
NSS = Network Services System
      (Mail, File, Relay)
  G = CATENET Gateway
Probe = Monitoring Machine

Fig. 2:   Schematic of the UCL Interconnection System

Our connection to the DARPA CATENET is via a gateway to the Atlantic Packet Satellite Network (SATNET) [6]). Our local area networks are a component of the CATENET. SATNET is the main communication vehicle of a collaborative project between a number of national defence agencies; as a result of this funding British Telecom have insisted that we do not use this network for civil traffic using our services, unless that traffic is sponsored by Defence agencies. To accommodate this restriction we also use one or more X.25 virtual calls to a special gateway in the US to reach the ARPANET [7]. This connection carries the IP/TCP protocol packets that would otherwise use SATNET. We call this connection an IP tunnel, it uses PSS, IPSS and TELENET in the US; Fig. 1 shows this path.

3. NAMING, ADDRESSING AND ROUTING

The position of UCL, astride several networks, not only provides wide connectivity for UCL hosts, but also allows UCL to act as a relay agent between differing systems. Both these activities require the communication facilities at UCL to be able to accommodate diverse sets of protocols, addressing structures and routing strategies.

For computer mail, UCL belongs to two separate mail domains: the DARPA CATENET system and the JNT system (Joint Network Team of the Computer Board and SERC) in the UK. Other mail domains which connect into these systems can also be reached from UCL. The largest single one reachable directly is USENET [8], a collection of computers using the UUCP protocols running on the UNIX operating system. The extent of USENET is truly world wide, and is reached via a series of store and forward paths at the Host level. Mail constitutes the widest address space dealt with by UCL. While mail addresses can be viewed as belonging to a single, flat, address space, the use of different message transfer protocols result in the need for mail relays to provide connectivity between them. UCL acts as a mail relay between the DARPA and JNT domains, and is used also as a relay to the USENET one.

At the network level, UCL belongs to at least four addressing domains: PSS, JANET, the DARPA Catenet, and the USENET network. Addresses on PSS conform to the X.121 format, with UCL represented as a 12 digit network address and 2 digits for local call demultiplexing. JANET uses separate network addressing, distinct from X121, without a demultiplexing ability. JANET hosts can, however, use the extended addressing capabilities provided by the Network Independent Transport Service (NITS) [9]. The DARPA Catenet uses a hierarchical 32-bit host address structure, within which UCL has defined its own internal addressing. USENET uses a complex source-routing host address structure; these host addresses are mapped directly at each stage into PSTN telephone numbers or Packet Data Network Host numbers.

The interconnection facilities at UCL aim to shield applications as much as possible from such network diversity. This is achieved in three ways: the use of host names at the user interface rather than actual addresses; the use of a standard transport interface for all destination networks (Clean and Simple); and the separation of network-specific functions into distinct network access machines.

For applications such as mail, file transfer and terminal traffic to be able to use host names rather than direct addresses, a table-based name translator mechanism [10] is used which produces the three components of an address needed locally within UCL: the network to be accessed, the address of the remote host on its network, and the service required. The first determines which network access machine is to be used; the latter two are supplied to the access machine for addressing and/or routing purposes, and possible forwarding.

The network access machine for the DARPA Catenet has two routes available to it: via SATNET and via the IP tunnel. The information necessary for correct routing, namely the user's affiliation, is initially available only at the user or application level. This has to be preserved across any name-to-address translation and must be passed down to the network level. In the absence of any other facility in the DARPA protocols, this information is embedded in the source address [11].

4. SERVICES

We offer three primary services to external customers, these are terminal access, file transfer and computer mail. The terminal access service uses its own resources and hosts, though these are all small micro-computers. The file and mail services use general purpose hosts, and run as applications on those hosts. All of the services use the shared network access machines to provide and obtain network connections. We provide accounts for external customers on a dedicated computer called the Network Service System (NSS). These accounts allow customers to stage file transfers and access computer mail where their local facilities are not suitable. The NSS is also used as the collection point for incoming mail.

4.1 Terminal Access

The terminal access facilities enable a user at a terminal, or host, to access an interactive host on the other side of the Atlantic. We find UK users logging into ARPANET computers as well as US users logging into JANET and PSS connected computers. The users host, or terminal connection, would be using quite different protocols to those used by the remote host, at all levels.

The terminal protocol translation is carried out on a dedicated machine called the Terminal Gateway (TG, currently a LSI11/23 with no backing store) attached to a Cambridge ring (see Fig. 2). Each connected network is represented by a separate process, whose task is to interface with the network using IPCS on one hand and translate the appropriate protocol on the other. The TG allows access, via the Cambridge Ring, from local terminal multiplexors.

In order to avoid the "N-Squared" problem of translating between N different protocols, it was decided to define a canonical internal protocol into which all the terminal protocols are converted on input and output. We have to deal with five terminal protocols; X.29 over PSS and JANET, ITP which is a protocol defined specifically for use on JANET, TELNET which is used on the DARPA Catenet, a null protocol for interaction with local terminals, and another protocol internal to UCL for accessing local hosts over the Cambridge Ring. Due to the large differences between the protocols, it is not easy to convert protocol messages from one terminal protocol to another. Hence a simple subset of messages was defined, in particular ECHO/NO-ECHO is available which is necessary for providing host password privacy.

As one of the access paths to the US is over IPSS it is necessary to provide as economical a service as possible. Thus the preferred mode of operation of the TG is forwarding of data a line at a time with local echo.

Access control is necessary on this system, therefore it is required for users to "Login" to the TG before making a connection to a host computer. Login verification is done using the normal password file of a service host at UCL. A backup password file is kept on the TG in memory. Call accounting information is sent to the service host and stored in a database in order to produce accounting information.

4.2 File Transfer

We provide a facility at UCL whereby users may transfer files between local UCL hosts, UK hosts and hosts on the ARPANET. The file transfer protocol largely used within the UK community is the Network Independent File Transfer Protocol (NIFTP), whilst the DARPA community has a completely different file transfer protocol (FTP). Neither of these protocols provides for the user to specify addresses containing routes, in the way that the mail protocols do. To use these protocols directly for transfers between the UK and the ARPANET, a user must stage the file transfer at a host which implements both the protocols. We encourage customers to use the mail facilities to transfer files to take advantage of the automatic relaying. However, mail systems do not always offer all the facilities needed for file transfers (e.g. binary transfers) hence we provide facilities for users to access the file transfer protocols directly.

The NIFTP system at UCL [12] is a spooled system in contrast to the majority of ARPANET FTP interfaces which do the transfer in real time. The spooling system allows us greater flexibility to use resources which are scarce, for example to do transfers when the IPSS Tunnel is open for terminal traffic etc. We have arranged that our NIFTP and FTP interfaces are unified to provide both spooled and real-time facilities.

Access control is provided by using the standard facilities of the UCL host. The NIFTP protocol includes parameters for user identification, for an incoming request these are checked before the transfer is accepted. For outgoing requests the user's permissions to use a particular network can be checked. In the case of the ARPANET, this mechanism is used to determine if the transfer is to occur over SATNET or IPSS.

4.3 Computer Mail

UCL provides two major types of mail service. The basic service enables local users to send messages, and to manipulate received messages. A selection of interfaces are provided, so that users may choose a system appropriate to their needs. The second service is to provide a connection to remote message systems, which takes both the form of remote services for local users, and increasingly of relayed services. The major relaying service is between the UK Networks, and the DARPA CATENET, as discussed in sections 2 and 3. Another function performed by UCL is the handling of network mailing lists [13].

Within the Department we operate a number of Message Transfer Protocols, two of which are particularly important for the services provided. The first is the JNT Mail Protocol, which is based on a UCL proposal, and is now used extensively in the UK Academic Community, with NIFTP (see Section 4.2) used to perform the data transfer [14]. The second is SMTP (Simple Mail Transfer Protocol) used within the DARPA CATENET. A less important third one is the UUCP philosopy used in USENET. UUCP is much more than a simple protocol. It presupposes that it is possible to route messages over a concatenation of Packet Data Network and PSTN paths, by store and forward techniques. Many hosts throughout the world have adopted the UUCP protocols, and their collection is called USENET. While USENET does not provide the same level of reliability or error notification as JNT or ARPA mail, it does provide additional connectivity on a world-wide basis to Hosts using the UNIX operating system. We both run the UUCP PSTN protocols ourselves, and use other systems to relay our mail via their systems. Message level relay allows protocol conversion for the services offered to be performed in a straightforward manner. Besides the functionality of the system within one machine, it is also important to be able to utilise the flexibility of the interconnection architecture. An additional problem faced by UCL is in the differences between the UK and DARPA CATENET address specification. Some solutions are discussed in [15].

Message Level relaying also introduces another set of addressing problems on top of those discussed in Section 3. In essence, the syntax of the present mail protocols allows the specification of a user mailbox to be translated into a source route consisting of a number of mail relay hosts. The translation algorithms of Section 3 will be applied to the first host of the source route to identify a message level relay. In some cases, it is desirable for the Message Handling System to map a domain specification into a source route when there is no direct connectivity with the domain specified. This will hide complex connectivity from the user, and allow simple specification of mailboxes on remote networks.

The local mail system at UCL is not described here, suffice to state that it provides the needed functionality for a message relay. All the host machines at UCL run identical systems and derive connectivity information from a common set of tables. This consists of many thousands of lines of user and host mappings, and so for efficiency reasons, this information is loaded into a database on each machine which is accessed by the mailsystem. Some tables are updated manually, and others (such as the DARPA host tables) are automatically extracted from remote network databases. To allow machine dependencies, and connectivity to the various front end machines to be adjusted independently, a dynamic tailoring schema is used to allow variable mappings. This set up allows the various mail protocols to be used over different transport protocols in a simple manner. The JNT Mail Protocol implementation is also structured to allow protocol layering at the file transfer level, to allow dynamic configuration with alternative mail and file transfer protocols.

4.4 Protocol Structures Supported

The gammut of external protocols used is of clear significance. Figure 2 has already shown direct gateways to PSS, JANET, SATNET and the PSTN. The actual protocols used internally have been designed to suit our own purposes - they are not relevant externally, and change as we extend our internal technology. However, the protocols supported externally are relevant to others. It is convenient to divide these at the transport level. At first sight, the network access level (Interface to level 3 of the ISO structure) would seem the right point. Unfortunately at the network level some of our networks (the PSTN and the DARPA CATENET) use the connectionless service, whilst JANET and PSS provide the connection-orientated service. However at the transport level, all the networks use a connection-oriented service, hence we have chosen this as one of the protocol conversion levels [16]. A summary of the external protocols we use currently up to the transport level is shown in Fig. 3. Here the protocols used have been mentioned specifically earlier in this section. The common interface to the transport level is provided almost always by the Clean and Simple interfaces mentioned in Section 4.1.

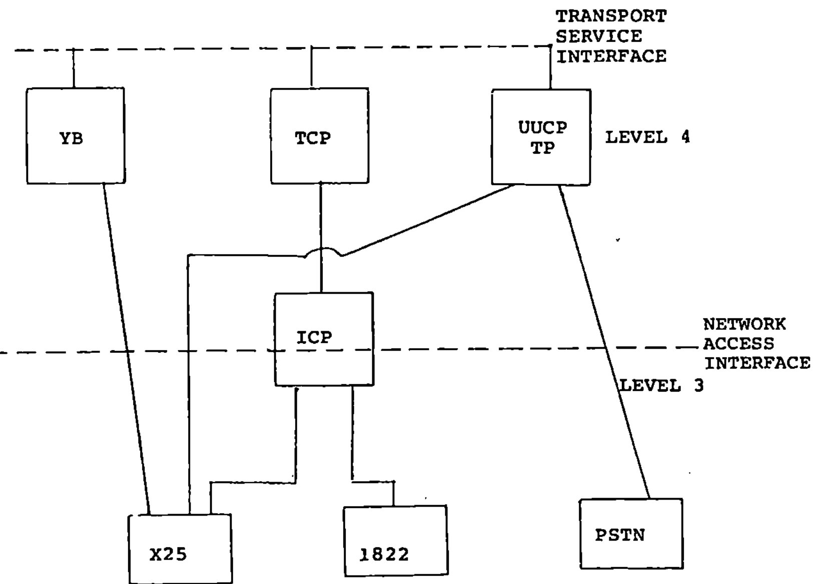

Fig. 3: The Protocols Used at the Transport Levels and below

For the high level functions, the specific protocols have also been mentioned already in the text. A schematic of their interconnection is shown in Fig. 4 for terminal facilities, and in Fig. 5 for file transfer and electronic mail.

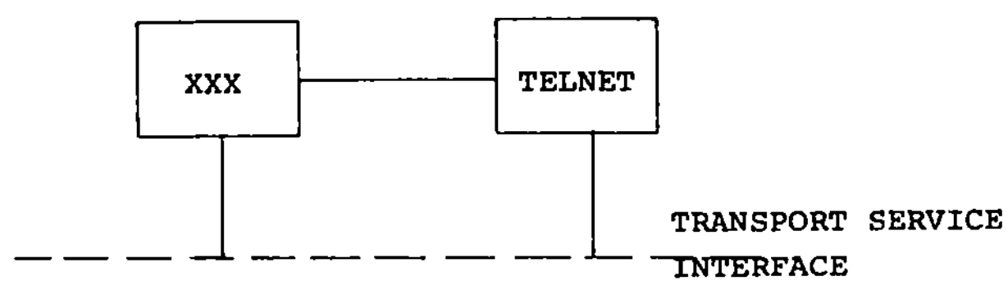

Fig. 4: Protocol Translation for Terminal Traffic

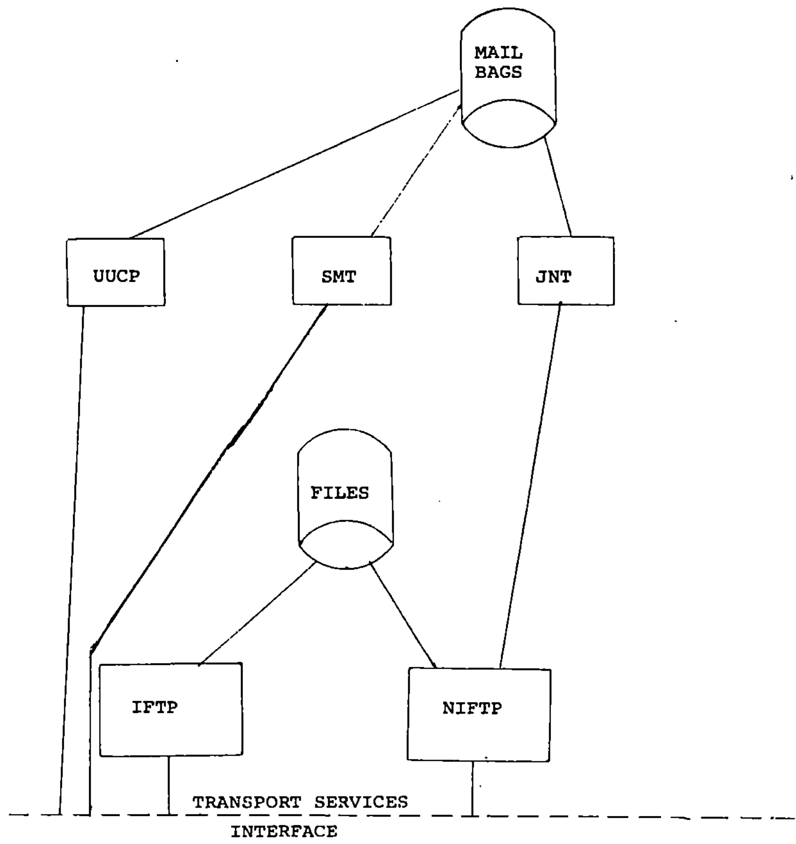

Fig. 5: The Protocols Used for Bulk Traffic

5. MANAGEMENT FACILITIES

The interconnection facility has considerable management problems, there are four areas in which we have provided support for managing the service:

(1) Daily operation - detecting failures and reconfiguring the resources.

(2) Access control and accounting.

(3) Flexibility for growth and parallel development.

(4) Performance measurement.

The interconnection service is run without operators. All of the network access machines and the Terminal Gateways are automatically rebooted when they crash. All the connected networks, and consequently the network access machines, are monitored so all crashes are detected and logged.

Access control is applied in both the Terminal Gateway and NSS by traditional password mechanisms. Access control is also applied to mail relayed through UCL by checking the sending and receiving addresses. Apart from limiting access to authorised users, we need to know who is using the facilities and how much resource they are consuming so that we can enforce charging. Data collection is extensive, but policy implementation is kept simple and in many cases manual, so that managerial decisions can be quickly effected but not triggered automatically.

The service has grown and developed over a number of systems. Use of the LAN and network access machine architecture has allowed test systems to run in parallel with production systems, and the easy reconfiguration of resources by adjusting tables. We now have two X.25 access machines, two Terminal Gateways and one TCP access machine. We can quickly reconfigure these to almost any combination required if any other component fails. There is room for further expansion of machines as the load increases or to introduce new services.

As part of our requirement to understand and improve the functions of the system, some facilities have been built in to measure throughput and delay at different levels. For example we regularly echo packets from distant gateways. Of course this is used to detect malfunctions; also, however, the time for the echo to return gives a measure of load and bottlenecks. This can be used, for example, to open additional X25 circuits over the packet switched networks or to arrange the increase of the speed of access lines. Where remote hosts implement high level echos, or provide time stamps, these can be used to monitor high level performance. Examples are relaying messages by different routes (eg. UUCP, SATNET or or PSS). From the time of the messages, we can determine whether the multiple relay host via UUCP result in unacceptable performance, or whether certain of the complex relay facilities are more expeditious. From the difference in delays in message arrival via SATNET and IPSS, we can determine if the underlying networks are overloaded. This overload may be the load on transmission or gateways; it may also be an excessive request for use of the transmission paths, leading to longer queue times in the UCL (or remote) spooling mailers.

6. LIAISON

To enable customers to make full use of the inter-connection facilities described above we also provide an **Internet Liaison** service.

The liaison provision is the difference between a 'service' to others and the mere provision of access to our facilities; personal contact is very important. Other members of the research group provide technical help but all contact with external customers is channelled through a single liaison person. Liaison can be contacted by an advertised telephone number, letter, or computer mail to a special mailbox. The tasks performed by liaison include:

- provision of technical information on using the facilities,
- dealing with specific questions or difficulties,
- ensuring adequate notification of forthcoming changes or breaks in service,
- helping potential customers apply for formal permission to use the service and provision of early support,
- writing and distribution of a monthly newsletter,
- ensuring all users have up to date applications and removing old users from the system.

On average some 10 queries are dealt with each working day from a

user population of approximately 150 people, on both sides of the
Atlantic.

Our system is continuously expanding, additional services are being
offered, and poorer ones occasionally being replaced. While the USER
notices help keep users informed of the changes and addresses, the
documentation can easily become unwieldy. We try to hold as much HELP
information on-line, to deal with this problem; it is very difficult
to keep our **written** documentation for users up to date, and
distributed in a streamlined form.

7. FUTURE EXTENSIONS AND INTERNATIONAL STANDARDS

It must by now be obvious that only in the use of Packet and Public
Switched Telephone Networks, and one terminal protocol, are we
compatible with the emerging international standards. To a large
extent this is inevitable, because of our interconnection role; we
cannot progress more rapidly than our interconnection partners.
However we are committed to move towards international standards to
the extent this is feasible and economic.

The British JANET is currently considering an "INTERCEPT" Strategy
to move away from its "Coloured Books" service of protocols to the
international standard ones. The progress will be slow. Many of the
older computers will never be equipped with the newer standards, and
one will have to wait until those systems are replaced. Even when
incorporation of new standards is feasible, it will be expensive and
take a long time to occur - after the standards are really accepted.
For this reason, we expect that the major motivatation for us to adopt
the new standards, other than research interest, is the desire to
increase our connections with new communities. This was our real
reason to add UUCP to our repertoire; it is no international standard,
but a very prevalent pragmatic one.

We are moving towards the international standards services in
several directions. We already are implementing the X400 Message
protocol [17]. One reason is to allow access from Teletex terminals;
a second is to allow X400 interconnected with Canadian Academics [18],
the ESPRIT community [19] and other European partners. Technically we
could expect to provide an X400 relay service also within a year or
so; we do not yet know whether we will provide a real service. We
already have provided a Teletex service interface in a dedicated

server on our LAN [20]. We have not provided a service here because
we are more interested in Computer Based Message Services (CBMS) than
a terminal-terminal Teletex Service; however we are interested in the
X430 provision of access to our CBMS from Teletex terminals. We are
also exploring whether the widely accepted KERMIT procedures [21] are
a suitable service interface to our system.

A good guide to the emerging ISO standards is given in [22]. We
are interested in the ISO Transport (TP) for several reasons. Of
course it is fundamental to X400 and Teletex. It will be required for
the international File Transfer Access (FTAM) when it is defined. The
British JANET plans for FTAM are still undefined, so that aspect of
Fig. 8 is shown with a question mark. Finally we have a research
interest in the interoperatability between TP and TCP.

The move towards different universal terminal protocols we find
interesting, and a subject for research. Its relevance to relay
service is not clear to us at present. Not only are the relevance of
those terminal protocols still under, but also whether they will be
taken up by JANET or ARPANET has not even been raised.

We will certainly attach some of the newer network services
experimentally. We already have ordered two lines to the UK
Integrated Data Access (the Experimental ISDN) [23]. This will allow
us to understand how its features can be integrated into our system;
whether it will be part of any UCL interconnection service is
premature to say. Finally, as part of a pure research programme [2],
we will be attached to British Telecom 2 Mbps Megastream facilities.
Again its relevance to the subject of this paper is unclear.

Thus the service extensions we will be exploring over the next
couple of years are illustrated in Figs 6 - 8. We would hope that
some of the existing ones can be stopped when new ones are introduced
- otherwise the software maintenance would become excessive.

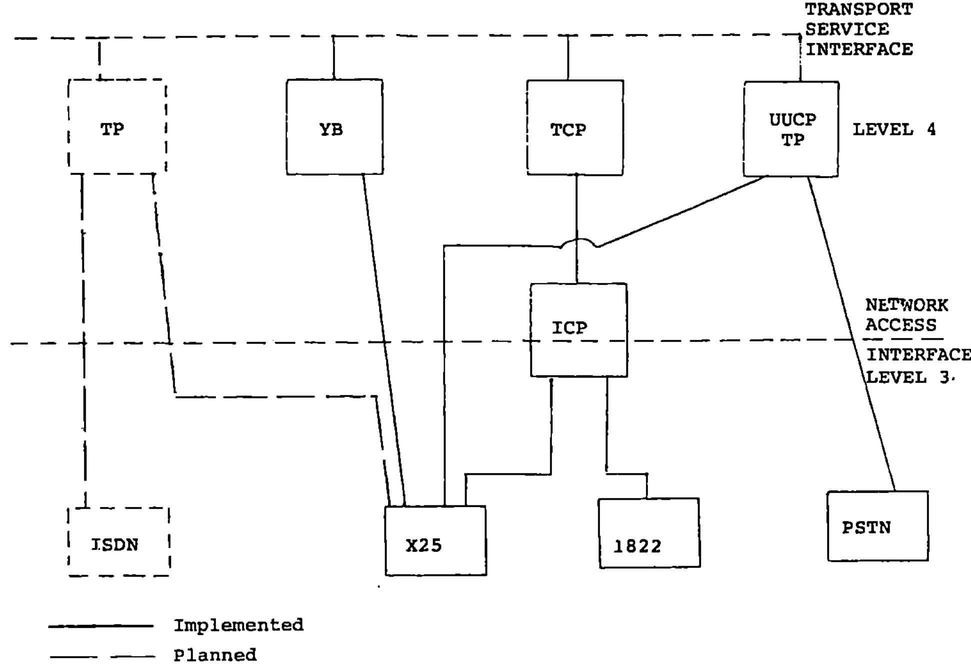

Fig. 6: The Protocols used at the Transport Level and Below

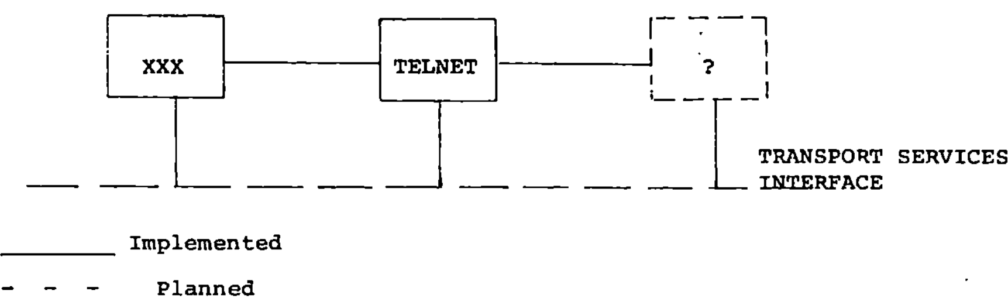

Fig. 7: Proposed Protocol Translation for Terminal Traffic

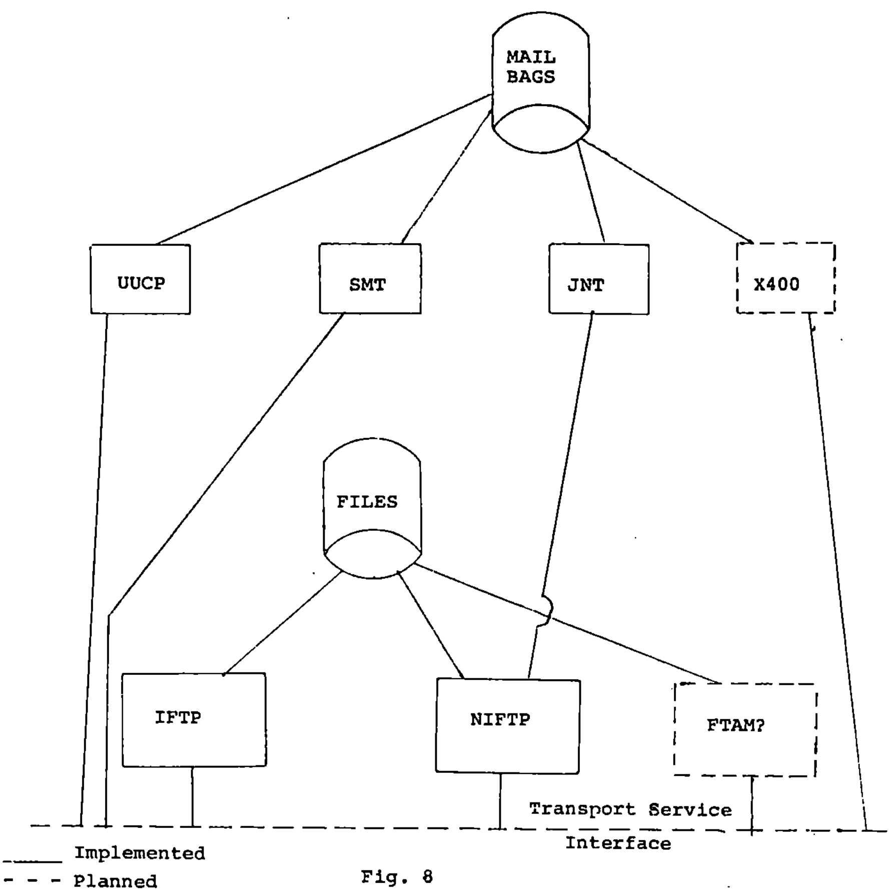

Fig. 8:   Proposed Protocols for Bulk Traffic

8.   FUTURE EXTENSIONS AND MANAGEMENT

Concurrently with protocol implementation, we intend an ongoing and high priority activity in Network Management and Control. As the new services are introduced, the need for monitoring, reconfiguration and performance measurement becomes ever more essential. As the interconnection alternatives increase, and the changes become better understood, automated methods of cost reduction will assume an increasing role. Regulatory issues must be considered, of course; in this context, the international aspect of our activity is of particular relevance [24].

Probably paramount in this management area is, however, that of Directory Services. Many of these will be distributed - for example the location, addresses and routes to specific services or individual. Some will be local - for example the authorisation, authentication, billing and facilities of the Users of our relays. Data Base Management facilities will be essential; we do not yet know to what extent the service requires different Data Base Management Systems. In some areas CCITT and IFIP are taking the initiative already; where we can we will participate in, and follow, their recommendations. Where they are not being treated yet in standard ways, we will make our own developments.

We are greatly helped in these newer areas by two initiatives - the UK Alvey programme [25] and the ESPRIT one [19]. We participate in both these programmes with industrial partners, who are initially interested in incorporating International standards and recommendations. We would wish to follow them in any case; our industrial partners would force us to even, if we wished something different.

9. SUMMARY

The Department of Computer Science is now offering a network interconnection service between the US and the UK for scientific users. The primary users of this service are research groups cooperating with co-workers on either side of the Atlantic. Most of the US groups use hosts on the ARPANET.

This service has grown from the very early provision of dial-in access on an experimental system to the provision of sophisticated terminal, file and mail relaying facilities. The problems of network inter-connection encountered in the provision of these facilities have been tackled through research; however the research has been directed by the service requirements.

This service has never remained static and frozen. It has been necessary to evolve the service facilities underlying protocols, hardware, management facilities and liaison services. To provide a technical and institutional architecture to allow this evolution has proved a demanding but feasible task. While further work is required both to improve the services and keeping with the external developments, there is no sign that the underlying architecture cannot meet these challenges.

While the emergence of international standards should make an interconnection service like the UCL one unnecessary in the future, there is no sign that this will occur within the next five years in the UK. It may well be politic to run such a service in a different organisation than that of an academic research group in Computer Science; the range of services will not be providable without both a service operator and a development group, with a strong interest in international computer communications services.

10. ACKNOWLEDGEMENTS

This work represents the efforts of many people. A large part of the paper is based on [26] which was authored by Cole, Daniel, Higginson, Kille, Lawson, Lloyd and myself. The hardware development and maintenance relied on Gamble and Munoz. The research and monitoring rely also on the work of Bacarisse, Brink, Knight, Ma and Wilbur. Even this list is restricted arbitrarily to those who contributed to the work and were at UCL on September 1, 1984; the work really has built on dozens of people at UCL and our collaborators in other places over the last decade.

Finally, the development work described would not have been possible without financial support from the UK Atomic Energy Authority, British Library, British Telecom, Department of Trade and Industry, Ministry of Defence, Science and Engineering Research Council and the US Defence Advanced Research Projects Agency. The Service activities are now supported almost exclusively by the Ministry of Defence for their category of users, the Science and Engineering Research Council for academic ones, and the Department of Trade and Industry (through the Alvey Directorate) for their contractors. The support of all these bodies is gratefully acknowledged.

References

(1) Kirstein, P., et al, "The UNIVERSE Project", Proc ICCC 82, pp442-447, North Holland, 1982.

(2) Rosner, R., "Towards OSI among UK Universities", Proc ICCC 82, pp607-611, North Holland, 1982.

(3) Postel, J.B., et al, "Recent developments in the DARPA Internet Program", Proc ICCC 82, pp975-980, North Holland, 1982.

(4) Postel, J.B., et al "The ARPA Internet Protocol", Computer Networks, 5(1981), pp261-271, November 1981.

(5) Braden, R., et al, "A Distributed Approach to the Interconnection of Heterogeneous Computer Networks", Proc. Communications Architectures and Protocols Symp, (SIGCOMM 83) Austin Texas, pp254-259, ACM, March 1983.

(6) Jacobs, D.M. et al: "Packet Satellite Network Design Issues", Proc. Nat. Telecom Conference, IEEE, pp45.2.1-12, 1979.

(7) Cole, R., "User Experience & Evaluation of International X.25 Services", Proc. Telecoms Today Conf. ONLINE, pp107-118, 9 USENET 1984.

(8) Nowitz, D.A. and M.E. Lasks "A Dial-up Network of UNIX Systems", UNIX Programmers Manual, 7th Edition, Bell Laboratories, August 1978.

(9) Post Office SG3, "A Network Independent Transport Service", Joint Network Team, Rutherford Appleton Laboratory, February 1980.

(10) Cole, R., et al, "International net face problems handling mail and file transfer", Data Communications, vol 12,6, pp175-187, McGraw-Hill, June 1983.

(11) Cole, R., and P. Lloyd, "A Flexible Architecture for Protocol Studies in a Multi-network Environment", Proc. IFIP 83, pp401-406, North Holland 1983.

(12) Higginson, P.L., and R. Moulton, "Experiences with Use of the UK Network Independent File Transfer Protocol on Several Networks", Proc ICCC 82, pp913-918, North Holland, 1982.

(13) Kille, S.E., "Integration of Electronic Mail and Conferencing Systems", second IFIP on Message Handling Systems, pp259-268, North Holland, 1984.

(14) Kille, S.E., (editor), "JNT Mail Protocol", Joint Network Team, Rutherford Appleton Laboratory, March 1984.

(15) Kille, S.E., "The Interconnection of Multiple Internetwork Mail Systems Using Different Addressing Strategies", IFIP WG6.5 Workshop, Versailles, North Holland Publications, April 1982.

(16) Braden, R. and R. Cole, "Some Problems in the Interconnection of Computer Networks", Proc ICCC 82, pp969-974, North Holland, 1982.

(17) CCITT.5/VII, "Recommendations X400, Message Handling Systems Model - Service Element", ITU, Geneva, 1983.

(18) Neufeld, G.W., "EAN a Distributed Message System", Proc CIPS National Meeting, Ottawa, pp144-149, May 1983.

(19) "Call for Proposals for the Informtion Exchange Systems of the European Strategic Programme for Research and Development in Information Technology", CEC, Brussels, ITTTF/2443/84-EN, 1984.

(20) Tuck, W. et al, "The Teletex Experiment on Project Universe" Proc ICCC 84, pp359-363, North Holland, 1984.

(21) da Cruz, F., and B. Catchings, "File Transfer Protocol for Universities", Byte Magazine, 4, 6, pp255-278, 1984.

(22) McCrum, W.A., "International Standards for Data Communications", Proc ICCC 84, pp554-559, North Holland, 1984.

(23) Wedlake, J., "The Use of X Series Interfaces in an ISDN", Proc ICCC 84, pp619-625, North Holland, 1984.

(24) Kirstein, P.T. "The Interconnection of Message Handling Systems in the Light of Current CCITT Recommendations, Proc 2nd IFIP Conf. on Message Handling Systems, pp227-234, North Holland, 1984.

(25) Barber, D., "Net Working Expectations and the Fifth Generation", Proc Networks 84, ONLINE, pp451-464, London, 1984.

(26) Cole, R., et al: "Network Connection Facilities at UCL", Proc. ICCC 84, pp742-746, North Holland, 1984.

NETWORKS-INTERCONNECTIONS - AN ARCHITECTURAL REFERENCE MODEL

B. Butscher (HMI, Berlin, Germany)

L. Lenzini   (CNR, Pisa, Italy)

R. Morling   (PCL, London, UK)

C.A. Vissers (Twente Univ., Enschede, Netherlands)

R. Popescu-Zeletin (HMI, Berlin, Germany)

M. Van Sinderen (Twente Univ., Enschede, Netherlands)

Edited by R. Popescu-Zeletin

        Hahn-Meitner-Institut

        1000 Berlin 39

        Glienicker Str. 100

        Germany

Abstract: One of the major problems in understanding the different
approaches in interconnecting networks of different technologies is
the lack of reference to a general model. The paper develops the rationales for a reference model of network interconnections and focus
on the architectural implications for the definition of the services
and protocols in the ISO/OSI environment.

This work has been sponsored by the Commission of the European Communities; the programmes in informatics: COST 11bis and EWICS.

1. INTRODUCTION

A major technical challenge associated with the use of data communication technologies is how to interconnect different networks which have been designed at different times for different environments and applications. Today network designers are faced with heterogeneity of networks just as they were previously faced with heterogeneity of computers in a single network. Different types of networks are available and in operation from long-haul networks (X.25 public networks, satellite networks, etc.) to local area networks. The diversity of available products makes their interconnection to be at least as difficult as interconnecting heterogeneous computers.

In order to solve the network interconnection problem in a global way, it is necessary to refer to a general model for interconnection. Since the OSI Reference Model (OSI/RM) /ISO83/ was developed to define the architecture of interconnected systems, designing network interconnection in a global way means establishing the implications and necessary extensions in the OSI/RM in order to achieve this scope. Note that the model refers to <u>systems interconnections</u> and does not preclude the interconnection of networks.

The adopted structuring technique in terms of layers permits each layer to wrap the layers below and so to isolate them from the layers above. This technique allows the development of products at each layer separately making them independent of the other layers. In other words, the internal functioning of one network in the interconnected environment is not visible from outside of it from a certain layer above; only the external result is visible. The concept of <u>external visibility</u> permits heterogeneity to be reconciled with compatibility, by restricting the commonality to external relations and leaving room to solve the heterogeneity internally /GIE79/.

From an <u>internal point of view</u>, considerable problems have to be solved in interconnecting networks of widely different technologies, conceived in different environments, at different times, for different purposes, etc. This paper focuses on the network interconnection problem in the framework of the OSI/RM. It outlines the different approaches already available or proposed by different organizations and develops an architectural model for the interconnection of heterogeneous networks in the above mentioned framework.

2. Existing approaches

The different approaches reported in literature may be classified in two categories:

- approaches relying on available subnetworks without referencing to a global solution. They do not rely on the ISO/RM even if they are based on a layered architecture.

- approaches which define global architectural solutions within the ISO/RM framework.

In the first category the ARPANET IP architecture and the XEROX PUP Internetwork Architecture are the best known solutions, the second category comprises the NBS and ECMA proposals for the internetwork architectures. The ARPANET and the Xerox PUP design architectures rely on the uniformity of the internetwork packet format which can be encapsulated for transmission through individual subnetworks in their own formats. On top of the interconnection layer the homogeneity of the layers above is presupposed in the interconnected networks. The service offered by the ARPA IP and XEROX PUP layer is a global connection less type of service. The interconnection level in both architectures enhance/de-enhance the services offered by each subnetwork and offer a global uniform service for the layers above.

The second category of proposals pertains to the ISO/OSI RM model and defines the architecture of the layer 3 (network layer). The OSI/RM /ISO83 states that subnetwork interconnection has to be solved in layer 3, since layer 2 provides the access mechanism in a particular subnetwork and layer 4 has an end-to-end significance.

The basic service of the Network Layer is to provide the transparent transfer of all data submitted by the Transport Layer. The Network Layer is assumed to contain functions necessary to mask the differences in the characteristics of different transmission and network technologies. The service provided at each end of a network access point shall be the same even in the case of a network interconnection spanning several subnetworks where each subnetwork offers dissimilar services.

The first three sublayers in layer 3 identified by ECMA and NBS proposals have the same functionality. The first sublayer, is termed Subnetwork Services Sublayer /NBS/ or Communication Services Sublayer /ECMA/.
They are peculiar to the subnetworks to be interconnected. They include the functions needed to support the direct interactions between a
pair of entities, using a particular subnetwork type. These services
can be connectionless or connection oriented type and are subnetwork
specific. The second sublayer performs any functions necessary to map
the type of service provided by any specific subnetwork into that type
chosen for subnetworks interconnection. This type of service can be
richer (enhanced), poorer (de-enhanced) or equivalent to the type of
service provided by each individual subnetwork to be interconnected.
This implies that, in general, before being interconnected, subnetworks
have then to be "wrapped" by a functional shell which matches them to an
agreed common type of service (enhanced service).

The main difference between NBS and ECMA is that in the former the
type of service required by the internet sublayer does not necessarily
coincide with the type of service provided by the Global Network. In
the ECMA proposal they are equivalent.

The services provided by the third sublayer to the transport layer are
on a Global Network wide basis whereas the services of the second sublayer are defined over individual component subnetworks only. Using
the sublayering technique the subnetworks are then concatenated in such
a way as to provide the appearance of one single uniform network.

In the NBS approach a fourth sublayer is introduced (internet interface sublayer), which is motivated for the cases where the type of service provided by the global network service does not match the transport layer requirements, the different service types and protocols.

The constraints, the different peculiarities of each proposal in both
categories have motivated us towards the definition of a reference model for network interconnection. The model developed by the COST 11bis
network interconnection subgroup and its service and protocols implications are described in detail in /COST84/.

The following chapters describe the rationales and the design principles which lead us to the proposed model.

3. Scope of network interconnection and design considerations

One of the fundamental objectives in any subnetworks interconnection strategy are solutions for the required adaptations between subnetworks. In this section it is assumed that each subnetwork has a layered structure not necessarily conforming to that of the ISO/RM, and that these structures are not (necessarily) identical.

In general terms, the scope of interconnection appears from the user requirement to extend his resources over the Global Network. This scope implies the existence of the same type of user applications in the interconnected environment.

As a consequence, the network designer has to determine by a top-down analysis the level of homogeneity in the extended environment. This determines the level from which the external compatibility will be visible and from which internally the interconnection problem has to be solved.

The mechanism used to isolate layers in the OSI/RM are the services offered by a layer to the layer above. Note that services are defined as the set of capabilities offered at the boundary of a layer to a user in the next higher layer and that they are an abstraction which is independent of any particular interface implementation. The service does not define how this behaviour is realized in the layers below and how its functions are distributed among layers and systems.

Taking a closer look, we have to observe that a service user must adapt to the behaviour of the service provider in order to use it properly. It follows that, for example, the Network Service does not only define the (complete) behaviour of the Network Service Provider but defines also some of the behaviour of the Transport Entities.

Generally:

- a service completely defines the behaviour of the service provider and defines the boundaries of the behaviour of the users of the service.

Note that in the rest of the section the internetworking problems are considered in a general way without referring to a particular layer. The above observation is depicted in the fig. 1, where the SAP's (Service Access Points) model the amount of specification that applies to User as well as to Provider:

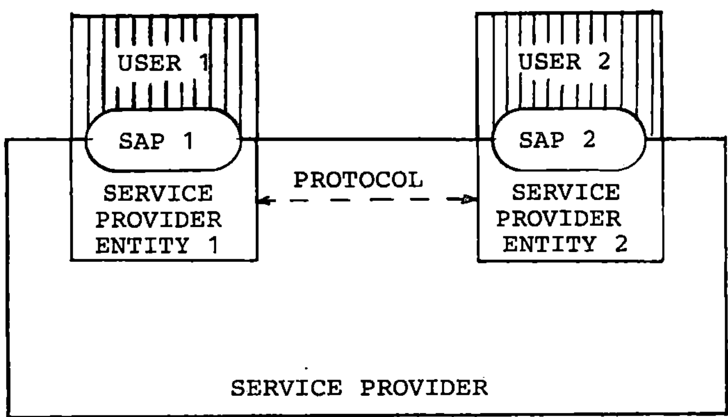

Figure 1. System structure for a Service

In this model we have extended the concept of Service Access Point by attributing to it also the requirements of sequencing of primitives, local negotiation, agreed behaviour, etc. We call this the 'Abstract Interface' of User and Provider and is located within one Open System pertaining to two adjacent layers.

Thus a service can be conceived as consisting of abstract interfaces which are common to Users and Provider, and the communication function between the abstract interfaces, which is exclusive to the Provider.

First conclusion

The above considerations bring us to the conclusion that one aspect of the understanding of the problem of internetting is the notion of a defined common activity of User Entities which is subjected to decomposition.

The decomposition of this common activity shows us which portion of the behaviour of User Entities is defined by the Service Provider. Through this approach we introduce a top-down treatment rather than a bottom-up approach of interconnecting subnetworks of arbitrary different technologies and trying to reason about the resulting structure.

Second conclusion

The entities in a layer are defined while knowing on what service they are placed on and what service has to be provided.

This clearly is a bottom up reasoning. The definition of (the entities within) a layer is based on the assumption that the Service, as seen by an entity in one Open System, partly defines the behaviour of the peer entity which resides in the other Open System in the same layer. It is obvious that heterogeneous subnetworks cannot be simply concatenated.

The first conclusion shows us that the interconnected subnetworks do not result from a decomposition of a defined interaction of the users on top of these different subnetwork. Instead, the users have a different view of their interaction, and thus cannot interact according to their orginal specification (fig. 2).

The second conclusion shows us that the Global Network, resulting from interconnecting two different subnetworks, is in principle undefined, since the entities in the service provider are designed to provide different services and use different protocols.

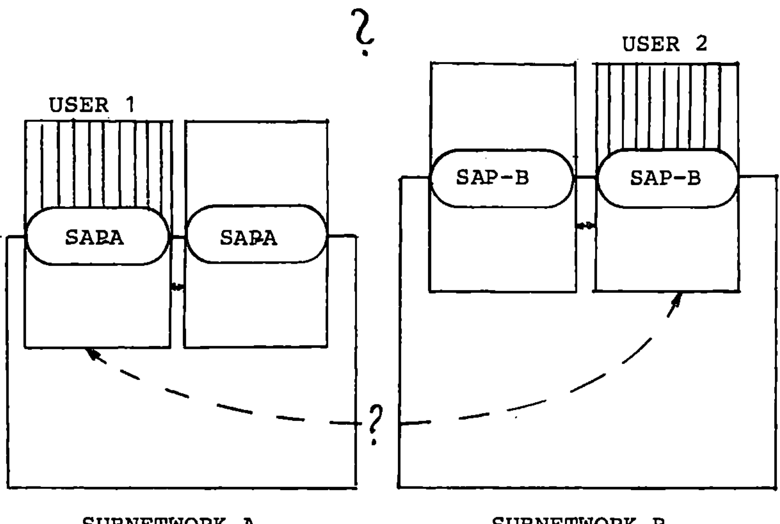

Figure 2. Interconnecting subnetworks of different technologies

The interconnection problem can be stated as follows:

- How can User Entities which have complete different views of their
  interaction, harmonize these views so that they can interact in a defined way, and
- How can we define a Global Service, starting from incompatible subnetwork services, which incorporates the harmonized view of the entity
  interaction?

By following the reasoning developed above, a general model may be defined.

Let us consider two dissimilar subnetworks we wish to interconnect to
form one Global Network. Each subnetwork will be made up of OSI layers
1, 2 and its network layer which will be designated as sublayer 3A.

These layers provide their services to the original subnetwork users.
However, since the subnetwork services are different with respect to
type, functionality and/or quality, users of different subnetworks have
a different view of their interaction with the subnetwork. According
to the first clause of the interconnection problem these views should
be harmonized to make the interconnection useful.

The user's view of its interaction is determined by the abstract interface common to both user and provider. Thus, both users and provider of
either subnetwork must be modified in order to allow proper interconnection.

The first step to interconnection is to separate the users with their
SAPs from the original service providers and to define a Global Network
service provider which wraps and hides both subnetworks.

We now come to the second clause of the interconnection problem stated
in this section, namely "How can a Global service be defined starting
from the incompatible subnetwork services?" A communication function
is required between the abstract interfaces of the Global Network.

We cannot concatenate the communication services of the two subnetworks. This is because it is not possible to make a direct, two-step
mapping between two abstract interfaces pertaining to two dissimilar
subnetworks since their protocols will be different. Thus both of the

subnetworks must be wrapped in such a way that these differences will be invisible to the user of the subnetwork.

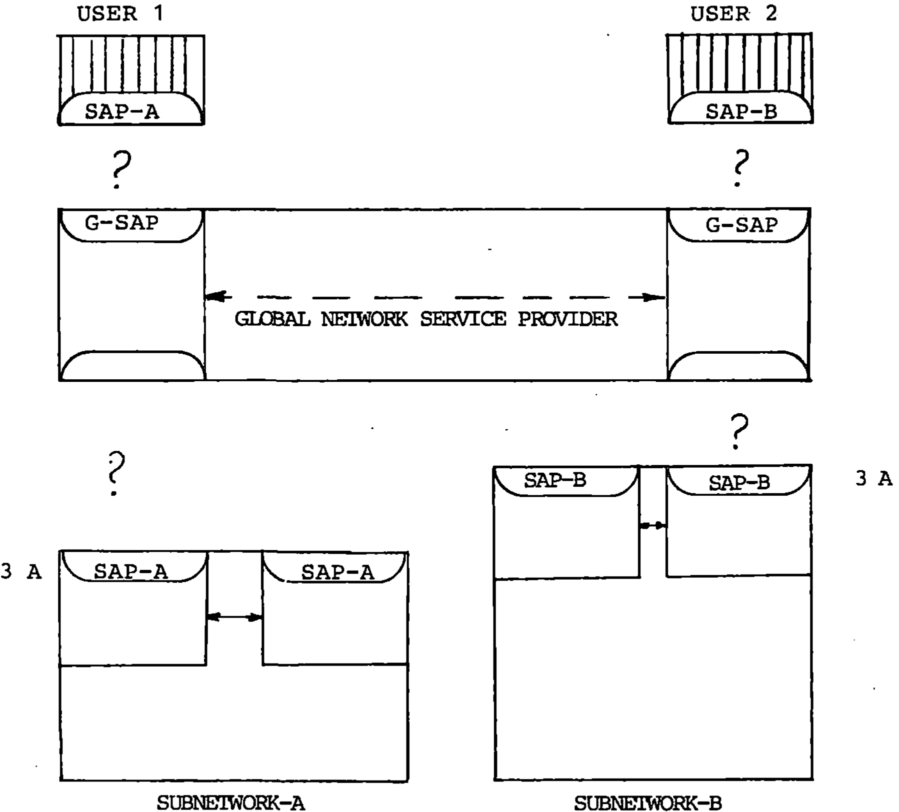

Fig. 3

This process we refer to as enhancement or de-enhancement. The first term is used if a subnetwork service is 'enhanced', the latter if a subnetwork service is 'reduced'. The sublayer which results from the (de-)enhancement process will be denoted as the 3B sublayer (Figure 4).

The services provided by the 3B sublayers of both subnetworks are now compatible. This means that the services are of the same type and provide the same functionality. The global communication function can now easily be accomplished by adding another sublayer above sublayer 3B which takes care of functions such as routing, forwarding, stripping/ embedding of data units leaving/entering a subnetwork, etc.

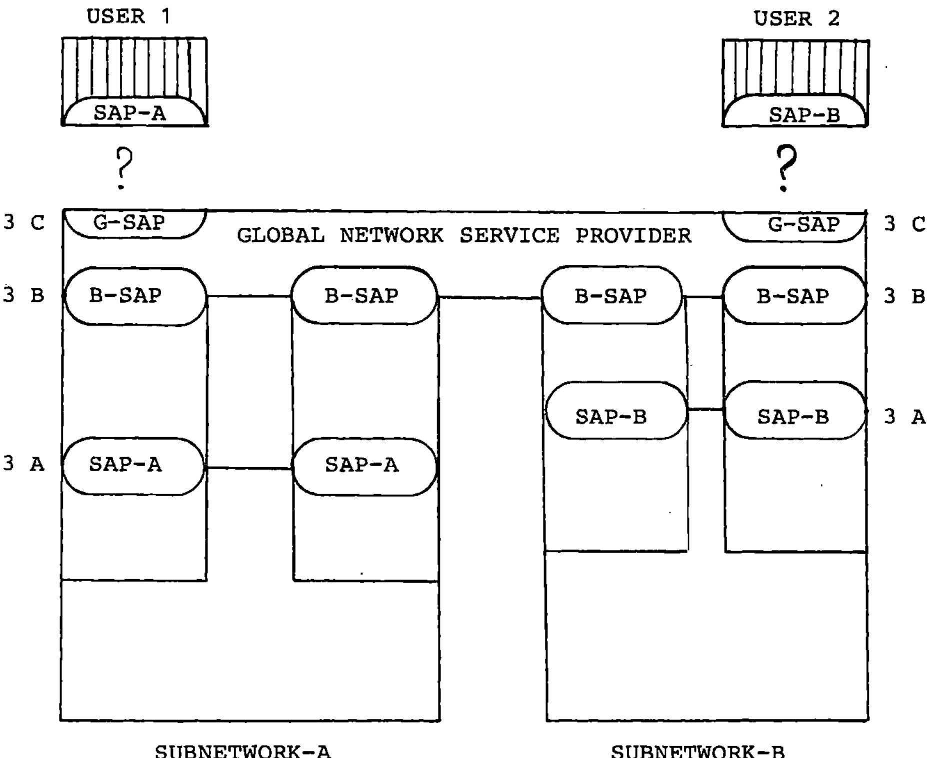

Fig. 4

These functions mask the existence of more than one subnetwork from the
users of its service. The inter-subnetwork adaptions are performed by
the inter-subnetworkgateway. Thus the service provided by this sublayer
has a global scope and the protocol governing its functions may have
end-to-end as well as hop-by-hop aspects. In this context a hop is defined as a communication path through a single subnetwork provided by
the sublayer 3B service. For this reason we refer to this protocol as
an end-hop-end protocol (as distinct from end-to-end protocol), and
the sublayer itself is named the 3C sublayer.

At this point we may or may not fall short of the Global Network service, previously defined as the desired service to be provided to the
Global Network service users (i.e. transport entities). If the former
is the case, we still have to add some functionality in order to meet
the service requirements of these entities. The sublayer holding the
extra functions is the fourth sublayer we might distinguish when sub-
structuring the Network Layer for interconnection purposes, and will

be denoted as the 3D sublayer (fig. 5).

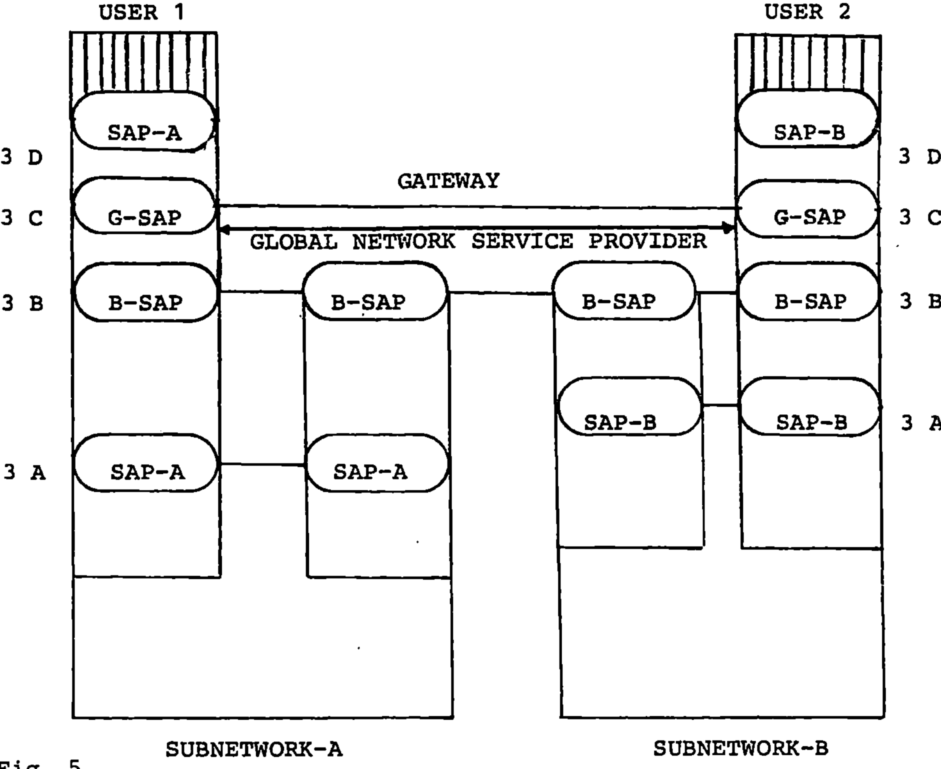

Fig. 5

4. The reference model for network interconnection

The interconnection model described above is depicted in Figure 6 (comprising 3 subnetworks).

4.1 Sublayer Functionality

The Global Network designer is faced with two sets of constraints. One set is the services provided by the existing un-enhanced (3A) subnetworks. The other set is the service required by the Global Network user. These services, will, in general, be of different type. It will be necessary, in these cases, to change the type of services in one of the sublayers. This could be done on a hop-by-hop basis in which case the change of service type will be within sublayer 3B. Alternatively the change of the service type could be performed on an end-to-end basis in which case it will be within the 3D layer. It is pointless to change the type of service within the 3C sublayer because to do so would effectively decompose the 3C sublayer again into smaller sublayers.

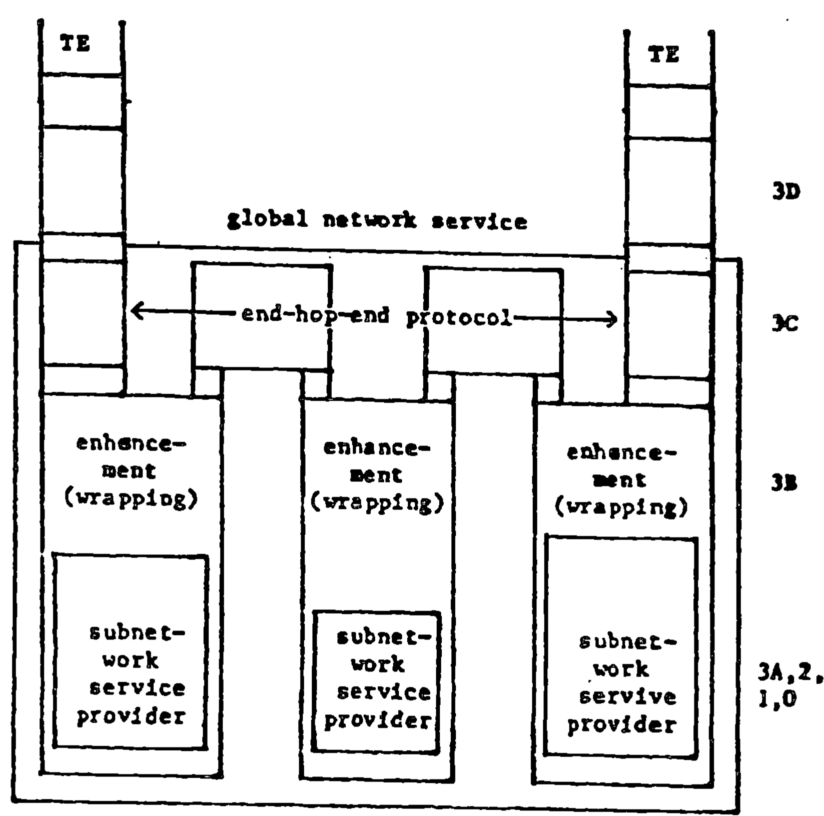

Figure 6

However, these new sublayers will be the same as the existing 3B and
3D sublayers (which may change the type of service) and a new 3C sublayer (which takes care of interconnection and will <u>not</u> change the type
of service).

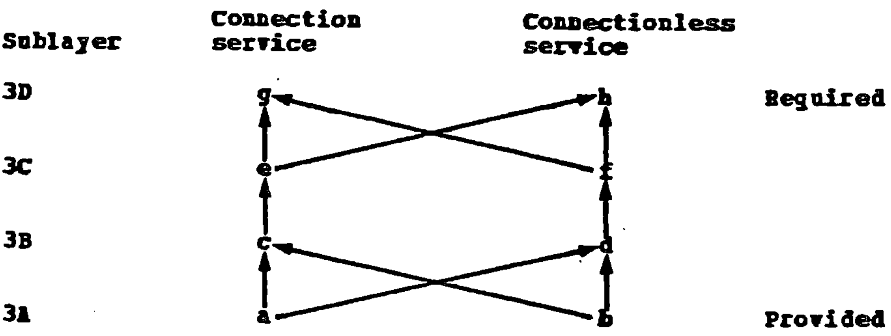

Figure 7 shows the mapping possibilities for the choice of connection
oriented service and one of the connectionless services. Note that there
are two possible pathways for each combination of service type provided
by the 3A sublayer and required from the 3D sublayer.

4.2 <u>Sublayer 3B enhancement/de-enhancement</u>

The functionality of the 3B sublayer may be different for each subnetwork participating in the Global Network, i.e. some 3A sublayers need
enhancement of service, other dehancement, and all probably in different amounts. In addition since the 3B sublayer protocols have local
significance, there is no relation between the 3B sublayers.

However, the result is a set of compatible services offered by all 3B
sublayers involved in the Global Network.

We call two services compatible if they offer:

- the same set of service primitives and their same temporal ordering,
- a non-exclusive value ranges for their parameters.

3 A type of service	3 B type of service	Observations
Connection-less	Connection oriented	enhancement: -provide establishment/ termination phase -fragmenting -reassembly -addressing space -sequencing -etc.
Connection oriented	Connection oriented	enhancement/deenhancement towards: -same set of primitives and temporal ordering -etc.
Connection-less	Connection-less	changes: -same set of primitives and temporal ordering -fragmenting -reassembly -etc.
Connection oriented	Connection-less	changes: -deenhancement towards the connectionless set of primitives -source/destination address -reassembly -fragmenting to suit the 3 A connection requirements -etc.

Table 8 summarizes the possible enhancement/de-enhancement steps to be performed in the 3 B sublayer.

4.3 Sublayer 3C Interconnection

If the sublayer 3B is supposed to provide a <u>compatible</u> set of services, sublayer 3C provides a <u>single, consistent</u> service at its boundary. The type of service provided by sublayer 3C is connection or connectionless and its scope is global.

The entities involved in providing the 3C service are situated in the end systems and inter-subnetwork gateways and they communicate using the end-hop-end protocol. The functions provided in the 3C sublayer are:

- routing and forwarding;
- updating information (e. g. routing etc.);
- error reporting;
- 3C management functions etc.:

4.4 Sublayer 3D User Requirements Enhancement

The main scope of this sublayer is to map the user requirements to the services provided by the sublayer 3C. The specification of this sublayer is defined by the global user requirements and the service provided by the sublayer 3C.

5. Conclusions

The general model developed above was a pre-requisite for the study of the networks interconnections developed by the COST 11bis/EWICS group. The understanding of the protocol and service implications of internetting and of the architectures of the available solutions was eased by referencing them to a general model.

The internetting rationales can be applied for the interconnection of networks without refering to a particular layer, although the model described above focues on the ISO/OSI layer 3.

The reader is encouraged to refer to /COST 84/ for details of the model
and for the services and protocol specifications of the different sublayers within the model.

Note that /COST 84/ describes global connection <u>and</u> connectionless types
of services and the related protocols since we strongly believe that
for different environments and requirements, <u>both</u> service and protocol
types have to be considered.

Note also the model offers an architectural reference for any specific
implementation and allows a comparison between different architectural
solutions.

6. <u>References</u>

/COST84/   B. Butscher, L. Lenzini, R. Morling, C. A. Vissers,
            R. P. Zeletin, "Network interconnection: Principles, Archi-
            tecture and Protocol Implications"
            COST11bis/EWICS Report CEC 1984

/ECMA/   ECMA
            "Network Layer principles" Final Draft
            ECMA/TC24/82/18 January 1982

/GIE79/   M. Gien, H. Zimmermann
            "Design principles for Network Interconnection"
            6th Data Communication Symp. Pacific Grove Cal 1979

/ISO83/   ISO/TC97/SC16
            "Open Systems Interconnection-Basic Reference Model"
            ISO DIS 7498

/NBS/   NBS
            "Features of internetwork Protocol"
            ICST/HLNP-80-8, July 1980

<u>DFN-EARN Gateway für den Message-Verbund</u>

G. Schulze
Gesellschaft für Mathematik
und Datenverarbeitung mbH
Bereich Darmstadt
Rheinstr. 75
D-6100 Darmstadt

G. Müller
IBM Deutschland GmbH
Wissenschaftszentrum Heidelberg
Tiergartenstr. 15
D-6900 Heidelberg

Kurzfassung

Da der EARN und DFN Nachrichtenverbund jeweils unterschiedliche Kommunikationsprotokolle benutzen, ist ein Gateway
notwendig, wenn DFN und EARN Teilnehmer gegenseitig Nachrichten austauschen wollen. Der Gateway verhält sich wie
ein Nachrichten Transfer Knoten im Sinne des CCITT Message
Handling Modells und bedient DFN-seitig die CCITT Nachrichtenaustausch Protokolle der Serie X.400 und EARN-seitig IBM
spezifische Kommunikationsprotokolle.

Der vorliegende Beitrag führt kurz in den DFN und EARN Nachrichtenverbund ein, gibt einen Überblick über die Funktionen
des Gateways, beschreibt die benutzten Kommunikationsprotokolle und zeigt auf, in welche Richtungen zukünftige Erweiterungen angestebt werden.

Abstract

Since the DFN and EARN message handling systems are using
different communication protocols, a gateway becomes
necessary in order to exchange messages between users of
both systems. The gateway behaves as a message transfer
agent in the sense of the CCITT message handling model and
performs at the DFN side the CCITT defined message handling
protocols of the series X.400 and at the EARN side IBM
specific protocols.

This paper introduces briefly the DFN and EARN message
handling system, gives an overview of the functionallity of
the gateway and the used protocols and mentions possible
future extensions.

Einleitung

Das Deutsche Forschungsnetz (DFN) ist ein öffentlich gefördertes Projekt zur Unterstützung überinstitutioneller Datenkommunikation für den
deutschen Wissenschaftsbereich und mit diesem kooperierende Partner.
Hauptziel ist die Bereitstellung von Kommunikationsdiensten und entsprechender Infrastrukturen. Ein wesentlicher Bestandteil des Gesamt-
vorhabens ist der Versand von Nachrichten über einen Message-Verbund.

Viele Benutzer sind schon heute in zahlreiche Message-Systeme eingebunden. Die vorhandenen Message-Systeme sind entweder zentral oder
dezentral organisiert und herstellerspezifisch oder projektspezifisch
im Hard- und Softwarebereich konfiguriert. Es spricht vieles dafür,
daß der Trend zur verstärkten Bereitstellung und Akzeptanz lokaler

Message-Systeme sowohl für die organisationsinterne als auch -externe Kommunikation anhält, sofern dafür gesorgt wird, daß auch Message-Systeme unterschiedlicher Bauart miteinander kommunizieren können.

Das DFN hat sich daher entschieden, für die Verbindung von Message Systemen die ISO und CCITT Standards bzw. Empfehlungen aus dem Bereich der Message Handling Systems (MHS) einzusetzen und entsprechende Realisierungen mit öffentlichen Mitteln zu fördern.

In einer von DFN in Auftrag gegebenen Studie, (1), wurden diverse vorhandene Message Systeme beschrieben und miteinander verglichen und ein Satz von Funktionen für den DFN Message Verbund empfohlen. In einer weiteren Studie, (2), wurde untersucht, wie der Verbund auf der Basis der CCITT Empfehlungen realisiert werden soll und in (3) sind weitere Festlegungen für die Implementation getroffen worden.

Im europäischen Wissenschafts- und Forschungsbereich wurde zu Beginn des letzten Jahres mit Unterstützung von IBM das European Research Network (EARN) aufgebaut und in Betrieb genommen. Dieser Verbund besitzt auch Zugangsmöglichkeiten in nordamerikanische Message Systeme. Zur Kommunikation untereinander wird eine IBM spezifische Technik eingesetzt und es können nur solche Rechner als Netzknoten eingesetzt werden, die sich an diese Konventionen halten.

Gegen Ende des letzten Jahres bestand der Verbund europaweit aus etwa 80 Knoten, innerhalb der Bundesrepublik waren etwa 35 Knoten installiert. Allerdings waren zu diesem Zeitpunkt noch nicht alle vorgesehenen Knoten bzw. Verbindungen voll betriebsbereit.

Aufgrund der technischen Inkompatibilität des IBM Kommunikationsprotokolls und der für DFN vorgesehenen MHS Protokolle ist zur Zeit kein Nachrichtenaustausch zwischen EARN und DFN Systemen möglich. Das Ziel des DFN-EARN Gateways ist es, diese Barriere abzubauen und einen technischen Übergang zwischen EARN und DFN zu schaffen, damit EARN und DFN Teilnehmer auch untereinander Nachrichten austauschen können.

Message Handling Systems und DFN

In der letzten Studienperiode (1980 - 1984) wurde vom CCITT u.a. der Bereich der Message-Handling-Systems (MHS) bearbeitet. Das derzeitige Ergebnis ist in den Empfehlungen der Serie X.400 niedergelegt, (4) und (5). Parallel dazu hat auch die ISO das Thema aufgegriffen und die Erarbeitung entsprechender Standards dem TC97/SC18 aufgetragen. Zu diesem Themenkomplex hat die IFIP Arbeitsgruppe WG 6.5 einige grundsätzliche Vorarbeiten geleistet, die in die Normungsarbeit eingeflossen sind.

Die Standards enthalten Festlegungen, die sich auf die Ebenen 6/7 des OSI Referenzmodells beziehen. Sie beschreiben ein Modell des MHS und die Dienste und Protokolle für den Transfer von Nachrichten zwischen den Endbenutzern eines MHS.

Das MHS Modell, siehe Bild 1, enthält als wesentliche Bestandteile die User Agents (UA) und Message Transfer Agents (MTA). Ein UA stellt aus Benutzersicht die lokale Schnittstelle zum MHS dar und aus Systemsicht einen Repräsentanten des Benutzers. Der UA hat letztlich die Funktion eines elektronischen Briefkastens.

Die MTAs bewerkstelligen den Nachrichtentransport und sind für die vorübergehende Zwischenspeicherung und Weitervermittlung von Nachrichten verantwortlich.

Die Auslieferung einer transportierten Nachricht an den UA des Empfängers kann auf Wunsch quittiert werden. Eine solche Auslieferungsquittung (delivery notification) wird durch den letzten MTA dann erzeugt, wenn die eigentliche Nachricht an den Empfänger UA erfolgreich übergeben werden konnte. Eine negative Auslieferungsquittung wird erzeugt, wenn ein MTA die Nachricht aus irgendwelchen Gründen nicht ausliefern konnte.

Falls beim Absenden einer Nachricht eine Empfangsbestätigung (receipt notification) verlangt war, erzeugt der Empfänger UA eine solche, nachdem die Nachricht erfolgreich an den Teilnehmer übergeben wurde.

Für DFN wird darüberhinaus eine Gruppenkommunikation und ein Auskunftsdienst gefordert. Diese Dienste sind im derzeitigen MHS nicht vorgesehen. Zu diesem Zweck wurden für den DFN Verbund verschiedene Verfahren untersucht, wie solche Erweiterungen in das augenblickliche MHS eingebettet werden können, (3).

Der EARN Verbund

Der Kern des EARN besteht aus einem Verbund von Knotenrechnern, die über festgeschaltete Leitungen miteinander verbunden sind. Als Kommunikationsvehikel wird das Remote Spooling Communication Subsystem (RSCS) eingesetzt. Nähere Einzelheiten über RSCS sind in (6) und (7) zu finden. Das RSCS kann vom Benutzer Dateien annehmen, sie zwischenspeichern und bei gegebener Gelegenheit an einen Nachbarknoten weitervermitteln. Die Datei wird schrittweise solange weitervermittelt, bis sie vom Zielsystem an den Benutzer übergeben werden kann.

Ein Datentransfer über einen solchen Verbund von RSCS Knoten ist insofern quasi öffentlich, als ein Teilnehmer jeden anderen Teilnehmer unmittelbar erreichen kann. Er braucht nicht erst irgendwelche Benutzerberechtigungen für das fremde System zu beantragen und sich dann jeweils bei jedem Transfer zu authorisieren, sondern er kann, sobald ihm nur der Name eines Teilnehmers (Benutzerkennung) bekannt ist, diesem sofort eine Datei übermitteln. Hiervon sind natürlich die Zugangsberechtigungsverfahren zum lokalen Benutzersystem nicht betroffen, sie bleiben unverändert bestehen.

Für den EARN Verbund wurde vereinbart, daß oberhalb des RSCS Transfersystems die sogenannten NETDATA Formate zu benutzen sind. Dies kann als eine oberhalb von RSCS liegende nächsthöhere Protokollschicht angesehen werden, die für die Übermittlung versandspezifischer Angaben (wie z.B. Absendezeit, Dateibenennung im Absendersystem) und für das Packen der Daten verantwortlich ist. Auf dieser Ebene kann auch eine Empfangsbestätigung angefordert und erzeugt werden.

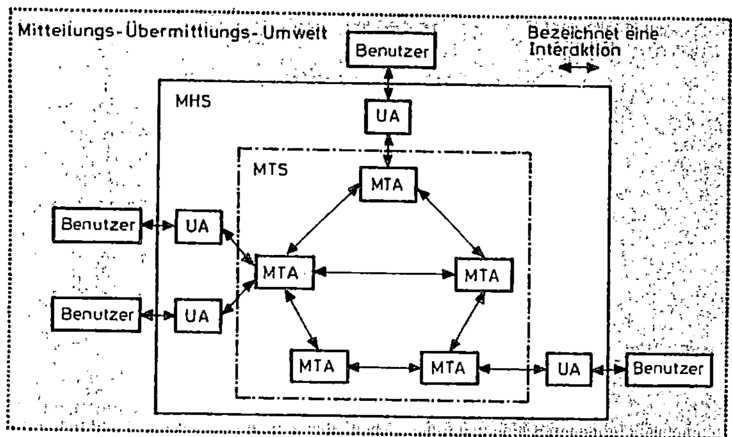

User   *Benutzer: der Mensch als Benutzer des Systems*

UA   *User Agent: End-Systemteil*
Darunter wird die Systemkomponente verstanden, die der Mensch anspricht,
wenn er von Leistungen des Systems Gebrauch machen will.

MTA   *Message Transfer Agent: Transfer-Systemteil*
Darunter wird die Systemkomponente verstanden, die einem Benutzer über ein
End-Systemteil Transferfunktionen zu anderen End-Systemteilen und
Benutzern ermöglicht. MTAs stellen den UAs Dienste der
Mitteilungs-Transferschicht zur Verfügung.

MTS   *Message Transfer System: Mitteilungs-Transfer-System*

MHS   *Message Handling System: Mitteilungs-Übermittlungs-System*

Bild 1: Modell des Message Handling Systems   (Quelle: (4))

In der VM/SP Betriebssystemumgebung, (8), stehen dem Benutzer die CMS SENDFILE und RECEIVE Kommandos, (9) und (10), zur Verfügung, die dieses NETDATA Format erzeugen bzw. verarbeiten.

Überblick über den DFN-EARN Gateway

Der Gateway stellt einen technischen Übergang zwischen dem DFN und EARN dar. EARN seitig ist er über RSCS und DFN seitig über DATEX-P erreichbar, siehe Bild 2.

Funktional kann er im Sinne der MHS Architektur als ein MTA angesehen werden, der auf Nachrichtentransfer Ebene empfangene Nachrichten solange zwischenspeichert bis er sie an einen Nachbarknoten übergeben kann bzw. im Falle einer nicht durchführbaren Weitervermittlung eine negative Auslieferungs Quittung erzeugt und diese in Richtung Absender auf den Weg schickt.

Die Realisierung des Gateways wird in Kooperation zwischen der GMD in Darmstadt und dem WZH der IBM in Heidelberg durchgeführt. Sie basiert auf Implementationen von S.70 und S.62 Protokollen, die ebenfalls gemeinsam von der GMD und dem WZH durchgeführt wurde, (11), (12), (16) und (17). Die Entwicklung erfolgt in der VM/SP Betriebssystemumgebung und wird zunächst so ausgelegt, daß der Gateway an einer zentralen Stelle, geplant ist die GMD in Darmstadt, betrieben werden kann, (18).

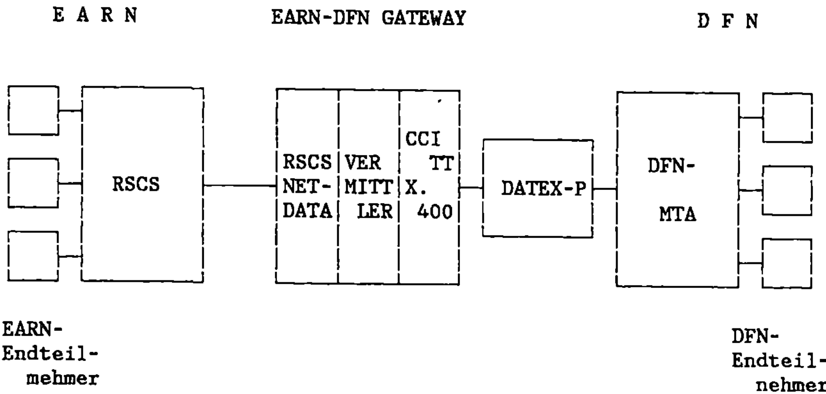

RSCS ... Remote Spooling Communication
Subsysytem
MTA ... Message Transfer Agent

Bild 2: Einordnung des Gateways

Es wurde nicht eine unmittelbare Protokollumsetzung zwischen dem EARN
und DFN Protokoll vorgesehen, sondern eine etwas allgemeinere Lösung.
Diese soll es ermöglichen, daß in späteren Erweiterungen zwischen mehr
als nur zwei Protokollen umgesetzt werden kann. Der Gateway soll daher
als zentrale Komponente einen Vermittler enthalten und an den Rändern
zum jeweiligen Netz die entsprechende Protokollbedienung vornehmen.

Ein Überblick über die funktionalen Komponenten ist in Bild 3 enthalten. Das Bild enthält bereits auch solche Komponenten, die für eine
spätere Ausbauversion beabsichtigt sind.

Die Protokollinstanzen bedienen das zugehörige Protokoll und formen
daraus Gateway intern einheitliche Auftragselemente. Diese werden
anschließend an den Vermittler übergeben. Der Vermittler ermittelt
seinerseits aus dem Empfängernamen den Netztyp, den Protokolltyp und
eine zugehörige Adresse. Im DFN Fall ist die Adresse die Transportadresse des nächsten MTA und im EARN Fall besteht sie aus RSCS Knoten-
und Benutzerkennung des Empfängers (/userid/ AT /nodeid/). Die Auftragselemente werden dann an die entsprechende Protokollinstanz
abgegeben, von ihr in die zugehörigen Protokollelemente umgewandelt
und dann zum nächsten Knoten verschickt.

Um bereits zu Beginn einen möglichst großen Benutzerkreis
anzusprechen, wurde zunächst DFN seitig vereinbart, vorhandene Message
Systeme auf der Basis von Teletex kompatiblen Protokollen zu koppeln.
Dieses Verfahren ermöglicht es auch, daß bereits Teilnehmer an Teletex
kompatiblen Systemen oder Stationen den Message Dienst in Anspruch
nehmen können. Da die verwandten Protokollelemente in einer menschlich
lesbaren Form vorliegen, kann der Dienst bereits in Anspruch genommen
werden, auch wenn noch nicht der volle Funktionsvorrat auf dem betrof-

fenen System implementiert wurde. Die Bearbeitung der übermittelten Nachrichten, wie z.B. Verteilen, Weiterleiten, und die Erzeugung von Bestätigungen kann in solchen Fällen zunächst manuell erfolgen.

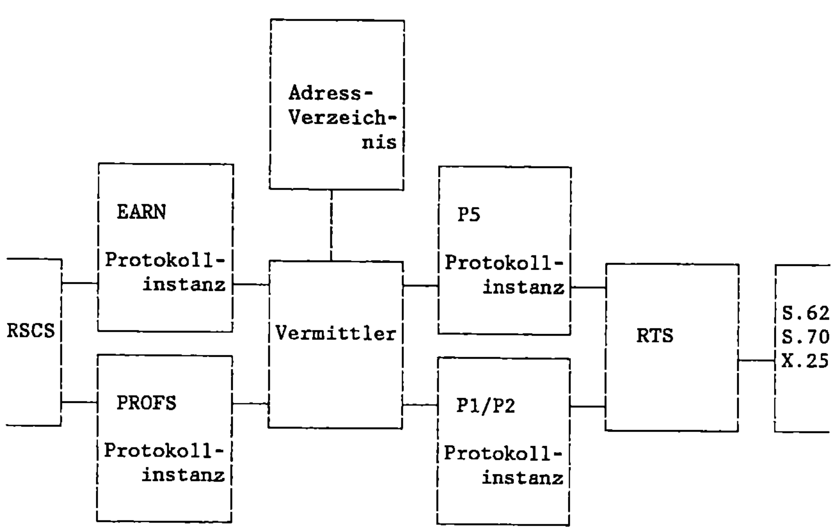

Bild 3: Funktionale Komponenten des Gateways
(in einer möglichen Ausbaustufe)

Außerdem können dann schrittweise mehr Funktionen implementiert und benutzt werden, ohne daß die Partner davon berührt werden.

Die derzeitige Realisierung basiert auf dem CCITT MHS-Protokoll P5 entsprechend der Empfehlung X.430, sie benutzt lediglich eine Untermenge der in X.430 vorgesehenen Protokollelemente. Für eine spätere Version ist vorgesehen, den Gateway um das Protokollpaar P1/P2 entsprechend den Empfehlungen X.411 und X.420 zu ergänzen. Mit dieser Erweiterung ist es dann möglich, aus DFN Sicht den Verkehr mit EARN wahlweise über einen der beiden Protokolle abzuwickeln als auch innerhalb des DFN einen Übergang zwischen den beiden Protokolltypen zu schaffen. Längerfristig soll der Nachrichtentransfer mit den Protokollen P1/P2 abgewickelt werden sobald eine ausreichende P1/P2 Infrastruktur vorhanden ist. Die P5 Protokollkomponente kann dann auch so erweitert werden, daß ein in Sinne von X.430 vollständiger Dienst angeboten werden kann.

In Bild 3 ist ferner dargestellt, wie der Anschluß an das IBM Bürosystem PROFS (Professional Office System) erfolgen kann. In einer solchen Ausbaustufe würde dann ein Übergang zwischen allen Kombinationen der vier Protokolltypen technisch möglich sein.

Der Gateway aus der Sicht von DFN

Aus der DFN Sicht stellt der Gateway einen MTA im Sinne des CCITT MHS Modells dar. Er repräsentiert einen Zwischenknoten, der über Datex-P angerufen werden kann oder der von sich aus über Datex-P einen DFN Knoten anwählt. Er fungiert sowohl in Richtung EARN als auch in Richtung DFN als ein Weitervermittler.

Oberhalb der X.25 Protokollschichten werden auf den Ebenen 4 und 5 des OSI-Modells die Protokolle S.70 und S.62 dazu benutzt, einen Reliable Transfer Service (RTS) in Anlehnung an die CCITT Empfehlung X.410 zu realisieren. Mit Hilfe des RTS können Dateien (z.B. Teletex Dokumente, Message Transfer Protokollelemente) sicher von einem Knoten zum nächsten übermittelt werden, s. Bild 4.

Oberhalb des RTS wird, wie bereits erwähnt, in der ersten Version eine Untermenge des Protokolls P5 entsprechend der Empfehlung X.430 benutzt.

Das der Empfehlung X.430 zugrunde liegende Modell sieht vor, daß ein Benutzer mittels eines Teletex Gerätes (TTX) eine Teletex Access Unit (TTXAU) Dienstleistung in Anspruch nehmen kann. Dieser TTXAU stellt im wesentlichen einen UA des Message Handling Modells dar, der aber in der Lage ist, mit einem Teletex Benutzer zu kommunizieren. Ihm können Aufträge übermittelt werden, Nachrichten zu versenden, eingegangene Nachrichten aufzuzeigen, bestimmte empfangene Nachrichten dem Benutzer zu übermitteln und Empfangsquittungen zum Absender zurückzuschicken. Außerdem kann die TTXAU auf Wunsch Nachrichten und Quittungen sofort nach Erhalt (d.h. ohne explizite Aufforderung) an den TTX Benutzer übermitteln.

Im Sinne des X.430 Modells verhält sich der Gateway beim Nachrichtenversand und Empfang von Quittungen wie die TTX Seite und umgekehrt beim Empfang von Nachrichten und der Übermittlung von Quittungen wie die TTXAU Seite. Durch diese symmetrische Benutzung einer Untermenge von X.430 wird es ermöglicht, daß sich beide Seiten wie ein MTA verhalten können und nicht eine der Seiten immer ein Endsystem darstellen muß. Eine detailierte Beschreibung ist in (3) bzw. (15) enthalten.

Im Bild 5 ist diese Situation grafisch verdeutlicht. Mit dem SEND Auftrags Element werden gleichzeitig Dokumente übergeben, die dann von der empfangenen Instanz an einen lokalen Benutzer übergeben oder an einen weiteren MTA weitervermittelt werden. Das SEND Element besteht aus einem oder mehreren Teletex Dokumenten. Ein Teil des ersten Dokuments oder das ganze erste Dokument ist nach vorgegebenem Schema strukturiert und enthält alle relevanten MHS Parameter in lesbarer Form. Die hier sichtbaren Parameter beziehen sich sowohl auf die Message Transfer (MT) als auch auf die Interpersonal Messaging Ebene (IPM) und entsprechen bezüglich Typ und Wertevorrat den Parametern der Protokolle P1 und P2 entsprechend den Empfehlungen X.411 und X.420.

Der nachfolgende Teil des Dokumentes bzw. der Dokumentenfolge enthält den Inhalt der Nachricht. Für die Anfangsversion wird davon ausgegangen, daß die Nachricht entweder ein druckbarer Text mit dem Teletex

Zeichenvorrat oder dem Zeichenvorrat der internationalen
Referenzversion (IA5) oder aber eine transparente Datei sein kann.

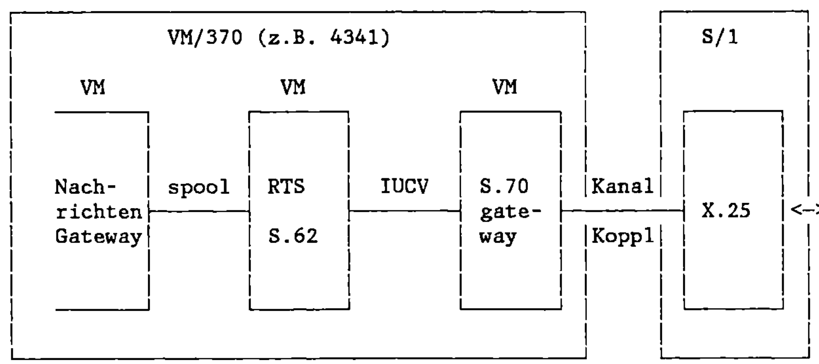

IUCV ... Inter User Communication
Vehicle
VM ..... Virtual Machine

Bild 4: Implementationsstruktur der OSI Protokolle

Das DELIVERY STATUS NOTIFICATION Element besteht aus einem Teletex
Dokument, das nur Parameter enthält, die sich auf die Auslieferung
einer Nachricht beziehen. Es wird mit diesem Element entweder die Aus-
lieferung einer Nachricht an den UA des Empfängers bestätigt oder
Gründe für eine Nichtauslieferung an den Absender übergeben. Letzteres
kann auch die Tatsache enthalten, daß die Nachricht durch irgendeinen
MTA nicht weitervermittelt werden konnte. Inhaltlich entspricht das
Element der P1 Protokoll Einheit DELIVERY REPORT aus der Empfehlung
X.411.

Das RECEIPT STATUS NOTIFICATION Element besteht ebenfalls aus nur
einem Teletex Dokument, das empfangsspezifische Angaben in lesbarer
Form enthält. Es entspricht dem IPM STATUS REPORT Protokoll Element
(SR-UAPDU) des P2 Protokolls der Empfehlung X.420.

Der Gateway aus der Sicht von EARN

Aus EARN Sicht erscheint der Gateway wie ein normaler EARN
Endbenutzer. Ein EARN Teilnehmer adressiert den Gateway so, als ob
dieser ein anderer EARN Teilnehmer sei, also über die Benutzer- und
Knotenkennung des Gateways. Der Gateway erwartet nun aber, daß die
übermittelten Dateien eine bestimmte Struktur haben, die der Gateway
für seine Zwecke interpretieren können muß. Die neue Struktur kann als
neue Protokollebene angesehen werden, die oberhalb der IBM NETDATA
Schicht angesiedelt ist.

Auf dieser neuen Protokollschicht sind alle Funktionen vorhanden, die
durch die IBM SENDFILE/RECEIVE Dienste nicht erbracht werden können.
Die zu übermittelnde Datei muß daher mit einem Vorspann versehen wer-

den, der die neue Gateway spezifische Struktur haben muß. Es wurde als
sehr sinnvoll angesehen, diese Struktur ähnlich X.430 zu gestalten.
Das bietet die Möglichkeit, daß alle notwendigen Parameter in lesbarer
Form vorliegen und daher auch manuell bearbeitet werden können. Auf

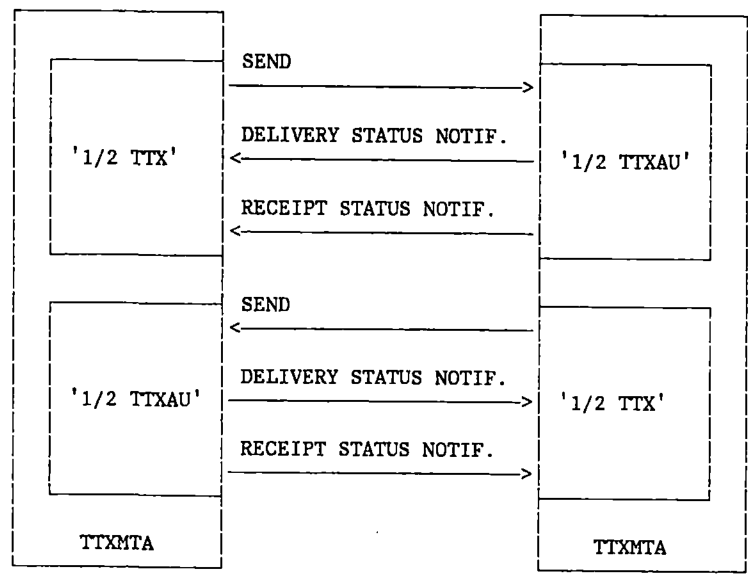


```
TTX ...... Teletex
TTXAU .... Teletex Access Unit
TTXMTA ... Teletex Message
           Transfer Agent
```

Bild 5: Untermenge der benutzten X.430 Protokoll Elemente

diese Weise ist es möglich, daß auch über bereits vorhandene EARN
Benutzerschnittstellen der DFN-MHS Dienst verfügbar wird.
Darüberhinaus können je nach Bedarf und Verfügbarkeit weitere lokale
Dienste bereitgestellt werden, die benutzerfreundlichere Schnittstellen anbieten. Beispiele hierfür wären Erweiterungen an den derzeitigen SENDFILE/RECEIVE Prozeduren oder der Einsatz von PROFS, sobald
der in Bild 3 dargestellte Anschluß bereit steht.

Implementationsstruktur

Abbildung 3 zeigt die funktionalen Komponenten des Gateways und setzt
sie in Beziehung zum RSCS Subsystem und dem RTS Transfer Dienst. Sie
zeigt die generelle Struktur für eine erweiterte Version, in der auch
bereits das Protokoll P1 für die MTA-MTA Kommunikation entsprechend
X.411 eingebunden ist und EARN seitig die Anbindung von PROFS vorgenommen wurde.

Dateien, die vom RTS empfangen wurden, werden der Protokollinstanz P5
bzw. P1/P2 zusammen mit einigen zusätzlichen Informationen übergeben.
Diese Komponente wiederum erwartet, daß die Dateien Protokollelemente

auf der MT Ebene darstellen, das sind im Falle von X.430 sogenannte Aktions Elemente und im Falle von X.411 Nachrichten Protokoll Einheiten (MPDUs) des Protokolls P1. Da eine Protokolleinheit u.U. aus mehreren Dateien bestehen kann, werden erforderlichenfalls die empfangenen Dateien vor der Verarbeitung zu solchen Einheiten verkettet.

Für den Fall, daß ein empfangenes Protokoll Element weiter verarbeitet werden kann, wird erforderlichenfalls ein entsprechendes Gedächtnis angelegt, das Protokollelement auf die Gateway interne Darstellung umgeformt und dann dem Vermittler übergeben. Im anderen Fall wird die Nachricht ignoriert und u.U. mit einer negativen Auslieferungs Quittung beantwortet.

Der Vermittler ermittelt aus dem Empfängernamen der Nachricht die Adresse der Instanz, die als nächste die Nachricht zur eventuellen Weitervermittlung oder Auslieferung an den Empfänger übernehmen kann. Die Adresse stellt im DFN Fall die Transportadresse (incl. Datex-P Adresse) des nächsten MTA dar und im EARN Fall die Benutzer- und Knotenkennung des Teilnehmers. Neben der Adresse werden noch einige weitere Eigenschaften, wie z.B. Netztyp und Protokolltyp aus dem Empfängernamen abgeleitet. Die Beziehungen zwischen Empfängernamen, Adresse und sonstigen Eigenschaften können z.T. aus dem Adress Verzeichnis (Directory Services) ermittelt oder auch aus dem Namen des Empfängers (Recipient O/R Name) direkt abgeleitet werden.

Der Empfängername ist entsprechend CCITT Empfehlung X.400 struktiert. Solange aber noch keine ausreichenden Directory Dienste verfügbar sind, muß der O/R Name des Empfängers in einer vorgegebenen Weise benutzt werden, um aus ihm die Adresse ermitteln zu können.

Der Vermittler übergibt dann das Auftragselement, mit Adresse und sonstigen Merkmalen versehen, an die entsprechende MT-Protokoll Komponente. Diese Instanz wandelt daraufhin das Auftragselement in die jeweils erforderliche Struktur um und übergibt, je nach Netztyp, die Nachricht entweder dem RSCS System oder dem RTS zur Übermittlung an die nächste Instanz.

Mit der oben beschriebenen Gatewaystruktur kann der Nachrichtenaustausch zwischen EARN und DFN erfolgen. In der ersten Version soll nur die oben beschriebene Untermenge von X.430 implementiert werden. Diese Version ermöglicht den Verkehr mit solchen Teilnehmern, die Teletex kompatible Datex-P Endeinrichtungen besitzen oder die bereits über X.430 Implementationen auf ihren lokalen Nachrichtensystemen verfügen. Zur letzteren Gruppe gehören z.B. das von der GMD in Birlinghoven entwickelte Konferenzsystem KOMEX, (13), und das von der Europäischen Kommission in Auftrag gegebene Committee Support System (CSS), (14).

In einer Nachfolgeversion soll die P1/P2 Komponente in der dargestellten Weise mit in den Gateway aufgenommen werden. Diese Version ermöglicht dann auch den Nachrichtentransfer von und zu EARN über Installationen, die die MTAs und UAs entsprechend X.411 und X.420 implementiert haben. Darüberhinaus würde aufgrund der gewählten Gatewaystruktur dann auch ein Übergangsdienst zwischen P1/P2 und P5 bei Bedarf bereitgestellt werden können.

Nach Ausbau und Verfügbarkeit von mächtigeren Directory Diensten könnten auch benutzerfreundlichere Namen zur Adressierung der Empfänger benutzt werden. Hierbei sind nicht nur gatewayinterne Dienste relevant sondern vielmehr auch externe Auskunftsstellen. Diese können in solchen Fällen vom Gateway abgefragt werden, in denen die Auskunft zur Durchführung der Vermittlung benötigt wird.

<u>Zusammenfassung</u>

Die erste Version des Gateways wird den DFN-EARN Übergang auf der Basis der zuvor beschriebenen Untermenge von X.430 schaffen. Für die Adressierung der Teilnehmer gelten solange noch besondere Vorgaben, wie noch keine quasi öffentlichen (gatewayübergreifende) Adressverzeichnisse existieren. In einer Nachfolgeversion soll der Gateway um die Protokolle ensprechend den Empfehlungen X.411 und X.420 erweitert werden, damit dann auch der EARN Zugang über P1/P2 mölich wird. Diese Erweiterung könnte auch DFN seitig als Übergang zwischen P1/P2 und P5 benutzt werden.

Es ist ferner beabsichtigt, daß EARN seitig an der Benutzerschnittstelle weitere Hilfsmittel bereitgestellt werden können, um die X.430 ähnlichen Dateien bequemer zu handhaben. Eine Möglichkeit wäre die Anbindung des IBM Bürosystems PROFS.

Der Gateway soll zunächst an einer zentralen Stelle für den Betrieb bereitgestellt werden. Längerfristig wäre es aber möglich, daß die Software auf mehreren EARN Knoten installiert wird und auf diese Weise immer näher an die Endsysteme herangeführt wird. Diese Lösung wäre eine Möglichkeit, den EARN Verbund in der jetzigen Form abzulösen.

<u>Literaturverzeichnis</u>

(1)   D. Conrads, et. al.
      Studie I:
      Funktionalität und Bewertung von Message Systemen
      Universität Düsseldorf, April 1984

(2)   M. Bogen, et. al.
      Studie II:
      Konzept für einen DFN-Message Verbund auf der Basis
      von CCITT-Message-Handling-Systems
      Universität Düsseldorf, Juni 1984

(3)   P. Kaufmann, et. al.
      Der DFN-Message-Dienst
      Pflichtenheft für die Implementierung
      Universität Düsseldorf, Oktober 1984

(4)   W. Tietz
      Stand der internationalen Normung im Bereich
      "Message-Handling"
      Nachrichtentechnische Zeitschrift Bd. 37 (1984)
      H. 1, S. 20-26

(5)   CCITT Empfehlungen der Serie X.400
      Message Handling Systems
      CCITT Genf, 1984

(6)   IBM Virtual Machine Facility/370:
      Remote Spooling Communications Subsystem (RSCS)
      User's Guide
      IBM, GC20-1816-2

(7)   IBM VM/370
      Remote Spooling Communications Subsystem
      Networking,
      Allgemeiner Überblick
      IBM, GC20-1379-0

(8)   IBM Virtual Machine Facility/370:
      Introduction
      IBM, GC20-1800-9

(9)   IBM Virtual Machine/
      System Product:
      CMS Command and Macro Reference
      IBM, SC19-6209-1

(10) EARN
      Pocket Reference Summary
      IBM, 1984

(11) E. Giese
      Recommendation for the GILT Session Service Interface
      GMD Arbeitspapier Nr. 31
      GMD Darmstadt, 1983

(12) L. Eckstein
      Benutzungsanleitung der Ebene 4 Prozedurschnittstelle
      in Pascal Programmen
      GMD Arbeitspapier Nr. 90
      GMD Darmstadt, 1984

(13) U. Pankoke
      Experience with the KOMEX system as an inhouse CBMS
      IFIP WG 6.5 Working Conference on Computer based
      Message Services
      Nottingham, Mai 1984

(14) Committee Support System
      Band 1 und 2
      CAP GEMINI Nederland, ICL European Institutions und
      Barco Video Communications, Januar 1984
      (zu beziehen über die EG Brüssel)

(15) G. Schulze
      Message Handling Systems:
      Protocol for Teletex-MTA, (2xP5/2)
      GMD Darmstadt, Oktober 1984
      (dieses Dokument ist auch Bestandteil von (3))

(16) Holliday, R.,Kropp, D., Mueller, G., Schulz, W.
      End User Applications in Open Systems - Project Description,
      IBM HDSC Techn. Report TR 84.06.007, June 1984

(17) Schulz, W., Reinhardt, K.
      Access from VM/CMS to the DATEX-P network via the Series/1,
      IBM Heidelberg SC Internal Working Paper, August 1984

(18) Philipp, R., et.al.
      EARN/DFN Gateway - Design Overview,
      IBM Heidelberg SC Internal Working Paper, September 1984

Synchronisation und Recovery in verteilten Datenbanken
Ein implementiertes Beispiel aus der Praxis

Ewald Maria Mund
Mund Software
Buchholzstr. 13
CH-3066 Stettlen/Bern

Abstract:

Durch ingenieurmässige Anwendung von Konstruktionsmethoden wie
dem Entity-Relationship-Model und Petri-Netze einerseits, sowie
durch integrierende Benutzung flexibler Betriebssysteme, homogener
Rechnerverbundsysteme und Programmiersprachen der vierten Generation andererseits ist es heute möglich, bestimmte Typen verteilter
Datenbanksysteme mit Synchronisation und Recovery wirtschaftlich
zu realisieren.

The use of well-structured modelling methologies like the Entity-
Relationship Model and Petri-Nets together with the integration
of operating systems, homogen computer-networks and fourth generation programming languages, allow us today to create a limited
synchronized distributed data-base with recovery in a cost effective way.

I. Einleitung

Dieser Vortrag verfolgt das Ziel, anhand eines existierenden Beispieles dem Konstrukteur von Informationssystemen einen leichten, in der Praxis bewährten Einstieg in die Implementierung von einfachen verteilten synchronen Datenbanksystemen zu vermitteln. Insbesondere wird gezeigt, dass bei ingenieurmässiger Anwendung von klaren Konstruktionsmodellen sowie Software- und Kommunikationswerkzeugen kostengünstige Lösungen erstellt werden können.

II. Zur Aufgabe

Für die Firma Du Pont de Nemours in Genf hatte ich die Aufgabe, ein rechnergestütztes Preislisten-Verwaltungssystem zu entwickeln. Ohne auf die Aufgabe selbst tief einzugehen, handelt es sich hierbei um ein System, bei dem an verschiedenen Verkaufsniederlassungen in Europa Preislisten erstellt und modifiziert werden. Diese Preislisten werden zur Zentrale (Genf) geschickt, dort zur Kenntnis genommen, gelegentlich geändert und anschliessend zur Verkaufsniederlassung zurückgesandt. Ganz allgemein befindet sich je eine ganze Kopie einer Preisliste sowohl in der Zentrale in Genf als auch in mindestens einem Knotenrechner der Verkaufsniederlassungen. Soweit zur Aufgabe selbst.

II.1 Vereinfachungen und Generalisierung der Aufgabenstellung

a) Wie oben erwähnt, befindet sich immer eine vollständige Kopie einer Liste sowohl in der Zentrale als auch in mindestens einer Verkaufsniederlassung.

b) Gleichzeitige globale Zugriffe, die eine oder mehrere Preislisten modifizieren, sind sowohl innerhalb einer Verkaufsniederlas-

sung als auch im gesamten Niederlassungsnetz nicht notwendig und
daher auch nicht implementiert. Dies stellt eine wesentliche Vereinfachung der Aufgabenstellung dar, weil damit die Implementierung
eines vollsynchronisierten Zugriffes auf verteilte Daten einschliesslich globalen Sperrens und Dead-Lock Vermeidung und Erkennung entfällt. Generell haben wir es mit einem verteilten Informationssystem
zu tun mit folgenden einschränkenden Eigenschaften:

1. von jedem Objekt (im Sinne von Entity) des Informationssystems
   befindet sich je eine vollständige Kopie in dem als "Zentrale"
   bezeichneten Rechner als auch in mindestens einem anderen
   Rechnerknoten im Netz.
2. Transaktionen an einem oder mehreren Objekten des Informationssystems können nur an einem definierten, pro Objekt(e) durchaus verschiedenen Rechner durchgeführt werden. Gleichzeitige
   Zugriffe auf das gleiche Objekt auf verschiedenen Rechnern im
   Netz sind durch die Anwendung nicht vorgesehen. Beispiele:
   Alle Informationssysteme, die nicht zur Klasse von Reservie-
   rungssystemen gehören wie, verteilte elektronische Post, verteilte Auskunftsysteme, Bildschirmtext? etc.
3. Die Synchronisation, d.h. die Verteilung der durch eine Transaktion modifizierten Objekte muss nicht online ($<$5 Sekunden)
   erfolgen. Zeitspannen bis zu 30 Minuten sind tollerierbar.

III. <u>Lösungsziel</u>

Als Lösung für obige Aufgabe wurde eine soweit wie möglich applika-
tionsunabhängige Methode entwickelt, die auf jede andere Aufgabe
ähnlicher Struktur anwendbar ist. Insbesondere wurde bei der Entwicklung der Software darauf geachtet, dass eine klare Trennung zwischen
Synchronisationsebene und Anwendungsebene gegeben ist. Wie später
noch gezeigt wird, haben wir eine Methode entwickelt, die es uns
erlaubt, die eigentliche Applikationssoftware sowohl zentral als
auch auf verteilten Rechner zur Ausführung zu bringen, ohne hierbei
Aenderungen an eben dieser Applikationssoftware vornehmen zu müssen.

IV. Randbedingungen

IV.1 Hardware

Sowohl in der Zentrale als auch in den verschiedenen Verkaufsniederlassungen befinden sich Systeme der Familie HP/3000 verschiedener
Leistungsfähigkeit. Alle diese Systeme werden durch das gleiche Betriebssystem MPE V verwaltet. Datenbank, Uebersezter etc. sind jeweils die gleichen.

IV.2 Software

Datenbanksystem IMAGE/3000

Auf allen Systemen werden die Daten in IMAGE/3000 Datenbanken abgelegt. Im weiteren gehen wir davon aus, dass IMAGE/3000 im Zusammenspiel mit der Applikationssoftware nach einem Programm- oder Systemabsturz sowohl physische- als auch logische Konsistenz der Daten
garantiert. Die Synchronisationssoftware kann also nach einem Pro-
gramm- oder Systemabsturz davon ausgehen, dass nach Wiederanlauf
des Systems die Daten auf den einzelnen Rechnern im Netz konsistent,
wenn auch nicht synchron, vorhanden sind.
Zur Klärung: unter synchronem Informationssystem verstehen wir somit
bei unserem System, dass nach einem gewissen Zeitraum ($<$ 30 Min.)
sämtliche lokale Aenderungen an Preislisten auch an den entsprechenden Kopien im Netz durchgeführt wurden und somit überall die gleichen
Preislisten mit gleichem Inhalt vorzufinden sind.

Rechnerverbundsystem DS/3000

DS/3000 erlaubt HP/3000 Systeme zu einem homogenen Rechnerverbund
zusammenzuschliessen. Dieses Verbundsystem unterstützt die wesentlichen Transportsysteme wie Wähl- und Standleitungen, X.25- und X.21
Netzwerke. Die Funktionen Virtuelles Terminal, Remote File Access
und Transfer, Remote Date-Base Access und Transfer sowie Device
Sharing des Presentation-Layers sind voll unterstützt; dies in allen
Programmiersprachen. Der programmatische Zugriff auf remote Dateien
und Datenbanken erfordert somit KEINEN! zusätzlichen Softwareaufwand.

TRANSACT/3000

Sowohl die Applikationssoftware als auch die Software zur Synchronisation wurden in TRANSACT/3000 geschrieben, um den Codieraufwand
wesentlich zu reduzieren. TRANSACT/3000 kann man als eine Programmiersprache der vierten Generation betrachten und arbeitet in Verbindung mit einem Dictionary als Meta-Datenbank.

IV.3 Topologie des Datentransportsystems

Für unser Preislisten-Verwaltungssystem werden acht HP/3000 Rechner
in sieben verschiedenen Ländern Europas mittels eines privaten DU
PONT X.25 Datentransportnetzes verbunden. Jeder Rechner kann somit
mit jedem anderen Rechner im System eine Verbindung aufnehmen. Für
unsere Anwendung ist aber lediglich die Verbindungsmöglichkeit der
einzelnen europäischen Verkaufsniederlassungen mit der Zentrale in
Genf notwendig.

V. __Entwicklungsmodelle und Methoden__

Statisches Modell

Bei der Modellierung und Konstruktion des Informationssystems benutzten wir das Entity-Relationship Modell. Jedem Objekt (Entity)
wurde bei dessen Erstellung eine einheitliche und eineindeutige Objekt-
Nummer zugeteilt. Die Erzeugung dieser Objekt-Nummer ist Aufgabe der
Applikationssoftware. Aus der Struktur der Nummer kann man den Ob-
jekt-Typ erkennen. Die Objekt-Nummer ist zum Teil für den Anwender
nicht sichtbar. Dieser greift i.a. auf die Objekte mittels deren
logischen Schlüssels zu. Die verschiedenen Objekt-Typen werden in
sog. Base- und Dependant Segmenten representiert. Die Segmente
selbst befinden sich in dritter Normalform. Um Transaktionen verteil-
len und synschronisieren zu können, stellt jede Transaktion ebenfalls
ein Objekt dar mit einer einheitlichen, eineindeutigen Objekt-Nummer.
Die Erzeugung dieser Objekt-Nummer ist ebenfalls Aufgabe der Anwendungssoftware. Inhalt des Objektes "Transaktion" sind im wesentlichen
die Objekt-Nummern der durch die Transaktion modifizierten Objekte.
Aus der Objekt-Nummer einer Transaktion oder eines ihrer Attribute
geht der Rechnerknoten eindeutig hervor, auf dem die Transaktion
ursprünglich ausgeführt wurde.

Dynamisches Modell

Petri-Netze eignen sich hervorragend, dynamische und parallele Vorgänge darstellbar zu machen. In unserem Falle geht es darum, die
durch eine Transaktion modifizierten Objekte zu den entsprechenden
Rechnerknoten zu kopieren. Dies ist ein dynamischer und paralleler
Vorgang. Eine Transaktion ist somit genau dann synchronisisiert, wenn
alle die durch eine Transaktion modifizierten Objekte vollständig
auf die entsprechenden Rechnerknoten kopiert wurden, einschliesslich
der Transaktion als Objekt selbst.

VI. Synchronisation

Wenn immer Objekte durch eine Transaktion modifiziert wurden, müssen
die entsprechenden Objekte zu anderen Rechnern kopiert werden. Auch
das Erstellen eines Objektes selbst betrachten wir als eine Transaktion. Die Synchronisation der Transaktion   d.h. die Verteilung
der modifizierten Objekte dieser Transaktion zur Zentrale und den
anderen Knotenrechnern kann durch vier Zustände gesteuert werden.

VI.1 Transaktionszustände

Zustand LOKAL
In diesem Synchronisationszustand befindet sich eine Transaktion,
wenn diese (noch) nicht synchronisiert werden soll. Beispiele hierfür sind das Erstellen von Objekten etc. Wir nennen diese Transaktion dann auch eine lokale Transaktion.

Zustand BEREIT
In diesem Zustand wartete eine Transaktion und die damit verbundenen
Objekte auf den Transport zur Zentrale. Wohin die Objekte kopiert
werden ,   (ausser zur Zentrale selbst) steht in einer sog. Destinations-Tabelle. Die Anwendungssoftware ist verantwortlich für den Inhalt dieser Tabelle. Diese Destinations-Tabelle kann objekt- oder auch
nur transaktionsabhängig sein.
Die Anwendungssoftware ist dafür verantwortlich, dass nach Beendigung
einer Transaktion diese sich selbst in den Zustand "BEREIT" überführt.
Mit diesen beiden Funktionen ist das Zusammenspiel zwischen Anwendersoftware und Synchronisationssoftware vollständig beschrieben.

Abbildung 1: Teil-Petri-Netz für Synchronisation

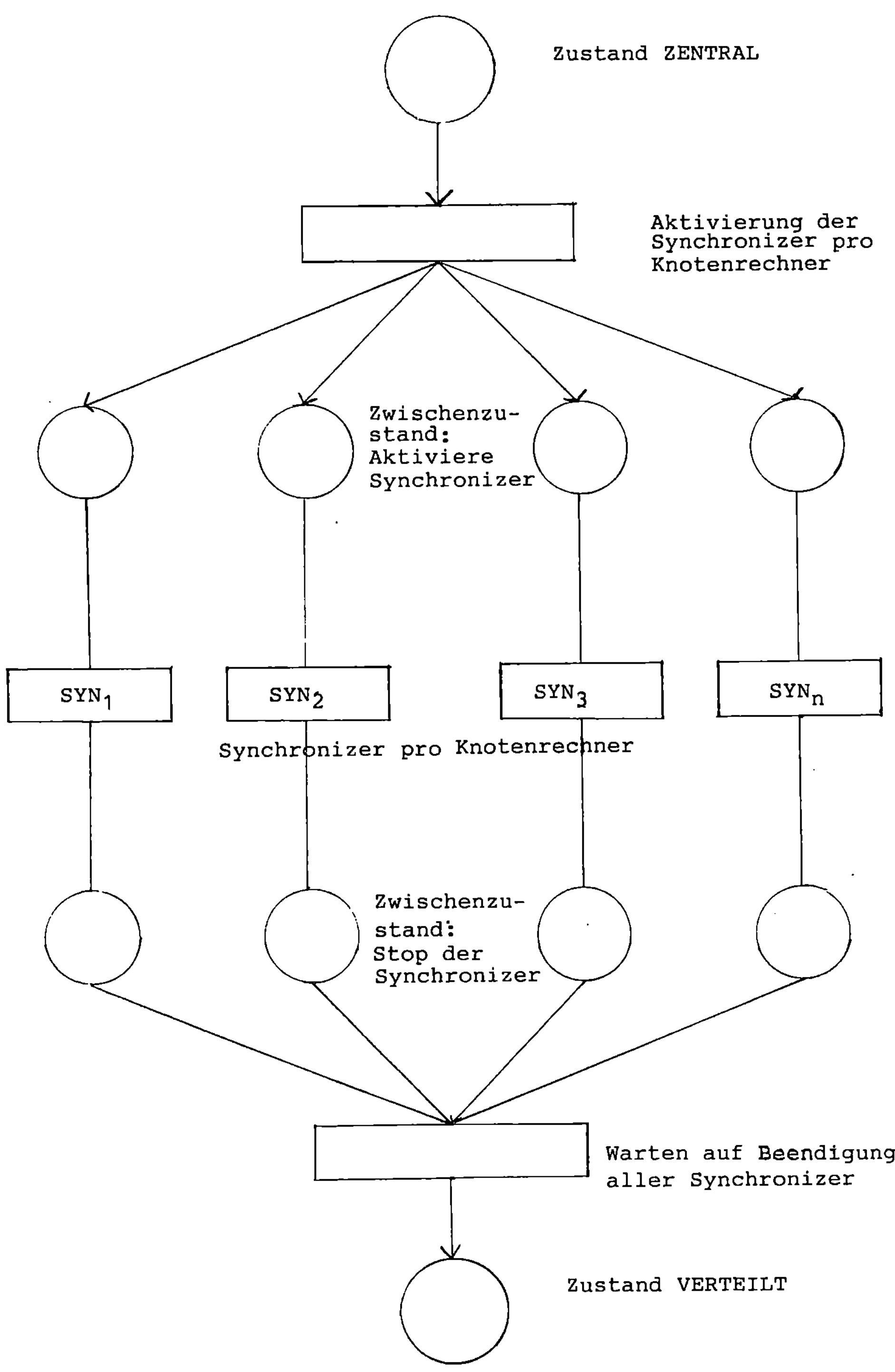

Zustandsübergang NICHT EXISTENT - LOKAL

Die Durchführung dieses Ueberganges obliegt dem Anwendungsprogramm.
Eine Transaktion befindet sich im Zustand "LOKAL", solange die Transaktion (logisch) nicht abgeschlossen wurde.

Zustandsübergang LOKAL - BEREIT

Nach Beendigung der Transaktion führt sich diese (das Anwenderprogramm) selbst in diesen Zustand über.

Zustandsübergang BEREIT - ZENTRAL

Wie oben erwähnt, ist dies die Aufgabe eines Prozesses mit dem Namen
Transporteur. Dieser Prozess hat im wesentlichen die Aufgabe, die
Transaktion und die modifizierten Objekte, versehen mit einem einheitlichen, zentralen Zeitstempel, von dem Knotenrechner zur Zentrale
zu kopieren. Pro Knotenrechner existiert ein Transporteur.

Zustandsübergang ZENTRAL - VERTEILT

Wenn alle sog. Synchronizer die Objekte einer Transaktion von der
Zentrale zu dem für sie zuständigen Knotenrechner kopiert haben,
wird die Transaktion in den Zustand "VERTEILT" überführt.

Anmerkung 1
Bei Zustandsübergängen gilt aus einsichtlichen Gründen die Regel,
dass die Zustansüberführung der kopierten Transaktion im Zielrechner
(Zentrale oder Knotenrechner) vor der Zustandsüberführung der zu
kopierenden Transaktion des Quellrechners (Zentrale oder Knotenrechner) vorzunehmen ist. Bei Verletzung dieser Regel kann bei Leitungs-
und Rechnerzusammenbruch ein ordentliches Recovery nicht vorgenommen werden.

Anmerkung 2
Sowohl der Transporteur als auch Synchronizer sind Prozesse, die in
der Zentrale gestartet werden und , technisch gesehen, auf die Datenbanken in den Niederlassungen remote zugreifen. Dies erleichtert die
Analyse der verschiedenen Recovery-Fälle.

VII. <u>Recovery</u>

Um das Wiederanlaufen des verteilten Informationsystem nach Zusammenbruch einer oder mehrer Knotenrechner einschliesslich Zentrale einfach steuern zu können, werden drei Datenbankzustände eingeführt.
Zum Verständnis sei nochmals darauf hingewiesen, dass nach einem
Programm- oder Rechnerzusammenbruch, das Datenbanksystem IMAGE/3000
in Zusammenspiel mit der Anwendungssoftware die physische und logische Konsistenz der Datenbank garantiert. Es ist jedoch möglich, dass
Transaktionen auf Grund eines Fehlers des Hintergrundspeichers etc.
verloren gingen.

VII.1 Datenbankzustände

Zustand NICHT SYNCHRON

Wann immer das Betriebssystem eines Knotenrechners oder der Zentrale
gestartet wird (IPL), setzt ein Anwendungsprozess mit dem Namen Starter den Zustand der Datenbank unmittelbar auf "NICHT SYNCHRON". Dies
bedeutet, dass wir nicht wissen, in welchem Zustand bezüglich der
Verteilung der einzelnen Transaktionen sich das isolierte Informationssystem befindet.

Zustand SEMI SYNCHRON

Dieser Zustand wird bei der Behandlung der einzelnen Recoveryfälle
genau erklärt.

Zustand SYNCHRON

In diesem Zustand befindet sich jede einzelne Datenbank, wenn wir davon ausgehen können, dass sich auch die anderen Datenbanken im Netz
in diesem Zustand befinden. Dies ist der normale Zustand der Appli-
kation. Anwender(programme) dürfen nur dann auf das System zugreifen,
wenn es sich in diesem Zustand befindet.

VII.2 Analyse der Recoveryfälle

Fall 1: Kommunikationsunterbruch

Sowohl der Transporteur als auch der Synchronizer stellen den Ausfall
des Transportnetzes fest. In den Knotenrechnern der Niederlassungen
und in der Zentrale werden die nicht vollständig kopierten Transaktionen als unvollständig angesehen und durch ein automatisches Back-
out herausgenommen. Sowohl die Datenbank in der Zentrale als auch
diejenigen der Niederlassungen befinden sich somit in einem logisch
konsistenten Zustand, wenn auch nicht vollständig synchron. Dies ist
jedoch in unserem Falle vernachlässigbar, da zum einen das Transportsystem i.a. nach wenigen Minuten wieder zur Verfügung steht und zum
anderen unsere Anwendung asynchrone Zeitintervalle, die nicht über
4 Stunden hinweggehen, toleriert. Ist dieses Zeitintervall überschritten, können die einzelnen Systeme kontrolliert angehalten werden, bis
die Synchronisation wieder gegeben ist.

Fall 2: Ausfall eines Knotenrechners

Der Ausfall eines Knotenrechners wird durch den Transporteur, bzw.
Synchronizer festgestellt. Diese starten einen sog. Mini-Recoverer.
Dieser wartet in der Zentrale, bis er eine Verbindung mit der Niederlassung (Starten einer remote Session) aufnehmen kann. Der Mini-
Recoverer sucht in der Niederlassung nach der jüngsten Transaktion
(Zeitstempel) im Zustand "SYNCHRON" und vergleicht diese mit der
jüngsten Transaktion im Zustand "SYNCHRON" in der Zentrale. Eventuel
werden dann Transaktionen, die dem Knotenrechner fehlen, zurückkopiert.
Nach Beendigung dieses Vorganges gibt der Mini-Recoverer die Datenbank in der Niederlassung wieder frei, indem er die Datenbank des
ausgefallenen Knotenrechners in den Zustand "SYNCHRON" überführt und
die Prozesse Transporteur, Synchronizer für diesen Knotenrechner re-
aktiviert. Während dieses Recoveryvorganges dürfen Anwender sowohl in
der Zentrale als auch in den anderen Knotenrechnern ungestört weiter-
arbeitet.

Fall 3: Ausfall der Zentrale

Nach Ausfall der Zentrale wird den noch aktiven Niederlassungen erlaubt, weiter zu arbeiten, denn die Datenbanken der Knotenrechner befinden sich weiterhin im Zustand "SYNCHRON".
Nach Starten des Betriebssystems (IPL) der Zentrale setzt der Prozess
"Starter" die Datenbank der Zentrale in den Zustand "NICHT SYNCHRON"

und aktiviert anschliessend pro Knotenrechner den Retransporteur
Prozess.
Die Retransporteure kopieren aufgrund der Zeitstempel die entsprechenden Transaktionen im Zustand "VERTEILT" von den Knotenrechnern zur
Zentrale. Nachdem ein Retransporteur seine Arbeit erledigt hat, setzt
er eine entsprechende Zustandsmarke. Sind alle Zustandsmarken gesetzt,
(gleiches Petri-Netz wie Abbildung 1) wird die Datenbank der
Zentrale in den Zustand "SEMI SYNCHRON" versezt. Alle Transaktionen
befinden sich nun wieder in ihrem früheren Zustand "VERTEILT". Da
möglicherweise in diesem Zeitraum ein oder mehrere Knotenrechner
ebenfalls einen Zusammenbruch erlitten haben könnten, wird die Zentrale nicht gleich wieder zur Weiterverarbeitung freigegeben. Damit
kommen wir zum

Fall 4: Ausfall der Zentrale und Knotenrechner gleichzeitig

Nach Beendigung des "Retransporteurs" werden sog. Resynchronizer-
Prozesse pro Knotenrechner gestartet. Diese kopieren nun fehlende
Transaktionen von der Zentrale zu den Niederlassungen, falls not-
wendig, ebenfalls mit Hilfe des Vergleichs von Zeitstempeln. Nach
Beendigung dieser Arbeit setzen die jeweiligen Resynchronizer die
Datenbanken in den Knotenrechnern in den Zustand "SYNCHRON", in der
Zentrale lediglich eine Zustandsmarke. Haben alle Resynchronizer
ihre Zustandsmarke gesetzt, so wird die Datenbank der Zentrale in
den Zustand "SYNCHRON" versetzt und anschliessend die Kontrolle an
die Prozesse Transporteur, Synchronizer pro Kontenrechner übergeben.

VIII. <u>Ausblick</u>

Um den Softwareaufwand in Zukunft für ähnliche Anwendungen zu minimieren, ist ein standartisiertes Verfahren denkbar. Die Strukturen von
Transaktionen und Objekten sind bei diesem Verfahren in einem Dictionary abgelegt. Anwenderprogramme sowie die für die Synchronisation
notwendigen Prozesse halten sich an ein gemeinsames Synchronisations-
protokoll, das im wesentlichen die in diesem Artikel beschriebene
Synchronisationsmethode beinhaltet. Ein weiterer Schritt wäre die
Integration dieses Verfahrens in ein Datenbankmanagementsystem.

IX. Schlussfolgerung

1. Ohne Hilfe von konzeptuellen Modellen und Methoden sind Anwendungen dieser Komplexität in einer verteilten Umgebung nicht zu beherrschen.

2. Ohne Hilfe eines (homogenen) Rechnerverbundsystems wie DS/3000
   als auch eines komfortablen Betriebssystems ist die Implementierung von Synchronisation und Recovery wirtschaftlich nicht möglich.

3. Eine geeignete höhere Programmiersprache der vierten Generation
   kann eine gewaltige Reduktion des Codieraufwandes und somit der
   Gesamtkosten erbringen.
   (Die Synchronisationssoftware für 10 verschiedene Objekt-Typen
   wurde von einer Person in SECHS Monaten erstellt!).

4. Rom wurde nicht an einem Tage erbaut! Auch wenn das in diesem
   Vortrag beschriebene Verfahren kein globalen Zugriff und Sperren
   auf verteilte Objekte erlaubt, zeigt es, dass auch SCHRITTWEISES
   Vorgehen auf diesem Gebiet machbar und wirtschaftlich vertretbar
   ist.

Literatur

/1/ Dadam, P; Synchronisation in verteilten Datenbanken:
            Ein Ueberblick, Teil 1 und 2. Informatik Spektrum
            August, Oktober 1981.

/2/ Flavin, M; Fundamental Concepts of Information Modeling.
            A Yourdon Press Monograph.

/3/ Goos, G; Programmkonstruktionen.
            Fakultät für Informatik, Uni Karlsruhe.
            Vorlesungsskript 1975-76.
            Kapitel 7.6 Petri-Netz

/4/ HP/3000 System Manuals.
        DS/3000 Reference Manual.
        IMAGE/3000 Reference Manual.
        TRANSACT/3000 Reference Manual.

<u>THYNET, ein Kommunikationssystem für den betrieblichen Einsatz</u>

H.J. Haubner
Fraunhofer-Institut für Informations- und Datenverarbeitung (IITB),
Karlsruhe
R. Körner und K. Kruppa
Thyssen Stahl AG, Duisburg

Abstract

Modern distributed production processes require data communication
between all local and central sub-systems to have an optimal control,
even for very complex production processes. But open system interconnections are not yet available for the present generation of process control
computers. THYNET was developed as a communication system for industrial
applications to link heterogeneous hosts.

The communication network is a packet switching system, with flexible
configuration consisting of minicomputers as switching nodes connected
by means of private lines as well as the public network DATEX-P. Starting
with the user's demands this paper deals with main structures and components of the system and operational experiences.

Zusammenfassung

Moderne, verteilte Produktionsprozesse erfordern einen Informationsfluß
zwischen allen lokalen und zentralen Subsystemen, um auch sehr komplexe
Produktionsabläufe optimal steuern zu können. Offene Systeme sind für
die heute im Einsatz befindliche Generation der Betriebs- und Prozeßrechner erst in Ansätzen vorhanden. Für diesen Zweck wurde THYNET als flexib-
les Kommunikationssystem für heterogene Rechner entwickelt.

Es besteht aus einem beliebig vermaschten Netz von Netzknoten und Nachrichtenwegen, wie betriebseigene Leitungen oder öffentliche Datennetze.
In diesem Beitrag werden ausgehend von den Benutzeranforderungen die
wesentlichen Strukturen des Systems sowie betriebliche Erfahrungen behandelt.

<u>Zusammenfassung:</u>

Die Entwicklung der Datenverarbeitung in Industriebetrieben ist gekennzeichnet durch einen ständig wachsenden Informationsbedarf und die Forderung nach einem schnellen Informationsfluß für die Prozeß-, Produktions-
und Betriebsablaufsreuerung. Dies erfordert eine integrierende und produktionssynchrone Bereitstellung und Auswertung dezentral anfallender
Daten. Als Voraussetzung hierfür sind Kommunikationsmöglichkeiten zum
Austausch von Daten auch zwischen weit auseinderliegenden heterogenen
Datenquellen und -senken zu schaffen.

THYNET ist ein gemeinsam von der Thyssen Stahl AG und dem Fraunhofer-
Institut für Informations- und Datenverarbeitung entwickeltes und er-
probtes verteiltes paketvermittelndes Kommunikationssystem zum Aufbau von
beliebig vermaschbaren Rechnernetzen. In dessen Konzeption fanden gerade
betriebliche Forderungen nach Heterogenität der Kommunikationspartner,
hohe Zuverlässigkeit, flexible Einsatzmöglichkeiten und transparente
und einfache Unterhaltung und Betrieb Berücksichtigung.

In diesem Beitrag werden zunächst ausgehend von den Benutzerforderungen
die wesentlichen Strukturen des Systems erläutert werden. Hierauf aufbauend erfolgt eine Analyse der betrieblichen Erfahrungen.

1. Ausgangssituation

Eine zusammengefaßte und produktionssynchrone Bereitstellung und Auswertung aller dezentralen Prozeß-, Produktions- und Betriebsdaten beispielsweise zur Optimierung des Rohstoff- und Energiebedarfs, Produktqualitäten
und -kosten, sowie von Beständen und Lieferterminen wird in den Industriebetrieben zu einem immer dringender verlangten Instrumentarium. Als Voraussetzung hierfür sind Kommunikationsmöglichkeiten zum Austausch von
Daten zwischen auch weiträumig verteilten heterogenen Datenquellen und

-banken zu schaffen. Die Integration von einzelnen Anwendersystemen zu diesem Zweck in einem Verbund wird jedoch gerade in Industriebetrieben durch eine Vielzahl von historisch gewachsenen Rechnertypen mit unterschiedlichen Kommunikationsschnittstellen erschwert /1/.

Die Öffnung solcher geschlossenen Einzelsysteme wird durch die Normung des Architektenmodelles für offene Kommunikationssysteme durch das ISO-Referenzmodell und durch die Normung von Kommunikationsprotokollen angestrebt. Solche genormten Schnittstellen sind für die heute im Einsatz befindlichen Rechnergenerationen erst im Ansatz vorhanden. Ihre Nachrüstung ist aus Kapazitäts- und/oder Kostengründen oder wegen fehlender Verfügbarkeit bei älteren Systemen schwierig /2/.

Auch führt der Einsatz dedizierter Einzelverbindung bei wachsender Zahl der Rechner zu einer immer unübersichtlicheren Kommunikationsstruktur. Der Einsatz von Systemen wie sie beispielsweise in überregionalen Netzen Verwendung finden, würde hier eine kosten- und leistungsmäßige Überdimensionierung bedeuten. Diese Überlegungen führten zwangsweise zur Konzeption und Entwicklung von THYNET.

2. Benutzeranforderungen an ein betriebliches Kommunikationssystem

Aufgrund dieser Ausgangssituation erfolgte die Entwicklung insbesondere für einen flexiblen und kostengünstigen Einsatz bei breitem Anwendungsspektrum gerade für den industriellen Bereich. Die Konzeption erfolgte mit Blickrichtung auf die bekannten interregionalen Netze wie ARPA-Netz, DATEX-P, TRANSPAC usw .

Es sollten deren Vorteile wie

- effektive Leistungsausnutzung durch Multiplexen von Datenströmen mittels Paketvermittlung und

- hoher Sicherheitsgrad durch Vermaschungsredundanz verbunden mit Wegesteuerung und Datenflußkontrolle

genutzt werden.

Daneben sollten aber insbesondere folgende betriebliche Anforderungen erfüllt werden /3/:

- Anschluß im Prinzip beliebiger vorhandener Rechner und Terminals verschiedener Generation, Größe und Hersteller über flexible und frei programmierbare modulare Schnittstellen im Kommunikationssystem mit einfacher Kommunikationsschnittstelle,

- weitgehende Entlastung der Teilnehmer von Kommunikationsaufgaben durch einfache Kommunikationsmöglichkeiten mittels Einzelnachricht (Telegramme).

- Verwendung vorhandener betriebseigener Leitungen unter teilweiser Einbeziehung öffentlicher Leitungen und Netze,

- stufenweise kontinuierliche Erweiterung des Verbundes im laufenden 24 h-Betrieb,

- Transparenz der Funktionen und Abläufe, so daß für die Betreiber eine unkomplizierte Unterhaltung sowie Wartung und Testmöglichkeiten von einer zentralen Instanz möglich sind.

- Beschaffungs- und Unterhaltungskosten müssen in einem vergleichbaren Rahmen zu den überwiegend Klein- und mittleren Rechnern zuordenbaren vorhandenen Rechnern und Geräten liegen.

3. Strukturen und Komponenten

Das Gesamtsystem bildet einen heterogenen Rechnerverbund, der sich aus den beiden Teilsystemen Kommunikationssystem und den daran angeschlossenen Teilnehmern zusammensetzt. Das Kommunikationssystem ist ein beliebig vermaschbares Netz aus Netzknoten und Nachrichtenwegen (Bild 1).

3.1 Netzwerk

Das Netz arbeitet nach dem Prinzip der Paketvermittlung. Übertragen werden sog. Telegramme oder Nachrichten, deren maximale Länge 512 Bytes beträgt. Jedes Telegramm enthält einen Kopf mit eindeutiger Absender- und Empfängerangabe sowie weiteren Kontrolldaten und die eigentlichen Nutzdaten.

Eine Wegesteuerung verbunden mit einer Ende-zu-Ende-Kontrolle garantiert die geforderte Zuverlässigkeit. Weitere wichtige Merkmale sind eine einfach handhabbare Bedienung und ein transparentes Überwachungs- und Bediensystem. Alle Funktionen sind fernbedienbar ausgelegt, so daß sie von einem Netzknoten systemweit angesprochen werden können.

3.2 Programmierung

Das Programmsystem von ca. 41 kW ist in der höheren Echtzeitprogrammier-
sprache PEARL realisiert. Es ist in 14 getrennt übersetzbare Module mit
insgesamt 11 Tasks gegliedert. Der interne Datenverkehr zwischen diesen
Tasks erfolgt über ein in Software realisiertes bus-ähnliches Auftrags-
und Kommunikationssystem. Zusätzlich werden ein PEARL-Betriebs- und
Laufzeitsystem von ca. 7 kW verwendet. Hinzu kommt die in Assembler realisierten Programme der Prozedurprozessoren, die je nach Anwendung zwischen 2 und 4 kW zusätzlich 2 kW Datenspeicher betragen.

3.3 Netzknoten

Jeder Netzknoten besteht aus einem Hauptprozessor mit 48 k Wörtern Ar-
beitsspeicher und mikroprogrammierten Befehlsatz. Ein zusätzlicher mikroprogrammierter Co-Prozessor (AUR) übernimmt die Sonderfunktionen des
Überwachens, Urladens, Dumpens und Startens im Initialzustand des Haupt-
rechners /4/.

Der E/A-Bereich besteht aus dem Stationsbedienfeld, über dessen V24-
Schnittstelle, Tastatur bzw. Anzeigenelemente lokale Bedienein- bzw. Be-
dienausgaben sowohl manuell als auch rechnergesteuert möglich sind, sowie
maximal 24 Prozedurprozessoren (PP), die in Gruppen zu je 8 über Inter-
ruptsammler (IS) und Businterfarce (BI) angekoppelt sind (Bild 2).

3.4 Prozedurprozessor

Die Baugruppe PP ist ein autarker, frei programmierbarer Ein-Platinen-
Rechner, der speziell für den Anschluß der netzinternen Leitungen zwischen Netzknotenrechner als auch zur Realisierung der Teilnehmeranschlüs-
se entwickelt wurde. Der PP ist für synchrone vollduplexfähige serielle
Übertragung mit Geschwindigkeiten bis zu 19 200 bit/s ausgelegt. Aufgabe
des PPs ist die Abwicklung der auf dem jeweiligen Anschluß (Netzknoten-
Netzknoten bzw. Netzknoten-Teilnehmerrechner) eingesetzten Datenübertra-
gungsprozedur. Zur Unterstützung eines schnellen Datentransfers zwischen
PµP und PP ist im PµP ein mikroprogrammierter Bytegruppentransfer realisiert.

3.4 Verwendete Kommunikationsprinzipien

Die Kommunikation basiert auf dem Prinzip des Nachrichtenverbundes. Hier-

unter sei verstanden, daß Teilnehmer und Kontrollinstanzen im Netz untereinander über Einzelnachrichten oder Sequenzen von Nachrichten miteinander kommunizieren. Eine Nachricht (Telegramm) besteht aus einem
Nachrichtenkopf mit Sender- und Empfängeradresse sowie weitere Kontroll-
informationen undden eigentlichen Nutzdaten (Bild 3). Die Nutzdaten werden transparent übertragen, währenddessen die Nachrichtenköpfe zur Wegesteuerung und Datenflußkontrolle verwendet werden.
Zur Sicherstellung der Transportfunktion im Netz wird zwischen zwei
Netzknoten eine Punkt-zu-Punkt-Kontrolle durchgeführt, die den ordnungs-
gemäßen Telegrammverkehr zwischen zwei benachbarten Netzknoten sicher-
stellt. Hierfür wird HDLC gemäß CCITT-Empfehlung X.25, Level 2, verwendet.

3.5 Ende-zu-Ende-Kontrolle

Dieser ist zur Erhöhung der Zuverlässigkeit für jeden Telegrammverkehr
zwischen einem Empfänger- und Sendepaar von Nachrichten einer Ende-zu-
Ende-Kontrolle überlagert.

Für die nur innerhalb des Netzes zwischen Kontroll- und Bedieninstanzen
auszutauschenden Nachrichten besteht diese aus einem einfachen Quittungs-
verfahren nach dem Hand-shake-Prinzip mit Quittungsfensterbreite = 1.
Für den Teilnehmerverkehr besteht für die Ende-zu-Ende-Kontrolle wahlweise die Möglichkeit von Einzeltelegramm-Übertragung mit Quittungsfensterbreite = 1 (Datagrammverkehr mit Quittung) oder von einer ver-
bindungsorientierten Mehrfach-Telegramm-Übertragung mit Quittungsfensterbreite = 7. In beiden Fällen wird der gesamte Telegrammweg innerhalb des Netzes überwacht. Die Telegramme bleiben dabei solange in dem
Netzknoten der Sendeteilnehmer gespeichert, bis die Übertragung zum
Zielknoten als gesichert gilt bzw. die Übertragung definiert abgebrochen wird.

3.6 Teilnehmerspezifische Anpassung

Um die gefordete flexible Ankopplung von Teilnehmern sicherzustellen,
werden in den den Teilnehmern zugeordneten Prozedurprozessoren die teil-
nehmerspezifischen Protokolle auf dieses neutrale netzinterne Kommuni-
kationsprinzip abgebildet und umgekehrt. Letztere wurden bewußt so einfach gehalten, daß die Abbildung auch bei einfachen teilnehmerspezi-
fischen Protokollen wie z.B. die hier i.A. vorhandenen BCS-ähnlichen
Prozeduren erbracht werden können.

4. Betriebserfahrungen

Das vorgestellte System THYNET ist seit 1981 bei der Thyssen Stahl AG
in Betrieb und wurde kontinuierlich auf den heutigen Ausbau von 8 Netzknoten mit ca. 30 Teilnehmern erweitert. Wegen der geforderten Verfügbarkeit sind davon fast alle Teilnehmer mit Doppelanschlüssen ausgerüstet
(Bild 1).

Die Inbetriebsetzung und der Probebetrieb erfolgten mittels eigens entwickelten Simulatoren als einfache Datenquellen und -senken. So konnten
bereits im Vorfeld des endgültigen Betriebes wichtige Erfahrungen gesammelt werden.

4.1 System-Management

Das System ist so ausgelegt, daß weitgehend alle Funktionen automatisch
ablaufen. Trotzdem war es ein weiteres Entwicklungsziel, dem Betreiber
ein effektives Instrumentarium zur Verfügung zu stellen, mit dessen
Hilfe eine transparente und zentral geführte Kontrolle und Steuerung
(System-Management) möglich ist.

Die Überwachungs- und Bedienfunktionen sind für alle Netzknoten fernbedienbar, d.h. sie können von einem im Prinzip beliebigen Netzknoten
angesprochen werden. Hierfür wird aber i.a. ein Netzknoten mit Sonderfunktion, der als Minimalausstattung mit einem Bediengerät, bestehend
aus Terminal, Drucker und Floppy-Laufwerk, gesteuert über ein Mikrocom-
putersystem, ausgerüstet ist, eingesetzt. Bei komplexeren Verbundstruk-
turen oder aus Gründen der Ausfallsicherheit können auch mehrere Netzknoten im Verbund mit Sonderfunktion ausgestattet sein.

Alle angesprochenen Funktionen werden über spezielle Steuer- und Kontroll-
telegramme abgewickelt, die zwischen dem zentralen Teil und den lokalen
Teilen des Überwachungs- und Bediensystems zusammen mit dem Teilnehmer-
telegrammverkehr über die Leitungen zwischen den Netzknoten ausgetauscht
werden.

Während die Kernfunktionen des Systems weitgehend unverändert bleiben,
wurde die Managementfunktion aufgrund der Erfahrung auf den im folgenden
beschriebenen Funktionsumfang erweitert.

4.2 Die Überwachungsfunktionen

Das Überwachungssystem dient dazu, schnell und zielgerichtet auf Störungen reagieren zu können. Hierzu wird in jedem Netzknoten eine Ablauf-
überwachung durchgeführt, bestimmte Ereignisse, wie z.B. Ausfall und
Wiederanlauf von Netzleitungen oder Telegrammfehler registriert und an
den Netzknoten mit Sonderfunktionen gemeldet.

Das Überwachungssystem kann selektiv für einzelne bzw. für alle Netzknoten zugeschaltet werden. Die Meldungen enthalten Uhrzeit, Netzknoten,
Meldenummer, Kanal und Art des Ereignisses und werden am Netzknoten mit
Sonderfunktion zusammen mit Datum und Protokollierungszeitpunkt protokolliert. Insgesamt sind derzeit ca. 150 verschiedene Meldungen implementiert.

4.3 Bedienfunktionen

Das Bediensystem in den Netzknoten wird vom Bediengerät über eine einfache Kommandosprache angesprochen. Es enthält zunächst die notwendigen
Funktionen für das Urladen aller Netzknoten (Kaltstart) wie auch für das
Nachladen einzelner Netzknoten im laufenden Betrieb mit den Programmda-
teien von den Disketten des Bediengerätes. Die Funktion des Laders in
dem urzuladenden lokalen Netzknoten übernimmt der AUR. Dieser ist nach
Spannungseinschaltung bzw. nach $P\mu P$-Ausfall oder Grundstellung aktiv.

Der Ladevorgang ist nach dem Prinzip des "Down-Line-Loadings" konzipiert,
d.h. er wird wie alle Bedienvorgänge über spezielle Telegramme vom Netzknoten mit Sonderfunktion zu den urzuladenden Netzknoten über die Netzleitungen abgewickelt. Ein urgeladener Netzknoten wird nach Beendigung
des Ladevorganges automatisch gestartet. Nach erfolgtem Start koppelt
sich der Netzknoten selbsttätig an den Verbund an. Eine weitere Bedingung
ist hierbei nicht erforderlich.

Auf diese Weise werden die Netzknoten sukzessive geladen und gestartet,
wobei sich der Verbund jeweils nach dem Start eines Netzknotens rekon-
figuriert. Dadurch wird erreicht, daß auch ein teilkonfigurierter Verbund
bereits voll funktionsfähig ist. Das Laden weiterer Netzknoten erfolgt
dann unter Ausnützung aller Sicherheitsmechanismen wie Wegesteuerung
und Punkt-zu-Punkt-Kontrolle des teilkonfigurierten Verbundes.

Um beim Umkonfigurieren oder Erweitern des RDC-Datenverbundes ein Nachladen einzelner Netzknoten zu vermeiden, können über das Bediensystem
im laufenden Betrieb Konfigurations- und Betriebsparameter gelesen und
verändert werden (z.B. Wegetabelle, s. Bild 4).

Über den gleichen Mechanismus kann weiterhin auf umfangreiche Tabellen
für Fehler- und Telegrammverkehrstatistiken zugegriffen werden (Bild 5,
6,7). Durch deren Auswertung können frühzeitig Fehlertrends oder System-
engpässe erkannt und entsprechende Gegenmaßnahmen ergriffen werden.

4.4 Ferndiagnosefunktionen

Speziell für Test- und Wartungszwecke ist in das Bediensystem ein Fern-
diagnosesystem integriert. Es dient dazu, Fehler und Störungen in den
lokalen Netzknoten, die mit den bisher geschilderten Mitteln nicht ausreichend behoben werden können, gezielt vom Netzknoten mit Sonderfunktion zu lokalisieren und zu diagnostizieren. Es umfaßt Funktionen für

- Programmverfolgung (Tracefunktion) auf PEARL-Ebene im laufenden Betrieb
  und

- Speicherabzug (Dumpfunktion) nach Rechnerausfall.

Die Tracefunktion ist tabellengesteuert über das Bediensystem selektiv
sowohl funktionsbezogen als auch kanalbezogen im laufenden Betrieb fernbedienbar, zu- und abschaltbar. Zusätzlich kann diese Funktion ständig
als auch nur im Fehlerfall zugeschaltet werden. Diese Trace-Information
wird dabei am Netzknoten mit Sonderfunktion ausgegeben.

Mittels der Dumpfunktion des Bediensystems können nach Ausfall von lokalen Netzknoten gezielt Arbeitsspeicherbereiche und Registerbereiche, wie
z.B. die internen Hardware-Fehlerregister, am Netzknoten mit Sonderfunktionen gelesen und so post mortem die Ausfallursachen analysiert werden.

4.5 Zuverlässigkeit

Für den betrieblichen Einsatz ist ein Höchstmaß an Zuverlässigkeit gefordert. Diese wird zum einen durch die Zuverlässigkeit der einzelnen
Hard- und Software-Komponenten, zum anderen durch die Struktur des Verbundes sowie besonderer konstruktiver Maßnahmen erreicht.

Durch die beliebig vermaschbaren Netzknoten kann durch entsprechende
Konfiguration des Kommunikationssystems erreicht werden, daß bei Ausfall
von Netzknoten bzw. Leitungen zwischen diesen im Netz selbst eine hin-
reichende Anzahl redundanter Nachrichtenwege zwischen zwei Kommunikationspartnern vorhanden sind.

Die Auswahl eines Nachrichtenweges erfolgt durch eine dezentrale Wegesteuerung, die in jedem Netzknoten installiert ist. Dieser ermittelt für
jedes Telegramm adaptiv aus den Leitungszuständen. d.h. betriebsbereit,
ausgefallen, überlastet und der Ziel- und Quelladresse einen Teilweg zum
nächsten Netzknoten.

Die Leitungszustände wie betriebsbereit, ausgefallen sowie den Belastung-
en werden laufend gemessen und für die Wegesteuerung benutzt. Die Wegesteuerung ist abhängig vom Ziel- und Empfangsknoten und erlaubt so eine
gezielte Steuerung des Datenflusses. Sie beinhaltet einen vorgegebenen
Primärweg und bis zu 2 vorgegebene Ersatzwege (Bild 4). Die angegebenen
Wege 1,2,.. entsprechen den Kanälen 17,18.. in Bild 1. Die Wegesteuerung
ist zentral über die Managementfunktions-Bedienung im laufenden Betrieb
steuerbar und kann so bei Konfigurationsänderung oder durch Lastanpassung
geändert werden.

Diese einfache Art der Wegesteuerung bietet gegenüber komplexerer Verfahren für den vorliegenden Anwendungsfall folgende Vorteile:

- Transparenz: bei der i.a. vorhandenen Netzgröße sind die Nachrichtenwege a priori bekannt.

- Steuerbarkeit: Datenflüsse können von außen sehr leicht beeinflußt
  werden.

- Geringe Systembelastung: Es werden für die Wegesteuerung keine zusätzlichen Telegramme ausgetauscht.

Um eine vergleichbare Zuverlässigkeit für die Teilnehmer-Anschlüsse an
den Netzknoten zu erreichen, können diese gleichzeitig an zwei Netzknoten
angeschlossen werden. Dadurch wird insgesamt erreicht, daß bei einer angemessenen Vermaschung des Netzes mindestens der Ausfall einer Komponente
toleriert wird, ohne die Funktionsfähigkeit des Kommunikationssystems
zu beeinträchtigen. Die Anpassung an Ausfälle erfolgt automatisch durch
das System.

Neben dem Ausfall von Komponenten werden weiterhin kurzfristige Störung-
en der Übertragung zwischen Netzknoten toleriert. Hierzu dienen zum einen
das eingesetztes Punkt-zu-Punkt-Protokoll-HDLC, das durch sein Fehler-
prüfverfahren CRC und die Numerierung von Datenblöcken den bestmöglichen
Schutz gegen Verfälschung oder Verlust von Telegrammen auf den Punkt-zu-
Punkt-Verbindungen zwischen den Netzknoten bietet.

Um den Verlust von Telegrammen weitgehend unmöglich zu machen, ohne die
Teilnehmer mit einer Ende-zu-Ende-Kontrolle zu belasten, wurde innerhalb
des Kommunikationssystems zwischen den Ziel- und Quellknoten weiterhin
eine Ende-zu-Ende-Kontrolle integriert. Sie sichert zum einen die Sequenz
von Telegrammstapeln zwischen zwei Teilnehmern, außerdem stellt sie
sicher, daß zwischen Quell- und Zielknoten keine Telegramme verloren
gehen. Hierzu werden die Telegramme in den Quellknoten gespeichert, bis
eine entsprechende Quittung vom Zielknoten eintrifft.

Durch die laufende Überwachung und die damit verbundene Analyse und Be-
hebung lokaler Fehler, die fast ausschließlich auf den Ausfall einzelner
Hardware-Komponenten zurückzuführen waren, liegt die Verfügbarkeit des
Netzes bei annähernd 100 %. Aufgetretene Störungen waren größtenteils
auf den Neuanschluß von Teilnehmern zurückzuführen.

4.6 Weiterreichende Managementfunktionen

Die bisher geschilderten im Netz realisierten Mechanismen stellen Basisfunktionen dar. Auf diese wurden zur Erhöhung des Komforts weiterreichende
Managementfunktionen aufgesetzt. Diese wurden auf einem Tandemrechner,
der anstelle des einfachen Bediengerätes angeschlossen wird, realisiert.
Hierzu gehören

- Generieren der Wegetabellen in Abhängigkeit von der aktuellen Netzkonfiguration,

- Einbringen von aktuellen Parametersätzen wie z.B. Wegetabellen in das
  Programmsystem beim Urladen von Netzknoten,

- Zyklische Abfrage und Auswertung von Verkehrs- und Störungsstatistiken
  und deren Langzeitdokumentation,

- Aufschalten von definierten Verkehrsströmen für Durchsatz- und Bela-
  stungsmessung.

Die Statistiken werden als Übersicht- und Detailstatistiken geführt. Es
werden damit z.B. der Durchsatz an vermittelten und aufgrund von Fehlern
nicht vermittelbaren Telegrammen an Teilnehmer, Netzkanäle und an Über-
wachungsprogramme getrennt nach Telegrammarten in Telegrammen und in
Portionen (Telegrammsegmente = 40 Bytes) angezeigt (Bild 5).

Zum anderen können die pro Teilnehmer- und Netzkanal aus- und einge-
gangenen Telegramme sowie die dabei aufgetretenen Fehler detailliert
ausgegeben werden. Dies kann als Schnappschuß (Bild 6) oder in seinem
zeitlichen Verlauf erfolgen (Bild 7).

5. Ausblick

Im praktischen Betrieb hat sich besonders die Transparenz des Systems
bewährt, durch die es möglich ist, letztlich fast alle auch die dezentralen Abläufe nach außen zentral sichtbar zu machen. So konnten bei
Neuanschlüssen von Teilnehmern, was in einem heterogenen System oft
problematisch ist, auftretende Fehler schnell lokalisiert und behoben
werden.

Neben dem beschriebenen innerbetrieblichen Einsatz ist es auch möglich
abgesetzte, in räumlich getrennten Betriebsbereichen installierte Teilnetze über DATEX-P zu koppeln (Bild 8). Dies erfolgt über den Basis-
dienst DATEX-P10. Die Kommunikation zwischen zwei Teilnehmern in zwei
verschiedenen Teilnetzen erfolgt dabei in gleicher Weise wie innerhalb
eines Teilnetzes.

Literatur:
/1/ Wissel, H.: Datenkommunikation im technischen Bereich, eine Voraus-
        setzung zur integrierten Produktionssteuerung, FhG-Be-
        richte 2-83.

/2/ Haubner,H.-J.; Borchers, G.: RDC Datenverbundsystem mit DATEX-P-Ver-
        bindung, FhG-Berichte 3-82.

/3/ Haubruck, A.; Früchtenicht, H.W.: THYNET, ein Rechnernetz für indu-
        strielle Anwendungen, Intern. Kongreß für Datenverarbei-
        tung und Informationstechnologien, Berlin 1981.

/4/ Heger, D.; Steusloff,H.; Syrbe, M.: Echtzeitrechnersystem mit ver-
        teilten Mikroprozessoren, Forschungsbericht DV 79-01,
        BMFT 1979.

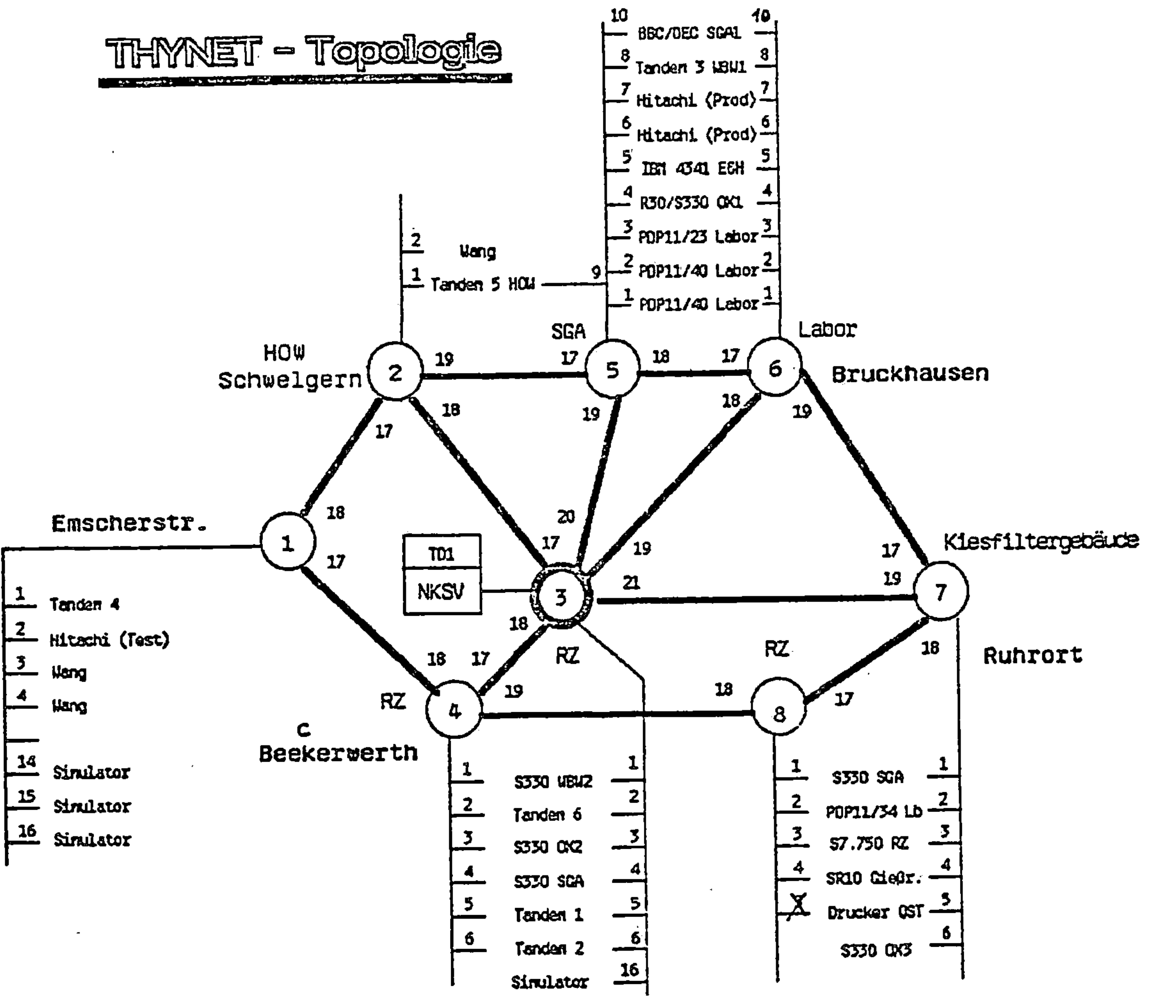

Bild 1: THYNET-Topologie

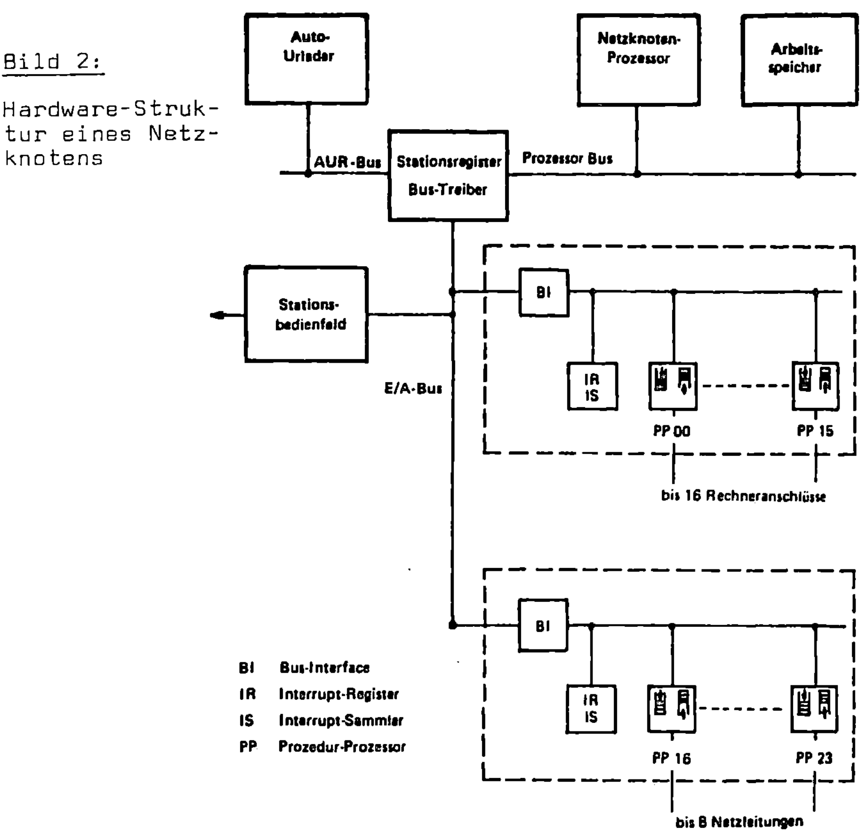

Bild 2:

Hardware-Struktur eines Netzknotens

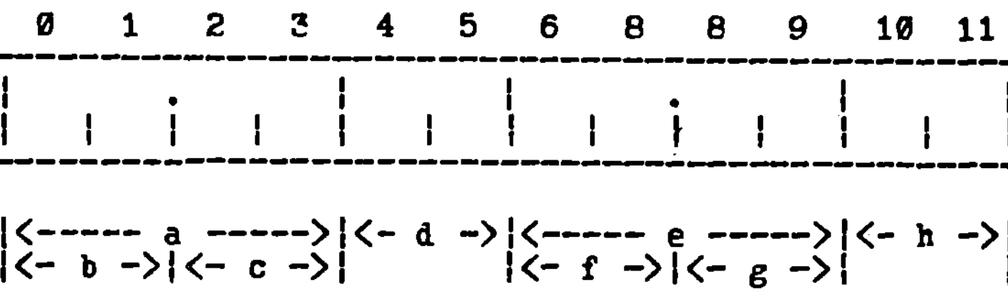

Bild 3:

Telegramm-struktur

Beschreibung der einzelnen Kopfelemente:

a   Adresse des Empfaengers mit

   b   Adresse des Vermittlungknotens und
   c   Adresse des Teilnehmers am Knoten;

d   Telegramm-Art

e   Adresse des Senders mit

   f   Adresse des Vermittlungknotens und
   g   Adresse des Teilnehmers am Knoten;

h   Telegramm-Status

```
KNOTEN  03              14.11.1984-16:55:39

TABELLE  2  (WEGE)
SKN -> EKN 1      2      3      4      5      6      7      8      9

    1      ----   143    ----   253    431    354    532    253    ----
    2      253    ----   ----   253    435    345    532    253    ----
    3      124    142    ----   215    413    345    532    253    ----
    4      143    143    ----   ----   413    354    534    534    ----
    5      125    125    ----   251    ----   352    532    523    ----
    6      214    142    ----   251    412    ----   521    251    ----
    7      214    143    ----   214    413    341    ----   214    ----
    8      124    143    ----   214    413    345    534    ----   ----
    9      ----   ----   ----   ----   ----   ----   ----   ----   ----
```

Bild 4: Beispiel Wegetabelle

```
KNOTEN  02              14.11.1984-15:07:37

TABELLE  1  (STATISTIK-UEBERSICHT)                TELE:       PORT:

VERMITTELT    DATEN-TELE  AN  TR-KAN              12138       27703
VERMITTELT    DATEN-TELE  AN  NE-KAN              42116       50969
VERMITTELT    UEBWA-TELE  AN  NE-KAN              35018       40010
VERMITTELT    UEBWA-TELE  AN  MONITOR             00376       00468
VERMITTELT    TELE        AN  LEIT-ZE             37078       37844
VERMITTELT    GESAMT                              52844       17510
VERNICHTET    DATEN-TELE  AN  TR-KAN              00000       00000
VERNICHTET    DATEN-TELE  AN  NE-KAN              00000       00000
VERNICHTET    UEBWA-TELE  AN  NE-KAN              00000       00000
VERNICHTET    UEBWA-TELE  AN  MONITOR             00000       00000
VERNICHTET    GESAMT                              00000       00000
LOWLITE  (IN  1000  DURCHLAEUFEN)     31064
```

Bild 5: Übersichts-Statistik

```
KNOTEN  03              14.11.1984-16:07:54

TABELLE   14/15  (VERMITTELTE  TELE)

KANAL    TELE       PORT
01       02760      08280
02       00507      01582
03       07618      11756
04       01127      05631
05       00126      00454
17       11435      17419
18       05346      10943
20       13465      32592
21       01263      01263


TABELLE   16/17  (EMPFANGENE   TELE)

KANAL    TELE       PORT
01       02928      03095
02       00508      00512
03       07995      11684
04       02913      20613
05       00385      00892
17       14631      14849
18       42899      33908
19       49045      49441
20       25023      30410
21       12755      18765


TABELLE   19       (LP-STATISTIK HDLC)

KANAL     17     18     19     20     21
          00003  00000  00000  00000  00000  CTS-INAKTIV
          00001  00001  00000  00000  00002  DCD-INAKTIV
          00001  00000  00002  00000  00000  CRC-FEHLER
          00000  00000  00000  00000  00000  PORTIONS-LE
          00000  00000  00000  00000  00000  PORTIONS-NR
          00000  00001  00000  00000  00000  SIO-FEHLER
          00000  00000  00000  00001  00000  FIFO-FEHLER
          00001  00000  00000  00000  00000  TIMEOUT-SIO
          00001  00000  00000  00000  00000  T1-TIMEOUT
```

Bild 6: Detail-Statistik

```
STATISTIK   KNT01
            ALLG.-STATISTIK-DATEN
            VMDTT VMDTN VMUTN VMUTM VMTEL VNDTT VNDTN VNUTN VNTEL
10.02.-00:00 36359 30979 00121 01398 34542 00000 00036 00000 00215
10.02.-01:00 37257 31411 00123 01402 37636 00000 00036 00000 00215
10.02.-02:00 38149 31845 00125 01406 40722 00000 00036 00000 00215
                                       "
                                       "
                                       "
10.02.-22:00 17013 11977 00936 03193 55119 00000 00069 00000 00412
10.02.-23:00 17897 12401 00938 03197 59141 00000 00036 00000 00215
10.02.-24:00 18794 12830 00942 03203 61249 00000 00036 00000 00215
```

Bild 7: Beispiel Statistikauswertung

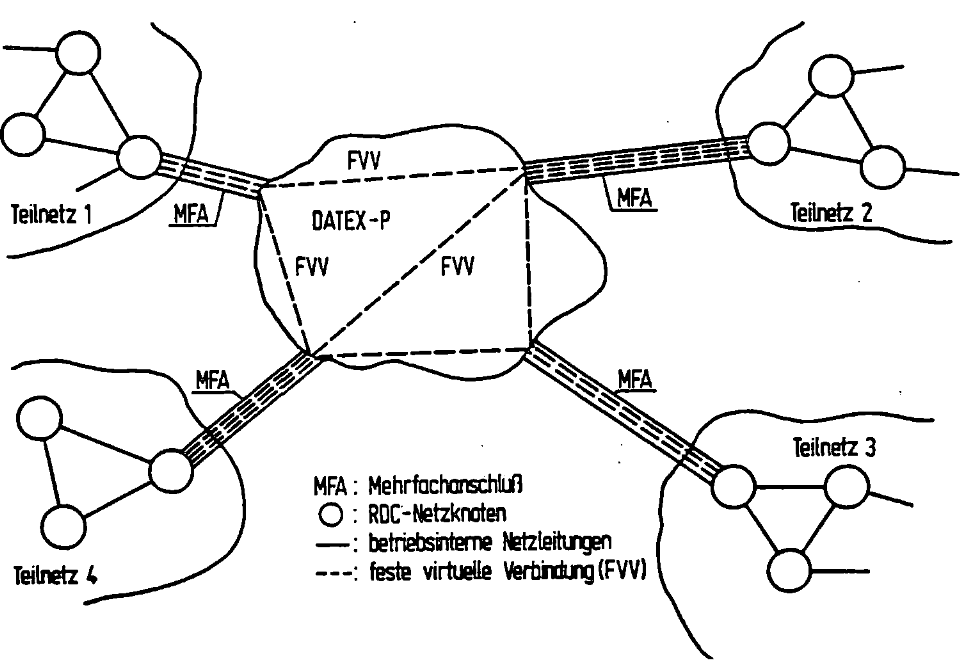

Bild 8: Teilnetz-Integration mittels DATEX-P

<u>Praktische Realisierung eines verteilten, fehlertoleranten</u>
<u>Fertigungsautomatisierungssystems</u>
<u>sowie Aspekte der Kommunikation</u>
M. Rudolf
Fraunhofer-Institut für Informations- und Datenverarbeitung (IITB)

<u>Zusammenfassung:</u>

Besondere Probleme bei der Automatisierung im Bereich der Fertigungstechnik stellen die Berücksichtigung der Individualität der Fertigung
jedes einzelnen Werkstückes, der komplexen, zwischengeschalteten Trans-
portvorgänge sowie die Auslegung auf schnelle Instandhaltung ausgefal-
lener Anlagenkomponenten und der damit zusammenhängenden Störungserken-
nung und -analyse dar.

In diesem Beitrag wird das im IITB entwickelte und in der Automobilin-
dustrie eingesetzte Fertigungsautomatisierungssystem kurz vorgestellt.
Dieses System dient der Überwachung, Bedienung und Dokumentation des
Fertigungsprozesses. Das Systemkonzept basiert auf einem verteilten,
busgekoppelten, fehlertoleranten Prozessrechnersystem mit den Komponenten "Intelligente Unterstationen" zur Erfassung, Regelung und Steuerung,
dem Zentralwartensystem zur Prozessführung, -beobachtung und Bedienung
über Bildschirme und dem Anlagenüberwachungssystem zur Speicherung von
aufbereiteter Produktionsinformation in einer Echtzeitdatenbank.

Auf das Kommunikationssystem (LAN) für die Datenübertragung zwischen
allen beteiligten Stationen, einem Lichtleiterringbussystem, sowie auf
einige Aspekte der Kommunikation wird besonders eingegangen.

1. Problemstellung

Die Automatisierung umfangreicher diskreter Prozesse im Bereich der
Fertigungstechnik stellt sich wegen der <u>Individualität der Fertigung</u>
jedes einzelnen Werkstückes mit <u>komplexen</u> zwischengeschalteten <u>Trans-</u>
<u>portvorgängen</u> und der notwendigen Auslegung auf <u>schnellste Instandset-</u>
zung als besonders schwierig dar. Die klassische "signalorientierte"
Technik im Bereich Messen, Steuern, Regeln ist hier nicht mehr ausreichend. Es müssen qualitative Automatisierungsfortschritte gesucht wer-

den, die "informationsorientiertes" Arbeiten ermöglichen. Letztere Forderung bedeutet, daß Informationen bereitzustellen sind, die aus einer
logischen Verknüpfung von vielen Einzelsignalen über spezielle Abhängigkeiten aus dem meist komplexen Produktionsablauf gebildet werden müssen.
Diese Informationsverdichtung ermöglicht eine Optimierung des Normalbe-
triebes und eine bessere Beherrschung von Störungszuständen.

Letztere Forderung stellt jedoch auch hohe Ansprüche an die Erfassungs-
und Verarbeitungsgeschwindigkeit von Daten und Informationen in den jeweiligen Rechnern und - bei verteilten Systemen - insbesondere an die
Komponenten Geschwindigkeit und Sicherheit der Datenerfassung bei der
Rechner-Rechner-Kommunikation.

2. Systemkonzept

Das vom IITB entwickelte Fertigungsautomatisierungssystem, eingesetzt
in den Automobilwerksbereichen Rohbau, Lackierung und Montage der
Daimler-Benz AG, Bremen, dient der Überwachung, Bedienung und Dokumen-
tation des Fertigungsprozesses und liefert damit die Grundlage für die
Fertigungssteuerung. Aus den verdichteten Informationen lassen sich wei-
terhin Massnahmen zur vorbeugenden Wartung der Anlagen und zur gezielten
Fehlerbehebung ableiten. Es hat die folgende Lösungskonzeption (vgl.
Bild 1):

Eine zentrale Steuerwarte sammelt und dokumentiert den Betriebszustand
der einzelnen Fertigungsanlagen; Anlagen und zugehörige Steuerungen sind
von der zentralen Steuerwarte aus fernbedienbar. Die verdichteten Informationen werden zusammen mit den fallweise gewünschten Einzelsignalen
auf Farbbildschirmen dargestellt. Durch Bedienung über dieselben Bild-
schirmgeräte (Lichtgriffel) kann direkt in den Produktionsablauf einge-
griffen werden.

Das Systemkonzept basiert auf einem verteilten, busgekoppelten, fehlertoleranten Prozessrechnersystem mit Sichtgerätebedienplätzen. Dieser
Systemaufbau dient zur

o dezentralen Signalerfassung und Informationsermittlung über viele intelligente Unterstationen (IUS) in den Produktionsanlagen,

o zentralen Auswertung der Signale in einem Anlagenüberwachungsrechner
  (AÜR),

o zentralen Überwachung und Führung des Produktionsprozesses in einem
  Prozessführungsrechner (PFR).

Dieses System erfüllt damit die folgenden Anforderungen an Automatisie-
rungssysteme:

- Zentrales Anfahren, Überwachen, Steuern des Produktionsprozesses

- Anschliessbarkeit bestehender Anlagensteuerungseinrichtungen verschie-
  denster Art (inhomogenes System)

- Mensch-Maschine-Schnittstelle, ausgelegt für schnell erlernbare, feh-
  lerarme Bedienungen in der Zentralwarte

- Leichte Änderung oder Erweiterung des Fertigungsautomatisierungssystems
  (Flexibilität, Open End Design)

- Hohe Zuverlässigkeit, leichte Wartbarkeit durch umfangreiche Fehler-
  diagnose

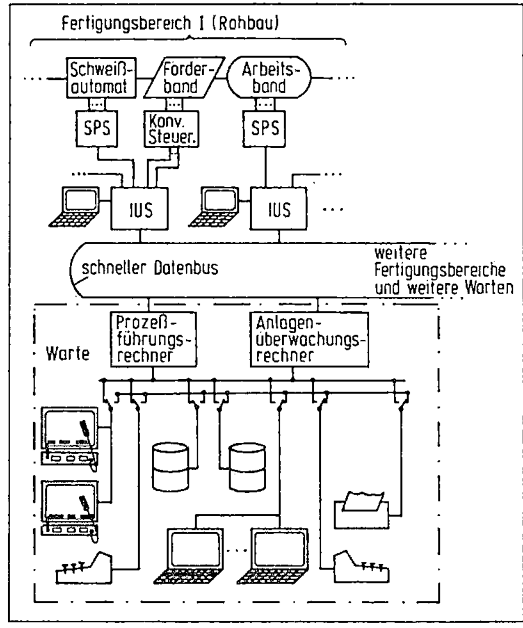

Bild 1: Grobstruktur des Fertigungsautomatisierungssystems

- Fehlertoleranz in Auslegung der Funktionen, der Prozessdatenerfassung,
  der Stationen sowie der Rechner-Rechner-Kommunikation.

3. Einzelkomponenten

Das Fertigungsautomatisierungssystem besteht aus den Komponenten

       o Intelligente Unterstationen (IUS)
       o Prozessführungsrechner (PFR)
       o Anlagenüberwachungsrechner (AÜR)
       o und dem Kommunikationssystem.

Eine Übersicht über das Programmsystem stellt Bild 2 dar.

Intelligente Unterstationen

Die Intelligenten Unterstationen, die "vor Ort" in den jeweiligen Pro-
duktions- und Fertigungsbereichen verteilt sind, haben folgende Aufgaben:

- Erfassung der binären und analogen Einzelsignale vom Produktionspro-
  zess

- Durchführung der Datenerfassung über diverse unterschiedliche Steuerungen mit seriellem oder parallelem Anschluss (inhomogenes System)

- Ausgabe von Schaltbefehlen, Stellbefehlen, sowie digitalen Informationen (z.B. Zählerstände)

- Weiterreichen vorverarbeiteter Informationen zu AÜR, PFR oder Nachbar-
  stationen über das Kommunikationssystem

- Durchführung spezieller Funktionen wie digitale Regelung, Dynamische
  Sollvorgabe für Arbeitsbänder, Abstandssteuerung, Anlagenhochlauf mit
  Hochlaufüberwachung, Pausenschaltungen etc.

- Verarbeitung und Durchführung von Befehlen und Bedieneingriffen vom
  PFR (z.B. manuelles Schalten)

- Überwachung von zugeordneten Nachbar-IUS'en und Übernahme deren Funktionen bei Ausfall (Rekonfiguration)

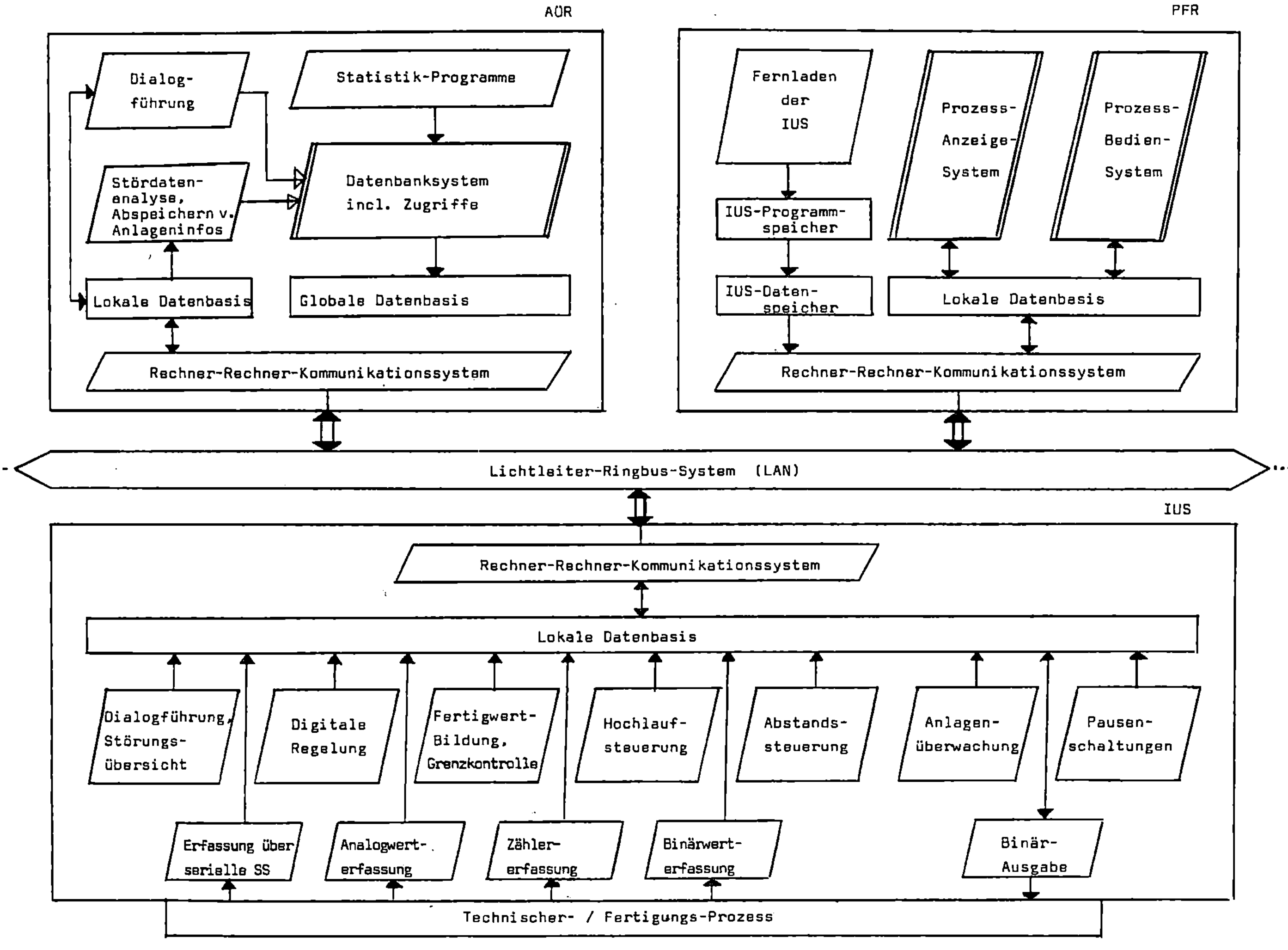

Bild 2: Übersicht über das Gesamt-Programmsystem

Einige Kenndaten:

In Bremen sind zur Zeit 10 IUS'en am Ring installiert, zusammen werden
ca. 11.000 Binärsignale, ca. 1.200 Analogsignale, ca. 160 "serielle"
Dateneingänge erfaßt und es erfolgt eine Informationsverdichtung in ca.
1.000 "Logischen Anlagen" im Anwendersinne.

Die IUS'en sind in der Hochsprache PEARL implementiert, die Rechner sind
Siemens R10 mit 512 KW Speicherausbau.

Prozessführungsrechnersystem

Das Prozessführungsrechnersystem (PFR) dient zur zentralen Prozessbeob-
achtung und -bedienung. Als Darstellungs- und Bediensystem wird ein im
IITB entwickeltes Ein-/Ausgabe-Farbbildschirmsystem (EAF) u.a. mit folgenden Eigenschafter eingesetzt:

o Darstellung sämtlicher Informationen über Bildschirm in Form von Kur-
  ven, Balken, Werten, Texten, Störeinblendungen oder Zustandscodierungen
  in Bildern des Betreibers (s.u.)

o Bedienung der Produktionsanlagen ausschließlich über Bildschirm, Lichtgriffel, Fliessbildern und dynamisch eingeblendeter virtueller Funktionstastatur mit Benutzerführung (vgl. Bild 3, 4)

o Protokollierung relevanter Ereignisse, Bedienungen sowie von Störungen

o Anschlussmöglichkeit von mehreren Bedienplätzen

o Durch den Anwender selbst anlegbare und parametrierbare Bilder, Signale,Verknüpfungen (Weiterentwicklung der Produktionsanlagen!) in der
  wahlweise zuschaltbaren ON-LINE-Betriebsart "Parametrierung".

Der PFR überwacht in der Zentrale ebenfalls den AÜR (gegenseitig) und
übernimmt bei Ausfall der Partnerstation deren Funktionen bei gewisser
Lastreduktion.

Einige Kenndaten:

Das Programmsystem für den PFR (Fortentwicklung von /1/) ist in Assembler realisiert, der Rechner ist eine Siemens R30 mit 512 KW Speicherausbau.

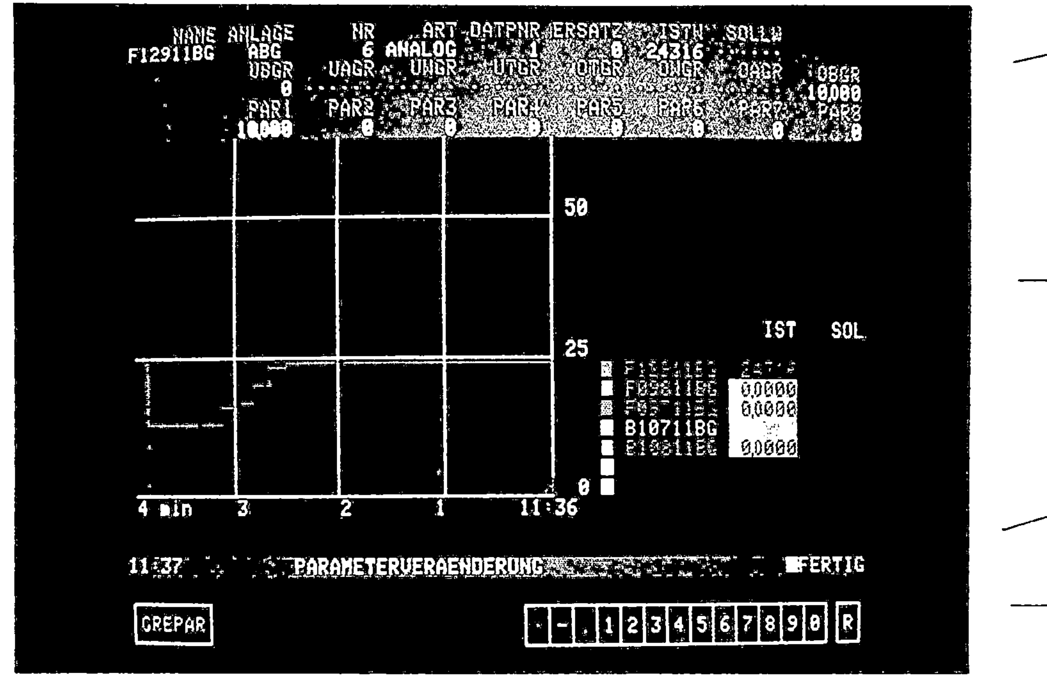

Dynamisch eingeblendete
Anlageninformationen
hier:
Sollvorgaben, Grenz- und
Parameteränderungen

In Fliessbildern einge-
gliederte Kurvendarstel-
lungen
hier:
Vorübergehende Darstel-
lung in Zeitlupenfunk-
tion (4 min) eines Band-
geschwindigkeitsverlaufes

dynamisch eingeblendete
Dialogführungszeile, da-
runter Störungsmeldezeile

situationsangepasste vir-
tuelle Funktionstastatur
hier:
Grenz-/Parameteränderung

Bild 3: Beispiel einer Bedienung am EAF-System
hier: Veränderung von Bandgeschwindigkeitsparametern

Momentan stehen 64 rollbare Fliessbilder mit jeweils 8-fachem Bildschirm-
inhalt zur Verfügung (weiter ausbaubar), die Bildaktualisierung ist $\leq$ 2
sec, die Bildwechselzeit ist $\geq$ 1 sec, die Datenerfassungszeit liegt zwischen 1 und 4 sec.

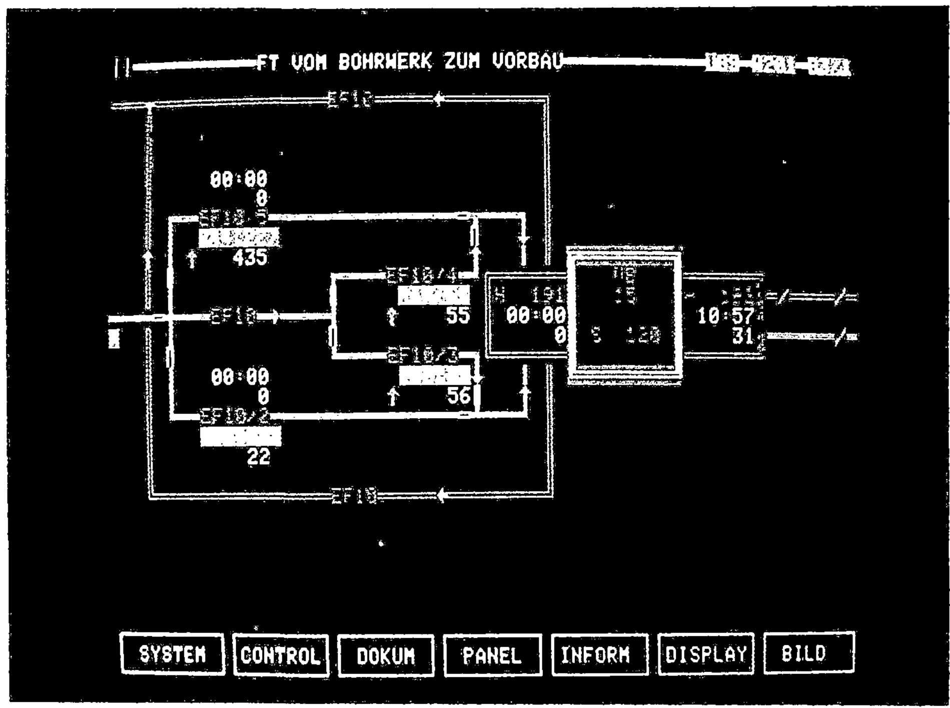

Bild 4: Bildausschnitt aus dem Bereich Rohbau:
Fördertechnik mit Puffern, Hebern, Bändern etc.

Anlagenüberwachungsrechner

Der zentrale Anlagenüberwachungsrechner (AÜR) dient

o zur Speicherung von aufbereiteten Anlagen- und Produktionsinformatio-
nen in einer Echtzeitdatenbank,

o zur interaktiven Auswertung dieses Datenbestandes an diversen Bedienplätzen sowie über Kommunikation an den IUS'en,

o zur automatischen und fallweise abrufbaren Erstellung von Protokollen,
  von Übersichten und Statistiken.

PFR und AÜR als Rechnerkomponenten der Zentralwarte können - wie schon
erwähnt - ebenfalls rekonfigurieren, d.h. bei Ausfall eines Rechners die
fremden Funktionen des jeweils anderen zusätzlich bei gewisser Lastreduktion übernehmen.

Die Programme des AÜR einschließlich der Echtzeitdatenbank sind in PEARL
realisiert und kommen auf einer Siemens R30 mit 1024 KW Speicherausbau
zum Einsatz.

Das Kommunikationssystem übernimmt die Datenübertragung zwischen den beteiligten Stationen mittels eines Lichtleiterringes. Nähere Einzelheiten
hierzu sind in den nachfolgenden Abschnitten aufgeführt.

4. Lichtleiterring mit Koppelstationen als Lokales Netz (LAN)

Für das lokale Netz war eine schnelle, leistungsfähige Kommunikation mit
hoher Teilnehmerzahl gefordert, die die gesamten Produktionsbereiche miteinander verbindet und die neben den Aspekten von niedrigen Realisie-
rungs-, Installations- und Betriebskosten u.a. den Forderungen genügen
sollten wie

- geringer Empfindlichkeit gegenüber unterschiedlichen Belastungsfällen
  und Ausbaustufen

- Fehlertoleranz bei Teilnehmerstörungen und Leitungsunterbrechung

- Selbstdiagnose mit Übermittlung der Diagnose.

Hierfür wurde eine Fortentwicklung des IITB-Lichtleiterringes /2/ als
Kommunikationssystem eingesetzt.

Hauptkennzeichen sind:

o die Übertragung erfolgt über einen Lichtleiter (Glasfaser) und ist damit unempfindlich gegen Störeinflüsse der "Fabrikumwelt" wie z.B. elektromagnetische Einflüsse

o Koppelstationen dienen zum Anschluss je einer Rechnerstation an den

Ring

o die Struktur besteht aus einem Ringbussystem mit dezentraler Steuerung,
  Vorzugs-Kommunikationsrichtung und aktiver Ankopplung aller Teilneh-
  merrechner (IUS, PFR, AÜR). Damit ist ein gleichzeitiger Kommunika-
  tionsbetrieb aller im Ring befindlichen Stationen ermöglicht; der
  Durchsatz ist dadurch wesentlich höher als bei üblichen Kommunikationssystemen mit passiver Ankopplung. Bei Unterbrechungen des Ringes wird
  mittels eines "Pendelverkehrs" die Kommunikation auf den jeweiligen -
  nun nicht mehr ringförmigen - Restübertragungsstrecken gewährleistet
  (Fehlertoleranz).

Einige Kenndaten:

o Momentan sind 14 Stationen in Bremen angeschlossen, technisch möglich
  wären bis zu 256

o Länge des Gesamtringes im Werk DB-Bremen: 3351 m

o Kerndurchmesser einer Glasfaser: 100 μm

o Bruttogeschwindigkeit der Datenübertragung auf dem Lichtleiter: 1,25
  Mio. Bit/s

o Nutzdatenrate unter Berücksichtigung sämtlicher Kommunikations-Level
  einschliesslich Netzbetriebssysteme, Betriebssysteme und Anwender-
  Kommunikationssoftware: 10 KW/s.

5. Einige Aspekte der Kommunikation

In diesem Abschnitt werden kurz die Gliederung und Aufgabenverteilung
der Kommunikationsprogramme des IITB für dieses Projekt vorgestellt,
desweiteren werden im Detail nur diejenigen Konzeptions- und Realisie-
rungspunkte angesprochen, die als einzelne und/oder in ihrer Kombination das Besondere des realisierten Kommunikationssystems bzw. Fertigungsautomatisierungssystems ausmachen.

5.1 Schichtenmodell der Kommunikationsprogramme

Angenähert an das ISO-Referenzmodell sind die Kommunikationsprogramm-
schichten konzeptioniert (vgl. Bild 5):

Netzbetriebssystem (NBS)

Das Netzbetriebssystem fungiert als Treiberprogramm für die spezielle
Lichtleiter-Kommunikation /3/ und entspricht dem ISO-Level 4. Es stellt
für die nachfolgende Schicht 5 zwei prinzipielle Kommunikationsfunktionen zur Verfügung:

- Implizite Kommunikation über globale Objekte

- Explizite Kommunikation durch Intertaskkommunikation.

Die implizite Kommunikation ist ein quasi-DMA auf dem gesamten Adress-
raum der am Ring befindlichen Stationen. Sendeseitiger Auftraggeber ist
immer eine Anwendertask, empfangsseitig wird der Auftrag (Lesen, Schreiben) immer vom NBS allein ausgeführt.

Die Intertaskkommunikation hingegen benötigt empfangsseitig eine korrespondierende Empfängertask mit zugehörigem Puffer, die den Auftrag entgegennimmt, interpretiert und entsprechend weiterverarbeitet (Mailboxing,
"PORT's").

Die Funktionen des NBS sind dabei in allen Rechnern identisch und unabhängig vom benutzten Rechnertyp, der Einbettung in die verwendeten Programmiersprachen wie PEARL; Assembler etc. und den jeweiligen Anwendungsfunktionen.

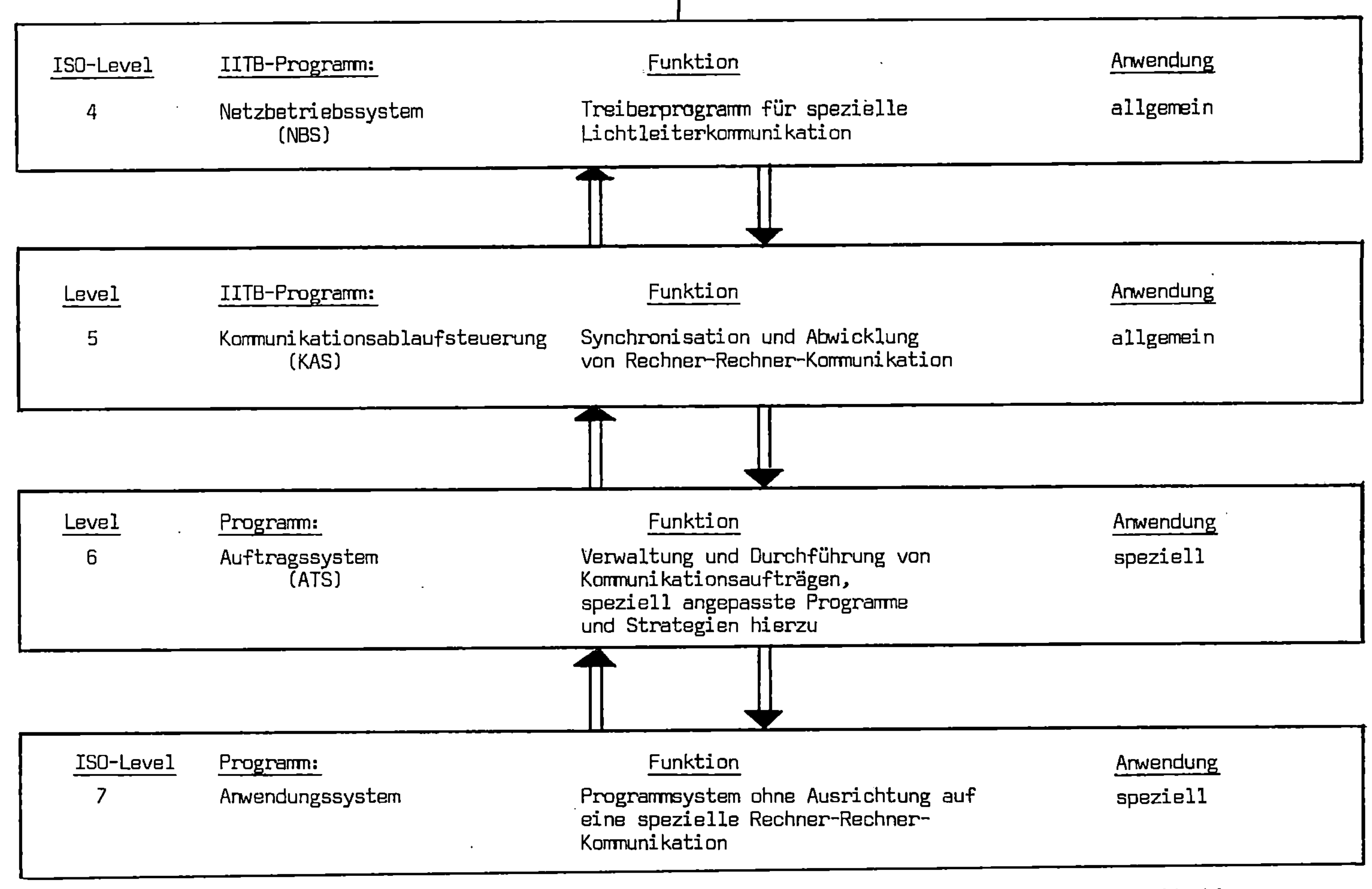

Bild 5: Grobrasterung der Programmschichten in Bezug auf die Rechner-Rechner-Kommunikation

Für alle im Ringbussystem befindlichen Rechner bzw. Programmsysteme ist
es somit unerheblich, in welcher Programmiersprache das Partnersystem
implementiert ist bzw. von welchem Rechnertyp der Zielrechner ist.

Zusätzlich zu den Kommunikationsfunktionen bietet das NBS noch einen Anschluß an die Diagnose- und Statusmeldungen des Lichtleiterringes.

Kommunikationsablaufsteuerung (KAS)

Die Kommunikationsablaufsteuerung entspricht dem Level 5 und stellt für
die nachfolgende Schicht zwei globale Funktionen zur Verfügung:

- explizite Kommunikation durch Intertaskkommunikation

- Synchronisation vom Kommunikationsabfolgen (Sessions) über Rechner hinweg.

Eine wesentliche Nebenaufgabe ist hierbei die Transformation logischer
Parameter (z.B. Zielprozess) in physikalische Parameter (z.B. aktuell
zuständige physikalische Station in Abhängigkeit vom Ringkonfigurations-
zustand). Ebenso zählt hierzu die dynamische Eröffnung logischer Kanäle
mit anschliessendem schnellen physikalischen Zugriff.

Die Funktionen der Kommunikationsablaufsteuerung (KAS) sind dabei wie
die des NBS

- unabhängig vom Rechnertyp

- unabhängig von der Programmiersprache

- unabhängig von den Anwendungsfunktionen.

Auftragssystem

Das Auftragssystem entspricht dem Level 6 und hat die Aufgabe, in Kenntnis der Anwendungsanforderungen einerseits und unter Benutzung der
Schnittstellen der Kommunikationsablaufsteuerung andererseits eine opti-
male und verklemmungsfreie Übertragung zu ermöglichen.

Folgende Aufgaben sind dabei zu bewerkstelligen:

- Festlegung der zu übertragenden Nutz- und Informationsdaten

- Portionierung der Aufträge

- Zuordnung von Auftragskennzeichnungen zu Aufträgen

- Festlegung von Übertragungsprinzipien für Datentransporte wie z.B.
  "Sende"- oder "Hol"-Prinzip

- Koordinierung von Rendez-Vous-Mannövern sende- und empfangsseitig

- Organisation des Empfangs von Aufträgen

- Organisation der Initiative von Übertragungssequenzen

- Verwaltung von Auftragssequenzen

- Erstellung von Tasking-Modellen unter Berücksichtigung der für die
  Kommunikationsauftragsentgegennahme durch das NBS beschränkte Taskan-
  zahl

- Organisation von verklemmungsfreien Abwicklungen auch beim Verkehr von
  N zu 1 Stationen

- Berücksichtigung von Reaktionen auf Anzeigen der KAS wie z.B. durch
  Wiederholung bzw. Zurückreichen der Anzeigenkennung an die Anwendungsschicht 7.

Im Gegensatz zum NBS und zur KAS sind die Programme des Auftragssystems
in jedem der Rechner unterschiedlich und dem zugehörigen Anwendungsauf-
gabenspektrum angepasst.

5.2 Ringdiagnose, Rekonfigurationseinleitung

Das Lichtleiterringbussystem ist in der Lage, sowohl den Status von Kop-
pel-Stationen als auch den (Ausfall-) Status zugehöriger Stationen zu
diagnostizieren. Diese Ereignisse und Zustände sowie auch Ringzustände
wie Normalbetrieb, Pendelverkehr, Eckstation etc. werden allen am Ring
beteiligten Stationen als Statusinformationen übersandt und dienen beispielsweise den IUS'en als Anlass für eine IUS-IUS-Rekonfigurationsein-
leitung.

5.3 Fernladefunktion

Bei dem Fertigungsautomatisierungssystem sind die verteilten Vor-Ort-
Rechner (IUS) aus Kosten-, Wartungs-, Ausfallsicherheits- und Fernsteu-
erungsgründen nicht mit Peripheriespeicher ausgestattet. Aus diesem
Grunde ist es notwendig, die Urladefunktionen für die IUS über das Kom-

munikationssystem durchzuführen.

Dies wurde durch ein 2-stufiges Verfahren gelöst:

- zunächst wird der "Urlader" selbst als Rumpfsystem in die Station fern-
  geladen,
- danach unternimmt das Rumpfsystem die vollständige Übertragung der
  restlichen Speicherinhalte, das neu geladene Programmsystem wird fern-
  gestartet und läuft an.

5.4 Dynamische Ankopplung

Starre Adressierungen von Kommunikationsaufträgen würden bei Ausfall von
Stationen diese bzw. eventuelle Ersatzstationen nicht mehr erreichen
können. Durch die Konzeption der logischen Adressierung von Stationen
durch das Auftragssystem jedoch und das Ersetzen der logischen Station
durch die momentan zuständige physikalische Station, durchgeführt durch
die Kommunikationsablaufsteuerung, können Ersatzstationen oder aber wieder neu geladene Normalstationen flexibel adressiert werden, wobei dies
dem Auftragssystem verborgen bleibt.

5.5 Telegrammstruktur

Die Rechner-Rechner-Kommunikation wird in der Schicht des Auftragssystems
einerseits über die zugehörigen und für jeden Kommunikationsstrang festgelegten Ports (siehe 5.1) und andererseits über feste Telegrammsteuer-
strukturen mit Telegramm- und Auftragskennungen abgewickelt.

Die Struktur der Steuerdaten ist hierbei fest definiert, die Telegramm-
und Auftragskennungen sind jedoch - wie die Ports - beliebig erweiterbar
(OPEN-END-DESIGN).

Je nach Abstimmung mit Laufzeit- oder Task-Synchronisationserfordernis-
sen können (neue) Telegramme dabei über weitere Ports "angeflanscht"
oder aber über vorhandene "Kanäle" mittels Multiplexer parallel betrieben werden.

5.6 Heterogene Rechner, Programmiersprachen

Wegen der verwendeten Funktionen des NBS (5.1) und den Prinzipien der

Telegrammstrukturierung (5.5) ist es möglich, am Lichtleiterring unterschiedliche Rechnertypen zu installieren (im IITB installiert: Siemens R30, R10, S310 sowie IITB-RDC-Stationen), deren Programme wiederum in verschiedenen Programmiersprachen erstellt sind (Assembler, PEARL ...) und die von unterschiedlichen Betriebssystemen (Assembler-Betriebssystem ORG, PEARL-Betriebssystem BAPAS ...) gemanagt werden können.

Besonders ist zu erwähnen, daß das NBS in der Lage ist, die Kommunikation in einem Mischsystem (z.B. bei Rekonfiguration) mit mehreren Betriebssystem- und Anwenderprogramm-"Welten" in einer Station zu gestatten /3/.

5.7 Adressieren fremder LAN's

Bei dem vorgestellten Projekt existiert ein singuläres, lokales Netz, mehrere - auch fremde - LAN's kommen hier nicht zum Einsatz. Wegen der Verwendung des Kommunikationssystems jedoch auch für andere Anwendungen wurde die Ausrichtung auf die Adressierung fremder LAN's bereits berücksichtigt, realisiert und somit kurz vorgestellt.

Bei der Adressierung von Stationen über Netze hinweg wird sowohl bei homogenen als auch bei heterogenen Netzen nach dem Prinzip der logischen Adressierung und der Weiterreichung von Aufträgen über Koppelstationen auf der Ebene des Kommunikationsauftragssystems (ATS) verfahren.

Es wurde diese Methode gewählt, um zum einen die Belastung der jeweiligen unterlagerten ISO-Level der singulären LAN's zu minimieren und zum anderen aus Gründen der Flexibilität und des Open End Designs, übergeordnete Adressierungen mit dynamischer Wegesteuerung und dynamisch sich ändernden Stationszuordnungen auf die relevanten Programme, nämlich die Auftragssysteme zu konzentrieren, insbesondere auch unter dem Aspekt der fallweisen Auswirkung auf die Datenerfassungsstrategien. U.a. können somit neue Stationen in LAN X ohne Änderungen in LAN Y im laufenden Betrieb installiert bzw. in Betrieb genommen werden.

5.8 Integrität von Daten

Für jedes verteilte Prozessautomatisierungssystem ist die Integrität der Daten eine lebensnotwendige Forderung.

Hierzu zwei Anwendungsbeispiele:

(a) Die chronologische Reihenfolge von Datensätzen einer Anlage soll
    unbedingt gewahrt werden.

(b) Störungsmeldungen beim <u>Übergang</u> eines Binärelementes in den Stör-
    zustand dürfen unter keinen Umständen "verloren gehen", die Ände-
    rung einer eventuellen Zustandsdarstellung allein reicht nicht aus.

Fehler unter Verletzung von (a) können zu unrichtigen Statistiken (Archiv) und Störaussagen führen (Überholvorgänge); bei Fehlern nach (b)
könnten wichtige Störhinweise an das Wartenpersonal und deren sofortige
Reaktionen darauf unterbleiben.

Die Integrität von Daten wird bei der Konzeption durch

- spezielle Synchronisationsmechanismen,

- durch das totale Quittungsprinzip des NBS (Anzeige positiv oder nega-
  tiv) und

- der daher resultierenden Transparenz der Kommunikation

gewahrt.

Das NBS bietet das vollständige Quittungsprinzip, d.h. sowohl abgebroche-
ne Übertragungsversuche als auch vollständig durchgeführte Transfers werden zurückgemeldet. Somit weiß das ATS, ob ein Telegramm definitiv den
Adressaten erreicht hat oder definitiv nicht.

Das ATS kann folglich entsprechend den Anforderungsprofilen abgestuft
reagieren und z.B.

o Anzeigen bei relativ unwichtigen oder zyklischen Datentransfers (z.B.
  Analogwertaktualisierung) ignorieren,

o bei Transfers von gewisser Wichtigkeit nach Anzeigen mehrfach wiederholen und dann nicht mehr

o oder aber bei wichtigen Informationen (z.B. Telegramme nach (b)) die
  Zusendung derselben solange verwalten, bis ein Transfer durchgeführt
  worden ist.

6. Trends, Spezielle Problemstellungen, Lösungsansätze

Der Trend in der Fertigungsautomatisierung führt zu einem integralen
System, das alle Bereiche wie

o Anlagensteuerung, -regelung, - Koordinierung

o Fertigungsüberwachung

o Wartungsdisposition

o Materialdisposition, Zuführung

o Dokumentation, Statistik, Planung, Simulation

o bis zum Bürobereich

zusammenfassen wird. Es werden somit weitere Bereiche automatisiert oder
aber bereits bestehende "Automatisierungsinseln" verknüpft.

Diese Integration führt zu einem immensen Zuwachs der zu verarbeitenden
Daten in der Datenerfassung, Datenverarbeitung, Informationshaltung und
dem Informationstransport zwischen den Rechnern. Bisherige Prozessrech-
nersysteme mit einfachen LAN's (Strecke, Ring) werden aber zunehmend
nicht mehr in der Lage sein, den resultierenden Anforderungen bezüglich
der zur Funktionserfüllung notwendigen Speicherkapazität und insbesondere des Echtzeitverhaltens gerecht zu werden. Eine entsprechend notwendige Verteilung von Funktionen durch den Einsatz von noch mehr Prozessrech-
nern führt jedoch zwingend zu einem extensiven, nicht mehr aktzeptablen
Engpass beim (singulären) Kommunikationssystem.

Es sind somit Fortentwicklungen der bisherigen Systemkonzeptionen notwendig, die u.a. die Hinzunahme beliebig vieler Prozessrechner erlauben.

Lösungsansätze bestehen in der

o Erweiterung der Hierarchiestruktur der beteiligten Prozessrechner und
   ihrer Funktionen durch Einschaltung von Zwischenebenen (Datenkonzen-
   tratoren, Vorrechner)

o Erweiterung der Struktur des Kommunikationssystems durch die Hinzunahme weiterer LAN's (Wabenstruktur?, mehrere homogene Netze?, heterogene
   Netze?).

Zwangsläufig ergeben sich für das Gesamtkommunikationssystem komplexe

Aufgabenstellungen:

o Untersuchungen zur funktionsorientierten optimalen Festlegung von Anzahl und Strukturen einzusetzender LAN's sowie deren Auslegung auf Rekonfiguration und Fehlertoleranz

o Definition einer logischen Adressierung zum Durchschalten von KS-Trans-
  fers bis zum augenblicklichen Zielrechner und zurück.

o Dynamische, evt. lastabhängige Wegesteuerung

o Dynamische Selektion von Datentransfermengen

o Erweiterung um zusätzliche Transportfunktionen wie Weiterreichen von
  Fremdaufträgen, Fernladen, Ferndiagnose, Kopieren über LAN's hinweg.

7. Einsatz, Inbetriebnahme, Betriebserfahrungen

Die Einsatzmöglichkeiten für solche Systeme sind nicht an die Fertigungstechnik gebunden, wegen der Vielfalt von Einzelsignalen wie etwa in der
Automobilfertigung und der dort notwendigen Informationsverdichtung wird
ihr Einsatz jedoch in der Fertigungstechnik besonders wichtig. Das IITB
hat ähnliche Systeme in der Stahlindustrie (Einsätze bei Thyssen), in
der Chenischen Industrie (Einsätze bei Bayer) und im EVU-Sektor seit
langem im Einsatz.

Speziell das Fertigungsautomatisierungssystem für Daimler-Bremen (Produktion des Typs 190) befindet sich momentan (Stand 12/84) in der End-
Inbetriebnahmephase:

o Die Basisfunktionen des Systems u.a. mit Fernladen, Prozessdatenerfassung -darstellung und -bedienung (Schalten etc.) wurden in II/84 in Betrieb genommen und seitdem genutzt.

o Zusatzfunktionen wie Anlagenhochlaufsteuerung mit manuellem Start
  sowie die Zählerfunktionen, die Abstandssteuerung und die Bandgeschwin-
  digkeitsregelung wurden in III/84 installiert.

o In IV/84 wurden die Funktionen "Dynamische Ankopplung", die Steuerung
  von Schaltungen sowie der automatische Start des Anlagenhochlaufs über
  Pausenmodelle integriert.

o In I/85 werden die Rechner-Rechner-Rekonfiguration sowie Restfunktionen
  in Betrieb genommen werden.

Für eine endgültige Beurteilung ist es noch zu früh, es zeichnet sich
jedoch jetzt schon ab, daß das System in der Lage ist, letztendlich zu
besseren Produktdurchlaufzeiten und zur Kostenreduktion im Fertigungs-
bereich beizutragen.

Literatur:

/1/ Laubsch, Rudolf: Ein/Ausgabe-Farbbildschirmsystem (EAF-System) mit
     Doppelbedienplatz zur zentralen Führung eines verteilten Automati-
     sierungssystems, FhG-Berichte (IITB) 2-80.

/2/ R. Bähre,D. Heger, F. Sänger: Der RDC-Ring, ein fehlertolerantes,
     dezentral- und ereignisgesteuertes Lichtleiter-Kommunikationsnetz,
     GI-NTG-Fachtagung'83,Informatik-Fachberichte,Springer-Verlag Berlin,
     S 517 ff.

/3/ G. Bonn et. al.: "Funktionen und Schnittstellen des Kommunikations-
     systems", IITB/Technische Notiz Nr. 10 vom 14.09.1982, Projekt "Ver-
     teiltes, fehlertolerantes Produktionswartensystem mit Einbezug einer
     Echtzeitdatenbank".

<u>NRW Jobverbund im Deutschen Forschungsnetz</u>
- Erfahrungen und Probleme beim Betrieb -

J. Knop
Universität Düsseldorf

1. Einleitung
2. Jobverbund in NRW
    2.1   Teilnehmer und Systeme
    2.2   FEP-Methode
    2.3   Die Betriebskonfiguration
    2.4   Protokolle und Schnittstellen
3. Bewertung des Konzepts aus betrieblicher Sicht
4. Erfahrungen und Probleme im Betrieb
    4.1   Zentrale Betreuung
    4.2   RJE-Dienst und Ergänzung durch Dialog-Dienst
    4.3   Hauptrichtungen des Verkehrs
    4.4   Nutzung
5. Accounting
6. Verbesserungsmöglichkeiten

<u>1. Einleitung</u>

Im deutschen Hochschulbereich wurden in den vergangenen
Jahren eine Reihe von vom Bundesminister für Forschung und
Technologie geförderten Projekten im Bereich des Rechnerverbundes
durchgeführt.
Ziel dieser Projekte war, unter anderem über theoretische
Arbeiten und pilotmäßige Implementierungen Beiträge und
Impulse zur Einführung von "offenen" Kommunikationssystemen
im Sinne der ISO-Open Systems Interconnection zu geben und
die Ergebnisse für die Etablierung von praktischen Verbund-

Betriebskonfigurationen nutzbar zu machen. Diese Arbeiten
waren zum Teil eng verknüpft mit der Benutzung neuer Postdienste
im Netzbereich, insbesondere mit der Nutzung des X.25-Dienstes
von DATEX-P.
Dieser Beitrag gibt einen Überblick über eine größere Betriebskonfiguration für einen Job-Verbund, an der im Rahmen des
Deutschen Forschungsnetzes fünf Hochschulen in Nordrhein-Westfalen beteiligt sind, und die technisch auf den Ergebnissen
vorangegangener, vom BMFT geförderter Projekte an der Universität
Düsseldorf basiert (zum Beispiel /1/).
Im folgenden werden die Einsetzbarkeit der an der Universität Düsseldorf entwickelten Rechneranschlußmethode in einem
größeren Betriebsprojekt und die Erfahrungen beim Betrieb
des Verbundes diskutiert.

2. Das Jobverbundprojekt der Hochschulen in NRW

2.1 Teilnehmer und Systeme

Im Oktober 1983 wurde ein vom BMFT im Rahmen des Deutschen
Forschungsnetzes gefördertes Kooperationsprojekt der
Rechenzentren der Hochschulen in Aachen, Bielefeld, Bochum,
Düsseldorf und Köln begonnen, das die Einbindung der vorhandenen
DV-Systeme in einen gemeinsamen Verbund und dessen Integration
in das DFN zum Ziel hat. Die Koordinierung des Projektes
liegt bei der Universität Düsseldorf.

Folgende Systeme sind in den Verbund zu integrieren:

Hochschule	System	Betriebssystem
TH Aachen	CD 175	NOS 1.4
U. Bochum	CD 205	NOS 2.1
U. Bielefeld	TR 440	BS3
U. Düsseldorf	TR 445	BS3
	Siemens 7.760	BS2000
U. Köln	CD 76/72	NOS/BE

Im DFN ist dieses Projekt in dem Teilbereich "Basis-DFN"
angesiedelt. Dies bedeutet, daß - im Gegensatz zu DFN-Entwick-
lungsprojekten - hier der betriebliche Einsatz von vorliegenden Konzepten und Pilotimplementierungen durchzuführen ist.

2.2 Die FEP-Methode

Als Grundkonzept für den Anschluß von Hostsystemen an ein
offenes Verbundnetz wird die sogenannte FEP-Methode (Front-End
-Processor) eingesetzt. Hierbei wird zwischen Host und Verbundnetz ein (kleiner) Rechner geschaltet, der zum Verbundnetz
die dort erforderlichen Verfahren (Verbundprotokolle) abhandelt,
zur Hostseite die extern verfügbaren Systemschnittstellen
(systemspezifische DFÜ-Protokolle) bedient und die erforderlichen Übergänge bzw. Abbildungsprobleme zwischen diesen beiden
Protokollhierarchien behandelt.

Derartige Pilotimplementierungen für den Anschluß eines
TR445-Rechners (mit DUET-Vorrechner) mit Hilfe eines FEP-
Rechners vom Typ Dietz 621 auf der Basis der sogenannten
PIX-Protokolle (X.25, Message-Link, RJE) wurden am Rechenzentrum
der Universität Düsseldorf mit Förderung durch den BMFT
entwickelt. Die PIX-Protokolle (PIX war ein früheres Kooperationsprojekt von einzelnen BMFT-Projekten) werden als Startlösung
für das DFN eingesetzt.

Das folgende Bild zeigt die Struktur des Dietz 621-FEP mit
den wesentlichen Modulen.

```
           +============+=======+ ============+==========+===========+
           |  Job       |  RJE   |             |          |           |
           |  Map in    |  Job   |             |          |           |
           |    I       |  Empfaenger |        |          |           |
           +============+  +============+       |          |           |
           |  Output    |  RJE   |             |          |           |
  +=====+  |  Map in    | S |  Output    |      |          |         |+======+
  |Mi-  |  |    II      | P |  Empfaenger |  Message  |  X.25   || Z 80 |
  |kro  +===+============+ O +============+       |  Paket=  +  HDLC  |
  |MSV2|  |  Job       | O |  RJE   |      Link   |  ver=   ||        |
  +=====+  |  Map out   | L |  Job   |             |  mittlung |+======+
           |   III      |  Sender |            |          |           |
           +============+  +============+       |          |           |
           |  Output    |  RJE   |             |          |           |
           |  Map out   |  Output |            |          |           |
           |    IV      |  Sender |            |          |           |
           +============+=======+ ============+==========+===========+
```

Bild 1

2.3 Die Betriebskonfiguration

Die in Bild 1 dargestellte FEP-Struktur wird für den Anschluß
der im NRW-Projekt vorhandenen CDC-und TR440-Systeme eingesetzt.
Hierzu war die vorhandene Dietz-Software für den Anschluß
der CDC-Systeme zu modifizieren.
Außerdem waren spezifische Anpassungen bei dem Anschluß
der einzelnen Systeme in den Rechenzentren hinsichtlich
der unterschiedlichen Betriebsystemversionen erforderlich.
Diese Arbeiten wurden von den beteiligten Einrichtungen
gemeinsam (unter Federführung durch die Universität Düsseldorf)
im ersten Teil des Projektes durchgeführt, so daß ab Sommer
1984 folgende Betriebskonfiguration vorlag.

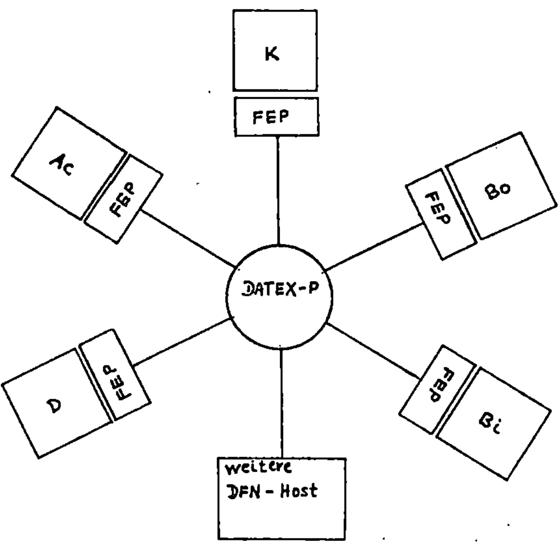

Bild 2

Übergänge zu anderen Hostsystemen im DFN mit RJE-Service
(z.B. BERNET, Universität Stuttgart, Universität Kiel, u.a.)
ist möglich.

Erfahrungen mit dieser Betriebskonfiguration werden in den
nachfolgenden Kapiteln dargelegt.

Im Rahmen der Fortführung des NRW-Projektes ist wichtig
zu erwähnen, daß die derzeit eingesetzten Dietz 621-Systeme
durch Systeme NORD 100 der Firma Norsk Data-Dietz abgelöst
werden. Diese Maßnahme ist erforderlich, da zum einen wegen
der Übernahme der Firma Dietz durch die Firma Norsk Data
der langfristige Einsatz vom Dietz 621-System problematisch
und somit insbesondere auch die angestrebte Einbeziehung
weiterer Hochschulen des Landes NRW auf Basis dieser Systeme
keine vernünftige Perspektive darstellt und zum anderen
die leistungsfähigeren NORD-Systeme für längerfristig zu
erwartende höhere Durchsatzanforderungen geeigneter sind.
Bei dieser Ablösung wird das PIX-Message-Link-Protokoll
durch das Standard-Transport-Protokoll S.70 (bzw. ISO Transport-
Klasse 0) ersetzt.

2.4 Protokolle und Schnittstellen

Der Host-unabhängige Teil der FEP-Software umfaßt X.25,
das Message Link-Transportprotokoll, PIX-RJE-Sender und
-Empfänger und die Schnittstelle zu dem Host-unabhängigen
Teil, in dem die Abbildung eines FEP-internen Job-/Output-
Standards auf den Standard der emulierten RJE-Station des
jeweiligen Host vorgenommen wird (vgl. Abb.).
Während X.25 Level 1 und 2 (HDLC LAP-B) von Dietz auf einem
eigenen Prozessor (MPUI, Mikro-Prozessor Universal-Interface)
geliefert wird, ist X.25 level 3 als eigenständiger Prozeß
auf dem Dietz 621 am Rechenzentrum der Universität Düsseldorf
entwickelt worden. Die Transport Layer-Realisierung umfaßt
einen selbständigen Prozeß mit Hauptaufgaben in der Verbin-
dungsaufbau- und -abbauphase und einem Unterprogrammpaket
mit der Benutzerschnittstelle für die Anwendungsprogramme
oberhalb des Transport Layer, das in das Betriebssystem
des FEP-Rechners eingebunden ist, und Aufrufe für die Services
und lokale Administrationsfunktionen des Transport Layers
in Form eines Betriebsystemeinsprungs verfügbar macht.
Die Schnittstelle zum Host-abhängigen Teil besteht aus einem

für das RJE-System entwickelten Spool-System und dem RJE-
Memory. Das Spool-System unterstützt die langfristige transparente Speicherung von großen Dateneinheiten - hier Job
oder Output-Daten -, organisiert in Warteschlangen. Das
System ist so realisiert, daß die Zugriffsroutinen zum
Schreiben in bzw. Lesen aus bestimmten Spool-Warteschlangen
und einige Hilfsroutinen ebenfalls als Betriebssystemaufrufe
zur Verfügung stehen. Schreiben und Lesen der vollständigen,
allein relevanten logischen Dateneinheiten ist in beliebigen
(und unabhängigen) Teilportionen möglich, aber stets rein
sequentiell. Im RJE-Memory, als File mit Direktzugriff auf
Sätze fester Länge organisiert, werden RJE-spezifische Parameter und Verarbeitungszustände von Netjobs gehalten,
auf die netz- und hostseitige Programme des FEP-Systems
Zugriff haben müssen.
Die Funktionen des PIX-RJE-Protokolls - Job-Send, Job-Receive,
Output-Send, Output-Receive - sind in zwei Programmen, FEP-
Sender und FEP-Receiver zusammengefaßt. Beide Programme
benutzen die genannten Betriebsystemschnittstellen des
Message Link und des Spool-Systems und Routinen für einen
koordinierten Zugriff auf das RJE-Memory.
Der Implementierungsaufwand für die einzelnen Komponenten
ist in folgender Tabelle zusammengfaßt:

Impl.- Sprache	Aufwand	Aufwand Protokollimpl.: Aufwand System-Einbettung
X.25	Assembler	18 MM	70 : 30
ML	Assembler	36 MM	30 : 70
RJE	PASCAL	24 MM	35 : 65
Spool	Assembler	12 MM	

3. Bewertung des Konzepts aus betrieblicher Sicht

Für den Benutzer ist es im Allgemeinen nicht sichtbar, ob
die Verbundsoftware im FEP realisiert wird, oder ob es sich
um eine Host-Lösung bzw. eine Realisierung im systemeigenen
Vorrechner handelt. Erschwert ist evtl. die Integration
weiterer Dienste wie Dialog, Message oder File-Transfer.
Auch ergänzende Serviceleistungen zum RJE-Betrieb wie Statusinformationen, Cancelfunktionen etc. könnten schwer zu integrieren sein. Für die Benutzer entstehen Funktionsverluste
im Bereich der Jobsteuerung, der Jobüberwachung, der Ein-
griffsmöglichkeiten und der Formatsteuerung bei Verwendung
eines Netzstandards gegenüber den Möglichkeiten, die ein
bestimmtes Hostbetriebssystem bieten kann. Der Standard
kann immer nur einen (kleinsten) gemeinsamen Nenner der
verschiedenen möglichen Services anbieten. Zusätzliche Ver-
luste entstehen beim FEP-Konzept durch die Verwendung von
System-Remote-Batch-Stationen - 8418 Station bei TR/Siemens
und 200UT/Hasp-Workstation bei den CDC-Anlagen. Die 8418-
Station hat z.B. keine zusätzlichen RJE-Funktionen. Der
TR-Benutzer informiert sich im parallelen Dialog über den
Status seines Jobs. Alle RJE-Stationen sind natürlich nur
für Job-In- und Output-Out-Verkehr ausgelegt. Bei den entgegen-
gesetzten Verkehrsrichtungen werden Umsetzprogramme benötigt,
durch die z.B. die Formatsteuerung (zumindest bei sog. "Host-
erzeugten-Output") eingeschränkt ist.

Aus Sicht der beteiligten Rechenzentren ist der Einsatz
des FEP-Konzepts im Vergleich zu einer Host-integrierten
Lösung zum derzeitigen Zeitpunkt (und auch in den nächsten
Jahren) vorzuziehen, und zwar solange bis verbindlich standardi-
sierte Verbundprotokolle für RJE vorhanden sind und insbesondere
solange diese nicht durch den Hersteller unterstützt werden.
Es folgt eine Zusammenstellung von Vor- und Nachteilen des
FEP-Einsatzes aus betrieblicher Sicht.

<u>Vorteile:</u>

- durch die Vermeidung von Host-Implementierungen von Ver-
  bundprotokollen mit entsprechenden Verbesserungen, Fehler-
  verfolgung, etc. wird eine Beeinträchtigung des auf dem
  Host ablaufenden normalen Benutzerbetriebs vermieden;
  diese Vorgänge spielen sich vielmehr außerhalb des Hosts
  an externen und gesicherten Systemschnittstellen ab, an
  denen ein Fehlverhalten des FEP i.a. keine Host-internen
  Störungen bewirken kann. Die Installation des Job-Verbun-
  des an den fünf Hochschulen hat aus Sicht des Betriebs
  keine lokalen Einschränkungen mit sich gebracht (keine
  gesonderten Testzeiten am Host-System, Störungen, etc.)
- durch die (auch hardwaremäßige) Separierung von Verbund-
  und Host-Aktivitäten wurde eine Fehlerlokalisierung und
  -behebung erleichtert
- durch den Einsatz einer verbundbezogenen Spool-Möglichkeit
  im FEP wurde der Verbundverkehr von Wartungszeiten, Aus-
  fällen, etc. des Hosts unabhängig; hierbei sind allerdings
  die Störzeiten des FEP-Systems zusätzlich zu berücksichtigen.
- durch den Einsatz einer einheitlichen FEP-Lösung, hier
  Dietz 621, wird der Aufwand für die Betreuung von Verbundsoftware reduziert;
- durch die Abspaltung des FEP-Verbundteils kann auch organi-
  satorisch eine bessere Zuordnung von betrieblichen Verant-
  wortlichkeiten erreicht werden;
- in vielen Pilot-Realisierungen von Host-integrierten Ver-
  bundlösungen wird der Host durch die DFÜ-Aufgaben CPU-
  mäßig stark belastet, was nicht nur aus betrieblichen,
  sondern auch aus wirtschaftlichen Gründen inakzeptabel
  ist;
- das Einbringen neuer Standard-Protokolle (z.B. S.70) kann
  durch Entwicklung und Test an einer Stelle für alle beteiligten Hochschulen gleichzeitig erfolgen.

Nachteile:
- durch die Benutzung recht einfacher Systemschnittstellen
  (z.B. Station 8418 mit MSV2) wird ein Verlust von Funktionen
  in Kauf genommen, die im Verbundbereich vorhanden sind,
  wie z.B. simultane Verbindungen u.ä..
  So kann z.B. ein sehr umfangreicher JOB den Eingangskanal
  zu einem Host evt. für längere Zeit blockieren.
- die im NRW-Projekt gewählten Systemanschlüsse sind in
  ihrem Durchsatz im allgemeinen auf Raten bis max. 9600
  bit/sec. (evt.19.2 kb) beschränkt; obwohl die derzeit
  genutzte Leistung von 4800 bit/sec. für den derzeitigen
  Verkehr ausreichend ist, könnten an dem Bochumer CD205-Rechner,
  dem meistgenutzten im Verbund, längerfristig evt. Kapazitäts-
  engpässe auftreten
- generell ist der Dietz 621-Mini-Rechner (mit der recht
  umfangreichen FEP-Software) im Hinblick auf den Daten-
  durchsatz doch recht beschränkt, insbesondere da für die
  CD UT200-Anschlüsse keine moderne Interface-Hardware/
  Software, wie sie für den X.25-Anschluß und den TR 440-
  MSV-Anschluß vorliegen, eingesetzt werden kann.
- bei der derzeitigen Diskussion, weitere DFN-Dienste (wie
  Dialog, Message) zusätzlich zu dem RJE-Service für die
  Hochschulen in NRW zur Verfügung zu stellen, werden größere
  Nachteile dieser FEP-Lösung sichtbar; einfacher DFN-File-
  Transfer ließe sich dagegen leicht integrieren.

4. Erfahrungen und Probleme im Betrieb

4.1 Zentrale Betreuung

Die einheitliche Lösung beim FEP-Einsatz (ausschließlich
Dietz 621-Systeme) reduzierte den Aufwand der Inbetriebnahme und der Betreuung und Wartung der FEP-Verbund-Software
im wesentlichen auf die speziellen Gegebenheiten am jeweiligen Host. Hierbei waren eine Reihe von Detailproblemen zu

testen und zu lösen, die - abgesehen von den grundsätzlichen
Unterschieden des TR440- und CDC-Anschlusses in unterschiedlichen Betriebssystemversionen der CDC-Rechner (NOS 1.4,
NOS 2.1, NOS/BE) und der TR440-Rechner (lokale Erweiterungen
an BS3 wie z.B. Spoolkomponenten) bestanden.
Zur praktischen Durchführung dieser Aufgaben wurde in allen
Dietz-Systemen zuerst mit der X.25-Software eine X.28/X.29
und eine Teletex-Software (S.70,S.62) installiert. Hierdurch
wurde ermöglicht
       a) der Dialogzugriff zwischen den Dietz-Systemen
       b) die Übertragung von Teletex-Files zwischen Dietz-
          Systemen

Die X.28/X.29-Software erlaubte, von Düsseldorf aus die
Anpassungen der Verbundsoftware an die externen Host-Systeme
im jeweiligen vorgeschalteten Dietz-System direkt durchzuführen,
wobei durch die Teletex-Software Programm-Files zu diesen
Systemen sicher transferiert werden konnten.
Diese Methode erleichterte die Aufgaben beträchtlich und
vermied die Notwendigkeit von zu häufigen Dienstreisen;
sie ist auch für die längerfristige Wartung (und evt. Erweiterung) der Verbundsoftware eine große Hilfe.
Zu erwähnen ist ferner, daß sich sowohl das parallele Senden
als auch das parallele Empfangen von mehreren Jobs und
Outputs mit Hilfe des Spool-Systems im Betrieb als nützlich
erwiesen hat, obwohl der Host-Eingang nur sequentiell
organisiert ist.

4.2 RJE-Dienst und Ergänzung durch Dialog-Dienst

Zusätzlich zu dem RJE-Verbund stellte die Universität Bochum
mit gesonderten Eingängen eine Dialogmöglichkeit an ihrem
CDC-System über Datex-P mit X.28/X.29 zur Verfügung.
Hierbei war interessant zu beobachten, daß viele Benutzer
der TH Aachen, die mit einem internen Terminalnetz DATEX-P

Zugang haben, das Arbeiten auf dem Bochumer Rechner so organisieren, daß

- im Dialog die Verarbeitungsaufträge erzeugt und
  gestartet werden, während
- mit dem RJE-Dienst die Ergebnisse der Aufträge von
  Bochum nach Aachen zurückgeschickt wurden.

Dieses Benutzerverhalten sollte beim Ausbau des Verbundes
berücksichtigt werden.

4.3 Hauptrichtungen des Verkehrs

Die folgende Skizze zeigt die Hauptrichtungen der externen
Nutzung.

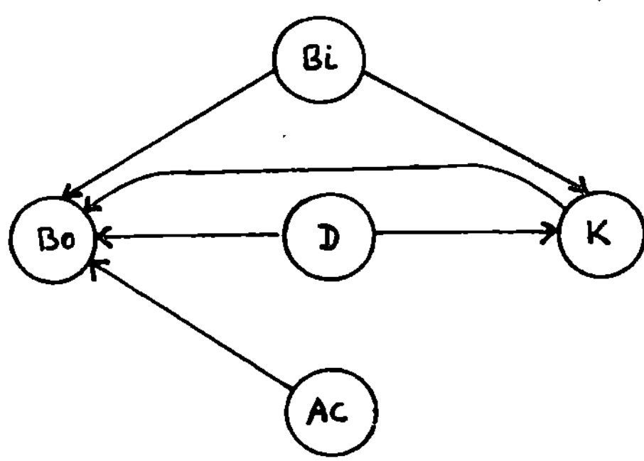

<u>Bild 3</u>

Erste Erfahrungen zeigten, daß im wesentlichen die Rechenkapazitäter
externer Systeme genutzt werden (insbesondere wenn die lokalen
Hosts stark ausgelastet sind). Die Nutzung von Spezial-Software
durch externe Benutzer spielt in der bisherigen Betriebsphase
keine größere Rolle, was evt. auf nicht genügender Kenntnis
extern verfügbarer Möglichkeiten beruht. Hieraus könnte
gefordert werden, daß die Projekte - nach Lösung der technischen
Anschlußprobleme - in den folgenden Betriebsphasen des Verbundes
stärker diesen Aspekt berücksichtigen sollten.

4.4 Nutzung

Die folgende Skizze gibt einen ersten Überblick über die
Nutzung des Verbundes. Da, wie erwähnt, ein Teil der Outputs
über Dialogzugang erzeugt werden, gibt die Tabelle die Zahl
von Job-Input- und Job-Output-Transfervorgängen im Netz
wieder.

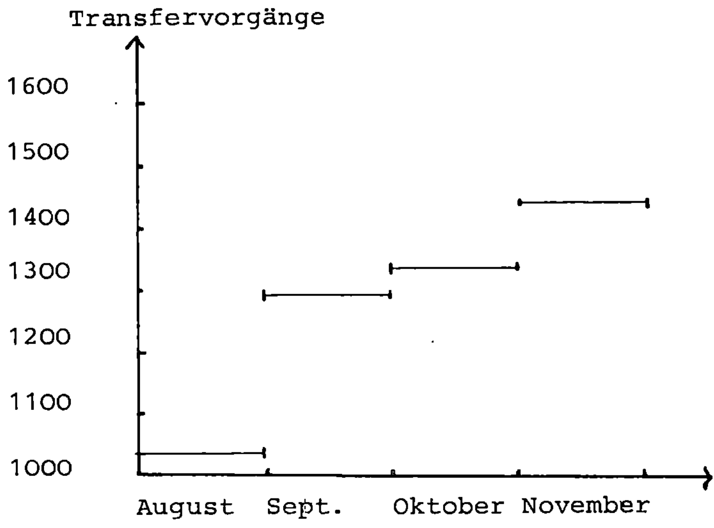

Bild 4

Hiervon wurden

      ca. 80 % auf dem Bochumer CDC-Rechner
      ca. 15 % auf dem Kölner CDC-Rechner

durchgeführt. Einige wenige Jobs wurden auf den anderen
Systemen gerechnet (wobei auch Testjobs zu berücksichtigen
sind).
Für die Nutzung des Verbundes durch die einzelnen Hochschulen ergeben sich folgende erste Anteile (als Anhaltspunkt):

Hochschule	Anteil	Nutzung der
TH Aachen	ca. 25 %	CD205 in Bochum
Uni. Bielefeld	ca. 45 %	CD205 in Bochum
		CD Cyber in Köln
Uni. Bochum	0 %	-
Uni. Düsseldorf	ca. 10 %	CD205 in Bochum
Uni. Köln	ca. 20 %	CD205 in Bochum

Erwähnenswert ist noch (in Verbindung mit der kombinierten
Anwendung von RJE- und Dialog-Diensten (s. 4.2), daß der
Umfang von Job-Input typischerweise im Bereich weniger
K-Bytes liegt (2-4K), während der Umfang der Outputs typischerweise bei 20-100 K-Bytes liegt. Hieraus läßt sich für
die Steuerung des Betriebs sofort ableiten, daß die Übertragung von Outputs möglichst in die Nachtstunden zu schie-
ben ist (billigste DATEX-P-Tarifklasse), während es für
den Job-Input keine Einschränkung geben sollte.

5. Accounting

Folgende Kosten müssen mit den Benutzern abgerechnet werden
- die herkömmlichen Kosten zur Ausführung der Jobs und zum
  Drucken der Outputs
- die Verbundkosten
Die herkömmlichen Kosten werden mit Hilfe der Benutzerkennung
in dem Rechenzentrum, in dem sie entstehen, abgerechnet.
Die Verbundkosten beinhalten
- die DATEX-P-Gebühren
- die Host-Leistungen
- die FEP-Leistungen

Die Kosten für den Rechner (Host und FEP) werden je nach
Kostengruppe aufgeteilt in Betriebskosten und Selbstkosten.
Zur Abrechnung der Verbundkosten gibt es im Prinzip zwei
Möglichkeiten
- Abrechnung der Kosten am Ort der Entstehung unter der
  Benutzerkennung
oder
- Sammlung aller Verbundkosten zu einem Netjob (also die
  Kosten für das Suchen des Jobs und der zugefügten Outputs)
  bei der Station oder dem Rechner, bei dem der Job (oder
  Output) ursprünglich initiiert wurde.
Im NRW-RJE-Verbund wird die zweite Alternative aus folgenden
Gründen favorisiert:
Die Verbundkosten werden auf den FEPs gesammelt. Dort ist
es aber u.U. technisch schwierig, den Erzeuger eines
Druckerprotokolls zu ermitteln (keine Benutzernummer im
Druckerprotokoll).
Außerdem bietet die Abrechnung nach Netjobs für den Benutzer
z.B. für seine Statistik etc. Vorteile.
Technisch wird folgender Ablauf gewählt:
An einer Station oder einem Host werden den Benutzern Netz-
benutzerkennungen vergeben. Normalerweise sind das die ge-
wöhnlichen Benutzerkennungen. Bei der Ausgangsstation wird
die Verknüpfung zwischen NETJOBID und Netzbenutzerkennung
hergestellt. Job und zugehöriger Output werden bei allen
Verbund-spezifischen Aktivitäten stets von der NETJOBID
begleitet. Da aus der NETJOBID die Ausgangsstation erkenn-
bar ist, können alle Aktivitäten zu einen NETJOBID mit der
Ausgangsstation abgerechnet werden und dort wieder den Netz-
benutzern in Rechnung gestellt werden.
Zur Zeit realisiert sind die Verknüpfung von NETJOBID und
Benutzerkennung sowie die Sammlung der DATEX-P-Kosten auf
dem FEP zu einer NETJOBID.

6. Verbesserungsmöglichkeiten

Wie bereits erwähnt, sind die beteiligten Rechenzentren
der Ansicht, daß bis zur Bereitstellung einer auf ISO bzw.
CCITT-Standards basierenden, herstellerunterstützten Lösung
das gegenwärtige FEP-Konzept geeignet ist, den Verbund in
einer Übergangszeit von (mittelfristig) einigen Jahren zu
betreiben (s. hierzu auch Kapitel 3).
Es gibt jedoch eine Reihe von Verbesserungsmöglichkeiten
an den derzeitigen Konzepten und an der Hard- und Software-
konfiguration der FEP-Lösung (sowohl an der derzeitigen
Dietz 621, als auch an der demnächst eingesetzten NORD 100),
die zukünftig zu untersuchen und eventuell zu lösen sind.
Aus Benutzersicht liegt die wohl wesentlichste Ergänzung
bei einem verbundweiten Nachfrage- und Informationsdienst,
der Informationsmöglichkeiten über das aktuelle Stadium
der Bearbeitung eines Jobs im Netz zur Verfügung stellt.
Beispiele für solche Informationen sind:
- Job mit seinen Netzparametern lokal akzeptiert für eine
  Verbundverarbeitung
- Job transferiert zum Zielhost
- Job am Zielhost in der Input/Execution/Output-Queue
- Output transferiert zum Output-Host
- gesamter Netz-Job beendet, etc.
Es bleibt zu diskutieren, wie solch ein Informationssystem
technisch zu realisieren ist. Denkbar sind hier einfache
Protokollelemente, mit denen solche Status-Information zu
zentralen oder auch verteilten Monitorkomponenten übertragen
werden. Ein entsprechendes Konzept liegt im ersten ISO Draft
Proposal zum Job-Transfer-Bereich vor /2/.
In diesem Zusammenhang ist auch zu sehen, daß den Benutzern
im Verbund Informationen über Betriebsdaten der im Verbund
vorhandenen Ressourcen zur Verfügung gestellt werden (Wartungszeiten, Ausfälle, Betriebssystemversionen und -änderungen,
u.ä.). Die technische Realisierung zum Beispiel mit entsprechendem Zugriff oder dezentral mit Kommunikationsunterstützung
(oder ähnliches) wäre festzulegen.

Die Steuerung von Transfer- und Verarbeitungsvorgängen innerhalb einer initiierten Netz-Job-Aktivität ist aus wirtschaftlichen Gründen (z.B. Nutzung günstiger DATEX-P-Tarifklassen)
oder auch aus Benutzersicht (Einflußnahme auf aktuelle Abläufe) wünschenswert.
Im Hinblick auf die derzeitige FEP-Hard- und Software sollte
angestrebt werden, die DÜ-Anschlüsse (Netz und Hosts) gänzlich
in intelligente DMA-fähige Interfaces zu verlagern, da nur
so eine merkliche Erhöhung der Durchsatzraten zu erzielen
ist.
Verbesserungsmaßnahmen im Softwarebereich müßten Messungen
von einzelnen FEP-Systemkomponenten vorausgehen; zu konzentrieren wäre sich hier zum Beispiel auf den Einfluß von
Plattenzugriffen, der Interprozeß-Kommunikation, u.ä..
Es ist zu erwarten, daß hier unterschiedliche Maßnahmen
bei den FEP-Typen Dietz 621 und NORD 100 erforderlich sind.

Referenzen

/1/ Rechnerverbund mit Minirechnern Dietz 621;
    B.Cappel, E.Dregger-Cappel, A.Haag, H.Kerutt, J.Knop,
    W.Müller, R.Speth, P.Wallerath, H.Zschintzsch;
    Bericht zum Förderungsprojekt DV 081 5614 des BMFT;
    September 1981
/2/ OSI- Job Transfer and Manipulation Concepts and Services;
    ISO DP 8831; ISO/TC97/SC21/N210

    OSI-Specification of the Basic Class Protocol for Job
    Transfer and Manipulation; ISO DP 8832; ISO/TC97/SC21/N211;

<u>**AUFBAU, BETRIEB UND WEITERENTWICKLUNG EINES NETZWERKES**</u>
<u>**AUS ANWENDERSICHT**</u>

R. Göldner
Rechenzentrum der Finanzverwaltung des
Landes Nordrhein-Westfalen
Düsseldorf

Einleitung

ISO hat mit dem Architekturmodell den Weg zu offenen Systemen gewiesen.
Es beschreibt in 7 Schichten die für die Datenfernverarbeitung notwendigen Funktionen. Die Schichten haben nach [1] folgende Bezeichnungen:
1. Bitübertragung, 2. Sicherung, 3. Vermittlung, 4. Transport, 5. Kommunikationssteuerung, 6. Darstellung und 7. Verarbeitung. Die Normung der
zugehörigen Protokolle und Dienste ist in einem fortgeschrittenen Stadium [2]. Für den Anwender und Betreiber eines privaten Netzwerkes sind
in erster Linie die Schichten 6 und 7 von Bedeutung. Der vorliegende Beitrag versucht, die Problematik privater Netze aus Anwendersicht am Beispiel des "Kommunikationssystems der Finanzverwaltung des Landes Nordrhein-Westfalen" darzustellen, das auf der Grundlage eines am ISO-Modell
orientierten Betriebssystems realisiert wurde. Es wird der Aufbau und Betrieb des Netzes mit den Mitteln des Anwenders aufgezeigt und ein Ausblick auf die weitere Entwicklung gegeben.

Definition "Anwendersicht"

Der Anwender sieht das Netzwerk als reines Transportvehikel, das es ihm
ermöglicht, Daten mit einem entfernten Partner auszutauschen (Bild 1).
Aus der Datenverarbeitung wird damit die Datenfernverarbeitung, bei der
sich aus der Sicht der Anwendung zwei grundsätzliche Nutzungsarten unterscheiden lassen:

        - die Mensch-Maschine-Beziehung und
        - die Maschine-Maschine-Beziehung.

Die Mensch-Maschine-Beziehung findet üblicherweise in Form eines Dia-
loges statt, bei dem der Bearbeiter von seinem Bildschirmplatz mit dem
Rechner über ein Netzwerk verkehrt. Die ausgetauschten Daten (Transaktionen) sind in der Regel sehr kurz.

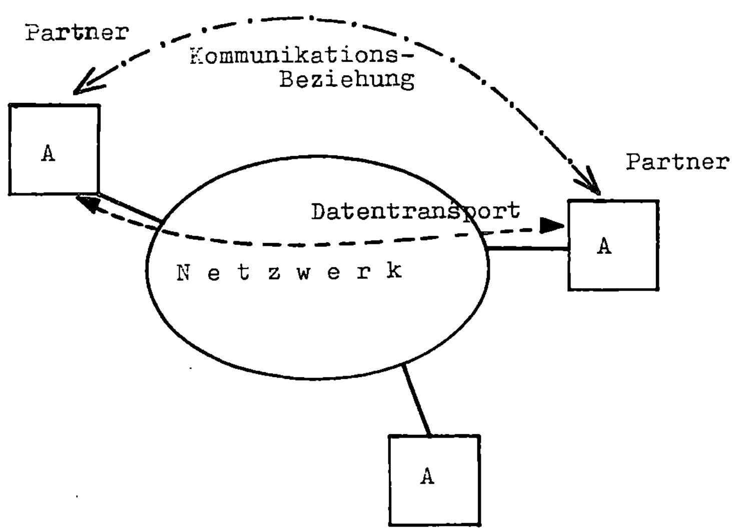

Bild 1   Netzwerk aus Anwendersicht als Transportvehikel
         für den Datenaustausch zwischen zwei entfernten
         Partnern
         A: Anwender

Typisch für die Recher-Rechner-Beziehung ist eine andere Art des Datenaustauschs - die Dateiübertragung (File Transfer). Hierbei werden komplette Dateien von einem Rechner zu einem anderen Rechner über ein Netzwerk übertragen und dann am Zielrechner verarbeitet. Die Stapelfernver-
arbeitung (Remote Job Entry) und das Fernladen (Down Line Load) sind nur
Sonderfälle der Dateiübertragung, bei denen die übertragene Datei durch
das empfangende System lediglich entsprechend interpretiert werden müssen.

Außer den genannten Zweier-Beziehungen sind auch 1:N-Beziehungen möglich.
Beispiel hierfür ist das Rundspruchverfahren (Broadcasting), bei dem ein
Anwender an mehrere Partner über das Netzwerk eine Mitteilung versendet,
die vom Netzwerk automatisch vervielfältigt wird.

Anwendersicht eines Netzwerkes soll heißen, daß alle Leistungen auf der Grundlage der Dialog- oder Transaktionsschnittstelle und der Dateiübertragungsschnittstelle der vorhandenen Betriebssysteme realisiert werden.

Für den Anwender. existieren im Grunde genommen nur verteilte Systeme mit lokaler Anwendungssoftware., die zum Zweck des Datenaustauschs über Leitungen verbunden werden müssen. Hier gibt es zwei Möglichkeiten: festgeschaltete Leitungen (HfD) oder digitale öffentliche Netzwerke wie Datex-L und Datex-P. Die Entscheidung hängt in der Regel von der Datenmenge, der für die Datenübertragung zur Verfügung stehenden Zeit und den Kosten ab und stellt eine keinesfalls triviale Optimierungsaufgabe dar.

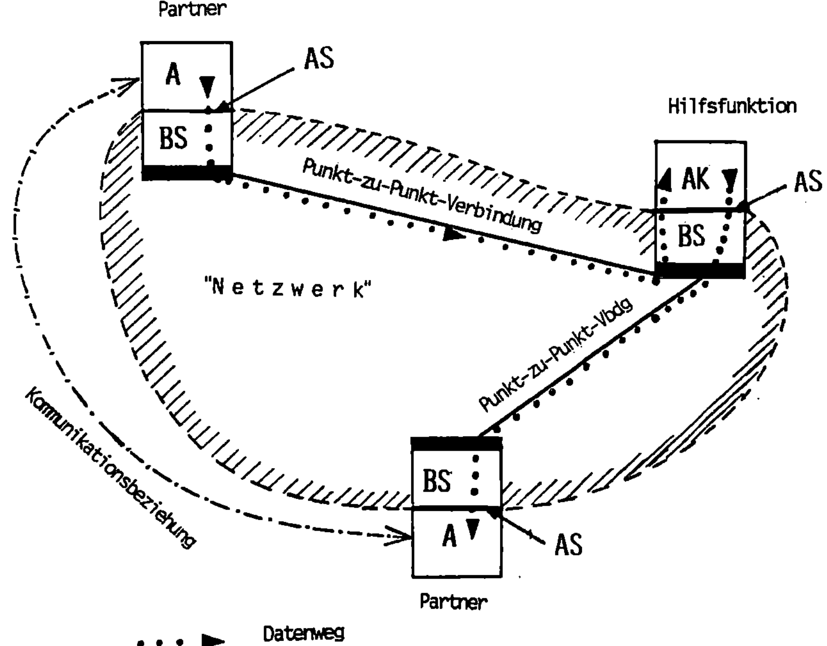

Bild 2   Knotenfunktion als Anwendungssoftware
A Anwender mit lokaler Software
AK Anwender mit Knotensoftware
BS Betriebssystem
AS Anwenderschnittstelle zum Netzwerk

Entscheidet sich der Anwender für HfD-Leitungen, dann hat er die Vermittlungsfunktion eines Netzwerkes selbst zu erbringen und eigene Netzknoten zu installieren. Diese zur Kanalisierung der Datenströme notwendigen Netzknoten stellen allerdings nur eine Hilfsfunktion dar, die zusätzlich zur lokalen Anwendungssoftware in den verteilten Systemen zu installieren ist (Bild 2).

Ein auf diese Weise konstruiertes Netzwerk besteht logisch gesehen nur
aus Punkt-zu-Punkt-Verbindungen und Anwendungssoftware zur Weitervermittlung der Daten.

Die Dialogschnittstelle und die Dateiübertragungsschnittstelle dieses
Netzwerks ist entweder bei verteilten Systemen desselben Herstellers
durch das Betriebssystem gegeben oder bei heterogenen verteilten Systemen durch Emulation z.B. des "Industriestandards" BSC. Selbstverständlich sind hier - soweit realisiert - auch die nationalen EHKPs einsetzbar
bzw. später, genormte Protokolle, wenn die europäischen DV-Hersteller
ihre Absichtserklärung wahr gemacht haben, ihre Betriebssysteme durch
Einbau der ISO-Protokolle für die Kommunikation zu öffnen [3].

**Aufgaben und Konstruktionsidee des betrachteten
realen Netzwerks**

Grundlage der weiteren Ausführungen ist das "Kommunikationssystem der
Finanzverwaltung des Landes Nordrhein-Westfalen" genannte Netzwerk, über
das alle 107 Finanzämter (FÄ) des Landes mit dem Rechenzentrum (RZF) in
Düsseldorf verbunden sind. Im RZF werden zentral für alle FA Daten gespeichert, Steuerberechnungen durchgeführt und Steuerbescheide gedruckt,
sowie die Kontenführung der Steuerpflichtigen (Sollkonten) abgewickelt.

Die Datenerfassung wird in den FÄ selbst durchgeführt; zusätzlich ist
das Auskunftssystem für Kontenabfragen dezentralisiert.

Das Netzwerk besteht aus insgesamt 143 Minirechnern Bull System 6 (S6)
mit dem Betriebssystem GCOS MOD400/DSS, die über HfD 4800 und 9600
Bit/sec sternförmig in zwei Hierarchieebenen gekoppelt sind (Bild 3). In
jedem FA ist ein S6 als FA-Rechner mit bis zu 24 Bildschirmplätzen für
die Datenerfassung und das Auskunftssystem installiert. In 16 FA sind
zusätzlich insgesamt 32 S6 als Netzknoten installiert. Diese Netzknoten
enthalten zusätzlich die dezentrale Datenbank des Auskunftssystems. Im
RZF sind 4 gekoppelte S6 als Netz-End-Rechner installiert. Tab. 1 zeigt
die Daten des Netzwerkes auf einen Blick.

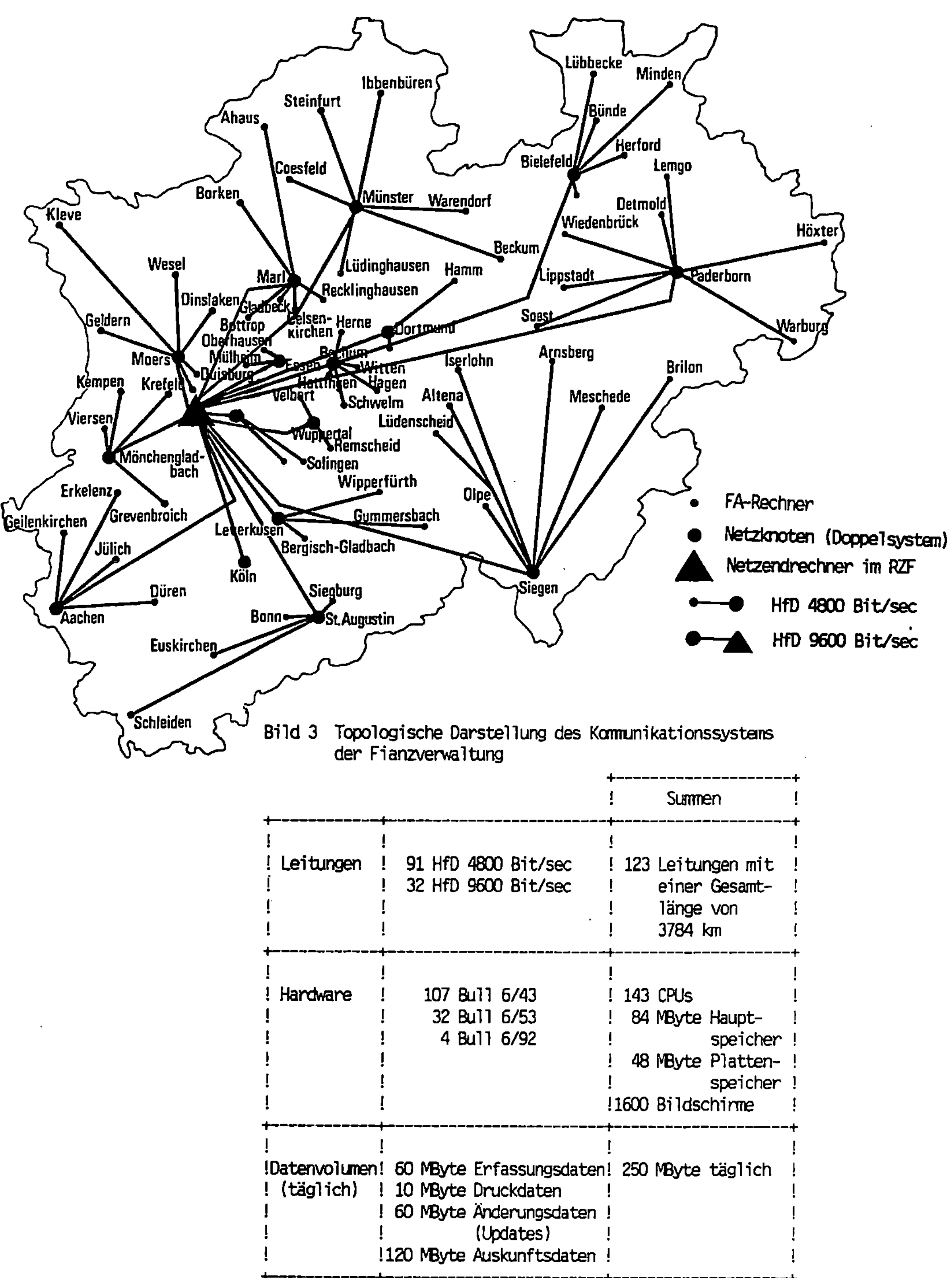

Bild 3 Topologische Darstellung des Kommunikationssystems der Fianzverwaltung

			! Summen
! Leitungen	!	91 HfD 4800 Bit/sec 32 HfD 9600 Bit/sec	! 123 Leitungen mit einer Gesamt- länge von 3784 km
! Hardware	!	107 Bull 6/43 32 Bull 6/53 4 Bull 6/92	! 143 CPUs 84 MByte Haupt- speicher 48 MByte Platten- speicher !1600 Bildschirme
!Datenvolumen! ! (täglich)	!	60 MByte Erfassungsdaten! 10 MByte Druckdaten 60 MByte Änderungsdaten (Updates) !120 MByte Auskunftsdaten	250 MByte täglich

Tab. 1 Technische Kennzahlen des Netzwerkes

Die Konstruktionsidee des Netzwerkes ist im Bild 4 dargestellt.

Der Aufbau des Netzwerkes

Das für die Realisierung der Anwendungssoftware zur Verfügung stehende
Betriebssystem GCOS MOD 400/DSS zeigt in seinem Netzwerkteil große Ähnlichkeit mit dem ISO-Architekturmodell. Dies ist dem Konstrukteur der
Netzwerksoftware Charles Bachmann zu verdanken, der lange Zeit Chairman
der ISO-Arbeitsgruppe TC97/SC16 war. Natürlich sind nur die Schichten 1
bis 3 identisch mit den Normen. Die höheren Schichten weisen lediglich
funktionale Ähnlichkeit auf.

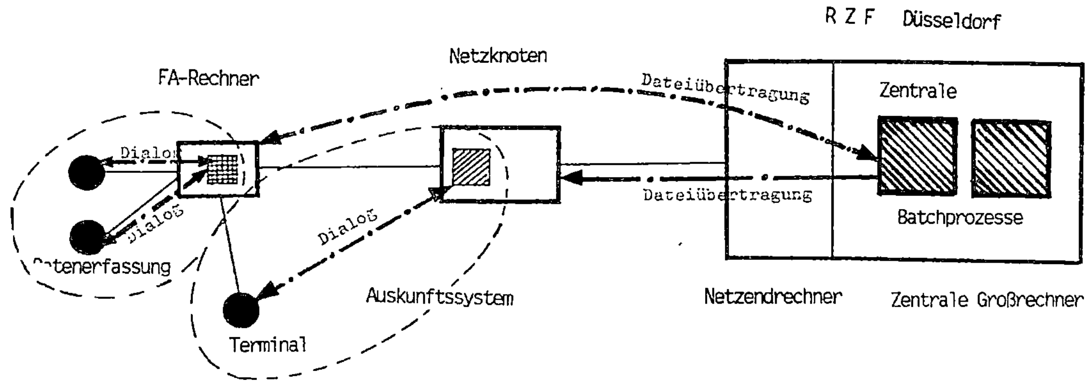

Bild 4   Konstruktionsprinzip des Netzwerkes:
         Auslagerung der Datenerfassung in die FA-Rechner und
         Auslagerung der Auskunftsdateien in die Netzknoten
         Verbindung mit den zentralen Batchprozessen mit
         Dateiübertragung

Die konsequente Realisierung der Netzwerksoftware in den ISO-Funktions-
schichten hat für den Anwender große Vorteile. So kann im Kommunikationssystem der Finverw, in Schicht 3 wahlweise für HfD, Datex-P oder
Datex-L konfiguriert werden, wobei lediglich die Netzwerkparameter geändert werden müssen.

Schicht 4 umfaßt die im ISO-Modell definierten Transportleistungen in
Form Bull-spezifischer Dienste und Protokolle.

Schicht 5 und 6 existieren nicht als eigenständige Funktionen. Sie sind
in der "Distributed Transactional Facility (DTF)" und in der "File
Transfer Facility (FTF)" enthalten, die dem Anwender auf Schicht 7 einerseits eine Schnittstelle für Dialog- oder Transaktionsbetrieb und andererseits eine Schnittstelle für Dateiübertragung zur Verfügung stellen.

Sowohl im DTF- als auch im FTF-Modul sind dieselben Mechanismen für die
Kommunikationssteuerung (Schicht 5) enthalten. Als Zugriffspunkte für
die Anwendungsprozesse dienen adressierbare Speicherbereiche, Mailboxes
genannt. Zwischen den Mailboxes der beteiligten Systeme wird eine Sitzung, Logical Connection genannt, aufgebaut, unterhalten und abgebaut.

Als Funktionen der Darstellungsschicht in DTF und FTF kann die Zeichen-
kompression und das Verstecken des Mailbox-Mechanismus vor den Anwen-
dungsprogrammen durch das Bereitstellen entsprechend einfacher Schnittstellen bzw. Makroaufrufe angesehen werden.

Die realisierten Anwendungen lassen sich klassifizieren in lokale Anwendungen und Netzwerkanwendungen.

Von besonderem Interesse aus der Anwendersicht des Netzwerkes sind die
Netzwerkanwendungen wie Durchschleusen von Daten und Vervielfältigung
von Daten. In Bild 2 ist die Knotenfunktion des Durchschleusens, also
des Vermittelns von Daten dargestellt. Nach der Philosophie des ISO-Modells handelt es sich hierbei um eine typische Funktion der Schicht 3.
Die Hersteller sehen dafür nur eine Unterstützung mit spezieller Hardware vor. Üblicherweise werden die Vorrechner der Hosts angeboten, die
auch abgesetzt (remote) und vernetzt betrieben werden können. Beispiele
sind IBM 37x5, Siemens DUET und Bull DATANET.

Dezentrale Systeme bieten in der Regel nicht die Möglichkeit des Vermittelns mit Funktionen des Betriebssystems. Der Anwender ist gezwungen, in
den Knoten die Daten bis auf die Anwendung (Schicht 7) durchzureichen,
dort zwischenzuspeichern und dann erst weiterzusenden.

Im Kommunikationssystem der Finverw wurde das Problem dadurch gelöst,
daß bei durchzuschleusenden Daten auf FTF verzichtet wird und Dateien in
kleinen Abschnitten - Stapel genannt - mit DTF übertragen werden. Der
Stapel ist gleichzeitig Wiederaufsetzpunkt für Restart bzw. Recovery.

Die Stapel werden durch ein Anwendungsprogramm - den Input-Output-Queue-Manager (IOQM) - durchgeschleust, der Einträge der Empfangswarteschlange in die Sendewarteschlange überträgt. Durch Vereinbarung eines Mehrfachziels (Multiple Destination) kann der IOQM Nachrichten vervielfältigen, indem sie nacheinander in verschiedene Sendewarteschlangen eingetragen werden. FTF wird nur bei Verbindungen zwischen maximal 2 Systemen eingesetzt.

Netzwerküberwachung

Der Betrieb eines Netzwerkes setzt in einem unvergleichlich höherem Maße Überwachung und Kontrolle voraus als dies beim Betrieb eines normalen Rechenzentrums notwendig ist. Ursache dafür ist die Tatsache, daß die Teilsysteme eines Netzwerkes für das Rechenzentrum "unsichtbar" sind und allenfalls über Leitungsverbindungen zugänglich sind. Erschwerend kommt hinzu, daß es sich immer um eine Vielzahl von Teilsystemen handelt.

Objekt	Verfahren	Maßzahl	Auswertg.	A
DÜ-Verbindung (Postleitung)	Messung	Ausfallzeit/ Betriebszeit	Vergleich gegen Sollwert oder Erfahrungswert	
Hardware der dezentralen Systeme(CPU, Peripherie,...)	Messung	Ausfallzeit/ Betriebszeit		
Anwendungssysteme (Software)	Messung	Ausfallzeit/ Betriebszeit		x
	Zählung	Anzahl der Starts bzw. Restarts		x

Tab. 2 Überwachung der Verfügbarkeit eines Netzwerkes und ihre Bedeutung aus Anwendersicht
A: relevant für Anwendersicht

Die wichtigsten Merkmale für den Betrieb eines Netzwerkes sind Verfügbarkeit, Durchsatz und Sicherheit.

Die Überwachung dieser Merkmale läßt sich anhand verschiedener Objekte
und unterschiedlicher Messungen durchführen. Die Tab. 2-4 enthalten Zu-
sammenstellungen nach Merkmalen, Objekten und Kontrollverfahren mit den
zugehörigen Maßzahlen und ihrer Auswertung.

Das Meßverfahren besteht überwiegend darin, daß Zeiten gemessen und Mengen gezählt werden, die dann auf Einhaltung von Soll- oder Erfahrungs-
werten geprüft werden. Die Meßhäufigkeit und die Schnelligkeit, mit der
auf Normabweichungen reagiert werden muß, ist abhängig von der speziellen Aufgabe, die das Netzwerk zu erfüllen hat.

Objekt	Verfahren	Maßzahl	Auswertg.	A
DÜ-Verbindung (Postleitung)	Zählung	Anzahl der Blockwieder- holungen	Vergleich gegen Sollwert oder	
	Zählung	Fehlerrate (Anzahl feh- lerhafte Zei- chen / Anzahl übertragene Zeichen)	Erfah- rungswert	
Logische Verbindung (zwischen 2 An- wenderprozessen)	Zählung und Messung	effektive Übertragungs- rate(Zeichen/ sec)		
Warteschlangen	Zählung und Messung	Anzahl be- legte Plätze in der Zeit- einheit		
Prozessor (CPU)	Messung	Auslastung (CPU-Zeit/Ge- samtzeit,auf- gegliedert nach Priori- tätsebenen)		
Anwendungssystem (Software)	Messung	Reaktionszeit		x

Tab. 3   Überwachung des Durchsatzes eines Netzwerkes und ihre Bedeutung aus Anwendersicht
A: relevant aus Anwendersicht

Während Verfügbarkeit und Durchsatz durch meßbare Größen dargestellt
werden können, ist die Sicherheit eine Boolesche Größe. Sicherheit ist
entweder gegeben oder nicht gegeben. Die Maßzahlen der zugehörigen Objekte müssen stets ohne Ermessensspielraum auf Identität geprüft werden.

Objekt	Verfahren	Maßzahl	Auswertg.	A
Logische Verbindung (zwischen 2 Anwenderprozessen)	Transport-kontrolle	Satzprüfzeizeichen	Identitätsvergleich	x
		Satzzahl		x
	Quittierung	ja/nein-Zustand	Abfragen	x
	Plausibilitätsprüfung	diverse	Vergleich mit div. Prüfnormalen oder Intervallen	x
Hardware der dezentralen Systeme(CPU, Peripherie,...)	Prüfprogramme	diverse	Vergleich gegen Sollwert	
Anwendungssysteme (Software)	Zulässigkeitskontrolle	Versionsnummer	Identitätsvergleich	x
Datenbestand	Zulässigkeitskontrolle	Versionsnummer	Identitätsvergleich	x

Tab. 4 Überwachung der Sicherheit eines Netz-
        werkes und ihre Bedeutung aus Anwendersicht
        A: relevant aus Anwendersicht

Wenn wir nun die Netzüberwachung aus Anwendersicht betrachten, dann
stellen wir fest, daß durch Überwachungsmaßnahmen auf der Verarbeitungs-
schicht, Netzstörungen genau so gut erkannt werden können wie durch
hardwarenahe oder betriebssystemspezifische Maßnahmen. Genauer gesagt,
es werden diejenigen Störungen erkannt, die Auswirkungen auf die Anwendung haben. Für das Rechenzentrum als Netzwerk-Steuerungsinstanz ist es

wichtig, ob die dezentralen Anwendungen funktionieren oder nicht und ggf.
welche Datei- und Softwareversionen bzw. welches Aktualitätsdatum dezentral im Einsatz sind. In den Tab. 2-4 sind die für die Netzüberwachung
aus Anwendersicht erforderlichen Kontrollverfahren in der letzten Spalte
angekreuzt.

Selbstverständlich kann auf die anderen Kontrollverfahren keinesfalls
verzichtet werden. Sie stellen jedoch in erster Linie Hilfsmittel zum
Einkreisen von Fehlern im Störungsfalle dar, müssen aber nicht unbedingt
zur laufenden Netzüberwachung eingesetzt werden.

```
            Uhrzeit------>

            7    8    9    10   11   12   13   14   15
Netzknoten  I----I----I----I----I----I----I----I----I----I
   010      I  **I****I****I****I****I** *I****I****I****I     97,0%
            I    I    I    I    I    I    I    I    I    I
   019      I  **I****I****I****I****I****I    I    I    I     64,7%   A
            I    I·   I    I    I    I    I    I    I    I
   020      I o*I****I****I****I****I****I****I****I**ooI     91,1%
            I    I    I    I    I    I    I    I    I    I
   030      I  **I****I****I****I****I****I****I****I****I    100,0%
            I    I    I    I    I    I    I    I    I    I
   039      I  **I****I****I****I****I****I****I****I****I    100,0%   '
            I    I    I    I    I    I    I    I    I    I
   040      I  **I****I****I**,**I****I    I    I  **I***oI     67,6%   B
            I    I    I    I    I    I    I    I    I    I
   049      I   *I****I****I****I****I****I****I****I**ooI     91,1%
            I    I    I    I    I    I    I    I    I    I
   050      I  **I****I****I****I****I****I****I****I****I    100,0%
            I    I    I    I    I    I    I    I    I    I
   059      I  **I****I****I****I****I****I****I****I****I    100,0%
            I    I    I    I    I    I    I    I    I    I
   060      I  **I****I****I****I****I*oooIo***I****I**ooI     82,3%
            I    I    I    I    I    I    I    I    I    I
   069      I  **I****I****I****I****I****I** *I*   I ***I     85,2%   B
            I    I    I    I    I    I    I    I    I    I
   070      I  **I****I****I****I****I****I****I****I*** I     97,0%
            I    I    I    I    I    I    I    I    I    I
   079      I    I    I  **I****I** *I*** I****I ***I**  I     61,7%   C
            I    I    I    I    I    I    I    I    I    I
   100      I   *I****I****I****I****I****I****I****I**  I     91,1%
            I----I----I----I----I----I----I----I----I----I
            7    8    9    10   11   12   13   14   15
```

Dienstzeit 7.30 - 16.00

Bild 5   Ausschnitt aus der Verfügbarkeitstatistik des
          Auskunftssystems vom 1.6.84

```
*   Auskunft voll verfügbar    A) Leitungsstörung
o   Auskunft abgeschaltet      B) Neuaufbau Datenbank
-   Auskunft fehlerhaft        C) Hardware-Störung
    DÜ-Verbindung gestört
```

444

Diese Aussagen gelten für das hier vorgestellte Netzwerk, das Kommunikationssystem der Finverw. zu dessen Charakteristiken es zählt, daß der Ausfall eines dezentralen Systems für einen Zeitraum bis zu 1 Tag verkraftet werden kann. Netzstörungen müssen daher auch nicht präventiv erkannt werden.

Die in Tab. 2-4 angekreuzten Kontrollverfahren sind im RZF alle eingesetzt. Als Beispiel ist in Bild 5 ein Ausschnitt aus der Verfügbarkeitsstatistik des Auskunftsystems dargestellt. Alle 15 Minuten wird im Netzknoten ein Zugriff auf die Auskunftsdatenbank durchgeführt. Der dabei erkannte Zustand

- Auskunft voll verfügbar oder
- Auskunft abgeschaltet oder
- Auskunft fehlerhaft oder
- DÜ-Verbindung gestört

```
+--------------------------------------------------------------------+
! FA-Nr.! Anzahl   !        !        !        !                      !
!(Netz- ! Abfragen! 4 sec  ! 7 sec  ! 13 sec ! Bemerkungen          !
!knoten)!         !        !        !        !                      !
!==================================================================!
! 305   !   634   ! 89.8   ! 97.9   ! 98.9   ! Direktanschluß!
! 335   !   566   ! 59.4   ! 92.4   ! 96.8   !                      !
! 331   !   246   ! 60.6   ! 96.8   ! 98.4   !                      !
!  -------------------------------------------------------------- !
! (010) !  1446   ! 72.9   ! 95.6   ! 98.0   ! 6/92                 !
!                                                                    !
! 359   !   416   ! 63.7   ! 78.4   ! 86.8   ! Direktanschluß!
! 307   !   649   ! 41.0   ! 66.7   ! 80.4   !                      !
! 318   !   289   ! 49.5   ! 76.1   ! 83.7   !                      !
! 319   !   414   ! 46.9   ! 73.7   ! 85.2   !                      !
!  -------------------------------------------------------------- !
! (040) !  1768   ! 49.1   ! 72.6   ! 83.6   ! 6/53                 !
!                                                                    !
! 301   !   144   ! 45.1   ! 95.8   ! 100.0  !                      !
! 308   !   243   ! 56.4   ! 82.3   ! 91.0   !                      !
! 320   !   246   ! 61.8   ! 86.6   ! 91.0   !                      !
!>>340  !   484   ! 29.0   ! 51.8   ! 63.3   !                      !
!  -------------------------------------------------------------- !
! (049) !  1117   ! 44.3   ! 71.8   ! 80.1   ! 6/53                 !
!                                                                    !
!>>204  !   879   ! 41.0   ! 68.1   ! 78.2   !                      !
!  212  !   658   ! 59.6   ! 88.0   ! 93.9   !                      !
!  -------------------------------------------------------------- !
! (229) !  1537   ! 49.0   ! 76.6   ! 84.9   ! 6/92                 !
!                                                                    !
+--------------------------------------------------------------------+
```

Tab. 5 Ausschnitt aus der Reaktionszeitenststistik
des Auskunftssystems vom 26.3.1984
FA-Nr. Finanzamtsnummer
(...)   Netzknotennummer
6/..   Prozessortyp des Netzknotens

wird gespeichert und, soweit die DÜ-Verbindung zum RZF intakt ist, zum
RZF übertragen und dort an einem Bildschirm angezeigt. Zum Dienstschluß
in den FÄ wird eine komplette Übersicht für alle 16 Netzknoten(doppel)-
systeme zentral ausgedruckt.

Tab. 5 zeigt als weiteres Beispiel einen Ausschnitt aus der täglichen
Reaktionszeitenstatistik des Auskunftssystems. Aufgeführt sind jeweils
die Prozentsätze für die Reaktionszeiten kleiner gleich 4 sec, 7 sec und
13 sec. Aus der Tab. 5 lassen sich z.B. anhand des 4 sec Wertes verschiedene Aussagen ermitteln. Sinkt der Wert unter 45%, so ist erfah-
rungsgemäß eine Fehleranalyse erforderlich: beim FA 340 lagen Leitungs-
störungen vor und beim FA 204 Hardware-Störungen.

Unterschiedlich leistungsfähige Prozessoren (6/92 - 6/53) der Netzknoten
haben ebenso Einfluß wie die Tatsache, ob ein FA-Rechner über Leitung
4800 Bit/sec) oder Direktanschluß (19200 Bit/sec) angeschlossen ist.

Die Weiterentwicklung des Netzwerkes

Im Zuge der Weiterentwicklung des Netzwerkes sind zwei Maßnahmen
von grundlegender Bedeutung geplant:

        - die Verlagerung der Auskunftsdatenbank von den
          Netzknoten in die FA-Rechner und
        - die Erweiterung der dezentralen Rechnerkapazität
          im FA durch Personal Computer (PC).

Mit der Verlagerung der Auskunftsdatenbank in die FA-Rechner entfällt
der Dialogbetrieb über das Netzwerk. Durch die gleichzeitige Abkehr von
HfD-Leitungen und den Übergang auf Wählleitungen des Datex-P-Neztes ent-
fallen die heutigen Netzknoten.

Die Netzwerkproblematik ergibt sich dann im Inhaus-Bereich, in dem PCs
an den Arbeitsplätzen der Sachbearbeiter über ein internes Netzwerk mit
dem FA-Rechner kommunizieren werden.

Dem liegt folgende Situation zugrunde: die Anzahl der Auskunftsterminals
ist heute auf 6 begrenzt. Erforderlich sind jedoch bis zu 80 Terminals
pro FA, um die Informationsbedürfnisse der Sachbearbeiter befriedigen zu
können. Das bisher übliche Verfahren wäre, den vorhandenen FA-Rechner
hardwaremäßig hochzurüsten, unintelligente Terminals anzuschließen und
die Software zu optimieren. Statt Mittel und Aufwand in herkömmliche
Technik zu investieren, bietet sich als Alternative die Beschaffung von
PCs an. Mit der so gewonnenen Kapazität können über die Informations-
zugriffsfunktion hinaus zusätzliche Leistungen z.B. aus dem Bereich der
Bürokommunikation [4] realisiert werden. Es ergeben sich jedoch noch
weitere Möglichkeiten. So ist z.B. daran gedacht, die Einkommensteuer-
berechnung, die heute als Batchverfahren im RZF abgewickelt wird, in den
PCs als Dialogverfahren am Arbeitsplatz des Sachbearbeiters zu realisieren [5].

Als Hauptprobleme aus Anwendersicht ergeben sich hierbei:

- die Absicherung der Richtigkeit bzw. Zulässigkeit
  der eingesetzten PC-Software,
- die Gewährleistung der Funktionsfähigkeit
  der dezentralen Verarbeitung und
- die Sicherung der am PC gespeicherten Daten.

Ein Hauptproblem des PC-Einsatzes ist zweifellos die Absicherung, daß nur
die offiziell genehmigte Software eingesetzt wird. Infolgedessen scheiden
Diskettenlaufwerke als Massenspeicher für den PC aus. Da aus Kostengründen nicht an jedem PC ein Festplattenlaufwerk installiert werden kann,
lassen sich nur Mehrplatzsysteme mit zentralisiertem Plattenspeicher realisieren. Die Mehrplatzsysteme können entweder aus einem Cluster-Prozessor mit unintelligenten Terminals oder mit intelligenten Terminals, also
PCs ohne eigenen Massenspeicher, bestehen. Im letzteren Fall benötigt
man schnelle DÜ-Verbindungen, um z.B. in annehmbarer Zeit Programme vom
Massenspeicher des Cluster-Prozessors in den PC laden zu können.

Die Funktionsfähigkeit der PC-Software muß durch automatisierte Vertei-
lung der jeweils gültigen Versionen gewährleistet werden. Zusätzlich
ist die Versionsnummer im laufenden Betrieb regelmäßig abzuprüfen. Die
Betriebsbereitschhaft der PC-Hardware ist hierarchisch zu überwachen.
Im FA ist jeder einzelne PC zu überwachen, im RZF sind nur summarische
Angaben pro FA von Interesse.

Die Sicherung der an den PCs bzw. dem Cluster-Prozessor gespeicherten
Daten kann ohne Problem am FA-Rechner durchgeführt werden.

Die Arbeitsteilung zwischen zentralem Großrechner und PC's stellt offensichtlich einen allgemeinen Trend dar. Denkt man folgerichtig weiter, so
ergibt sich, daß die hier diskutierte Klasse von privaten Netzwerken ersetzt werden wird durch mehr oder weniger einfache Verbindungsinstituti-
onen zwischen lokalen Netzen und zentralen Großrechnern.

Literaturhinweise

1. Burkhardt, H.J. (Bearb.): Kommunikation offener Systeme - Eine Einführung. GMD-Spiegel Sonderheft (1981)

2. Rosenberg, H.J.: OSI-Standards für die Transport- und Kommunikationssteuerungsschicht verfügbar. Ang.Inf. 6,296(1984)

3. Presseinformation 18/84 der BULL-Gruppe vom 19.03.1984: Zwölf führen-
   de Firmen der EDV-Industrie erklärten gegenüber der EG-Kommission, daß
   sie einheitliche Normen für offene Datenkommunikation ab 1985 einführen wollen unter schrittweiser Einbeziehung der CCITT- und ISO-Standards.

4. Höring, Bahr, Struif, Tiedemann: Interne Netzwerke für die Bürokommunikation. R.v.Decker's Verlag G.Schenk Heidelberg 1983.

5. Göldner, R.: Dialogverarbeitung beim Einsatz von Personalcomputern -
   Planungen und Voruntersuchungen in der Finanzverwaltung Nordrhein-
   Westfalen. Veröffentlicht im Tagungsband des 21. Erfahrungsaustauschs
   ADV Bund/Länder/Kommunaler Bereich am 29. und 30.03.1984 in Kassel,
   herausgegeben vom KGRZ Kassel und dem Senator für Inneres Berlin

Jürgen Hänle   DANET GmbH, Darmstadt

Das Petrus-Testsystem
für Zulassungs- und Abnahmeprüfungen von
Bildschirmtext- und Teletex-Einrichtungen

1.   Zusammenfassung

PETRUS ist ein universelles Testsystem der Deutschen Bundespost
für Zulassungs- und Abnahmeprüfungen von Kommunikationseinrich-
tungen im Telematik-Bereich, speziell bei Bildschirmtext und
Teletex. Die Anforderungen, Voraussetzungen und realisierten
Konzeptionen in den verschiedenen Testbereichen werden dargestellt, von Erfahrungen und möglichen Weiterentwicklungen wird
berichtet.

2.   Einführung

Die PETRUS-Testsysteme decken heute (Dez.1984) drei Anwendungsgebiete ab, nämlich

o   Testsystem EHKP-TS für Funktions- und Abnahmetests
    von Bildschirmtext-Vermittlungsstellen (Btx-Vst) und
    Bildschirmtext-Externen Rechnern (Btx-ER)

o   Testsystem Ttx-TS für Zulassungstests von Teletex-
    Endgeräten (Ttx-EG)

o   Testsystem Btx-T-TS für Zulassungstests von Bild-
    schirmtextterminals (Btx-T)
    (In Entwicklung)

Das EHKP-Testsystem prüft die Kommunikationsschnittstelle
zwischen einer Btx-Vst und einem Btx-ER, definiert durch
die Kommunikations- und Präsentationsprotokolle EHKP4 und

EHKP6 (Bild 1). (Die ebenfalls relevante X.25-
Schnittstelle gehört nicht zum Testbereich).

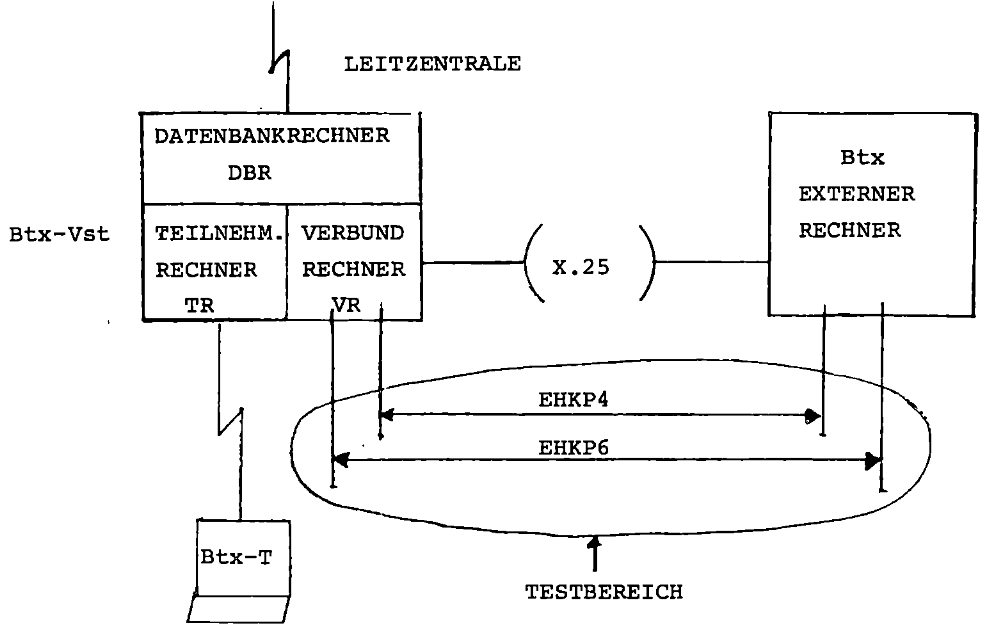

BILD 1: TESTBEREICH DES EHKP-TESTSYSTEMS

Das EHKP-TS dient sowohl zum Funktionstest der Btx-Vst,
genauer: der Verbundkomponente, als auch zur Durchführung
der Anschlußzulassung Externer Rechner im Btx. Es ist
installiert auf einer PDP11/44 unter dem UNIX-Betriebssystem
IS 3 (Bild 2).

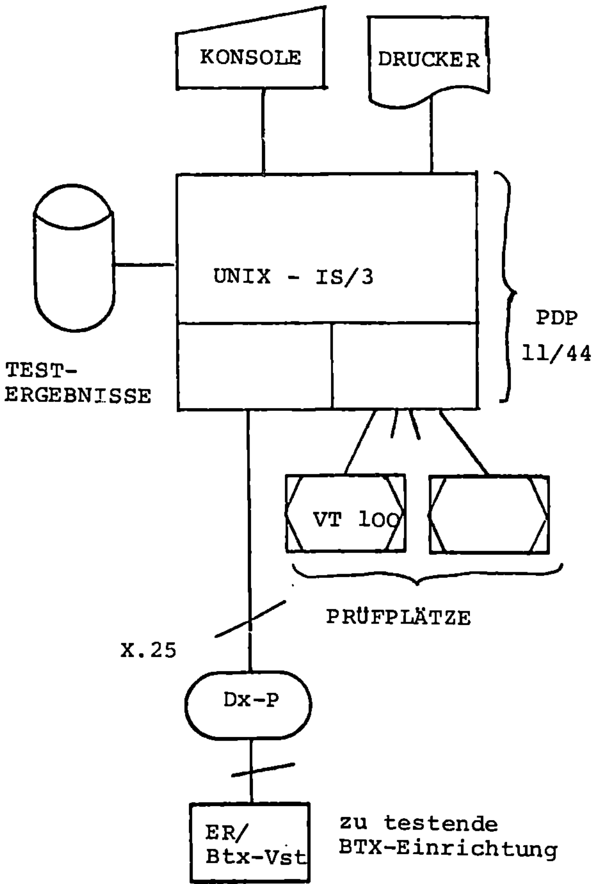

Bild 2: PETRUS - EHKP TESTSYSTEM

Das Ttx-Testsystem dient der Zulassung von Teletex-Endgeräten;
Testziel ist die Überprüfung der normgerechten Implementierung
der einschlägigen Kommunikationsprotokolle gemäß Teletex-Rahmen-
werte, nämlich HDLC, Netz- und Transportprotokoll (T.70),Session-
und Dokumentenprotokoll (T.72) sowie der Datenpräsentation
gemäß T.61. Das Ttx-Testsystem ist auf einer VAX 11/750 unter
UNIX-IS/3 installiert (Bild 3).

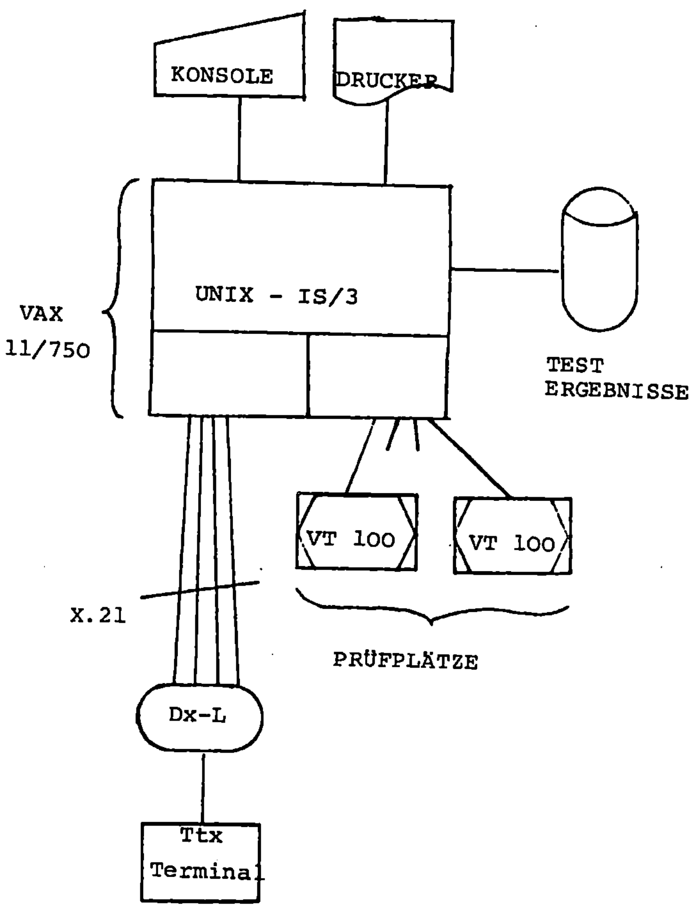

Bild 3: PETRUS -Ttx TESTSYSTEM

Das Btx-Terminal Testsystem ist für die Zulassungstests von
Bildschirmtextendgeräten gedacht. Es umfaßt den Test der Datenübertragung (Link Level Protocol), des sogenannten "Bulk Transfers" und der Datenpräsentation (Decoder Test) (Bild 4). Dieses
System ist im Aufbau begriffen; über seine Hardwarebasis ist
noch nicht endgültig entschieden.

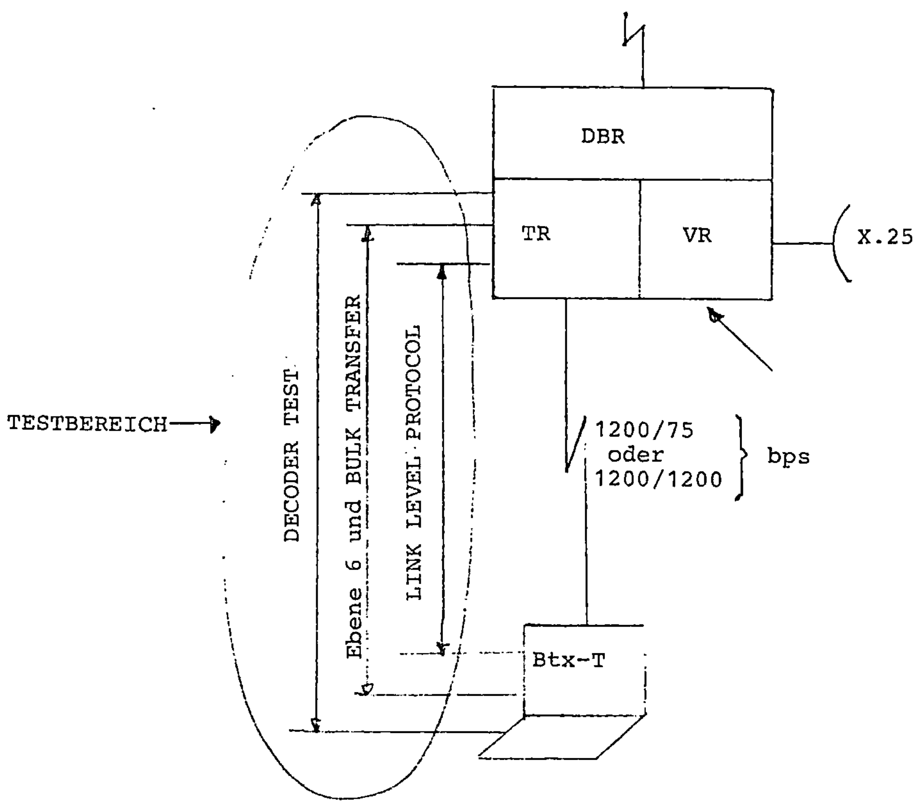

BILD 4: TESTBEREICH DES Btx-TERMINAL-TESTSYSTEMS

Der Teil "Decoder Test" ist aber bereits realisiert. Er besteht aus ca. 120 Btx-Seiten, die aus dem Btx-Netz abrufbar sind. Jede dieser Btx-Seiten beinhaltet eine bestimmte zu testende Funktion der Datenpräsentation (z.B. Scrolling, Invokierung von Zeichensätzen, Attributsteuerung, etc.)

Der Test besteht aus dem visuellen Vergleich des Soll-Bildes mit dem Ist-Bild. Eine automatische Abwicklung solcher Tests wäre - wenn überhaupt - nur mit einem unvertretbar großen Aufwand an Hard- und Software machbar.

3. Die verschiedenen Konzeptionen der Testsysteme

3.1 EHKP-Testsystem

Zwei wichtige Voraussetzungen sind hierbei gegeben, nämlich:
(Bild 5)

- jeder Externe Rechner verfügt über einen
  vordefinierten Testresponder oberhalb EHKP4

  und

- über eine sogenannte Referenzanwendung (RA)
  oberhalb EHKP6.

Btx - EXTERNER RECHNER

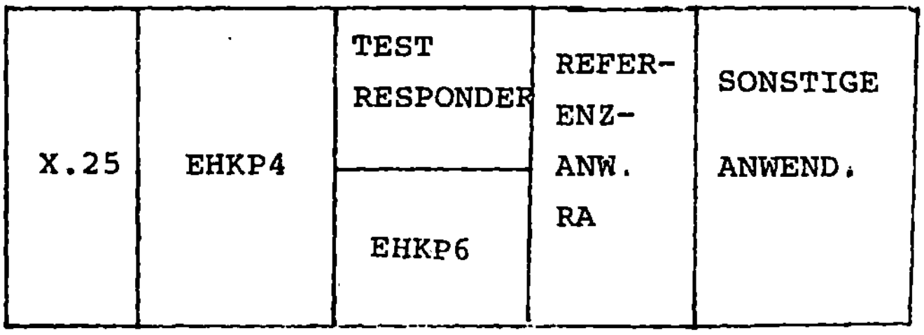

X.25	EHKP4	TEST RESPONDER ——— EHKP6	REFER- ENZ- ANW. RA	SONSTIGE ANWEND.

Bild 5: FUNKTIONSBAUSTEINE EINES Btx-ER

Diese Funktionsbausteine sind nur für die Durchführung von Zulassungstests erforderlich und nicht für den Btx-Betrieb. Der Testresponder ist dabei ein sogenannter "standardisierter" Be-

nutzer der EHKP4-Funktionsschicht. An ihn können über das Transportprotokoll "Anweisungen" gegeben werden, wie "Sende Informa-
tionseinheiten", "Breche Transportverbindung ab und baue neue
auf", etc. Mit seiner Hilfe läßt sich die EHKP4-Funktionsschicht
im Externen Rechner in zum Austesten gewuenschte Zustände
bringen.

Desgleichen ist die Referenzanwendung eine standardisierte Btx-
Anwendung, bestehend aus einigen wenigen Btx-Seiten (Begrüßungs-
seite, Datensammelseite, Abschiedsseite, etc.). Die RA ist Benutzer der EHKP6 und erlaubt das Austesten der Schicht 6- und
Schicht 7-Funktion.

3.1.1 Externer Rechner Test

Bei diesem Test emuliert der PETRUS-Rechner eine Btx-Vst
(Bild 6)

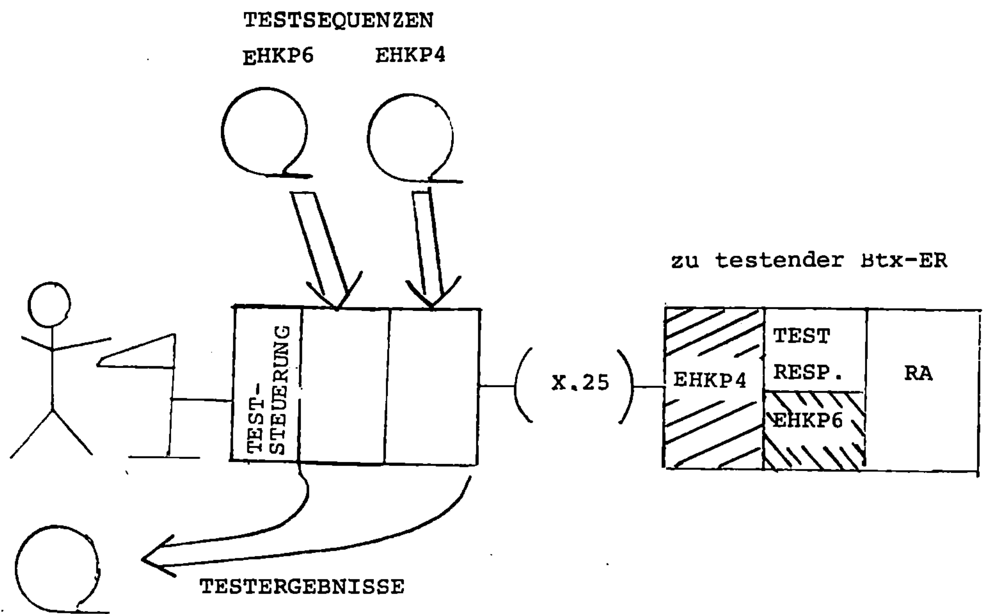

BILD 6: EXTERNER RECHNER TEST

Zum Test der EHKP4 benutzt PETRUS vordefinierte Testszenarien, das sind Folgen von EHKP4 protocol data units, die die EHKP4-Funktionsschicht - unter Zuhilfenahme des Testresponders - in einen ganz bestimmten Zustand bringen und dann dessen Reaktion beobachten. Die Testszenarien können dabei - im Sinne des EHKP4 - protokollkonforme und protokollwidrige Kommunikationsabläufe abwickeln. Die Testszenarien repräsentieren die einzelnen Testfälle; sie sind in einer Datei vordefiniert und werden vom Testingenieur am Testrechner abgerufen. Die Testszenarien sind erweiterbar. Das Ergebnis jedes Tests wird protokolliert und ist vom Testingenieur auszuwerten.

In prinzipiell der gleichen Weise wird EHKP6 getestet. Die entsprechenden Testszenarien bedienen sich einer (oder mehrerer) Referenzanwendungen, um beim ER bestimmte EHKP6-Funktionen zu veranlassen.

An den PETRUS-Testrechner kann auch ein Btx-Terminal angeschlossen werden und über dieses eine Teilmenge obiger Tests wie eine normale Btx-Anwendung getestet werden. Die Testergebnisse werden dann zum Btx-Terminal in Form von Btx-Seiten geliefert. Dies gibt dem Betreiber (Entwickler) eines Externen Rechners die Möglichkeit, einen Konfidenztest seines Externen Rechners über ein gewöhnliches Btx-Terminal zu fahren.

3.1.2 Btx-Vst-Test

Dieser Test betrifft insbesondere die EHKP4- und EHKP6-Funktionsschicht in der Verbundrechnerkomponente der Btx-Vst. Der Testrechner PETRUS, der einen Externen Rechner emuliert, kann dabei im sogenannten "manual Mode" (Bild 7) oder "automatic mode" (Bild 8) betrieben werden.

Beim "manual mode" kommuniziert der Tester zum einen mit der RA im Testrechner über ein Btx-Terminal und veranlaßt andererseits über das PETRUS-Terminal die Aussendung ganz bestimmter EHKP4- und EHKP6-Testsequenzen in Abhängigkeit der am Btx-Terminal gemachten Eingaben.

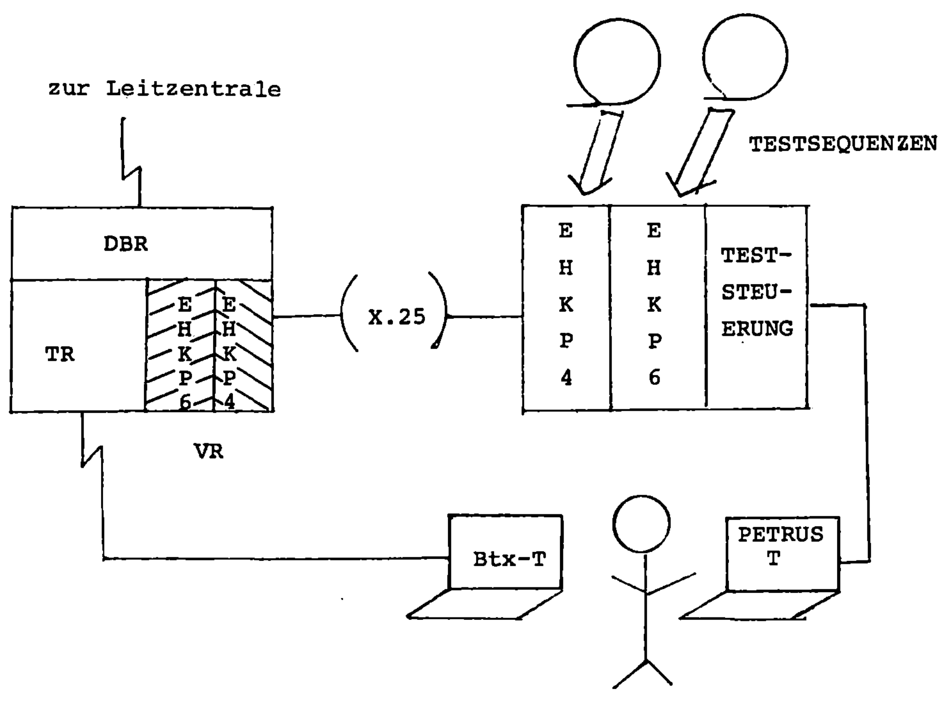

Bild 7: Btx-Vst im 'manual mode'

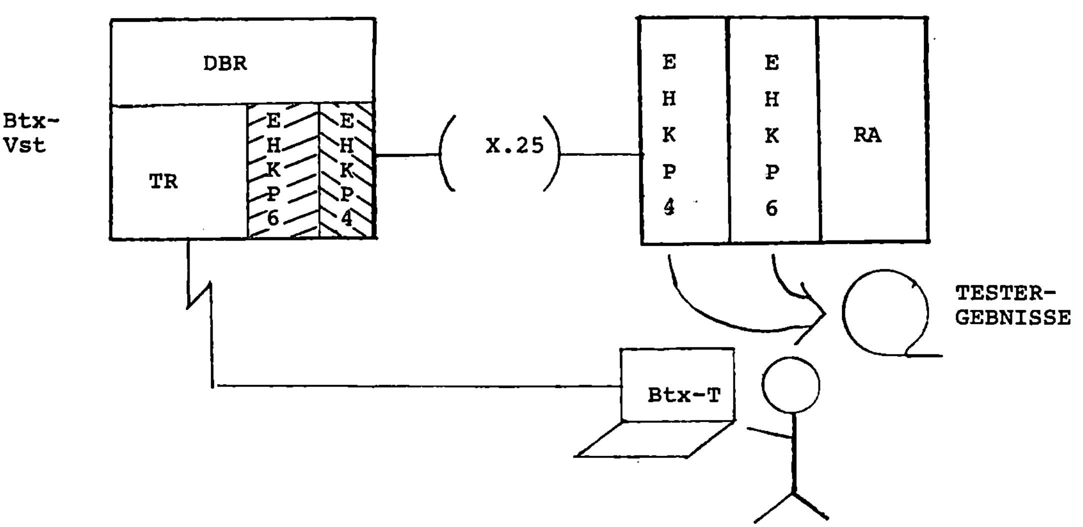

Bild 8: Btx-Vst Test im 'automatic mode'

Beim "automatic mode" kann der Tester nur über das Btx-Terminal die RA im Testrechner ansprechen. Der Testrechner antwortet wie ein normaler Externer Rechner, zeichnet aber zusätzlich alle Kommunikationsvorgänge auf. Bei diesem Test können keine bewußt protokollwidrigen Vorgänge herbeigeführt werden. Dieser Test deckt also nur den "Gut"-Fall ab.

3.2 Teletex-Testsystem

Im Gegensatz zum Btx-Dienst sind bei Teletex keine automatischen Testresponder definiert worden. Als "Testresponder" auf der Seite des Teletex-Endgerätes (Prüfling) dient der Mensch, sprich Bediener des Endgerätes, der lediglich die Aussendung und die Bereitschaft zum Empfang eines Dokumentes veranlassen kann. Infolgedessen kann eine Protokollschicht im Prüfling nicht unabhängig von den darüberliegenden Protokollschichten getestet werden, sondern jeder Test erfolgt innerhalb einer intendierten Dokumentenübertragung von oder zum Prüfling (Bild 9).

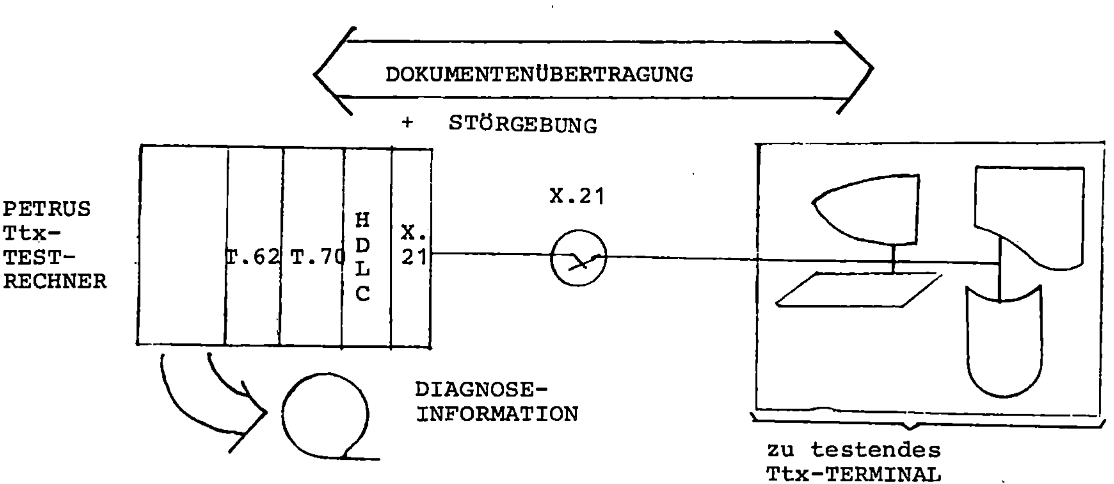

BILD 9: TESTANORDNUNG BEI PETRUS - Ttx-TEST

Das gewählte Testkonzept sieht wie folgt aus:

Der PETRUS-Rechner enthält eine vollständige Referenzimplementation der Teletex-Protokolle, kann also wie ein normales Ttx-Endgerät Dokumente senden und empfangen. Gewisse optionale Eigenschaften eines Ttx-Endgerätes sind zusätzlich für jede Testsession parametergesteuert vorgebbar. Außerdem überprüft das Testsystem die Richtigkeit aller Kommunikationsvorgänge und diagnostiziert protokollwidrige Vorkommnisse in Form von Diagnosemeldungen in einer Diagnosedatei. Die Diagnosetiefe ist variabel und z.B. einstellbar auf eine umfassende Aufzeichnung aller kommunikationsrelevanten Ereignisse. Am Ende einer Testsession versucht der Testrechner, die Diagnosedatei als Ttx-Dokument dem Pruefling zu übermitteln. Bei Mißerfolg wird es lokal ausgedruckt. Mit diesem System können bereits die "Gut"-Fälle einer Ttx-Kommunikation getestet werden, und zwar auch gesteuert vom Prüfling ohne manuelle Hilfe am Testrechner.

Um auch protokollungewöhnliche oder protokollwidrige Kommunikationsvorgänge zu testen, verfügt der Testrechner über eine Reihe von Testdriver, die während einer Dokumentenübertragung vorprogrammierte "Störungen" auf den verschiedenen Protokollebenen erzeugen und die Reakton des Prüflings darauf überwachen (Bild 10). Für jeden Testdriver ist definiert, wer Initiator der Verbindung ist, welches Dokument von wo nach wo zu übertragen ist und welche Protokollebene damit getestet wird, d.h. auf welche Ebene sich die "Störungen" beziehen. Die Testdriver können auch lokal vom Testrechner-Terminal gestartet werden. Sie werden (später) auch vom Bediener des Prüfling gestartet werden können, indem dieser bei Beginn einer Testsession ein sog. Testanforderungsdokument sendet, in welchem der zu startende Testdriver spezifiziert ist.

Mit dieser Einrichtung soll also der Bediener beim Prüfling alle verfügbaren Tests durchführen können ohne manuelle Unterstützung beim Testrechner.

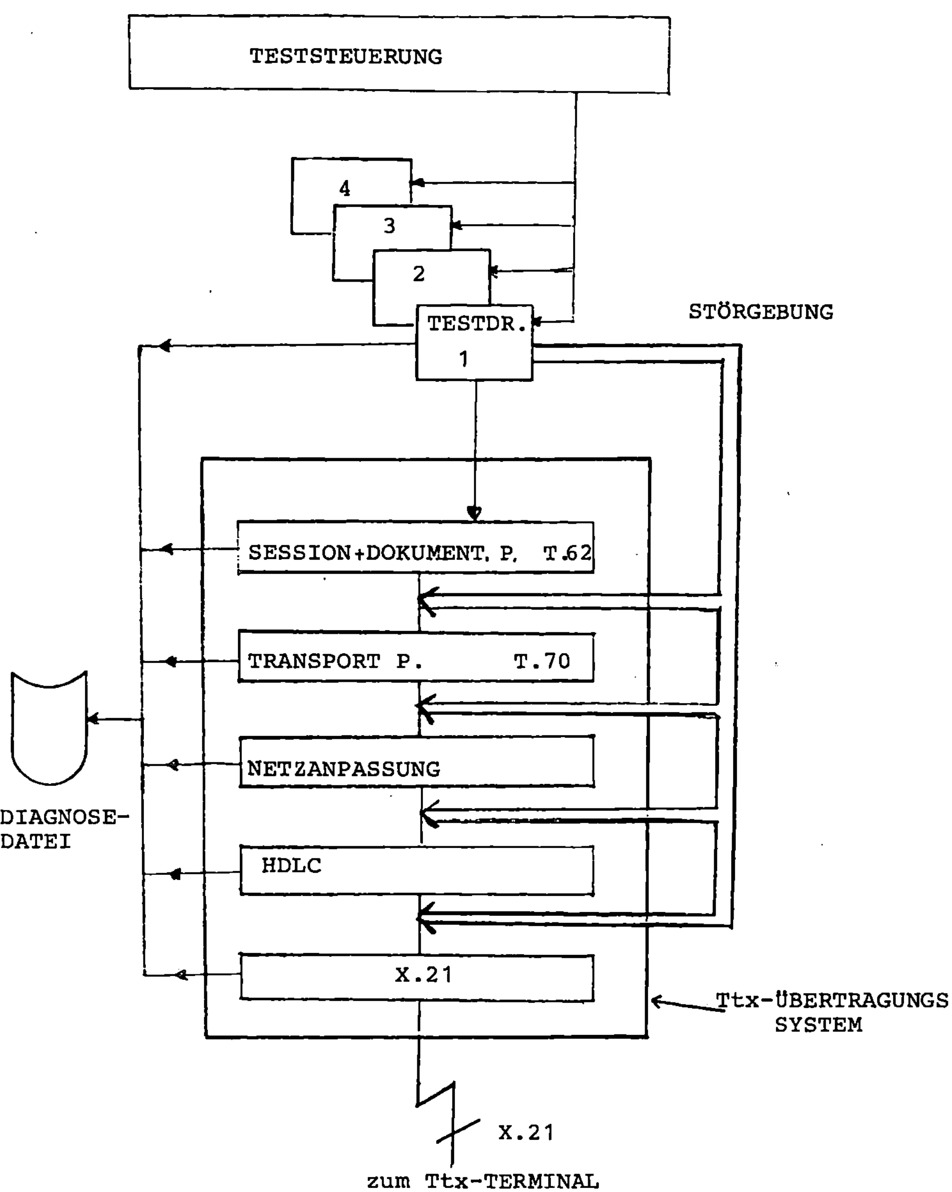

BILD 10: FUNKTIONSBAUSTEINE DES Ttx-TESTSYSTEMS

4. Erfahrungen und Weiterentwicklungen

Das PETRUS-EHKP-Testsystem hat sich sicher als sehr nützliches,
ja unerläßliches Instrument erwiesen. Dies zeigt sich auch darin,
daß die Inanspruchnahme des Testrechners weit größer war und ist
als ursprünglich geplant, so daß ein Testrechner speziell für
diese Anwendung reserviert werden mußte. Die Weiterentwicklung
besteht

o   in der Vervollständigung der Testfälle,

o   in der Verbesserung der Benutzerschnittstelle am
     Testrechner (weitgehend automatische Testabwick-
     lung ohne Interaktion)

     und

o   im Zurverfügungstellen eines (eingeschränkten)
     Zugangs zum Testsystem für den Externen Rechner-
     Betreiber über ein normales Btx-Terminal.

Das PETRUS-Teletex-Testsystem teilt sich seinen Einsatz mit
einem bereits seit längerem benutzten Testsystem, basierend
auf einem Mikrorechner (IDA 11). Der Unterschied zwischen
den beiden Testsystemen besteht darin, daß das IDA 11-
Testsystem zwar eine zeitgleiche Beobachtung der Kommunikationen mit allen Möglichkeiten der Interaktion erlaubt,
dafür aber ein ständiges Operating während einer Testsession
erfordert, während das PETRUS-Testsystem von Anfang an
konzeptionell auf automatische Testabwicklung ausgerichtet ist, mit der Möglichkeit, Tests über den normalen
Ttx- Anschluß vom Prüfling aus zu initiieren. Solange beim
PETRUS- Testsystem die Menge der Testdriver noch nicht
vollständig ist, haben beide Testsysteme sinnvolle und
nicht deckungsgleiche Einsatzgebiete. Langfristig aber
sollte sich der Einsatz des "manuellen" Testsystems auf
besonders schwer diagnostizierbare Testfälle beschränken.

Communication in a Software Development Environment

by

Herbert Weber

Universität Dortmund

Abstract

The paper presents distributed software development environment DDS (Distributed Development System). It therefore describes first the software development methodology that is enforced by the system it explains the components of the system and discusses the communication architecture of the system.

Zusammenfassung

Das vorliegende Papier beschreibt die verteilte Softwareentwicklungsumgebung DDS (Distributed Development System). Es enthält zunächst eine Beschreibung der Softwareentwicklungsmethodologie, die durch die Benutzung der verteilten Entwicklungsumgebung erzwungen wird, es beschreibt die Komponenten des Systems und diskutiert die Kommunikationsarchitektur des Systems.

1. Introduction

The DDS is a complete environment for the cost effective <u>industrial production and subsequent modification of high-quality software.</u> It supports software development professionals residing at the same or at geographically distant locations in the joint development of a software system. As such it supports the development or new software components and the construction of systems by <u>re-using prefabricated components</u> in accordance to a given Software Development Methodology (see section 2 of the paper).

In order to serve in that task the DDS provides a Software Development Expert (see section 3 of the paper) to each software development professional. The Software Development Expert guides the professional in his/her task; it supports him/her in the development of all kinds of descriptions that are needed to be produced in the software development process; it supports him/her with an information repository that contains all information about the system under development; it supports him/her with verification and validation capabilities that are needed to be applied to guarantee the correctness of the systems; it provides him/her with evaluation capabilities that support the early evaluation of the system during its development to guarantee the expected performance characteristics of the system; it provides him/her with communication capabilities to support his/her communication with other professionals working on the development of the same system or on other systems, with the information repository and the verification and validation capabilities.

2. The Software Development Methodology (SDM)

The Software Development Methodology (SDM) is aimed at providing an instrument for the constuction of <u>modular software systems.</u> Modularity is meant to provide the means for the development of selfcontained software components and for the construction of systems out of (maybe prefabricated) components in an industrial software production process. Thus, the goal to enable the modular construction of software, is the driving force in the specification of all methods that pertain to the software development methodology.

The software development methodology is based on a number of interrelated models for the construction of complex software. The most fundamental principle, the software development methodology is based on, is called Software Model (SM).

The other models introduced below all adhere to that fundamental principle. The software model and the other models together constitute the basis for the software development methodology. All of the models support in their particular way the construction of modular systems. The interrelationship between the models within the software development methodology may be depicted as follows:

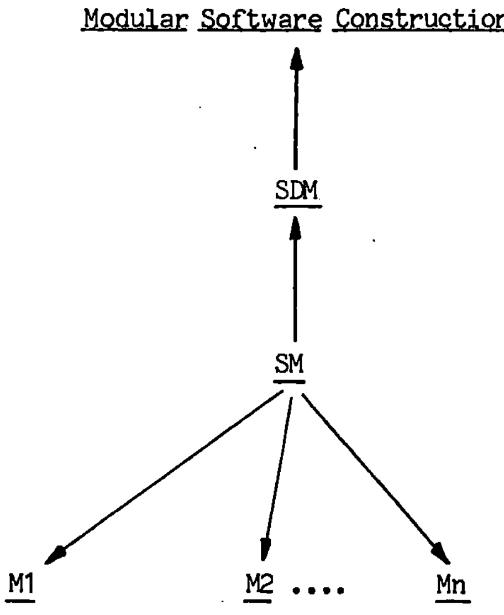

All the models have been developed for the reasons explained below.

2.1 Software Model (SM)

The Software Model is applied in order to achieve a most suitable structuring of software. An <u>object-oriented</u> <u>software</u> <u>structuring</u> principles serves as that fundamental software model. Object-orientation refers to the construction of selfcontained software components through the encapsulation of data objects of a certain type. It therefore embodies good software engineering practices like

- data and procedural abstraction
- hierarchical composition and decomposition of components
- information hiding
- decision postponement.

For our purposes only object-oriented software structuring provides the means for a truly modular construction of software.

On the basis of that principle we expect to achieve a higher degree

- of reliability, robustness and fault tolerance,
- of software modifiability
  and
- of simplicity and comprehensibility.

It, thus, provides the means to achieve a higher quality of the software products and a higher productivity in the development of software.

We expect the object-oriented structuring principle to be applied not only in the coding but on all stages of the software development process from early requirements descriptions through design and implementation. Modularization may therefore be achieved not only on the code level but also for requirements and designs of complex software systems. This general object orientation is the basis for a

- software life model
  (described in section 2.2)
- software description model
  (described in section 2.3)
- verification and validation model
  (described in section 2.4)

- project management model
  (described in section 2.5)
- evaluation model
  (described in section 2.6)
- re-usability model
  (described in section 2.7)

The justification for an object-oriented software model and the advantages of its rigid use are explained in detail in a separate paper.

2.2 Software Life Model (SLM)

The Software Life Model is applied in order to achieve a most suitable structuring of the software evolution process that streches from the first ideas about a software system through its development and use to its final abandoning. We do not call the model a life cycle model to indicate the principal difference between conventional life cycle models that support the structuring of the software development and maintenance process in consecutive phases and our model that supports the <u>structuring of an evolution process.</u> In our view evolution takes place in all phases of development and use of a software system. A justification for the model and a detailed description of its advantages can be found in the aforementioned separate paper.

2.3 Software Description Model (SDM)

The Software Description Model is applied in order to achieve a most suitable description of software and its properties. As such the model supports first the development of a complete description through its <u>incremental completion.</u> The description starts out with a simple description of only some basic characteristics of the system under development and terminates with a detailed description of all its static and dynamic properties. The incremental completion approach allows to "devide and conquer" in order to enable the handling of the complexity of software descriptions. It also supports the modifyability of descriptions.

Secondly, the Software Description Model provides only one <u>uniform notation</u> to be applied in the description of software. Rather than providing different notations to describe different characteristics of software as in conventional software descriptions our SDM is aimed at avoiding the difficulties that arise in the use of multiple notations.

Instead of providing different notations to express different properties and characteristics our software description model allows to describe systems uniformly with different degrees of formality. Language constructs with intended identical semantics can be expressed either informally (i. e. in natural language) or formally by using a pseudo-code (with a precisely defined syntax) or by using a formal specification language (with both precisely defined syntax and semantics). Informal descriptions are meant to be produced in the early stages of the development with only little precise knowledge available about the components under development. They will be turned into formal descriptions in pseudo-code and/or formal specifications later on for maybe only some of the components of the system. Informal descriptions are not meant to be executable, whereas pseudo-code description shall be executable. Formal specifications will be used in formal verifications of the correctness of the software components (see verification and validation model in section 2.4).

Attempts to specify formally and also execute the resulting formal specification will be supported in the software description model.

The first two descriptions will be produced for all components of a system whereas the third one may be produced where necessary. All three descriptions serve as the only kind of documentation that will be produced for a system. This documentation will be necessarily developed during the software development process and not in the aftermath. It is, thus, a realistic documentation for the developed system.

The justification for providing a software description model with the aforementioned principal characteristics can once again be found in the aforementioned paper.

2.4 Validation and Verification Model (VVM)

The Validation and Verification Model is applied to systematize

- the validation of a maybe only partially developed system description against initially given requirements and constraints; as such it supports the compatibility of the system description with preformulated requirements

- the validation of a partially developed system description
  against its description at previous refinement levels; as
  such it supports the consistency in incremental completions
  of the system description

- the validation of changed system description against previous versions of that descriptions; as such it supports
  consistency in evolutionary modifications of the system and
  its respective descriptions

- the verification of the compatibility of independently
  developed components; as such it supports the consistency of
  the distributedly developed system.

The Verification Model is also used to verify the correctness of a completed
software system description or of a completed software component description.
The uniform notation for software descriptions has therefore been chosen to
also allow a mathematically sound specification (see our explanations about the
different degrees of formality of the choosen uniform notation in section 2.3)
whose correctness can be formally verified by a suitable verification system.

2.5 Project Management Model (PMM)

The Project Management Model is applied to ensure a consistent and up-to-date
management of a software development project. The model is aimed at
guaranteeing the supply of all necessary information to all parties
participating in the project and it is aimed at providing the necessary
supervisory mechanisms for the management of the work conducted by the involved
parties. As such the model enforces

- the recording of the history of the project
- the structuring of the project through mile stones and check points
- the application of an information distribution mechanism to ensure the
  mutual up-to-date information of all participating parties on system parts
  with "external" relationships, to guarantee the consistency and acceptance
  of those parts on a global level
- the application of voting mechanisms to ensure a mutual agreement between
  the parties involved in the project on certain development steps of common
  interest.

In order to fulfill these functions the PMM correlates the respective entities of the Software Life Model, the Software Description Model, the Validation and Verification Model and the Evaluation Model. It also ensures an appropriate allocation of recources and an appropriate co-ordination of the work.

2.6 Evaluation Model (EM)

The Evaluation Model provides a prototyping strategy and technique. As such it is meant to support the early evaluation of a system by its expected users. Prototyping is thought of as an early execution of system description that have been developed only until a certain refinement level. The expected user performs tests of these still "incomplete" systems with their execution. The incomplete description may then be corrected and refined.

We thus avoid with this approach the writing of "throw away code". An appropriate system that interprets those "incomplete" descriptions of systems and simulates their execution has been developed and its use in practical prototyping is under way (see evaluation model section 2.6).

The Evaluation Model is also applied to evaluate the performance of alternative designs of a system. It is meant to be applicable during the entire design process from early stages with only a very coarse design until the end with an ultimately refined design. In order to serve in that capacity the model guides in early executions of incompletely developed design descriptions that describe systems as they appear after their development until a certain degree of refinement.

Therefore the EM is meant to be the same here as the one that is used in order to enable early prototyping.

In addition to enabling the validation of a system through prototypes the EM also enables with performance estimates and/or measures the analytic determination of the respective "execution costs" of alternative designs or their comparisons in simulation experiments. A system for the evaluation of alternative designs has been implemented and used successfully in the evaluation of special purpose hardware/software system for the management of data bases (i. e. multicomputer data base machines).

2.7 Re-Usability Model (RUM)

The Re-Usability Model is applied to systematize the construction of software systems by re-using prefabricated components. Software components need to fulfill certain prerequisited in order to be re-usable in a software environment for which they have not been developed originally.

Therefore, provisions have to be taken during their design to fully specify all the expected characteristics of the possible new environments the components may be used in later on. The RUM is therefore based on the modularity principle introduced in section 2. In order for modules to be re-usable we need to enhance them with a specification of all the expected characteristics of possible new environments the modules may be used in later on. A module concept that enables to specify those additional features is based on the well known concept of parameterized abstract data types. The resulting module concept has been called strictly modular systems concept. Early experiments in the construction of strictly modular system have already been conducted.

3. The Software Development Expert (SDE)

The Software Development Expert is a uniform tool system that supports one or a number of maybe geographically separated software development professionals in a common software development task. It consists logically of a number of functional units that either support in the construction, evaluation or modification of complex software system. The "expert support" provided by the SDE to the software development professional is based on an "intelligent" data base and a number of guiding functions embodied in the SDE. The functional units of the SDE maybe physically distributed over a computer communication network in a way most suitable to the application environment. The functional units of the SDE are

(I)   an _Interface Support Unit_ that provides a uniform interface to the user of the system, e. g. the system development professional. It embodies a number of capabilities to guide the user in the development of descriptions of software modules in accordance to the software description model (SDM) as described in section 2.3.

(II)   A _Data Base Support Unit_ that serves as the host for the information repository that contains all necessary information about the system under

development. The information repository is meant to contain information that is globally available to all parties involved in the software development process and privat information that is only locally available for one Software Development Expert or just one professional working on a SDE. The Data Base Support Unit, thus, provides means to differentiate between different types of information and provides appropriate data base management functions for the respective type of information.
The principal types of information that are kept in the information repository and must be administered by the Data Base Support Unit are

- the various forms of module descriptions
- all management information that is needed to perform version control and procurement, project control and life model support and control
- mail-information for common use.

For the administration of this information the Data Base Support Unit provides appropriate retrieval and modification operations.

(III) A **Validation and Verification Support Unit** that will be used to verify the consistency of a developed description of a software module. It thus embodies the verification model described in section 2.4.

A software module description that has been produced with the support of the Interface Support Unit will be forwarded to the Validation and Verification Support Unit. It will here be checked for its internal consistency (i. e. validation against given consistency constraints and against previous versions) and for its external consistency (i. e. verification against restriction imposed by other modules of the system). On the basis of these checks the Validation and Verification Support Unit decides on the admittance of the module description under development to the global information repository.

(IV) A **Software Life Model Support Unit** that serves as the controller of the progress of the project in accordance to the software life model as described in section 2.2.

As such it imposes a discipline on all parties involved in the project for the organization of all activities in the conduct of a project: it enforces an order on all activities related to the project, enables possible parallel activities within a project and controls a proper sequencing of all activities conducted by all parties involved in the project.

The Software Life Model Support Unit also provides a bookkeeping facility that enforces the recording of all information about the progress made in the development process for its use by the Project Management Unit (see section 3 (VI)).

(V) An _Evaluation_ _Support_ _Unit_ that serves in the early execution of only partially developed system description for their evaluation by the expected users of the system in accordance to the prototyping strategy as described in section 2.6.

The application of the Evaluation Support Unit presupposes the description of the system under development in pseudo-code (see the software description model explained in section 2.3). It does, however, only presuppose the completion of the system description up to a certain level of refinement in the software development process. Early prototyping may in fact start after the pseudo-code description of·the top-most module of the whole system.

The Evaluation Support Unit also serves in the determination of the performance characteristics of the system under development. It is especially valuable in the comparison of those characteristics for alternative designs. The evaluation is based on measures or estimates for the performance of system components. It determines then the overall performance characteristics of the entire system on the basis of the performance characteristics of the components. The Evaluation Support Unit may be applied for the determination of the system performance from the beginning of a project (for a very incomplete development of the system delivering only very coarse information about the system performance) through arbitrary many refinement levels (delivering increasingly refined results) to the end of the development process for the ultimate refinement of the system under development. It thus supports the collateral design and performance evaluation refinement over the entire software development process.

(VI) A _Project_ _Management_ _Support_ _Unit_ that supports and guides management in its supervisory function for a software development project. As such the Project Management Support Unit serves as an interface system for the management to insert manpower and resource allocation plans and to insert time frames with milestones and deadlines. The Unit guides the management in that activities and assures that all this will be fixed in accordance to the Software Life Model.

The Unit also provides the interface for obtaining information about the status of the project, the status of work reached for certain components and the status of certification reached for the various system components.

The Project Management Support Unit also serves as a "request repository" containing also requests for management information or management decision on certain problems related to the development task. Conversely, it also supports the selective information distribution that provides information to those who need it. The functions embodied in the unit also enable the initiation of a voting mechanism between project participant for joint decision making.

4. The SDE Communication Architecture

The functional units of the Software Development Expert may all be located centrally at a single host or may be arbitrarily distributed to support the joint development of software by geographically separated professionals. The latter problem will be addressed with the development of a suitable communication architecture that enables the communication of all professionals with all functional units for their respective task.

A communication architecture does not yet refer to a physical distribution of the software development expert to a real computer communication network. The communication architecture identifies only logical communications needs and services between the different functional units of the Software Development Expert. The logical communication services are needed to be mapped into the (physical) services provided on a real computer communication network. An appropriate physical distribution depends on the particular software development setting or even on a particular project. The DDS is in fact thought of as an "Open System Development Environment" that may be adjusted to the particular needs and circumstances. The logical services proposed in the sequel will correspond to those proposed as standard application services in the ISO/OSI framework to enable their simple mapping into the physical services of a real open system. The following lists the needed communication services in the proposed communication architecture:

(I)   User - Interface Support Unit:

This communication enables the editing of software description of any kind and the initialization of the communication between the Interface Support Unit and all other functional Units. It also serves in the user directed monitoring of the software development activities.

The communications between a user (e. g. software development professional) and the Interface Support Unit via a terminal will be based on a local or remote terminal access service (e. g. X3, X28, X29).

(II) Interface Support Unit - Data Base Support Unit:

This communication enables the transfer of edited software component description into/from the information repository independent of whether it is private to one software developer or common to the entire community of developers.

The communication between the Interface Support Unit and the Data Base Support Unit will be based on a file transfer and access service (e. g. FTAM).

Another type of communication between the Interface Support Unit and the Data Base Support is needed to enable the transfer of textual unformatted information for the monitoring of user directed activities and for the co-ordination of activities between different users of the DDS.

This kind of communication between the Interface Support Unit and the Data Base Support Unit will be based on a message handling service (e. g. MAIL).

(III) Interface Support Unit - Validation and Verification Support Unit:

This communication enables the transfer of newly edited module descriptions of any kind for their validation and verification by the Validation and Verification Support Unit.

This communication will be based on a file transfer and access service (e. g. FTAM).

(IV)   Interface Support Unit - Software Life Model Support Unit:

This communication enables the transfer of monitoring information for the initialization and control of activities in the development of a software component and of the entire software system.

This communication will be based on a message handling service (e. g. MAIL).

(V)   Interface Support Unit - Evaluation Support Unit:

This communication enables the transfer of newly edited module descriptions in Pseudo-Code from/to the Interface Support Unit to/from the Evaluation Support Unit for their evaluation in early prototyping and for the evaluation of their performance.

This communication will be based on a file transfer and access service (e. g. FTAM).

(VI)   Interface Support Unit - Project Management Support Unit:

This communication enables the transfer of requests for management information and decision from the Interface Support Unit to the Project Management Support Unit as well as the transfer of requests and responses from the Project Management Support Unit to the Interface Support Unit.

This communication will be based on a message handling service (e. g. MAIL).

The communication between the Interface Support Unit and the other units may be depicted as in Fig. 1:

User
Interface Support Unit
Project Management Support Unit
Data Base Support Unit
Validation and Verification Support Unit
Software Life Model Support Unit
Evaluation Support Unit

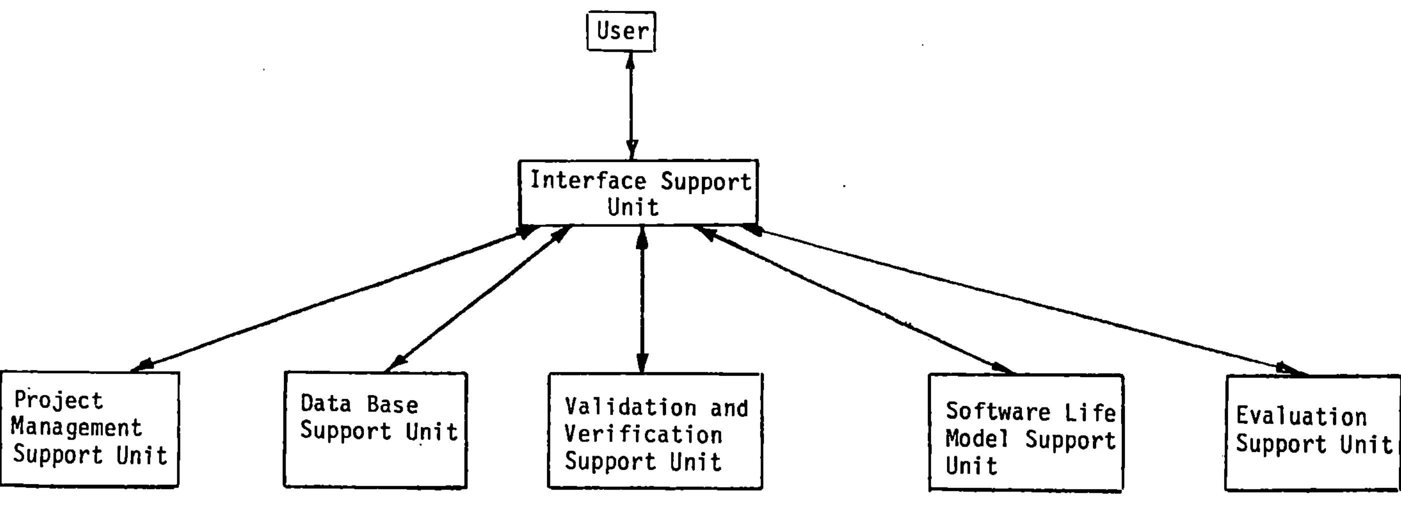

Fig. 1

A number of other services will be needed to enable the communication between some other functional units of the SDE.

(VII)   Data Base Support Unit - Validation and Verification Support Unit:

This communication enables the transfer of any kind of descriptions of software system components (e. g. modules and module compositions) between the Data Base Support Unit and the Validation and Verification Support Unit for the validation and verification of those components.

The communication will be based on a file transfer and access service (e. g. FTAM).

(VIII) Data Base Support Unit - Evaluation Support Unit:

This communication enables the transfer of pseudo-code descriptions of software system components for their evaluation.

The communication will be based on a file transfer and access service (e. g. FTAM).

(IX)   Data Base Support Unit - Project Management Support Unit:

This communication enables the transfer of unformatted textual information for the supervision of the project by the project management.

The communication will be based on a message handling service (e. g. MAIL).

(X)   Data Base Support Unit - Life Model Support Unit:

This communication enables the transfer of bookkeeping information for its storage in the Data Base Support Unit and for its later use by the users and managers.

The communication will be based on a message handling service (e. g. MAIL).

A complete picture of the Communications Architecture for the SDE takes then the following form as in Fig. 2.

For their allocation to hardware components of the DDS the functional units of the Software Development Expert

- may be assigned in total to one computer system
- may be split into different sub-units that may be assigned to different computers
- may be duplicated an arbitrary number of times with each incarnation of the unit assigned to a different computer.

(XI)   In a typical setting the Interface Support Unit will be duplicated an arbitrary number of times with each incarnation provided to a different group of software development professionals. Each of the incarnations may be allocated to a different computer system.

A configuration of the SDE that supports the duplication of the Interface Support Unit requires some additional communication connection between the different incarnations of the Interface Support Unit. The communication service that may be appropriately used to enable this communication between different incarnations of the Interface Support Unit is a message handling service (e. g. MAIL) refered to as service number XI in the rest of the paper.

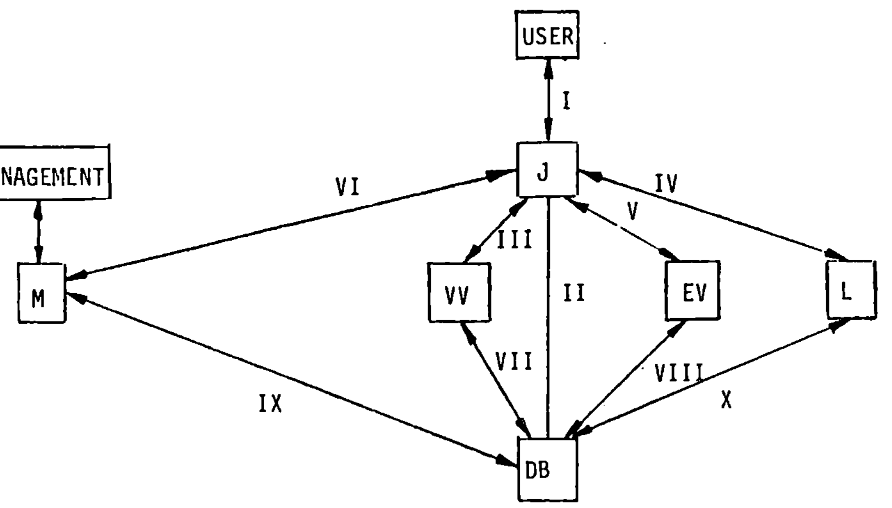

J: Interface Support Unit

VV: Validation and Verification Support Unit

EV: Evaluation Support Unit

L: Life Cycle Support Unit

DB: Data Base Support Unit

M: Management Support Unit

The number I to IX identify the communication mechanism described in the respective paragraph above.

Fig. 2

(II') The Data Base Support Unit may be first split into an "interface part" and a "common part". The interface part may be duplicated an arbitrary number of times and each incarnation of that part may be assigned to a different computer. The interface part of the Data Base Support Unit provides then the local information repository available to all professionals using the same Interface Support Unit. The remaining common part of the Data Base Support Unit provides the common information repository for all professionals working on the same task. This kind of a configuration also requires then additional communication connections among all the interface parts and the common part of the Data Base Support Unit. The services that are needed to enable that communication may be of quite a different nature depending on the requirements:

- it may be a MAIL and FTAM service if no global consistency requirements are needed to be maintained between the interface parts and the common part of the data base,

- it must be a distributed data base system service if global consistency constraints are needed to be maintained.

Since the Data Base Support Unit in the DDS is ultimately meant to guarantee the global consistency of all data in the data base the communication service that will be finally needed is a distributed data base system. For the moment we will call that service II' for the rest of this paper.

(III') The Validation and Verification Support Unit may be similarily split into an "interface part" and a "common part". The interface part may be duplicated an arbitrary number of times and each incarnation allocated to a different computer. It provides then services in the validation and verification that can be accomblished locally without any reference to the total systems knowledge (e. g. simple syntax driven consistency checks for the modules that are under development at that site). The common part of the Validation and Verification Support Unit provides then the common validation and verification services needed to validate and verify the entire system under development. The additional FTAM service needed will be called III' in the rest of the paper.

(IV') The Life Model Support Unit may be equally split and duplicated for its "interface part" to guide in the local development of modules of the entire system. The common part in contrast guides in the overall development by all parties involved in the development project.

For a configuration of the SDE that supports a number of incarnations of the interface part and the common part of the Life Model Support Unit additional communication connections are needed among the incarnations of the interface part and the common part. The additionally needed MAIL service will be called IV'.

(V'+VI') The Evaluation Support Unit and the Management Support Unit may be allocated in total to their respective computer system. Their connection to each of the incarnations of the interface part of the above functional units must be provided in addition to the communicaton connections provided so far. The required additional MAIL and FTAM services will be called VI' and V' respectively. The communications architecture of the SDE that full- fills all the communication needs identified above may then be depicted as shown in Fig. 3).

A complete picture of the Communications Architecture for the SDE takes then the following form as in Fig. 2.

For their allocation to hardware components of the DDS the functional units of the Software Development Expert

- may be assigned in total to one computer system
- may be split into different sub-units that may be assigned to different computers
- may be duplicated an arbitrary number of times with each incarnation of the unit assigned to a different computer.

(XI)   In a typical setting the Interface Support Unit will be duplicated an arbitrary number of times with each incarnation provided to a different group of software development professionals. Each of the incarnations may be allocated to a different computer system.
A configuration of the SDE that supports the duplication of the Interface Support Unit requires some additional communication connection between the different incarnations of the Interface Support Unit. The communication service that may be appropriately used to enable this communication between different incarnations of the Interface Support Unit is a message handling service (e. g. MAIL) refered to as service number XI in the rest of the paper.

5. The DDS Architecture

The DDS Architecture incorporates three major types of components:

        (I)   workstations

        (II) servers

        (III) communication networks.

The first two types of components are meant to host the functional units or subunits (e.g. interface part or common part) of the Software Development Expert. The third type is meant to provide the communication connections and the needed services for the communication between the units and subunits of the SDE.

The workstations serve as the front-end hardware systems to the DDS. They may be either single-user or multi-user systems that all support an appropriate operating system - preferabely UNIX.

They are meant to host - depending on the special setting in the application environment - the "Interface parts" of the Software Development Expert like

- the interface Support Unit

- the interface part of the Data Base Support Unit

- the interface part of the Validation and Verification Support Unit

- the interface part of the Software Life Model Support Unit.

A workstation in a typical setting of a usual application environment may serve in the following manner:

- one or a number of software development professional work with the Interface Support Unit in the editing of module descriptions; they generate management information in that task that will be temporarily stored at the workstation

(II') The Data Base Support Unit may be first split into an "interface part" and a "common part". The interface part may be duplicated an arbitrary number of times and each incarnation of that part may be assigned to a different computer. The interface part of the Data Base Support Unit provides then the local information repository available to all professionals using the same Interface Support Unit. The remaining common part of the Data Base Support Unit provides the common information repository for all professionals working on the same task. This kind of a configuration also requires then additional communication connections among all the interface parts and the common part of the Data Base Support Unit. The services that are needed to enable that communication may be of quite a different nature depending on the requirements:

- it may be a MAIL and FTAM service if no global consistency requirements are needed to be maintained between the interface parts and the common part of the data base,

- it must be a distributed data base system service if global consistency constraints are needed to be maintained.

Since the Data Base Support Unit in the DDS is ultimately meant to guarantee the global consistency of all data in the data base the communication service that will be finally needed is a distributed data base system. For the moment we will call that service II' for the rest of this paper.

(III') The Validation and Verification Support Unit may be similarily split into an "interface part" and a "common part". The interface part may be duplicated an arbitrary number of times and each incarnation allocated to a different computer. It provides then services in the validation and verification that can be accomblished locally without any reference to the total systems knowledge (e. g. simple syntax driven consistency checks for the modules that are under development at that site). The common part of the Validation and Verification Support Unit provides then the common validation and verification services needed to validate and verify the entire system under development. The additional FTAM service needed will be called III' in the rest of the paper.

(IV') The Life Model Support Unit may be equally split and duplicated for its
"interface part" to guide in the local development of modules of the
entire system. The common part in contrast guides in the overall
development by all parties involved in the development project.

For a configuration of the SDE that supports a number of incarnations of
the interface part and the common part of the Life Model Support Unit
additional communication connections are needed among the incarnations
of the interface part and the common part. The additionally needed MAIL
service will be called IV'.

(V'+VI') The Evaluation Support Unit and the Management Support Unit may be
allocated in total to their respective computer system. Their
connection to each of the incarnations of the interface part of the
above functional units must be provided in addition to the
communicaton connections provided so far. The required additional MAIL
and FTAM services will be called VI' and V' respectively.
The communications architecture of the SDE that full-
fills all the communication needs identified above may
then be depicted as shown in Fig. 3).

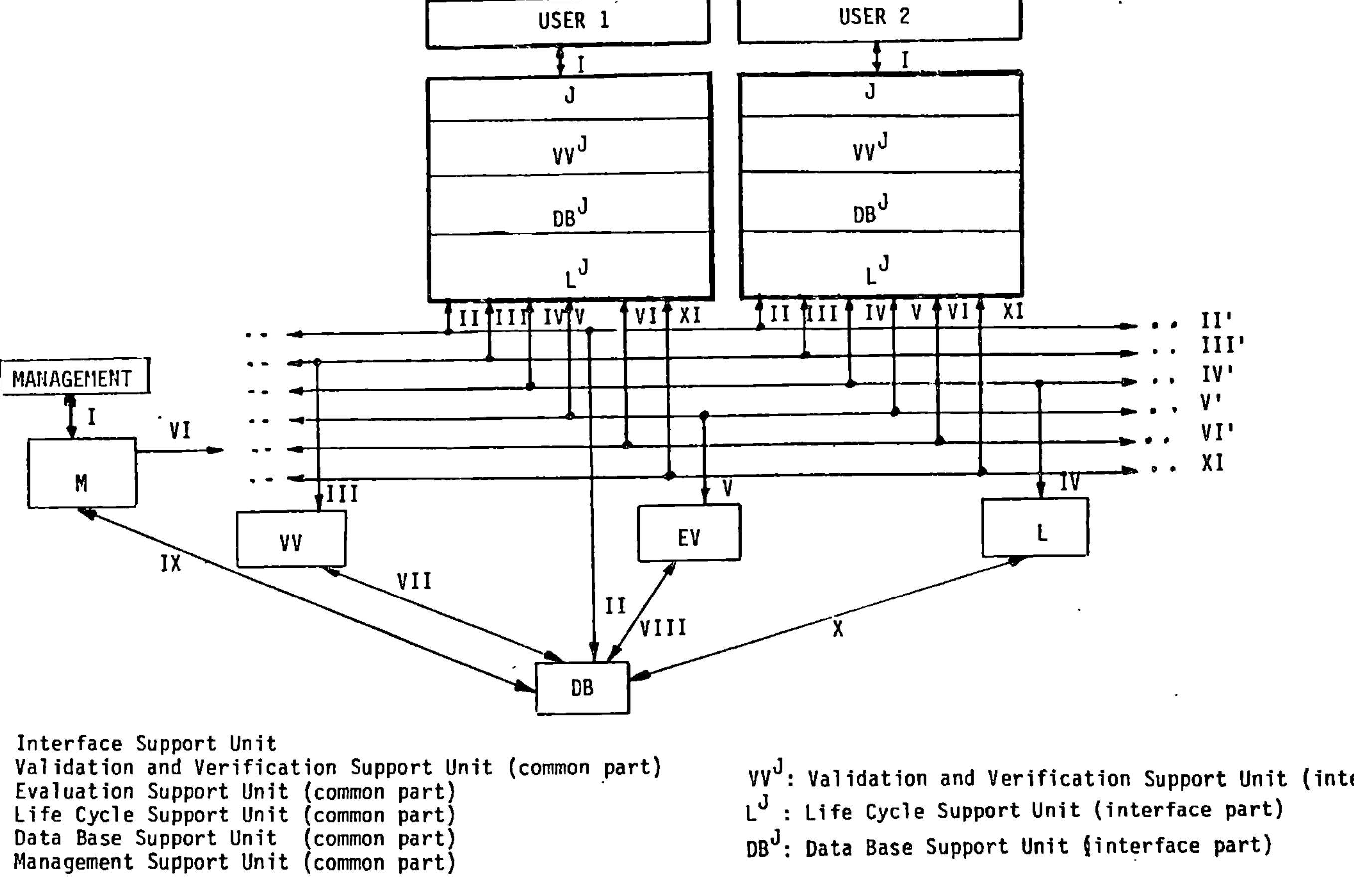

J: Interface Support Unit
VV: Validation and Verification Support Unit (common part)
EV: Evaluation Support Unit (common part)
L: Life Cycle Support Unit (common part)
DB: Data Base Support Unit (common part)
M: Management Support Unit (common part)

VV^J: Validation and Verification Support Unit (interface part)
L^J : Life Cycle Support Unit (interface part)
DB^J: Data Base Support Unit (interface part)

The number I to XI and II' to IV' identify the communication mechanism described in the respective paragraph above.

Fig. 3

5. The DDS Architecture

The DDS Architecture incorporates three major types of components:

(I) workstations

(II) servers

(III) communication networks.

The first two types of components are meant to host the functional units or subunits (e.g. interface part or common part) of the Software Development Expert. The third type is meant to provide the communication connections and the needed services for the communication between the units and subunits of the SDE.

The workstations serve as the front-end hardware systems to the DDS. They may be either single-user or multi-user systems that all support an appropriate operating system - preferabely UNIX.

They are meant to host - depending on the special setting in the application environment - the "Interface parts" of the Software Development Expert like

- the interface Support Unit

- the interface part of the Data Base Support Unit

- the interface part of the Validation and Verification Support Unit

- the interface part of the Software Life Model Support Unit.

A workstation in a typical setting of a usual application environment may serve in the following manner:

- one or a number of software development professional work with the Interface Support Unit in the editing of module descriptions; they generate management information in that task that will be temporarily stored at the workstation

- the interface part of the Software Life Model Support Unit guides each professional individually in his/her development task

- the interface part of the Validation and Verification Support Unit enables the simultaneous validation and verification of the description under development by each professional allocated to that workstation

- the interface part of the Data Base Support Unit allows a temporary storage and administration of partially or completely developed module descriptions produced by the software development professionals at that workstation.

Servers are computer systems that are dedicated to support one functional unit of the Software Development Expert. In a typical setting one may find

- a data base server

- a validation and verification server

- an evaluation server.

They are primarily necessary to host those functional units that are jointly used by all software development professionals and which can not be supported entirely on a workstation.

The Data Base Server is dedicated to host the common part of the Data Base Support Unit. It is meant to provide the computational capabilities for an efficient administration of the common information repository.

The Validation and Verification Server is dedicated to host the common part of the Validation and Verification Support Unit and is meant to provide sufficient computational capabilities to enable the efficient validation and verification of the total system under development.

The Evaluation Server is dedicated to host the Evaluation Support Unit and is meant to provide sufficient computational capabilities to enable the efficient evaluation of the system under development.

The communication networks that provide the facilities for the interconnection of the various functional units and subunits may be themselves be of very

different architectures.
One may think of local Area Networks - that only interconnect some functional units - interconnected via gateways to Wide Area Networks etc. The many opportunities for the connection that may be chosen result in a great variety of Distributed Development Systems with basically the same functionality but taylored to the special communication needs in a special application setting. It is, however, assumed that all computer networks supporting the DDS should provide the standard services identified above.

One particular type of network that may be very appropriately used to support the DDS is the ESPRIT INFORMATION EXCHANGE SYSTEM.

6. Conclusion

The paper contains a very coarse description of a new type of software development environment. It supports uniform specifications and renders itself on the basis of this uniform specification technique into a monolithic environment. The environment is composed of a number of support units some of them are meant to support users with the aid of expert knowledge maintained in those support units. All support units are interconnected in a taylored communication network that supports standard communication services. The paper presents work in progress. The described features of the DDS are, therefore, subject to changes.

7. Acknowledgement

This paper has been outlined in a number of brainstorming sessions by the following group of people:

Radu Popescu-Zeletin (Hahn-Meitner-Institut)

Silke Seehusen (University of Bremen)

Dieter Schuh (University of Bremen)

Herbert Weber (University of Dortmund)

Some helpful comments have been provided by Fritz Voigt (Hahn-Meitner-Institut).

Client/Server Model of Distributed Processing

Liba Svobodova
IBM Zurich Research Laboratory
Rüschlikon, Switzerland

Abstract: The client/server model of distributed computing is a structuring concept which can be identified in many different distributed systems. Clients and servers are active modules communicating through messages. Systems composed of server machines and personal workstations as their clients exemplify this concept. The paper focuses mainly on two issues: server design and communication protocols. File servers are used as an example in the server design overview. The communication issues are discussed in the framework of the OSI Reference Model and protocols.

1. INTRODUCTION

The notion of a client/server model of distributed computing gained widespread use in the context of local-area networks where separate server machines are used to support work performed on workstations. Workstations are rather powerful personal computers in their own right, but they need to communicate, and, above all, they still do need to share resources, specifically:

- information
- complex services
- expensive hardware

The individual workstations should retain a high degree of autonomy, that is, it should be possible for the owner to insert and remove his workstation from the network, without affecting other users. In order to achieve such an autonomy, the shared resources should not be controlled by the workstations; rather, the shared resources should be managed by highly-available, carefully-administered machines, which are usually called servers. Such a combination of client and server machines has been claimed to be the most suitable scenario for the office environment. Many research efforts of this sort have been conducted in the past decade. The projects at Xerox PARC [SWIN79, STUR80, MITC82a, BIRR82, SCHR84, BROW84] and the University of Cambridge [NEED82, DION80, CRAF83] are particularly notable. Two large projects co-sponsored by IBM have been started in the recent past to deploy this sort of system in a university environment: an ITC (Information Technology Center) project at Carnegie-Mellon University [SATY84], and project Athena at MIT [BALK84]. Each of these systems is intended to support in a few years thousands of personal computers on multiple connected local-area networks.

An alternative scenario for distributed computing, sometimes called "integrated distributed computing", assumes that each node[1] controls some resources (usually information), which

---

[1] A node stands here for a single processor or a cluster of processors with memory and possibly various storage and input/output devices. A node is associated with a single network address.

can be used from any other node in the system. Thus, the principal difference between this scenario and the one described above is that there is no distinction between a server machine and a client machine: each machine is both a server and a client. Many of these projects are based on UNIX:[2] LOCUS [POPE81, WALK83], Newcastle Connection [BRBR82], and Berkeley Distributed Unix [FERR84] are just a few well-known examples. Another example of extending an existing operating system is VMDF [FUSI81]. An alternative approach is to start with a new operating system, specifically designed to support distributed computing [LAZO81, RASH81, LEAC83, CHERI84a]. Finally, distributed database systems such as R* [WILL82, LIND84] also belong to the category of integrated systems.

It should be added that many of the integrated systems are designed so that it is possible to configure some nodes as servers and others as clients only. On the other hand, many designs that fall into the workstation/server category provide a uniform interface to local and remote services, thus giving the client an image of an integrated system. Perhaps the main distinction between these two types of system is where one starts: in the first case, the initial design emphasis is on server design, and the integration is added as a layer between the client and the server. In the other scenario, uniform mechanisms for being able to access resources on any node from any node forms the basis; server and client machines can be configured by restricting these general access possibilities.

Yet another scenario, rarely described in scientific literature, is built around a general-purpose host. Such systems represent a natural evolution from a central time-shared facility towards a distributed system. Intelligent workstations make it possible to off-load many tasks, especially those dealing with the user interface, while the host continues providing access to databases and services developed in the past. The IBM VM/PC product is a good example of this trend [IBM83].

1.1. Generalized client/server model

The workstation/server scenario is the most obvious demonstration of the client/server model of distributed computing, but the notion of servers and clients can be identified in all three scenarios described above.[3] In a broader sense, a *server* is simply a subsystem that provides a particular type of *service* to a priori unknown clients. The definitions given below are adapted from those given by Mitchell [MITC82b].

- A *service* is a software entity running on one or more machines.[4]

- A *server* is an instance of a particular service running on a single machine.

- A *client* is a software entity that exploits services provided by servers. A client can but does not have to interface directly with a human user.

Although clients could run on the same physical processor as the server, such a cohabitation is assumed to be an exception rather than a rule; as a default, clients and servers must be designed so that their communication can be carried over the connecting network.

---

2   UNIX is a trademark of Bell Laboratories.

3   The notion of a client and a server (User Agent and Service Agent) appears also in the CCITT System Model for Message Handling System [CCITT83] and Directory Systems [CCITT84].

4   This software entity might control some specific hardware resources, such as a disk or a printer.

Thus, the client/server model is more a communication and operating system concept than a particular hardware configuration. This concept is also finding its way into programming languages designed for distributed environments [LISK79, LISK83, LISK84].

The rest of this paper reviews the basic design issues in the client/server model. Section 2 discusses various options and trends in server design. Section 3 focuses on the communication protocols. Section 4 addresses the programming aspects.

2. SERVER DESIGN

Various types of servers have been built or proposed to provide:

- Control of shared resources:

    disk server
    print server
    computing server
    file server
    database server

- Management services:

    directory server
    authentication server
    resource allocation server
    boot server
    timing server

- User-oriented services:

    "yellow-pages" server
    mail server

The servers in the first two groups implement functions that are crucial to the correct and efficient operation of the whole system, including synchronization and scheduling of accesses to shared resources, checking the rights of the clients to use a particular resource, determining where to find the resource and how to communicate with the server managing the resource. In the last group are servers that provide higher-level, user-oriented services and applications; many other application servers could be imagined, both for an office and a factory environment.

This section discusses the server to hardware processor mapping, partitioning of function between the server and the client machine, and the issues of transparency. A brief case study of file servers is included to demonstrate some design trade-offs.

2.1. Server machines

Physical separation of servers from ordinary clients[5] has distinct advantages:

- it is easier to ensure reliable and secure operation and install access control mechanisms for one or a few server machines than for many separately administered general-purpose nodes

- a server machine can be designed as a "closed" system, using a simpler and consequently more efficient operating system than what is required to support a user-programmable system

- the server machine can be tailored (e.g., via special hardware) for the type of service it performs

As an extreme, each of the servers listed above could run on a separate machine; this approach has been pursued in the Cambridge Distributed System [NEED82]. In this system, even the client machines are in fact shared servers: they are general-purpose processors allocated to users from a processor bank controlled by a Resource Manager, which is a server running on its own machine [CRAF83], and loaded with the appropriate client software from a file server running on yet another machine. Another extreme would be to put all servers needed in a particular installation on the same physical machine. Especially for small systems (small number of workstations), this might be the only feasible arrangement. The Cambridge Distributed System engineered a different solution: individual small servers run on separate processors which share, however, the network attachment and the physical housing, including the power supply.

2.2. Internal structure of a service

The complete service as needed by the application-level clients cannot in general be provided only by a remote server. Some component is needed in the client machine to provide an interface between the client and the remote server. Such a component, which is, in the sense of the layered OSI architecture [OSI81], above the basic communication protocols, usually includes some of the mechanisms needed to achieve transparency (see Section 2.3.), and might include other mechanisms such as a cache to improve performance, or mechanisms needed to tailor the service to a particular client.[6] How to divide the service between the server proper and these local service components is one of the major design problems. Of course, the problem of modularization is inherent to any system design. However, a distributed system presents some additional complications:

- the dividing line implies in general a remote operation

- the client population may not be homogeneous

Thus, the decision how to structure the service must take into consideration above all performance and generality.

---

[5]   Some clients are servers themselves, such as a directory server might be a client of a file server. For such a client, it might be in fact better to share hardware with the server which provides it with the basic resources.

[6]   This is not the only possible view, though. Mitchell, for example, defines this type of component to be the actual client of the remote server [MITC82b].

Finally, a service can run on multiple machines, that is, a service can consist of multiple servers. In general, such a provision is considered for two reasons:

Availability. One of the main advantages of a personal workstation is not being dependent on a complex system that is often down because of a crash, or because of maintenance. However, if the processing being done on a workstation needs assistance from servers, then to sustain this advantage, the service must always be available, that means, at least one server operational and accessible via the network from the workstation.

Incremental growth. It should be possible to increase the capacity of a service gradually, by adding servers, without changes to the client interface. Additional mechanisms are needed in order to achieve these properties and to hide the distributed nature of the service from the clients.

2.3. Transparency

Transparency means hiding certain aspects of the system from the clients. In a distributed system, several forms of transparency are desired:

location transparency. local and remote resources can be accessed in exactly the same way, without having to specify the actual physical location

replication transparency. if several copies of a resource exist for the reasons of reliability or faster access, the clients see only a single logical resource, which is accessed in the same way as if it had only a single copy

concurrency transparency. the clients do not have to be concerned about other clients which might try to access the same resources concurrently

failure transparency. the system hides the different failure characteristics of the accesses to remote resources.

Transparency can be implemented using general mechanisms independent of the semantics of the service; the best example of this approach are atomic transactions, which provide concurrency and failure transparency [GRAY78, SVOB83]. However, it has been argued that atomic transactions might be too expensive, and that in many cases a much less expensive solution can be found if one takes advantage of the properties of the application. The latter approach was applied, for example, in Grapevine, the distributed mail system developed at Xerox PARC [BIRR82, SCHR84].

Besides these forms of transparency which focus on the functional aspects, some projects strive to achieve *performance transparency* as well. Performance transparency requires that the response time for a remote request be on the same order of magnitude as if the request were handled locally. This goal is particularly important in the integrated distributed system scenario or where a server is used to provide some basic resource such as a disk. Performance transparency has been achieved with specialized protocols on local-area networks [PAXT79, POPE81, CHERI83b, CHERI85], but in more general networks and interconnected networks such a goal is likely to be unattainable.

2.4. Example: file servers

To get a better appreciation of the problems discussed in this section, it is instructive to study some actual designs. File servers are by far the most common type of servers built thus far; an analysis of a number of experimental file servers can be found in [SVOB83]. There seems to be no general agreement on the exact function of a file server; some designs are driven by the goal of generality, some by the need to remain compatible with a particular existing software. We shall exclude the latter, and review the designs of the "universal" file servers for heterogeneous environments.

A file server might need to support:

- filing systems
- virtual memory
- database applications

First, a universal file server should be able to support several distinct filing systems since different categories of clients might desire different organization and protection of their data or simply the ability to access files on a remote server direct by their internal names, without having to use a higher-level filing system.

Second, different application areas require different modes of access:

File storage and retrieval. Clients submit and retrieve entire files. This mode of access can be exploited to provide an extension to a local filing system, or for down-loading programs to client machines.

Page-level access to files. Clients can retrieve and update individual pages. This mode of access can be extended to provide large virtual memory for clients that do not have their own, secondary storage devices.

Byte-level access. Clients can retrieve and update fields of variable lengths at random. Such access is needed to support database systems.

Thus, on one hand, a universal file server might have to support fast bulk transfer of entire files, on the other hand, efficient random access to small fields within a file. Alternately, a whole file can be cached by the service component on a client machine which then provides the other two types of access. Similarly, page-level access to a server can be converted to a record-level access by the local component. Finally, it might be better to relax the requirement of universality; for example, the ITC project [SATY84], which assumes that a typical workstation does not have its own permanent storage device, includes separate file servers and paging servers.

A file server also ought to protect clients' data from being improperly modified or destroyed by events beyond the clients' control. Such protection includes, besides control of access to shared data, backup of both private and shared files, and an automatic recovery (rollback) of a file to a consistent state if an update of the file cannot be completed because of a failure of the server, the client, or the communication between the two. Such an automatic recovery, combined with control of concurrent accesses in case of shared data, forms the basis of what has become known as atomic transactions. For many applications it might be sufficient to be able to update a single file atomically while the respective transaction has an exclusive access to it. However, in a database system, a transaction may have to include several files, while allowing other transactions concurrent access to the same files. Such multi-file transactions could be supported by the server, or built from more elementary transactions on the client side [PAXT79, ACCE80].

3. COMMUNICATION MODEL

Perhaps the most distinguishing feature of the client/server model compared to just ordinary modular design is communication. This section reviews how a client and a server can communicate across a general-purpose network. Typically, it will be a high-speed local-area network [METC76, WILK79, BUX82].

For the discussion of the client/server communication issues, the seven layers of the OSI Reference Model [OSI81] can be compressed to three sets, providing:

Node-to-node connectivity. This layer includes everything needed to exchange information between physical nodes. It encompasses OSI layers 1, 2 and 3, that is, the Physical, Data Link, and the Network layers.

End-to-end transport. This layer provides service-independent communication channels between specific software entities in distinct physical nodes. It encompasses OSI layers 4 and 5, that is, the Transport and the Session layers.

Client/server communication. This layer hides the remoteness of the actual server, possibly uses other services needed to locate the server and establish a connection with it, and chooses how the information representing requests and responses is to be transferred between the client and the server. It encompasses the OSI layers 6 and 7, that is, the Presentation and Application layers. In local-area networks, there is a general tendency to use the simplest protocols possible, in particular, to minimize use of connected mode in the first two sets of layers. First, a connectionless (or datagram) protocol is used whenever possible. Second, when a connection must be established, it is maintained only for the time needed to satisfy a request.

3.1. Connection-oriented vs connectionless protocols

According to the OSI Reference Model, a connection-oriented protocol has three distinct phases: connection establishment, data transfer, and connection release. The Protocol Data Units (PDU's) sent on a connection are sequenced, so that missing PDU's can be detected and duplicates suppressed. A connection-oriented protocol contains also flow control mechanisms. A PDU at a particular layer may carry the entire data unit passed to it by the next higher level, but the latter, called Service Data Unit, or SDU, may need to be segmented by the layer providing the service.[7] By sequencing the PDU's in layer n, SDU's of the layer n+1 sent on the same connection are guaranteed to be delivered in the right order, and when an SDU cannot be delivered, the problem is at least detected.

In a connectionless protocol, an SDU is transmitted as a self-contained message which bears no relation to SDU's sent to the same destination before or after. The term connectionless (datagram) protocol has been used in local-area networks mostly for protocols where a whole application-level SDU is sent in a single packet (layer 1 PDU) [MITC82a, JANS83]. However, the recent specifications of connectionless mode for OSI protocols allow for the possibility of sending bigger units without an explicit connection establishment. Internally, the respective layer must establish sort of a connection in order to be able to reassemble the individual segments at the other end [OSI83].

---

[7] The 'service' defined by the OSI Reference Model has nothing to do with the 'service' provided by the servers which are the topic of this paper.

In the network and data link layers, a connectionless protocol has become a *de facto* standard;[8] further, ECMA [ECMA82] has been actively promoting such an architecture within the OSI standardization efforts. However, while ECMA promotes use of a full Transport protocol Class 4, many distributed systems built around a local-area network use very specialized end-to-end transport protocols, based on a network-level datagram or "light-weight" connections; some of these protocols are quite similar to the connection-less mode defined for OSI.

While some impressive performance results have been reported with specialized protocols [CHERI83b, CHERI84, POPE81], this approach has also its drawbacks, the difficulty to accommodate more complex networks being the most important one. A gateway can be used to translate the special light-weight protocols of a particular local-area network to general-purpose protocols available on other systems [BRAD83, CHERI83a], but this solution still assumes that any node attached to the same local-area network supports the same specialized protocols, that is, in fact, such a network is internally a closed system. Further, the end-to-end implications of such a conversion are not yet fully understood. Alternatively, one could start with general-purpose protocols and try to optimize the implementation; a recent experiment at MIT with implementing TCP/IP [POST81a, POST81b] protocols on the IBM PC testifies that such an approach is feasible [SALT84].

3.2. Client/server protocols

One of the major problems in designing distributed systems has been how to deal with the uncertainties caused by component failures and imperfect communication. To prevent a possibly infinite wait, it is a standard practice to set a timeout on a reply to a remote request. However, when such a timeout expires, it is not possible to deduce how far the request has progressed. Any of the following could have happened:

- Execution of the request never started: the communication subsystem never delivered the request message to the server or the server crashed before the service could be invoked.

- The server crashed while executing the request.

- Execution of the request terminated, but the reply was not delivered to the client: either the server crashed before the reply could be sent, or the communication subsystem failed to deliver the reply message.

- Execution of the request still continues: the timeout was set too short even for the best case, or execution is unusually slow, for example, because of high load, an internal maintenance, or server recovery.

When a timeout expires, the client has two options: abandon the request, or retry. When the request was sent as a datagram, not supported by a reliable connection at any level below, retrying at the request level is the only possible recovery from packet losses due to transmission errors or lack of buffers. However, if the request is still executing or the reply

---

was lost, a retransmission creates another problem, and that is duplicate requests.[9] To prevent unexpected effects, the following approaches have been used:

All operations provided by the server are repeatable. An operation is repeatable if executing that operation with the same parameters several times has the same final effect as if that operation had been executed exactly once. Thus, a duplicate request can be simply re-executed. This approach has the following disadvantages

- re-execution of a request consumes additional resources; if the timeout occurs because the server machine is overloaded, it will only worsen the situation

- this approach restricts the type of operations that can be offered by a server

- re-execution of a request might result in a duplicate reply to the client

Duplicates are detected and suppressed. The basic technique for detecting duplicate requests is to assign an identifier to each request; a retransmitted request must bear the same identifier as the original request. Such identifiers must be used in addition to any sequencing employed by the underlying protocols, since when a client resends a request, it appears to the lower layers as a new unit. The request identifiers, however, do not have to be assigned strictly sequentially, since this type of protocol does not have to check for *missing* requests. The server must reliably log the identifier, and check each request against this log file. The amount of checking can be reduced if a retried request is explicitly marked as a retry.

When a duplicate request is received while the original request is being executed or is waiting for execution, some form of acknowledgement should be sent back to prevent another retransmission or an abortion of the operation on the client side. If the execution of the request has already completed, the reply must be resent. Thus, the server must reliably log a reply as long as there is a chance that the client might send a duplicate. The choice of the identifiers can have a significant impact on how long this uncertainty might last. Identifiers which are guaranteed to increase monotonically also in the presence of client's crashes and restarts are the best choice; such identifiers can be generated from a clock [SHRI82].

Combined approach. If making the server operations repeatable is not too restrictive, the two approaches can be combined. The advantage of this scheme is the relaxation of the conditions under which replies can be discarded; in particular, the log does not have to survive server crashes.

For some type of service, the request/reply can be viewed as a basic unit of work, but for other services the client must establish a sort of connection with the server at the service level. Multi-request transactions supported by some file servers are a good example [SVOB83]. If a client fails in the middle of a transaction, the server must be able to free the resources tied up by that client. Client/server protocols based on connection-oriented transport and link protocols can benefit from the mechanisms provided by these protocols to detect network problems and hardware crashes [LIND84], but to cover also higher-level software problems, a timeout mechanism is needed at this level as well.

---

[9] A request or a reply could also be lost because of a problem in the higher-level software. Thus, even when the client/server communication protocol is built on top of a highly-reliable connection-oriented protocol, some end-to-end recovery is still needed [SALT81].

4. PROGRAMMING MODEL

In the most general sense, the client/server model is a way of structuring distributed programs. A client and a server are 'active' modules, communicating via messages. A client or a server module resides on a single node in its entirety, but multiple modules can share a node. A server module encapsulates a resource or a collection of resources, and provides a set of well-defined operations that other modules can call. Server modules include the necessary synchronization, protection, and crash recovery mechanisms.

To aid in the construction of such distributed systems, a programming system combining a high-level programming language and operating system features is needed [SVOB79, LISK79]. Argus is an example of a programming system designed specifically to support this model [LISK83, LISK84], including synchronization, protection, and crash recovery.

It is widely agreed, that it is desirable to provide a procedural interface to remote services. The mechanism has become known as a *remote procedure call* (RPC). Various semantics of an RPC have been defined and studied [SPEC82, NELS81], but distributed systems projects favor semantics similar to local procedure calls with exception-handling mechanisms. Essentially, the remote call is either executed exactly once, in which case exactly one reply is returned to the caller, or an exception is signaled to the caller.

An implementation of an RPC embodies the request/reply protocol discussed in Section 3.2. Consequently, all of the problems of dealing with duplicate requests and duplicate replies must be solved in the implementation. However, there is more to it. If a call terminates abnormally, it is necessary to ensure that indeed the call had no effect. In case the server has performed or still is acting on the request, the effects have to be undone. *Atomic actions* are a general mechanism for solving this problem. Integration of atomic actions with RPC's to support building of resilient distributed systems was studied in [SVOB84]. The RPC's in Argus are also based on atomic actions [LISK84].

Extensions to a basic RPC include runtime binding, which may transparently call on a name server, and possibly undergo a security check before the call is passed to the server requested by the calling client [BIRR84]. This powerful RPC is the quintessence of the client/server model of distributed computing.

5. CONCLUDING REMARKS

The client/server model of distributed processing is a structuring concept, which ought to be supported by the operating system and the programming languages and tools. The client/server model is often reflected directly in the hardware base as server machines supporting client personal computers.

An important component in the client/server model is the communication architecture. Various light-weight transport protocols and protocols for a remote procedure call have been implemented and studied in the experimental client/server systems. Current activities in the standardization bodies include specifications of similar protocols; it will be interesting to see whether these protocols become part of common operating systems.

REFERENCES

ACCE80   Accetta, M., Robertson, G., Satyanarayanan, M., and Thompson, M. "The Design of a Network-Based Central File System," Tech. Rep. CMU-CS-80-134, Dep. of Computer Science, Carnegie-Mellon Univ., Pittsburgh, PA., Aug. 1980.

BALK84   Balkovich, E., Lerman, S., and Parmelee, R., "Project Athena, A Joint Academic-Industry Experiment in Computers in Education," *Proc. IEEE Frontiers in Education 14th Annual Conference*, Philadelphia, PA., Oct. 3-5, 1984, pp. 466-471.

BIRR82   Birrell, A., Levin, R., Needham, R., Schroeder, M., "Grapevine: An Exercise in Distributed Computing," *Commun. ACM 25*, 4 (April 1982), pp. 260-273.

BIRR84   Birrell, A.D. and Nelson, B.J., "Implementing Remote Procedure Calls," *ACM Trans. Computer Systems 2*, 1 (Feb. 1984), pp. 39-59.

BRAD83   Braden, R., Cole, R., Higginson, P., and Lloyd, P., "A Distributed Approach to the Interconnection of Heterogeneous Computer Networks," *Proc. ACM SIGCOMM'83 Symposium on Communications Architectures and Protocols*, Austin, Texas, March 1983, pp. 254-259.

BROW84   Brown, M.R., Kolling, K., and Taft, E.A., "The Alpine File System," Xerox Palo Alto Research Center, Feb. 1984 (submitted to ACM-TOCS).

BRBR82   Brownbridge, D.R., Marshall, L.F., and Randell, B., "The Newcastle Connection or UNIXes of the World Unite!," *Software Practice and Experience 12*, 1982, pp. 1147-1162.

BUX82   Bux, W., Closs, F., Janson, P., Kümmerle, K., Müller, H.R., and Rothauser, E.H., "A Local-Area Communication Network Based on a Reliable Token-Ring System," *Proc. Int. Symp. Local Computer Networks*, Florence, Italy, April 1982, pp. 69-82.

CCITT83   CCITT Draft Recommendations on Message Handling, Nov. 1983.

CCITT84   "Directory Systems Working Document," Interim Rapporteur on Directory Systems, CCITT Study Group VII, June 1984.

CHERI83a   Cheriton, D.R., "Local Networking and Internetworking in the V-System," *Proc. 8th Data Communications Symposium*, Cape Cod, Massachusetts, Oct. 1983, pp. 9-16.

CHERI83b   Cheriton, D.R. and Zwaenepoel, W., "The Distributed V Kernel and its Performance for Diskless Workstations," *Proc. of the 9th ACM Symposium on Operating Systems Principles*, Bretton Woods, New Hampshire, Oct. 1983, pp. 129-140.

CHERI84a   Cheriton, D., "The V Kernel: A Software Base for Distributed Systems," *IEEE Software 1*, 2 (April 1984), pp. 19-43.

CHERI85   Cheriton, D.R. and Roy, P.J., "Performance of the V Storage Server," *Proc. 1985 ACM Computer Science Conference*, New Orleans, Louisiana, March 12-14, 1985 (to appear).

CRAF83   Craft, D.H., "Resource Management in a Decentralized System," *Proc. 9th ACM SIGOPS Symposium on Operating Systems Principles*, Bretton Woods, New Hampshire, Oct. 1983, pp. 11-19.

DION80   Dion, J. "The Cambridge File Server," *ACM SIGOPS Operating Syst. Rev. 14*, 4 (Oct. 1980), pp. 26-35.

ECMA82   "Final Draft, ECMA Standard on Local Area Networks (token ring)," ECMA Working Paper, Dec. 1982.

FERR84   Ferrari, D., "The Evolution of Berkeley UNIX," *IEEE Distributed Processing Technical Committee Newsletter 6*, SI-2, June 1984, pp. 3-6.

FUSI81   Fusi, A. and Sommi, G., "Distributed Virtual Systems," *Proc. 2nd Int. Conf. Distributed Systems*, Paris, France, April 1981, pp. 41-49.

GRAY78   Gray, J.N. "Notes on Database Operating Systems," *Lecture Notes in Computer Science 60*, Springer-Verlag, 1978, 393-481. Also Tech. Rep. RJ2188, IBM San Jose Research Laboratory, San Jose, California, Feb. 1978.

IBM83   "IBM Virtual Machine/Personal Computer User's Guide, International Business Machines Corporation, Personal Computer Professional Series 6936733, Nov. 1983.

JANS83   Janson, P., Svobodova, L., and Maehle, E. "Filing and Printing Services on a Local Area Network," *Proc. 8th Data Communications Symposium*, Cape Cod, Massachusetts, Oct. 1983, pp. 211-220.

LAZO81   Lazowska, E.D., Levy, H., Almes, G., Fischer, M., Fowler, R., and Vestal, S., "The Architecture of the Eden System," *Proc. 8th ACM Symp. Operating Systems Principles*, Asilomar, California, Dec. 1981, pp. 148-159.

LEAC83   Leach, P.J., Levine, P.H., Douros, B.P., Hamilton, J.A., Nelson, D.L., and Stumpf, B.L., "The Architecture of an Integrated Local Network," *IEEE Journal on Selected Areas in Communications SAC-1*, 5, Nov. 1983, pp. 842-857.

LIND84   Lindsay, B., Haas, L.M., Mohan, C., Wilms, P.F., and Yost, R.A. "Computation and Communication in R*: A distributed database manager," *ACM Trans. Computer Systems 2*, 1 (Feb. 1984).

LISK79   Liskov, B., "Primitives for Distributed Computing," in *Proc. 7th ACM Symp. Operating Systems Principles*, Asilomar, California, Dec. 1979, pp. 33-42.

LISK83   Liskov, B., "Guardians and Actions: Linguistic Support for Robust, Distributed Programs," *ACM Transactions on Programming Languages and Systems 5*, 3 (July 1983), pp. 381-404.

LISK84   Liskov, B., "Overview of the Argus Language and System," Programming Methodology Group Memo 40, M.I.T. Laboratory for Computer Science, Cambridge, Massachusetts, Feb. 1984. pp. 381-404.

METC76   Metcalfe, R.M. and Boggs, D.R., "Ethernet: Distributed Packet Switching for Local Computer Networks," *Commun. ACM 19*, 7 (July 1976), pp. 395-404.

MITC82a   Mitchell, J.G. and Dion, J., "A Comparison of Two Network-Based File Servers," *Commun. ACM 25*, 4 (April 1982), pp. 233-245.

MITC82b   Mitchell, J.G. "File Servers for Local Area Networks," Lecture Notes, Course on Local Area Networks, University of Kent, Canterbury, England, March 1982, pp. 83-114.

NEED82   Needham, R.M. and Herbert, A.J. *The Cambridge Distributed Computing System*, Addison-Wesley Publ. Co., Reading, Massachusetts, 1982.

NELS81   Nelson, B.J., "Remote Procedure Call," Technical Report CSL-81-9, XEROX Palo Alto Research Center, Palo Alto, Calif., May 1981. (Also CMU-CS-81-119, Carnegie-Mellon University, PhD Thesis.)

OSI81   "Data Processing — Open Systems Interconnection — Basic Reference Model," *Computer Networks 5*, 2 (April 1981), pp. 81-118.

OSI83   "Information Processing Systems — Open Systems Interconnection — Addendum to the ISO 7498 Covering Connectionless-mode Transmission, ISO/DP8524, 1983.

PAXT79   Paxton, W.H. "A Client-Based Transaction System to Maintain Data Integrity," in *Proc. 7th ACM Symp. Operating Systems Principles*, Asilomar, California, Dec. 1979, pp. 18-23.

POPE81   Popek, G., Walker, B., Chow, J., Edwards, D., Kline, C., Rudisin, G., and Thiel, G. "LOCUS: A Network Transparent, High Reliability Distributed System," *Proc. 8th ACM Symp. Operating Systems Principles*, Asilomar, California, Dec. 1981, pp. 169-177.

POST81a   Postel, J.B., Internet Protocol, DARPA Internet Program Protocol Specification, Sep. 1981.

POST81b   Postel, J.B., Transmission Control Protocol, DARPA Internet Program Protocol Specification, Sep. 1981.

RASH81   Rashid, R. and Robertson, G., "Accent: A Communication Oriented Network Operating System Kernel," *Proc. 8th ACM Symp. Operating Systems Principles*, Asilomar, California, Dec. 1981, pp. 64-75.

SALT81   Saltzer, J.H., Reed, D.P., and Clark, D.D., "End-to-End Arguments in the System Design," in *Proc. 2nd Int. Conf. Distributed Systems*, Paris, France, April 1981, pp. 509-512.

SALT84   Saltzer, J.H., Clark, D.D., Romkey, J.L., and Gramlich, W.L., "The Desktop Computer as Network Host," M.I.T. Laboratory for Computer Science, submitted to *IEEE Journal on Selected Areas in Communications*, Sep. 1984.

SATY84   Satyanarayanan, M., "The ITC Project: An Experiment in Large-Scale Distributed Personal Computing," *Proc. Networks 84 Conference*, Madras, India, October 1984. (Also Report CMU-ITC-035, Information Technology Center, Carnegie-Mellon University, Pittsburgh, PA.)

SCHR84   Schroeder, M.D., Birrell, A.D., and Needham, R.M., "Experience with Grapevine: The Growth of a Distributed System," *ACM Trans. Computer Systems* 2, 1 (Feb. 1984), pp. 3-23.

SHRI82   Shrivastava, S.K. and Panzieri, F. "The Design of a Reliable Procedure Call Mechanism," *IEEE Trans. on Computers C-31*, 7 (July 1982) pp. 692-697.

SPEC82   Spector, A.Z., "Performing Remote Operations Efficiently on a Local Computer Network," *Commun. ACM 25*, 4 (April 1982), pp. 246-260.

SVOB79   Svobodova, L., Liskov, B.H., and Clark, D.D., "Distributed Computer Systems: Structure and Semantics," MIT/LCS/TR-215, Laboratory for Computer Science, Massachusetts Institute of Technology, Cambridge, Mass., March 1979.

SVOB83   Svobodova, L., "File Servers for Network-Based Distributed Systems," Technical Report RZ1258, IBM Zurich Research Laboratory, Rüschlikon, Switzerland, Oct. 1983.

SVOB84   Svobodova, L., "Resilient Distributed Processing," *IEEE Trans. on Software Engineering SE-10*, 3 (May 1984), pp. 257-268.

STUR80   Sturgis, H.E., Mitchell, J.G., and Israel, J.E. "Issues in the Design and Use of a Distributed File System," *ACM SIGOPS Operating Syst. Rev. 14*, 3 (July 1980), 55-69.

SWIN79   Swinehart, D., McDaniel, G., Boggs, D., "WFS: A Simple Shared File System for a Distributed Environment," *Proc. of the 7th ACM Symposium on Operating Systems Principles*, Asilomar, California, Dec. 1979, pp. 9-17.

WALK83   Walker, B., Popek, G., English, R., Kline, C., and Thiel, G., "The LOCUS Distributed Operating System," *Proc. of the 9th ACM Symposium on Operating Systems Principles*, Bretton Woods, New Hampshire, Oct. 1983, pp. 49-70.

WILL82   Williams, R., Daniels, D., Haas, L., Lapis, G., Lindsay, B., Ng, P., Obermark, R., Selinger, P., Walker, A., Wilms, P., and Yost, R. "R*: An Overview of the Architecture," *Proc. 2nd Int. Conf. on Databases: Improving Usability and Responsiveness*, Jerusalem, Israel, June 1982, pp. 1-27.

WILK79   Wilkes, M.V., Wheeler, D.J., "The Cambridge Digital Communications Ring," *Proc. of the NBS Local Area Communications Network Symposium*, Boston, Massachusetts, May 1979.

Btx-Anwendungen mit Großbanken am Beispiel automatisierter
Kontoführung

Hartmut Bressel
Projekt Bildschirmtext
Dresdner Bank AG
Frankfurt am Main

Grundlage für eine verläßliche Finanzdisposition sind aktuelle Informationen. International bestehen seit Jahren Balance-Reporting-Systeme, mit denen die Konten weltweiter Bankverbindungen abgefragt werden können. Auf nationaler Ebene ist dies komfortabel und kostengünstig durch die Kopplung von Bildschirmtext und Personal-Computern
möglich. Durch den Einsatz entsprechender PC-Software kann das Lesen
und Speichern von Umsätzen und Salden von Bankkonten sowie das Senden
von Zahlungsaufträgen voll maschinell abgewickelt werden.

Base for a reliable cash-management-arrangement are actual information.
On the international level has existed for some years balance-reporting systems by which the accounts of the worldwide banklink can be obtained. On the national side it can be achieved comfortable and
with minimizing of fees by coupling interactive videotex and personal-computer. With this application provided by banks it is possible
to read and store turnovers and balances of bankaccounts as well as
to send money transfers fully automatically.

Eine verläßliche Finanzdisposition bedarf aktueller Informationen, auf die auch schnell zugegriffen werden kann. International bestehen seit einigen Jahren Balance-Reporting Systeme, die Bankkunden die Möglichkeit geben, die Konten ihrer weltweiten Bankverbindungen mit Hilfe grenzüberschreitender Nachrichtennetze abzufragen. Als Beispiel sei hier das Internationale Cash-Management-System der Dresdner Bank "drecam" genannt.

Auf nationaler Ebene war hier bis vor kurzem nichts entsprechendes vorhanden. Jedoch bot sich gerade das Bildschirmtext-System (Btx) als eine komfortable und vor allem kostengünstige Möglichkeit an, schnell zu den gewünschten Informationen bei den Banken zu gelangen. Dieses nationale Cash-Management-System der Dresdner Bank mit dem Namen "drebit" (Dresdner Bank Bildschirmtext) soll hier vorgestellt werden.

Eine technische Grundvoraussetzung, um mit "drebit" arbeiten zu können, ist das Vorhandensein eines Personal-Computers (PC's) sowie eines Btx-Anschlusses im Unternehmen des Kunden.

Folgende drei Konfigurationsmöglichkeiten stehen zur Auswahl, um "drebit" ablaufen lassen zu können:

- Zum einen kann der Personal-Computer über einen Adapter mit einem bereits vorhandenen Btx-Gerät gekoppelt werden (siehe Anlage 1). Diese Konfiguration bietet den Vorteil, daß der Kunde mit zwei klar abgegrenzten Systemen arbeiten kann. D.h., neben "drebit" kann auch weiterhin Btx betrieben werden. Bei dieser Lösung ist auch der Einsatz eines nicht Btx-fähigen PC's möglich. Der Nachteil dürfte in der umständlichen Handhabung durch die parallele Bedienung beider Geräte sowie in der sehr umfangreichen Hardware liegen.

- Als zweite Konfigurationsmöglichkeit kann der Kunde seinen PC mit einem Btx-Decoder koppeln (s. Anlage 2), was ihm den Vorteil bringt, daß kein zweiter Bildschirm mehr notwendig wird. Hinzu kommt, daß bei Neuanschaffung ein Btx-Decoder preisgünstiger ist, als ein entsprechendes Btx-Gerät mit bereits integriertem Decoder. Mit dieser Konfiguration kann er nicht mehr selbstständig Btx betreiben. Es sind nur "drebit"-Anwendungen möglich, auch können aufgrund des fehlenden Bildschirms die einzelnen Btx-Dialogschritte optisch nicht mehr verfolgt werden.

- Als dritte, wohl auch beste Konfigurationsmöglichkeit bietet sich
die integrierte Lösung in Form eines Btx-fähigen PC's an, der neben
den PC-Funktionen auch für den Btx-Betrieb geeignet ist (siehe Anlage 3).

Dadurch sind Btx- sowie PC-Anwendungen sowohl gemeinsam, als auch
getrennt möglich. Ein weiterer Vorteil liegt in der besseren Handhabung eines einzigen Gerätes. Ein Nachteil ist allerdings, daß beide Systeme nicht zeitgleich bedient werden können.

Die Funktionsweise von "drebit" ist zu unterscheiden zwischen offline-,
den sogenannten Verwaltungsfunktionen und online-, den sogenannten
Btx-Dialogfunktionen (s. Anlage 4). Die Verwaltungsfunktionen bein-
halten:
- Btx-Teilnehmer-Kennungen
- Telekonto-Zugangskennungen
- Verwaltung von Transaktionsnummern
- Kontoumsätze/-Auszüge
- Erfassung und Verwaltung von Überweisungen.

Die Btx-Dialogfunktionen beinhalten

- Verbindungsaufbau-Routinen
- Überweisungen senden
- Kontoumsätze/-auszüge lesen.

Hinzu kommt eine Sonderfunktion

- Valutensalden

Zu den Funktionen im einzelnen:

Da bekanntlich für den Zugang zum Btx-System die teilnehmerbezogenen
Kennungen benötigt werden, ist es vorab erforderlich, sich über die
Funktion "Btx-Teilnehmerkennungen" die entsprechenden Kennungen verschlüsselt im PC abzuspeichern. Die gleiche Vorgehensweise gilt für
die kontobezogene Kennung (PIN), die über die Funktion Telekonto-Zu-
gangskennung ebenfalls verschlüsselt im PC abgelegt werden kann. Auf
diese Dateien wird dann beim automatischen Verbindungsaufbau zur Btx-
Vermittlungsstelle und der automatischen Anwahl des Externen Rechners
der Bank vom System her zugegriffen. Auf diese Prozedur wird im Anschluß an den folgenden Abschnitt noch detailliert eingegangen.

Zunächst zu einigen Funktionen:

- Transaktionsnummern

  Die Sicherungsnummern (Transaktionsnummern), die dem Kunden für seine elektronische Unterschrift im Btx-System für z.B. auszuführende
  Überweisungen von der Bank zugestellt werden, können ebenfalls über
  die entsprechende Funktion erfaßt und verschlüsselt abgespeichert
  werden. Die Verwaltung wird dann vom System übernommen. Bei Geschäftsvorfällen, wie z.B. Ausführen von Überweisungen, werden die
  Transaktionsnummern nicht wie im Btx-System manuell in die kontrol-
  lierte Überweisungsmaske eingesetzt, sondern beim Absenden der Überweisungen automatisch den einzelnen Geschäftsvorfällen zugeordnet.
  Der Verbrauch wird ebenfalls vom System überwacht.

- Kontoumsätze/-auszüge

  Über die Funktion "Kontoauszüge lesen" werden im Btx-Dialog die Kon-
  to-Umsätze sowie die Kontosalden aus dem Externen Rechner der jeweiligen Bank rechnergesteuert gelesen. Hierbei sind bis zu 99 Bank-
  anbindungen möglich.

  Die Verbindung zur Btx-Zentrale wird per Programm unter Zugriff auf
  unsere angelegten Dateien automatisch hergestellt; die Anwahl des
  Externen Rechners erfolgt ebenfalls automatisch. Nach dem Lesevor-
  gang der Umsätze und Salden geht der PC im Btx-Programm automatisch
  auf die Leitseite der Post zurück, liest entweder Umsätze und Salden
  weiterer Banken oder unterbricht die Verbindung zur Btx-Zentrale automatisch.

  Da die Tagesauszüge der einzelnen Banken ein unterschiedliches Erscheinungsbild haben, werden anschließend offline die erfaßten Umsätze zu einem einheitlichen, institutsneutralen Tagesauszug umfor-
  matiert und wahlweise über einen Printer in einem für alle Banken
  einheitlichen Erscheinungsbild ausgedruckt. In jedem Fall aber zur
  weiteren Bearbeitung auf externen Speichern, z.B. Disketten, abgespeichert. Auf diesem Weg hat der Disponent die Salden und Kontoum-
  sätze aller -an Btx angeschlossenen- nationalen Bankverbindungen
  seines Unternehmens schnell als Grundlage für eine verläßliche Finanzdisposition vorzuliegen.

- Überweisungen

    Über die entsprechende Verwaltungsfunktion können offline die auszu-
    führenden Einzel- bzw. Sammelüberweisungen erfaßt werden. Der Mas-
    kenaufbau der Überweisungen entspricht dem des Bundesverbandes Deut-
    scher Banken. Durch die wählbaren Ausführungstermine einmalig, täg-
    lich, wöchentlich, monatlich, quartalsweise, halbjährlich oder jähr-
    lich können die Überweisungen auch als Daueraufträge verwaltet und
    ausgeführt werden. Die Ausführung der vorbereiteten Überweisungen
    erfolgt zeitgesteuert zum jeweiligen Ausführungstermin. Nach Absen-
    den der Überweisungen wird ein "Protokoll der ausgeführten Überwei-
    sungen" erstellt. Über fehlerhaft erfaßte Überweisungen fällt zu-
    sätzlich ein Fehlerprotokoll an, anhand dessen die entsprechenden
    Überweisungen nochmals überarbeitet und neu zur Ausführung gebracht
    werden können. Für den Zahlungsverkehr sind hier Verbindungen zu
    anderen Anwenderprogrammen z.B. Lohn- und Gehaltszahlungen vorge-
    sehen (s. Anlage 5).

Das kontoführende System 'drebit' kann durch ein Zusatzmodul "Valutensalden" erweitert werden. Mit diesem Programmteil wird nach dem Ausle-
sen der Salden und Umsätze zusätzlich noch der aktuelle valutarische
Saldo je Buchungstag ermittelt. Der Kunde kann sich den valutarischen
Saldo pro Bankkonto, pro Bankverbindung oder auf die Gesamtsumme aller
aufgerufenen Bankkonten anzeigen oder ausdrucken lassen. Für den Fi-
nanzdisponenten ergibt sich hierdurch der Vorteil, daß er den valutarischen Saldo, den er für seine Disposition bisher sehr zeitaufwendig manuell ermitteln mußte, zusammen mit den Kontosalden und -um-
sätzen exakter und schneller als bisher vorliegen hat. Diesen Zeitge-
winn kann er für eine intensivere Vorbereitung seiner Geld- bzw. An-
lagegeschäfte nutzen.

Die täglichen Valutensalden werden auf externen Speichern (z.B. Disketten) fortgeschrieben und jeweils mit den aktuellen Umsätzen je Buchungstag aktualisiert.

D.h., daß neben dem valutarischen Saldo per Buchungstag auch valutarische Salden bis zu 30 Tagen -ausgehend vom Bankbuchungstag- in der Zukunft fortgeschrieben werden, in dem die bis dahin kumulierten Valutenumsätze SOLL und HABEN dann schon enthalten sind (siehe Anlage 6).

Um die Informationen der Konten der nationalen Bankverbindungen auch auf internationaler Ebene nutzen zu können, besteht in "drebit" über ein Umsetzprogramm eine Schnittstelle zu internationalen Cash-Management-Systemen z.B. drecam (s. Anlage 7). Über diese Schnittstelle können die nationalen Kontosalden bzw.- umsätze in internationale Systeme überspielt werden, wodurch die internationalen Kontodaten kostengünstiger um die entsprechenden nationalen Angaben für eine weltweite Finanzdisposition ergänzt werden können.

Anlage 1

OREBIT

P R I N Z I P U N D T E C H N I K
(KOPPLUNG VON PC MIT BTX-GERAET)
=====================================

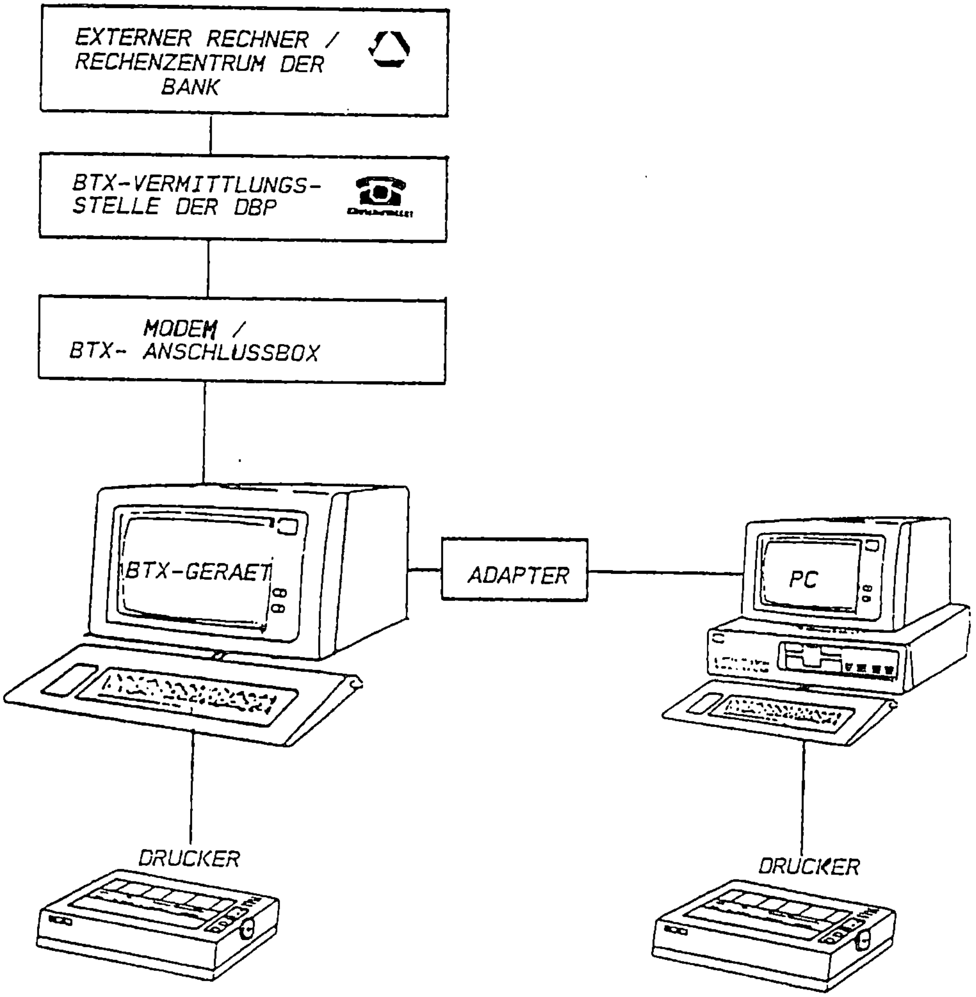

<u>Anlage 2</u>

DREBIT

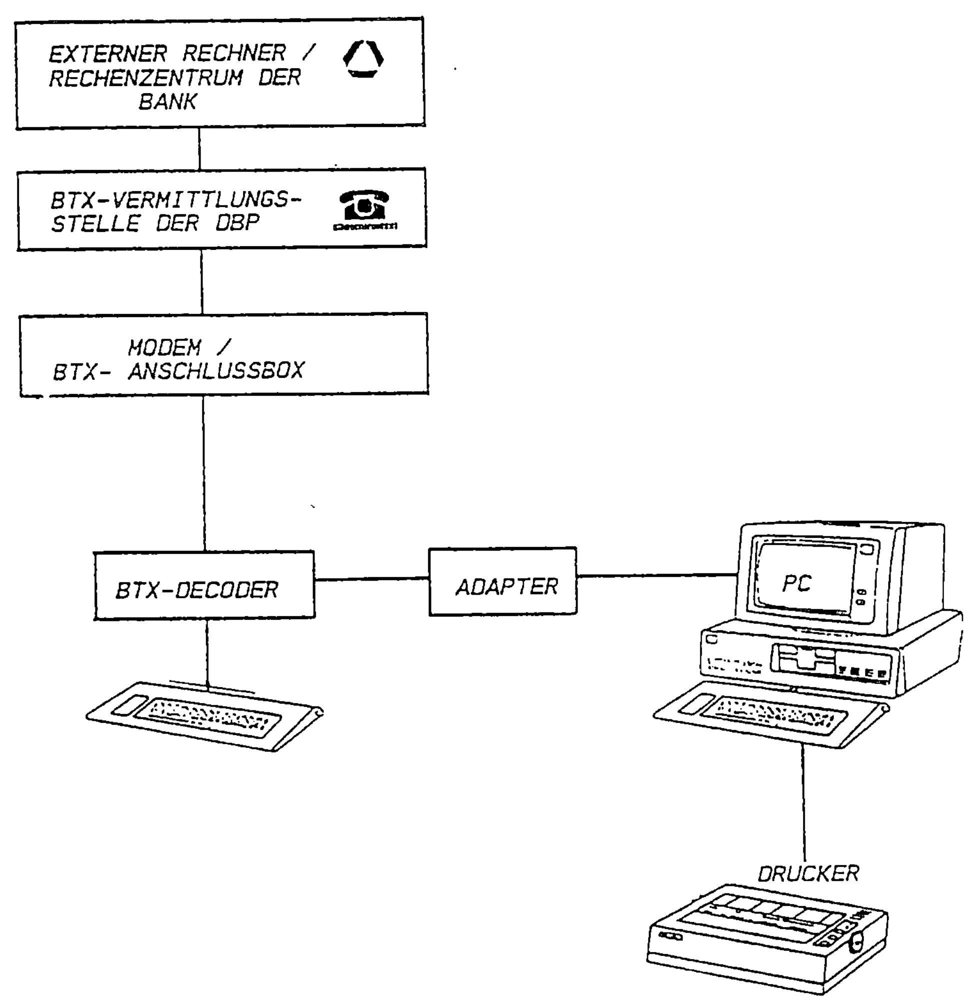

<u>Anlage 3</u>

DREBIT

P R I N Z I P   U N D   T E C H N I K
(BTX-FAEHIGER PC)
=====================================

BTX-FAEHIGER PC
MIT

DOPPELFLOPPY-LAUFWERK
ALTERNATIV:
FESTPLATTE UND FLOPPY

<u>Anlage 4</u>

DREBIT

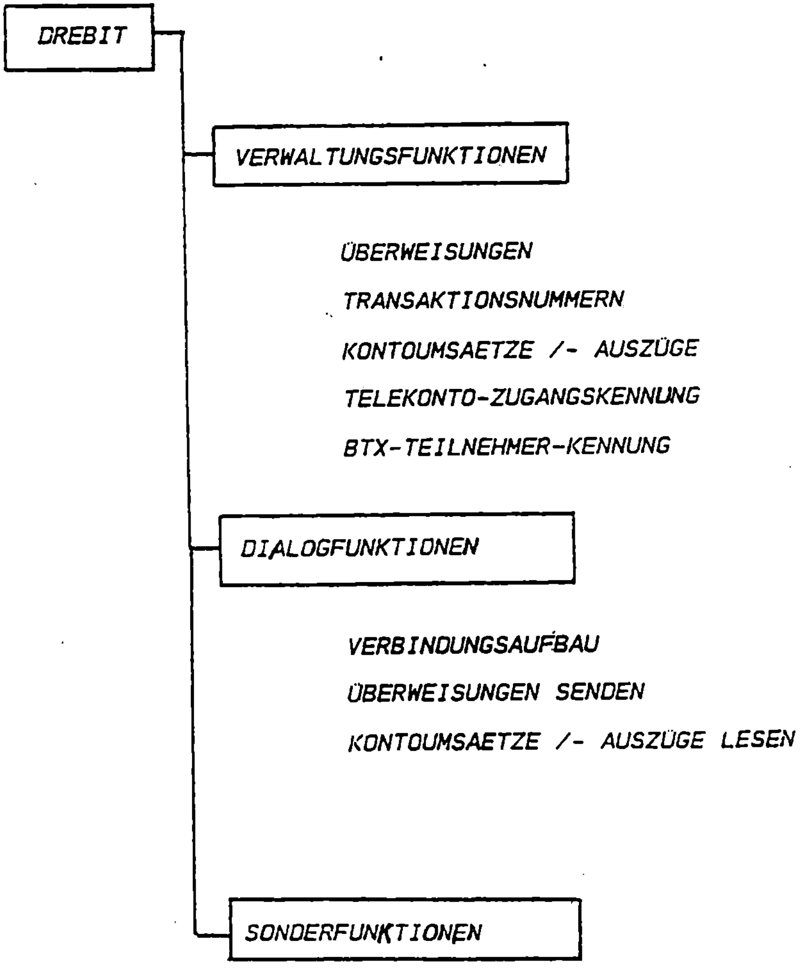

F U N K T I O N S U M F A N G
==============================

Anlage 5

DREBIT

S C H N I T T S T E L L E Z U A N D E R E N
P C - P R O G R A M M E N
(Z.B. LOHN- U. GEHALTSABRECHNUNG)
===

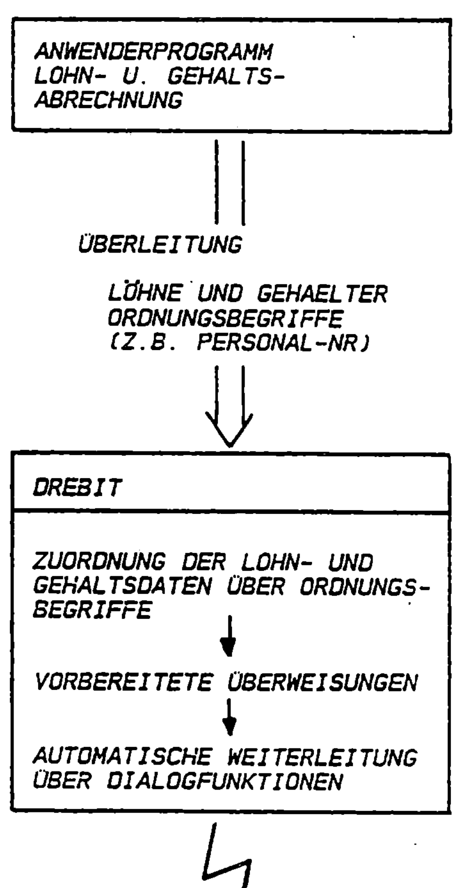

Anlage 6

DREBIT

V A L U T E N S A L D E N
FUNKTION FORTSCHREIBUNG
==========================

ARBEITSTAG 25.1.   (BUCHUNGSTAG 24.1.)

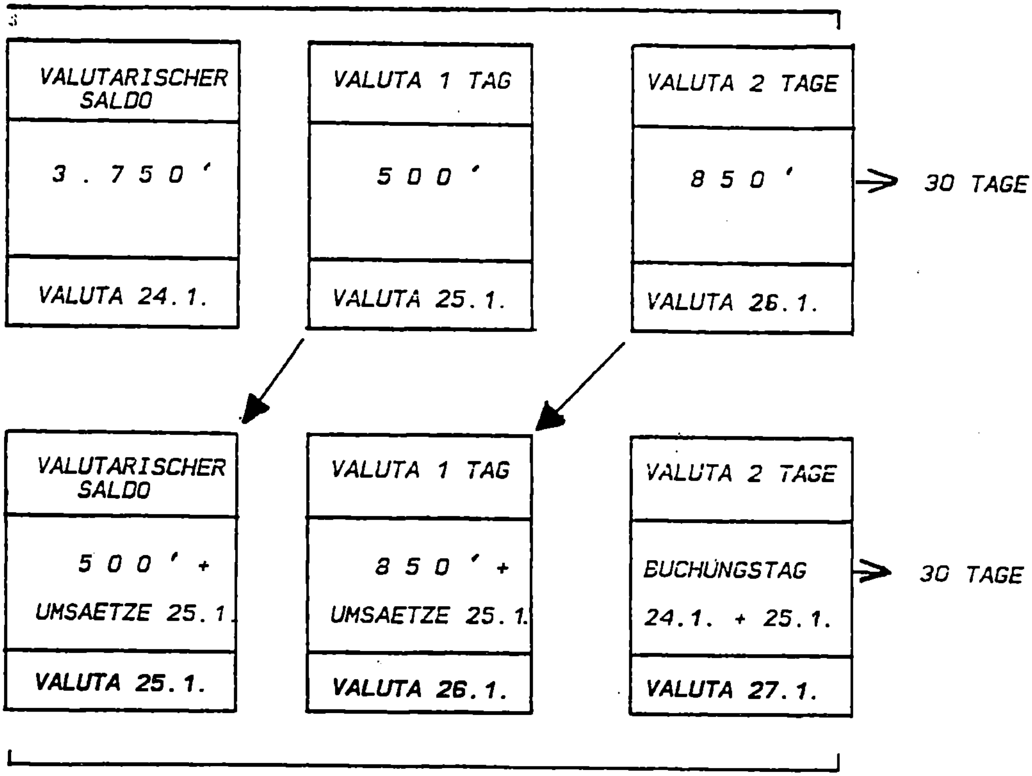

VALUTARISCHER SALDO	VALUTA 1 TAG	VALUTA 2 TAGE	
3 . 7 5 0 '	5 0 0 '	8 5 0 '	⇒ 30 TAGE
VALUTA 24.1.	VALUTA 25.1.	VALUTA 26.1.	

VALUTARISCHER SALDO	VALUTA 1 TAG	VALUTA 2 TAGE	
5 0 0 ' + UMSAETZE 25.1.	8 5 0 ' + UMSAETZE 25.1.	BUCHUNGSTAG 24.1. + 25.1.	⇒ 30 TAGE
VALUTA 25.1.	VALUTA 26.1.	VALUTA 27.1.	

ARBEITSTAG 26.1. (BUCHUNGSTAG 25.1.)

<u>Anlage 7</u>

DREBIT  BILDSCHIRMTEXT

S C H N I T T S T E L L E   Z U   D R E C A M
==

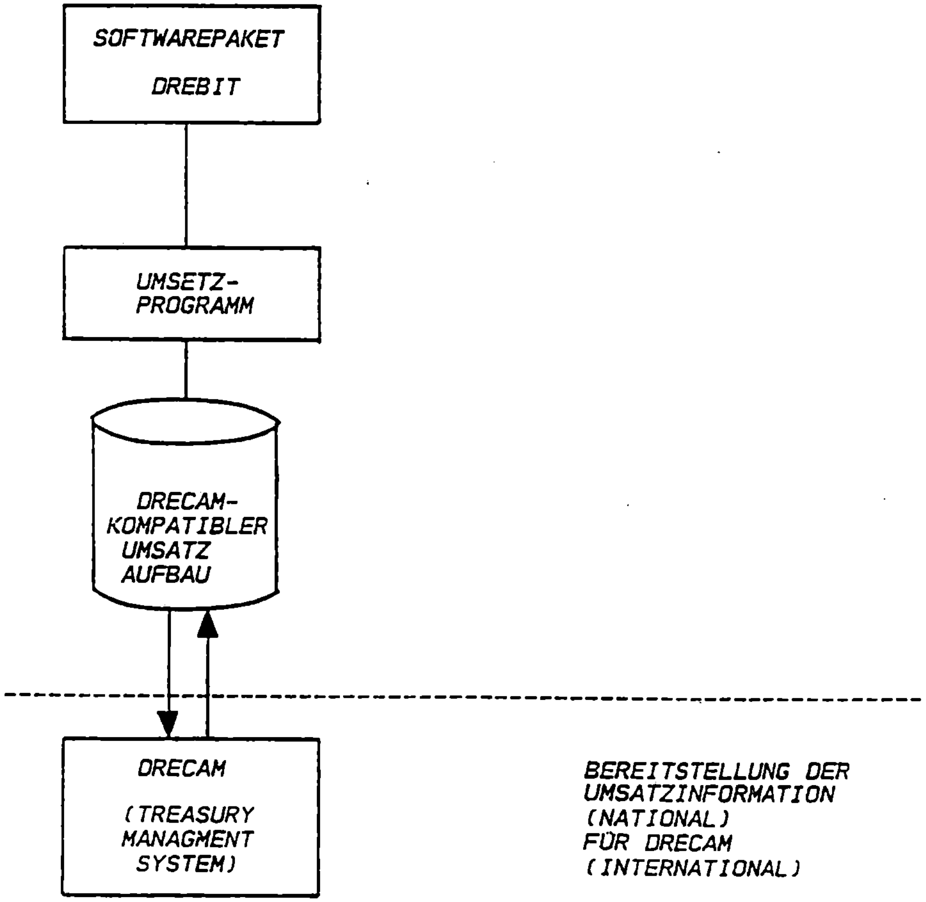

Z U M   B E T R I E B   D E S   D F N

Klaus Truöl

Zentrale Projektleitung des DFN - Berlin
Ges. f. Mathematik und Datenverarbeitung - Darmstadt

Dezember 1984

Inhalt

1. Ziele für ein DFN
2. Dienste im DFN
    2.1 Basisdienste
    2.2 Dienste zum Nachrichtenaustausch
    2.3 Graphik- und weitere spezialisierte Dienste
3. Protokollarchitektur des DFN
4. Technische Struktur des DFN
5. Realisierungsstand
6. Organisationsformen des DFN
7. Betriebskonzept des DFN
    7.1 Bilaterales Kommunikationsmodell
    7.2 MHS-Kommunikationsmodell
    7.3 Protokoll-Konformitätstest
    7.4 Softwareverteilung und -wartung
    7.5 DFN-Informationssystem
8. Literatur

Zusammenfassung

Es werden die Ziele für Aufbau und künftigen Betrieb eines Deutschen
Forschungsnetzes DFN geschildert sowie die Kommunikationsdienste
dargestellt, die im DFN realisiert und angeboten werden. Die wesentlichen Konzepte zum Betrieb eines solchen Forschungsnetzes werden
geschildert. Grundlage hierfür sind bilaterale Beziehungen zwischen
Anbieter und Nutzer von Kommunikationsdiensten und Rechner-Resour-
cen. Für ein zentrales DFN-Management verbleiben wichtige Aufgaben

wie Nutzerberatung, Informationsangebot, Protokolltest, Software-
verteilung und Softwarewartung, Organisation eines verteilten Nach-
richtendienstes.

Abstract

The goals for building up a German Research Network DFN and the concepts for the future operation of it are discussed. DFN offers a set
of communication services which are treated briefly. The basic concepts of DFN operation are the two-way relations between the provider and the user of communication services and hard-/software resources. There are some tasks remaining for a central DFN management, namely advice and information of users, protocol test and certification, software distribution and maintenance, organization of a
distributed message service.

1. Ziele für ein DFN

Das Deutsche Forschungsnetz DFN wird geschaffen, um ein Instrumentarium an Kommunikationsdiensten für Anwender aus dem deutschen Wissenschaftsbereich zur Verfügung zu stellen. Hierzu gehören die Hochschulen, die Großforschungseinrichtungen, die öffentlich geförderten
Forschungseinrichtungen sowie die Forschungsstellen der Industrie.
In enger Kooperation mit Hardware- und Softwareherstellern und basierend auf international abgestimmten Dienst- und Protokollspezifikationen wird für diesen Nutzerkreis die verstärkte Möglichkeit zur
Kommunikation und zur Kooperation mit Hilfe von Telekommunikations-
medien geschaffen.

Das DFN ist für wissenschaftliche Nutzer und Nutzergruppen konzipiert. Die Idee für ein DFN ist aus dem Bedarf dieser Anwenderklasse
entstanden, und das Verbundnetz wird mit der Akzeptanz der Anwender
leben und wachsen. Es ermöglicht dem Anwender die Nutzung der Kapazitäten entfernter wissenschaftlicher Rechenzentren, und es unterstützt den Zusammenschluß kooperativer Wissenschaftlergruppen. Damit
ist eine Steigerung der wirtschaftlichen Nutzung von Rechnern und
Programmen im Wissenschaftsbereich verbunden. Der dezentrale Zugriff
auf kostspielige Betriebsmittel an Hard- und Software, z. B. auf
Vektorrechner oder spezielle periphere Geräte erweitert das instru-

mentelle Spektrum für den einzelnen Wissenschaftler. Auch für Wissenschaftler aus den nicht-technischen Disziplinen, z. B. den Geisteswissenschaften, eröffnet der computergestützte Nachrichtenverbund neue Dimensionen der Kooperation.

2. Dienste im DFN

Die DFN-Zielvorstellungen, für wissenschaftliche Nutzer Kooperationen durch

* Geräteverbund
* Programmverbund
* Datenverbund
* Nachrichtenverbund

zu ermöglichen, werden durch ein Angebot an Kommunikationsdiensten realisiert. Diese werden im folgenden kurz skizziert.

2.1 Basisdienste

Die für das DFN realisierten Basisdienste für

* Dialog
* Remote Job Entry
* File Transfer

geben die Grundlage für den Zugriff auf Rechner anderer Forschungseinrichtungen, sei es im Dialog oder im Stapelbetrieb, sowie für den Austausch von Dateien. Dadurch wird die Basis für die angestrebte Unterstützung von Kommunikation und Kooperation geschaffen.
D i a l o g : Ein einfacher zeilenorientierter Dialogzugriff auf Rechner über das öffentliche Transportnetz Datex-P ist gemäß den CCITT-Empfehlungen X.3/X.28/X.29 realisiert. Dieser Dienst gestattet es, von asynchronen zeilenorientierten Terminals aus über eine PAD-Einrichtung (PAD = Packet Assembly/Disassembly) und das Datex-P-Netz zu einem entfernten Zielrechner eine Verbindung herzustellen. Diese PAD-Funktion kann sowohl als eine private oder posteigene Hardware-PAD-Einrichtung realisiert sein oder auch als Software-PAD. Letzterer ist eine Softwarekomponente in einem Host, welche auf einer X.25-Implementierung aufsetzend die 'Packet Assembly/Disassembly'-Funktionen gemäß X.29 nachbildet. Über einen Software-PAD ist die Dialogfähigkeit für alle Terminals des Hosts gegeben; er kann zusätzlich für eine sehr einfache Realisierung eines Filetransfers erweitert werden. Im letzeren Fall werden Eingaben an den Zielrechner

nicht vom Terminal sondern aus einer Datei abgerufen und umgekehrt
Ausgaben von dem Zielrechner nicht auf das Terminal sondern in eine
Datei geleitet.

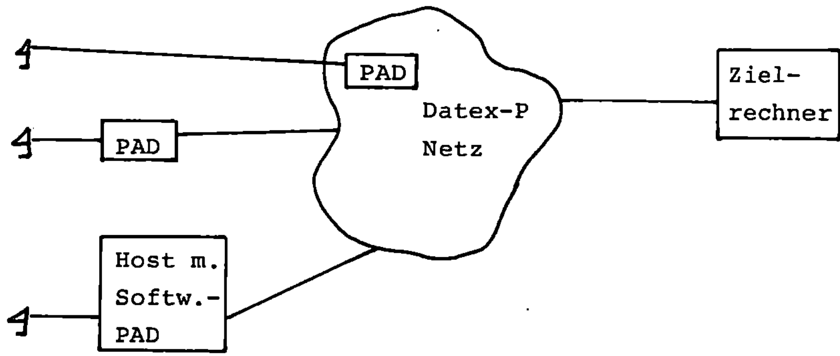

Für Dialogfunktionen allgemeinerer Art, z. B. für seiten- und for-
matorientierten Dialog, fehlen zur Zeit noch die international standardisierten Spezifikationen für ein v i r t u e l l e s   T e r -
m i n a l. Von Fachinformationszentren ist ein Bedarf formuliert
worden, der sich zunächst an dem nationalen Zwischenprotokoll EHKP6
für den Bildschirmtext-Dienst orientiert. Hierfür sind spezielle
Implementierungen für TRANSDATA-Netze vorgesehen.
R e m o t e   J o b   E n t r y :   Der RJE-Dienst ermöglicht es,
über das Netz einen Stapelauftrag in der Kommandosprache des Zielrechners zu schicken, den Auftrag dort ausführen zu lassen und das
Ergebnis zum initiierenden Rechner zurück oder zu einem dritten
Rechner weiterzuschicken.

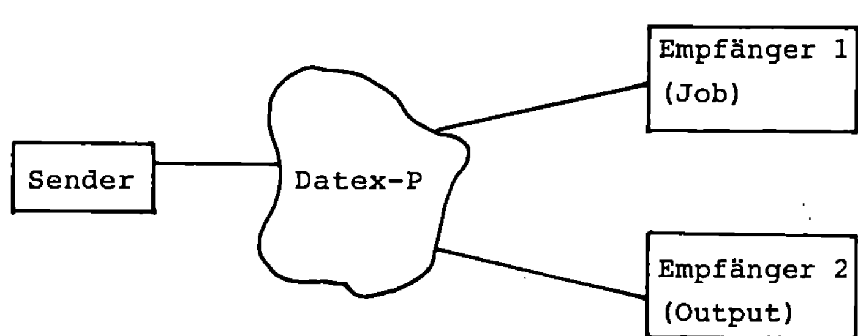

Da die Spezifikationen des JTM-Protokolls (Job Transfer and Manipulation) zur Zeit noch nicht verabschiedet und auch nicht stabil
sind, wird ein im PIX-Projekt erarbeitetes RJE-Protokoll als natio-

nale Zwischenlösung implementiert.

F i l e   T r a n s f e r :   Die Übertragung einer Datei von dem lokalen Speicher eines Rechners in den Speicher eines entfernten Rechners und umgekehrt ist ein wichtiger Dienst zur Unterstützung des Datenaustausches zwischen kooperierenden Wissenschaftlern. Auch hier ist die Situation dadurch gekennzeichnet, daß die FTAM-Spezifikationen der ISO noch nicht stabil genug sind. Es wird daher auch wieder als Zwischenlösung eine vereinfachte Fassung des für BERNET in Berlin spezifizierten RDA-Protokolls (Remote Data Access) zugrunde gelegt.

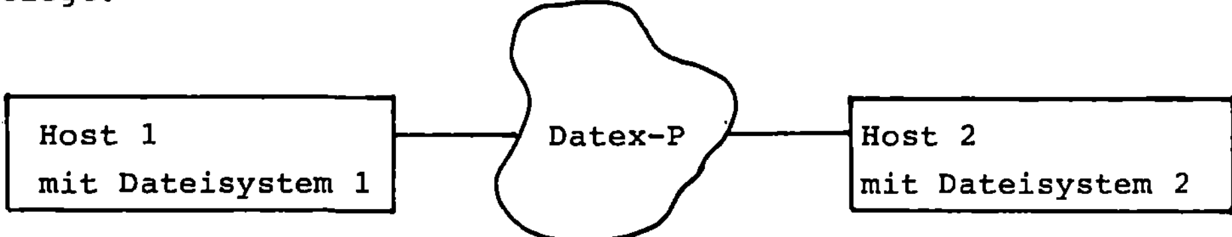

2.2 Dienste zum Nachrichtenaustausch

Eine der wichtigsten und am meisten verbreiteten Anwendungsformen der computerunterstützten Kommunikation ist der Nachrichtenverbund (Electronic Mail, Electronic Messaging) und darauf aufbauend rechnergestützte Konferenzführung. Hierzu gehören Funktionen wie Definition von Verteilerlisten, Anfordern von Empfangsbestätigung und Antwort, Ablage von Briefen zur Wiedervorlage, Archivverwaltung usw. Im DFN wird das zentrale Messagesystem TELEBOX der Deutschen Bundespost eingesetzt.

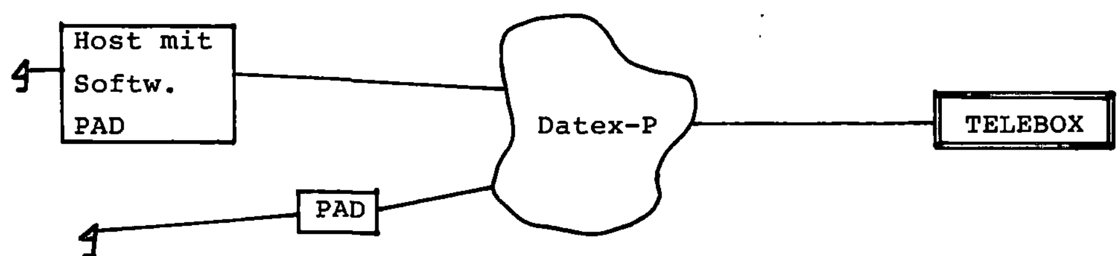

Darüber hinaus sind Vorbereitungen getroffen und es werden Projekte beginnen, in enger Kooperation mit der DBP ein dezentrales Messagesystem MHS zu realisieren. Dieses wird auf den CCITT-Empfehlungen der Serie X.400 für 'Message Handling Systems' beruhen und auch Übergänge in das Teletexnetz sowie in ausgewählte private Message-Systeme ermöglichen.

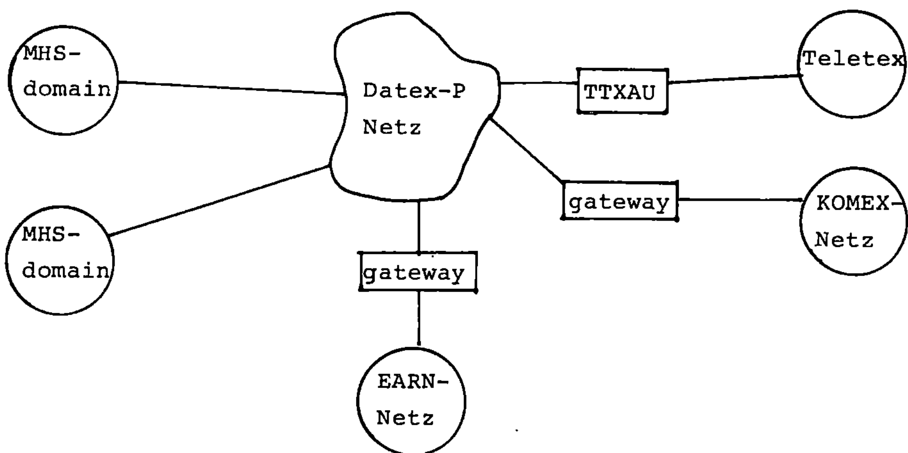

Aufgrund der international standardisierten MHS-Protokolle bzw. über Gateway-Realisierungen wird hier verstärkt die Möglichkeit zur internationalen Kommunikation geboten.

2.3 Graphik - und weitere spezialisierte Dienste

Die graphische Datenverarbeitung läßt einen zunehmenden Einsatz in vielen Anwendungsbereichen erkennen. Damit wächst auch der Bedarf, graphische Informationen in einem Netz zu übermitteln, auf Spezial-hard- und -software im Graphikbereich zuzugreifen. Die zu realisierenden Dienste können den folgenden Bereichen zugeordnet werden:

* GKS-orientierter Dialog und Filetransfer
* Übertragung produktbeschreibender Modellierungsdaten
* Dokumentenübertragung von Text und Graphik

Die Konzipierung und Realisierung weiterer anwendungsbereichspezifischer Dienste ist vorgesehen. Dies entspricht der allgemeinen Zielvorstellung, mit dem DFN einen Kommunikationsmarkt für Nutzergruppen zu schaffen.

3. Protokollarchitektur im DFN

Der erklärte Grundsatz für den Aufbau des DFN ist es, den Implementierungen für die Kommunikationsdienste nur international verabschiedete und akzeptierte Spezifikationen für Dienste und Protokolle zugrundezulegen. Dadurch soll die Kommunikationsmöglichkeit nicht nur beschränkt auf den engeren DFN-Bereich gegeben sein, sondern darüber hinaus der Anschluß an internationale Netze problemlos oder

zumindestens mit geringem Aufwand realisierbar sein. Nationale Zwi-
schenlösungen sind nur zeitlich befristet eingeplant, wenn die internationale Normen noch nicht abgeschlossen ist (z. B. FTAM, JTM).
Auch an den Bestrebungen zur europaweiten und weltweiten Harmonisierung der Normung, der Festlegung von Optionen und Freiheitsgraden in
den Normenspezifikationen, ist das DFN beteiligt und wird sich ihnen
anschließen.

Einen Überblick über die Protokollarchitektur im DFN - wie sie im
DFN-Protokoll-Handbuch festgelegt ist - und ihre Einordnung in die
ISO - Architektur für Offene Kommunikationssysteme gibt das folgende
Bild.

<u>Schichten</u>

6 - 7	X.29- Dialog	RJE	FT	VT- EHKP6	Graphik- Dienste	MHS CCITT X.400ff
5						CCITT T.62 (ISO Layer 5, BAS)
4		CCITT T.70 (ISO Transport Class 0)				
1 - 3		CCITT X.25				

In dieser Übersicht sind die ISO-Dienste für FTAM, JTM und das Virtuelle Terminal nicht aufgenommen, da sie zukünftigen Weiterentwicklungen vorbehalten sind.

4. Technische Struktur des DFN

Das DFN ist ein offenes, heterogenes Netz für die deutsche Wissenschaft, das

        * auf international standardisierten und abgestimmten
          Diensten und Protokollen aufsetzt und das

        * die öffentlichen Transportnetze der DBP benutzt.

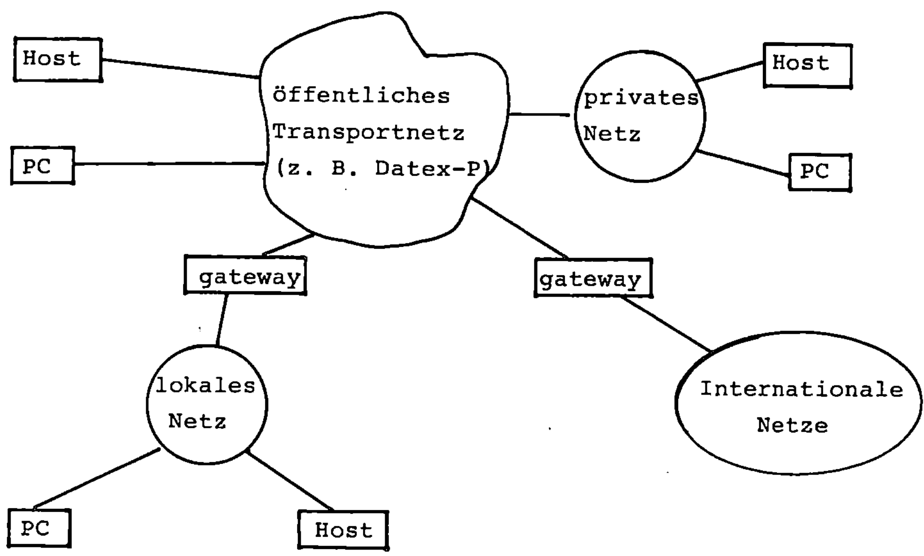

Zum DFN gehört damit kein eigenes Leitungsnetz; es gibt keine Netzknoten, kein DFN-Netzkontrollzentrum. Über öffentliche Transport-
netze werden Personal-Computer, Rechner, private Netze (z. B. Unter-
vermittlungen) und auch lokale Netze unterschiedlicher Architekturen
verbunden. Probleme, die damit zu bewältigen sind, liegen insbesondere im Bereich des Anschlusses einzelner Systeme, der Kopplung von
'Wide Area'-Netzen mit 'Local Area'-Netzen und der konzeptionellen
Beherrschung verteilter Message-Systeme.

5. Realisierungsstand

Betriebssysteme im DFN, für die zur Zeit Kommunikationsbausteine
verfügbar sind bzw. für die solche entwickelt werden, sind:

CDC NOS/BE	Siemens R30 ORG
CDC NOS	DEC PDP11 RSX
IBM MVS	DEC10 TOPS10
IBM VM	UNIX (AT&T)
IBM SNA	UNIX (Berkeley)
Siemens BS2000	ND 100
Siemens BS3000	UNIVAC OS1100
Siemens TRANSDATA	

Das DFN konnte aus früheren Projekten bzw. auch aus Herstellerpro-
dukten bereits eine Reihe von Kommunikationsdiensten übernehmen und
damit mit einer <u>0-ten Protokollgeneration</u> starten. Hierzu gehören
die folgenden Dienste.

X.29:	CDC/NOS BE, PDP11 RSX, VAX11 VMS, DIETZ 621, IBM MVS, BS2000, BS3000,TR440, UNIVAC 1100
Software-PAD:	PDP11 RSX, VAX11 VMS, DIETZ 621, HP 1000, BS2000, Siemens R30
RJE:	CDC NOS/BE, CDC NOS, PDP11 RSX, DIETZ 621, BS2000, TR440, UNIVAC 1100
File Transfer:	PDP11 RSX, VAX11 VMS, BS3000

Allerdings setzen die Dienste RJE und FT dieser 0-ten Generation auf
dem Transportprotokoll 'Message Link' auf, einer Entwicklung aus dem
früheren PIX-Projekt.

Die Kommunikationsdienste der <u>ersten Protokollgeneration</u>, der ersten
wirklich DFN-spezifischen Protokollgeneration sind gemäß der in Kapitel 3 skizzierten Protokollarchitektur aufgebaut. X.29-Dialog und
Software-PADs können als Herstellerprodukte bzw. aus früheren Projekten beibehalten werden. Die übrigen Dienste werden zur Zeit auf
das Transportprotokoll T.70 (ISO Transport Class 0) umgestellt bzw.
für viele der o. g. Systeme neu entwickelt. Fertigstellung und Einsatz sind ab Anfang 1985, z. T. auch erst in 1986 geplant. Insbesondere die Realisierung eines verteilten Message-Dienstes wird erst
1986 zum Einsatz kommen.

Die <u>zweite DFN-Protokollgeneration</u> wird insbesondere auch für File
Transfer und Job Transfer internationale Protokollspezifikationen
benutzen und verstärkt Herstellerprodukte einsetzen. Produkte hierzu
werden nicht vor 1987 verfügbar sein.

6. Organisationsformen des DFN

Das deutsche Forschungsnetz DFN ist ein Begriff für eine Einrichtung, die je nach Blickwinkel und nach Interessenlage unter verschiedenen Sichtweisen betrachtet und damit auch beschrieben werden kann.

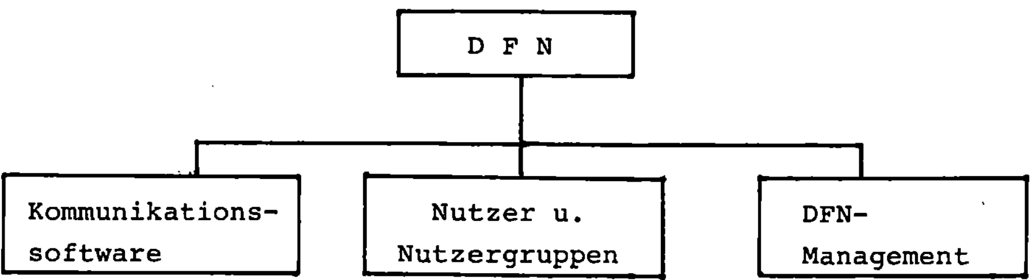

1.Sicht: Kommunikationssoftware

Das DFN ist eine Menge von Kommunikationssoftware, die auf den unterschiedlichen, heterogenen Rechnern der dem DFN angeschlossenen
Mitgliedsinstitutionen implementiert ist. Sie bietet damit der
'wissenschaftlichen Community' unter Benutzung öffentlicher Trans-
portnetze die Möglichkeit zur Kommunikation und Kooperation.

2. Sicht: Nutzer und Nutzergruppen

Das DFN ist eine Menge von Nutzern und - nationalen und internationalen - Nutzergruppen, die sich im DFN zusammenschließen, um Unterstützung für ihre wissenschaftliche Kooperation zu finden. Das DFN
wird für den Kommunikationsbedarf solcher Nutzer und Nutzergruppen
geschaffen.
Einige Nutzergruppen, die sich derzeit schon gebildet haben, sind
aus den Bereichen

* Schaltkreisentwicklung (Pasinger Kreis)
* Hochenergie- und Plasmaphysik (HEPNET)
* Verteilung der Methodenbank RSYST
* Schiffsbau
* Künstliche Intelligenz
* Jobverbund Nordrhein-Westfalen
* Bibliothekswesen, Informationsdatenbanken

3. Sicht: DFN als Projekt und als Management-Aufgabe

Für den Aufbau und den Betrieb eines solchen Netzes sind gewisse Ma-
nagement-Aufgaben zu organisieren und durchzuführen. Hierzu gehören
Projektbetreuung für den Netzaufbau, Organisation des Betriebes (s.
Kap. 7), Betreuung und Beratung der Nutzer, Unterstützung der Bildung neuer Nutzergruppen, Koordinierung des Ausbaus weiterer Dienste.

Zu diesem Zweck wurde ein 'Verein zur Förderung eines Deutschen Forschungsnetzes' e. V. gegründet. Er betreut den Aufbau des DFN, verwaltet die Fördermittel des BMFT und wird den Betrieb organisieren.

Dieser Verein hat die folgende Struktur:

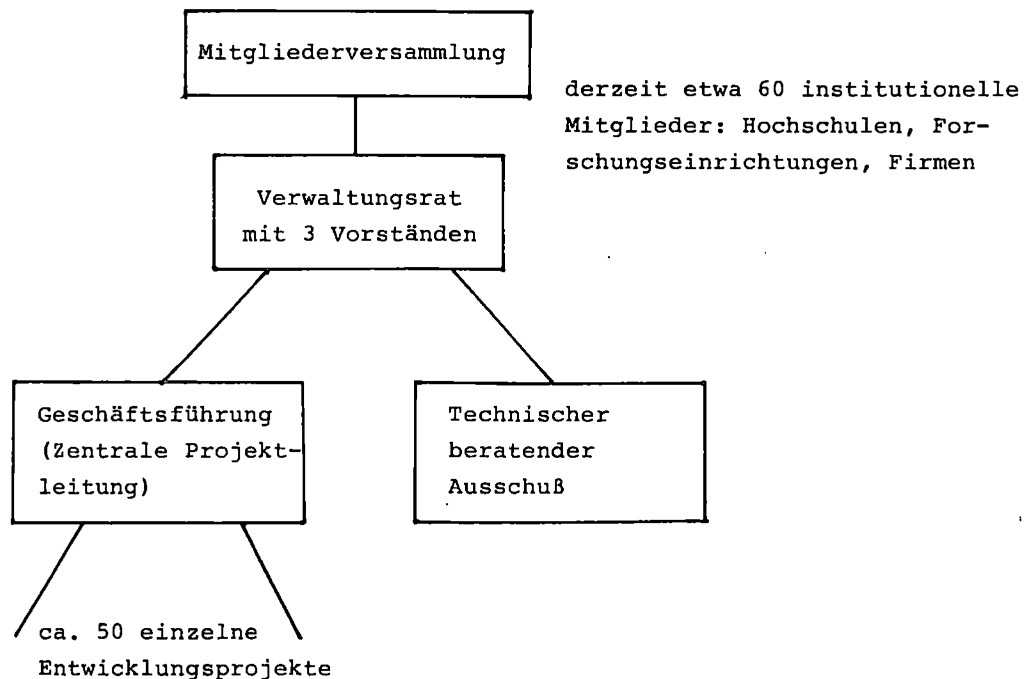

7. Betriebskonzept des DFN

Das Deutsche Forschungsnetz befindet sich im Aufbau. Die ersten Softwareprodukte der 1. Protokollgeneration werden 1985 zum Einsatz kommen. Es sind daher rechtzeitig Überlegungen anzustellen und Konzepte zu entwickeln, um einen reibungslosen und vor allem nutzerfreundlichen und auf die speziellen Bedürfnisse der Nutzer abgestimmten Betrieb des DFN sicherzustellen. Das Grundprinzip hierbei ist die Selbststeuerung des Netzes aufgrund von Angebot und Nachfrage, aufgrund des natürlichen Kooperationsverlangens der wissenschaftlichen Nutzergruppen. Zentrales Management im DFN soll auf das notwendige Minimum beschränkt bleiben. Im folgenden werden die Konzepte zum Betrieb des DFN kurz skizziert. Dabei liegt die Betonung auf dem künftigen Betrieb, nicht auf dem Aufbau des Netzes und der Betreuung der Entwicklungsprojekte. Letztere erfordern naturgemäß ein erhöhtes Maß an zentraler Koordinierung.

7.1 Bilaterales Kommunikationsmodell

Software für Kommunikationsdienste wird zentral im DFN und für DFN entwickelt. Diese Produkte sind anschließend auf den verschiedenen autonomen Rechnern und Rechnersystemen der DFN-Mitgliedsorganisationen installiert und verfügbar. Damit kann das Angebot und die Nutzung dieser Dienste durch bilaterale Anbieter/Nutzer-Beziehung geregelt werden. Hierzu gehören einerseits das Angebot der Dienste, die Bereithaltung von Kapazitäten, die Vergabe von Benutzerkennzeichen und Zugriffsberechtigungen, die Kostenverrechnung für in Anspruch genommene Leistungen; auf der anderen Seite die Nutzung der Dienste, die Beantragung von Zugangsberechtigungen und die Anerkennung der beim jeweiligen Zielsystem geltenden Betriebs- und Gebührenordnung. Das Grundprinzip ist eine verursacherorientierte und nutzungsorientierte Abrechnung auf zweiseitiger Basis. Eine DFN-zentrale oder gar einheitliche Regelung dieser Fragen würde einen hohen Verwaltungsaufwand bedeuten und vor allem zu sehr in die Autonomie der Rechenzentren und der einzelnen Länderverwaltungen eingreifen. Insbesondere sind auch die Gebührenstrukturen und Modalitäten der Kostenverrechnung bei den DFN-Mitgliedern zu unterschiedlich, um in ein einheitliches Modell überführt werden zu können. Übertragungskosten sind hierbei unproblematisch, da sie dem Verursacher von der DBP in Rechnung gestellt werden.

7.2 MHS-Kommunikationsmodell

In einem verteilten Nachrichtenvermittlungssystem, wie es gemäß CCITT X.400 ff für DFN realisiert werden soll, ist der Betrieb nicht mehr in einfacher Form nach dem Verursacherprinzip zwischen Anbieter und Nutzer organisierbar. Hier sind zusätzliche vertragliche Absprachen zwischen MHS-Domains und des lokalen MHS-Nutzers mit der für ihn zuständigen Domain erforderlich. Diese beziehen sich z. B. auf ständige Verfügbarkeit, auf die Bereithaltung von Speicher für die Aufnahme von Nachrichten, auf die Verrechnung von Sekundärkosten (Weiterleitung von Nachrichten an die Empfänger, automatisches Rücksenden von Empfangsbestätigungen, d. h. Bearbeitungsleistungen des empfangenden Message-Transfer-Agenten).
Absprachen des Benutzers mit seiner lokalen Domain:
Es wird ein domainspezifischer pauschaler Preis je gesendete Nach-

richteneinheit festgesetzt. Dieser fixe Kostensatz kann variieren je nach Zieldomain (Tarifautonomie der einzelnen MHS-Domains!) und abhängig sein von gewünschten Leistungsoptionen (z. B. mit oder ohne automatische Empfangsbestätigung). In die Kalkulation für einen solchen fixen Kostensatz gehen ein

* Kosten für die lokale Nachrichtenverteilfunktion
  (lokale MTA-Funktion)
* Übertragungskosten
* Sekundärkosten in der Empfänger-Domain
  (Empfänger-MTA-Funktion)

Absprache zwischen MHS-Domains:
Sekundärkosten für die Bearbeitung empfangener Nachrichten werden pauschaliert je Nachrichteneinheit erfaßt und der sendenden Domain in Rechnung gestellt. Diese von der Sender-Domain zu tragenden Kosten sind dort bereits bei der Kostenkalkulation für die lokalen MHS-Nutzungsgebühren berücksichtigt.

Die Implementierung eines solchen Modells wird einer gewissen Zeit des Anlaufs und des Sammels von Erfahrungen bedürfen. Es ist daher ein etwa 1-jähriger Probebetrieb unter zentraler DFN-Koordination vorgesehen. Auf die Probleme der Kostenverrechnung bei Gruppenkommunikation und bei Gateways zu anderen Netzen kann in dem Umfang dieser einführenden Darstellung nicht eingegangen werden.

7.3 Protokoll-Konformitätstest

Das DFN soll Kommunikation in einem offenen heterogenen Netz ermöglichen. Damit ist die Verpflichtung verbunden, die Korrektheit, Zuverlässigkeit und Verträglichkeit mit den Normen und Standards des DFN-Protokollhandbuchs zu gewährleisten. Durch Nutzung der Test- und Abnahmedienste der DBP (X.25, künftig gegebenenfalls MHS) sowie durch Einrichtung und Betrieb eines Protokoll-Testlabors bei der GMD in Darmstadt als zentralem DFN-Dienst wird diese Protokollkonformität überprüft und zertifiziert. Das bedeutet, ein Prüfling, die Implementierung eines Kommunikationsdienstes, muß bei einer Kommunikation mit einer Test-Implementierung korrektes Protokollverhalten zeigen und insbesondere auch auf vom Tester beabsichtigte fehlerhafte Protokolldaten vorschriftsmäßig reagieren. Nur auf diese Weise zertifizierte Produkte werden im DFN zum Einsatz kommen.

7.4 Softwareverteilung und -wartung

Neben der Zertifizierung der Protokollkonformität ist als wichtige
zentrale Aufgabe im DFN die dienst- und funktionsspezifische Abnahme
von entwickelten Produkten im softwaretechnologischen Sinne identifiziert. Dies bedeutet Überprüfung auf Konformität mit dem Pflichtenheft, auf korrekte Diensterbringung, auf sinnvolle Einbettung in
das Betriebssystem, auf eine modulare und übersichtliche Systemstruktur sowie auf Vollständigkeit der Dokumentation. Weiterhin gehören hierzu die Verteilung der Software an DFN-Mitglieder. Instal-
lationsberatung, Verfolgung von Fehlern und Organisation der Wartung. Diese Aufgaben sind betriebssystemspezifisch. Es wird daher
für jedes im DFN vertretene Betriebssystem eine Referenzmaschine
eingerichtet, die für die jeweilige Systemumgebung zentral die geschilderten Arbeiten durchführt.

Das bedeutet, Softwareprodukte im DFN durchlaufen vor ihrem Einsatz
im DFN die Instanzen Protokoll-Testlabor und - betriebssystemspezifische - Referenzmaschine. Insbesondere ist hier eine enge Kooperation zwischen dem Protokoll-Testlabor und den Referenzmaschinen vorgesehen.

7.5 DFN-Informationssystem

Die Akzeptanz der Netzdienste im DFN wird sehr stark davon abhängen,
wie leicht der Zugang zu diesen Diensten ist und wie flexibel ein
Interessent Auskunft über angebotene Netzdienste, ihre Verfügbarkeit, über Kosten und Adressen erhält. Im Falle von länger dauernden
Netzstörungen sollten die Benutzer zentral informiert werden können.
Für diesen Zweck wird ein - aus Sicherheits- und Verfügbarkeitsgrün-
den - auf zwei BS2000-Anlagen der FU in Berlin laufendes zentrales
Informationssystem aufgebaut. Es wird auf einer mit SESAM-DRIVE realisierten Datenbank beruhen. Der Benutzer kann über X.29-Zugang auf
dieses Informationssystem zugreifen, um aktuelle Informationen über
das DFN zu erfragen.

8. Literatur

(1)   Deutsches Forschungsnetz - Kurzbeschreibung
          DFN, Berlin 1984

(2)   Deutsches Forschungsnetz - Dienste und Architektur
DFN, Berlin 1985

(3)   Deutsches Forschungsnetz - Konzept für den Betrieb
DFN, Berlin 1985

(4)   Dienste im Deutschen Forschungsnetz
DFN, Berlin 1985

(5)   K. Görgen, H. Parslow, U. Viebeg, S. Vollmer
Protokoll-Testlabor im DFN
DFN, Berlin 1985

(6)   K. Birkenbihl, K. Kröger, F. Limburger
Abnahme, Pflege und Wartung von DFN-Produkten
DFN, Berlin 1985

Zur Einbettung von lokalen Netzwerken
im Deutschen Forschungsnetz - DFN

Wulfdieter L.-Bauerfeld
DFN - Verein, Berlin

1 Einleitung

Das Konzept des im Gesamtprojektplan /DF84.1/ beschriebenen Deutschen Forschungsnetzes - DFN sieht den Verbund von für den Wissenschaftsbereich typischen Rechensystemen ("Hosts") vor. Diese werden als Teilnehmer an ein herstellerunabhängiges Datenübertragungsnetz, zur Zeit überwiegend an das paketvermittelnde DATEX-P der Deutschen Bundespost /BP79/, angeschlossen. Nach dem ISO-Referenz-Modell "Zur Kommunikation offener Systeme" /IS83.1/ werden diese Teilnehmer "Systeme" genannt und sind auf Schicht 3 dieses Kommunikationsmodells, der Vermittlungsschicht, eindeutig identifizier- und adressierbar. Den Nutzern der Systeme stehen zur Kommunikation die DFN-Basisdienste zur Verfügung, die durch die im DFN-Protokoll-Handbuch /DF84.2/ festgelegten und zum Teil international standardisierten Protokolle vermittelt werden. Solche "öffentlichen" Netze mit u.U. weit voneinander entfernten Teilnehmern heißen "Wide Area Networks (WAN)".

Der Zusammenschluß von lokal dicht beieinander stehenden Rechensystemen über ein privates Datenübertragungsnetz mit im allgemeinen höherer Übertragungsrate und anderer -technik als beim WAN heißt "Local Area Network (LAN)". Die internationale Normung beschränkt sich derzeit bei LANs mit IEEE 802 /IE83/ auf die Schichten 1 und 2 des ISO-Referenzmodells. Lokale Netze bilden neben den "traditionellen" Hosts eine neue Gruppe, die ebenfalls für eine regionale oder überregionale Kommunikation an das WAN angeschlossen werden soll. Die Kopplung an ein öffentliches Netz bzw. auch der Zusammenschluß mehrerer LANs unterschiedlicher Technik geschieht über spezielle Rechner, in denen die verschiedenen Protokolle aufeinander abgebildet ("Mapping") werden.

Im DFN sollen grundsätzlich die Basis-Dienste wie Dialog - der Zugriff zeilenorientierter Terminals zu dedizierten Systemen, File Transfer - der Transport von Dateien eines Systems auf ein anderes und Remote Job

<u>Entry</u> - der Transport von Aufträgen von einem System an ein anderes
mit Ausgabesteuerung zu einem beliebigen Dritten, allen Nutzern gleichermaßen zur Verfügung stehen. Damit wird für die Basisdienste kein
Unterschied zwischen "Host-to-Host-", "Host-to-LAN-" oder "LAN-to-LAN-
eSrvices" gemacht.

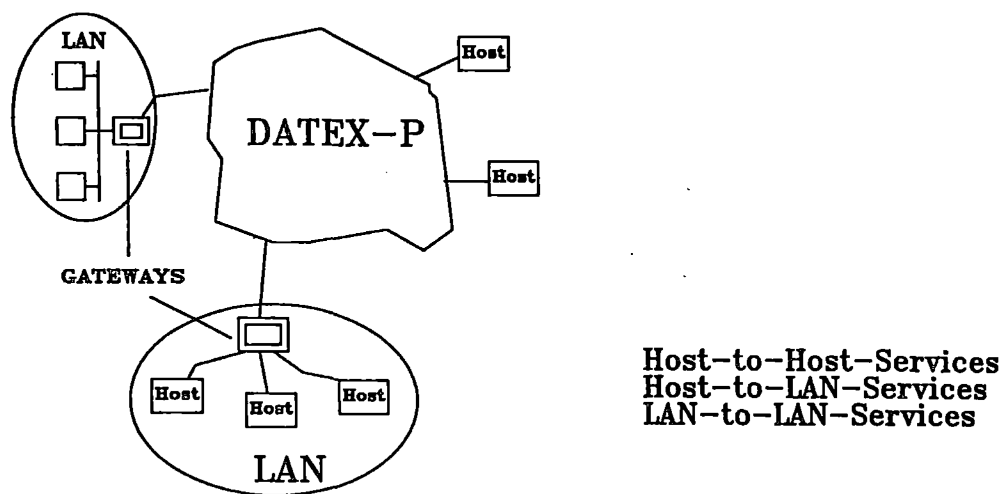

Unterschiede ergeben sich vielmehr für die Kopplungsarchitektur, die
Adressierung und die Protokollübergänge zwischen LAN und WAN, da verschiedene Betrachtungsweisen möglich sind. Zur Einbettung von lokalen
Netzwerken in das DFN sind mehrere Aspekte zu untersuchen, um zu einer
geschlossenen und möglichst aufwandsarmen Vorgehensweise zu gelangen.
Ausgehend von der allgemeinen Zugangs-Problematik im folgenden Kapitel
wird in Verbindung mit einer aus dem ISO-Referenzmodell hergeleiteten
Sichtweise dargelegt, wie sich der Anschluß von lokalen Netzen strukturieren läßt.

Der Zusammenschluß von Rechensystemen zum DFN wird immer unter dem Hintergrund gesehen, welche Rechen- bzw. Betriebssysteme in einem <u>Infra-
strukturprojekt</u> für die Wissenschaft relevant sind und mit Basisdiensten versehen werden sollen. Auch die Einbettung von LANs in das DFN
muß unter diesem Aspekt gesehen werden. Die Vielfalt von unterschiedlichen Konzepten läßt sich in hohem Maß durch die Verfügbarkeit von
LANs eingrenzen. Auch aus den Anwendungsgebieten resultiert, welche
Protokolle im LAN und im WAN einheitlich sind oder sein können und welche Architektur als Folgerung ausgewählt werden muß. Typische Einsatzgebiete von lokalen Netzen, wie sie für die Bürokommunikation und -au-
tomatisierung propagiert werden, werden bereits von der einschlägigen
Industrie abgedeckt und daher im DFN nicht vorrangig verfolgt.

2   Das allgemeine Gateway-Problem

Bei der Verbindung zweier verschiedener Netzkonzepte mit unterschiedlichen Protokoll-"Welten" sollten vom Standpukt des Nutzers aus diese so wenig wie möglich verändert werden, um eine Nutzung verteilter Anwendungen ungestört zu gestatten. Das läßt sich nur dann erreichen, wenn beide Netze für sich als Subnetze betrachtet werden und der <u>Zugang</u> von einem Netz zum anderen über einen <u>"Gateway-Rechner"</u> erfolgt.

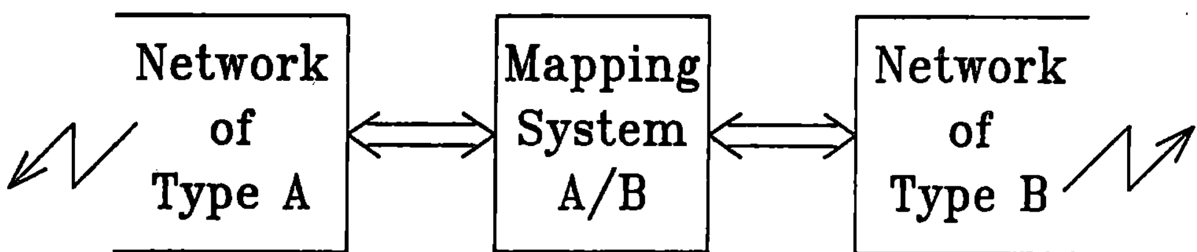

Ein Gateway-Rechner sollte so konstruiert sein, daß jedes Subnetz dieses <u>Mapping-System</u> für Protokolle nur als Fortführung seiner eigenen Netz-Architektur ansieht. So ist SNATCH /EG83/ eine Gateway-Lösung, die die Netzkonzepte zweier verschiedener Hersteller konsequent auch für höhere Protokolle umsetzt. Aufgabe dieses speziell konstruierten Mapping-Systems ist jedoch nicht die Anbindung weiterer Netzkonzepte oder die "Offene Kommunikation" von verschiedenartigen Systemen. Dazu muß jedes Netzkonzept (z.B. vom Typ A oder B) grundsätzlich auf ein "neutrales Netz" abgebildet und die einzelnen Mapping Systeme über dieses neutrale Netz (vom Typ X) miteinander verbunden werden. Das ISO-Referenz-Modell und die daraus abgeleiteten Standard-Protokolle stellen ein solches (hersteller-) neutrales Konzept dar.

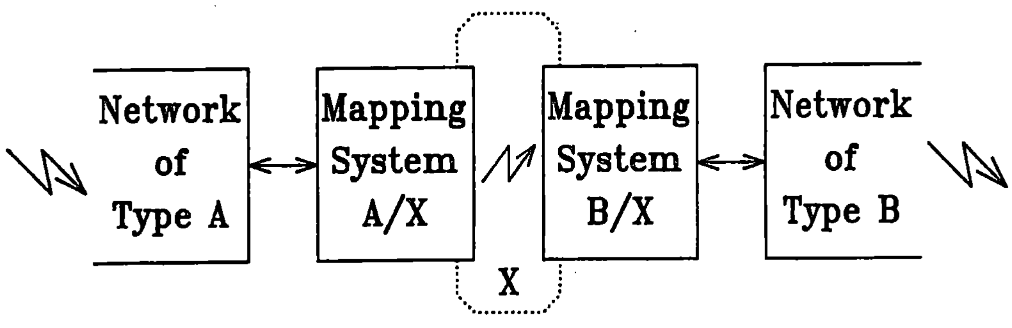

Network of Type X

Ein Mapping-System ist aus drei Moduln zusammenzusetzen, damit ein Anwender T(A) mit einer Anwendung U(X) kommunizieren kann. (Die Begriffe "Anwender" und "Anwendung" sind hier nicht notwendigerweise in Zusammenhang mit der "Anwendungsschicht" des ISO-Modells zu sehen.)

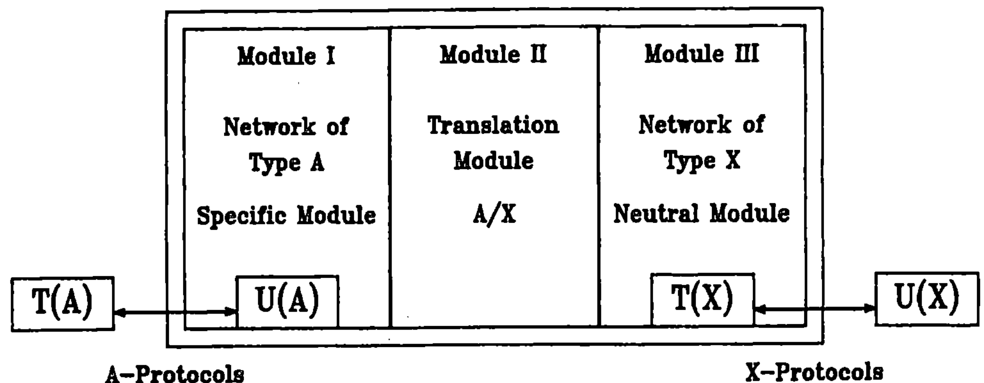

T(n) Terminal im Netz n

U(n) Anwendungsprogramm im Netz n

Das auf der ISO-Architektur und allgemeinen (zumindest DFN-weiten) Standards aufbauende Deutsche Forschungsnetz ist ein ideales neutrales Kommunikationsnetz, um auch lokale Netze mit unterschiedlicher Architektur, Technik und Anwendung miteinander und mit anderen Hosts zu verbinden. Um unterschiedlichen Anwendungsfällen für LANs einerseits und andererseits der Forderung gerecht zu werden, Nutzern in LANs und in einzelnen Hosts gleichermaßen und möglichst aufwandsarm das Angebot von Basis-Diensten zur Verfügung zu stellen, muß das Abbildungsproblem gemäß der Schichten-Architektur des ISO-Referenzmodells untersucht werden, um dann die geeignete Gateway-Architektur herleiten zu können.

3 Unterschiedliche Gateway-Architekturen

Als verbindlich gelten die im DFN-Protokoll-Handbuch niedergelegten Übereinkünfte zur Zeit nur für Hosts, die zur Kommunikation an die WAN-Technik im DFN angeschlossen werden. Innerhalb eines lokalen Netzes kann ein Betreiber eine beliebige Protokoll-Welt nutzen. Um jedoch als Kommunikationspartner am DFN teilnehmen zu können, muß ein Mapping-System zum Anschluß des LAN so konstruiert werden, daß die oben skizzierten Abbildungsmodule in jeder der Kommunikationsschichten implementiert werden, die im LAN und WAN unterschiedlich sind. Ein Mapping-System, welches als Spezialfall im LAN und WAN zueinander völlig inkompatible Protokolle aufeinander abbildet, heiße "Level-7-Gateway".

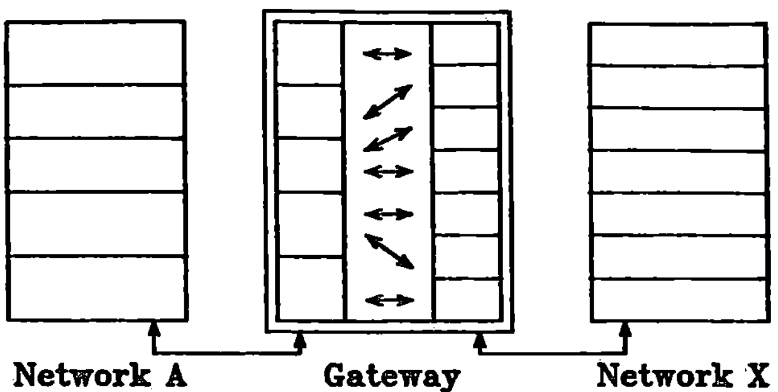

Network A   **Gateway**   **Network X**

Verbindung von zwei inkompatibel strukturierten Netzen

Eine Eins-zu-Eins-Abbildung der einzelnen Protokoll-Elemente und auch der von den einzelnen Kommunikationsschichten erbrachten Dienste wird dabei nicht immer möglich sein, d.h. auf der Anwendungsebene könnten bestimmte Teile eines DFN-Dienstes nicht im LAN bzw. umgekehrt angeboten werden. Da im DFN jedoch lediglich die Einhaltung bestimmter Protokolle und nicht die Spezifikation von Diensten fest vorgeschrieben ist, sollten solche Fälle nur zu Einschränkungen für Benutzer, aber nicht zu Mapping-Systemen führen, die den Protokoll-Spezifikationen des DFN nicht genügen.

Im anderen Extrem wird eine Abbildung schon auf der untersten Ebene nur aufgrund unterschiedlicher physischer Verbindungen das Gateway-Problem zufriedenstellend lösen. Alle darüberliegenden Protokolle sind sowohl auf den direkt am WAN als auch auf den im LAN angeschlossenen Systemen identisch und müssen nicht neu implementiert werden. Ein solches Mapping-System heiße "Level-1-Gateway".

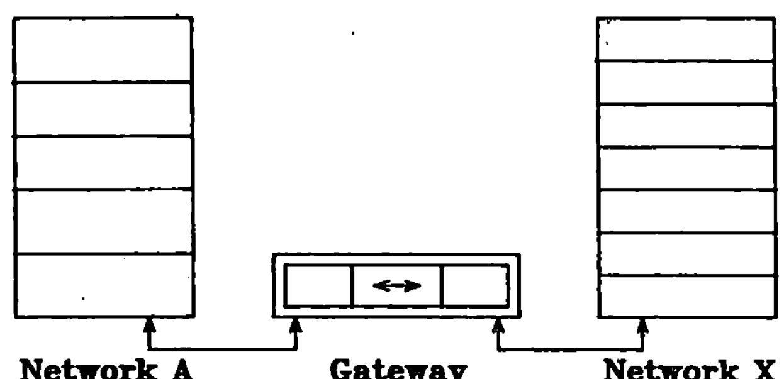

Network A   **Gateway**   **Network X**

Verbindung von zwei nur auf der Bitübertragungsschicht inkompatiblen Netzen

Innerhalb des DFN wird man meist eine dazwischenliegende Lösung finden müssen, um die Vorteile einer bestimmten LAN-Technik mit Implementierungen von höheren Diensten, die übernommen und innerhalb des LAN Verwendung finden können, zu vereinen. So soll z.B. eine File-Transfer-Implementation gemäß DFN-Protokoll-Handbuch auch im lokalen Netz eingesetzt werden und gleichzeitig die Verbindung jedes LAN-Endsystems nach "draußen" zu anderen DFN-Partnern ermöglichen.

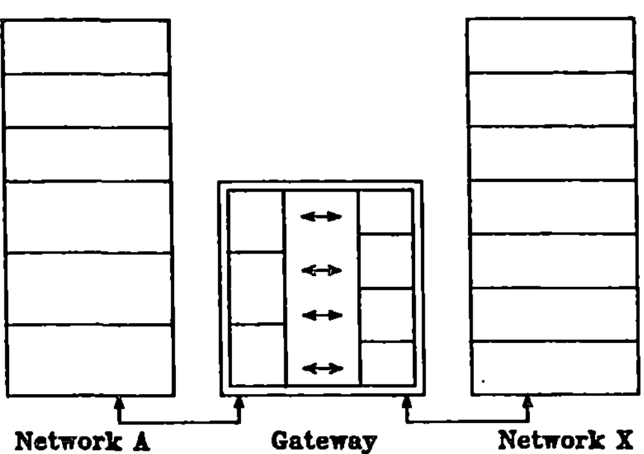

Verbindung von zwei Netzen auf der Schicht n

Typische Schichten, bis zu denen Abbildungsfunktionen durchgeführt werden müssen, sind dann die Schichten 3 oder 4. Sind die Systeme als Endteilnehmer eines LANs auf der Vermittlungsschicht (Schicht 3) adressierbar, d.h. können die WAN-Adressen eindeutig auf Systeme im LAN abgebildet werden, wird das LAN als "Subnetz" des WAN betrachtet. Das entsprechende Mapping-System heißt "Level-3-Gateway". Wird das LAN als ein Endteilnehmer betrachtet, bei dem z.B. erst auf der Transportschicht (Schicht 4) unterschiedliche "Prozesse" unterschieden werden, die durchaus auf unterschiedlichen Systemen im LAN ablaufen können, ist das LAN ein "verteiltes System", und nur über einen "Level-4-Gateway" erreichbar. (Der Begriff "System" ist dann im Sinne des ISO-Modells allerdings nicht mehr eindeutig anwendbar /s.a. Ba84/).

Einige DFN-Architekturen /DFN84/ gehen von einem "angeglichenen" Transportdienst im LAN aus. Um einem über ein LAN angekoppelten Rechensystem oberhalb der im DFN durch X.25 /CC80.1/ realisierten Vermittlungsschicht den DFN-Transportdienst anbieten zu können, müssen u.U. kompliziertere Architekturen entworfen werden, wenn man auf ein bestimmtes, nur innerhalb des LAN notwendiges Transportsystem nicht verzichten will. Im folgendem Beispiel wird ein X.25 Support-Layer definiert, der auf dem LAN-Transport-Protokoll ISO Class 4 /IS83.2/ von und zum Mapping System X.25-Protokollelemente so transferiert, als würde das Rechensystem direkt am DATEX-P angebunden sein. Ohne jede Mitwirkung des Gateways wird darüber dann das DFN-Transport-Protokoll (T.70, ISO Class 0 /IS83.2/) vermittelt, auf dem die Anwendungsdienste aufsetzen.

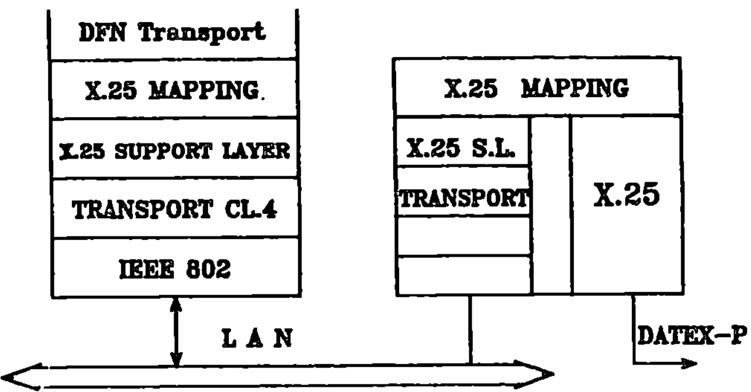

Andere Unterscheidungen sind entweder wegen des' im WAN verwendeten Zu-
gangs-Protokolls X.25 (für Level 2) oder wegen der nicht konsequent
ver.olgten Hierarchie von Diensten im ISO-Modell (für Level 5 und 6)
bei Gateway-Rechnern nicht anwendbar.

4   Gateway-Architekturen im DFN

4.1 Protokoll-Architektur des DFN und Einsatzgebiete von LANs   .

Neben der Betrachtung der einzelnen Kommunikationsschichten des ISO-Re-
ferenz-Modells ist die mögliche Architektur eines Mapping Systems auch
dienstabhängig zu untersuchen. Im DFN wird nach den ISO- bzw. CCITT-
Spezifikationen oberhalb der Vermittlungsschicht zwischen dem Dialog-
Protokoll für asynchrone Terminals (X.3,X.28,X.29 /CC80.1/) und dem
Transport-Protokoll für andere höhere Dienste unterschieden. Auf der
durch T.70 realisierten Schicht 4 basieren dann der File Transfer und
der Remote Job Entry, die als höhere Dienste die Schichten 5 - 7 (Kommunikationssteuerung, Datendarstellung und Anwendung) einschliessen.

Die typischen Nutzungsarten eines lokalen Netzes im Wissenschaftsbereich gehen über die Kommunikation mit anderen DFN-Nutzern mit Hilfe
der Basis-Dienste hinaus und lassen sich ohne Anspruch auf Vollständigkeit wie folgt umreißen:

- Datenerfassung zur Experimentsteuerung und -Auswertung
- Erstellung und Verwaltung von aus Text und Graphik bestehenden
  Dokumenten
- Mehrfache Nutzung von Betriebsmitteln wie CPUs, Massenspeichern,
  Druckern etc. nach dem Server-Prinzip durch Übermittlung von
  Transaktionen

Vor allem in Laborumgebungen sind derartige Einsatzgebiete durch das nutzerspezifische, LAN-interne Bedürfnis nach Realzeitverhalten, d.h. nach einer garantierten Übermittlungszeit zwischen zwei oder mehreren Kommunikationspartnern, geprägt. Innerhalb des DFN als Infrastruktur- projekt werden aber nicht die Entwicklungsarbeiten für spezielle LANs in einem der beschriebenen Einsatzgebiete durchgeführt, sondern - ana- log zum Zusammenschluß von als Produkt verfügbaren Hosts - verstärkt Anschlußmöglichkeiten von als Produkt verfügbaren lokalen Netzen unter- sucht und implementiert. In der Regel wird davon ausgegangen, daß ein Nutzer entweder schon ein lokales Netz betreibt oder eines betreiben wird, welches sich auf dem Markt durchgesetzt hat.

4.2   WAN-Untervermittlungen

Problemlos ist der Zugang vom LAN zum WAN und umgekehrt, wenn im LAN die gleiche Protokoll-Welt für alle Dienste wie im WAN "gefahren" wird. Sind lediglich bestimmte Eigenheiten spezifischer Link-Protokol- le, unterschiedliche Leitungsgeschwindigkeiten oder die Adresse von "Vielfachteilnehmern" auf Einzeladressen abzubilden, läßt sich im DFN als Level-1-Gateway zu einem entsprechenden LAN eine X.25-Untervermitt- lung einsetzen.

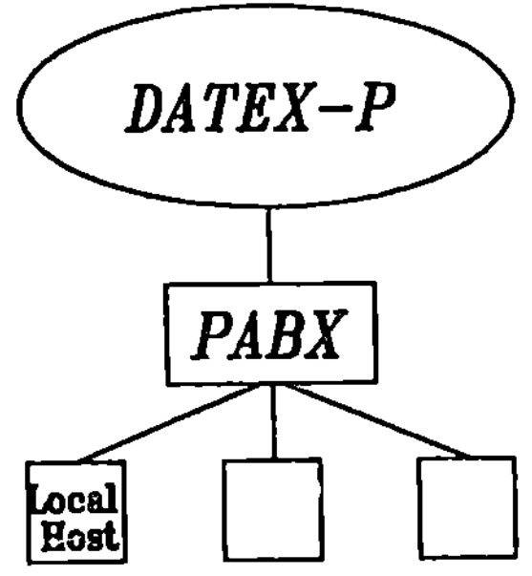

PABX:
Private Automatic Branch Exchange
(Private autom. Untervermittlung)

- Angleichung von Leitungs-
  geschwindigkeiten
- Interpretation von Endadressen
- Umsetzung von Link-Protokollen

Die Untervermittlung agiert dann sowohl als zentraler Knoten des loka- len Netzes als auch als Gateway-Rechner zwischen LAN und WAN /BH83/. Der Dialog- und der Transport-Dienst und die höheren Protokolle sind im WAN und LAN identisch, wobei im LAN ein Dienst durchaus "reichhalti- ger" sein kann /s.a. DFN84.2: Definition des einfachen Filetransfers/. Typische Anwender derartiger LANs, für die die Kopplungsprobleme an die heutige Protokoll-Welt des DFN praktisch gelöst sind, sind diejeni- gen wissenschaftlichen Einrichtungen, die keine hohen Anforderungen an

Übertragungsgeschwindigkeiten oder spezielle Nutzer-Dienste wie Real-Zeit-Applikationen stellen und an ihren Endgeräten die zur Zeit noch relativ aufwendigen X.25-Kopplungsbausteine implementieren können.

4.3   LANs als Subnetze des WAN

Wird im LAN - was meist der Fall ist und ein "echtes" LAN charakterisiert - aufgrund anderer Technik auf den unteren Schichten eine andere Protokollwelt als im WAN eingesetzt, muß ein (zumindest) Level-3-Gateway den Zugang zwischen beiden Netzen ermöglichen. In IEEE 802 werden oberhalb des LAN-spezifischen, technik-abhängigen "Medium-Access-Sublayer (MAC)" für den darüberliegenden "Logical Link Control Sublayer LLC" zwei Protokolle vorgeschlagen, die datagramm- (Typ 1) oder verbindungs-orientiert (Typ 2) sind. Beide sind unabhängig voneinander, können aber in einem Endteilnehmer gemeinsam existieren, da sie anhand eines "Command Codes" voneinander unterschieden werden. Allerdings wird ein Mapping-System, welches die verbindungsorientierte Vermittlungsschicht des WAN bis in das LAN hineinziehen soll, notwendigerweise auf ein bisher nicht standardisiertes Protokoll im LAN abbilden müssen, damit beispielsweise eine Adressierung gemäß X.121 /CC80.2/ wie in einem privaten X.25-LAN durchgeführt werden kann.

LAN als Subnetz eines globalen Netzes

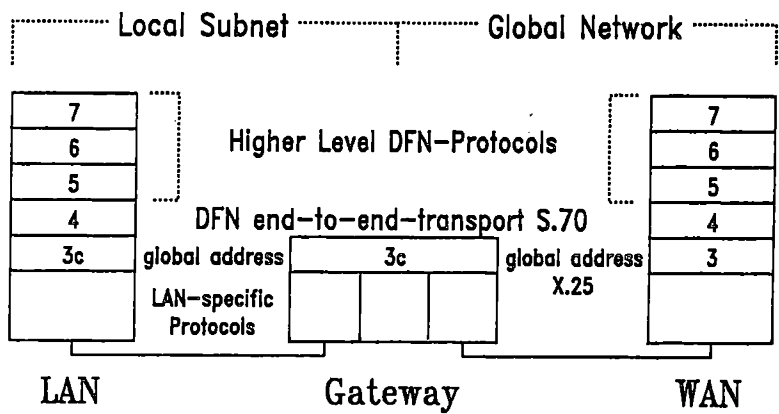

Zur Realisierung wäre im LAN ein aus Link Layer Typ 2 und X.25-Level-3 Protokoll bestehender Vermittlungsdienst eine vom architekturellen Standpunkt aus unproblematische Alternative, diese wird aber zu sicherlich nicht akzeptablen Leistungseinbußen im LAN führen /VD84/. Ein wesentlich leistungsfähigerer Implementierungsvorschlag geht von einem

Dienst aus, der die Funktionen des Link-Layers und der Vermittlungsschicht zusammen erbringt und direkt auf der MAC-Schicht im LAN auf-
setzt. Typische Anwender solcher LANs propagieren entsprechende .Gate-
way-Architekturen genau dann, wenn die Dienste eines WAN (z.B. X.29,
Telematikprotokolle) auch in den Endgeräten des LAN unterschiedslos
zur Verfügung stehen sollen. Abweichungen von zur Normung anstehenden
Vorschlägen können aber nur dann in Kauf genommen werden, wenn die potentiellen Entwickler nicht aus dem für das DFN typischen Wissenschaft-
lerbereich, sondern aus dem Herstellerbereich stammen.

4.4   Produkt-orientierte Gateways

In den meisten Fällen werden zur Erfüllung der oben skizzierten Aufgaben bereits in käuflichen Produkten implementierte Protokolle wie z.B.
XNS (Xerox Network System) genutzt werden können, die wegen ihrer en-
gen funktionalen Anlehnung an Normungsvorschläge einerseits und ihrer
weiten Verbreitung und Implementierung bei Zweitherstellern andererseits einen weltweiten "Quasi-Standard" darstellen. Diese Hersteller-
Produkte bieten in der Regel auf der Transportschicht einen solchen
Quasi-Standard an, auf dem dann Hersteller-spezifische oder vom Benutzer entwickelte Anwendungsprotokolle aufsetzen können. Zugang zu
diesen Netzen erfolgt in der Regel über einen Level-4-Gateway.

LAN als verteiltes Endsystem eines globalen Netzes

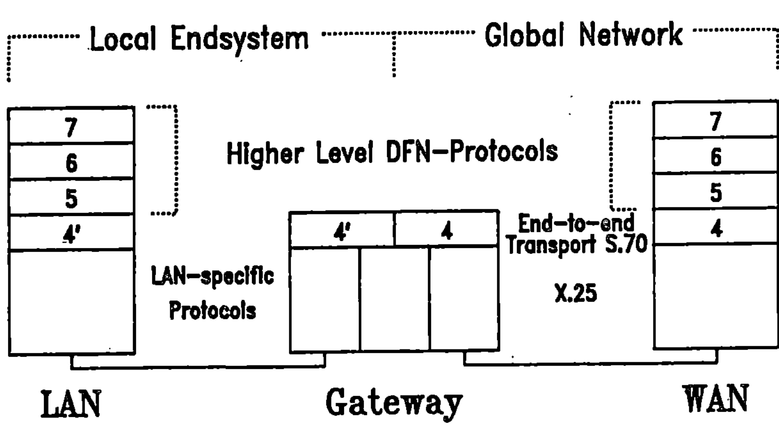

Auf dem "Internet Transport Protocol (ITP)" von XNS können auch spezielle, zur Steuerung und Überwachung abgestimmte Realzeit-Protokolle implementiert werden, da sowohl datagramm- als auch transaktions- bzw.
verbindungsorientierte Transportdienste unterstützt werden. Will man

die vom Hersteller gelieferte Transport-Software nicht verändern, muß
die Angleichung an den im WAN vorgeschriebenen Transportdienst über
eine "Transport-Enhancement- (TE-)" Schicht im Level-4-Gateway erfolgen, da z.B. das "Sequenced Packet Protocol" aus XNS sich nicht direkt
auf die T.70-Schicht des WAN abbilden läßt.

Eine TE-Schicht stellt keinen allzu großen Implementierungsaufwand dar
und liesse sich bei etwaigen Änderungen im WAN- oder LAN-Protokoll verändern, ohne für daß für darauf aufsetzende Dienste im LAN die entsprechende Schnittstelle variiert /He84/. Derartige "Quasi-Standards", die
den Einsatz einheitlicher LAN-Protokolle mindestens bis zur Transportschicht garantieren und damit die Beschaffung, Installation und Wartung innerhalb des DFN rationalisieren, sind bisher nur auf der Basis
von ETHERNET vorgeschlagen worden.

Der im WAN verwendete X.29-Dialog baut direkt auf X.25 auf und verlangt daher bei Level-4-Gateways eine Sonderlösung. Durch Einführung
einer Art "Network-Enhancement" auf der Vermittlungsschicht (s. 4.3)
muß die LAN-weite Nutzung entsprechender Hard- und Software überhaupt
erst ermöglicht werden. Innerhalb des LAN sollte daher ein Dialog-
Dienst auf der Anwendungsschicht angeboten werden, wobei für diesen
Dienst der Nutzer die einheitliche Sicht des gesamten DFN verliert.

4.5   Anwendungsspezifische Gateways

Der Einsatz des im WAN verwendeten X.29-Dialog ergibt im LAN ein grund-
sätzliches Problem: Da X.29 nur zeilenorientiert ist, würde sein Einsatz würde die Einschränkung in der Funktionalität für die LAN-Benut-
zer nach sich ziehen, falls mit einem entsprechenden Protokoll zum
"Full-Screen-Editing" formatorientierte Terminals LAN-weit genutzt
werden könnten.

Die durchgängige Möglichkeit, einen Dialogdienst sowohl im WAN als
auch in angekoppelten LANs erreichen zu können, kann über ein anwen-
dungsspezifisches Mapping-System erreicht werden. Zum Dialog-Verkehr
über das WAN muß der Nutzer die entsprechende Funktion im Gateway-Rechner über das LAN zunächst anwählen. Als aufwandsarme Lösung für einen
solchen Gateway, der auf standardisierten Protokollen im WAN einerseits und auf standardisierten Schnittstellen im LAN andererseits beruht, sei hier für den DFN-weiten Dialog auf die Installierung eines

PAD-Bausteins an einem LAN verwiesen /LZ84/. Auf diese Weise wird nach einem Prädialog DFN-Nutzern von "außen" der Teilnehmer-Zugriff auf Hosts im LAN bzw. nach Anwahl des PAD als Endsystem im LAN DFN-Nutzern von "innen" der Zugriff auf entfernte Hosts über das WAN ermöglicht.

Auch die Implementierung eines "Virtual Terminal"-Protokolls zur Möglichkeit der DFN-weiten Nutzung formatorientierter Terminals /GE84/ erfolgt über einen anwendungsspezifischen Gateway-Rechner sinvollerweise dann, wenn Änderungen in einem vorhandenen Netzkonzept und in etablierten Anwenderprogrammen vermieden werden können. Auch wenn im vorliegenden Beispiel Herstellerarchitekturen (SNA/IBM, Transdata/Siemens) eingebunden werden können, die nicht unbedingt typisch für lokale Netze sind, zeigen die zu lösenden Probleme und die resultierende Architektur solcher Ankopplungen dieselbe Strukturierung wie bei LANs. Durch dieses anwendungsspezifische Gateway-Konzept werden anderenfalls entweder in der Anwender- oder in der Betriebssystem-Software notwendige Änderungen vermieden. Die Hineinführung des Protokolls in die Endsysteme wäre zumindest mit einem höheren Organsisations- und Wartungsaufwand verbunden, der zudem außerhalb des Verantwortungsbereiches eines Herstellers bleibt. Hier bleibt die Wartung für Software im Endsystem beim Hersteller, da sein Netzkonzept genutzt wird. Ein hoher Anteil des notwendigen Aufwands im Mapping-System ist durch die Implementierung eines (noch nicht DFN-standardisierten) Protokolls bedingt.

Eine andere Art anwendungsspezifischer Mapping-Systeme und damit Level-7-Gateways kann die Realisierung von sogenannten Server-Konzepten darstellen. Soll ein solcher z.B. "File-" oder "Print-Server" als besonderes Betriebsmittel eines LAN auch DFN-weit benutzt werden, stellt er das Bindeglied zwischen einem der Basisdienste des DFN (File Transfer oder Remote Job Entry) und einem möglicherweise von der Funktionalität ähnlichen, im verwendeten Protokoll davon verschiedenen Dienst im LAN dar. Den lokalen Nutzern mag dann aufgrund des LAN-spezifischen Server-Protokolls ein wesentlich reicherer Dienst zur Verfügung stehen, ohne daß der Zugriff von "außen" dadurch benachteiligt wird. Bei einem solchen Server sind auch die synchrone Integration in einen DFN-iDenst ("on the fly") oder die asynchrone Ankopplung ("staging") über ein spezielles Steuerprotokoll zu unterscheiden.

Als Beispiel seien zwei File Transfers vom WAN zum LAN, im synchronen und im asynchronen Modus einander gegenübergestellt.

Asynchroneous Operation Mode

Example:
File Transfer from WAN to LAN

    1. Confirm WAN–Connection

    2. Perform File Transfer in WAN

    3. Store File internally

    4. Disconnect WAN–Connection

    5. Open LAN–Connection

    6. Perform File Transfer in LAN

    7. Disconnect LAN–Connection

    8. Delete File internally

requires:

– no change of communication software

– execution request for LAN file transfer

– resources for file storage

Synchroneous Operation Mode

Example:
File Transfer from WAN to LAN

    1. Confirm WAN–Connection

    2. Open LAN–Connection

    3. Perform File Transfer

      from WAN to LAN

      componentwise

    4. Disconnect LAN–Connection

    5. Disconnect WAN–Connection

requires:

– change of communication software

– 1:1 mapping of file transfer

– availability of three hosts

---

5   Protokoll-Portierungen von WAN-Hosts zu LAN-Hosts

Allgemein gilt im DFN, daß die Ankopplung einzelner Hosts ebenso wie nutzerspezifischer privater Netze an die vereinbarte Protokoll-Welt so aufwandsarm wie möglich und so herstellernah wie nötig erfolgen soll. Im Sinne einer Minimierung des Aufwands sollten Betreiber von LANs daher für ihre Rechensysteme Software aus anderen Projekten des DFN oder direkt von Herstellern übernehmen können.

Allgemein sind aber - außer beim Level-7-Gateway - die jeweils ersten, über den Mapping-Moduln liegenden WAN und LAN gemeinsamen Protokolle die Nutzer des tieferen, im WAN und LAN unterschiedlich erbrachten Dienstes. Bei der Auswahl der LAN-Technologie muß daher daraufhingearbeitet werden, daß <u>bis zur Transportschicht</u> möglichst <u>die sich in der Standardisierung befindlichen Protokolle</u> in den verschiedenen LANs eingesetzt werden, um die Entwicklung von höheren Anwendungsprotokollen zu vereinheitlichen und zumindest die Kopplung von LANs als verteilte Endsysteme zu ermöglichen. Insbesondere sollten einersseits die Festlegungen für LANs gemäß IEEE 802berücksichtigt werden, aber ist andererseits auch auf die Koexistenzmöglichkeit von besonderen Anwendungsdiensten auf dieser Transportschicht zu achten.

Es ist dann gewährleistet, daß die Portierung der höheren Protokolle von WAN-Implementierungen auf LAN-Hosts bzw. zwischen unterschiedlichen LAN-Techniken auch bei identischen Rechensystemen nur geringe Anpassungsarbeiten erfordern wird, auch wenn die von der unteren Schicht erbrachten Dienste nicht identisch sind.

In den oben erwähnten Beispielen ist bei Level-1- und Level-3- Gateways die direkte Übernahme von Basis-Diensten gewährleistet, unabhängig, ob das betrachtete Rechensystem direkt am DATEX-P angeschlossen oder aber über ein LAN und einen Gateway mit der WAN-Technik verbunden ist. In allen Fällen setzen File Transfer und Remote Job Entry auf die sowohl im LAN- als auch im WAN-Anschluß vorhandene und identische Transportschnittstelle mit End-zu-End-Signifikanz auf. Auch der hier genutzte Transportdienst T.70 kann bei gleichen Rechensystemen vermutlich auch zwischen LAN und WAN portiert werden. Dieser Dienst setzt im WAN direkt auf X.25, im LAN auf einer Mapping-Schicht, die sich aber in ihrer Funktionalität nicht von einer WAN-X.25-Schnittstelle unterscheiden wird. Ein entscheidende Vereinfachung kann hier aber nur die Standardisierng eines "Global Network Layer" /BE83/ bringen, damit das LAN als integriertes Subnetz eines globalen Netzes, der zum DFN zusammengeschlossenen Hosts, betrachtet werden kann.

6   Schlussfolgerungen

Da innerhalb des DFN keine LAN-Techniken entwickelt werden, ist das Problem der Einbettung von LANs im DFN auf die Zugangsproblematik reduziert. Verstehen wir unter einem LAN eine beliebige Netzarchitektur, so ist diese Architektur an ein neutrales Kommunikationsnetz anzuschließen. Es ist wichtig festzuhalten, daß damit auch der Verantwortungsbereich des DFN definiert ist, der unmittelbar bei der Mapping-Software endet. Die Verantwortung für das LAN bleibt damit in den Händen des Betreibers, anders als bei einem öffentlichen Netz hat er im allgemeinen nach außen keine Betriebsverpflichtung. Dafür bleiben sowohl die Zertifikation der LAN-spezifischen Mapping-Module als auch die Wartung und Anpassung an eventuelle Protokoll-Änderungen im LAN in seinem Aufgabenbereich.

7   Literaturhinweise

Ba84   L.-Bauerfeld,W.
          Zur Spezifikation von Protokollen und Diensten
          in: Spezifikationstechniken im DFN - Tagungsband zum
              Arbeitstreffen am 28. - 30. Nov. 1983 in Darmstadt
              ZPL-DFN, Berlin (Jan.1984), pp. 1.1 - 1.13

BE83   Burkhardt,H.J., Eckert,K.J., Krecioch, T.H.
          A Universal Network Independent Interface Based on the OSI
          Network Service - Some Basic Thoughts
          Computer & Standards
          1 (1983), pp. 5 - 16

BH83   Butscher,B., Henken,G., Lausch,B., de Meer,J.
          Private lokale X.25-Netze mit Untervermittlung
          Handbuch der modernen Datenverarbeitung
          HMD 111, Forkel-Verlag (Mai 1983), pp. 13 - 26

BP79   Deutsche Bundespost
          DATEX-P Handbuch
          FTZ Darmstadt (1979)

CC80.1   CCITT
          Data Communication Networks
          Services and Facilities, Terminal Equipment and Interfaces
          Recommendations X.1 - X.29
          Yellow Book, Vol. VIII - Fasc. VIII.2 (1980)

CC80.2   CCITT
          Data Communiation Networks
          Transmission, Signalling and Switching, Network Aspects,
          Maintenance, Administrative Arrangements
          Recommendations X.40 - X.180
          Yellow Book, Vol. VIII - Fasc. VIII.3 (1980)

DF84.1   Deutsches Forschungsnetz - DFN
          Gesamtprojektplan Version 2.1.1
          ZPL - DFN, Berlin (April 1984)

DF84.2   Deutsches Forschungsnetz - DFN
          Protokollhandbuch Version 1
          ZPL - DFN, Berlin (Juni 1984)

DF84.3   Deutsches Forschungsnetz - DFN
          Zur Architekur von Kopplungen von "Local AREA Networks" und
          "Wide Area Networks" im Deutschen Forschungsnetz
          ed: ZPL - DFN, Berlin (Juli 1984)

EG83   Einert,D., Glas,G.
          The SNATCH Gateway: Translation of Higher Level Protocols
          Protocols & Standards
          Journal of Telecommunications Networks
          3 (1983), pp. 83 - 102

EG84   Einert,D., Glas,G., Klinger,P., Mayerhofer,L., Pawlik,W.
          Vanselow,W.
          Pilotmäßige Implementierung des DFN-VT für IBM-Systeme
          (DFN-SNA-Gateway) und UNIX-Systeme (System III)
          DFN-Pflichtenheft (Dez. 1984)

IS83.1   International Organisation for Standardisation - ISO
          DIS 7498
          Information Processing Systems - Open System Interconnection
          Basic Reference Model
          (April 1983)

IS83.2   International Organisation for Standardisation - ISO
          DIS 8073
          Information Processing Systems - Open System Interconnection
          Connection Oriented Transport Protocol Specification
          (Sep. 1983)

IE83   IEEE
          Project 802
          Local Network Standards
          Draft D (June 1983)

He84   Henckel,L.
          Kopplung von Local Area Networks (LANs) mit dem HMINET2/DFN
          Wide Area Network (WAN)
          DFN-Pflichtenheft
          HMI Berlin (Juni 1984)

LZ84   Lackner,H., Zorn,W.
          LINK - Lokales Informatiknetz Karlsruhe
          Elektr. Rechenanl. 26 (1984) H.5, pp. 254 - 260

VD84   Verband Deutscher Maschinen- und Anlagenbau VDMA
          Fachgemeinschaft Büro- und Informationstechnik
          System Architecture for Local Area Networks, Network Layer
          VDMA, Frankfurt (May 1984)

Die Graphischen Protokolle

im Deutschen Forschungsnetz DFN

P. Egloff

Hahn-Meitner-Institut, Berlin

1. Einleitung

Mittel der Graphischen Datenverarbeitung werden zunehmend bei der
Entwicklung komplexer technischer Produkte eingesetzt. Dadurch steigt
auch der Wunsch von Anwendern, graphische Informationen im Rahmen
solcher Anwendungen über ein Rechnernetz auszutauschen, um sie
"local" oder "remote" weiterverarbeiten zu können. Der Austausch
graphischer Information kann dabei interaktiv, aber auch über
Dateitransfer erfolgen. Im folgenden Beitrag werden drei
Entwicklungen vorgestellt, die im Deutschen Forschungsnetz DFN zum
einen den Dienst eines graphischen Dialogs, zum anderen den Dienst
eines Bildaustauschs bieten und zum dritten die dezentrale Erstellung
technischer Dokumente ermöglichen werden. Auf das weite Gebiet des
Austauschs produktdefinierender Daten und des Modellierens in
Rechnernetzen wird in diesem Beitrag nicht eingegangen.

2. Das Basis-Referenzmodell und Graphische Protokolle

Um eine Kommunikation zwischen sogenannten offenen Rechnersystemen –
das sind über ein Kommunikationssystem verbundene Rechenanlagen
verschiedener Hersteller – zu ermöglichen, wurde in der International
Organization for Standardization ISO ein Architekturmodell
entwickelt, welches die Kommunikationsbeziehungen in sieben Schichten
strukturiert /ISO83/: das sogenannte Basis-Referenzmodell. Für einige
dieser Schichten sind von der ISO und der CCITT (Comite Consultatif
International Telegrafique et Telefonique) Normempfehlungen
erarbeitet worden /CCITT81/. Diese Empfehlungen beinhalten
Protokollvereinbarungen der Schichten 1 bis 5. Für die Schicht 6 –
die Darstellungsschicht – und die Schicht 7 – die Anwendungsschicht –
liegen bisher keine verbindlichen Normungsempfehlungen für
Kommunikationsprotokolle vor.

Im Rahmen der Festschreibung des ISO-Architekturmodells wurde der Begriff des Kommunikationsprotokolls definiert. Dabei gilt, daß die Funktionalität einer Schicht durch die Kooperation der Instanzen (entities) dieser Schicht geliefert wird; mehrere solcher Instanzen bilden zusammengefaßt ein Teilsystem, welches eine Komponente des Kommunikationssystems darstellt. Die Regeln zur Kommunikation zwischen einzelnen Partner-Instanzen (peer entities) werden Protokoll genannt. Dabei gehören diese Partner-Instanzen zur selben Schicht, liegen aber auf verschiedenen Rechnersystemen vor. Ein Protokoll definiert die Syntax und Semantik der gemeinsamen Nachrichten sowie die Reihenfolge, in der diese ausgetauscht werden, um einen bestimmten Dienst zu erbringen. Es ist offensichtlich, daß unterschiedliche Protokolle in einer Schicht unterschiedliche Dienste zur Verfügung stellen, auch wenn sie nur einen Dienst in der darunterliegenden Schicht benutzen.

Die sogenannten Graphischen Protokolle sind den Schichten 7, 6 und 5 oder einer Untermenge davon zuzuordnen, je nachdem welche Dienste in einem Kommunikationssystem angeboten werden /CAMP84,GI83/. Falls ein Protokoll für die Schicht 5 – die Kommunikationssteuerungsschicht – existiert, legen Graphische Protokolle nur Dienste in den Schichten 6 und 7 fest, da auf den Mechanismen der Kommunikationssteuerungsschicht aufgebaut werden kann. Falls auch für die Schicht 6 ein Protokoll realisiert ist, welches zum Beispiel auf der Empfehlung X.409 der CCITT "Presentation Transfer Syntax and Notation" /CCITTX409/ aufbaut, sind die zu treffenden Vereinbarungen für einen graphischen Informationsaustausch der Schicht 7 zuzuordnen.

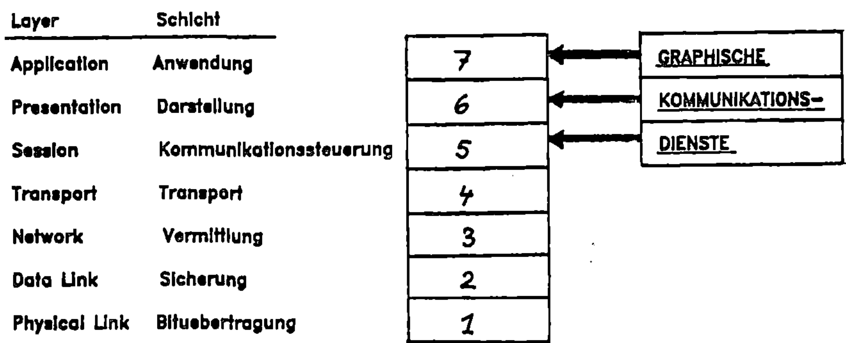

Bild 1: Das Basis-Referenzmodell und Graphische Protokolle

Das Ziel der Entwicklungsarbeiten, die im Rahmen des Aufbaus des Deutschen Forschungsnetzes DFN auf dem Gebiet der Graphischen Datenverarbeitung durchgeführt werden, ist die einheitliche Entwicklung von Softwarebausteinen, die in Rechnernetzen den Austausch von graphischer Information bzw. von graphischen Funktionen in zwei Betriebsarten ermöglichen:

- interaktiver graphischer Informationsaustausch, d.h Dialogfunktion

- passiver graphischer Informationsaustausch, d.h. Transfer von Bildern und Dokumenten (Dokument=Kombination von Bildern und Text).

Die durchzuführenden Aufgaben lassen sich dabei in vier Stufen wie folgt strukturieren:

1. Es muß netzeinheitlich eine Funktionalität festgelegt werden, nach der graphische Funktionen bzw. Bilder beschrieben werden können. Dabei muß sichergestellt werden, daß dem Stand der Technik entsprechend mit den dann festgeschriebenen Funktionen sämtliche herkömmlichen Bilder bzw. graphischen Interaktionen beschrieben bzw. ausgeführt werden können.

2. Für die unter Punkt 1 festgelegte Funktionalität muß eine netzeinheitliche (=netzneutrale) Darstellung vereinbart werden. Nur mit einer solchen neutralen Kodierung ist es möglich, graphische Information und graphische Interaktionen in einem Rechnernetz auszutauschen bzw. durchzuführen.

3. Um den Austausch graphischer Information möglichst ökonomisch abwickeln zu können, muß die entsprechend Punkt 2 kodierte Information komprimiert werden. Dazu gehören einerseits Verfahren zur Komprimierung der Kodierung selbst, andererseits aber auch die Definition und netzeinheitliche Festlegung bzw. Beschreibung von höheren graphischen Funktionen; dabei wird beispielsweise an Funktionen wie Kreis, Ellipse (oder generell Kegelschnitte), Splines, Beschreibung von Flächen u.s.w. gedacht.

4. Um eine weitere Minimierung der Menge der auszutauschenden graphischen Information bzw. Interaktionen in speziellen Kommunikationssystemen - wie z.B. in lokalen Netzen - zu erreichen, muß zwischen Sende- und Empfangssoftware Information ausgetauscht werden, die zu einem Intelligenzabgleich der beiden Systeme führt. Als Einsatzbeispiel sei hier der "dumme" Zeichenautomat mit geringer Auflösung genannt, dem die graphische Information zur Verarbeitung nicht in der archivierten hohen Auflösung übermittelt zu werden braucht.

Diese vier Aufgabenstufen lassen sich entsprechend den Strukturierungsmechanismen des Basis-Referenzmodells einteilen. Dieses führt zu folgender Zuordnung:

Aufgabenstufe 1:
Definition der
netzeinheitlichen Funktionalität   => Anwendungsschicht (7)

Aufgabenstufe 2:
Definition der
netzeinheitlichen Kodierung   => Darstellungsschicht (6)

Aufgabenstufe 3:
Entwicklung von
Komprimierungsalgorithmen   => Schichten 6 und 7

Um die Aufgaben der Stufe 4 einordnen zu können, muß festgelegt
werden, welcher Schicht oder welchen Schichten eine
"Vorab-Kommunikation" zuzuordnen ist, die vor dem eigentlichen
Informationsaustausch Parameter für diesen Austausch festlegt.

3. Das GKS-orientierte Kommunikationssystem

Um einen interaktiven graphischen Informationsaustausch in einem
Rechnernetz zu ermöglichen, wurde das sogenannte GKS-orientierte
Kommunikationssystem entworfen /HMI84a/. Dieses basiert auf dem
internationalen Standard für das Graphische Kernsystem GKS /ISO82/
und soll dem Anwender über einen graphischen Dialog ein interaktives
Arbeiten an graphischen Arbeitsstationen (workstations) ermöglichen,
wobei verschiedenartige Ein- und Ausgabemöglichkeiten bestehen: für
die Ausgabe Bildschirm und Zeichentisch, für die Eingabe Fadenkreuz,
Tastatur oder graphisches Tablett. Der graphische Dialog schließt
auch die alphanumerische Ein- und Ausgabe des Anwenderprogramms ein.

Ziel des GKS-orientierten Kommunikationssystems ist die Verteilung
von Softwarebausteinen bestimmter Anwendungen auf verschiedene über
ein Rechnernetz gekoppelte Rechnersysteme. Dazu ist es nötig,
innerhalb des GKS eine Schnittstelle festzulegen, über die die
Kommunikation zwischen den sogenannten Kernfunktionen des GKS und den
"workstation"-abhängigen Funktionen abgewickelt wird. Diese
Schnittstelle ist das "workstation interface" WSI /DIN84/.

Ein verteiltes GKS soll es dem Anwender ermöglichen, mit graphischer
Anwendersoftware interaktiv in offenen Netzen zu arbeiten. Außerdem
soll ihm im Netz vorhandene graphische Peripherie zur Verfügung
stehen, die lokal nicht vorhanden ist. Das im GKS vorgesehene
Konzept, mit einem Anwendungsprogramm beliebig viele graphische
Geräte gleichzeitig betreiben zu können, darf durch die Verteilung
auf mehrere Rechner nicht eingeschränkt werden.

Zur Realisierung des graphischen Dialogs müssen die graphischen
Kommunikationsdienste folgende Funktionen ermöglichen:

- Auf- und Abbau der Verbindung zwischen GKS-Kern und Arbeitsstation
  (verbindungsorientierter Dialog) und die Verwaltung mehrerer
  Arbeitsstationen, d.h. eine 1:n Kommunikation zwischen Kern und
  Arbeitsstationen;

- Abbildung des Ein-/Ausgabeverhaltens der WSI-Schnittstelle auf

Kommunikationsprimitive zur Interprozesskommunikation zwischen Anwendungsprogramm und GKS-Arbeitsstation;

- fehlerfreie Auslieferung der transparent übertragenen Daten und Operationen an der Schnittstelle;

- Sicherstellung der Synchronisation zwischen den kommunizierenden Prozessen.

Die Workstation-Schnittstelle WSI ist für verteilte graphische Anwendungen gegenüber anderen möglichen Schnittstellen wie Treiberschnittstelle oder GKS-Funktionsschnittstelle am besten geeignet, da sie allen Anforderungen des graphischen Dialogs genügt: Anhand der GKS-Funktionen kann sie eindeutig definiert werden, die Datenraten sind gegenüber der direkten Übertragung einzelner Primitive stark reduziert und die Peripherie im gesamten Netz kann benutzt werden.

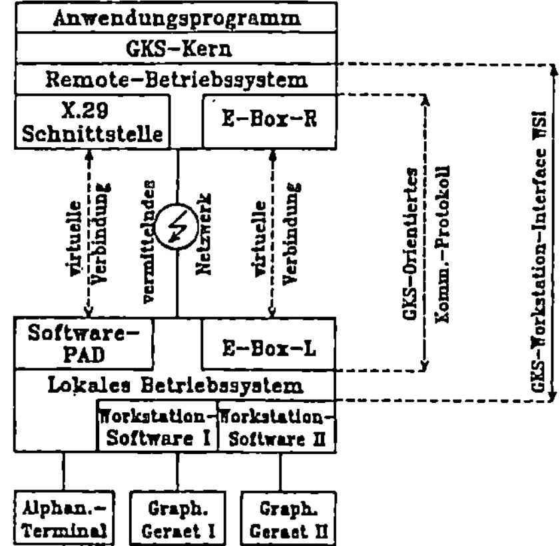

Bild 2: Das GKS-orientierte Kommunikationssystem

In einem verteilten System sind GKS-Kern und Workstations auf verschiedenen Rechnern installiert. Bild 2 stellt eine mögliche Verteilung der Softwarebausteine auf zwei Rechnersysteme dar. Dabei stehen die Bausteine E-Box-R und E-Box-L für die Verwaltung und Abwicklung der graphischen Kommunikation zwischen GKS-Kern und verschiedenen Arbeitsstationen. Bild 3 verdeutlicht die innere Struktur des Bausteins E-Box-R. Als zusätzliche Moduln werden beim "Kern-Rechner" und beim "Workstation-Rechner" ein Remote bzw. Local Workstation Controller (RWC/LWC) und ein Graphic Communication Controller (GCC) benötigt. Die Funktionalität des GCC richtet sich nach der Art des verwendeten Netzes (LAN oder WAN) bzw. nach den vom Netz zur Verfügung gestellten Diensten.

<u>Remote Workstation Controller (RWC)</u>
Die Schnittstelle zwischen Remote Workstation Controller (RWC) und GKS-Kern entspricht der Funktionalität des WSI. Der RWC hat die

Aufgabe, die Remote-Verbindungen zu verwalten, den Auf- und Abbau von Verbindungen zu veranlassen sowie das "Multicasting" zu realisieren. Unter "Multicasting" wird verstanden, daß graphische Primitive, die an mehrere Workstations eines Rechners gerichtet sind, nur einmal übertragen werden. Für die Durchführung der Multicasting-Funktion stellt der RWC zwei Steuerfunktionen zur Verfügung.

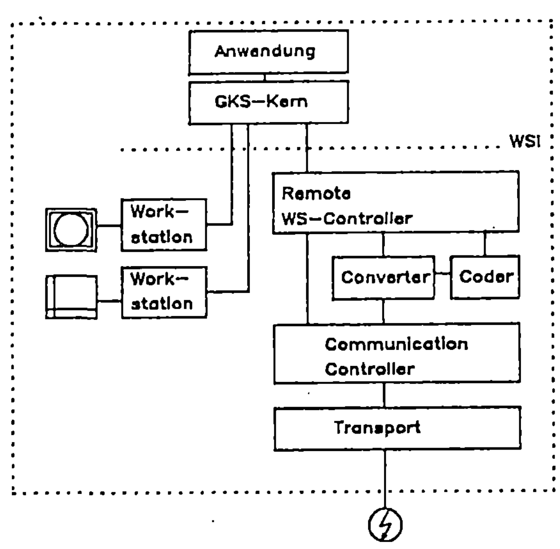

Bild 3: Struktur der Kommunikationsbausteine

Der Graphic Communication Controller (GCC)

Der Graphic Communication Controller (GCC) baut die Datenstruktur auf, in der die graphische Information für die Workstations auf dem entfernten Rechner enthalten ist. Er veranlaßt mit Hilfe der Transportdienste die Übertragung der Daten. In der Prozedurschnittstelle zwischen GKS-Kern und Workstation sind Ein- und Ausgabeparameter gemischt, der Funktionsaufruf enthält Daten, die für die Workstation bestimmt sind, und Parameter, die von der Workstation geliefert werden sollen.

Der Local Workstation Controller (LWC)

Der Local Workstation Controller (LWC) verwaltet die Workstations am lokalen Rechner. Falls WSI-Funktionen für mehr als eine Workstation am lokalen Rechner bestimmt sind, wird jeweils ein Aufruf für die entsprechende Workstation erzeugt.

Den in Kapitel 2 erläuterten Strukturierungsmechanismen entsprechend ist die Definition des WSI und die Funktionalität des RWC/LWC der Schicht 7 zuzuordnen. Die Festlegung der Datenstrukturen im GCC erfolgt in der Schicht 6.

4. Der Transfer von Bildern
===============================

Zum passiven graphischen Informationsaustausch gehört der Transfer
von Bildern /HMI84b/. Unter Bildern wird hier die Erzeugung bzw.
Darstellung zweidimensionaler graphischer Information verstanden,
deren Funktionalität in sogenannten graphischen Metafiles
(Bilddateien) beschrieben ist. Für solche Metafiles existieren zur
Zeit zwei Normempfehlungen:

- der Metafile des GKS: GKSM
- der Computer Graphics Metafile: CGM

Im Rahmen der Standardisierung des Graphischen Kernsystems GKS ist
auch der GKSM definiert worden. Inhalt und Format des GKSM sind im
Anhang des GKS-Dokuments enthalten /ISO7942/ und stellen nur eine
Empfehlung für die Darstellung graphischer Information dar. Der GKSM
ermöglicht die geräteunabhängige Speicherung zweidimensionaler Bilder
sowie die Protokollierung einer GKS-Sitzung. Erstellung,
Interpretation, Weiterverarbeitung und Speicherung müssen über das
GKS erfolgen.

Neben dem GKSM gibt es den Computer Graphics Metafile CGM /ISO5-88x/
als eigenständigen Standard. Begründet ist diese Zweigleisigkeit in
der Tatsache, daß viele Anwendungen die geräteunabhängige Speicherung
und den Transport graphischer Daten benötigen ohne das GKS zu verwenden, da z.B. Rechenanlagen zu klein sind, spezielle Anwendungsgebiete
vorliegen oder das GKS in etablierte Anwendungssysteme nicht
integriert werden kann.

Der CGM definiert eine Datenstruktur, die eine peripherie-, rechner-
und systemunabhängige Speicherung graphischer Daten ermöglicht. Die
Funktionalität des CGM entspricht einer Untermenge des geplanten
Standards des Computer Graphics Interface CGI /CGI84/, welches die
Basis für die Definition eines virtüllen graphischen Terminals sein
wird. Der CGM beinhaltet saemtliche Elemente, die notwendig sind, um
Bilder unabhängig voneinander zu beschreiben. Ferner ist die Möglich-
keit gegeben, nicht standardisierte graphische Elemente einzufügen.

Getrennt von der Beschreibung der Funktionalität legt der Standard
drei mögliche Kodierungsarten für den CGM fest:

- Zeichenkodierung (character encoding):
  Jeder Funktion ist eine feste Bitkombination (in den meisten Fällen
  1 Byte, bei einigen Ausnahmen 2 Bytes) zugeordnet. Für Parameter
  mit Integer- oder Real-Werten sind spezielle Formate vorgesehen.

- Binärkodierung (binary encoding):
  Die einzelnen Funktionen werden zu Klassen zusammengefaßt, wobei
  jede Klasse durch eine binär kodierte Integer-Zahl gekennzeichnet
  wird.

- Klartextkodierung (clear text encoding):
  Die Klartextkodierung wird in einer für den Menschen lesbaren Form
  in Textdateien erstellt.

Die einfachste Möglichkeit, einen CGM zu übertragen, ist, den
herkömmlichen File-Transfer-Dienst zu verwenden und die Datei - je
nach Kodierungsart - als Textdatei oder transparente Datei zu
transferieren. Dieses hat unter anderem den Nachteil, daß die
Übertragung kostenintensiv bzw. langsam sein kann, was insbesondere
für eine klartextkodierte Datei gilt. Die Zeichen- und Binärkodierung
erzeugen Dateien mit geringerem Umfang, doch ist nicht
sichergestellt, daß ein in dieser Kodierung vorliegender CGM mit dem
transparenten File-Tranfer-Dienst übertragen werden kann.

An einen spezifischen CGM-File Transfer werden Anforderungen
gestellt, die vom herkömmlichen File Transfer nicht erfüllt werden
können:

- Die Datenstruktur bzw. Datentypen sind abhängig von der
  Semantik. Sie können daher nur mit Kenntnis der Semantik
  interpretiert und umgesetzt werden.

- Da der CGM in drei unterschiedlichen Kodierungen vorliegen kann,
  muß davon ausgegangen werden, daß beim Sender und Empfänger
  unterschiedliche Kodierungen realisiert sind; daher sollen diese
  für die Übertragung in eine netzeinheitliche Repräsentation
  überführt werden.

- In der Initialisierungsphase des Verbindungsaufbaus sollen
  Informationen ausgetauscht werden, um unterschiedliche
  Funktionalitäten der CGM-Systeme bei Sender und Empfänger
  festzustellen und somit gegebenenfalls nur eine Teilmenge über
  das Netz auszutauschen.

Um eine - wie in Kapitel 2 dargestellt - netzeinheitliche
Repräsentation der Daten zu erzeugen, wird eine vierte Kodierung -
die sogenannte Transferkodierung - entworfen werden, die die Aufgabe
hat, das Datenvolumen so weit wie möglich zu minimieren. Diese
Optimierung ist für die Senkung der Übertragungskosten relevant, da
bei der Kostenberechnung vorwiegend die Anzahl der Datensegmente zugrunde liegt. Dieses rechtfertigt auch die durch die Erzeugung der
Transferkodierung entstehenden erhöhten Konvertierungskosten.

Der folgende Entwurf eines CGM-Transfers basiert zum Teil auf dem im
DFN verfügbaren RDA(Remote Data Access)-Protokoll /RDA84/. Eine
Einbettung in FTAM /ISO16-1669/ ist noch nicht berücksichtigt. Die
Dienste des CGM-Transfers werden in vier Phasen eingeteilt:

- Aufbauphase
- Vereinbarungsphase
- Datentransferphase
- Abbauphase

Anhand der Vereinbarungsphase soll die Abwicklung eines Elementardienstes des CGM-Transfer erläutert werden, wobei die Dienste CONNECT, GIVENEG (GIVE NEGOTIATION RIGHT), NEGDAT (NEGOTIATION DATA), CONVERT und DISCONNECT eingesetzt werden.

<u>Vereinbarungsphase</u>:

        GIVENEG.request
        GIVENEG.indication
        GIVENEG.response
        GIVENEG.confirmation

        NEGDAT.request
        NEGDAT.indication
        NEGDAT.response
        NEGDAT.confirmation

Durch die in der Vereinbarungsphase ausgetauschte Information soll die Übertragung überflüssiger Daten vermieden werden. Diese Information ist in einem dem Metafile vorangestellten Header enthalten. Dazu gehoeren z.B.:

- die Genauigkeit von Koordinatenwerten
- die Genauigkeit von Farbwerten
- die Verwendung bestimmter Zeichensätze
- die Verwendung bestimmter Textfonts
- die Verwendung von nicht standardisierten
  Werten in Indextabellen

Es gibt mehrere Möglichkeiten automatisch festzulegen, ob ein Metafile, der eine größere Funktionalität beinhaltet als der Empfänger verarbeiten kann, übertragen wird oder nicht:

- Der Sender entscheidet, ob es sinnvoll ist, den Metafile trotzdem zu übertragen;
- der Empfänger entscheidet, ob er den Metafile erhalten möchte, nachdem ihm mitgeteilt wird, welche Funktionalität der Metafile beinhaltet;
- es gibt eine pauschale Festlegung, ob ein Metafile gesendet wird oder nicht.

Es muß noch untersucht werden, ob auch der Anwender die Möglichkeit haben soll, den Informationsaustausch in einer Vereinbarungsphase zu beeinflussen.

Die Festlegung der Funktionalität eines graphischen Metafile gehört zu den Aufgaben, die der Schicht 7 zugeordnet werden. Die Definition der Transferkodierung ist der Schicht 6 zuzuordnen. Die Aktionen der Vereinbarungsphase können nach Kapitel 2 nicht eindeutig zugeordnet werden.

5. Dezentrale Erstellung von Dokumenten

Unter einem Dokument wird in diesem Zusammenhang die einheitliche Erstellung, Verarbeitung, Darstellung und der Transfer von Text und graphischer Information verstanden /HMI84c/. Einführend soll der Unterschied von logischer und Layout-Struktur solcher Dokumente verdeutlicht werden.

Ein Dokument kann auf zweierlei Weise in Komponenten unterteilt werden. Zum einen kann ein Dokument aus "abstrakten Objekten" bestehen, das sind z.B. Kapitel, Absätze, Sätze, Wörter. Zum anderen kann ein Dokument aus "konkreten Objekten" bestehen. Diese konkreten Objekte sind für die zweidimensionale Darstellung geeignet z.B. in Form von Seiten, Spalten, Textblöcken und Zeichnungen. Ein Textobjekt gilt als abstrakt, wenn es unter dem Aspekt der inhaltlichen Vermittlung gesehen wird. Ein Textobjekt gilt als konkret, wenn die den Text bildenden Symbole auf einer Seite betrachtet werden.

Der Vorgang des "Formatierens" bildet die abstrakten Objekte eines Dokuments auf die konkreten Objekte ab /FUR82/. Das Formatieren eines Absatzes kann beispielsweise das Setzen des Textes im Blocksatz mit beidseitigem Randausgleich innerhalb festgelegter Ränder bedeuten. Die Operationen des "Editierens" definieren Abbildungen von abstrakten Objekten auf abstrakte Objekte und von konkreten Objekten auf konkrete Objekte. Beispielsweise bildet das Editieren vom Text logische Textobjekte aufeinander ab, da sprachliche Informationen modifiziert werden. Konkrete Objekte werden aufeinander abgebildet, wenn z.B. schon formatierte Objekte eines Dokuments (Tabellen, Bilder) interaktiv neu positioniert werden.

Die Gesamtheit der abstrakten Komponenten eines Dokuments ergibt zusammen mit deren logischen Beziehungen untereinander die "logische Struktur" des Dokuments. Unter dem Begriff "Layout" wird eine Anordnung von Schrift und Bildern auf einer Fläche verstanden. Das Layout entspricht also dem endgültigen Aussehen eines Dokuments. Unter "Layout-Struktur" ist die Gesamtheit der konkreten Komponenten eines Dokuments zusammen mit deren Beziehungen zur Ausgestaltung untereinander zu verstehen /ISO18-3-283/. Der Autor eines Dokuments hat die logische Struktur im Auge, wenn er etwa ein Unterkapitel eröffnet, um einen Gedankengang in das umfassende Schema eines Kapitels einzugliedern, dagegen hat er die Layout-Struktur im Auge, wenn er festlegt, den Text zweispaltig zu formatieren. Beim Formatieren wird die logische Struktur eines Dokuments in die Layout-Struktur überführt.

Im Rahmen des DFN wird ein Dokumentendienst entwicklet, der dem Anwender die Möglichkeit bietet, dezentral Dokumente oder Teile davon zu erzeugen, zu modifizieren und formatiert darzustellen. Da die Realisierung eines idealen Dokumentensystems sehr zeitaufwendig ist, werden an dieser Stelle zunächst Untermengen mit eingeschränkter

Funktionalität dargestellt.

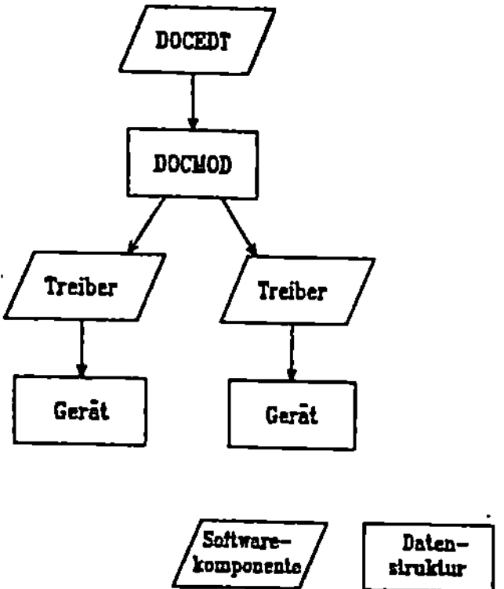

Bild 4: Aufbau eines idealen Dokumentensystems

In einer ersten Entwicklungsstufe werden möglichst viele bereits
bestehende Softwarekomponenten verwendet. Für die Texterstellung werden bereits existente Text-Editoren eingesetzt. Unabhängig davon werden die darzustellenden Bilder mit Hilfe von GKS als GKSM bzw. CGM in
Klartextkodierung erzeugt und später durch den Text-Editor in die
Textdatei integriert.

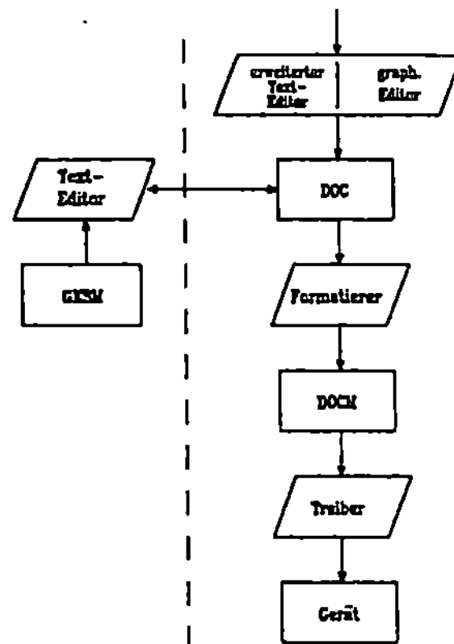

Bild 5: Dokumentenverarbeitung im DFN

Die zunächst unterschiedlichen Datenstrukturen für Text und Graphik
werden in einer Datei über eine einheitliche Kodierung vereint. Da
das nicht der Philosophie des in Bild 4 dargestellten idealen Doku-
mentenmodells (DOCMOD) entspricht, muß hier anfangs und für Anwender,
die nur mit Untermengen des Dokumentensystems arbeiten wollen, eine
weitere Datenstruktur eingeführt werden. In den Dokumenten-Editor
DOCEDT ist zur Ausführung von WYSIWYG-Funktionen (What You See Is
What You Get) der sogenannte Formatierer integriert. Dadurch enthält
DOCMOD sowohl logische als auch Layout-Strukturen. Dieses Konzept

läßt sich bei der Verwendung von bestehenden Editoren und
Formatierern nicht realisieren. Hier werden getrennte Datenstrukturen
für formatierte (DOCM) und unformatierte Dokumente (DOC) benötigt,
wie es in Bild 5 dargestellt ist. Für die beiden Datenstrukturen DOC
und DOCM werden soweit wie möglich aktuelle Standardisierungs-
vorschläge eingesetzt.

Das Modell aus dem Normvorschlag ODA (Office Document Architecture)
/ISO18-3-283/ wird für die Datenstruktur DOC übernommen. Da in ODA
eine Syntax zur Definition dieses Modelles auf Anwenderebene fehlt,
bietet sich hier die Verwendung der Structured Generalized Markup
Language SGML an, wie sie in Teil VI von CLPT definiert ist
/ISO5-12-9x/. Mit SGML lassen sich aber nur die generische logische
Definition und die spezifische logische Struktur beschreiben. Die
Angabe der generischen Layout-Definition und der Logical-Layout-Be-
ziehungen läßt sich nicht in SGML realisieren. Diese Beziehungen sind
in ODA/ODIF selbst zur Zeit noch nicht geklärt. Daher ist die Zuord-
nung der vom Anwender definierten Markups zu den ausführenden
Prozeduren unklar. Wenn der Anwender sich z.B. ein logisches Objekt
‹heading› definiert, muß die ausführende Prozedur wissen, ob die
Überschrift mittig gerückt und mit welchem Zeichen unterstrichen werden soll. Für die Steuerung solcher Layout-Anweisungen können die
Variablen aus Teil V ("Formatting and Composition Functions") an
/ISO5-12-9x/ eingesetzt werden, die an die entsprechenden Prozeduren
des Formatierers als Parameter weitergereicht werden.

Um im Rahmen des DFN möglichst bald eine Datenstruktur DOC anbieten
zu können, die auch bei Weiterentwicklung von CLPT /ISO5-12-9x/ noch
verwendet werden kann, werden zunächst zwei Dokumentenklassen DFN-ein-
heitlich angeboten werden. Die generische logische Definition für
diese beiden Dokumentenklassen werden in SGML-Syntax spezifiziert.
Mit Hilfe dieser Markups kann der Anwneder zunächst über einen Text--
Editor und in den folgenden Entwicklungsstufen über spezielle
Editoren seine spezifische logische Struktur erzeugen. Die vom
Formatierer zur Erzeugung von DOCM aus DOC benötigten generischen
Layout-Definitionen werden durch einen festen Satz von Prozeduren
festgelegt. Falls in Zukunft vollständige SGML-Parser angeboten wer-
den, ist der Anwender in der Lage auch seine generische logische
Definition selbst zu bestimmen. Dazu müssen dann auch die dazu-
gehörenden Prozeduren in der Sprache des Formatierers entwickelt
werden.

Im folgenden wird zur Veranschaulichung ein Auszug aus der Dkalartion
der DFN-Dokumentenklasse "report" in SGML-Syntax wiedergegeben. Eine
Elementdeklaration kann aus folgenden Teilen in der angegebenen
Reihenfolge bestehen:

- Elementname
- Markup-Minimierung
- Typ des Elementeninhaltes
- Ausnahmen
- Attributdeklarationen (Name, Typ und Standardwert)

Zur Erhoehung der Übersichtlichkeit werden innerhalb der
Element-Deklarationen teilweise Entities verwendet. Die Deklaration
der erlaubten Markup-Minimierung besteht pro Element aus zwei
Einträgen: einem für das Metasymbol, das ein neues Element einleitet,
und einem für das beendende Metasymbol. Die Einträge haben folgende
Bedeutung:

> \- : Metasymbol muß in jedem Fall angegeben werden
> O : Metasymbol kann weggelassen (omit) werden

<u>"report"</u>
```
‹!ENTITY  % titlepage "(title & author & institution? &
                       date? & publisher?)" ›

‹!ENTITY  % graphic   "(gksm | cgm | subtitle)" ›

‹!ENTITY  % chapter_content
                      "(heading , paragraph ,
                      (paragraph | indent | FLATTER | table |
                      formula | list | %graphic;)*)" ›

‹!ENTITY  % emphasis  "(e1 | e2 | e3 | e4)" ›

‹!ELEMENT report          - -
                      (%titlepage; , headline? & footline? ,
                      preface? , chapter+, appendix*)
                      +(index, %emphasis;)
                      author    CDATA         UNKNOWN
                      version   NUMBER        UNKNOWN
                      gksm      ("NO"|"YES")  UNKNOWN
                      cgm       ("NO"|"CLEAR" UNKNOWN›


‹!ELEMENT title       - O      RCDATA ›
‹!ELEMENT author      - O      RCDATA ›
‹!ELEMENT institution - O      RCDATA ›
‹!ELEMENT date        - O      RCDATA ›
‹!ELEMENT publisher   - O      RCDATA ›

‹!ELEMENT headline    - O      RCDATA ›
‹!ELEMENT footline    - O      RCDATA ›

‹!ELEMENT preface     - O
                      (heading, paragraph+)        +(footnote) ›

‹!ELEMENT chapter     - O
                      (%chapter_content; , sub1*)  +(footnote) ›
‹!ELEMENT sub1        - O
                      (%chapter_content; , sub2*)  +(footnote) ›
```

.
.
.

```
< !ELEMENT cgm            - -        NDATA
                 sizex    NUMBER   REQUIRED
                 sizey    NUMBER   REQUIRED
                 orgx·    NUMBER   REQUIRED
                 orgy     NUMBER   REQUIRED ›

< !ELEMENT subtitle       - O        RCDATA ›

< !ELEMENT appendix       - O
                 (%chapter_content; , appl*)  +(footnote) ›
```

Gemäß der in Kapitel 2 dargestellten Strukturierung der Aufgaben ist die Definition der Metasprache und der damit zusammenhängende Aufbau der Datei DOC der Schicht 7 zuzuordnen. Die Struktur der vom Formatierer abhängigen Datei DOCM gehört in die Schicht 6.

6. Schlußwort

Die Zeitplanungen für die in den letzten drei Kapiteln beschriebenen DFN-Dienste sehen bis zum Herbst 1985 folgendermaßen aus:

- Das GKS-orientierte Kommunikationssystem soll in einer Prototyp-Version auf dem dann im DFN verfügbaren Transportprotokoll einsatzfähig sein.

- Für die Erzeugung und Interpretation des Computer Graphic Metafile CGM steht portable Software zur Verfügung, die einen CGM in Klartextkodierung oder in Characterkodierung verarbeitet.

- Für die dezentrale Verarbeitung von Dokumenten wird für die Dokumentenklasse "report" die Deklaration abgeschlossen sein; ein Parser setzt die Metasymbole von SGML in Prozeduraufrufe um, über die Makros eines Formatierers angeschlossen werden können. Der Anschluß des Formatierers TEX von D. Knuth wird vorhanden sein.

Es soll an dieser Stelle erwähnt werden, daß wesentliche Arbeiten an den Inhalten des Kapitels 3 von J. Bechlars, C.Egelhaaf und G. Schürmann von der Zentraleinrichtung für Datenverarbeitung der Freien Universität Berlin geleistet wurden. Beiträge zum Kapitel 4 wurden von M. Schulz, G. Foest und L. Hammerl (alle HMI) geleistet. An der Erarbeitung der Inhalte des Kapitels 5 haben C. Smith und A. Scheller (beide HMI) wesentlichen Anteil.

7. Literaturverzeichnis

/CAMP84/   P. Egloff: Graphische Datenverarbeitung
in Rechnernetzen , CAMP - Konferenz , Berlin
September 1984

/CCITT81/   CCITT - Empfehlung der V-serie und der X-Serie
Band I
Datenpaketvermittlung - Internationale Standards
4. erweiterte Auflage
R.V. Decker's Verlag, G.Schenk
Heidelberg-Hamburg 1981

/CCITTX409/   CCITT/SGVII: Draft Recommendation X.409
Message Handling Systems:
Presentation Transfer Syntax and Notation
Juni 1983

/CGI84/   Computer Graphics Interface
(Virtual Device Interface)
dpANS X3H3 84/45

/DIN84/   J. Bechlars, C. Egelhaaf, G. Schürmann
The GKS-Workstation Interface, Version 1.0
DIN-AK 5.9.4 / 15-84 , August 1984

/FUR82/   Document Formatting Systems:
Survey, Concepts, and Issues
R. Furuta, J. Scofield, A. Shaw
University of Washington
Computing Surveys, vol. 14, No. 3, Sept. 1982

/GI83/   P. Egloff : Graphische und Modell - Datenverarbeitung
im Rechnernetz DFN , Fachgespräch Graphische
Kommunikation, Jahrestagung der GI, Hamburg
Oktober 1983

/HMI84a/   J. Bechlars, Chr. Egelhaaf, G. Schürmann
GKS-orientiertes Kommunikationssystem
in offenen Rechnernetzen, DFN-Pflichtenheft G003-1
August 1984

/HMI84b/   P.Egloff, G.Foest, L.Hammerl, M.Schulz
Computer Graphics Metafile Transfer
in offenen Rechnernetzen, DFN-Pflichtenheft G003-2
August 1984

/HMI84c/   P.Egloff, A.Scheller, C. Smith
                Verarbeitung von Dokumenten in offenen Rechnernetzen
                DFN-Pflichtenheft G003-3
                August 1984 ·

/ISO7498/   ISO/DIS 7498
                Open System Interconnection
                Basic Reference Model
                April 1982

/ISO7942/   ISO/DIS 7942
                Graphical Kernel System
                Functional Description
                November 1982

/ISO5-88x/   ISO/TC97/SC5/N881 - N884
                Computer Graphic - Metafile for transfer and
                storage of picture description information
                Functional description and encoding , Part 1, 2, 3, 4
                July 1984

/ISO5-12-9x/   ISO/TC97/SC5/WG12/N96 - N101
                Text Interchange and Processing
                Part 1, 2, 3, 4, 5, 6
                April 1984

/ISO16-1669/   ISO/TC97/SC16/N1669
                File Transfer Access and Management
                Februar 1984

/ISO18-3-283/   ISO/TC97/SC18/WG3/N283
                Office Document Architecture
                Februar 1984

/ISO18-3-284/   ISO/TC97/SC18/WG3/N284
                Office Document Interchange Formats
                Februar 1984

/ISO18-3-285/   ISO/TC97/SC18/WG3/N285
                Document Profile
                Februar 1984

/RDA84/   R. Popescu-Zeletin, L. Henckel, G. Grolms
                Specification of the Remote Data Access System
                in the HMINET2 (reduced version)
                Hahn-Meitner-Institut
                Mai 1984

Der Directory-Dienst
im Message-Handling-System des
Deutschen Forschungsnetzes

Alfons Warnking, Horst Santo

Gesellschaft für Mathematik und Datenverarbeitung Bonn (GMD)
Institut für Angewandte Informationstechnik (F3)
Forschungsgruppe Gruppenunterstützungssysteme (GUS)
Schloß Birlinghoven
Postfach 1240
D-5205 St. Augustin 1

Zusammenfassung

Ein Message-Handling-System erfordert benutzerfreundliche Konventionen für die Namen von Sendern und Empfängern von Nachrichten. Ein Directory-Service unterstützt zum einen die Benutzer des Systems bei der Ermittlung korrekter Namen, zum anderen das Transfersystem bei der Ermittlung von Adressen und Routen. Dieses Papier beschreibt das Architekturmodell, das Datenmodell, die Dienste und Protokolle für einen auf internationalen Standards basierenden verteilten Directory-Dienst für das Message-Handling-System im Deutschen Forschungsnetz.

1. Einleitung

Internationale Standards für 'Message Handling Systeme (MHS)' stehen kurz vor ihrer Verabschiedung bzw. sind kürzlich verabschiedet worden [CCITT-1, ECMA-1, ISO-1]. Die Standards basieren auf Modellen, in denen jeder Benutzer innerhalb des Systems durch seinen 'User Agenten' (UA) repräsentiert wird, welcher als Sender/Empfänger von Nachrichten auftritt. Jeder Benutzer hat einen oder mehrere O/R-Namen (originator/recipient), darunter eine O/R-Adresse, die die Lokalisierung seines UA's innerhalb des MHS angibt. Das 'Message Transfer System' (MTS) besteht aus 'Message Transfer Agenten' (MTA's) und umfaßt eine oder mehrere Management Domains (MDs'). Letztere können unter privater (PRMD) oder öffentlicher (ADMD) Verwaltung stehen (Bild 1.1).

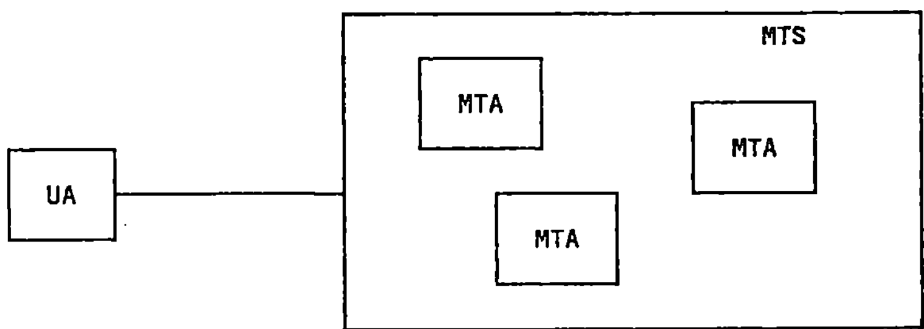

Bild 1.1: Funktionale Sicht des MHS

Alle Instanzen im Message-System benötigen für ihre Arbeit Verzeichnisse über Namen, Adressen, Eigenschaften und Beziehungen der Kommunikationsinstanzen. Die Aufgabe des Directory-Dienstes (Verzeichnisdienstes) ist, die Führung und Verwaltung der Verzeichnisse zu unterstützen. Die Dienstelemente beziehen sich auf die Erzeugung, die Änderung, das Lesen und das Löschen von Einträgen in der Directory-Datenbank.

Es gibt verschiedene Gründe, die Verwaltung der Verzeichnisse nicht in die Instanzen des Message-Systems zu integrieren. Zunächst steht ein eigenständiger Dienst per Definition verschiedenen Anwendungen zur Verfügung. Zweitens sind Directory-Funktionen im Gegensatz zum 'store and forward' Prinzip des Message-Systems dialogorientiert. Und drittens ergibt sich durch einen eigenständigen Dienst eine größere Flexibilität. Dazu gehört die Entlastung der Benutzer von technischen Identifikatoren für Sender/Empfänger durch benutzerfreundliche Namen oder ein netzwerkunabhängiger Zugang zu Diensten.

Da das MHS durch die kooperierenden Management Domains verteilt verwaltet wird, sollte auch der Directory-Dienst durch ein verteiltes System erbracht werden. Bei größeren Netzen läßt sich im Übrigen nur durch ein verteiltes System die nötige Effizienz und Zuverlässigkeit erreichen.

Im Rahmen der Konzeption eines MHS für das "Deutsche Forschungsnetz" wurde deshalb ein eigenständiger Directory-Dienst entworfen, der - obwohl zunächst nur lokal angeboten - als verteiltes System angelegt ist. Das dazu zugrundegelegte Architekturmodell wird in Abschnitt 2 beschrieben, die Formatspezifikation in Abschnitt 3. Abschnitt 4 behandelt die Dienste und Protokolle, Abschnitt 5 schließlich die Vorgehensweise im DFN.

2. Das Architekturmodell

Konzipiert man derzeit ein Directory-System, so sollte man darauf achten, kompatibel zu den kommenden internationalen Directory-Standards für offene Systeme (OSI) zu bleiben. Daher wurde die Konzeption im DFN eng an ein von der IFIP WG 6.5 vorgelegtes Modell angelehnt [IFIP-1], welches sich heute auch in den Arbeitspapieren der Normungsgremien wiederfindet.

In diesem Modell wird der Dienst durch 'Directory-Service-Agenten' (DSA) erbracht, die dazu ggf. miteinander kommunizieren. Den Klienten (MTA, UA, ...) des Services steht ein 'Client-Service-Agent' (CSA) zur Anforderung eines (Teil-) Dienstes und Abwicklung der Protokolle zur Verfügung (Bild 2.1). Im Rahmen des MHS ist ein CSA eine Komponente innerhalb des UA's oder MTA's.

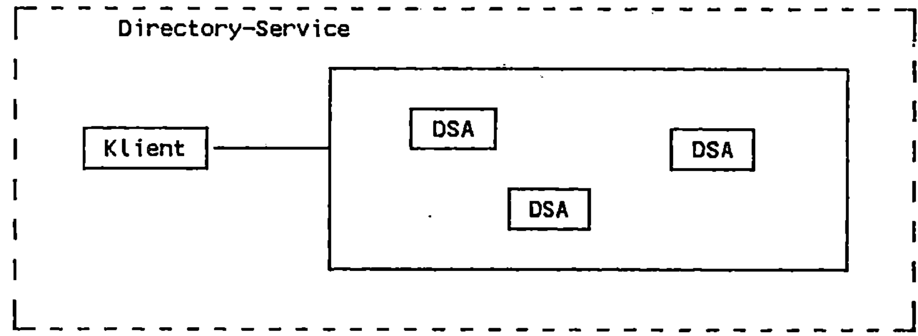

Bild 2.1: Funktionale Sicht des Directory-Dienstes

Bild 2.2 zeigt ein Schichtenmodell des Directory-Systems und ordnet die beteiligten Instanzen, Dienste, Protokolle und Interaktionen in dieses Modell ein. Bild 2.2 liegt voll in der Anwendungsschicht des OSI-Referenzmodells. Die Klientenschicht repräsentiert die Benutzer, die Directoryschicht die Erbringer des Dienstes.

Das Schichtenmodell ist unabhängig vom Schichtenmodell für Message-Handling-Systeme [CCITT-1]. Der Directory-Dienst stellt eine eigenständige Anwendung im Sinne des OSI-Referenzmodells dar.

Die 'Directory Access Entity '(DAE)' ermöglicht einem Klienten den Zugang zum Directory-Service bei verteilter Realisierung. Das Zugriffsprotokoll behandelt die Kommunikation zwischen DAE und DSA, die Formatspezifikation Absprachen über Aufbau und Inhalt der Directory-Einträge. Das Directory-System ist ein verteiltes System. Daher ist ein internes Directory-Protokoll erforderlich, welches die Kommunikation zwischen DSA's behandelt und u.a.Austauschformate zur Unterstützung der Verwaltung von Kopien der Directory-Einträge in dem verteilten System festlegt.

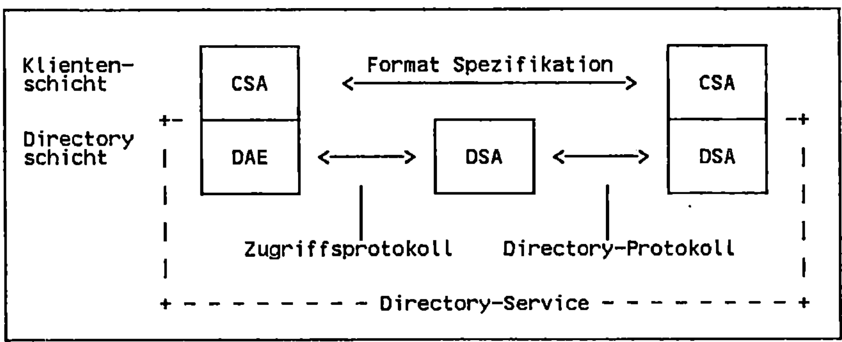

Bild 2.2: Schichtenmodell des Directory-Services

Eine Anfrage an den Service präsentiert der CSA über ein definiertes Protokoll den ihm nächstgelegenen DSA. Hat der DSA die gewünschte Information, liefert er sie zurück. Hat er sie dagegen nicht, bestehen verschiedene Möglichkeiten:

- Im einfachsten Fall muß er dem CSA eine Meldung der Art "Versuche es bei DSA XY" zurückgeben, auf den dann der CSA direkt zugreifen kann.

- Die zweite Möglichkeit ist, die Anfrage an einen weiteren DSA zu vermitteln, welcher die Antwort ggf. direkt an den fragenden CSA zurückgibt.

- Die dritte Möglichkeit besteht darin, daß der DSA selbst als Klient anderer DSA's auftritt, um die gewünschte Information zu erhalten.

Die beiden letzten Alternativen erfordern, daß der Klient die Entscheidung über den Ablauf seiner Anfrage behält (Kosten!). Dies gilt nicht für 'update-'Funktionen. Hier muß der erste DSA selbständig entsprechende Nachrichten an weitere betroffene DSA's schicken. Im DFN werden die Betreiber von DSA's bilaterale oder unilaterale Verträge über den Austausch der Informationen schließen. Das Directory-Protokoll muß deshalb den Austausch und die Konsistenzerhaltung von Directory-Einträgen (Kopien) in dem verteilten System unterstützen.

3. Die Formatspezifikation

Die Directory-Datenbank speichert Informationen über Objekte im Message-System. Dazu gehören Benutzer (bzw. ihre UA's), Organisationen und Organisationseinheiten, Rollen, MTA's, DSA's, Länder, Domains, verschiedene Formen von Benutzergruppen (Verteilerlisten, elektronische Zeitungen, Konferenzen,...) u.a. Im DFN wird die Datenbank zunächst Informationen über die oben unterstrichenen Objekttypen aufnehmen. Zu den Informationen gehören Namen (ggf. mehrere unterschiedlicher Form für ein gegebenes Objekt) und Eigenschaften der Objekte. Der Directory-Eintrag eines Objektes ist deshalb eine Zuordnung einer Eigenschaftsmenge zu einer Namensmenge:

$$\{Name1,...,Namek\} \longrightarrow \{ Eig.1,...,Eig.l\}$$

Darüberhinaus finden sich in der Directory-Datenbank noch Angaben zur Repräsentation diverser Rechte von Klienten auf den Directory-Einträgen und zur Konsistenzerhaltung von Kopien in dem verteilten System.

Die für die Klienten des Directory-Dienstes relevanten Daten beziehen sich also auf

- Namen,
- Eigenschaften und
- Rechte.

3.1 Namen

Namen sind Werkzeuge. Werkzeuge müssen angepaßt sein, an das, was mit ihnen 'bearbeitet' wird, und denjenigen, der das Werkzeug handhabt. Außerdem müssen Werkzeuge ihren Zweck erfüllen.

Namen werden in Message-Handling-Systemen von Menschen und Maschinen gehandhabt. Da Computer die Menschen bedienen sollen, ist Benutzerfreundlichkeit das höherrangige Ziel. Sie erfordert die Benennung eines Objektes durch natürliche Angaben, die mit einer Bedeutung verbunden sind und das Objekt beschreiben (keine Computeradressen). Die Anzahl der Namensteile sollte gering sein, wobei die Ordnung irrelevant und in der Schreibweise Alternativen erlaubt sein müssen. (Universität Dortmund, Uni Dortmund, UNIDO). Zur Aufhebung von Mehrdeutigkeiten, sollten möglichst keine künstlichen Konventionen erforderlich sein (nicht Müller1, Müller2 sondern Hubert und Peter Müller). Benutzerfreundlichkeit erfordert außerdem, daß ein Name statisch d.h. zeitunabhängig ist.

Diese Forderungen können mit descriptiven Namen erfüllt werden, die Charakteristiken des betreffenden Objekts angeben. Ein descriptiver Name ist eine Aussage, die für genau ein Objekt im Message-System wahr ist. Die Form der Aussage wird für das MHS im Deutschen Forschungsnetz eine Attributliste sein. Eine Attributliste ist eine ungeordnete Menge von Attributen. Attribute besitzen einen Attributnamen und einen Attributwert. Der Attributname bezeichnet die betreffende Charakteristik (Vorname, Zuname), der Attributwert den jeweiligen Wert (Peter, Müller).

Als Beispiel sind im folgenden einige für das DFN festgelegten Namensformen für Benutzer aufgelistet. Sie unterscheiden sich nach den Kontexten, aus welchen die jeweiligen Attribute stammen. Spitze Klammern deuten optionale Attribute an, Punkte Iterierungsmöglichkeiten.

```
organisatorisch:    <Landesbezeichnung>
                     Organisationsname
                    <Organisationseinheit>...

                    <Initialen>
                    <Vorname>                =  Personen-
                     Zuname                     name
                    <Generation>
                     (jun./sen.)
geografisch:         Landesbezeichnung
                    <Region>...
                     Stadt
                    <Straße>
                    <Hausnummer>
                    "Personenname"

architekturell:      Landesbezeichnung
                     Administration-Domain-Name
                     Private-Management-Domain-Name
                    <Organisationsname>
                    <Organisationseinheit>...
                    <"Personenname">
```

In einem Directory-Eintrag kann für jedes Attribut mehr als ein Wert registriert sein. Auch kann die Namensmenge eines Objektes Namen aus verschiedenen Kontexten beinhalten. Es ist dabei zu beachten, daß aus dieser Menge gemäß den derzeitigen CCITT-Empfehlungen [CCITT-1] nur zwei Formen – die architekturellen und die (oben nicht aufgeführte) terminalorientierte – als O/R-Namen im Transfersystem verwandt werden können. Diese treten im Rahmen des Directory-Services als ausgezeichnete Namen eines Objektes auf. Die organisatorische und die geografische Namensform sind jedoch im Rahmen eines Directory-Services ein wichtiges Hilfsmittel, um Informationen (insbesondere den O/R-Namen) zu erhalten.

Außerdem steht zu erwarten, daß nach Verabschiedung der Normen für einen Directory-Dienst Namen obiger Art im Übermittlungssystem möglich sein werden. ECMA hat bereits eine Obermenge der CCITT-Attribute standarisiert [ECMA-1].

Der Namensraum

Wie gesagt, ist der Zweck eines Namens die Unterscheidung eines Objekts von "allen restlichen". Um dies zu sichern, müssen Mittel zur Verwaltung und Beschreibung des globalen Namensraumes insbesondere für Sender/Empfänger bereitstehen. Ein nützliches Beschreibungselement ist ein gerichteter Graph mit genau einer Wurzel. Bild 3.1 gibt ein Beispiel.

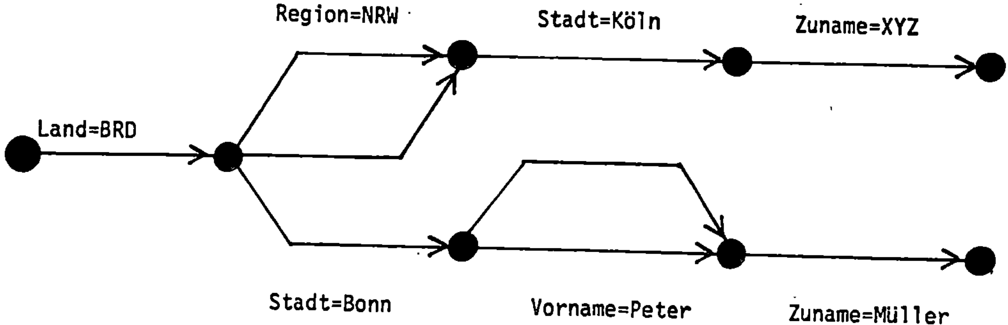

Bild 3.1: Ein sehr einfacher globaler Namensgraph

Zur Modellierung des Namensraums wird der Graph in bestimmter Weise interpretiert. Jedes Blatt des Graphen repräsentiert ein einzelnes benanntes Objekt, innere Knoten Namensautoritäten, eine Kante ein Attribut eines oder mehrerer Objekte. Die in Frage kommenden Objekte sind diejenigen, zu denen ein Weg von der Wurzel über die betreffende Kante führt.

Eine Kante ist mit dem Attribut markiert, das sie repräsentiert. (Name=Wert). Eine unmarkierte Kante repräsentiert das "Null-Attribut", ein fiktives Attribut das alle zugehörigen Objekte besitzen.

Zur Verwaltung des Namensraums gehören insbesondere Änderungen des Raumes und Existenzprüfungen von Namen. Die Zugriffskontrolle muß Änderungsversuche nicht autorisierter Benutzer abweisen. Ein Verfahren zur Existenzprüfung findet sich in [White-1].

3.2 Eigenschaften

Ein Name hat eine Bedeutung. Diese ist ihm zugeordnet. Die Bedeutung ist das Objekt, für welches der Name steht. Um das Objekt zu benutzen, benötigt man Kenntnisse über das Objekt. Um Klienten des Directory-Service dieses Wissen zumindest teilweise verschaffen zu können, wird der Namensmenge eines Objekts in der Directory-Datenbank eine Eigenschaftsmenge zugeordnet. Als Beispiel sind im folgenden einige Eigenschaften eines 'User Agenten' aufgelistet.

- Objekttyp:UA
- zugehöriger MTA
- UA-Klasse
- deliverable encoded information types (Facsimile, IA5-Text,...)
- maximum content length
- Paßwort
- Namen in anderen Netzen
- persönliche Angaben des Benutzers
- (Netzwerk- oder Transport) Adresse.

Soweit absehbar, werden die internationalen Standards eine Eigenschaft als ein Tripel aus Eigenschaftsnamen oder -identifier, Eigenschaftstyp und Eigenschaftswert auffassen. Der Eigenschaftstyp spezifiziert, ob der Wert atomar ist (Typ: item), oder aus einer Menge von Namen besteht (Typ: group).

Im Rahmen des 'Message-Handling-Systems' im DFN wird der Directory-Service die Verwaltung von Gruppen übernehmen. Zu Gruppen gehören einfache Verteilerlisten, geschlossene Kommunikationszirkel, elektronische Journale, Schwarze Bretter u.ä. Eine Gruppendefinition umfaßt deshalb nicht nur die Liste der Mitglieder, sondern diese müssen nach ihrem Verhalten und ihren Rechten differenziert werden. (Automatisches oder empfängergesteuertes Empfangen, Senden, Ändern der Gruppendefinition). Dazu muß das äußere Erscheinungsbild der Gruppen spezifizierbar sein (öffentlich, privat, abonnierbar,...). Schließlich sollen Gruppen ein gemeinsames Kommunikationsgedächtnis (Archiv) besitzen können. Zukunftsmusik sind Ablaufsteuerungen ('coordination procedures') im Rahmen eines MHS auf der Basis internationaler Standards.

Bild 3.2 zeigt als Beispiel die Definition eines 'elektronischen Journals', Bild 3.3 ein allgemeines Schema für eine Gruppendefinition, wie sie im DFN angestrebt wird.

Die Bedeutung der einzelnen Einträge in Bild 3.3 soll hier nicht im einzelnen erläutert werden. Trotzdem dürfte an der Namensgebung klar werden, daß die verschiedenen Einträge differenzierte Rechte für die Mitglieder und Nichtmitglieder der Gruppe spezifizieren.

```
Gruppenname:            der Name des Journals
Beschreibung:           eine detailliertere Ausfuehrung ueber
                        den generellen Inhalt, wie man abon-
                        niert, welche Kosten entstehen, ...
Mitgliedschaft:         offen fuer 'Empfangen' oder geschlossen
Sichtbarkeit:           sichtbar
Mitglieder:
  Leiter:               'Schreiben', 'Empfangen', 'Listen'
                        'Organisation'
  Editoren:             'Schreiben', 'Empfangen'
  Abonnenten:           'Empfangen'
```

Bild 3.2 "Elektronisches Journal"

```
Gruppenname:                     <O/R-Name>
Beschreibung:                    <freier Text>
Externe Erscheinung:
  Potentielle Mitgliedschaft:    (geschlossen
                                 beschraenkt
                                       auf <Gruppen-Liste>
                                       fuer <Zugriffsrechte>
                                 offen fuer <Zugriffsrechte>)
  Sichtbarkeit:                  (verborgen,
                                 sichtbar,
                                 transparent,
                                 schreibbar)
Mitglieder-Liste mit
  Eintraegen:                    <O/R-Name>, <Zugriffsrechte>

<Zugriffsrechte> ::= set of      (Schreiben, Empfangen,
                                  (Mitglieder) Listen,
                                 Organisieren)
```

Bild 2.4: Allgemeines Schema zur Gruppendefinition

Diese Rechte sind in der Directory-Datenbank auf den ersten Blick durch Rechte der Klienten auf dem Eintrag einer Gruppe zu repräsentieren. Nun kann der Directory-Dienst die Rechte nur bei Operationen (Erzeugen, Ändern, Lesen, Löschen) auf der Directory-Datenbank garantierten, nicht jedoch, daß der Übermittlungsdienst eine an eine Gruppe adressierte Nachricht den empfangenden Mitgliedern zustellt (und niemandem sonst) und dies nur dann, wenn der Sender eine entsprechende Schreibbe-rechtigung besitzt. Daher gehören (neben der Beschreibung der Gruppe und dem Objekttyp) die schreibenden und empfangenden Mitglieder zu den Eigenschaften einer Gruppe.

3.3 Rechte

Für die Manipulation des <u>Namensraumes</u> wird es im DFN zunächst nur zwei verschiedene Rechte geben. Das erste umfaßt alle lesenden Operationen auf dem Namensraum. Es wird allen Klienten automatisch gewährt und kann nicht entzogen werden. Das zweite Recht umfaßt zusätzlich die Erzeugung, die Veränderung und das Löschen von Namen. Dieses Recht besitzt in der ersten Realisierungsstufe nur der lokale DSA-Administrator.

Es gibt also insbesondere keine Möglichkeit, Namen zu verbergen und keine Unterstützungsfunktionen zur Verwaltung des Namensraumes.

Um die mit dem allgemeinen Schema zur Gruppendefinition (Bild 3.3) an die Verwaltung einer Gruppe festgelegten Anforderungen erfüllen zu können, unterscheidet der Directory-Dienst bzgl. der Eigenschaften sieben verschiedene Rechte. Diese beziehen sich zum Teil auf die gesamte Eigenschaftsmenge, zum Teil auf gegebene Eigenschaften eines Objektes.

Für die <u>Eigenschaftsmenge</u> eines Objektes existieren folgende Rechte:

Erzeugen: Das Recht zum Erzeugen von Eigenschaften in der Eigenschaftsmenge eines Objektes. Dieses Recht besitzt automatisch der Erzeuger des DS-Eintrages.

Listen: Das Recht zum Listen der Eigenschaftsnamen bzw. -identifier.

Mit den Rechten 'Erzeugen' und 'Listen' sind also insbesondere keine Rechte zum Lesen des Eigenschaftswertes einer Eigenschaft verbunden. Es lassen sich also Situationen modellieren, in denen beispielsweise auf bestimmte Eigenschaften nur ein Systemadmininistrator auf andere nur der durch den Eintrag in der Datenbank abgebildete Klient Zugriff hat.

Für jede <u>Eigenschaft</u> einer Eigenschaftsmenge existieren folgende Rechte:

Vergabe: Das Recht zur Vergabe der im folgenden aufgeführten Rechte. Dieses Recht besitzt automatisch der Erzeuger der Eigenschaft.

A/D-Self: Dieses Recht bezieht sich auf Eigenschaften des Typs 'group'. Es umfaßt die Operationen 'Add Self' und 'Delete Self'.

Löschen: Das Recht zum Löschen der Eigenschaft.

Ändern: Das Recht zum Ändern des Wertes eines Eigenschaft.

Lesen: Das Recht zum Lesen einer Eigenschaft.

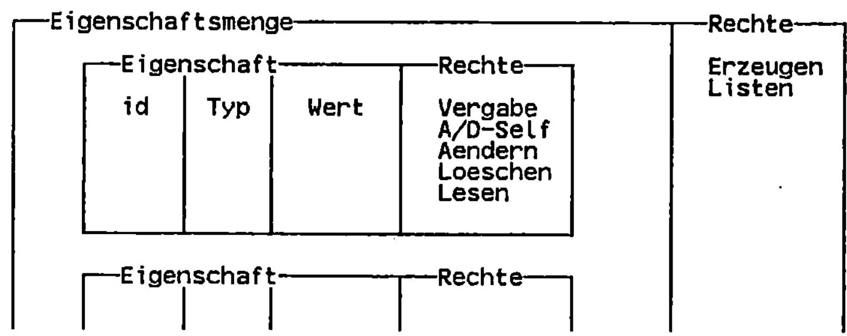

Bild 3.4:   Eigenschaften und Rechte

Der Besitz bestimmter Rechte impliziert den Besitz anderer Rechte.   Dafür gelten
folgende Regeln:

$$\text{Erzeugen} =====> \text{Listen}$$

$$\text{Vergabe} =====> \text{A/D-Self}$$

$$\text{Vergabe} =====> \text{Löschen} =====> \text{Ändern} =====> \text{Lesen}$$

Bei der Vergabe eines Rechtes läßt sich spezifizieren, ob dieses weitergewährt
werden darf. Entziehbar ist ein Recht nur durch denjenigen, der das Recht gewährt
hat.

Die Handhabung der Rechte sei am Beispiel einer 'elektronischen Zeitung' in der
Organisation ''HMI' dargestellt. Ein möglicher Name der Zeitschrift wäre:

Land='Deutschland',   ADMD=<Nullstring>,   PRMD='HMI',   Org.-Name='HMI',
Zuname:'HMI-News'.

Die einzelnen Rechte könnten/sollten folgende Besitzer haben:

a) Eigenschaftsmenge
   Erzeugen: der DSA-Administrator, der Chefredakteur
   Listen: die Mitarbeiter des HMI
   Damit können alle Mitarbeiter des HMI erfahren, daß es sich bei diesem Objekt
   um eine Gruppe handelt und welche Eigenschaften die Gruppe hat.

b) Eigenschaft: "Objekttyp: Gruppe"
   Typ: item
   Wert: Nullwert
   Vergabe: DSA-Administrator
   A/D-Self: nicht anwendbar
   Löschen: keiner
   Ändern: keiner
   Lesen: keiner
   Diese Eigenschaft muß jede Gruppe haben. Allerdings hat diese Eigenschaft kei-

nen Wert. Daher erhält auch niemand das Recht zur Manipulation dieser Eigenschaft.

c) Eigenschaft: "schreibende Mitglieder"
```
Typ:        group
Wert:       die Redakteure der Zeitschrift
Vergabe:    der Chefredakteur
A/D-Self:   keiner
Löschen:    keiner
Ändern:     der Vorstandsvorsitzende
Lesen:      die HMI-Mitarbeiter, die MTAs
```
Alle HMI-Mitarbeiter können erfahren, wer Artikel für die Zeitschrift schreibt. Der Vorstandsvorsitzende hat durchgesetzt, daß er (ihm genehmen) Mitarbeitern das Schreibrecht geben kann. Die MTAs müssen die Eigenschaft lesen können, um das Schreibrecht zu überprüfen.

d) Eigenschaft: "empfangende Mitglieder"
```
Typ:        group
Wert:       die Abonnenten der Zeitschrift
Vergabe:    der Chefredakteur
A/D-Self:   die HMI-Mitarbeiter
Löschen:    keiner
Ändern:     der/die Sekretärin/in der Redaktion
Lesen:      die MTAs
```
Jeder HMI-Mitarbeiter kann die Zeitschrift selbständig abonnieren oder sich dazu an den/die Sekretär/in wenden.

e) Eigenschaft: "Beschreibung der Gruppe"
```
Typ:        item
Wert:       Ein netter Text, der erläutert, daß die HMI-News eine wissenschaft-
            lich herausragende Zeitschrift ist, und wie wichtig es doch für jeden
            HMI-Mitarbeiter wäre zu abonnieren.
Vergabe:    der Chefredakteur
A/D-Self:   nicht anwendbar
Löschen:    keiner
Ändern:     keiner
Lesen:      die HMI-Mitarbeiter
```

Rechte können an eine oder mehrere Klientenklassen vergeben werden (einzelne Klienten treten dabei als Spezialfall auf). Die Beschreibung einer Klientenklasse besteht aus einer Liste von Attributen (im Sinne der Namenskonventionen) und Eigenschaftsnamen bzw. -identifiern. Ein Klient ist Mitglied der Klientenklasse, falls in seinem DS-Eintrag die Attribute (Name und Wert) und Eigenschaftsidentifier der Liste existieren. Die UA's der Mitarbeiter des HMI lassen sich beispielsweise mit der Klassenbeschreibung

Org.-Name=HMI,"Objekttyp:UA"

zu einer Klientenklasse zusammenfassen. Außerdem existiert die Klientenklasse 'JEDER', zu der jeder Klient gehört.

4. Dienste und Protokolle

Da der Directory-Service des MHS in einen globalen, verschiedene Anwendungen umfassenden allgemeinen Dienst aufgehen soll, dürfen die Dienstelemente nicht MHS-spezifisch ausgeprägt sein. Die Service-Elemente (s. Bild 4.1) sind daher Operationen, die an den allgemeinen Strukturelementen (Namen, Eigenschaften, Rechte) der Directory-Datenbank ansetzen.

```
Operationen auf Namen              Operationen auf Eigenschaften
   * Create object                    * List properties
   * Delete object                    * Delete property
   * List objects
     Bind                           Operationen auf 'items'
     Find distinguished name          * Add item property
     Add name                         * Retrieve item
     Add value                        * Change item
     Delete Name
     Delete value                   Operationen auf 'groups'
     Test equivalence                 * Add group property
     List equivalent names            * Retrieve members
     Check existence                  * Add member
                                      * Add self
Service_Kontrolle und Rechte          * Delete member
     Get controls                     * Delete self
     Grant                            * Is member
     Revoke
     Read rights
     Change password
     Read object description
```

Bild 4.1: Directory Operationen

Im Hinblick auf die kommenden internationalen Standards ist derzeit Zurückhaltung bei der konkreten Spezifikation der Dienste angebracht. Im DFN wurde deshalb zunächst eine Untermenge der in dem Normentwurf [ECMA-2]. spezifizierten Operationen ausgewählt (in Bild 4.1 mit einem Stern '*' markiert). Diese wurde um Elemente erweitert, die für die Behandlung einer differenzierten Namensstruktur und der Rechte erforderlich sind. Die Funktionalität umfaßt die Erzeugung, die Änderung und das Löschen von Strukturelementen. Beim Retrieval sind z. T. einfache boolsche und/oder reguläre Ausdrücke zulässig.

Alle Operationen sind eigenständig. Retrievaloperationen können beispielsweise nicht auf den Ergebnissen vorausgehender Anfragen aufbauen. Integraler Bestandteil aller Dienstelemente ist eine Zugriffskontrolle.

Im DFN werden in die Instanzen des MHS (UA, MTA) CSA's integriert werden, die MHS-spezifische Dienste unter Rückgriff auf diese Operationen realisieren. Dazu gehören Dienste

- zur Ver- und Entschlüsselung von Passworten

- zur lokalen Verwaltung der einem MTA zugeordneten UA's

- zum Lesen von Routinginformationen

- zur Definition von 'Defaults' für die Werte von Attributen.

Die Directory-Dienste müssen durch die zugehörigen Protokolle unterstützt werden. Bild 4.2 zeigt die Struktur der Protokolle für den Zugang zum DSA. Danach implementiert ein DSA den Service als 'Remote Operationen' in Anlehnung an [CCITT-2]. Das dort festgelegte Protokoll definiert vier 'Operation Protocoll Data Units' (OPDU's), die zum Aufruf, Rückgabe des Ergebnisses, Rückgabe einer Fehlermeldung oder Rückweisung einer OPDU dienen (Bild 4.3). Weiter wird die Nutzung unterliegender Schichten angesprochen.

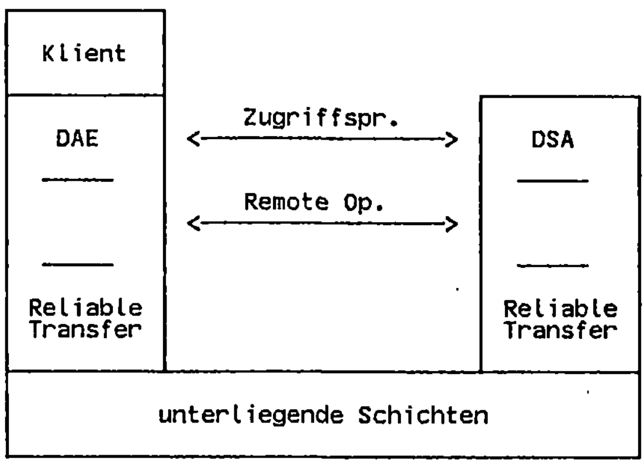

Bild 4.2: Protokollstruktur fuer den Zugang zum Dienst

Obiges Protokoll eignet sich nicht, falls zwischen CSA und DSA keine Teilnehmerverbindung im Sinne des OSI-Referenzmodells aufgebaut werden kann. Im Rahmen des MHS wäre es dann sinnvoll, den Directory-Service über "normale" Messages anzusprechen. Der DSA würde in diesem Fall als spezialisierter UA auftreten.

Für das Directory-Protokoll kann ein DSA als Klient eines weiteren DSA's auftreten. Besonders interessant ist dies für einen Funktionsverbund, wo mehrere DSA's einen Dienst gemeinsam erbringen. Für einen Informationsverbund zur gemeinsamen Verwaltung der verteilten Datenbank bietet sich das MTS, Filetransferdienste oder der Reliable Transfer-Service an. Dies hängt von der Topologie des DSA-Verbundes und der Menge und Häufigkeit der ausgetauschten Daten ab. Letzteres wird wesentlich durch das zugrundegelegte Verbundkonzept bestimmt, welches u. a. Gegenstand derzeitiger konzeptioneller Arbeiten für einen anwendungsunabhängigen Directory-

Dienst im DFN ist.

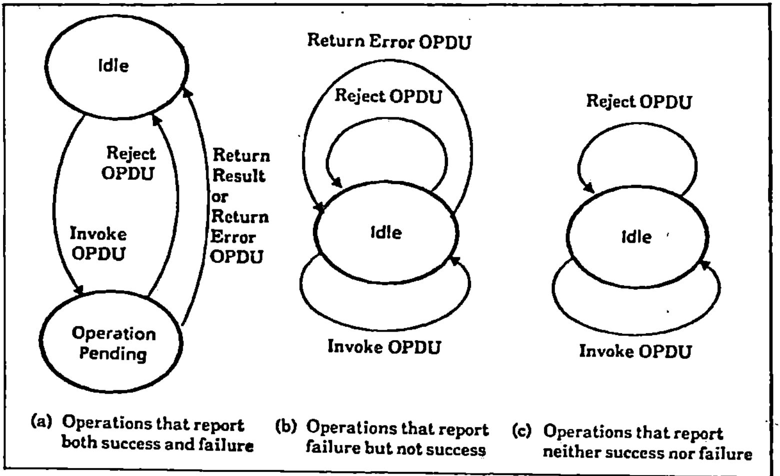

4.3 Zustandsdiagramme für Remote-Operationen [CCITT-2].

5. Vorgehensweise im DFN

Zu dem Zeitpunkt, wo dieses geschrieben wird, stehen viele (politische) Entscheidungen für ein MHS im DFN noch aus. Daher kann hier nur über die Vorschläge in [DFN-1] berichtet werden.

Danach wird die Realisierung des Directory-Dienstes in mehreren Stufen erfolgen:

Stufe 1

In der ersten Stufe wird ein einfacher verteilter Directory-Service mit Komponenten in jeder privaten Management Domain realisiert. In die Directory erfolgen nur MHS-relevante Eintragungen lokal erforderlicher Daten. Ein Protokoll zwischen den DSAs wird nicht realisiert. Benutzern fremder Domains wird Zugang zur Directory über die interaktive Geräteschnittstelle X3,X28,X29 ermöglicht.

Stufe 2

Stufe 2 gliedert sich in zwei Phasen. In der ersten Phase werden noch offene Probleme wie Authentisierung, Abrechnung, sowie die Integration anderer Applikationen in den Directory-Service diskutiert und an die zu erwartende Normung angepaßte Lösungen erarbeitet. Dies umfaßt auch eine Erweiterung des Directory-Dienstes durch DSA-Protokolle, Einbindung von Registrier- und Kopien-Diensten sowie die Erarbeitung einer Lösung der sich daraus ergebenden Datenkonsistenzprobleme.

In der zweiten Phase werden die erarbeiteten Lösungen realisiert.

Stufe 3

Nach einer am Ende der nächsten Studienperiode zu erwartenden internationalen Normung des Directory-Dienstes müssen eventuelle Abweichungen in den DFN-Protokollen an diese Normen angepaßt werden.

6. Schluss

Die Funktionalität des hier beschriebenen Dienstes ist eine Basis für weitere Entwicklungen. Beispielsweise werden für die Unterstützung von Gruppen in Zukunft Erweiterungen vorgenommen werden müssen. Für ein betriebsfähiges System ist es nötig, sofort einen solchen Basis-Dienst zur Verfügung zu stellen. Die hier vorgestellte Architektur des Systems dürfte es ermöglichen, konsistente Erweiterungen im Hinblick auf zukünftige Anforderungen vorzunehmen.

Literatur

CCITT-1   CCITT, Study Group VII: Message Handling Systems; System-Model-Service Elements, Draft Recommendation X.400.

CCITT-2   CCITT, Study Group VII: Message Handling Systems; Remote Operations and Reliable Transfer Server, Draft Recommendation X.410

CCITT-3   CCITT/SG VII: Directory Systems Working Document (Version 1), June '84

DFN-1   M. Bogen, u.a., Konzept für einen DFN-Message-Verbund auf der Basis von CCITT-Message-Handling-Systems, unveröffentlichte Studie im Rahmen des DFN, Juni 1984.

DFN-2   Arbeitsgruppe Message Systeme: Der DFN Message-Dienst, Pflichtenheft für die Implementierung, Okt. '84

ECMA-1   Standard ECMA-93: Distributed Application for Message Interchange (MIDA), Sept. '84

ECMA-2   ECMA/TC23/84/54, ECMA Directory Service Standard, Draft Proposal, May '84

ECMA-3   ECMA/TC23/84/110, Directory Service Standard, First Draft, Sept. '84

IFIP-1   IFIP WG 6.5 Working Paper, Naming and Directory-Services for Message Handling Systems, Version 4, July '83.

ISO-1   ISO/TC97/SC18/WG4, Functional Description of Message Oriented Text Interchange Systems, Draft Proposal.

ISO-2   ISO/TC97/SC16/N1957, Directory-Service for OSI Systems, Draft Service Specification, June '84

White-1   J.E. White, A User-friendly Naming Convention for Use in Communication Networks, Proceedings of the IFIP 6.5 Working Conference, Nottingham, England, May '84, Nottingham University

Time in Formal Protocol Specifications

Harry Rudin
IBM Zurich Research Laboratory
8803 Rüschlikon, Switzerland

Abstract: The need for formal, machine-readable specification of the protocols used in distributed computer systems is widely accepted. In these formal protocol specifications relatively little attention has been paid to the problem of including time — the time it takes to execute various portions of the protocol as well as explicit values for timeouts. In this tutorial, models for including time are reviewed. Techniques for analyzing these models are then discussed. A comprehensive bibliography is included.

I. INTRODUCTION

The need to unambiguously disseminate a computer-communication protocol description to its various implementers has led to the use of formal protocol specifications. In the case of teleprocessing-equipment manufacturers, for example, IBM describes its Systems Network Architecture in a formal language, FAPL. In the case of international standards bodies the International Standards Organization (ISO) has its Estelle and LOTOS languages and CCITT its SDL language. Failure to use a formal specification, i.e., to rely instead on a description in prose, invites incompatibility among the individual implementations of the protocol in the various system components.

A formal specification consists of two parts: a model (or abstraction) of the protocol mechanism and a language to describe the model. If the formal specification language is machine-readable (all the examples just mentioned are) then there are a number of benefits which may be had in addition to that of unambiguous dissemination (Rudin85). Some of these are automated validation (a check for logical self-consistency and completeness), automated verification (demonstration that the protocol fulfills its functional objective), prediction of performance (throughput or response time), partially automated generation of the code necessary for an implementation, and automated testing to demonstrate that the implementation does indeed conform to the protocol specification. Research is well underway and at least some encouraging results have been obtained in all these areas.

Why add time to the formal specification?

1) Often, it is a timeout which as a last resort guarantees the reliability of a protocol in the normal, error-prone environment. Checking (validating and verifying) a protocol to see that it performs its task should take these timeouts into account. If time cannot be taken into account, then, in many cases, one must accept the liability inherent in a significantly restricted check.

2) The success of a protocol depends to a large extent on its performance (responsiveness and throughput) capability. The inclusion of time in the formal specification opens the door to an estimate of the protocol's performance early in the design phase.

3) Most of the techniques used for automated protocol analysis, be it validation, verification, or performance prediction, rely at least in part on exhaustive exploration of a state space. The inclusion of time could potentially decrease the size of the space which needs to be explored, even though it may complicate the way in which this space is explored.

The hazard in adding time to the formal specification:

On the other hand, even without the inclusion of time, the methods known for carrying out the aforementioned analyses are impractical for some protocols. Adding another dimension — time — to the model can only increase the complexity of computation for the general case.

Where the balance of advantages and disadvantages will come to rest is not clear and will certainly vary from case to case. But, there seems to be no doubt that the inclusion of time will play an increasingly important role in protocol specification.

In Section II the most popular models used for formal protocol specification are discussed. In Section III the analysis of these models for purposes of validation and performance prediction are covered.

II. FORMAL DESCRIPTION TECHNIQUES

This section reviews much of the work on formal protocol specification which explicitly includes time. Three categories are discussed: Petri-net models, finite-state machine models, and algebraic models.

II.1 Petri-net models

While Petri nets have not yet been widely used as a formal means of disseminating protocol description to a community of implementers, they have been heavily used for the analysis of protocols. In fact, most of the work done which includes time in formal protocol specification uses Petri nets as a basis. PETERSON has written a very readable introduction to Petri nets (Peterson77).

Figure 1 shows a generic Petri-net transition with a number of input and output places. In Petri nets, tokens are used as an indication of momentary process status. Tokens reside in places. Consistent with Petri-net terminology, when all its input places have a token, a transition may fire, removing one token from each input place and adding a token at each of the output places. When a time specification is added to Petri nets, time delay can be associated either with transitions or with places. Almost all investigators associate the execution time or delay with the execution of a transition, rather than with the time tokens spend in places. An exception are the papers of WONG *et al.* (Wong84) and COOLAHAN and ROUSSOPOULOS (Coolahan83). Under some conditions, the one can be converted to the other (Sifakis79). For modeling timeouts, the association of delay only with places is insufficient.

There are many ways in which delay can be ascribed to transitions in Petri nets. Table I, further on, summarizes these as used by various authors.

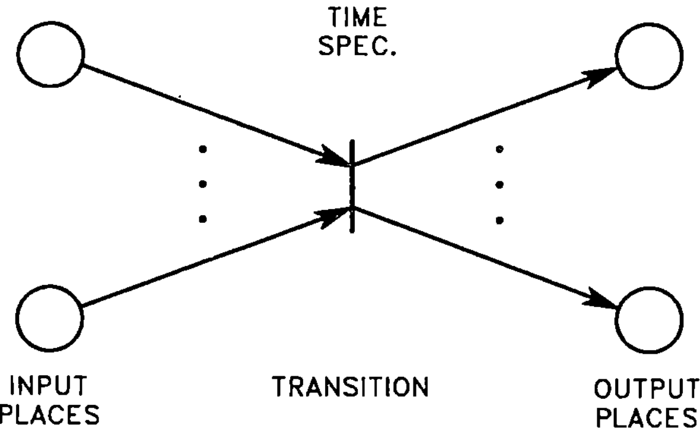

Fig. 1. Generic Petri net with time specification

The simplest form is simply a fixed delay as defined in (Ramchandani73) and (Zuberek80). As soon as there are sufficient tokens to enable the transition, it fires immediately, effectively consuming the tokens for a constant amount of time $\tau_{const.}$ and then delivering the tokens to the output places. Insofar as computer-communication protocols are concerned this is not a good model because there is no possibility for the arrival of a new event to change the event sequence established by the enabling of the transition. In Petri-net terminology this property, one undesirable for communication protocol modeling, is called persistence: Once a transition has been enabled, it can only become disabled by firing itself.

All the other models circumvent the above problem by using nonpersistent nets in a variety of ways. For example, in (Merlin76) the *firing* of a transition takes no time but a transition must be enabled for a minimum time $t_{min.}$ and possibly up to a maximum time $t_{max.}$. During the period over which the transition is enabled but has not been able to fire because of the specified delay, the transition may be disabled by the arrival of a new token which enables some other transition with a shorter delay specification. The firing of this second transition may remove the token which enabled the first transition. This means that the arrival of the new event may change the outcome of an event pattern established earlier. This is exactly the effect needed for modeling, say, a timeout. The expiration of a timer could then, for example, initiate a recovery procedure which would not occur under normal circumstances.

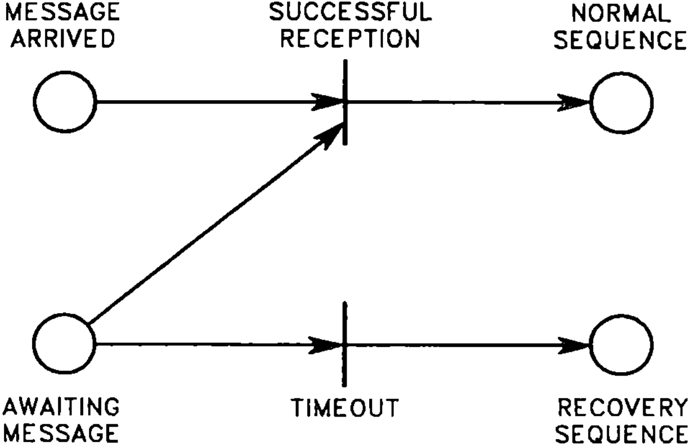

Fig. 2. Use of a Petri net with timeout

Figure 2 shows a simple timeout situation. Assume the AWAITING-MESSAGE place has a token, the TIMEOUT transition has a fixed delay, and the SUCCESSFUL-RECEPTION transition, a negligible delay. If the awaited message fails to arrive, the TIMEOUT will expire. The SUCCESSFUL-RECEPTION transition cannot be enabled, thus interrupting the normal sequence of operations. The recovery sequence is initiated in its stead.

Similarly, in (Molloy82) the firing of a transition takes no time but the transition remains enabled for a random interval $t_{rand.}$ which has a negative exponential distribution. Again a new event may influence the outcome if it arrives within this random interval.

Finally, in (Razouk84) there is an interval $t_{min.}$ during which a transition must be enabled if it is to fire. If the transition fires, then the firing will take a constant time interval $t_{const.}$. After this interval, the token(s) is delivered to the subsequent places.

Table I – Summary of Petri-net models which include time

$t = 0 \equiv$ time at which transition is enabled
  $t_1 \equiv$ earliest time at which token(s) can be removed from input place(s)
  $t_2 \equiv$ latest time at which token(s) can be removed from input place(s)
  $\tau \equiv$ time period required for firing the transition
  $t_3 \equiv$ earliest at which token(s) can be placed on output place(s)
  $t_4 \equiv$ latest time at which token(s) can be placed on output places(s)

Author	t_1	t_2	τ	t_3	t_4
RAMCHANDANI	0	0	$\tau_{const.}$	$\tau_{const.}$	$\tau_{const.}$
ZUBEREK	"	"	"	"	"
MERLIN	$t_{min.}$	$t_{max.}$	0	$t_{min.}$	$t_{max.}$
MOLLOY	0	$t_{rand.}$ $0 \leq t_{rand.} \leq \infty$	0	0	$t_{rand.}$ $0 \leq t_{rand.} \leq \infty$
RAZOUK	$t_{min.}$	$t_{min.}$	$t_{const.}$	$t_{min.} + t_{const.}$	$t_{min.} + t_{const.}$

II.2 Finite-state machine models

Relatively less work has been done with the inclusion of time in finite-state-machine (FSM) based models than with models based on Petri nets, although FSM-based models are more popular for formal protocol specification, per se. Most of the work, beginning with BEIZER in 1970 (Beizer70) associates an execution delay with the transition connecting two states. The states are used to model process status; transitions are used to model events. Examples of events are the offering (- sign) of a message by one process to another, capability to receive (+ sign) a message from another process, and timeouts. In FSM's, transitions may carry several labels: Event type, the probability of the transition being made when more than one is possible, and the execution-time or delay interval associated with the transition. An FSM fragment is shown in Figure 3.

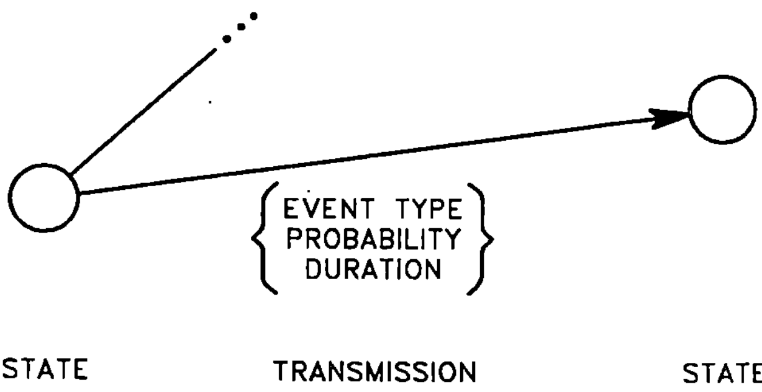

Fig. 3. Generic finite-state machine components

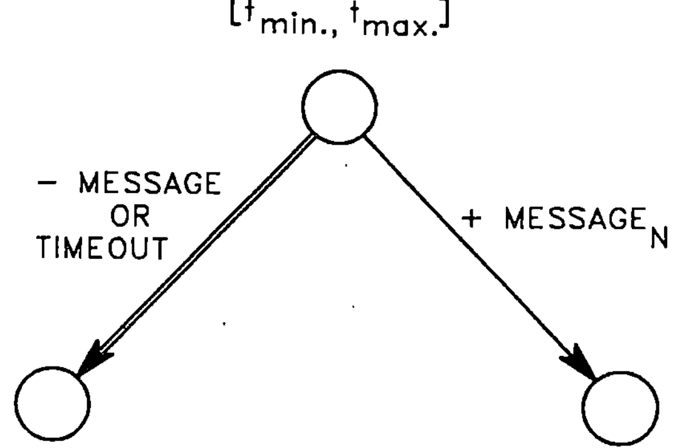

Fig. 4. Time limits associated with a state in an FSM model

In (Bolognesi84) the authors chose to associate a time duration with the state as shown in Figure 4. Here a state can be thought of as a process state in which an activity (such as processing or waiting) is going on. This activity continues until interrupted by the arrival of a message (a *passive* transition) $+ MESSAGE_N$ or, after an elapsed time of at least $t_{min.}$, an *active* transition (generation of a message or expiration of a timeout) occurs. If neither has taken place by time $t_{max.}$, the active transition must take place. Although only a single passive and a single active transition are shown in Figure 4, there could be several of each type. The specification has been called the CT-Graph (Coupled-Time Graph) model.

II.3 Algebraic models

The algebraic specification of protocols is an area of growing research interest at this time. Within ISO's effort on formal description techniques, Subgroup C is working on an algebraic language called LOTOS (Language for Temporal Ordering Specification) (Vissers83). LOTOS is related to MILNER's CCS (Calculus of Communicating Systems) (Milner80).

Insofar as the inclusion of time is concerned in an algebraic specification, the author is aware only of the following two examples:

AGGARWAL and KURSHAN (Aggarwal83) have defined a calculus for specifying and analyzing protocols called the Selection/Resolution model. Selections model potential alternatives in a process. The model provides for the inclusion of delay by assigning a holding-time interval for the duration of these selections.

In the work of NOUNOU and YEMINI (Nounou84), the various possible so-called τ-actions of MILNER's CCS (generally representing the exchange of messages between processes) are distinguished from one another and are assigned individual delay values or delay-value distributions.

So much for the formal specifications themselves. Now, on to what has been done with them.

III. ANALYSES WHICH INCLUDE TIME

This section discusses how models which include time may be used, first for protocol validation and second for performance analysis. These two topics are each organized according to the underlying models. For what follows, it is only fair to say that the author's particular interest is for techniques which pertain to computer-communication protocols and which have either been automated or evidently lend themselves to automation. There are other approaches to the various topics considered which have been published but which are not mentioned here.

III.1 Validation

Validation is a systematic examination of a protocol to check it for syntactic properties. In the case of Petri-net-based specifications, these properties include boundedness, being safe, being live, and freedom from deadlocks. Definitions of these properties may be found in (Diaz82). In the case of FSM-based specifications, the most usual properties checked are completeness and freedom from deadlocks. Validation is rather important, as a lack of completeness or a deadlock can cause a system to come to a complete halt. Further, experience has shown (Rudin85) that even skilled designers find it difficult to design error-free protocols.

Most of the work on validation has been in the absence of time constraints. Even without time, validation easily becomes unmanageable for complex protocols. But, since time often plays a key role in the recovery of a computer communication protocol, it is important to validate a protocol while including its time constraints.

Petri-net-based validation

In (Merlin76) a methodology is suggested for validating protocols and a small protocol is evaluated with the resulting discovery that the specification of two timeout values was inconsistent. SYMONS (Symons78) suggests using a reachability analysis and has used it to analyze a sample communication protocol which includes timeout.

WALTER (Walter83) uses a model similar to that in (Merlin76) and describes a method of analysis which will show whether or not timeouts are specified consistent with the rest of the protocol. The technique used is manual. MENASCHE and BERTHOMIEU (Menasche83) also use MERLIN's model but combined with an automated reachability analysis. In their paper, two protocols are analyzed. The first is the classic Alternating-Bit (A-B) protocol and the second is a protocol for a fault-tolerant, distributed communication system for real-time control applications. It was possible to demonstrate that both protocols have the properties sought for.

FSM-based validation

In 1978 ZAFIROPULO and WEST developed techniques for automated protocol validation wherein the formal specification of the protocol could contain the fact that timeouts were involved (Zafiropulo78) and (West78b). Successful validation of a protocol indicated that all was in order. Were validation unsuccessful, it would not be possible to separate actual from potential errors because *specific values* of timeouts could not be represented. The technique used was reachability analysis.

For the validation of a token-ring local-area-network protocol the above system was modified for synchronous operation and a trick — a circulating, countable phantom time pulse — was used to model a timeout (Rudin82a). The result is satisfactory (Rudin82b) but the modeling methodology was awkward.

Recently a more general approach (Bolognesi84) has been found driven from a CT-graph specification of the kind indicated in Figure 4. An efficient algorithm exists — a modification of a network flow assignment algorithm — when the communicating processes can be arranged in a ring topology. Results for an automated investigation of the A-B protocol are given (Bolognesi84).

Validation based on algebraic models

Based on their selection/resolution algebraic model, AGGARWAL and KURSHAN (Aggarwal83) analyze the A-B protocol. In this instance the inclusion of timing information allows them to substantially reduce the size of the state space which must be examined to do the validation. Specifically, the 88 states generated by the analysis without time constraints reduce to only 14 states with the inclusion of time constraints. The validation shows the timeout specification consistent.

NOUNOU and YEMINI (Nounou84) also examine the A-B protocol but under the assumption that delays, including timeouts, are distributed with a negative exponential distribution. Given this assumption it is possible to derive a value for the average timeout such that the probability that a message which has in fact not been lost is retransmitted is lower than a specified value.

III.2 Performance analysis

The analysis of a protocol's performance has long been a field of active investigation. The goal is usually estimation of throughput or response time. The method generally consists of first designing a model of the protocol and then using analytic techniques or simulation to determine performance. The model is nearly always derived heuristically from whatever description of the protocol is available.

Here we are concerned with techniques which avoid this heuristic derivation of a model suitable for performance analysis. The idea is to base the analysis direct on the formal, machine-readable specification. To do this one must add to the formal definition not only an estimate (or measurement) of execution time but also branching probabilities which specify the relative frequency of alternative paths. An example is the probability of message loss or generation of an error.

Performance simulation derived from the formal specification

While most of the work described here deals with a numeric or algebraic analysis driven direct from the formal description, there is an alternative. This would be first to derive a simulator and then to observe the simulator's performance. This was done in 1981 by BAUERFELD who described a software package called PODEX, capable of automatically generating a simulation, direct from a formal protocol specification of the FSM type (Bauerfeld81 and Bauerfeld83). The simulation is then run, and the performance results observed.

In a simulation, the response time observed for a large number of individual messages is averaged over the course of the simulation to produce an estimate of response time. An estimate for throughput is also based on the observation of a large number of messages. When analytic techniques are used, the calculation of response time is carried out by numerically or algebraically adding the component delays introduced along a path which represents the execution of the action for which the response time is desired. Throughput is usually found as the reciprocal of the delay around a cycle of events.

Petri-net-based performance analysis

The earliest work on performance analysis based on Petri-net models is that of RAMCHANDANI in 1973 (Ramchandani73). RAMCHANDANI calculates the maximum "computation rate" for cycles in an "event-graph" derived from the timed Petri-net description discussed above. A constant delay is associated with the firing of a transition.

In a refinement of the work of RAMCHANDANI, RAMAMOORTHY and HO find a more efficient algorithm for the calculation of the computation rate around a cycle (Ramamoorthy80). Unfortunately, at least for the kinds of protocols which generally exist in computer-communication systems, both of these analyses are restricted to the class of "persistent" Petri nets. In persistent Petri nets, once a transition has been enabled, it can only be disabled by firing said transition. As mentioned in Section II.1, this means that recovery actions, triggered for example by timeouts, cannot be properly modeled.

When the persistence property has been removed, making possible the modeling of timeouts, it has been shown that determining performance has, in general, the complexity of an NP-complete problem (Ramamoorthy80).

HAN in (Han78) also associates a constant delay to the firing of a transition. For systems represented by such a model, HAN calculates throughput in terms of transition-firing frequencies.

A constant delay associated with each transition is also used in the work of ZUBEREK (Zuberek80). Perhaps the largest contribution by ZUBEREK is his establishing of an "Instantaneous Description" of a Petri net in terms of vectors for: 1) the current marking, 2) the firing time remaining for the various enabled transitions, and 3) an indication (the "selector") of which transitions can simultaneously start firing. This system-state description paves the way for state exploration techniques. Two examples, both processor architectures, are explored. The underlying model — as stated above — is not suitable for protocols with timeouts.

MOLLOY takes quite a different approach. Assuming negative exponential distributions for delays, he can employ a Markov-chain analysis (Molloy82). As an example, a throughput analysis is made for the A-B protocol.

The Petri-net model used by RAZOUK has the advantages that timeouts can be modeled accurately, as discussed earlier (Razouk84). Using a state descriptor, like that of ZUBEREK, a "timed reachability graph" is found. From this, system performance can be derived. A modified A-B protocol serves as an example.

FSM-based performance analysis

Performance prediction based on an FSM model was first described by BEIZER in 1970 for predicting program running time (Beizer70). In retrospect it is interesting to note that he had no formal specification to work from; in fact his FSM is a heuristic model of the software code for which he wanted to make a performance estimate. He used a graphical reduction technique to reduce, stepwise, the directed graph representation of the FSM. At the end of this reduction process, the response time is represented by the delay associated with a single, equivalent arc.

Eight years later, but this time in an effort to estimate the performance of a computer-communication protocol, GOUDA derived the same graphical reduction procedure (Gouda78) independently.

The present author has calculated performance by automatically generating a tree rooted in the state from which response time for some action is to be measured (Rudin83). Component delays $D_{path\ i}$ along paths to leaves representing termination of the action are added, weighted by the probability P_i of path occurrence,

$$\text{Average delay} = \sum_{i=1}^{N} \{(P_i)(D_{path\ i})\}.$$

The calculation is terminated when the probability of uncovered paths drops below a specified threshold. The calculation is efficient when advantage is taken of a Markov property which allows the combining of equivalent states (Rudin84).

As an alternative to the above approach, KRITZINGER has used methods from traffic theory for handling networks of queues to achieve the same result (Kritzinger84a). A recent refinement of the same approach is dedicated to the analysis of protocols, taking into account their environment (Kritzinger84b). Specifically KRITZINGER is now able to take into account the fact that many protocol instances often compete with one another for the same implementation-processor cycles. In the Open Systems Interconnect architecture, for example, not only are there several different levels of protocols competing for processor cycles but also various numbers of active or open connections at each of these several levels. This work opens the door for the evaluation of a protocol embedded in its anticipated environment.

Algebraic-based performance analysis

This is a very recent notion and the author knows of only one investigation, that of NOUNOU and YEMINI (Nounou84). This work has been referred to above in connection with validation. In the context of performance evaluation, NOUNOU and YEMINI were able to calculate average response time, algebraically, for the A-B protocol.

Looking at the chronological development of analytic techniques, we see then that from a starting point where formal models were a heuristic abstraction — often not capable of accurately modeling computer-communication protocols — we have moved to a point where the formal protocol specification itself can be automatically analyzed. So far, the experiments with techniques for taking time into account have been limited to simple protocols, the most common being the A-B protocol.

IV. CONCLUSION

Work on the inclusion of time in formal computer-communication protocol specifications is still in its early days. Even so, significant results have been achieved on the fronts of validation and performance prediction. There are many open areas, particularly the application of existing techniques to more complicated protocols, improving the capabilities of existing techniques, and the further development and assessment of algebraic methods which may have substantial advantages for complex protocols. The author hopes that this tutorial paper and the bibliography it contains will help encourage work on the analysis of time in computer-communication protocols.

V. ACKNOWLEDGMENT

I thank my colleagues Tommaso Bolognesi, Pieter Kritzinger, and Liba Svobodova for many timely discussions and for their suggestions for this paper.

REFERENCES

Aggarwal83- S. Aggarwal and R. P. Kurshan, "Modelling elapsed time in protocol
     specification," Proc. Workshop on Protocol Specification, Testing, and Verification, III,
     Rüschlikon, Switzerland, May 1983, (North-Holland, Amsterdam, 1983) pp. 51-62.

Bauerfeld81- W. L. Bauerfeld, "Description, verification, and performance prediction
     of computer network protocols," Proc. INWG/NPL Workshop on Protocols, Volume 1,
     Teddington, England, May 27-29, 1981, pp. 253-269.

Bauerfeld83- W. L. Bauerfeld, "Protocol performance prediction," Proc. International
     Conference on Communications, IEEE, Boston, Mass., June 20-23, 1983,
     pp. 1311-1315.

Beizer70- B. Beizer, "Analytical techniques for the statistical evaluation of program running time," Proc. Fall Joint Computer Conference, 1970, pp. 519-524.

Bolognesi84- T. Bolognesi and H. Rudin, "On the analysis of time-constrained protocols by network flow algorithms," Proc. Workshop on Protocol Specification, Testing, and Verification, IV, Sky Top, Pennsylvania, June 1984, (North-Holland, Amsterdam, 1984).

Coolahan83- J. E. Coolahan, Jr. and N. Roussopoulos, "Timing requirements for time-driven systems using augmented Petri nets," IEEE Trans. Software Eng., Vol. SE-9, No. 5, Sept. 1983, pp. 603-616.

Diaz82- M. Diaz, "Modeling and analysis of communication and cooperation protocols using Petri-net based models," Computer Networks, Vol. 6, 1982, pp. 419-441.

Gouda78- M. G. Gouda, "Protocol machines: towards a logical theory of communication protocols," PhD. thesis, University of Waterloo, Jan., 1978 (see especially Chapter 11).

Han78- Y. W. Han, "Performance evaluation of a digital system using a Petri net-like approach," Proc. National Electronics Conference, Chicago, 1978, pp. 166-172.

Kritzinger84a- P. Kritzinger, "Analyzing the time efficiency of a communication protocol," Proc. Workshop on Protocol Specification, Testing, and Verification, IV, Sky Top, Pennsylvania, June 1984, (North-Holland, Amsterdam, 1984).

Kritzinger84b- P. Kritzinger, "A performance model of the OSI communication architecture," IBM Zurich Research Laboratory Research Report, Rüschlikon, Switzerland, Nov., 1984.

Menasche83- M. Menasche and B. Berthomieu, "Time Petri nets for analyzing and verifying time dependent communication protocols," Proc. of the Workshop on Protocol Specification, Testing, and Verification, III, Rüschlikon, Switzerland, May 1983, (North-Holland, Amsterdam, 1983) pp. 161-172.

Merlin76- P. Merlin and D. J. Farber, "Recoverability of communication protocols: implications of a theoretical study," IEEE Trans. Commun., Vol. COM-24, Sept. 1976, pp. 1036-1043.

Milner80- R. Milner, "A Calculus of Communicating Systems," Lecture Notes in Computer Science, (Springer Verlag, Berlin, 1980).

Molloy82- M. K. Molloy, "Performance analysis using stochastic Petri nets," IEEE Trans. Computers, Vol. C-31, Sept. 1982, pp. 913-917.

Nounou84- N. Nounou and Y. Yemini, "Algebraic specification-based performance analysis of communication protocols," Proc. Workshop on Protocol Specification, Testing, and Verification, IV, Sky Top, Pennsylvania, June 1984, (North-Holland, Amsterdam, 1984).

Peterson77- J. L. Peterson, "Petri Nets," Computing Surveys, Vol. 9, September 1977, pp. 223-252.

Ramamoorthy80- C. V. Ramamoorthy and G. S. Ho, "Performance evaluation of asynchronous concurrency systems using Petri nets," IEEE Trans. Software Eng., Vol. SE-6, Sept. 1980, pp. 440-449.

Ramchandani73- C. Ramchandani, "Analysis of asynchronous concurrent systems by timed Petri nets," Ph.D. Thesis, M. I. T., Dept. of E. E., AD-775618, July, 1973.

Razouk84- R. R. Razouk and C.V. Phelps, "Performance analysis using timed Petri nets," Proc. of the Workshop on Protocol Specification, Testing, and Verification, IV, Sky Top, Pennsylvania, June 1984, (North-Holland, Amsterdam, 1984).

Rudin82a- H. Rudin and C.H. West, "A validation technique for tightly-coupled protocols," IEEE Trans. Computers, Vol. C-31, July, 1982, pp. 630-636.

Rudin82b- H. Rudin, "Validation of a token-ring protocol," in Proc. Intl. Symposium on Local Computer Networks, Florence, April, 1982, (North-Holland, Amsterdam, 1982) pp. 373-387.

Rudin83- H. Rudin, "From formal protocol specification towards automated performance prediction," Proc. Workshop on Protocol Specification, Testing, and Verification, III, Rüschlikon, Switzerland, May 1983, (North-Holland, Amsterdam, 1983) pp. 257-269.

Rudin84- H. Rudin, "An improved algorithm for estimating protocol performance," Proc. Workshop on Protocol Specification, Testing, and Verification, IV, Sky Top, Pennsylvania, June 1984, (North-Holland, Amsterdam, 1984).

Rudin85- H. Rudin, "An informal overview of formal protocol specification," IEEE Communications Magazine, probably February, 1985.

Sifakis79- J. Sifakis, "Performance evaluation using nets," Net Theory and Applications, Lecture Notes in Computer Science No. 84, (Springer Verlag, Heidelberg, 1979), pp. 307-319.

Symons78- F. J. W. Symons, "Modeling and analysis of communication protocols using numerical Petri nets," Ph.D. Thesis at University of Essex, England, May, 1978.

Vissers83- C. A. Vissers, R. L. Tenney, and G. V. Bochmann, "Formal Description Techniques," Special issue on OSI, Proc. IEEE, Vol. 71, No. 12, December, 1983, pp. 1356-1364.

Walter83- B. Walter, "Timed Petri-nets for modelling and analyzing protocols with real-time characteristics," Proc. Workshop on Protocol Specification, Testing, and Verification, III, Rüschlikon, Switzerland, May 1983, (North-Holland, Amsterdam, 1983) pp. 149-159.

West78- C.H. West, "General technique for communications protocol validation," IBM J. Res. Develop., Vol. 22, July 1978, pp. 393-404.

Wong84- C. Y. Wong, T. S. Dillon, and K. E. Forward, "Analysis of timing aspects of communication computer systems using timed places Petri nets," Proc. Seventh International Conference on Computer Communication, Sydney, October 30 - November 2, 1984, pp. 585-590.

Zafiropulo78- P. Zafiropulo, "Protocol validation by duologue-matrix analysis," IEEE Trans. Commun., Vol. COM-26, Aug. 1978, pp. 1187-1194.

Zuberek80- W. M. Zuberek, "Timed Petri nets and preliminary performance evaluation," Proc. 7th Annual IEEE Symposium on Computer Architecture, 1980, pp. 88-96.

Eine Methode zur Beschreibung von Kommunikationsprotokollen

C. Andres, A. Fleischmann, P. Holleczek,
U. Hillmer, R. Kummer

Regionales Rechenzentrum
der Universität Erlangen-Nürnbeg

Zusammenfassung

In dem vorgestellten Spezifikationskonzept wird ein DFÜ-System
als ein System von Prozessen betrachtet, die mit Hilfe von Botschaften untereinander kommunizieren. Die Botschaften entsprechen
dabei den einzelnen Dienstprimitiven. Die einzelnen Instanzen bestehen aus ein oder mehreren Prozessen. Die Prozesse einer Instanz kommunizieren mit den Prozessen ihrer Partnerinstanzen mit
Hilfe des unterlagerten Dienstes. Die Eigenschaften dieses Dienstes werden ebenfalls durch Prozesse modelliert.
Die einzelnen Prozesse werden mit Hilfe eines erweiterten Zu-
standsmodells beschrieben. Die Reihenfolge in der die einzelnen
Botschaften gesendet oder empfangen werden, wird graphisch mit
Hilfe eines Übergangsdiagramms dargestellt. Die Bedeutung der
einzelnen Nachrichten wird mit Hilfe von Funktionen beschrieben.
Beim sendenden Prozeß wird damit definiert, was dieser Prozeß
beim Senden einer Nachricht meint und beim Empfänger wird damit
festgelegt, wie er die Nachricht versteht. Diese Funktionen können mit Techniken spezifiziert werden, wie sie in der sequentiellen Programmierung üblich sind.

Abstract

We describe a method for specifying data communication systems as
a system of processes which communicate with each other via messages. The messages represent service primitives of various
types. The individual entities consist of one or more processes.
The processes belonging to a single entity communicate with each
other using the services of the next lower level. The properties
of these lower level services which are significant for the next
higher level are also specified by processes. The individual
prozcesses are described using an extended state transition diagram. The sequence in which the messages are sent or received is
specified using a state transition diagram. The meaning of the
messages is defined by functions, which define the meaning of the
message for the sending process, and how it is to be processed by
the receiving process. These functions can be specified as sequential programs.

0. Einleitung

Bei der Implementierung von Kommunikationsprotokollen zeigt sich,
daß viele Normdokumente, z. B. das S.70-Protokoll nach CCITT, nicht
direkt als Spezifikation verwendet werden können. Die zum großen
Teil in Prosa verfassten Dokumente müssen zuerst noch in eine mehr
oder weniger formale Spezifikation umgearbeitet werden.
Am Rechenzentrum der Universität Erlangen-Nürnberg wurde eine Spezifikationstechnik entwickelt, die sich bei der Spezifikation und Implementierung von Kommunikationsprotokollen und Programmen zur Pro-
zeßautomatisierung als sehr hilfreich herausstellte. Sie wird am
RRZE insbesondere auch zur Spezifikation der Protokolle im Deutschen
Forschungsnetz (DFN) eingesetzt.
In den folgenden Kapiteln sollen dieses Spezifikationskonzept vorgestellt und erste Erfahrungen diskutiert werden.

1. Das Schichtenmodell und seine Kommunikationsmechanismen

Datennetzen wird versucht, bestimmte Aufgaben in Schichten zusammen-
zufassen, um damit einen hierarchischen Aufbau eines Netzes bzw. der
verwendeten Protokolle zu erreichen. Das Architekturmodell für offene Systeme der ISO umfasst die gesamte Hierarchie von der Kommunikation zwischen Anwenderprozessen bis hinunter zur Datenübertragung
auf Leitungen.
Mit Ausnahme der untersten Schicht bedient sich jede Schicht (n) der
Dienste der Schicht (n-1), führt eigene Funktionen durch und bietet
der Schicht (n+1) wiederum Dienste an. (Siehe dazu auch Bild 1.) Die
Dienste einer Schicht werden durch sogenannte Instanzen realisiert.

Eine Instanz der Schicht (n) kommuniziert direkt mit den Instanzen
der Schicht (n-1) und der Schicht (n+1), d. h. sie verwendet den
Service, der von der Schicht (n-1) zur Verfügung gestellt wird und
bietet der Schicht (n+1) ihren Service an. Die Kommunikation erfolgt
über sogenannte Schnittstellensignale (direkte Kommunikation).
Mit Hilfe der Dienste der darunterliegenden Schicht kommuniziert
eine Instanz mit ihrer Partnerinstanz. Das dabei resultierende 'Kommunikationsverhalten' wird allgemein als Protokoll bezeichnet (indi-
rekte Kommunikation). Bei der indirekten Kommunikation werden zwischen den Instanzen sogenannte Protokolldateneinheiten ausgetauscht.

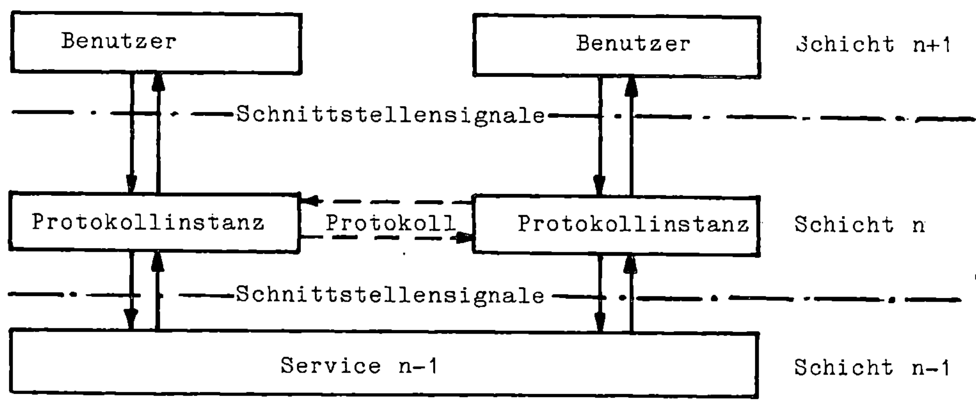

Bild 1: Schichten von Protokollen und Diensten (Service)

2. Aufgaben einer Protokollspezifikation

Wie oben dargestellt, handelt es sich bei einer Protokollspezifikation um die Beschreibung und Darstellung der direkten und indirekten Kommunikation in einer Protokollschicht. Außer der direkten und
der indirekten Kommunikation muß in einer Protokollspezifikation
auch der Aufbau der Protokolldateneinheiten festgelegt sein.
Die beiden Kommunikationsarten lassen sich auch unter einem anderen
Gesichtspunkt betrachten:
Unter der direkten Kommunikation versteht man das "lokale" Verhalten
einer Protokollschicht (n) auf einer Station im Zusammenwirken mit
den sie umgebenden Schichten, also vor allem mit der darüberliegenden (n+1) und der darunterliegenden (n-1) Schicht.

Unter der indirekten Kommunikation versteht man das "globale" Verhalten einer Protokollschicht (n) auf einer Station A im Zusammenwirken mit der Partnerschicht (n) auf einer Station B. Der dabei notwendige Austausch der Protokolldaten geschieht über die zur Verfügung stehende darunterliegende Verbindung und der dazugehörigen Dienste.

Daraus lassen sich bezüglich des Kommunikationsverhaltens folgende Forderungen an eine Spezifikationsmethode ableiten (/BOCH83/):
- Sie muß sowohl das globale als auch das lokale Verhalten einheitlich darstellen können; unterschiedliche Beschreibungen tragen nur zur Verwirrung bei.
- Da zwischen globalem und lokalem Verhalten Wechselbeziehungen bestehen und damit zwischen den beiden keine Trennung möglich ist, sollte die Methode auch beides im Zusammenhang darstellen können.

Zusätzlich zu diesen Anforderungen muß eine Spezifikationstechnik noch folgende Möglichkeiten haben:
- Es muß die Bedeutung von Schnittstellensignalen und deren Parameter beschrieben werden können.
- Ebenso muß es möglich sein die Bedeutung und den Aufbau der Protokolldateneinheiten eindeutig festzulegen.

Das im Folgenden vorgestellte Spezifikationskonzept versucht diesen Forderungen gerecht zu werden.

3. Das Spezifikationsmodell

Das hier vorgestellte Spezifikationskonzept beruht auf einem erweiterten Zustandsmodell wie es in ähnlicher Form in SDL /SDL83/ und ESTELLE /ESTE84/ verwendet wird. In diesen Spezifikationstechniken ist die Beschreibung des Kommunikationsverhaltens nicht strikt von der Beschreibung interner Aktionen getrennt, so daß es schwierig werden kann, z.B. nach der Beschreibung des Aufbaus einer Protokolldateneinheit oder der Reihenfolge der Schnittstellensignale beim Verbindungsaufbau zu suchen.
Außerdem kann in SDL bzw. ESTELLE die Bedeutung einer Nachricht (auch Ereignis genannt) für den Sender und Empfänger nicht explizit definiert werden.

Eine andere, zur Zeit sehr aktuelle Möglichkeit, Protokolle und Service zu definieren, beruht auf der direkten Beschreibung der zeitlichen Reihenfolge von Ereignissen, ohne explizit Zustände einzuführen (siehe dazu /LOT84/).

3.1. Das Schichtenmodell als System kommunizierender Prozesse

In diesem Spezifikationskonzept wird eine Protokollinstanz jeweils als eine Menge von Prozessen aufgefaßt ("Instanzprozesse"). Diese kommunizieren mit den Prozessen einer Partnerinstanz über einen dritten Prozeß ("Serviceprozeß"), der die Dienste der darunterliegenden Protokollschicht beschreibt bzw. das Verhalten des Verbindungssytems. Der Serviceprozeß ist somit eine Abstraktion aller darunterliegenden Schichten. Er beschreibt den Service, der von einer darüberliegenden Instanz vorausgesetzt wird.

Die Prozesse, die eine Instanz realisieren, können über gemeinsame Objekte verfügen oder über Botschaften miteinander kommunizieren. Die Instanzprozesse dagegen können mit dem entsprechenden Serviceprozeß nur über Botschaften kommunzizieren.
Die Dienste, die durch die Instanzprozesse der Schicht (n) und den Serviceprozeß (n-1) realisiert werden, können ebenfalls abstrakt als einzelner Prozeß beschrieben werden. Diese Beschreibung der Dienste der Schicht (n) durch einen Prozeß wird mit Serviceprozeß (n) bezeichnet.

Die Spezifikation besteht somit aus:
- **dem vorausgesetzten Serviceprozeß (n-1),**
- **den Instanzprozessen (n),**
- **dem Serviceprozeß (n).**

Zur Beschreibung des vorausgesetzten Serviceprozesses der Instanzprozesse und des zur Verfügung gestellten Services werden jeweils die gleichen Darstellungsmittel verwendet. Diese Darstellungsmittel sollen in den folgenden Abschnitten erläutert werden.

3.2 Darstellung der Spezifikationstechnik

Die im folgenden vorgestellte Technik /FLEI84/ benutzt bei der Beschreibung der Synchronisation und Kommunikation von Prozessen einen
Botschaftsmechanismus.

3.2.1 Kommunikationsstuktur

Die Beziehungen zwischen den Prozessen werden in der **Kommunikationsstruktur** aufgezeigt, d.h. dort wird beschrieben, .welche Prozesse an
einem Prozeßsystem beteiligt sind und welche Nachrichten sie austauschen. Graphisch werden die beteiligten Prozesse durch Rechtecke
dargestellt. Ein Pfeil, beschriftet mit dem Namen der Nachricht,
zeigt vom sendenden zum empfangenden Prozeß (B i l d   2).

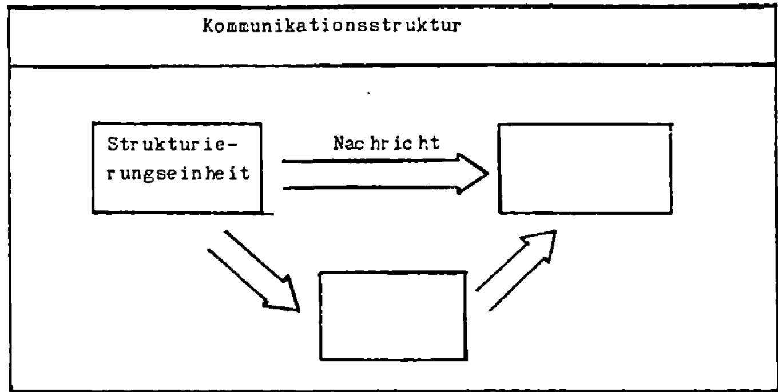

Bild 2: Beispiel einer einfachen Kommunikationsstruktur

Eine Nachricht kann mit Parametern versehen sein, die Kommunikation
zwischen den Prozessen erfolgt über diese Parameter.
Genauer müßte man in der Kommunikationsstruktur von Strukturierungseinheiten, statt von Prozessen sprechen. Strukturierungseinheiten
können einzelne Prozesse, Prozeßbündel oder Prozeßgruppen sein. Die
Prozesse können ihrerseits als elementare Objekte oder Prozeßtypen
definiert werden. Prozeßbündel bzw. Prozeßgruppen wurden eingeführt,
um eine zusätzliche Kommunikation über gemeinsame Objekte zu ermöglichen.
Jede Strukturierungseinheit wird in einem zweiten Schritt genauer
beschrieben.

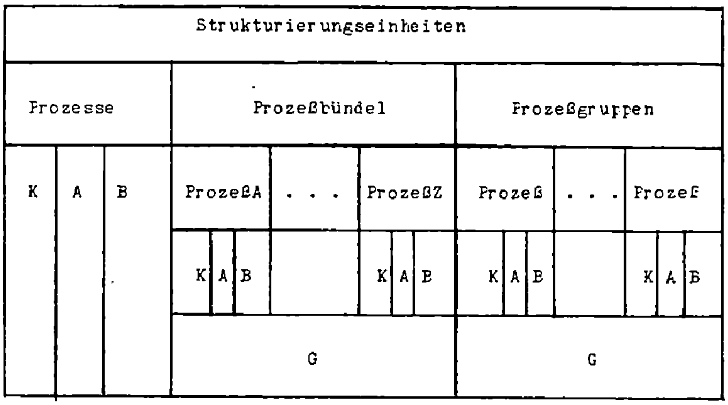

Strukturierungseinheiten						
Prozesse			Prozeßbündel		Prozeßgruppen	
K	A	B	ProzeßA . . . ProzeßZ		Prozeß . . . Prozeß	
			K A B ... K A B		K A B ... K A B	
			G		G	

K – Kommunikationsmaschine
A – Ablaufsteuerung
B – private Benutzermaschine
G – gemeinsame Benutzermaschine

Bild 3 : Arten von Strukturierungseinheiten und ihr Aufbau

3.2.2. Strukturierungseinheiten

3.2.2.1. Einzelprozesse

Die Struktur eines Einzelprozesses umfaßt die Kommunikationsmaschine, die Ablaufsteuerung und die private Benutzermaschine.
(B i l d 3)
Die **Kommunikationsmaschine** regelt das Senden und Empfangen von Nachrichten und veranlaßt dabei die Ausführung der entsprechenden Ausgabefunktion bzw. Eingabeoperation (näheres dazu folgt bei der Beschreibung der Benutzermaschine). Die Eigenschaften der Kommunikationsmaschine sind fest vorgegeben. Der Entwerfer des Systems kann lediglich angeben, ob die Botschaften synchron oder asynchron ausgetauscht werden sollen; d.h. es kann spezifiziert werden, ob Nachrichten über einen sogenannten Wartebereich empfangen werden sollen und wie groß dieser sein soll (Wieviele Nachrichten darin Platz haben, unabhängig von der Art und Anzahl ihrer Parameter). Ist einem Prozeß ein Wartebereich vorgelagert, so können Nachrichten, die an diesen Prozeß gesendet werden, darin abgelegt werden. Der sendende Prozeß wird nicht blockiert, wenn im Wartebereich noch ein Platz frei ist zum Abzulegen der Nachricht (asynchroner Nachrichtenaustausch).

Ist dem empfangenden Prozeß kein Wartebereich vorgelagert, wird der
Sender solange blockiert, bis ihm vom Empfänger die Nachricht abge-
nommen wird (synchroner Nachrichtenaustausch).

In der **Ablaufsteuerung** wird festgelegt, in welcher Reihenfolge ein
Prozeß Nachrichten sendet bzw. empfängt. Ein Prozeß tauscht nicht
nur Nachrichten mit anderen Prozessen aus, sondern führt auch interne Berechnungen durch. Aufgrund der Ergebnisse solcher Berechnungen
kann in verschiedene Teile der Ablaufsteuerung verzweigt werden.
In der Ablaufsteuerung gibt der Programmentwerfer somit an
- wann, welche Nachrichten an wen gesendet werden
- wann, welche Nachrichten von wem empfangen werden
- wann interne Berechnungen durchgeführt werden.
Die Ablaufsteuerung wird graphisch mit den Elementen Knoten und
Kanten dargestellt.
Erreicht ein Prozeß in der Ablaufsteuerung einen **Kommunikationsknoten** (Rechtecksymbol), möchte er Nachrichten senden oder empfangen.
Sollen Nachrichten gesendet werden, führen von einem Kommunikationsknoten Sendekanten (Doppelpfeile) weg. Die Sendekanten sind jeweils
mit dem Adressaten und dem Namen der zu sendenden Nachricht beschriftet. Kann die Nachricht gesendet werden, geht der Prozeß in
den Zustand über, zu dem die Sendekante führt.
Ein Prozeß erwartet Nachrichten, wenn er sich in einem Kommunikationsknoten befindet, von dem Empfangskanten (einfache Pfeile) weg-
führen. Die Empfangskanten sind jeweils mit dem Absender und dem
Namen der erwarteten Nachricht beschriftet. Trifft die Nachricht vom
angegebenen Quellprozeß (Absender) ein, wird in den Folgezustand
übergegangen.
Führen von einem Kommunikationsknoten mehrere Sende- bzw. Empfangskanten weg, so wird alternativ auf Nachrichten gewartet bzw. es werden alternativ Nachrichten gesendet.
In der graphischen Darstellung der Ablaufsteuerung werden interne
Berechnungen durch **Internknoten** (Ovale) gekennzeichnet. In diesen
Ovalen wird der Name der auszuführenden internen Berechnung eingetragen. Bei internen Berechnungen wird zwischen internen Operationen
und internen Funktionen unterschieden. Bei der Ausführung einer internen Operation werden die lokalen Daten eines Prozesses verändert,
die internen Funktionen dagegen fragen den Zustand der lokalen Daten
ab.

Führen von einem Internknoten sogenannte Operationskanten (Doppelpfeile) weg, wird eine interne Operation ausgeführt. Die Operationskanten sind mit den möglichen Resultaten der Operationsausführung beschriftet. Abhängig von den Resultaten wird in einen Folgezustand übergegangen.

Wird in einem Internknoten eine interne Funktion ausgeführt, so verlassen .diesen Zustand sogenannte Funktionskanten (einfache Pfeile). Die Funktionskanten sind mit den möglichen Ergebnissen der internen Funktion beschriftet.

Der Anfangszustand wird in der Ablaufsteuerung durch einen Pfeil gekennzeichnet, dessen Anfangspunkt an keinem Knoten liegt.

Mit der Ablaufsteuerung wird nur festgelegt, wann welche Aktionen ausgeführt werden, aber nicht, wie die einzelnen Aktionen definiert sind. Es gilt noch zu beschreiben, welche Parameter die einzelnen Nachrichten (Anzahl, Typ) haben, wie sich das Empfangen einer Nachricht auf den entsprechenden Prozeß auswirkt (außer daß in der Ablaufsteuerung in einen anderen Zustand übergegangen wird) bzw. wie beim Senden von Nachrichten der Wert der entsprechenden Parameter bestimmt wird.

Jeder Nachricht, die gesendet wird, wird eine Ausgabefunktion zugeordnet. Diese Funktion ermittelt aus dem internen Zustand eines Prozesses (Belegung der lokalen Variablen eines Prozesses) die Nachrichtenparameterwerte, ohne daß der interne Zustand verändert wird.

Jeder Nachricht, die empfangen wird, ist eine Eingabeoperation zugeordnet. Sie beschreibt, wie der Empfang einer Nachricht sich auf den internen Zustand eines Prozesses auswirkt.

Insgesamt müssen für eine vollständige Spezifikation eines Prozesses neben der Ablaufsteuerung folgende Funktionen und Operationen definiert werden:

- interne Funktionen
- interne Operationen
- Ausgabefunktionen
- Eingabeoperationen

Diese Funktionen und Operationen werden in der sogenannten **privaten Benutzermaschine** beschrieben. Die Benutzermaschine enthält außerdem die Daten eines Prozesses.

Die Benutzermaschine kann mit Techniken spezifiziert werden, wie sie
in der sequentiellen Programmierung üblich sind (umgangssprachliche
Beschreibung, pseudo-programmiersprachliche oder algebraische Spezifikation, Prädikatentransformation etc.).

Bisher wurden in der Spezifikationstechnik elementare Objekte beschrieben. Gibt es jedoch in einem System kommunizierender Prozesse
mehrere gleichartige Prozesse, d.h. Prozesse mit gleicher Ablaufsteuerung und gleicher privater Benutzermaschine, aber unterschiedlichen Kommunikationspartnern, so kann ein entsprechender **Prozeßtyp**
definiert werden. Objekte, deren Typ auf diese Art eingeführt wurde,
müssen im Anschluß an die Definition deklariert werden, wobei dann
die konkreten Kommunikationspartner festgelegt werden.

3.2.2.2 Prozeßbündel und Prozeßgruppen

Prozeßbündel und Prozeßgruppen bestehen aus mehreren Prozessen, die
über gemeinsame Objekte verfügen. Jeder einzelne Prozeß wird mit der
eben vorgestellten Methode, bestehend aus Kommunikationsmaschine,
Ablaufsteuerung und privater Benutzermaschine beschrieben. Die gemeinsamen Objekte werden zu einer gemeinsamen Benutzermaschine zusammengefaßt. Der Zugriff auf die gemeinsame Benutzermaschine kann
durch ressourcenorientierte Synchronisationskonzepte geregelt werden.
Jeder einzelne Prozeß eines Prozeßbündels kann Nachrichten zu anderen Strukturierungseinheiten senden bzw. von diesen empfangen. Die
einzelnen Prozesse einer Prozeßgruppe dagegen können nicht von außen
identifiziert werden. Prozesse einer Prozeßgruppe können demnach
weder als Absender noch als Adressat einer Nachricht auftreten. Nur
der Name einer Prozeßgruppe kann als Absender oder Adressat dienen.

4. Beispiel

Das Aussehen von Service- und Protokollspezifikationen die, mit dieser Technik beschrieben sind, soll an einem "akademischen" Beispiel
gezeigt werden.

4.1. Informelle Beschreibung des gewünschten Service

Zwei Programm auf verschiedenen Rechnern wollen Daten austauschen.
Dazu sind die beiden Rechner mit einer Leitung verbunden. Es soll
ein Übertragungssystem mit folgenden Eigenschaften entworfen werden:

- Sendet ein Programm Daten, so wird es solange blockiert, bis das
  Partnerprogramm die Daten angenommen hat.
- Ein Datenblock wird solange wiederholt, bis er fehlerfrei über-
  tragen werden konnte, d.h. bis kein Fehler mehr erkannt werden
  konnte.
- Es wird davon ausgegangen, daß die Verbindungsleitung nicht un-
  terbrochen wird.
- Ein Verbindungsaufbau bzw. -abbau ist nicht notwendig.
- Ein Prozeß, der Daten erwartet, wird solange blockiert, bis Daten
  vom Partner eintreffen.
- Als Datum können 16 Bytes gesendet werden.

Diese informell dargelegten Anforderungen an den Service des Über-
tragungssystems sollen nun in einem ersten Schritt formal dargestellt werden. Dazu wird mit den obigen Darstellungsmitteln zunächst
ein Modell des gewünschten Service erstellt, d. h., es soll der entsprechende Serviceprozeß konstruiert werden.
In einem zweiten Schritt soll ein Protokoll entwickelt werden, das
diesen Service realisiert.
Für eine Spezifikation nach dem dargestellten Konzept gibt es einen
konkreten Gliederungsvorschlag (siehe /FLEI84/). Diese Gliederung
wird hier nicht ausführlich dargelegt, sie ist hinreichend selbster-
klärend.

4.2. Formale Servicespezifikation

PROZESSYSTEM: verbindung

I. KOMMUNIKATIONSSTRUKTUR

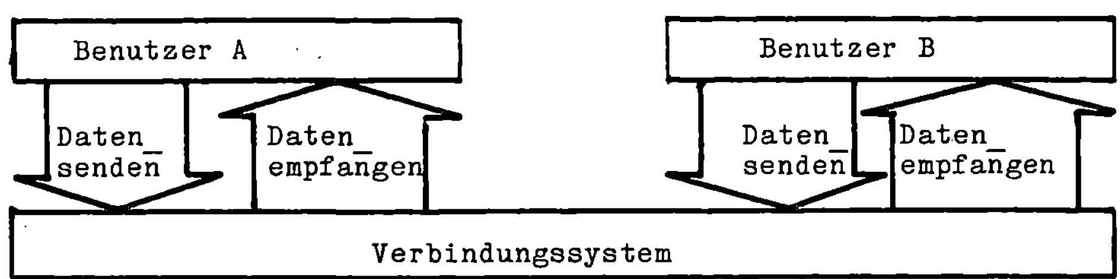

/* Der Service wird durch den Serviceprozeß 'Verbindungspro-
zeß' beschrieben. Das Beschreibung der Prozesse 'Benutzer
A' und 'Benutzer B' ist für die Beschreibung des Services
nicht notwendig */

II. TYPEN VON STRUKTURIERUNGSEINHEITEN

/* Keine */

III. ELEMENTE EINES PROZESSYSTEMS

1. DEKLARATIONEN

/* Keine */

2. DEFINITION WEITERER STRUKTURIERUNGSEINHEITEN

EINZELPROZESS: verbindungssystem
/* Durch diesen Einzelprozeß wird der Service des Verbindungssystems beschrieben, der den beiden Benutzern A
und B zur Verfügung steht */

1. KOMMUNIKATIONSMASCHINE
   WARTEBEREICH: keiner
2. ABLAUFSTEUERUNG

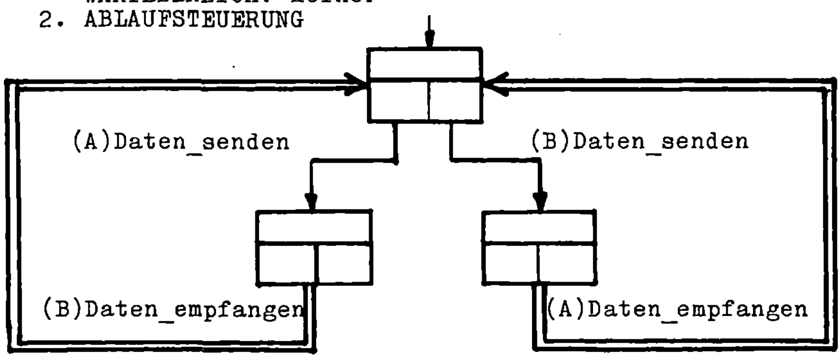

3. BENUTZERMASCHINE

3.1. AUSGABEFUNKTIONEN: Daten_empfangen
3.2. EINGABEOPERATIONEN: Daten_senden
3.3. INTERNE FUNKTIONEN: -
3.4. INTERNE OPERATIONEN: Fehler?

3.5. DEFINITION DER BENUTZERMASCHINE
3.5.1. DATEN
        speicher ARRAY (1..16) BIT(8)

3.5.2. EINGABEOPERATIONEN
        Daten_senden (daten ARRAY (1..16) BIT(8))
          EFFEKT: speicher := daten;

3.5.3. AUSGABEFUNKTIONEN
        Daten_empfangen (daten ARRAY (1..16) BIT (8))
          EFFEKT: daten := speicher;

ENDE verbindung

Mit dem oben definierten Kommunikationssystem werden die Eigenschaften des Verbindungssystems beschrieben, wie sie sich für die Benutzer darstellen.

4.3. Formale Protokollspezifikation

Es soll mit denselben Darstellungsmitteln ein Protokoll beschrieben
werden, das den obigen Service realisiert.
Das zu spezifizierende Protokoll setzt direkt auf der elektrischen
Verbindung auf, wobei eine Vollduplex-Leitung vorausgesetzt wird. Um
also das gewünschte Protokoll spezifizieren zu können, muß der vor-
ausgesetzte Service der darunterliegenden Schicht (hier die Leitung)
beschrieben werden.
Da das Protokoll symmetrisch ist, d.h. auf beiden Enden der Übertra-
gungsstrecke das gleiche Protokoll verwendet wird, liegt es nahe einen Prozeßtyp zu definieren, der das Protokollverhalten beschreibt.
Durch entsprechende Deklarationen werden dann Objekte dieses Typs
eingeführt die den Protokollinstanzen entsprechen.

PROZESSYSTEM

I. KOMMUNIKATIONSSTRUKTUR

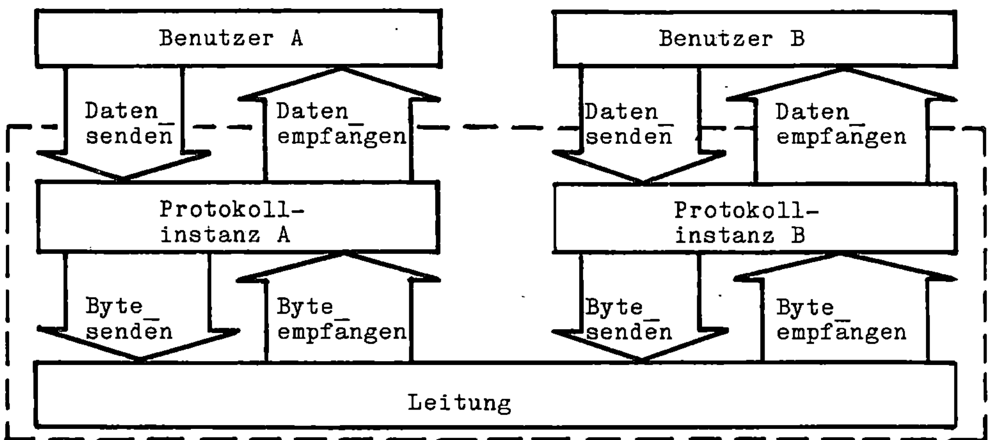

/* Die gestrichelt umrandeten Prozesse sollen gegenüber den
   Benutzern A und B ein Verhalten an den Tag legen, wie es
   weiter oben im Einzelprozeß 'Verbindungssystem' beschrieben ist. */

II. TYPEN VON STRUKTURIERUNG'SEINHEITEN

protokoll EINZELPROZESSTYP (benutzer, service)

/* 'benutzer' und 'service' sind formale Namen für Strukturierungseinheiten zur Absender- und Adressangabe (siehe
   dazu Abschnitt 3.4.) */

1. KOMMUNIKATIONSMASCHINE
   WARTEBEREICH: keiner

2. ABLAUFSTEUERUNG

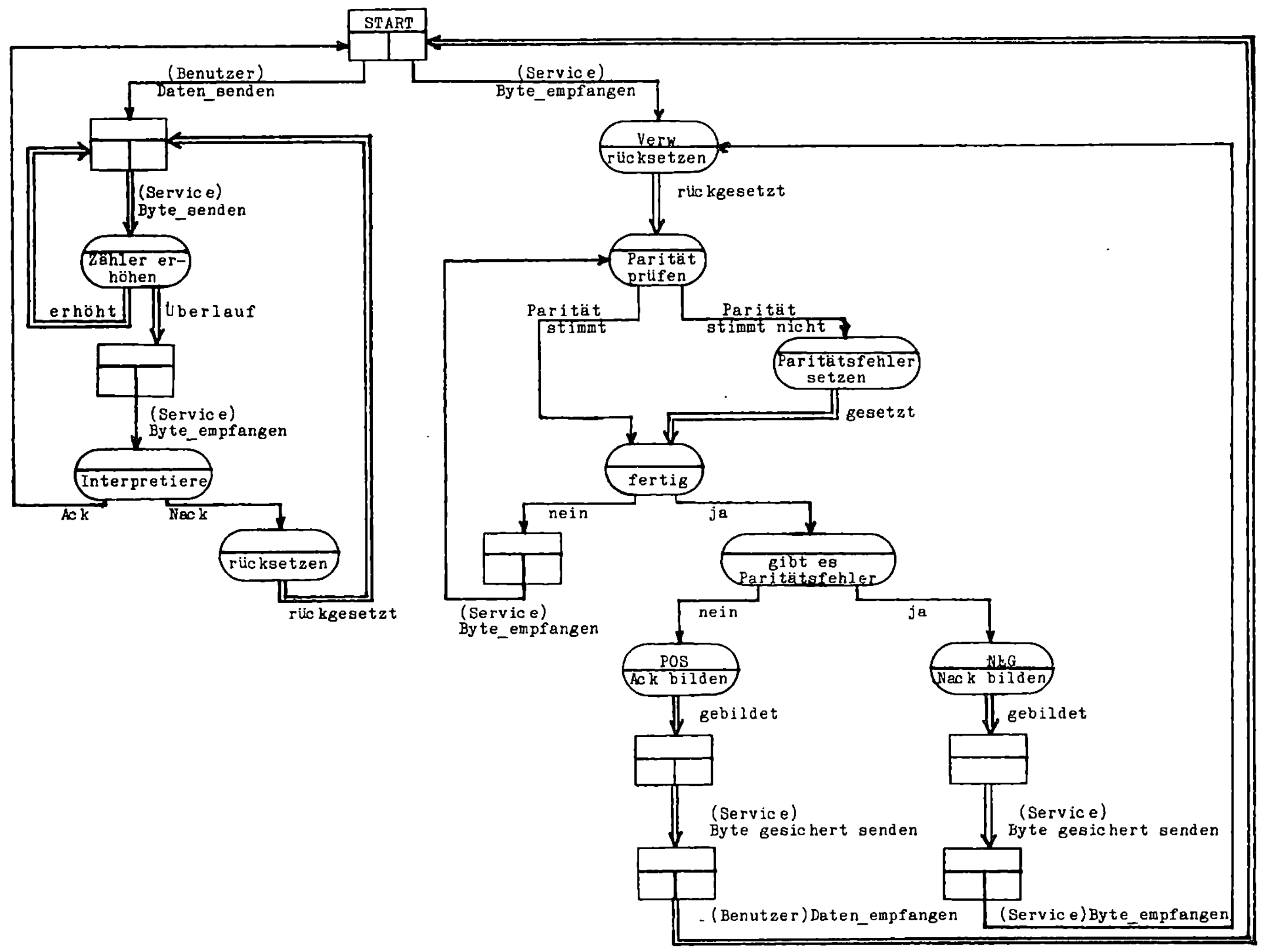

3. BENUTZERMASCHINE
   /* Wie bereits erwähnt, kann zur Beschreibung der Benut-
      zermaschine eine der Techniken verwendet werden, wie
      sie in der sequentiellen Programmierung üblich sind
      (Prosa, algebraische Spezifikation usw.). Aus Ein-
      fachheitsgründen wurde eine programmiersprachliche
      Notation gewaählt die weitgehend selbsterklärend ist.
   */

3.1. AUSGABEFUNKTIONEN: Byte_gesichert_senden,
                        Daten_empfangen
3.2. EINGABEOPERATIONEN: Daten_senden, Byte_empfangen
3.3. INTERNE FUNKTIONEN: Parität_prüfen, Fertig, Interpre-
                         tiere, Gibt es Paritätsfehler
3.4. INTERNE OPERATIONEN: rücksetzen, ACK_bilden,
                          NAK_bilden, Zähler_erhöhen,
                          Paritätsfehler_setzen
3.5. DEFINITION DER BENUTZERMASCHINE:

3.5.1. DATEN
       (parität, paritätsfehler) BIT(1);
       (empfspeicher, sendspeicher) ARRAY (1..16) OF BIT(8);
       zeiger FIXED;

3.5.2. EINGABEOPERATIONEN
       Daten_senden (daten ARRAY (1..16) BIT(8))
          EFFEKT: sendspeicher := daten;
                  zeiger := 1;
                  paritätsfehler := '0'B1 ;

       Byte_empfangen (byte BIT(9))
          EFFEKT: empfspeicher(zeiger) := BIT 1..8 OF byte;
                  parität := BIT 9 OF byte;
                  zeiger := zeiger + 1;

3.5.3. AUSGABEFUNKTIONEN

       Byte_gesichert_senden (byte BIT(9))
          EFFEKT: byte := sendspeicher(zeiger) CONCAT
                          EVEN_PARITY (sendspeicher
                                          (zeiger));
          /* Zur Datensicherung wird eine Parität verwendet.
             Durch die Operation EVEN_PARITY wird die gerade
             Parität des entsprechenden Feldelementes von
             sendspeicher berechnet und mit diesem Feldele-
             ment konkateniert. Normalerweise müßte die Ope-
             ration EVEN_PARITY ebenfalls in der Benutzerma-
             schine definiert werden, der Einfachheit hal-
             ber, wird darauf verzichtet. Das gleiche gilt
             für CONCAT */

       Daten_empfangen (daten ARRAY (1..16) BIT(8))
          EFFEKT daten := empfspeicher;

3.5.4. INTERNE OPERATIONEN

       Zähler_erhöhen
          EFFEKT: IF zeiger < 16
                     THEN zeiger := zeiger + 1; RETURNS erhöht
                     ELSE zeiger := 1   RETURNS überlauf
                  FIN;

```
rücksetzen
   EFFEKT: zeiger := 1;
           paritätsfehler := 'O'B1;
           RETURNS rückgesetzt

Paritätsfehler_setzen
   EFFEKT: paritätsfehler := '1'B1;
           RETURNS gesetzt

Ack_bilden
   EFFEKT: sendspeicher(1) := '00000110'B(8);
           zeiger := 1;
           RETURNS gebildet

Nack_bilden
   EFFEKT: sendspeicher(1) := '10010101'B(8);
           zeiger := 1;
           RETURNS gebildet
```

3.5.5. INTERNE FUNKTIONEN

```
Parität_prüfen
   IF parität EQ EVEN_PARITY(empfspeicher(zeiger))
    THEN RETURNS parität_stimmt
    ELSE RETURNS parität_stimmt_nicht
   FIN;


Gibt es Paritätsfehler
   IF paritätsfehler
    THEN RETURNS ja
    ELSE RETURNS nein
   FIN;

Interpretiere
   IF empfspeicher(1) EQ '00000110'B(8)
    THEN RETURNS  Ack
    ELSE RETURNS  Nack
   FIN;

Fertig
   IF zeiger EQ 16
    THEN RETURNS Ja
    ELSE RETURNS Nein
   FIN;
```

III. ELEMENTE EINES PROZESSYSTEMS

1. DEKLARATIONEN

```
/* Es werden die jeweiligen Instanzprozesse deklariert.
   Die formalen Namen 'benutzer' und 'service' werden
   durch die konkreten Namen 'benutzer a' bzw. 'benut-
   zer b' und 'leitung' ersetzt */
protokollinstanz a EINZELPROZESS (benutzer a, leitung)
                   TYP protokoll (benutzer, service)

protokollinstanz b EINZELPROZESS (benutzer b, leitung)
                   TYP protokoll (benutzer, service)
```

2. DEFINITION WEITERER STRUKTURIERUNGSEINHEITEN

leitung PROZESSGRUPPE
/* Diese Prozeßgruppe beschreibt den von den Protokollinstanzen vorausgesetzten Service */

A. PROZESSE
   /* Mit zwei Prozessen zur Modellierung der Voll-
      duplex-Leitung */
   PROZESS:
   /* Ohne Name, da in einer Prozeßgruppe die Prozesse
      nicht identifiziert werden */

   1. KOMMUNIKATIONSMASCHINE
      WARTEBEREICH: keinen

   2. ABLAUFSTEUERUNG

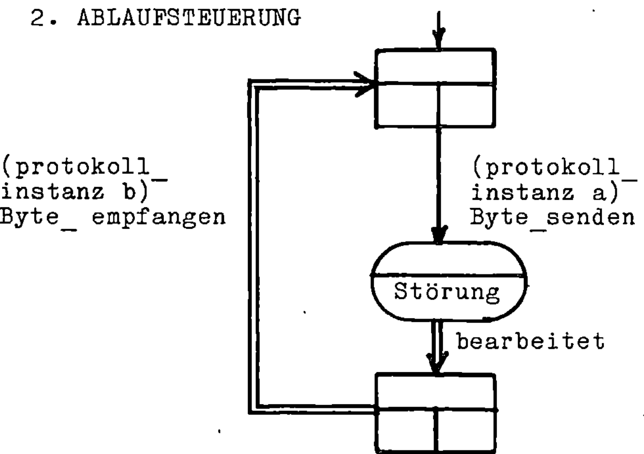

   3. BENUTZERMASCHINE
   3.1. EINGABEOPERATIONEN: Byte_senden
   3.2. AUSGABEFUNKTIONEN: Byte_empfangen
   3.3. INTERNE FUNKTIONEN : -
   3.4. INTERNE OPERATIONEN : Störung
   3.5. DEFINITION DER BENUTZERMASCHINE

   3.5.1. DATEN
          speicher BIT(9);

   3.5.2. EINGABEOPERATIONEN
          Byte_senden (datum BIT(9))
             EFFEKT: speicher := datum;

   3.5.3. AUSGABEFUNKTIONEN
          Byte_empfangen (datum BIT(9))
             EFFEKT: datum := speicher;

   3.5.4. INTERNE OPERATIONEN
          Störung
             EFFEKT: /* Mit der Störungswahrscheinlich-
                        keit wir der Inhalt von 'spei-
                        cher' geändert */
                     RETURNS bearbeitet

```
        PROZESS:
        /* Dieser Prozeß sieht analog dem oben beschriebenen
           Prozeß aus. Er beschreibt die Leitung in der Ge-
           genrichtung. Aus Platzgründen wurde auf die de-
           taillierte Beschreibung verzichtet
        */

    B. GEMEINSAME BENUTZERMASCHINE
        /*  Die beiden Prozesse der Prozeßgruppe haben keine
            gemeinsamen Objekte
        */
ENDE
```

5. Bisherige Anwendungen und Erfahrungen

Das hier vorgestellte Spezifikationskonzept hat sich im Laufe einiger Jahre entwickelt. Die bei den ersten Anwendungen festgestellten
Mängel wurden beseitigt, notwendige Erweiterungen eingeführt. Eine
grundsätzliche Überarbeitung führte dann zum heutigen Stand.
Die Ausgangsversion der vorgestellten Technik wurden in der Datenfernverarbeitung bei der Spezifikation und Implementation eines
File-Transferverfahrens zwischen einer CYBER 173 und einer R30 verwendet (/HART81/, /KRAG82/, /JORD83/). Um das Zeichnen der Ablaufsteuerung zu erleichtern, wurde ein graphischer Editor (/LEIN82/)
entwickelt.
Die überarbeitete und in diesem Artikel beschriebene Form der Technik wurde bei vor kurzem abgeschlossenen Arbeiten (/HERG83/,
/HOEF83/, /LIND83/ und /HOLL83/) eingesetzt. Außerdem wurde sie bei
der Erstellung des Pflichtenheftes für den Anteil des RRZE an den
Basisdiensten des Deutschen Forschungsnetzes (DFN) verwendet
(/AFHHKO/, /AFHHK1/, /AFHHK2/). Dabei wurden die Protokolle S.70,
X.29 und ein Subset des RDA (Remote Data Access) des Hahn-Meitner-
Instituts, Berlin, spezifiziert. Dabei konnte auch die Integration
dieser Protokolle in die Umgebung des ORG-300PV und von SINEC des
Siemens-Rechners R30 gut beschrieben werden (/ORG300/, /SINECE/).
Der Leistungsumfang und die Darstellungsmittel wurden dabei als aus-
reichend und zufriedenstellend empfunden. Als Vorteil wurde vor al-
lem die schnelle Umsetzbarkeit der Spezifikation in ein Programm gesehen.

Literaturverzeichnis

/AFHHKO/   Andres, Fleischmann, Hillmer, Holleczek, Kummer
           "Pflichtenheft für die Basis-Dienste des Erlanger DFN-
           Anschlusses"
           Interner Arbeitsbericht des Regionalen Rechenzentrums
           Erlangen Nr. 193

/AFHHK1/   Andres, Fleischmann, Hillmer, Holleczek, Kummer
           "Erste Ergänzung zum Pflichtenheft für die Basis-Dienste
           des Erlanger DFN-Anschlusses"
           Interner Arbeitsbericht des Regionalen Rechenzentrums
           Erlangen Nr. 201

/AFHHK2/   Andres, Fleischmann, Hillmer, Holleczek, Kummer
           "Zweite Ergänzung zum Pflichtenheft für die Basis-Dien-
           ste des Erlanger DFN-Anschlusses"
           Interner Arbeitsbericht des Regionalen Rechenzentrums
           Erlangen Nr. 202

/BOCH83/   G. v. Bochmann
           "Concepts for Distributed Systems Design"
           Springer Verlag 1983

/FLEI84/   A. Fleischmann
           "Ein Konzept zur Darstellung und Realisierung von ver-
           teilten Automatisierungssystemen"
           Dissertation an der Friedrich-Alexander-Universität Er-
           langen-Nürnberg, 1984

/HART81/   I. Hartmann
           "Entwicklung eines Systembausteins für Rechnerkopplungs-
           zwecke an der CYBER 173 des Rechenzentrums der Universi-
           tät Erlangen-Nürnberg"
           Diplomarbeit am IMMD IV der Universität Erlangen-Nürn-
           berg 1981

/HERG83/   G. Hergenröder
           "Erweiterung eines Großrechnerfiletransfers zum An-
           schluß mehrerer Kleinrechner"
           Laufende Studienarbeit am IMMD IV der Universität Erlan-
           gen-Nürnberg

/HOLL83/   P. Holleczek
           "Ein Terminalemulator und ein primitiver Filetransfer
           zum Anschluß von Kleinrechnern an die CYBER 173"
           Interne Notiz des Rechenzentrums der Universität Erlan-
           gen-Nürnberg 1983

/HOEF83/   G. Höfner
           "Entwicklung einer MSV2-Emulation"
           Studienarbeit am IMMD IV der Universität Erlangen, 1984

/SDL83/   SDL Draft Recommendation
           CCITT Temporary document Nr. 343, Working Party XI/3

/ESTE84/   ISO TC 97/SC 16/ WG 1-FDT Subgroup B
           A Formal Description Technique, based on an extended
           transitions model

/JORD83/   R. Jordan
           "Erstellung und Vergleich eines Filetransferprotokolls
           für Kleinrechner in verschiedenen Programmiersprachen"
           Diplomarbeit am IMMD IV der Universität Erlangen-Nürn-
           berg, 1983

/KRAG82/   G. Kragl
           "Entwicklung eines Sicherungsverfahrens für ein File-
           transferprotokoll zwischen Kleinrechnern und der CYBER
           173 des Rechenzentrums der Universität Erlangen-Nürn-
           berg"
           Diplomarbeit am IMMD IV der Universität Erlangen-Nürn-
           berg, 1982

/LEIN82/   M. Leinfelder
           "Ein graphisch-interaktives Hilfsmittel zur Erstellung
           von Transitionsdiagrammen"
           Diplomarbeit am IMMD IV der Universität Erlangen-Nürn-
           berg, 1982

/LIND83/   P. Lindlein
           "Entwicklung einer UT200-Emulation"
           Studienarbeit am IMMD IV der Universität Erlangen, 1984

/LOT084/   ISO/TC97/WG1-FDT Subgroub C, Draft tutorial document
           Temporal Ordering Specification Language

/ORG300/   "ORG-300PV" HANDBUCH, Beschreibung des Organisationspro-
           gramms der Siemens R30

/SINECE/   "SINEC 300, Version E" HANDBUCH,
           Produkt-Nr. P71100-B4030-X-X-35, 1983

Anschrift:

A. Fleischmann
c/o Regionales Rechenzentrum der
Universität Erlangen-Nürnberg
Martensstr. 1

8520 Erlangen

MODELLIERUNG VON OSI-KOMMUNIKATIONSDIENSTEN UND PROTOKOLLEN
MIT HILFE VON PRÄDIKAT-TRANSITIONSNETZEN

H. J. Burkhardt, H. Eckert, R. Prinoth
F2G3, Gesellschaft für Mathematik und Datenverarbeitung
Rheinstraße 75, D 6100 Darmstadt, FRG

Die in diesem Papier vorgestellte Modellierung von OSI-Kommunikationsdiensten und -Protokollen basiert auf einer Verfeinerung des OSI-Referenzmodells zu einem Kommunikationsdienstmodell und einem Kommunikationsprotokollmodell. Diese Verfeinerung ist Bestandteil eines systematischen Ansatzes, dessen Ziel es ist, den Entwurf von Kommunikations-
normen und die Realisierung entsprechender normkonformer realer
Kommunikationssysteme durch Etablieren einer Konstruktionslehre für
verteilte Systeme besser beherrschbar zu machen.
Die mathematische Basis des im GMD-Projekt PROSIT entwickelten Ansatzes
bilden Produktnetze, ein auf die formale Spezifikation von OSI-Diensten
und -Protokollen zugeschnittene Art von Prädikat-Transitionsnetzen.

1. Einführung

1.1 Der Modellierungskontext

Die in diesem Papier vorgestellte Modellierung von Kommunikationsdiensten und Protokollen ist als eine Verfeinerung des Referenzmodells
für die Kommunikation offener Systeme anzusehen (10).
Diese Verfeinerung bildet das Rückgrat eines systematischen Ansatzes,
dessen Ziel es ist, die Realisierung offener Systeme durch eine
Verbesserung von Methoden und Werkzeugen zu unterstützen, die einerseits der Spezifikation und Implementation von Kommunikationsdiensten
und -protokollen und andererseits dem Test und der Diagnose von Proto-
kollimplementation dienen.

Spezifikation von Kommunikationsdiensten und Protokollen zielt dabei auf
formale, problemorientierte und implementationsunabhängige Beschreibungen für Normungszwecke. Verifikation einer Protokollspezifikation gegen eine Dienstspezifikation dient dem Nachweis, daß ein

Protokoll den geforderten überlagerten Dienst erbringt und dabei den
unterlagerten Dienst korrekt benutzt.
<u>Implementation</u> eines Kommunikationsprotokolls wird verstanden als eine
Transformation der zugeordneten problemorientierten Beschreibung in eine
Implementationsbeschreibung (was die Abbildung auf reale Speicher und
Prozessoren beinhaltet), von der aus dann ausführbarer Programmcode erzeugt
werden kann.
<u>Test</u> dient dem Nachweis, daß eine Protokollimplementation konform mit
der problemorientierten Protokollspezifikation ist, von der sie abgeleitet wurde.
<u>Diagnose</u> stellt den Versuch dar, im Falle von Inkompatibilitäten
zwischen Systemen das die Inkompatibilitäten verursachende System durch
einen Vergleich des beobachteten Verhaltens der Protokollinstanzen mit
dem gemäß der problemorientierten Protokollspezifikation zu erwartenden
Problemverhalten zu lokalisieren.
Bild 1.1 verdeutlicht die verschiedenen Schritte, die wir innerhalb unseres Ansatzes zur Realisierung kompatibler Systeme basierend auf
Normen für notwendig halten.

1.2 Kennzeichnende Eigenschaften der Modellierung

Kommunikationsdienste und Protokolle zu modellieren, bedeutet, das OSI-
Referenzmodell zu einem Kommunikationsdienstmodell und einem Kommunika-
tionsprotokollmodell zu verfeinern.
Diese Modelle setzen die wesentlichen Konstrukte des Referenzmodells
sinnvoll zueinander in Beziehung und liefern damit eine allgemeinver-
bindliche Struktur für problemorientierte Spezifikationen von OSI-
Kommunikationsdiensten und -Protokollen.
Sowohl Dienste als auch Protokolle werden mit demselben formalen
Beschreibungsmittel spezifiziert, einer auf den Problembereich zuge-
schnittene Form von Prädikat-Transitionsnetzen, die wir Produktnetze
nennen.
Die Spezifikation eines Kommunikationsdienstes definiert wie üblich
eine Interaktion zwischen den Benutzern dieses Dienstes und seinem
Erbringer als Folgen von Dienstelementen an diskreten Interaktions-
stellen. Unsere Spezifikationen beschreiben jedoch nicht nur die
Sequenzen bezogen auf eine einzelne Verbindung, sondern die aller
Verbindungen; dadurch werden gegenseitige Abhängigkeiten zwischen
logisch unabhängigen Verbindungen erfaßt, die durch den Gebrauch gemein-

samer Betriebsmittel verursacht werden. Semantische Aspekte eines
Dienstes werden ausgedrückt durch die Modellierung von "Bewußtseins-
änderungen" der interagierenden Instanzen.

Die Spezifikation eines Kommunikationsdienstes wird als Vorgabe für die
Entwicklung des unterstützenden Protokolls benutzt. Deshalb ist die
Modellierung eines Dienstes darauf ausgerichtet, die Konstruktion von
Protokollspezifikationen aus Dienstspezifikationen zu unterstützen. Der
Modellierung von Kommunikationsprotokollen liegt ein Protokollmodell
zu Grunde, das aus der Überlagerung zweier Dienstmodelle entsteht und
Erbringerverhalten des Dienstes N mit Benutzerverhalten des Dienstes
(N - 1) korreliert.

Die Implementationsunabhängigkeit von Dienst- und Protokollspezifikationen, wie sie für Normungszwecke gefordert wird, wird in diesem
Ansatz dadurch erreicht, daß alle Interaktionen als nebenläufig
spezifiziert sind, die nebenläufig stattfinden können. Jede Implementation einer derartigen Spezifikation erscheint daher als eine
gültige Einschränkung von Nebenläufigkeit, d. h. als eine Sequentiali-
sierung, begründet durch die Abbildung auf reale Speicher und Prozessoren.
Gegenwärtig werden nur logische Beziehungen zwischen Interaktionen betrachtet, aber keine Beziehungen im Hinblick auf eine physikalische Zeit.
Dies hat zur Folge, daß Dienste und Protokolle nur bezüglich qualitativer
Aspekte beschrieben sind, z. B. dahingehend, daß Dateneinheiten verloren
gehen können, aber nicht in quantitativer Hinsicht, z. B. dahingehend,
wie oft dies geschieht.

2. Verfeinerung des OSI-Referenzmodells zu Modellen für
   Kommunikationsdienste und Protokolle

2.1 Ableitung und Verfeinerung des OSI-Begriffs Entität (entity)
Unseren Überlegungen liegt ein Szenario gemäß Bild 2.1 zu Grunde, in dem
ein Diensterbringer als Gegenüber einer Reihe von Dienstbenutzerinstan-
zen erscheint.
Als Aufgabe des Diensterbringers kann allgemein definiert werden,
Kommunikationsbeziehungen zwischen Dienstbenutzerinstanzen zu etablie-

ren, z. B. für eine einzelne Interaktion oder in Form von Verbindungen
für eine bestimmte Zeitdauer.
Eine Kommunikationsbeziehung zwischen Dienstbenutzerinstanzen basiert
immer auf zwei Arten von Interaktionen, einer Interaktion zwischen der
veranlassenden Dienstbenutzerinstanz und dem Diensterbringer und einer
anderen Interaktion zwischen dem Diensterbringer und der korrespondierenden Dienstbenutzerinstanz. Alle Interaktionen zwischen einer
einzelnen Dienstbenutzerinstanz und dem Diensterbringer finden an
spezifischen Interaktionsstellen unter Benutzung spezifischer Ausdrucks-
mittel statt, die Dienstelemente genannt werden.
Der Diensterbringer, der sich einer Dienstbenutzerinstanz als 'black
box' darstellt, ist selbst ein verteiltes System. Dieses System setzt
sich aus einzelnen Diensterbringerinstanzen zusammen. Zu jeder Dienstbenutzerinstanz ist eine Diensterbringerinstanz statisch und umkehrbar
eindeutig zugeordnet. Diensterbringerinstanzen kooperieren miteinander
gemäß den Regeln eines Kommunikationsprotokolls, um den geforderten
Dienst zu erbringen.
In Anlehnung an das OSI-Referenzmodell, nennen wir eine Interaktionsstelle, über die eine Dienstbenutzerinstanz mit der ihr zugeordneten
Diensterbringerinstanz kommunizieren kann, einen spezifischen Dienstzugangspunkt (Service Access Point: SAP).
Zur Modellierung verbindungsorientierter Kommunikation ist es notwendig,
statische Zuordnungen, die die Voraussetzungen für den dynamischen Auf-
und Abbau von Verbindungen bilden, von diesen dynamisch auf- und abge-
bauten Verbindungen selbst zu unterscheiden. Dies führt zu einer Verfeinerung von Dienstbenutzer- und Diensterbringerinstanzen, die einander
über einen spezifischen Dienstzugangspunkt zugeordnet sind, in jeweils
eine statische Instanz und dynamische Instanzen (siehe Bild 2.2).
Die dynamischen Instanzen werden durch die statischen Instanzen kreiert,
die zu diesem Zweck permanent existieren. Die Kreation dynamischer
Instanzen erfolgt während der Verbindungsaufbauphase und ihre Zerstörung
während der Verbindungsabbauphase.
Über einen spezifischen Dienstzugangspunkt kommuniziert genau eine
statische Dienstbenutzerinstanz mit genau einer statischen Diensterbringerinstanz. Jedem spezifischen Dienstzugangspunkt sind spezifische
Verbindungsendpunkte zugeordnet, die Betriebsmittel darstellen, die von
der statischen Diensterbringerinstanz während der Verbindungsaufbauphase
dynamischen Instanzen beiderseits einer Dienstschnittstelle zugeordnet

werden. Alle Instanzen, die mit einem spezifischen Dienstzugangspunkt
assoziiert sind, residieren im selben OSI-System.
Ein spezifischer Dienstzugangspunkt mit der globalen Adresse C ist somit
aufzufassen als eine Assoziation zweier statischer Instanzen gleichen
Namens in einem OSI-System: einer statischen Diensterbringerinstanz und
der mit ihr korrepsondierenden statischen Dienstbenutzerinstanz C. Entsprechend ist ein aktuell zugeordneter spezifischer Verbindungsendpunkt
(C,K) aufzufassen als die Assoziation zweier dynamischer Instanzen
gleichen Namens in einem OSI-System, d. h. einer statischen Diensterbringerinstanz (C,K) und der mit ihr korrespondierenden statischen Dienstbenutzerinstanz (C,K).
Daraus folgt, daß in einem System konzeptionell die folgenden Interaktionsstellen zwischen Diensterbringer und Dienstbenutzerinstanzen
existieren:
- für jeden spezifischen Dienstzugangspunkt eine Interaktionsstelle
  zwischen statischen Instanzen, die architekturell durch den Begriff
  des Dienstzugangspunktes selbst repräsentiert wird und über die die
  Dienstelemente zum Auf- und Abbau einer Verbindung ausgetauscht werden,
und
- für jede Verbindung eine Interaktionsstelle zwischen einer dynamischen
  Dienstbenutzerinstanz und einer dynamischen Diensterbringerinstanz, die
  architekturell durch den Begriff des Verbindungsendpunktes repräsentiert wird und über die die Dienstelemente, gültig in der Datentransferphase einer Verbindung, ausgetauscht werden.

Faßt man nun alle die Instanzen zu Äquivalenzklassen zusammen, die
identisches Kommunikationsverhalten zeigen, gelangt man zum OSI-Begriff
der Entität.
Zur Bildung von Äquivalenzklassen werden gemäß Bild 2.3 zusammengefaßt:
- Alle statischen Dienstbenutzerinstanzen zu einer statischen Dienstbenutzerentität,
- alle statischen Diensterbringerinstanzen zu einer statischen Diensterbringerentität,
- alle dynamischen Dienstbenutzerinstanzen zu einer dynamischen Dienstbenutzerentität,
und
- alle dynamischen Diensterbringerinstanzen zu einer dynamischen
  Diensterbringerentität.

Der OSI-Begriff 'Dienstzugangspunkt' bezeichnet danach die Kollektion
aller spezifischen Dienstzugangspunkte und der OSI-Begriff Verbindungsendpunkt die Kollektion aller spezifischen Verbindungsendpunkte.
Beide Begriffe definieren somit Beziehungen zwischen Äquivalenzklassen
von Instanzen; diese Äquivalenzklassen von Instanzen werden Entitäten
genannt.

2.2 Das PROSIT Kommunikationsdienstmodell

Das PROSIT Kommunikationsdienstmodell stellt eine Verfeinerung der
zuvor eingeführten Äquivalenzrelationen dar. Diese Verfeinerung basiert
auf der Unterscheidung zwischen aktiven und passiven Instanzen und
führt zu den Begriffen der aktiven und passiven Entitäten. Unter einer
aktiven Entität (Dienstbenutzer und -erbringer) wird eine Kollektion
aller rufenden Instanzen (Dienstbenutzer und -erbringer) und unter
einer passiven Entität (Dienstbenutzer und -erbringer) wird eine
Kollektion aller gerufenen Instanzen (Dienstbenutzer und -erbringer)
verstanden.
Diese Verfeinerung ist notwendig, um zusätzlich zu der Schnittstelle
zwischen Dienstbenutzer- und Diensterbringerinstanzen die Schnittstelle
zwischen OSI-Systemen sichtbar zu machen. Demzufolge wird ein Protokoll,
das aktive und passive Entitäten über Systemgrenzen hinweg zusammen-
bindet, als Korrelation von aktivem und passivem Kommunikationsverhalten interpretiert. Das resultierende Kommunikationsdienstmodell
setzt gemäß Bild 2.4 acht Entitäten zueinander in Beziehung:
- an einem Dienstzugangspunkt, eine aktive statische Dienstbenutzerentität zu einer aktiven statischen Diensterbringerentität und eine
  aktive dynamische Dienstbenutzerentität zu einer aktiven dynamischen
  Diensterbringerentität.
- am korrespondierenden Dienstzugangspunkt, eine passive statische
  Dienstbenutzerentität zu einer passiven statischen Diensterbringerentität und eine passive dynamische Dienstbenutzerentität zu einer
  passiven dynamischen Diensterbringerentität.
Der Aufbau einer Verbindung stellt sich in diesem Modell als Kreation
von vier korrespondierenden dynamischen Instanzen dar, jeweils eine
für jede dynamische Entität gemäß Bild 2.4, als Folge einer Interaktion
zwischen vier korrespondierenden statischen Instanzen, jeweils eine für
jede statische Entität gemäß Bild 2.4.

Der Abbau einer Verbindung stellt sich dementsprechend als die Zerstörung von vier dynamischen Instanzen dar, die zuvor kreiert worden sind.

2.3 Das OSI Kommunikationsprotokollmodell

Das OSI Kommunikationsprotokollmodell entsteht durch Überlagerung zweier Kommunikationsdienstmodelle an benachbarten Dienstschnittstellen (siehe Bild 2.5). Es enthält drei funktionale Ebenen, die hierarchisch aufeinander aufbauen:

- Die N-Dienstbenutzerebene, bestehend aus N-Dienstbenutzerentitäten des Modells für den N-Dienst,
- die N-Protokollebene, bestehend aus den N-Diensterbringerentitäten des Modells für den N-Dienst, den (N-1)-Benutzerentitäten des Modells für den (N-1)-Dienst, der Funktionalität des Abbildens von N-Diensterbringerverhalten auf (N-1)-Dienstbenutzerverhalten zusammen mit der Funktionalität der Überbrückung funktionaler Differenzen zwischen N-Dienst und (N-1)-Dienst; der OSI-Begriff N-Schicht entspricht dieser N-Protokollebene;
- die (N-1)-Diensterbringerebene, bestehend aus dem (N-1)-Diensterbringerentitäten des Modells für den (N-1)-Dienst.

Gemäß diesem Protokollmodell interagieren in einer Schicht N zwei Protokollentitäten, nämlich eine aktive N-Protokollentität und eine passive N-Protokollentität. Jede Protokollentität setzt sich dabei aus zwei N-Diensterbringerentitäten und zwei (N-1)-Dienstbenutzerentitäten zusammen.

Dieses verfeinerte Modell einer Protokollentität in der Schicht N macht es offensichtlich,

- daß eine N-Verbindung durch Zerstörung der sie repräsentierenden dynamischen N-Dienstbenutzer- und -erbringerinstanzen abgebaut werden kann, ohne gleichzeitig die unterstützende (N-1)-Verbindung abzubauen oder aber
- daß eine (N-1)-Verbindung durch Zerstörung der sie repräsentierenden dynamischen (N-1)-Benutzer- und -erbringerinstanzen abgebaut werden kann, ohne gleichzeitig die durch sie unterstützte N-Verbindung abzubauen.

3. Produktnetze - ein PROSIT-Beschreibungsmittel zur Spezifikation von Kommunikationsdiensten und -Protokollen

Spezifikationen von Diensten und Protokollen in PROSIT werden nach dem
in diesem Papier definierten PROSIT-Kommunikationsdienst - bzw. PROSIT-
Kommunikationsprotokoll-Modell strukturiert. Diese Struktur steckt den
Rahmen für weitergehende Unterstrukturierungen ab. Unterstrukturen
(Module) fassen Funktionsgruppen zusammen und legen die Kommunikation
mit "benachbarten" Moduln fest ((9), (13)). Beschreibung von Moduln
und ihre Einbindung in die Umgebung werden durch Benutzung des Begriffs
der verteilten Systeme formal gefaßt, wie sie in (12) und (16) eingeführt
sind. Verteilte Systeme sind dort definiert als Kollektion disjunkter
endlicher Automaten, deren jeweilige Zustandsübergangsfunktion formal
erweitert ist um den Aspekt der Kommunikation, wodurch das gesamte
System als Akzeptor einer formalen Sprache aufgefaßt werden kann (wobei
es sich in geeigneter Weise aus Teilakzeptoren zusammensetzt).
Netze (zur Def.von Netzen siehe (1)) sind ein geeignetes Beschreibungsmittel zur Darstellung verteilter Systeme, da sie Aspekte der Verteilung,
der Nebenläufigkeit, der Kommunikation und der lokalen Beobachtbarkeit
mit denen der formalen Sprache zu verbinden gestatten.
Um jedoch komplexe reale verteilte Systeme in adäquater Weise modellieren zu können (siehe 14,15), müssen den Netzen weitere Strukturen
hinzugefügt werden, um die dort auftretenden relevanten mathematischen
Objekte (Mengen, Operationen, Variablen, Funktionen, Prädikate, ...)
und ihre logischen Beziehungen zueinander beschreiben zu können. Hier
bietet es sich an, die Sprache der Prädikatenlogik 1. Stufe mit den
Netzen zu verknüpfen. Dabei sollte
- die Einheitlichkeit des Beschreibungsmittels erhalten bleiben,
- die mathematische Präzision nicht abgeschwächt werden,
- der Abstraktionsgrad der Modellierung durch die Anwendung und nicht
  durch das Beschreibungsmittel festgelegt werden.

Weiterhin hat es sich in der bisherigen Modellierungspraxis für verteilte
Systeme als zweckmäßig erwiesen, jeder Stelle in einem Netz eine eigene
Objekttypbeschreibung (Definitionsbereich) zuzuordnen, Objekten in
diesen Stellen eine Vektorstruktur zu geben und Operationen auf Komponenten dieser Vektoren zu beschränken. Hierdurch wird es möglich,

Syntax und Semantik einer Spezifikation aus rein lokaler Sicht zu prüfen, was für eine Modularisierung von Systemen (9) von großem Nutzen ist.

Diese Anforderungen werden erfüllt durch die <u>Produktnetze</u>, die in (7) und (8) eingeführt wurden. Produktnetze stellen eine Synthese aus Teilen der Netztheorie und der Prädikatenlogik dar:

- Jeder Stelle $s \in S$ wird eine Objektmenge D_s zugeordnet sowie für diese Objektmenge definierte Operationen.

- Jeder Kante $f \in F$ und jeder Transition $t \in T$ werden Zeichenketten zugeordnet (Kantenanschriften $K(f)$ bzw. Ausdrücke $P(t)$), deren syntaktische Regeln ähnlich den in der Sprache der Prädikatenlogik verwendeten sind.

- Die Zeichenketten werden auf der Basis der den Stellen zugeordneten Objektmengen konsistent interpretiert, ähnlich der Vorgehensweise in der Modelltheorie (formale Semantik der Netze).

Hierbei sind mit S, T und F die Stellenmenge, Transitionsmenge und die Flußrelation des dem Produktnetz zugrunde liegenden Netzes $N = (S,T,F)$ bezeichnet.

Der Name Produktnetz leitet sich aus der Struktur der Mengen D_s ab:
D_s ist eine nichtleere Menge von Objekten (Definitionsbereich).
Jedes Objekt $x \in D_s$ ist ein Vektor der Länge $n_s \in \mathbb{N}$.
n_s heißt Dimensionalität der Stelle s. Die einzelnen Vektorkomponenten sind durch nicht leere Mengen $M_1^S, \ldots, M_{ns}^S$ bestimmt (die Individuenbereiche heißen), so daß die Objektmenge D_s einer Stelle s beschrieben werden kann als nicht leere Teilmenge des Produkts der M_i^S:

$$D_s \subseteq \bigtimes_{i=1}^{n_s} M_i^s \quad , \quad D_s \neq \emptyset$$

Jedem Individuenbereich M_i^S ist eine (evtl. leere) endliche Menge G_i^S von Funktionsnamen zugeordnet.

Jedem Funktionsnamen $g \in G_i^S$ ist eine k-stellige (partielle) Funktion über M_i^S zugeordnet:

$$g : (M_i^S)^k \to M_i^S \quad , k \geq 1.$$

$K(f)$ ist – grob gesprochen – die formale Summe von Vektoren der Länge n_s, wobei jede Vektorkomponente eines Summanden ein Term ist. Terme sind hierbei wie üblich in der Prädikatenlogik definiert. Die Transitions-inschrift $P(t)$ ist ein prädikatenlogischer Ausdruck (4).

Beispiel 3.1:

Gegeben sei das Produktnetz $\tilde{N}$ (Abb. 3.1):

$\tilde{N}$

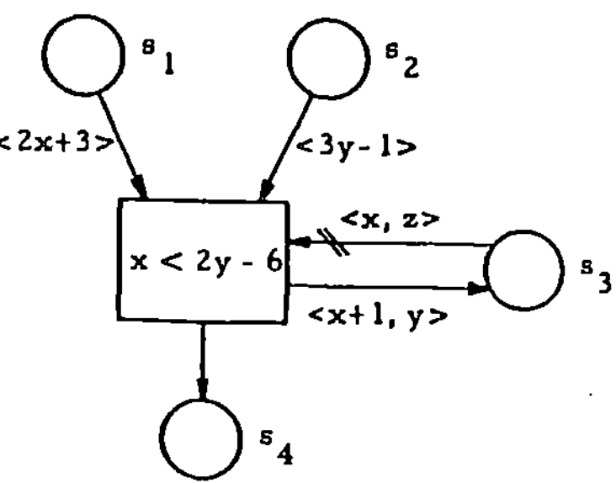

Abb. 3.1

1) Das dem Produktnetz $\tilde{N}$ zugrundeliegende Netz $N = (S,T,F)$ ist wie folgt gegeben:

$S = \{s_1, s_2, s_3, s_4\}$ (Stellenmenge)

$T = \{t\}$ (Transitionsmenge)

$F = \{(s_1,t),\ (s_2,t),\ (s_3,t),\ (t,s_3),\ (t,s_4)\}$ (Flußrelation)

$^\circ t = \{s_1, s_2, s_3\}$ (Eingangsstellen von t)

$t^\circ = \{s_3, s_4\}$ (Ausgangsstellen von t)

$f_1 = (s_1,t)$, $f_2 = (s_2, t)$ und $f_3 = (s_3,t)$ sind Eingangskanten von t,

$f_4 = (t,s_3)$ and $f_5 = (t,s_4)$ sind Ausgangskanten von t.

2) Jeder Stelle $s \in S$ wird ein Definitionsbereich D_s zugeordnet:

$$D_{s1} = D_{s2} = Z \quad (Z \text{ ist die Menge der ganzen Zahlen}).$$

$$D_{s3} = Z^2 \text{ und } D_{s4} = \{b\} \qquad (\{b\} \text{ ist eine einelementige Menge}).$$

Folgende Mengen von Funktionsnamen sind gegeben:

$$G_1^{s1} = G_1^{s2} = G_1^{s3} = \{+, -, \cdot\} \text{ und } G_1^{s4} = G_2^{s3} = \emptyset$$

Den Funktionsnamen "+", "-" und "$\cdot$" seien 2-stellige Funktionen von Z^2 nach Z zugeordnet, die wie üblich Addition, Subtraktion und Multiplikation bedeuten.

3) Jeder Kante $f \in F$ wird eine Kantenanschrift $K(f)$ zugeordnet:

$$K(f_1) \quad = \quad <2 \ x + 3> \qquad\qquad K(f_2) = <3y - 1>$$

$$K(f_3) \quad = \quad <x,z> \qquad\qquad\quad K(f_4) = <x+1,y>$$

$$K(f_5) \quad = \quad <b> \qquad\qquad (\text{Diese Kantenanschrift ist in Abb. 3.1}$$
$$\text{weggelassen})$$

Erläuterung zu $K(f_1)$:

Die Symbole 2 und 3 repräsentieren Konstante: $2 \in D_{s1}$ and $3 \in D_{s1}$.

x ist Variablenname und die Symbole + und $\cdot$ sind Funktionsnamen:

$+ \in G_1^{s1}$ und $\cdot \in G_1^{s1}$.

Die Zeichenkette + $(\cdot(2,x),3)$ ist ein syntaktisch korrekter Term.

Entsprechend ist $K(f_2)$ zu erläutern.

Erläuterung zu $K(f_3)$:

Die Struktur von $K(f_3)$ ist $\neg <$ Term 1, Term 2 $>$.

Das Symbol $\neg$ bedeutet, daß f_3 als Verbotskante zu interpretieren ist. In den Netzgrafiken dieses Papiers sind Verbotskanten durch das Symbol $\nrightarrow$ dargestellt. x und z sind Variablennamen. z kommt in keiner Anschrift einer Eingangskante vor, die nicht Verbotskante ist (z ist freie Variable).

$K(f_4)$ ist ein Termpaar und $K(f_5)$ besteht aus dem Term b (b ε D_{s4}).
Allgemein gilt für jede Transition t die Regel, daß in Kantenanschriften an Ausgangskanten von t und im Ausdruck $P(t)$ höchstens
solche Variablennamen vorkommen dürfen, die auch in Eingangskanten
von t, die nicht Verbotskanten sind, vorkommen.

4) Jeder Transition t ε T wird ein Ausdruck $P(t)$ zugeordnet:
$P(t): = x < 2y - 6$
Zugelassen sind prädikatenlogische Ausdrücke ohne Existenz- und
All-Quantoren.
Der Ausdruck dieses Beispiels ist von der Form 'Term $\overline{1}$ < Term $\overline{2}$'.
< ist Prädikatenname. Dem Prädikatennamen < ist eine 2-stellige Rela-
tion über Z zugeordnet. (< ist eine Abbildung von Z^2 nach $\{w, f\}$,
wobei w und f die aussagenlogischen Konstanten 'wahr' und 'falsch'
bezeichnen).
x ist Variablenname in Term $\overline{1}$ und y ist Variablenname in Term $\overline{2}$.
Term $\overline{1}$ und Term $\overline{2}$ sind Terme im Sinne der Termbildungsregeln für
Eingangskanten (hier etwa für f_1 oder f_2).
(Ende Beispiel 3.1).

Eine _Markierung_ in einem Produktnetz ist eine Abbildung
$$M : \bigcup_{s \varepsilon S} (D_s \times \{S\}) \to \mathbb{N}_0.$$

Mit $M(s) : D_s \to \mathbb{N}_0$ wird die Einschränkung der Markierung M auf
$D_s \times \{s\}$ für s ε S bezeichnet.
Für $M(s) : D_s \to \{0\}$ wird kürzer $M(s) = 0$ geschrieben.
Die Objekte x ε D_s sind Vektoren der Länge n_s.
Jeder Vektor x "kommt in s sooft vor", wie seine Vielfachheit h_x
= $M(x,s)$ angibt. Objekte mit der Vielfachheit 0 werden als "in s
nicht vorhanden" aufgefaßt.

In der zeichnerischen Darstellung werden Objekte x ε D_s in einen
Kreis eingeschlossen und in die betreffende Stelle s ε S abgelegt,
sofern ihre Vielfachheit > 0 ist. Wenn $|D_s| = 1$ gilt, dann ist das
Objekt in eindeutiger Weise der Menge D_s zugeordnet.
Schreibweise für die Elemente von $M(s)$:
$h_x <x> \varepsilon M(s)$ mit $h_x \varepsilon \mathbb{N}$ und x ε D_s.

Beispiel 3.2:

Eine Markierung M für das in Abb. 3.1 gegebene Produktnetz ist

$M(s_1)$ = {1<-1>, 1<4>} $M(s_2)$ = {2<5>, 1<8 >}
$M(s_3)$ = {1<-1,12>} $M(s_4)$ = 0

Graphische Darstellung:

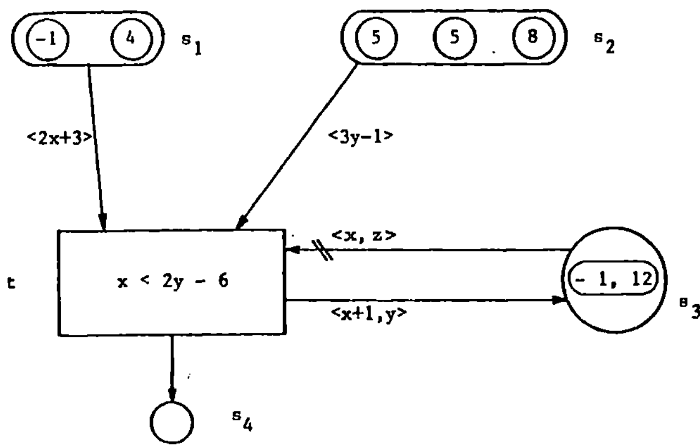

        Abb. 3.2
(Ende Beispiel 3.2).

Da alle in einem Produktnetz vorkommenden Zeichenketten (Kantenanschriften und Transitionsinschriften) als - aus netztheoretischer
Sicht - sinnvolle Einschränkung der Sprache der Prädikatenlogik
1. Stufe eingeführt wurden, und da jeder Stelle eines Netzes ein
Definitionsbereich zugeordnet wurde, der konsistent ist mit allen
"benachbarten" Kantenanschriften, ist der Interpretationsbegriff der
Prädikatenlogik in gleicher formaler Weise auf Produktnetze anwendbar
(formale Semantik von Produktnetzen).

Eine <u>Interpretation</u> δ_t eines Produktnetzes $\tilde{N}$ ist - grob gesprochen - eine Abbildung, die jeder Konstanten, jeder Variablen und jedem Term in der Anschrift K(f) an jede Kante $f \in F_t$, die mit einer Transition $t \in T$ inzidiert, einen Wert aus dem zugehörigen Individuenbereich M_i^s zuordnet:

Das Bild $\delta_t(z)$ liegt in M_i^s, falls z Konstante, Variable oder Term in der i-ten Komponente in K(f) ist, falls $f \in F_t$ gilt und $s \in S$ mit f inzidierende Stelle ist. Hierbei sind diejenigen Variablen von einer Wertsetzung durch δ_t ausgeschlossen, die nur in Anschriften von Verbots-kanten vorkommen (freie Variablen). Werte derartiger Variablen sind für jedes δ_t beliebig aus ihrem Individuenbereich wählbar. Der Transitions-inschrift P(t) wird durch δ_t letztendlich der Wert w oder f zugeordnet.

Der <u>Gültigkeitsbereich einer Variablen</u> ist bei Interpretationen δ_t beschränkt auf ihr Vorkommen in P(t) sowie in allen Anschriften K(f) von mit t inzidierenden Kanten $f \in F_t$.
Zu beachten ist, daß den Funktionsnamen partiell definierte Funktionen zugeordnet sind, was die Auswahl der δ_t einschränkt.

Beispiel 3.3:
In diesem Beispiel wird das in Beispiel 3.1 gegebene Netz mit einer Interpretation δ_t versehen.
- Konstanten sind die Zeichen 1,2,3,6 (aus Z).
- Variablennamen, die nicht ausschließlich an Verbotskanten vorkommen, sind x und y. Es werde V = {x,y} gesetzt.
- Funktionsnamen sind: x, -, ·
- < ist Prädikatenname
- Termmenge (ohne Terme, die die Variable z enthalten) ist
  τ = {2x + 3, 3y - 1, x, 2y - 6, x + 1, y, b}
- Menge der Transitionsinschriften:
  A = {x < 2y - 6}

Seien +, -, · und < wie gewöhnlich in Z definiert. Dann ist eine Inter-pretation δ wie folgt gegeben:

δ: {b} $\cup$ Z $\cup$ V $\cup \tau \cup$ A $\cup$ {w,f} $\rightarrow$ {b} $\cup$ Z $\cup$ {w,f}
derart, daß
1) $\bigvee$ z $\in$ Z: δ(z) = z δ(b) = b

2) $\delta(x) = -2$ $\delta(y) = 3$

       $\delta(2x + 3) = -1$ $\delta(3y - 1) = 8$

       $\delta(2y - 6) = 0$ $\delta(x + 1) = -1$

3) $\delta(w) = w$ $\delta(f) = f$

4) $\delta(x < 2y - 6) = w$ gilt.

(Ende Beispiel 3.3).

Es ist jetzt möglich, eine Aktivierungsbedingung für Transition $t \in T$, basierend auf einer Interpretation δ_t und der üblichen Aktivierungsbedingung für Stellen/Transitionsnetze (1), unter Berücksichtigung von Verbotskanten formal abzuleiten (8):
Kantenanschriften K(f) können als Verallgemeinerung von Kantengewichtungen in Stellen/Transitionsnetzen aufgefaßt werden.
In einem Stellen/Transitionsnetz (S,T,F) läßt sich eine Schaltregel für jede Transition $t \in T$ dadurch definieren, daß zunächst den Kanten des Netzes ein <u>Kantengewicht</u> zugewiesen wird (Abbildung $\beta : F \to \mathbb{N}$) und daß (für die aktivierte Transition) jeder Stelle $s \in {}^{\circ}t$ beim Schaltvorgang $\beta((s,t))$ Marken entzogen werden und jeder Stelle $s \in t^{\circ}$ $\beta((t,s))$ Marken hinzugefügt werden. Da hierbei Objekte in einer Stelle (Marken) nicht voneinander unterschieden werden, ist in der Schaltregel lediglich eine Aussage über die Anzahl der Marken gemacht, die bei jedem Schaltschritt für jede Stelle zu berücksichtigen sind.

Eine Schaltregel für Produktnetze muß neben der Kantengewichtung auch die Individualität der in Stellen befindlichen Objekte berücksichtigen. Dazu muß zunächst das <u>interpretierte Kantengewicht</u> δ_t (K(f)) für Produktnetze definiert werden (8). Anschließend läßt sich jedem interpretierten Kantengewicht δ_t(K(f)) einer Kante $f \in F_t$ eine Markierung M_f zuordnen. Sodann läßt sich einer Transition $t \in T$ in Bezug auf eine Interpretation δ_t eine <u>charakteristische Markierung</u> $M^{\delta t}$ zuordnen, die für die Aktivierung von t die Rolle einer "Schwellenmarkierung" spielt. $M^{\delta t}$ enthält für jede Eingangsstelle $s \in {}^{\circ}t$, die nicht mittels Verbotskante mit t verbunden ist, die durch δ_t (K(s,t)) induzierte Markenzahl und für alle anderen Stellen des gesamten Produktnetzes keine Marken ($\forall s \in S \setminus {}^{\circ}t : M^{\delta t}(s) = 0$).

Dabei gilt für alle mit Verbotskanten inzidierenden Stellen $M^{\delta t}(s) = 0$.
$M^{\delta t}$ drückt auf Netzebene gerade die Interpretation δ_t bzgl. der Eingangskanten von t aus.
Nachfolgend wird ein Beispiel für eine charakteristische Markierung gegeben.

Beispiel 3.4:
Aus der Interpretation δ des Beispiels 3.3 wird folgende charakteristische Markierung M^δ abgeleitet:

$$M^\delta(s_1) = \{1<-1>\} \quad M^\delta(s_2) = \{1<8>\}$$
$$M^\delta(s_3) = 0 \quad\quad M^\delta(s_4) = 0$$

(Ende Beispiel 3.4).

Mit Hilfe der charakteristischen Markierung läßt sich nun definieren, unter welchen Bedingungen eine Transition aktiviert ist.

Sei $\tilde{N}$ ein Produktnetz, M eine Markierung und sei δ_t eine Interpretation bezüglich einer Transition $t \in T$. Sei weiterhin $M^{\delta t}$ die zugehörige charakteristische Markierung. Transition <u>t heißt aktiviert unter M</u> (in Bezug auf δ_t), wenn jede Stelle $s \in {}^{O}t$ bezüglich der Markierung M mindestens die in der charakteristischen Markierung $M^{\delta t}$ abgeleiteten Objekte enthält:

$$\cdot \forall s \in S : M^{\delta t}(s) \leq M(s)$$

Weiterhin wird gefordert:
$\cdot\ \delta_t(P(t)) = w$
Schließlich soll für alle Stellen $s \in S$, die mit t durch Verbotskanten verbunden sind, gelten,
$\cdot$ daß sie keine Objekte enthalten, die durch Interpretation der an Verbotskanten stehenden Terme gemäß δ_t, bzw. durch beliebige Wertzuweisungen an die freien Variablen in Termen an Verbotskanten, identifiziert werden können (8).

Beispiel 3.5:

Transition t des Beispielnetzes $\tilde{N}$ ist aktiviert bzgl. der in Beispiel 3.2 gegebenen Markierung und· der aus Beispiel 3.3 abgeleiteten charakteristischen Markierung in Beispiel 3.4. Stelle s_3 enthält kein Element $(-2,a) \in D_{s3}$ mit $a \in Z$.

Bemerkung: Würde beispielsweise $M(s_3) = \{1<-1,12>,\ 1<-2,7>\}$ sein
statt $\{1<-1,12>\}$, dann wäre t nicht aktiviert.

(Ende Beispiel 3.5).

Aktivierte Transitionen können schalten. Durch das Schalten einer Transition $t \in T$ wird die Markierung M, unter der t aktiviert war, in eine (direkte) Nachfolgemarkierung M' überführt (M $[t>$ M'):

Sei t aktiviert unter der Markierung M bzgl. einer Interpretation δ. Dann wird durch das <u>Schalten von t</u> M überführt in die Nachfolgemarkierung M':

- $\forall\ s \in {}^{\circ}t \backslash t^{\circ}$: $M'(s) = M(s) - M^{\delta}(s)$
- $\forall\ s \in t^{\circ}$: $M'(s) = M(s) - M^{\delta}(s) + \delta(K(t,s))(s)$
-             $M'(s) = M(s)$ sonst

Beispiel 3.6:

Wenn im Beispielnetz $\tilde{N}$ die gemäß Beispiel 3.5 aktivierte Transition t schaltet, dann ergibt sich folgende Markierung M':

$$M'(s_1) = \{1<4>\} \qquad M'(s_2) = \{2<5>\}$$
$$M'(s_3) = \{1<-1,12>,\ 1<-1,3>\} \qquad M'(s_4) = \{1<b>\}$$

(Ende Beispiel 3.6).

4. Ein Beispiel

4.1 Allgemeine Einführung

Die vorgestellte Modellierungs- und Beschreibungsmethode soll
im folgenden an einem Beispiel veranschaulicht werden. Dasselbe Beispiel
wurde benutzt um eine Beweistechnik, basierend auf Computation Systems,
zu demonstrieren (siehe 6). Der Beschreibungstechnik des ISO Referenzmodells folgend, definieren wir zunächst den Dienst, der durch das
Protokoll zu erbringen ist, dann den Dienst, der vom Protokoll zu
benutzen ist und danach das Protokoll selbst.

Das Protokoll, das wir für Demonstrationszwecke gewählt haben, ist ein
HDLC ähnliches Sicherungsprotokoll, das die einseitige Datenübermitt-
lung zwischen zwei Benutzerinstanzen unterstützt.
Der Modellierung von Dienst und Protokoll liegen die folgenden Annahmen
zu Grunde:
- Zwei Verbindungsendpunkte (CEPs) identifizieren eindeutig eine etab-
  lierte Verbindung.
- Aufbau und Abbau von Verbindungen nicht modelliert.
- Benutzer der Dienste (sowohl des durch das Sicherungsprotokoll erbrachten Dienstes als auch des durch das Sicherungsprotokoll benutzten
  Dienstes) werden durch die entsprechenden Verbindungsendpunkte identifiziert.
- Der Diensterbringer korreliert jeweils zwei Verbindungsendpunkte, die
  lokal eine Verbindung darstellen.
Die formalen Definitionen der Dienste und des Protokolls des Beispiels
werden hier nur in graphischer, wenn auch in formal präziser Form angegeben.

4.2 Definition des erbrachten Dienstes

Der durch das Protokoll erbrachte Dienst ist in Bild 4.1 als Produktnetz definiert.
Der Zweck des Protokolls, wie er in Bild 4.1 zum Ausdruck kommt, ist
es, eine beliebige Anzahl von Dienstdateneinheiten von einem Produzenten
zu einem Konsumenten ohne Verlust und unter Erhalten der Reihenfolge zu
transferieren. Da für jede Dienstdateneinheit eine zwar beliebige, aber

endliche Länge angenommen werden kann, kann dieser Zweck im Dienst mit
Hilfe diskreter Objekte und Operationen modelliert werden. Der Dienst
wird durch vier Dienstelemente verfügbar gemacht, die im folgenden mit
ihren Parametern aufgelistet sind.

service primitive	abbreviated name	parameter	abbreviated name
PUT-request	PUTREQ	send-connection endpoint	S-CEP
		user-data	UD
PUT-confirmation	PUTCON	send-connection endpoint	S-CEP
GET-indication	GETIND	receive-connection endpoint	R-CEP
		user-data	UD
GET-response	GETRES	receive-connection endpoint	R-CEP

Eine korrekte Folge von Dienstelementen zum Transfer einer Dienstdateneinheit von einem spezifischen Produzenten (Absender) zu einem
spezifischen Konsumenten (Empfänger), welche einander durch eine
spezifische Verbindung zugeordnet sind, ist definiert durch PUTREQ,
GETIND, GETRES, PUTCON.
Bild 4.2 definiert nun nicht nur die erlaubten Folgen von Dienstelementen zwischen einem ausgezeichneten Sender-Empfänger-Paar, sondern
zwischen einer beliebigen Anzahl von Sender-Empfänger-Paaren. Um zu
klären, wie dies unter Anwendung der Beziehung zwischen Variablen und
deren Interpretation gemäß Kapitel 3 geschieht, betrachten wir die
PUT-Transition des Diagramms. Diese Transition koinzidiert mit drei
Kanten. Die Anschriften an zweien dieser Kanten stellen die zwei
Dienstelemente PUTREQ und PUTCON mit ihren Variablen dar. Die Anschrift
an der dritten Kante stellt die eine Dienstdateneinheit dar, die durch
die zwei Variablen S-CEP und UD festgelegt ist.
Mit der Stelle des Netzes, die mit der letztgenannten Kante koinzidiert,
sind alle Dienstdateneinheiten an allen sendeseitigen Verbindungsend-
punkten assoziiert zu denken. Diese Menge aller Dienstdateneinheiten
dient als Definitionsmenge gemäß Kapitel 3. Die Interpretation der
Kantenanschriften erlaubt mit derselben Topologie alle möglichen
Übertragungen von Dienstdateneinheiten zu beschreiben.

Dazu selektieren wir ein bestimmtes Objekt (Dienstdateneinheit), nämlich <s-cep, ud> mit einer Vielfachheit > 0, die die Existenz dieses
Objektes ausdrückt. Danach selektieren wir ein Objekt <PUTCON,s-cep,ud>,
das bezüglich des Verbindungsendpunktes mit der gewählten Dienstdateneinheit übereinstimmt, und das ebenfalls eine Vielfachheit > 0 besitzt.
Wenn beide Selektionen möglich sind, dann ist die Transition unter der
Interpretation von S-CEP als s-cep und UD als ud aktiviert. PUTCON ist
eine Konstante.
Wenn die Transition unter der gewählten Interpretation feuert, werden
beide Objekte aus den Ausgangsstellen entfernt und ein neues Objekt
<PUTREQ, s-cept, ud> in der Ausgangsstelle erzeugt. Das Schalten
aller anderen Transitionen ist in gleicher Weise definiert.
Auf diese Weise kann die Menge aller erlaubten Folgen von Dienstelementen als Folge von Markierungen bezogen auf die Stellen an der Dienstschnittstelle definiert werden, die von einer Anfangsmarkierung aus
erreicht werden können. Eine Anfangsmarkierung besteht dabei aus einer
Markierung von drei Stellen; sie repräsentieren die Menge der zu
übertragenden Nachrichten, die Menge der etablierten Verbindungen und
die dazu korrespondierende Menge der PUTCON-Dienstelemente.
Die erlaubten Folgen von Dienstelementen für eine ausgezeichnete Verbindung erscheinen als Folge von Markierungen, die auf bestimmte Verbindungsendpunkte Bezug nehmen.

4.3 Definition des benutzten Dienstes.

Der vom Protokoll benutzte Dienst ist durch das Produktnetz gemäß
Bild 4.2 definiert.
Bild 4.2 beschreibt einen Dienst zur Übertragung von Dienstdateneinheiten von einem Sender zu einem Empfänger, der reihenfolge-
erhaltend, aber verlustbehaftet ist. Dieser Dienst wird im Protokoll
zweimal benötigt, einmal zur Übertragung von PUTREQ in Hinrichtung und
zum anderen zur Übertragung von GETRES in Rückrichtung. Der Dienst für
jeweils eine Übertragungsrichtung wird durch zwei Dienstelemente
verfügbar gemacht, die in der folgenden Tabelle zusammen mit ihren
Parametern aufgelistet sind.

service primitive	abbreviated name	parameter	abbreviated name
DATA-request	DT-R	connection-endpoint	CEP
		user data	UD
DATA-indication	DT-I	connection-endpoint	CEP

Die erlaubten Sequenzen von Dienstelementen sind wiederum definiert
als Sequenzen von erreichbaren Markierungen im Produktnetz gemäß Bild
4.2.
Zur Erläuterung von Bild 4.2 wollen wir eine ausgezeichnete Verbindung
betrachten und für sie die erlaubte Menge von Folgen berechnen. Die
Beschränkung der Dienstdefinition auf eine Verbindung findet ihren
Ausdruck darin, daß nur ein Objekt als eine Interpretation von S-CEP
(Send-Connection Endpoint) und R-CEP (Receive-Connection Endpoint)
definiert wird. In der Startsituation enthält das Netz eine etablierte
Verbindung (modelliert durch das Objekt $\langle$s-cep,r-cep$\rangle$), einige Daten,
die vom Sender zum Empfänger zu übertragen sind (modelliert durch die
Objekte $\langle$s-cep, $ud_1\rangle$ ... $\langle$ s-cep, $ud_n\rangle$) und keine Objekte sonst.

In dieser Situtation ist die benutzerseitige Sendetransition aktiviert,
so daß ein DT-R-Element gestartet werden kann. Danach sind zwei
Transitionen aktiviert, die alternativ zu einer Beendigung dieses
Dienstelements führen:
- eine Transition, die den Verlust der Daten, enthalten in dem Parameter
  'user data' im DT-R-Element modelliert (diese Art von Datenverlust
  wird real durch Übertragungsfehler verursacht, die durch Auswerten
  von Prüfzeichenfolgen entdeckt werden).
- eine Transition, die die verlustfreie Übertragung von Daten modelliert
  und zum Anstoß eines DT-R-Elementes führt, dessen 'user data'
  identisch mit denen im DT-R-Element sind.
Beide Formen der Beendigung reaktivieren die Starttransition für ein
DT-R-Element, solange Nachrichten für die Übertragung vorliegen.

Im Falle eines erfolgreichen Datentransfers ist es offensichtlich, daß
der empfangende Benutzer ein angestoßenes DT-I-Element beenden kann, indem er die enthaltene Nachricht der Menge der bereits empfangenen Daten

hinzufügt. Wenn er dies jedoch nicht rechtzeitig tut, kann es passie-
ren, daß der sendeseitige Benutzer ein neues DT-R- Element startet,
daß der Erbringer die Daten von der Sender- zu der Empfängerseite
transferiert und dabei das zuvor übertragene Datum überschreibt.
(Überschreiben ist graphisch dargestellt durch eine Kante mit doppelter
Pfeilspitze). Diese Art von Nachrichtenverlust entsteht immer dann,
wenn der empfangende Benutzer DT-I-Elemente mit einer geringeren
Geschwindigkeit akzeptiert als der sendende Dienstbenutzer die entsprechenden DT-R-Elements aufbringt. Alle Modelle, die Kommunikations-
kanäle beschränkter Kapazität modellieren, müssen diesen Typ von
Nachrichtenverlust in Betracht ziehen. Es ist ferner zu beachten, daß
der sendende Dienstbenutzer keinerlei Wissen darüber hat, wie ein DT-R
beendet worden ist; Flußregelung mit Rückstaueigenschaften ist auf
dieser Dienstebene nicht verfügbar.

Aus Bild 4.2 kann leicht entnommen werden, daß der Benutzerdienst die
folgenden Eigenschaften hat:
- Für jedes DT-I-Element hat es
  ein DT-R- Element mit identischen Benutzerdaten gegeben.
- Jedes DT-R-Element resultiert entweder im Verlust der voll-
  ständigen Dienstdateneinheit oder in einem DT-I-Element.
- Für jede Verbindung kann zu jedem Zeitpunkt nur ein DT-R-Element
  aktiviert sein.
- Wenn während der Ausführung eines DT--Elemenets eine zuvor gesendete
  Nachricht überschrieben wird, so hat das den gleichen Effekt wie der
  Verlust einer Nachricht auf Grund von Übertragungsfehlern.
- Dienstdateneinheiten werden in der gleichen Reihenfolge empfangen,
  wie sie gesendet worden sind.

4.4 Definition des Protokolls

Das Protokoll ist durch das Produktnetz gemäß Bild 4.4 definiert. Das
Protokoll muß die funktionale Differenz zwischen dem Dienst, der benutzt
wird, und dem Dienst, der zu erbringen ist, überbrücken. Zu diesem
Zweck sind zwei Protokollentitäten zu konstruieren, eine Senderentität
und eine korrespondierende Empfängerentität (s. Bild 4.3). Gemeinsam
haben sie die Aufgabe, eine verlustfreie Datenübertragung trotz mög-

lichen Verlustes von Nachrichten und Quittungen im unterlagerten Dienst
zu gewährleisten.
Der Lösung dieses Problems liegt die folgende Strategie zugrunde:
Eine Senderinstanz (auf der PUT-Seite) wiederholt Benutzerdaten eines
PUTREQ-Elements solange, bis sie eine Quittung für diese Daten
empfängt. Nur infolge eines derartigen Quittungsempfanges aktiviert
die Senderinstanz das PUTCON-Element, das dem Dienstbenutzer
erlaubt, ein neues PUTREQ-Element zu initiieren. Für das Ausbleiben
einer erwarteten Quittung gibt es nun zwei Gründe:
- zum einen kann eine Dateneinheit verloren gegangen sein,
- zum anderen kann die Dateneinheit zwar empfangen worden sein, aber
  die Quittung darauf kann verloren gegangen sein.
Da eine Senderinstanz nicht zwischen diesen beiden Gründen entscheiden
kann, muß sie bei Ausbleiben einer Quittung in jedem Fall eine Nachricht
wiederholen; dies führt im zweiten betrachteten Fall zu einer Nachrich-
tenverdopplung. Daraus folgt unmittelbar, daß der Empfänger mit Nach-
richtenverdopplungen und der Sender mit Quittungsverdopplungen fertig
werden muß.
Eine grobe Analyse des Protokolls zeigt, daß die sendeseitige Dienstbenutzerinstanz schon ein neues PUTREQ initiieren kann, während die
sendeseitige Protokollinstanz noch duplizierte Quittungen auf die
vorausgehende Dienstdateneinheit abhandelt. Daraus folgt, daß das
Numerierungsschema, das im Protokoll zur Identifikation von Nachrichten
und Quittungen benutzt wird, erlauben muß, jeweils zwei Nachrichten
bzw. Quittungen voneinander zu unterscheiden.
Die sendeseitige Protokollentität enthält zwei Stellen: eine Stelle, in
der das Modulo 2 Numerierungsschema für Nachrichten niedergelegt ist,
und eine zweite Stelle, die Nachrichten für Wiederholungszwecke enthält.
Das Schalten von Transitionen in der sendeseitigen Protokollentität
wird durch die folgenden Ereignisse ausgelöst:

- Aufbringen eines PUTREQ Elementes durch eine Dienstbenutzerin-
  stanz,
- Ablauf des Wiederholungszeitgebers
- Empfang einer erwarteten Quittung
- Empfang einer nicht erwarteten Quittung,
- Empfang einer nachlaufenden Quittung.

Die empfangsseitige Protokollentität enthält ebenfalls zwei Stellen:
eine Stelle, die Empfangsfolgezähler modelliert, mit deren Hilfe entschieden werden kann, ob eine Nachricht die Originalnachricht ist
oder ein Duplikat einer zuvor empfangenen Nachricht darstellt, und
eine zweite Stelle mit Semaphor-Charakter, über die der Zugriff auf
den unterlagerten Dienst geregelt wird (ohne diese Semaphor-Variable
ist eine Modulo-2-Arithmetik nicht ausreichend).

Innerhalb der empfangsseitigen Protokollentität existieren drei unter-
schiedliche Transitionen, deren Schalten durch die folgenden Ereignisse
veranlaßt wird:
- Empfang einer Originalnachricht,
- Empfang des Duplikats einer Nachricht,
- Anstoßen eines GETRES-Elementes durch den Dienstbenutzer.

Die Startsituation für das Protokoll ist, jeweils bezogen auf eine
Verbindung durch die folgende Markierung des Produktnetzes (s. Bild 4.4)
modelliert:
- Die Verbindung durch (<s-cep,re-cep>)
- Die Menge der zu übertragenden Benutzerdaten durch (<s-cep,ud>)
- Sendefolgezähler auf 0
- Empfangsfolgezähler auf 0.

Für den Fall verlustfreier Übertragung kann Bild 4.4 leicht entnommen
werden, daß das Protokoll den geforderten Dienst erbringt. Dies bedeutet, daß die Folge von Dienstelementen an der oberen Schnittstelle
mit der durch die Dienstspezifikation vorgeschriebenen Folge übereinstimmt. Für den allgemeinen Fall, der auch nicht verlustfreie Übertragung mit einschließt, kann ein Beweis bezogen auf eine einzige Verbindung in (5) gefunden werden.

5. Zusammenfassung

Die PROSIT Modellierungs- und Beschreibungsmethode anzuwenden, heißt
- die Kommunikationsdienst und -protokollmodelle (s. Kap. 2)
  entsprechend der Funktionalität eines bestimmten OSI-Dienstes oder
  eines bestimmten OSI Protokolls zu Moduln zu verfeinern;

- für jeden einzelnen Modul die Topologie eines Produktnetzes zu
  entwerfen;
- die Topologien der verschiedenen Produktnetze zu einer übergreifenden
  Topologie zu verbinden, die die Struktur einer Entität und schließlich
  die Struktur des gesamten Kommunikationsdienst- oder -protokollmodells
  widergibt.

Die Beziehung zwischen dem Begriff einer Entität und der Instanz einer
Entität wird in dieser Spezifikationsmethodik durch Markierung der
entsprechenden Topologie hergestellt.

Durch Wahl einer entsprechenden Markierung kann sowohl eine einzelne
Instanz wie auch jede Kollektion von Instanzen modelliert werden. Durch
Wahl einer Markierung, die alle Instanzen innerhalb eines einzeln
OSI-Systems repräsentiert, läßt sich eine Spezifikation für eine proto-
typische Implementation gewinnen, die noch die ursprüngliche Neben-
läufigkeit enthält.

Diese nebenläufige Spezifikation kann durch eine Reduktion von Nebenläufigkeit auf Grund der Abbildung von Moduln auf reale Prozessoren in
eine Zustandsmaschinen-orientierte Spezifikation transformiert werden
(9). Diese Transformation überbrückt die semantische Differenz zwischen
der ursprünglichen Spezifikation, die die Relation zwischen Objekten
betont, und einer Spezifikation, die Ausführungsreihenfolgen eines
Prozessors in den Vordergrund stellt.

Die in diesem Papier vorgestellte Methodik wurde im Rahmen einer
industriellen Kooperation erfolgreich auf die Implementation von OSI-
Kommunikationsdiensten und -protokollen angewendet. Unsere Erfahrungen
lassen uns hoffen, daß sich die Konstruktion verteilter Systeme zu einer
wohlverstandenen Ingenieursdisziplin weiterentwickeln lassen wird.

Literaturverzeichnis

(1) Brauer, W. (ed.)
     Net Theory and Applications
     Proceedings of the Advanced Course on General Net Theory
     of Processes and Systems, Springer, 1979

(2) Burkhardt, H. J., Krecioch, T. H.
     Standards for the ISO-Reference Model for Open Systems
     Interconnection
     Computer Networks, Vol. 5, Nr. 3, May 1981
     North-Holland

(3) Burkhardt, H. J., Schindler, S.
     Structuring Principles of the Communication Architecture
     of Open Systems
     Computer Networks, Vol. 5, Nr. 3, May 1981
     North-Holland

(4) Chang, C.-L., Lee, R.
     Symbolic Logic and Mechanical Theorem Proving
     Academic Press, 1973

(5) Eckert, H.
     Ein mathematisches Verfahren zur automatisierten
     Verifikation von Kommunikationsprotokollen
     Dissertation, Universität Bonn, 1983
     (Ebenso: Berichte der GMD, Nr. 144, Oldenbourg 1984)

(6) Eckert, H., Prinoth, R.
     A computation-systems based method for automated proving
     of protocols against services
     Proceedings of the IFIP WG 6.1 Third International Workshop
     on Protocol Specification, Testing and Verification
     Zürich, 1983, North-Holland, 1983

(7) Eckert, H., Prinoth, R.
     Produktnetze - Definition eines PROSIT-Beschreibungsmittels
     (Vorversion), GMD-intern, 1983

(8) Eckert, H., Prinoth, R.
     Produktnetze - Definition eines PROSIT-Beschreibungsmittels
     Arbeitspapiere der GMD, Nr. 92, 1984

(9) Faul-Luers, E.; Prinoth, R.
     Ableitung von Implementationsvorgaben aus modularisierten
     Produktnetzen
     Arbeitspapiere der GMD, Nr. 123, 1984

(10) Information Processing Systems - Open Systems Interconnection -
     Basic Reference Model, ISO/IS 7498

(11) Linington, P.
     Fundamentals of the Layer Service Definitions and Protocol
     Specification
     Proceedings IEEE, 1983

(12) Prinoth, R.
An Algorithm to construct Distributed Systems from State-Maschines
Proceedings of the IFIP WG 6.1 Second International Workshop on
Protocol Specification, Testing and Verification
Idyllwild, 1982, North Holland, 1982

(13) Prinoth, R.
Modularisierung von Stellen/Transitionsnetzen
Arbeitspapiere der GMD, Nr. 26, 1983

(14) Burkhardt, H. J.; Eckert, H.
Transport-Protocol, Class 2, Formal Specification
GMD-F2G3, Projekt PROSIT, 1984

(15) Burkhardt, H. J.; Eckert, H.; Prinoth, R.
Modelling of OSI-Communication Services and
Protocols using Predicate/Transition Nets
Proceedings of the IFIP WG 6.1 Fourth International
Workshop on Protocol Specification, Testing and
Verification
Mt. Pocono, USA, 1984
North-Holland, 1984

(16) Starke, P. H.
Multiprocessor Systems and their Concurrency
Math. Found. of Comp. Sc., Praha, CSSR, 1984
Lecture Notes in Computer Science, Springer, 1984

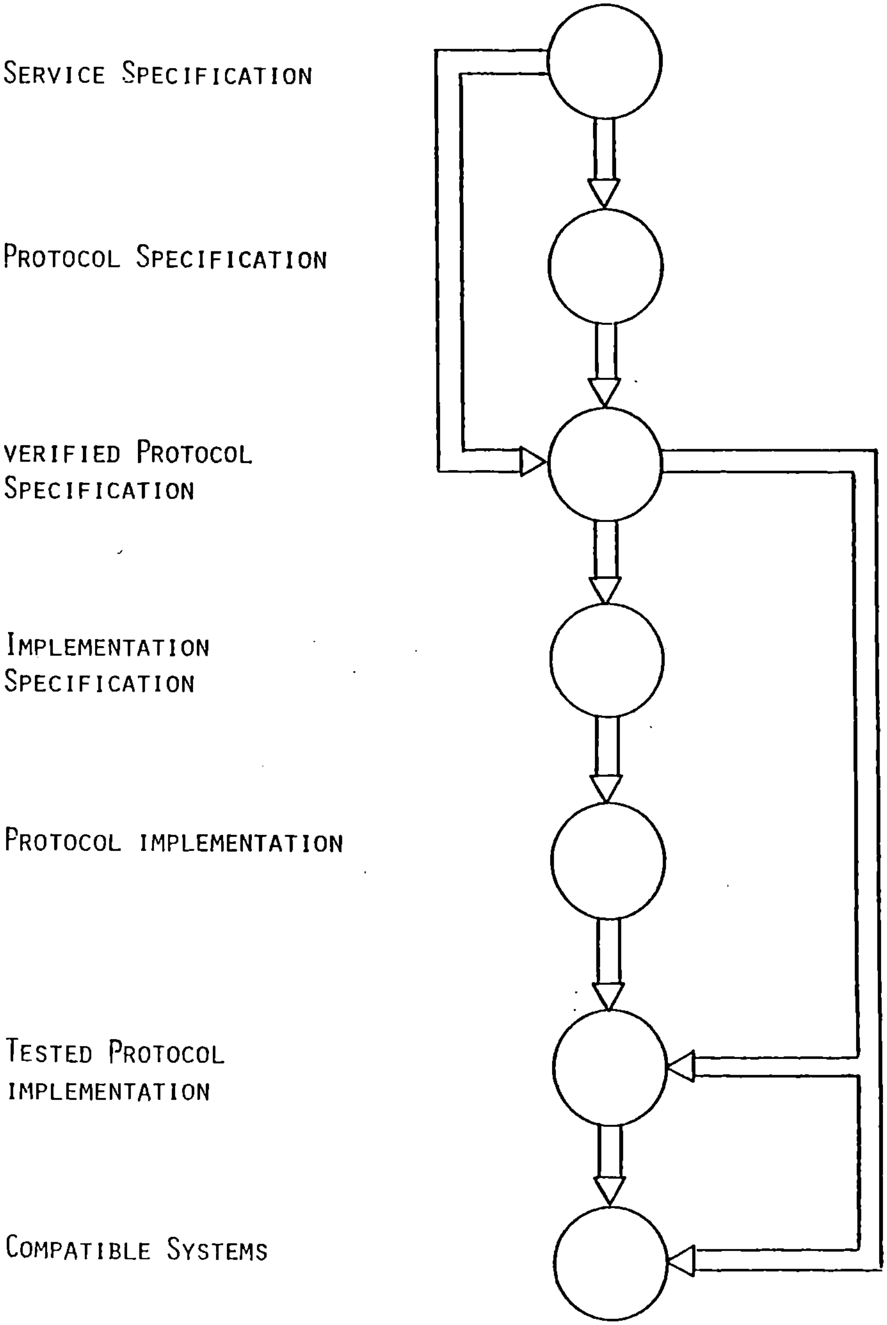

SERVICE SPECIFICATION

PROTOCOL SPECIFICATION

VERIFIED PROTOCOL SPECIFICATION

IMPLEMENTATION SPECIFICATION

PROTOCOL IMPLEMENTATION

TESTED PROTOCOL IMPLEMENTATION

COMPATIBLE SYSTEMS

Bild 1.1: Notwendige Schritte von der Norm zum kompatiblen System

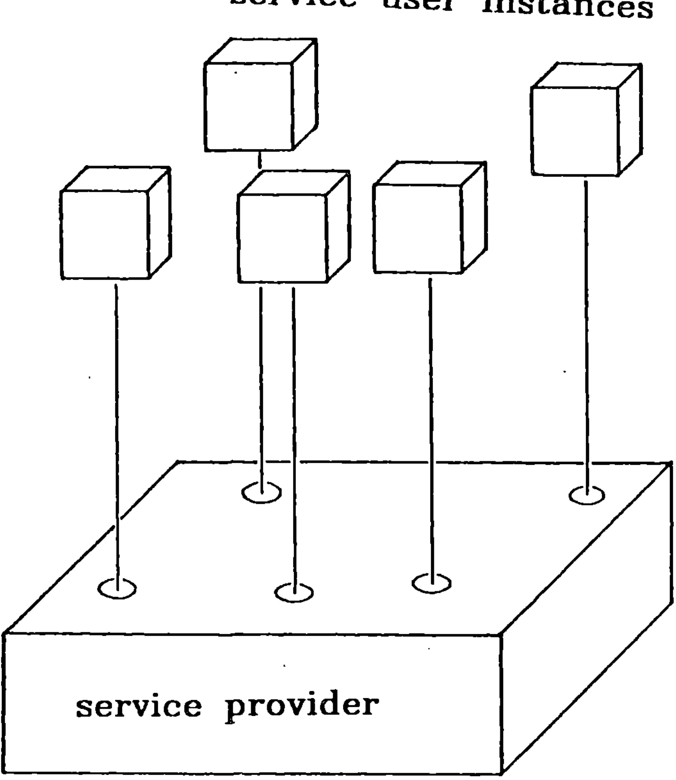

**Bild 2.1: Das allgemeine Szenario fuer einen Kommunikations-
dienst im Sinn des OSI-Referenzmodells**

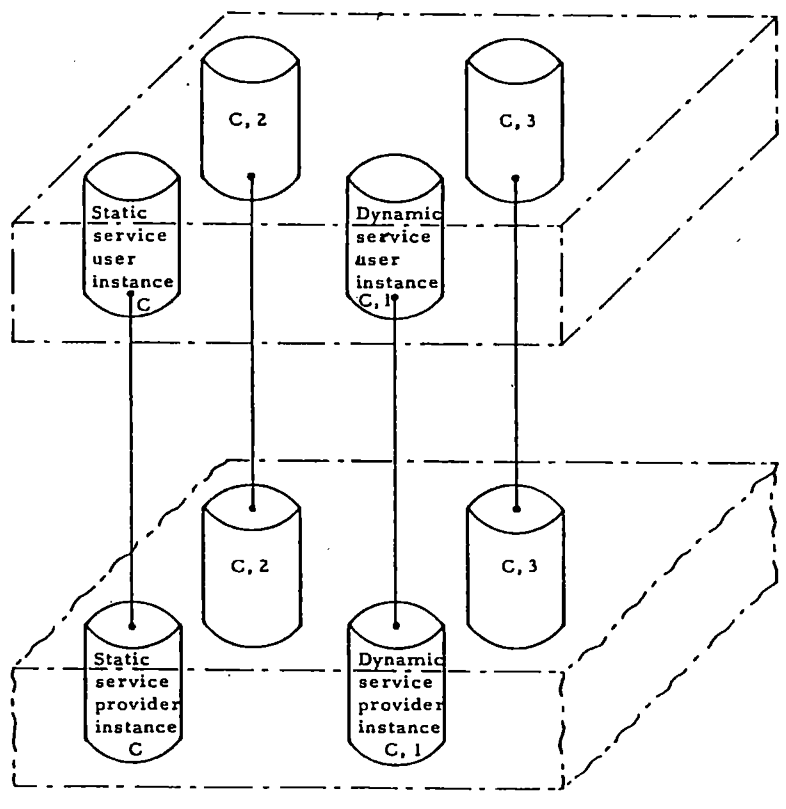

Bild 2.2: Die Verfeinerung von Dienstbenutzer- und Diensterbringerinstanz an einem spezifischen Dienstzugangspunkt mit der Adresse C in statische Instanzen
C und dynamische Instanzen C,K fuer K=1 bis 3

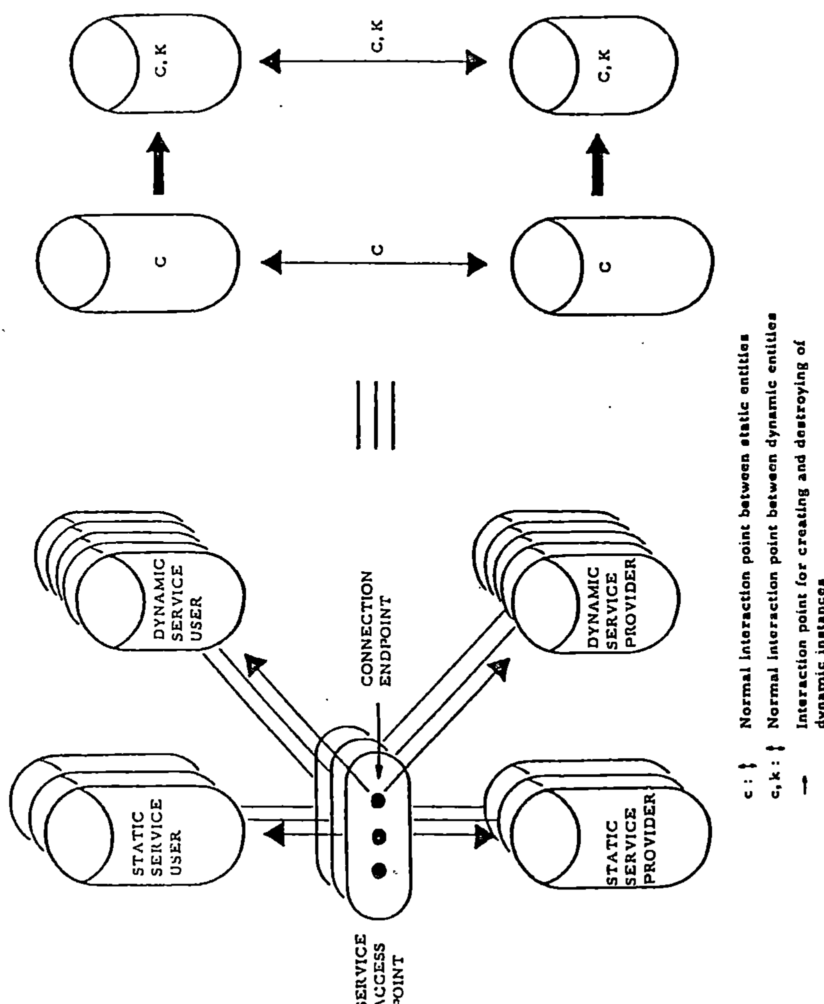

Bild 2.3: Das verfeinerte Modell eines Dienstzugangspunktes mit statischen und dynamischen Entitaeten

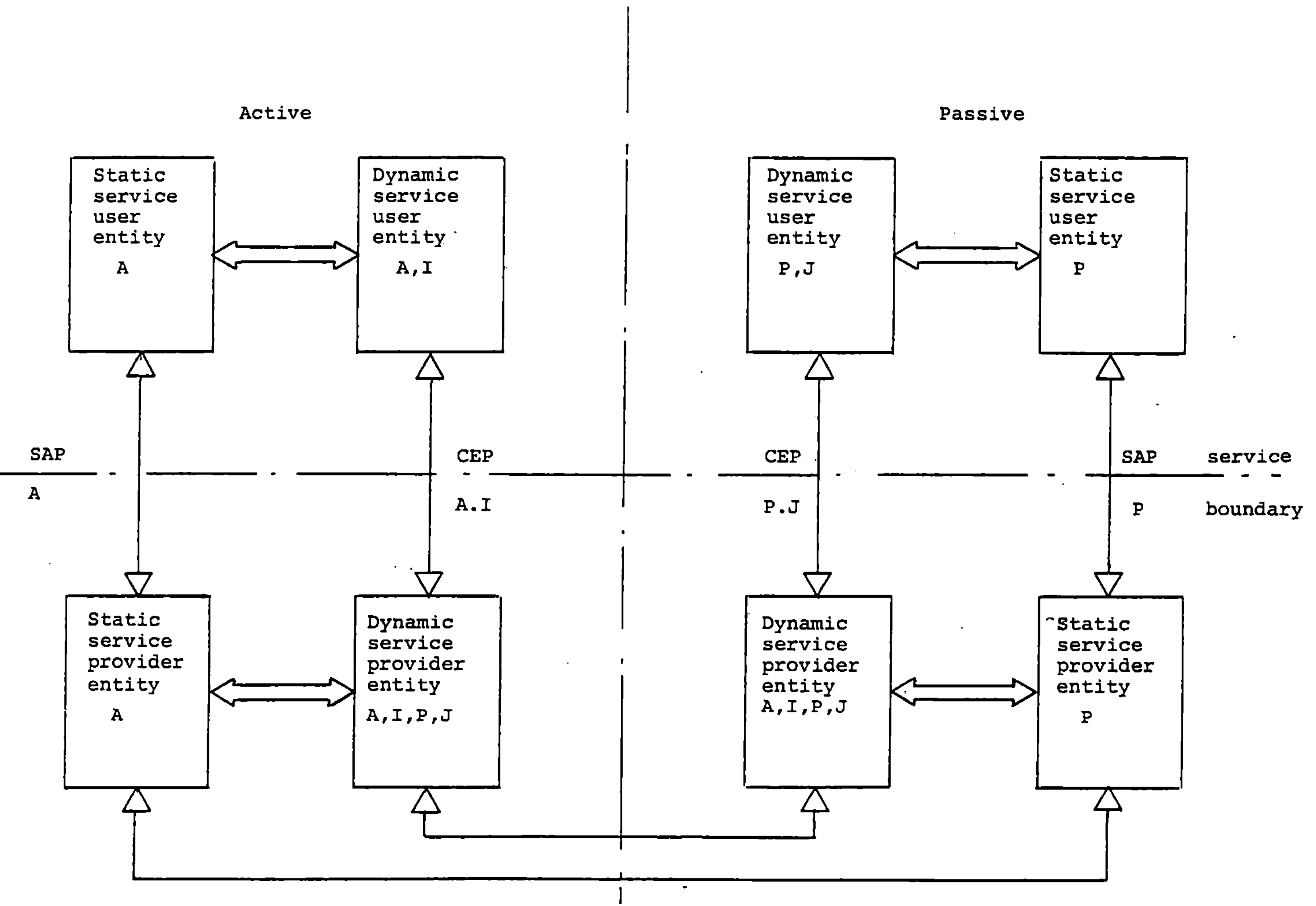

Bild 2.4: Das PROSIT Kommunikationsdienstmodell

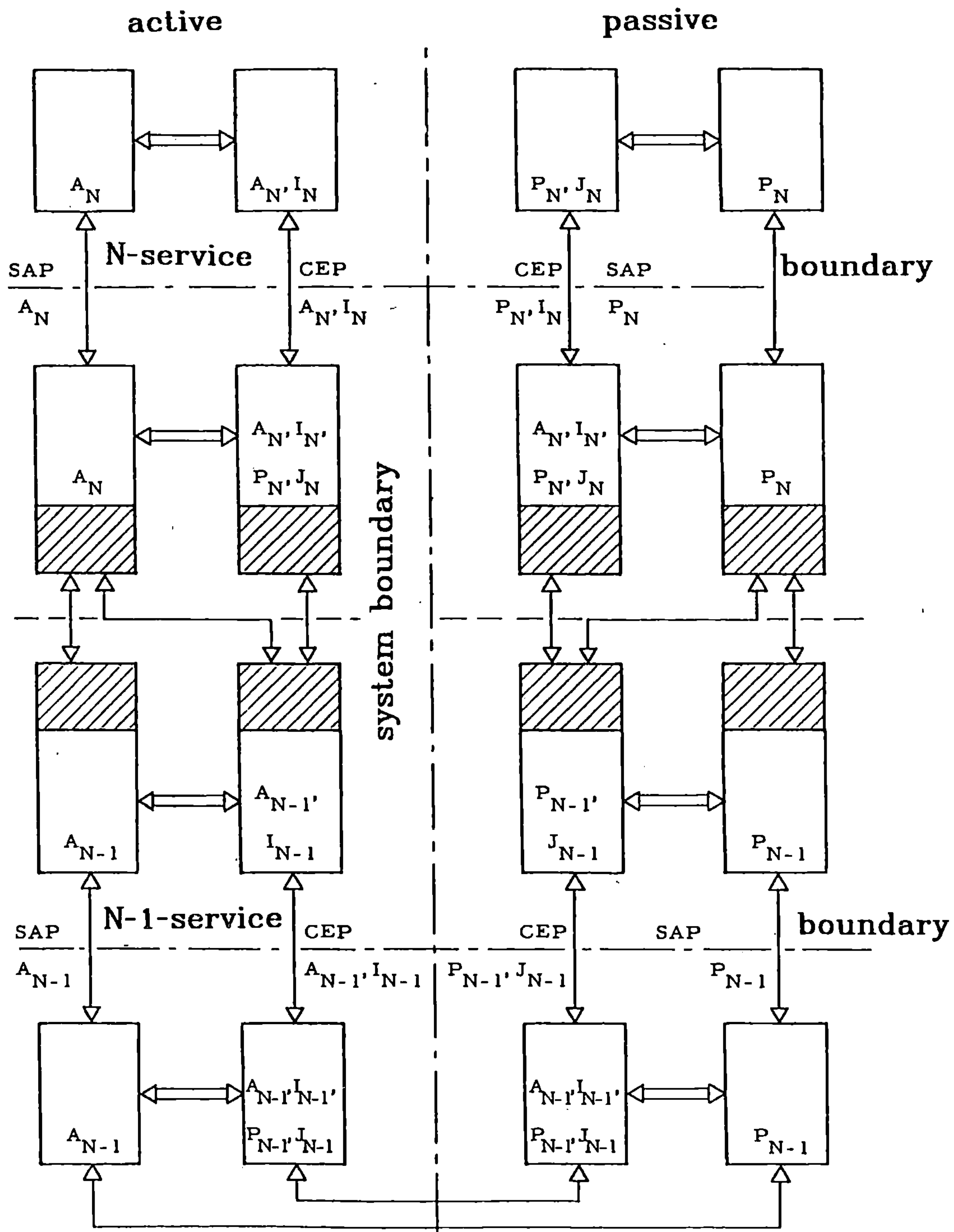

Bild 2.5: Das PROSIT Kommunikationsprotokollmodell

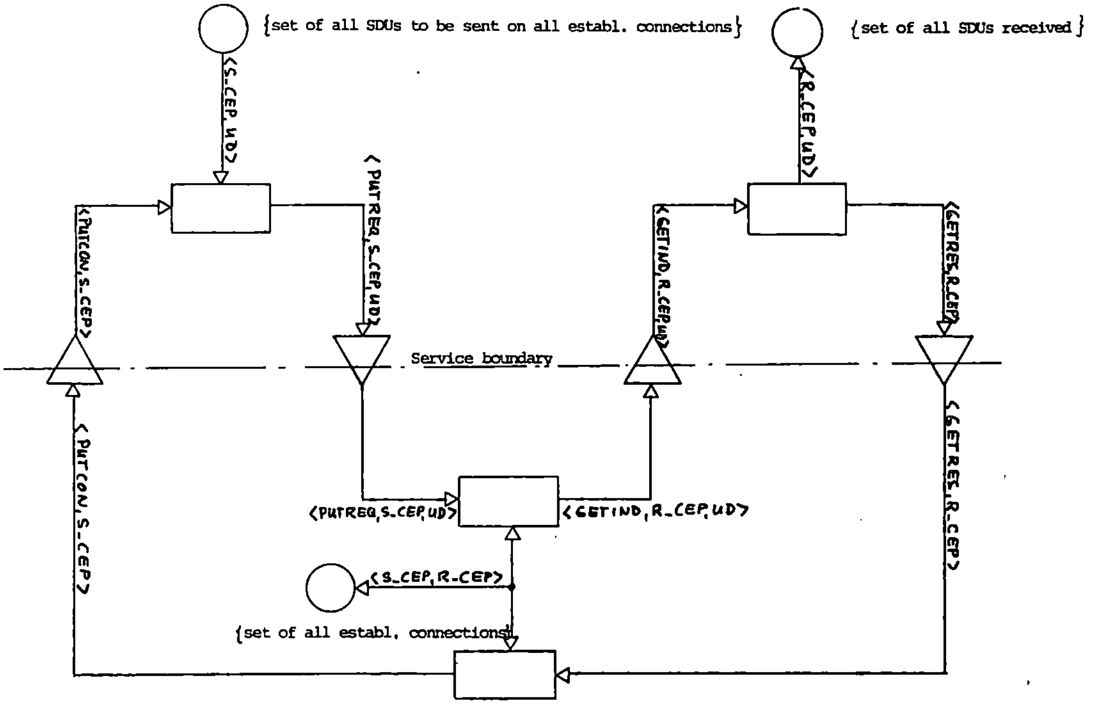

Bild 4.1: Der zu erbringende Dienst

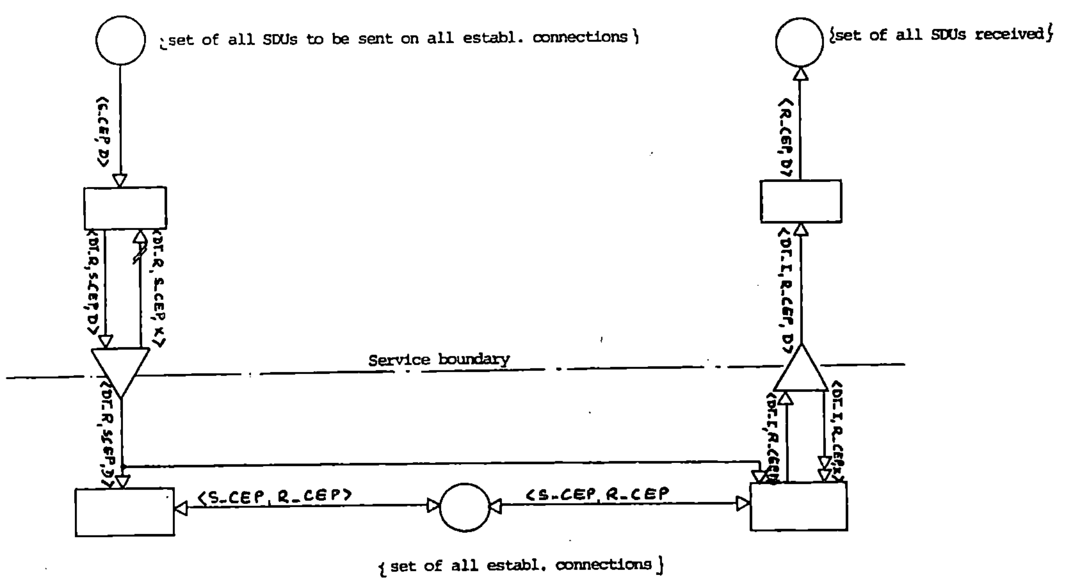

Bild 4.2: Der benutzte Dienst

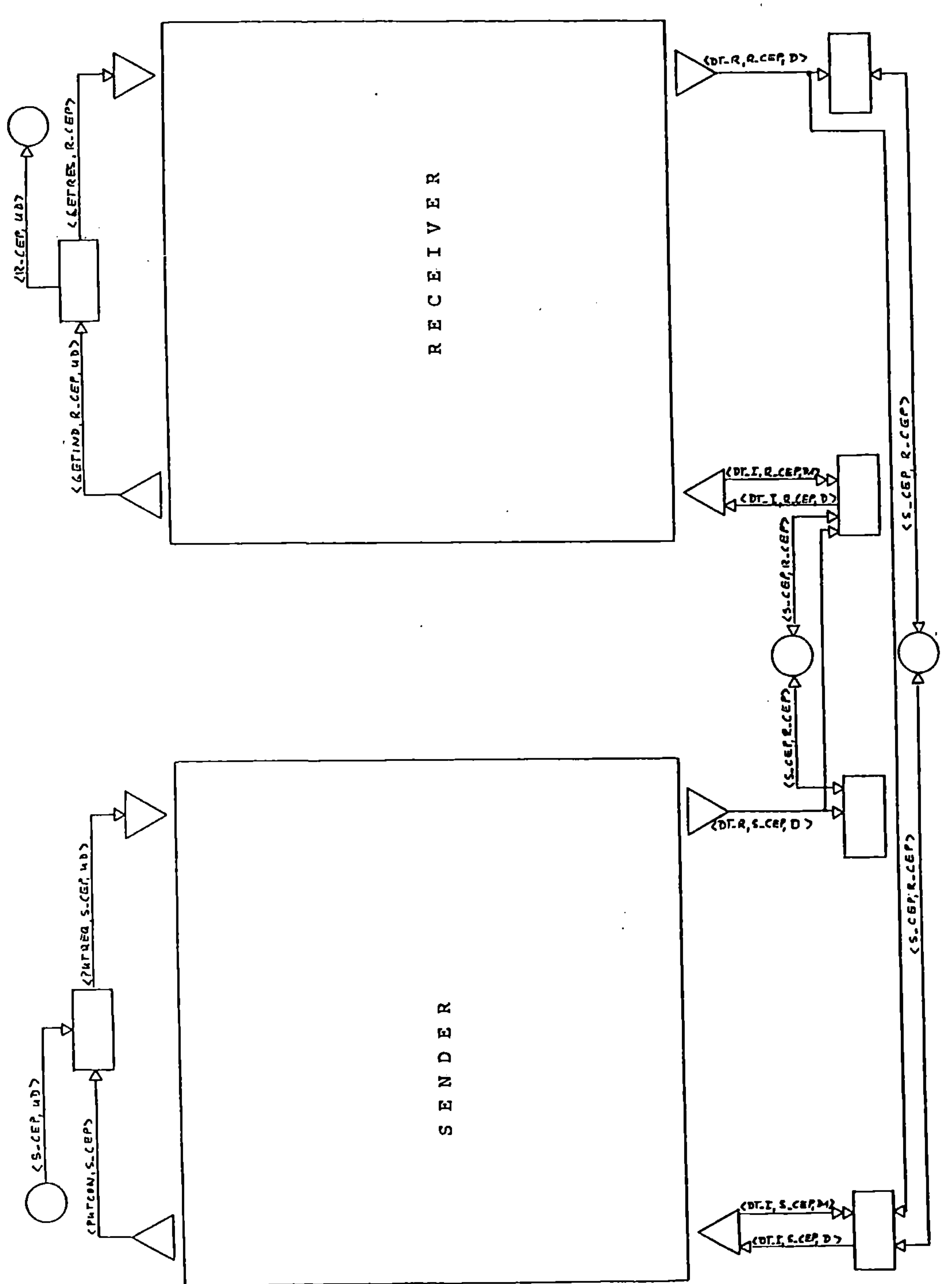

Bild 4.3: Berandende Dienste der Protokollinstanzen

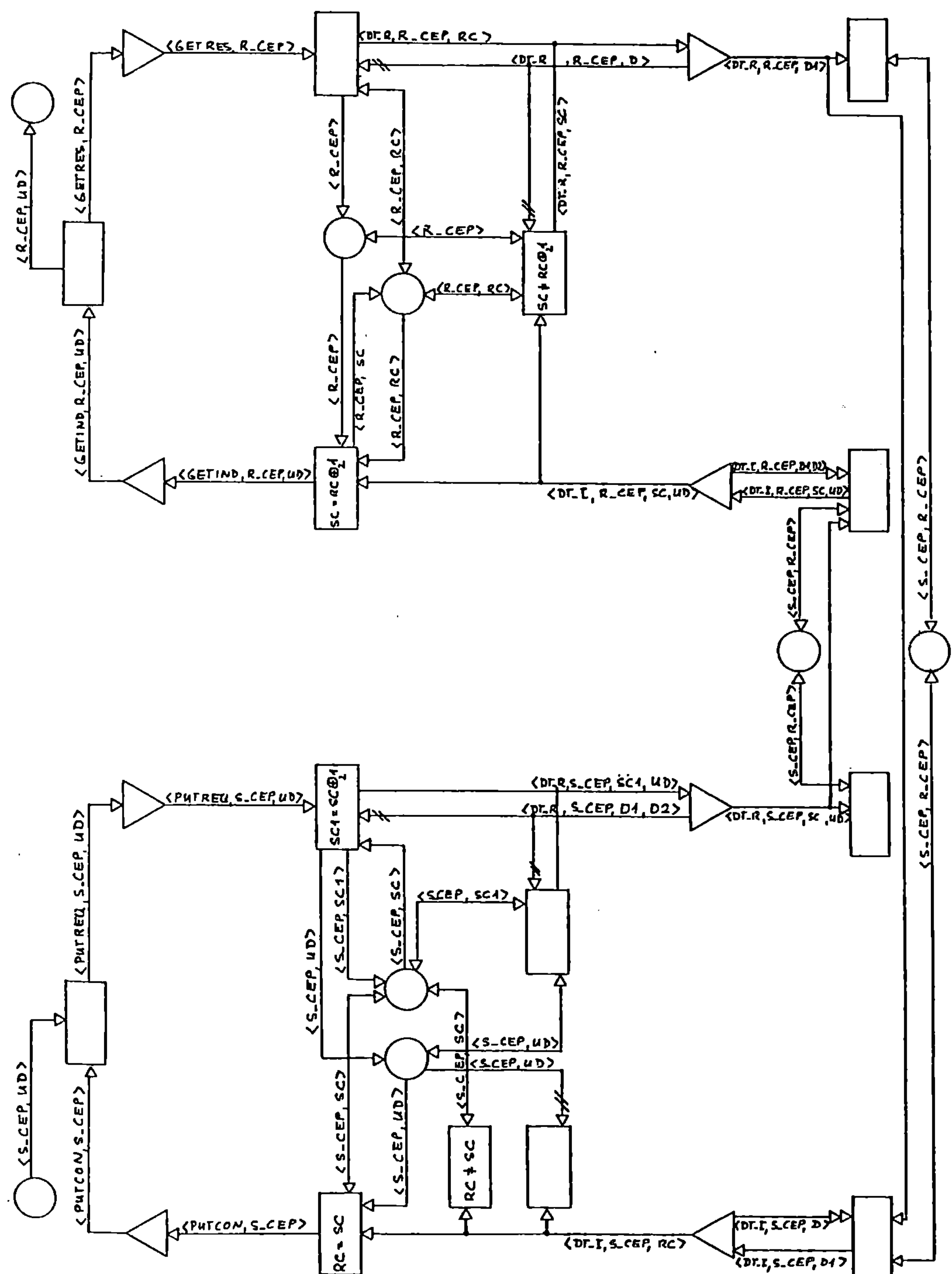

Bild 4.4: Das Protokoll

DESIGN – Eine verteilte Umgebung zur integrierten Entwicklung und Leistungsbewertung von Netzwerkanwendungen

M. Mühlhäuser, O. Drobnik
Institut für Informatik III
Universität Karlsruhe
Zirkel 2

D-7500 Karlsruhe

Kurzfassung

Die Entwicklung von Netzwerkanwendungen, d.h. von höheren Kommunikationsprotokollen und komplexen verteilten Anwendungen, erfordert entsprechende Hilfsmittel sowohl für die Softwareentwicklung als auch für die Leistungsbewertung. Das System DESIGN (Distributed Environment for Simulation, Investigation and Generation of Network Applications) integriert beide Bereiche. Es werden die Grundlagen zur Strukturierung von Netzwerkanwendungen und das Konzept für die Produktentwicklung in DESIGN erläutert. Die verschiedenen Laufzeitumgebungen, in denen eine Netzwerkanwendung während der Entwicklung ablaufen kann, werden besprochen. Ein Überblick über die Benutzerschnittstelle und die benötigten Entwicklungswerkzeuge schließt sich an.

Abstract

The development of network applications, e.g. of higher level protocols and complex distributed applications, requires adequate tools for the software production as well as for the performance evaluation. The system DESIGN (Distributed Environment for Simulation, Investigation and Generation of Network Applications) integrates both aspects. The fundamentals for structuring network applications and the concept for the application development with DESIGN are explained. DESIGN offers different run time environments to support different development phases. These runtime environments are discussed. The user interface and the development tools are outlined.

1.0 Einleitung

Nachdem die Standardisierungsbestrebungen im Bereich der transportorientierten Schichten von Kommunikationsdiensten vorangeschritten sind /IEE83/, wendet sich das Interesse höheren, anwendungsorientierten Schichten sowie den verteilten Anwendungen selbst zu /JEN82/; höhere Kommunikationsprotokolle und verteilte Anwendungen sollen im folgenden als **'Netzwerkanwendungen'** bezeichnet werden. Die zunehmende Komplexität dieser Netzwerkanwendungen erfordert den Einsatz neuartiger Methoden und Techniken aus den Bereichen Software-Engineering und Leistungsbewertung. Die Integration dieser beiden Bereiche bildet die Grundlage von DESIGN, einer verteilten Umgebung zur Entwicklung und Leistungsbewertung von Netzwerkanwendungen. Diese Arbeiten werden teilweise von der Deutschen Forschungsgemeinschaft gefördert. DESIGN soll als fortgeschrittenes Werkzeug in den unterschiedlichsten Netzen und Netzwerkarchitekturen eingesetzt werden können.

Die Komponenten zur Leistungsbewertung in DESIGN wurden nach folgenden
Anforderungen konzipiert:
- einfache Handhabung: weder Expertenwissen noch vertiefte mathematische Kenntnisse sollen erforderlich sein; Modelle sollen 'kostenlo-
  ses Nebenprodukt' der Softwareentwicklung sein; Benutzerführung und
  Funktionsumfang der Werkzeuge sollen höchsten Komfort bieten.
- frühzeitiger Einsatz: die Integration soll die Leistungsbewertung
  schon während der Programmentwicklung ermöglichen.
- höhere Effizienz: durch die Kombination mehrerer Leistungsbewertungsverfahren und die Entwicklung neuartiger Ansätze soll eine wesentliche Effizienzsteigerung erreicht werden.

Die Entwicklung der Software-Engineering-Komponenten von DESIGN besteht aus mehreren Schritten:
- Anpassung der Basis-Entwicklungswerkzeuge an die verteilte Programmierung,
- Einbringen fortgeschrittener Methoden zur integralen Unterstützung
  und weitgehenden Automatisierung des gesamten Entwicklungszyklus sowie
- Schaffung einer komfortablen und einheitlichen Benutzerschnittstelle.

Zum besseren Verständnis der getroffenen Entscheidungen soll zunächst
ein kurzer Überblick über existierende Ansätze aus beiden Bereichen
gegeben werden.

2.0 Stand der Entwicklung

2.1 Leistungsbewertung

Durch die Einführung neuerer Ansätze, wie BCMP-Theorie /BCM75/, Appro-
ximations- und Dekompositionstechniken (s. /COU77/) und rekursiver Be-
rechnungsverfahren /KMA82/ können wesentlich komplexere analytische
Modelle ausgewertet werden als zuvor. Dennoch konnten die grunsätzli-
chen Probleme der analytischen Modellierung verteilter DV-Systeme
nicht behoben werden: umfangreiche mathematische Kenntnisse sind erforderlich, die Lösbarkeit der Modelle ist oft nur durch starke Ab-
straktionen und unrealistische Voraussetzungen über Zwischenankunfts-
und Bedienzeitverteilungen zu erreichen, was teilweise zu falschen
Schlußfolgerungen führt /ZAH83, SG083/, und die Anwendung analytischer
Methoden bleibt i.a. auf die Untersuchung von einzelnen Protokollen
oder Phänomenen beschränkt /REI81, BUX81/. Wünschenswert wäre auch
eine Verknüpfung analytischer Methoden mit den beiden weiteren bedeu-
tenden Leistungsbewertungsverfahren, einerseits der Leistungsmessung
am realen System, vor allem zur Validation, und andererseits der Simulation, wodurch in 'hybriden' Modellen die Vorteile beider Modellie-
rungsverfahren ausgenützt werden könnten. Neuere Modellierungssysteme
wie RESQ /SAM77/ und COPE /BEI80/ erlauben nur die wahlweise, nicht
jedoch die gemeinsame Benutzung analytischer und simulativer Modellie-
rungstechniken.

Simulative Methoden gestatten es zwar, komplexe Modelle verteilter
DV-Systeme zu analysieren /FRC82, WOM83/, doch die bei Experimenten

auftretenden Laufzeiten lassen kaum mehr interaktive Benutzung der Systeme zu. Die erwähnten hybrid analytisch/simulativen Verfahren bilden eine Möglichkeit zur Laufzeitreduktion, kritisch ist hierbei jedoch das Problem, zur Laufzeit z.B. von einem analytisch modellierten Teilmodell in ein simulatives Teilmodell überzugehen. Eine andere Möglichkeit eröffnet sich durch den Einsatz der verteilten Simulation /PEA80, PWM79, BRY79, CHM81/, welche sich für die Modellierung verteilter Systeme besonders anbietet. Bislang wurde diese Modellierungsmethode für solche Zwecke jedoch noch nicht in größerem Umfang herangezogen.

Leistungsmessungen am realen System sind erst möglich, wenn ein meßbares System verfügbar ist. Um dennoch während der Entwicklungsphase von Softwaresystemen Leistungsmessungen zu ermöglichen, wurden spezielle verteilte Testumgebungen ('Testbeds') /COM82/ geschaffen, die mit aufeinander abgestimmten Komponenten für Entwicklung, Messung und Auswertung ausgestattet sind. Durch den stark verminderten Aufwand zur Leistungsmessung wird eine ständige Rückkopplung zwischen Entwicklung und Leistungsbewertung möglich. Die bekannten Testumgebungen sind allerdings auf lokale Netze beschränkt und erfordern großen Hardwareaufwand, da sie auf Hardwaremonitoren basieren. Die enge Kopplung zwischen Meß- und Entwicklungssystem beschränkt den Einsatz meist auf die eigens entwickelten (Mikro-)Rechnernetze der spezifischen Testumgebungen.

2.2 Software-Engineering

Software-Entwicklungswerkzeuge für sequentielle Programme haben teilweise einen sehr hohen Stand erreicht /RIF80/. Kennzeichen solcher fortgeschrittener Entwicklunssysteme sind beispielsweise umfassende Projektführung und -dokumentation, volle Unterstützung des Software-Entwicklungszyklus' vom Grobentwurf bis zur Implementierung und Wartung, Topdown Design, Rapid Prototyping, integrierte Benutzerschnittstellen für alle Werkzeuge mit gemeinsamer Syntax und gemeinsamen Daten, Anpassung an den Kenntnisstand des Benutzers, problemorientierte Entwicklung.

Demgegenüber sind noch kaum Entwicklungswerkzeuge für verteilte Systeme bekannt, am wenigsten für Netzwerkanwendungen in heterogenen Netzen. Hier ist bei einem Blick auf die Programmierpraxis, selbst bei Herstellern und Softwarehäusern, ein Mangel an **Basis-Entwicklungswerkzeugen** adäquater Ausprägung festzustellen: Compiler, Laufzeitsysteme, Testhilfen sind für sequentielle Programme konzipiert. Da sie die Module eines verteilten Programmes als isolierte Ablaufobjekte betrachten, können die Wechselwirkungen zwischen den Modulen nicht berücksichtigt werden. So überprüfen Compiler für sequentielle Programme nicht die Konsistenz von Modul-Interaktionen, zur Laufzeit führt der Fehlerabbruch eines Moduls nicht zum Abbruch der abgesetzten Programme, 'debugging' einzelner Module führt zu Synchronisationsverlust mit anderen Modulen, systemweite 'traces' sind nicht möglich, die Verteilung der Module sowie abgesetztes Übersetzen, Binden und Laden werden nicht unterstützt. In heterogenen Systemen erfordert der Übergang von

einem System zum anderen weitgehende Neuentwicklungen der Programme
und die Anpassung von Datenformaten etc. In neueren Ansätzen werden
einzelne dieser Probleme angegangen /KRD84/, integrale verteilte Pro-
grammierumgebungen existieren jedoch noch kaum.

Solange selbst die Basis-Entwicklungswerkzeuge für verteilte Programme
derart unzureichend sind, können fortgeschrittene integrale Entwicklungswerkzeuge, wie sie für die sequentielle Programmierung verfügbar
sind, keinen breiten Eingang finden.

3.0 Grundlagen von DESIGN

Das Konzept von DESIGN (Distributed Environment for Simulation, Investigation and Generation of Network Applications) ist von drei wesentlichen Ansätzen geprägt:
- Die Entwicklung einer Netzwerkanwendung findet von Anfang an in einem verteilten System statt, das dem Typ des Zielsystems entspricht.
  Dabei wird von Anfang an die Benutzung der vorhandenen Kommunikationsdienste – bezeichnet als **'Dienste des Informationsaustausches'**
  – durch die Netzwerkanwendung berücksichtigt.
- Von Anfang bis Ende der Entwicklung wird ein integriertes verteiltes
  Entwicklungssystem verwendet, das zur Erstellung des Grobkonzeptes,
  für die Leistungsbewertung, für schrittweise Verfeinerungen, zur
  Fehleranalyse etc., und schließlich als Laufzeitumgebung für die
  fertige Anwendung dient.
- Um ein Höchstmaß an Benutzerkomfort zu erreichen, wird soviel Arbeit
  wie möglich automatisiert, der theoretische Hintergrund vor allem
  der Leistungsbewertung bleibt dem Anwender im täglichen Umgang verborgen, und die Entwicklungsunterstützung orientiert sich speziell
  an den Anforderungen der Entwicklung verteilter Programme.

Das zentrale Konzept zur Strukturierung von Netzwerkanwendungen ist
die Funktionseinheit. Funktionseinheiten sind sequentieller Natur,
parallele Verarbeitung wird über mehrere kooperierende Funktionseinheiten realisiert. Entscheidend ist, daß der Entwickler während des
gesamten Entwicklungszyklus' auf demselben Satz Funktionseinheiten arbeitet, während die Laufzeitumgebungen wechseln können. In Rahmen der
schrittweisen Verfeinerung werden mehr und mehr Teile ausprogrammiert;
für noch nicht fertig entwickelte Teile stehen Operatoren zur Verfügung, um solche Teile grob zu beschreiben und somit den Ablauf der
Funktionseinheit in verschiedenen Laufzeitumgebungen bereits zu ermöglichen, während die Netzwerkanwendung zum größten Teil noch nicht ausprogrammiert ist.

Die Netzwerkanwendung setzt sich dann zusammen aus einer Menge von
Funktionseinheiten, deren Konfiguration und Zusammenspiel durch spezielle Funktionseinheiten beschrieben wird. Dabei kann die Netzwerkanwendung so konzipiert werden, daß pro Netzknoten und Netzwerkanwendung eine oder mehrere Funktionseinheiten gleichzeitig ablaufen, mehrere Inkarnationen desselben Typs pro Knoten sind erlaubt.

Während der Entwicklung kann die Netzwerkanwendung in unterschiedliche
Laufzeitumgebungen eingebettet werden, darunter 'verteilte Simula-

tion', '1:1-Simulation' und 'realer Dienst'. Eine komfortable Benutzerschnittstelle und eine Menge von Werkzeugen unterstützen die Entwicklung der Netzwerkanwendung während des gesamten Entwicklungszyklus.

4.0 Struktur einer Netzwerkanwendung

4.1 Funktionseinheiten

Eine Funktionseinheit wird in ihrem Ablauf beschrieben durch eine Folge von '**Rechenphasen**', die durch '**Interaktionspunkte**' voneinander getrennt sind. Dies ist in Abb. 1 dargestellt.

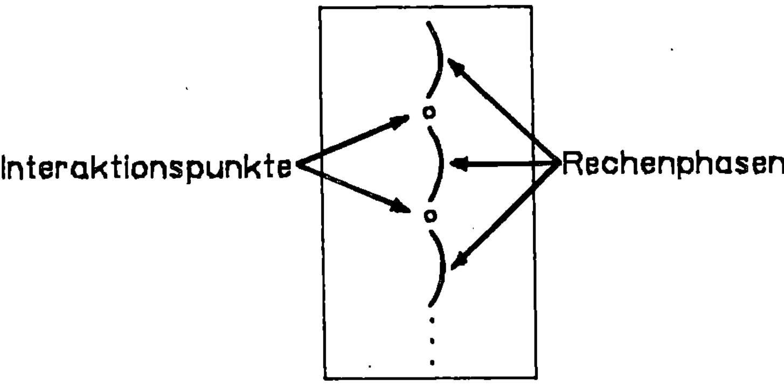

Abb. 1: Ablauf einer Funktionseinheit

Rechenphasen treten nach außen nur durch den Bedarf an Hauptspeicher und Prozessorzeit in Erscheinung. Es wird nur Code aus dem privaten Adressbereich der Funktionseinheit abgearbeitet.

Es werden Hilfsmittel angeboten, um in der Entwicklungsphase die noch nicht ausprogrammierten Teile von Rechenphasen beschreiben zu können, wodurch die Modellierung der Netzwerkanwendung möglich wird. Diese Beschreibungen betreffen zunächst Abschätzungen über den Ressourcenbedarf. Dazu gibt der Programmierer in Interaktion mit dem Präprozessor für Rechenphasen, wie auch für Interaktionspunkte, Abschätzungen über den Ressourcenbedarf an (bei Rechenphasen als CPU-/Hauptspeicherbedarf oder in benutzerorientierten Größen wie 'lines of code'). Anhand dieser Abschätzungen fügt ein Präprozessor am Ende jeder Rechenphase Anweisungen in den Programmtext ein, die zur Laufzeit für die Modellierung herangezogen werden.

Programmstücke, die noch ausprogrammiert werden müssen, können gesondert gekennzeichnet werden ('to_develop'). Diese Angaben werden von Werkzeugen der übergeordneten Projektführung verwendet, z.B. um einen schnellen Überblick über den Projektfortschritt zu ermöglichen.

Interaktionspunkte bezeichnen Punkte, an denen die Kontrolle von der Funktionseinheit auf andere Systemteile (andere Funktionseinheiten, Kommunikationsdienste oder das Betriebssystem) übergeht. Sie werden ausschließlich mit DESIGN-spezifischen Konstrukten beschrieben, um die

die Wirtssprache erweitert wird. Die Funktionalität dieser Spracherweiterung wurde so konzipiert, daß einerseits eine Portierung in praktisch jeden beliebigen Zielrechnertyp möglich ist, andererseits aber
auch alle wesentlichen Interaktionsmöglichkeiten zur Verfügung stehen.
Zwei wesentliche Bereiche sind hier zunächst zu nennen: Interaktionen
zwischen Programm und Betriebssystem (Zeitsteuerung, Ein-/Ausgabe) und
Interaktionen zwischen Programmen (Kommunikation, Synchronisation).

Sehr unterschiedlich in verschiedenen Betriebs- und Kommunikationssystemen sind die Möglichkeiten zur Zeitsteuerung, Synchronisation sowie
zur lokalen und abgesetzten Kommunikation. Daher mußte für DESIGN ein
einheitliches Konzept entwickelt werden; dieses sollte möglichst einfach und leicht zu handhaben, auf jedem Zielrechnertyp zu implementieren und möglichst im Funktionsumfang umfassend und universell sein.
Für die abgesetzte Kommunikation war zu beachten, daß verschiedene
Dienste des Informationsaustausches sehr unterschiedliche Funktionalität besitzen können und insbesondere Netzwerkanwendungen nach Abschluß
ihrer Entwicklung selbst Dienste des Informationsaustausches werden
können. Es gelang, Zeitsteuerung, lokale Kommunikation sowie Synchronisation und abgesetzte Kommunikation in einem einzigen Konzept zusam-
menzufassen, das auf dem gepufferten, optional zeitverzögerten Botschaftenaustausch fußt und mit den vier Operatoren **transmit, cancel,
get** und **test** auskommt.

Eine Funktionseinheit kommuniziert mit der Außenwelt über **Ports**.
Ports sind die Endpunkte von uni- oder bidirektionalen Punkt-zu-Punkt
Kommunikationspfaden, über die **Botschaften** ausgetauscht werden. Der
Begriff Botschaft umfaßt alle Arten von Information wie z.B. Ereig-
nis-, Status- und Synchronisationsmeldungen. Die innerhalb einer
Netzwerkanwendung verwendeten Botschaftentypen werden einmal festgelegt, Ports enthalten als Attribute die Namen der Botschaftentypen,
die an dem Port gesendet bzw. empfangen werden können. Jede Übertragungsrichtung eines Kommunikationspfades kann logisch als Botschaf-
ten-Warteschlange angesehen werden. **Transmit** dient dazu, Botschaften
in diese Warteschlange einzureihen. Solche Botschaften ermöglichen
sowohl Kommunikation als auch Synchronisation: 'transmit' enthält einen optionalen Zeitparameter, mit dessen Hilfe das verzögerte Eintreffen ('gültig werden') einer Botschaft erreicht werden kann. Da Kommunikationspfade auch zwei Ports derselben Funktionseinheit verbinden
können, ist auch die Zeitsteuerung möglich. Da vor allem bei Anwendungen in verteilten Systemen viele Zeitüberwachungsaufträge vor dem
Ablauf unnötig werden, ist eine Möglichkeit zum Zurücknehmen von Zeit-
aufträgen notwendig. Dies kann mit **cancel** geschehen. Zum Empfang von
Aufträgen wird **get** verwendet; optional kann der Empfangsport und/oder
der Botschaftentyp in 'get' vorgeschrieben werden. Liegen mehrere
Botschaften der gesuchten Art vor, so wird nach FIFO ausgewählt.
'Get' ist synchron implementiert, d.h. bei Fehlen einer entsprechenden
Botschaft wird bis zu deren Eintreffen gewartet – dies ist notwendig
zur Erhaltung der Sychronisationseigenschaft. Daher wurde es möglich
gemacht, die Warteschlange mittels **test** auf das Vorhandensein von Botschaften hin zu überprüfen. Bei 'test' kann analog zu 'get' nach Bot-

schaftentyp und Port unterschieden werden. Die Warteschlange selbst wird für den Programmierer nicht sichtbar, d.h. er kann sie außer über die vier genannten Operatoren nicht beeinflussen. Das Warteschlangenkonzept und die Operatoren müssen für jeden Zielrechner von DESIGN implementiert werden.

Die Netzwerk-Kommunikation über die verfügbaren Dienste des Informationsaustausches läßt sich durch das gewählte Konzept flexibel gestalten. In frühen Entwicklungsphasen der Netzwerkanwendung und in Fällen, wo keine besonderen Fähigkeiten der Dienste des Informationsaustausches explizit genutzt werden sollen, gibt es die Möglichkeit **'transparenter Kommunikation'**. Dabei bietet DESIGN, unter Ausnutzung der vorhandenen Dienste des Informationsaustausches, über Rechnergrenzen hinweg die gleichen Kommunikationsmöglichkeiten wie lokal, das heißt die Datenaustauschoperatoren können auch an abgesetzte Funktionseinheiten gerichtet werden. Bei der **'expliziten Kommunikation'** hingegen wird zwischen die Dienste des Informationsaustausches und die Funktionseinheiten eine Schnittstelle geschaltet, die den lokalen Kommunikationsdienst wie eine Funktionseinheit erscheinen läßt. Die Typen und Inhalte möglicher Botschaften sind dabei festgelegt und entsprechen dem Funktionsumfang des angebotenen Dienstes. In der Schnittstelle werden Botschaften der Funktionseinheiten in Aufrufe an die Kommunikationsdienste umgesetzt und Rückmeldungen und Anzeigen in Botschaften an die Funktionseinheiten verwandelt. Die Mächtigkeit des benutzbaren Dienstes schlägt sich dadurch nicht in den möglichen Schnittstellenaufrufen nieder, sondern nur in den verfügbaren Botschaftentypen, die wesentlich besser anpaßbar sind. Dies hat einen weiteren Vorteil: wird eine Netzwerkanwendung Teil der Dienste des Informationsaustausches, so enthält sie bereits die notwendige Schnittstelle.

Auch alle anderen Arten von Interaktionen, sowohl zur Verwaltung und Konfiguration von Funktionseinheiten, für die Ein-/Ausgabe etc., können auf den beschriebenen Botschaftenaustausch zurückgeführt werden, wobei allerdings vielfach zur Vereinfachung weitere DESIGN-spezifische Konstrukte angeboten werden. So wird die Ein-/Ausgabe abgebildet auf den Botschaftenaustausch zwischen einer Funktionseinheit und einer 'virtuellen' Funktionseinheit 'Disk' oder 'Terminal' o.a. An die Stelle der E/A-Aufrufe der Wirtssprache treten dabei DESIGN-spezifische Konstrukte.

Abb. 2 zeigt ein Beispielprogramm des Rumpfes ('Body') einer Funktionseinheit. Von den Deklarationen, die die Einbettung einer Funktionseinheit in eine Netzwerkanwendung beschreiben, soll hier nur die Port- und Botschaftenvereinbarungen dargestellt werden. Die zugehörige Netzwerkanwendung ist ein primitives verteiltes Datenbanksystem. Die dargestellte Funktionseinheit 'db_access' erhält über einen Port 'req' Transaktionen, die sie entweder lokal erledigt oder über einen Port 'net' zur netzweiten Bearbeitung weiterreicht. Bei lokaler Bearbeitung geht ein Ergebnis zurück an den Sender der Transaktion. Im Beispiel sind auch die Rechenphasen und Interaktionspunkte vermerkt.

```
SYSTEM db_access;
...
PORTS {    req:   -> transaction;
                  <- result;
           net:   <- transaction;
  };
...
MESSAGE mt TYPE transaction;
        mr TYPE result;


BODY

  ----------------------------------------------------------------------
         |   main () {
         |     INIT;
1        |     loopcount = 0;
|        |     .
1        |     .
         |     while (loopcount < reorgthreshold) {
    1    |       GET mt ON req;
         |       if ( act->node != mynode) {
2        |         act->processed++;
    2    |         TRANSMIT mt ON net;
         |       }
         |       else
3        |         TO DEVELOP "read/write";
    3    |       TRANSMIT mr ON req;
         |     }
4        |     TO DEVELOP "reorganize";
|        |     TO DEVELOP broadcast (400);
|        |     .
4        |     EXIT;
         |   }
  ----------------------------------------------------------------------

END SYSTEM;
```

Abb. 2: Beispielprogramm einer Funktionseinheit

4.2 Systemstruktur und Konfiguration

Zur integralen Betrachtung einer Netzwerkanwendung, um Modularität und
Top-Down-Entwurf zu ermöglichen, wurde das Konzept der Funktionseinheit erweitert: Eine Netzwerkanwendung wird dargestellt als eine Hierarchie von Systemen, wobei jedes System nur die Sicht auf seine direkten Subsysteme erhält. Wurzel, Subsysteme und Blätter werden durch
Funktionseinheiten realisiert. Eine Funktionseinheit, die Subsysteme
verwaltet, heißt administrativ, die Blätter des Systems heißen operationale Funktionseinheiten. Die für die direkt übergeordnete administrative Funktionseinheit sichtbare Schnittstelle einer Funktionseinheit besteht aus (Schnittstellen-) Ports - zusammen mit den auf dem
Port erlaubten Botschaftentypen - sowie frei wählbaren Formalparame-

tern. Die Verknüpfung der Schnittstellenports einer Funktionseinheit mit der Außenwelt wird von der direkt übergeordneten administrativen Funktionseinheit durchgeführt; für die subsystem-interne Verwendung der Schnittstellenports administrativer Funktionseinheiten gibt es zwei Alternativen: sie können von dieser Funktionseinheit selbst bedient werden ('serving') oder an die Schnittstellenports von Subsystemen weitergeleitet werden ('mapping'). Auf diese Weise kann eine administrative Funktionseinheit beispielsweise autonom entscheiden, ob Aufgaben - repräsentiert durch ankommende Botschaften - von der Funktionseinheit selbst erledigt werden oder ob und in welchem Umfang Subsysteme zu bilden sind, die zur Erledigung von Aufgaben herangezogen werden. Selbst die direkt übergeordnete Funktionseinheit ist nicht in der Lage zu erkennen, ob Subsysteme administrativ oder operational sind. Operationale Funktionseinheiten müssen naturgemäß all ihre Schnittstellenports selbst bedienen und brauchen dies nicht durch den 'serve'-Operator angeben. Zum Botschaftenaustausch zwischen einer administrativen Funktionseinheit und ihren Subsystemen besteht die Möglichkeit, daß die administrative Funktionseinheit weitere Ports vereinbart, die nicht Teil der Schnittstelle sind, sog. 'inner Ports', und diese mit Schnittstellenports der verwalteten Subsysteme verknüpft (ebenfalls Kommando 'connect'). Abb. 3 zeigt eine beispielhafte Systemstruktur. Zu beachten ist dabei, daß Funktionseinheit 1 nur die Schnittstelle zu den Funktionseinheiten 2 und 3 sieht, Funktionseinheit 2 nur die Schnittstelle zu Funktionseinheit 4 und die Funktionseinheiten 3 und 4 operationaler Natur sind.

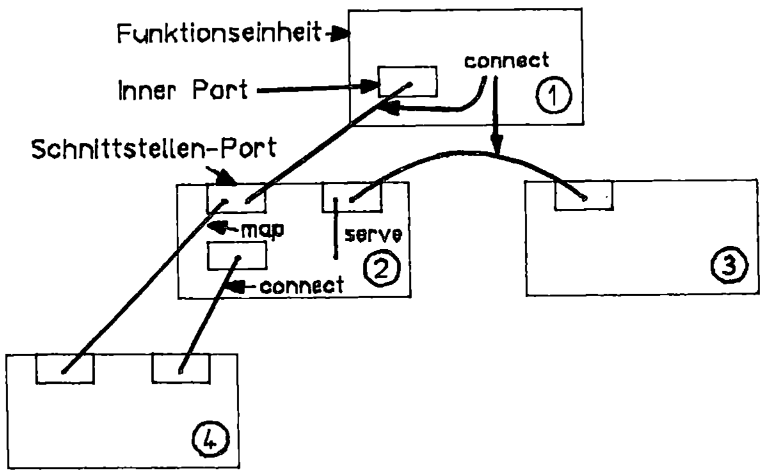

Abb. 3: Hierarchie von Funktionseinheiten

Den administrativen Funktionseinheiten kommt also vor allem Bedeutung bei der Konfiguration und Verwaltung einer Netzwerkanwendung zu. Dazu werden in DESIGN durch Sprachelemente und Laufzeitsystem neben der Installation von Funktionseinheiten bzw. Subsystemen und der Verknüpfung von Ports auch insbesondere **variable Topologien** und **dynamische Umkonfigurationen** unterstützt. Variable Systemgröße ist eine Grundvoraussetzung zur effizienten Programmierung von Netzwerkanwendungen: da meist die Zahl der benötigten Funktionseinheiten, die Systemstruktur, die Zahl der beteiligten Rechnernetzknoten etc. von Installation zu Installation einer Netzwerkanwendung variieren können, muß der Programmtext der Funktionseinheiten auch solche Unterschiede zu berücksichtigen erlauben. DESIGN bietet hier als wesentlichstes Hilfsmittel ein **Listenkonzept**, das es erlaubt, Knotenlisten, Subsystemlisten, Portlisten etc. zu verarbeiten, deren Inhalt und Länge erst zur Laufzeit bekannt werden. Darüberhinaus erlaubt DESIGN auch Netzwerkanwendungen, deren Topologie sich zur Laufzeit ändert, d.h. das Hinzukommen oder Wegfallen von Knoten, Funktionseinheiten etc. Als Hilfsmittel für solche dynamischen Umkonfigurationen stehen Operatoren zur Listenmanipulation zur Verfügung, außerdem Möglichkeiten zum Auflösen und Neueinrichten von Kommunikationspfaden. Wenn administrative Funktionseinheiten solche dynamischen Umkonfigurationen vornehmen, werden die betroffenen Funktionseinheiten über **Exceptions** benachrichtigt.

Neben dem Präprozessor ist eine große Zahl weiterer Hilfsmittel notwendig, um die Entwicklung und Leistungsbewertung von Netzwerkanwendungen zu gestatten und um denselben Satz von Funktionseinheiten, evtl. in verschiedenen Detaillierungsstufen, in den verschiedensten Experimentarten beobachten zu können. Diese Hilfsmittel sollen nachfolgend kurz beschrieben werden.

5.0 Laufzeitumgebungen

5.1 Verteilte Simulation

Hybride und simulative Experimente sind die hauptsächlichen Hilfsmittel, um in frühen Entwicklungsstadien zukünftige Netzwerkanwendungen im Ablauf beobachten zu können. Die Methode der verteilten Simulation wurde gewählt, da sie im Bereich der Modellierung verteilter Systeme neben den erwähnten allgemeinen Vorteilen zusätzliche Verbesserungen erwarten läßt; so bietet sich die 'natürliche' Partitionierung einer Netzwerkanwendung sowie der unterliegenden Dienste für die Verteilung der Modelle besonders an und ist geeignet, die Modularität des Gesamtmodelles zu unterstützen. Das Kernsystem für die verteilte Simulation wurde in /MÜD84/ vorgestellt. Es wurde in seiner Struktur und hinsichtlich der verwendeten Algorithmen und der Funktionalität speziell an die vorgesehene Verwendung in DESIGN angepaßt und bietet in einigen Punkten gegenüber den aus der Literatur bekannten Systemen wesentliche Vorteile. So ist bei der verteilten Simulation mittels DESIGN zu erwarten, daß eine große Zahl von Prozessen jeweils auf demselben Knoten abläuft. Die aus der Literatur bekannten Algorithmen berücksichtigen nicht die dadurch mögliche zentrale Verwaltung mehrerer Prozesse, welche insbesondere eine schnellere Synchronisation aller Prozesse ermöglicht. Zu diesem Zweck mußten jedoch die Synchronisationsalgorithmen wesentlich erweitert und verändert werden. Des weiteren zeigt die Literatur, daß bei schwankender Ereignishäufigkeit ein 'Beschleunigungsalgorithmus' den Synchronisationsalgorithmen zu überlagern ist, um das

Simulationssystem effizient zu gestalten. Die bekannten Beschleunigungsalgorithmen erzeugen bei stark vernetzten Prozessen jedoch eine sehr große Zahl von Synchronisationsnachrichten, die über das verteilte System zu senden sind. Da gerade bei der Anwendung der verteilten Simulation in DESIGN stark vernetzte Prozesse die Regel sein werden, mußte hier ein neuer Beschleunigungsalgorithmus gefunden werden.

Die Komponenten der Laufzeitumgebung zur verteilten Simulation sind:
- das Kernsystem: dieses bietet die grundlegenden Funktionen zum Aufbau simulativer Modelle und zur Durchführung von Experimenten.
- das Modellierungssystem: aufbauend auf dem Kernsystem wurde ein spezielles Modellierungssystem konzipiert mit Modellen für Kommunikations- und Betriebsysteme verschiedener Art und Ausprägung, für Ressourcen mit exklusivem und konkurrierendem Zugriff etc. Neben die Funktionseinheiten treten hier 'Systemeinheiten' und 'Betriebsmitteleinheiten', vom Simulations-Kernsystem gemeinsam als 'entities' gesehen.
- die Schnittstelle: hier erfolgt der weitgehend transparente Übergang zwischen Funktionseinheiten und Modellierungssystem.

Neben simulativen können auch **analytische Modelle** verwendet werden; für die besonderen Probleme beim Übergang von simulativen in analytische Modellteile müssen besondere Hilfsmittel bereitgestellt werden, deren Konzept gegenwärtig erarbeitet wird.

Als neuer Ansatz wird versucht, mit Methoden der **Regressionsanalyse** automatisiert Modelle spezieller Betriebs- und Kommunikationssysteme zu erstellen; vor allem die am meisten interessierende Schnittstelle zwischen Netzwerkanwendung und Diensten des Informationsaustausches soll mit dieser Methode modelliert werden. Dazu müssen umfangreiche Daten über das 'Schnittstellenverhalten' zwischen Funktionseinheiten oder auch anderen Anwenderobjekten und den zu modellierenden Systemteilen aufgezeichnet und mit möglichst viel Information über den Systemzustand in Verbindung gebracht werden. Da keine Erfahrungen über die Anwendung der Regressionsanalyse auf dieses Gebiet vorliegen, müssen über die Verwendbarkeit solcher Verfahren Untersuchungen angestellt werden; entsprechende Testsysteme sind im Rahmen der DESIGN-Entwicklung im Entstehen.

5.2 1:1-Simulation

Neben die verteilte Simulation mit ihren analytischen und regressionsanalytischen Komponenten tritt als Alternative die 1:1-Simulation. Dabei werden die Betriebsmittel im äquivalenten Umfang in Anspruch genommen, wie dies von der Netzwerkanwendung entsprechend den Abschätzungen über Ressourcenbedarf erwartet wird. Um ein solches Verhalten zu erreichen, müssen die vom Modell nicht direkt konsumierten Betriebsmittel künstlich belegt werden und Modell-Nachrichten, die von den Diensten des Informationsaustausches übermittelt werden sollen und kleiner sind als die realen, verlängert werden.

Die Verwendung von 1:1-Simulationen hat zwei Gründe. Zum einen können Leistungskenndaten, die durch simulative oder analytische Modellierung

gewonnen wurden, unvollständig, kaum interpretierbar oder nicht aussa-
gekräftig genug sein. In diesen Fällen kann es sehr hilfreich sein,
das zukünftige Programm ablaufen zu sehen, mit realen Benutzern zu
konfrontieren und in seinen Wechselwirkungen mit der Umwelt zu beobachten, wenn auch die von der Netzwerkanwendung gelieferten Antworten
in einem frühen Entwicklungsstadium kaum sinnvoll sein werden.

Der zweite, wesentlichere Grund für die Einführung der 1:1-Simulation
ist in der Einbettung der Netzwerkanwendung in eine sehr komplexe Umwelt zu finden. Netzwerkanwendungen stehen in Beziehung zu einer
Vielzahl von Protokollen und Schichten des Kommunikationssystems, anderen lokalen und kommunizierenden Benutzerprogrammen und zudem im
konkurrierenden Zugriff auf Betriebsmittel. Auch sehr komplizierte
und umfangreiche simulative oder analytische Modelle der gesamten Umwelt von Netzwerkanwendungen beseitigen nicht die Gefahr einer Akkumu-
lation von Fehlern der einzelnen Teilmodelle und somit stark unzurei-
chender Experimentergebnisse. Hier bietet die 1:1-Simulation besondere Vorteile. Da sie die Umwelt der Netzwerkanwendung voll nutzt und
auch real in Wechselwirkung mit konkurrierenden Programmen steht, wird
der mögliche Modellierungsfehler drastisch reduziert, nämlich auf fehlerhafte Abschätzungen des Betriebsmittelbedarfs der Netzwerkanwendung
selbst. Auch falsche Annahmen über das Benutzerverhalten können durch
den wahlweisen Austausch des Lastgenerators gegen reale Benutzer vermieden werden. Unter Zuhilfenahme von Softwaremonitoren kann somit
die 1:1-Simulation zur partiellen Validation der Simulationsmodelle
herangezogen werden, und zwar bereits zu einem Zeitpunkt, zu dem ein
Vergleich mit der fertig entwickelten Netzwerkanwendung noch nicht
möglich ist.

5.3 Realer Dienst

In den letzten Entwicklungsphasen wird die Netzwerkanwendung als realer Dienst ablauffähig. Damit ist eine erste Version der Netzwerkanwendung erstellt; zur Weiterentwicklung und Optimierung der Netzwerkanwendung stellt DESIGN ebenfalls Werkzeuge bereit (s.u.). Das Zu-
schneiden auf die Laufzeitumgebung des realen Dienstes wird wie bei
den anderen Laufzeitumgebungen über Präprozessor und Schnittstellen
vorgenommen. Die Aufrufe an Interaktionspunkten werden dabei dem realen Betriebssystem und den realen Diensten des Informationsaustausches
zugeführt. Eingebaut in die Schnittstellen befinden sich transparente
Softwaremonitoren, die bei Bedarf aktiviert werden können.

6.0 Benutzerschnittstelle

Die Benutzerschnittstelle orientiert sich an den in Kapitel 2 ausge-
führten Anforderungen an fortgeschrittene Softwareproduktionssysteme
sowie an den gemachten Erfahrungen mit umfangreichen Modellierungssy-
stemen. Eine wesentliche Rolle spielen dabei **graphische Schnittstellen.** Die verschiedensten Bereiche innerhalb DESIGN können durch umfangreiche graphische Unterstützung entscheidend verbessert werden.
Das sind zum einen Eingabe-Schnittstellen, z.B. für die Topologie-De-
finitionen im Rahmen der Systemdefinition von Netzwerkanwendungen, für

die Definition von Experimentkonfigurationen, aber auch für die Ablaufsteuerung in Form von Kommandos, wenn eine begrenzte und nicht zu komplexe Menge von Alternativen zur Auswahl steht. Ausgabeseitig hat sich beim Umgang mit umfangreichen Modellierungssytemen vor allem die farbgraphische Darstellung dynamischer Abläufe als hilfreich herausgestellt /WÜR82/.

Mit am verbreitetsten als Werkzeug der Leistungsbewertung ist die Graphik für die Datenaufbereitung. Die Aussagekraft solcher Graphiken läßt sich durch umfangreiche statistische Auswertungen wesentlich steigern, diese müssen jedoch weitgehend automatisiert vorgenommen werden, um ausreichende Akzeptanz zu gewährleisten.

In den letzten Jahren gingen die Versuche, komplexe Aufgaben des Softwaremanagements zu unterstützen, in die Richtung hochauflösender Graphikschirme mit mehreren 'Fenstern' (windows). Dadurch können alphanumerische und graphische Ausgaben mehrerer Prozesse gleichzeitig auf einem Schirm dargestellt werden, aktuell interessierenden Ausgaben kann dynamisch mehr Platz zur Verfügung geteilt werden, Zooming erlaubt detaillierten Einblick in Teilbereiche /SHE83/. Das Konzept von DESIGN berücksichtigt den Einsatz derartiger Hardware.

Eine möglichst **einheitliche Sicht** des Entwicklers auf alle verfügbaren Werkzeuge sollte durch die Benutzerschnittstelle ermöglicht werden. Dazu gehören neben einer einheitlichen Kommandosyntax auch vereinheitlichte Fehlermeldungen und -behandlungsmethoden sowie einheitliche Strukturen aller für den Benutzer sichbaren Schnittstellen, insbesondere der Hilfstexte. Durch eine **flexible Benutzerführung** müssen sich Menues, Kommandos, erklärende Texte etc. gut an den Kenntnisstand des Entwicklers anpassen lassen. Durchgriffsmöglichkeiten in Menue-Hierarchien, Kurzbezeichner, Einschränkung oder Erweiterung der Systemhilfen auf Tastendruck, benutzerspezifische Aufzeichnungen über Kenntnisstand, Fehlbenutzungsrate etc. müssen zur Unterstützung herangezogen werden.

Durch eine **problemorientierte Schnittstelle** können Aufgaben an die Entwicklungsumgebung in einer dem Benutzer geläufigen Form gestellt werden; in der Schnittstelle findet dann eine Umsetzung auf werkzeugorientierte Aktionsfolgen statt. Durch eine umfangreiche Gestaltung dieser Schnittstelle kann DESIGN in Richtung Expertensystem /HAW83/ weiterentwickelt werden.

7.0 Werkzeuge

Gemäß 2.2 sind zunächst die **Basis-Entwicklungswerkzeuge** zu schaffen. Dadurch werden die Voraussetzungen zur verteilten Programmierung und verteilten Administration von Funktionseinheiten geschaffen und die Grundlagen für die Implementierung der Laufzeitumgebungen für verschiedene Experimentarten gelegt. Teilaufgaben sind die Vereinheitlichung und zentrale Kontrolle der Administration, Übersetzung, Korrektur bis hin zu Binden und Installation sowie die Entwicklung verteilter Testhilfen und einer spezifischen Laufzeitfehlerbehandlung und eine einheitliche Schnittstelle für die Ein-/Ausgabe auf dem Endgerät des Entwicklers.

Auf diesen Basis-Entwicklungswerkzeugen baut das Konzept der Laufzeit-
umgebungen auf, zu deren benutzerfreundlichen Handhabung weitere Sy-
stemkomponenten erforderlich sind: mit der Vereinheitlichung der
Schnittstellen zum Betriebs- bzw. Kommunikationssystem im Konzept der
Funktionseinheiten wird es möglich, über **Präprozessor** und 'graphische
Topologiedefinition' zielmaschinen- und laufzeitumgebungs-spezifische
Versionen der Funktionseinheiten zu erstellen, anhand der 'graphischen
Experimentdefinition' die **Konfiguration** vorzunehmen, **Lastgeneratoren**
zu parametrisieren und Experimente durchzuführen.

Zur schrittweisen Verfeinerung wie auch zur Leistungsoptimierung der
Netzwerkanwendung muß deren Verhalten in den verschiedenen Laufzeitumgebungen beobachtet und analysiert werden. Zur Beobachtung dienen
Softwaremonitoren, die spezifisch für jede Laufzeitumgebung konzipiert
sind, ihre Meßergebnisse jedoch in weitgehend einheitlichen Datenformaten ablegen. Die Ergebnisse dieser Monitoren werden von der **stati-
stischen Aufbereitung** analysiert und einerseits über die graphische
Ausgabeschnittstelle dem Benutzer zur Interpretation zugeführt; andererseits werden die Ergebnisse aus der 1:1-Simulation und aus dem realen Dienst zur automatisierten **Validation** benutzt, um die in früheren
Entwicklungsstufen angegebenen Abschätzungen über Betriebsmittelbedarf
durch die gemessenen zu ersetzen und somit kalibrierte Modelle für
weitere Simulationsexperimente zur Verfügung zu haben. Durch verglei-
chende Experimente zwischen Simulation und realem Verhalten kann mit
diesem Werkzeug auch die Güte der Simulationsmodelle bestimmt werden.
Von einem Werkzeug zur **Optimierung** können Simulationen dann verwendet
werden, um nachträgliche Verbesserungen und Weiterentwicklungen der
Netzwerkanwendung zu testen, oder um Leistungsaussagen über die Verwendung derselben Netzwerkanwendung in anderen Umgebungen zu gewinnen
(Auswirkungen der Verwendung leistungsfähigerer Rechner etc.). Im
Rahmen der projektierten Entwicklung einer 'problemorientierten
Schnittstelle' von DESIGN ist auch vorgesehen, weitgehend automatisiert ganze Experimentserien zu erlauben, wobei das System auch
selbsttätig Parameteroptimierungen mit iterativer Anpassung von Ent-
wurfsvariablen durchführt.

Begleitend zum gesamten Entwicklungszyklus muß eine **übergeordnete Projektführung** den Entwicklungsstand der Netzwerkanwendung und ihrer Komponenten überwachen und protokollieren, vor Experimenten die Konsistenz des Entwicklungsstandes der beteiligten Module überprüfen, die
Koordination mehrerer Entwickler ermöglichen, etc.

Abb. 4 zeigt abschließend die wesentlichen Komponenten von DESIGN im
Zusammenhang. Die 'Datenbasen' enthalten im wesentlichen die Quell-
programme der Netzwerkanwendung, u.U. in verschiedenen Versionen, ab-
lauffähige Software von Funktionseinheiten, Daten zum Projektfortschritt, Experimentdefinitionen und Meßdaten über das Leistungsverhalten.

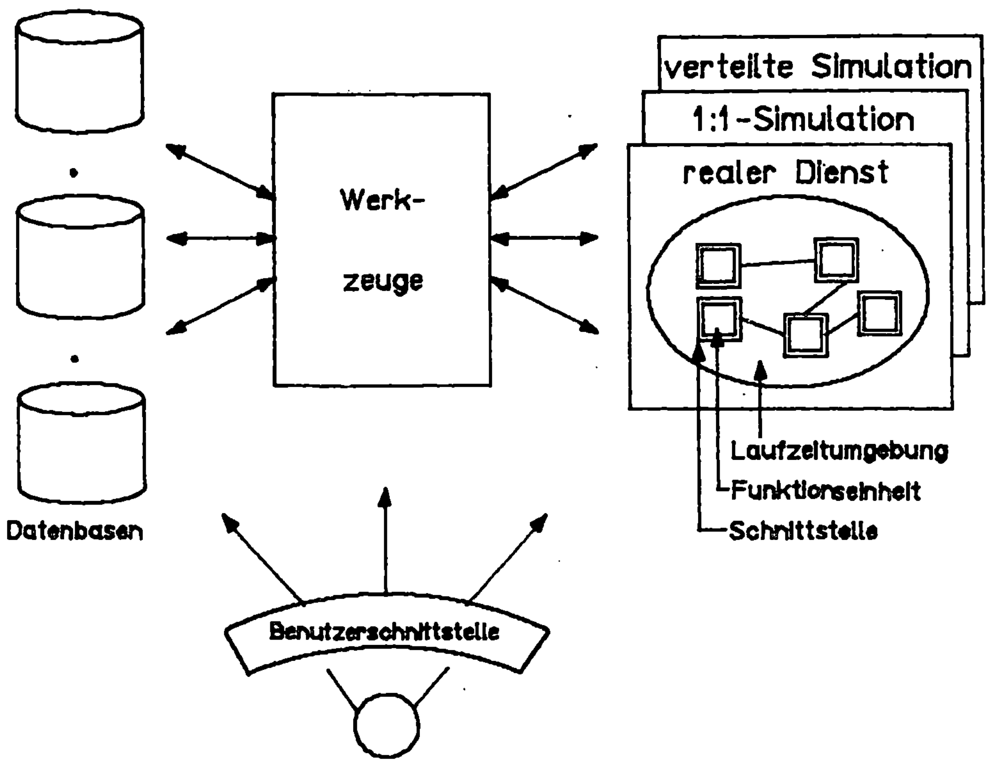

Abb. 4: Komponenten von DESIGN

8.0 Zusammenfassung

Die Problematik der Entwicklung und Leistungsbewertung von Netzwerkanwendungen wurde skizziert. Motiviert durch die zunehmende Bedeutung und Komplexität solcher Netzwerkanwendungen entstand das Konzept von DESIGN, einer verteilten Umgebung zur integrierten Entwicklung und Leistungsbewertung von Netzwerkanwendungen. Im Dezember 1984 sind die Spracherweiterung, das Kernsystem zur verteilten Simulation und die Basis-Entwicklungswerkzeuge zur verteilten Programmierung in der Implementierung befindlich, für Regressionsanalyse, statistisch-/graphische Aufbereitung und dynamische farbgraphische Darstellung liegen Pilotimplementierungen vor, die restlichen Komponenten sind noch in der Entwurfsphase.

9.0 <u>Literatur</u>

/BCM75/   Baskett, F., Chandy, K.M., Muntz, R.R., Palacios, F.G.
          Open, Closed and Mixed Networks of Queues with Different
          Classes of Customers
          J. ACM 22, 1975, 248

/BEI80/   Beilner, H.
          On The Construction of Computing System Simulators
          Second Summer School on Computer Systems Performance
          Evaluation, Urbino, June 1980

/BRY79/   Bryant, R.E.
          Simulation on a Distributed System
          Proc. IEEE, CH1445-6/79, 1979

/BUX81/   Bux, W.
          Local Area Subnetworks: A Performance Comparison
          IEEE Trans. Comm. Vol. 29, No. 10, Oktober 1981

/CHM81/   Chandy, K.M., Misra, J.
          Asynchronous Distributed Simulation via a Sequence of
          Parallel Computations
          Comm. ACM, Vol. 24, No. 11, 1981, 198 - 206

/COM82/   Berg, H.K. (guest editor)
          Distributed System Testbeds
          Special Issue of Computer, Oktober 1982

/COU77/   Courtois, P.J.
          Decomposability
          Academic Press, New York etc. 1977

/FRC82/   Franta, W.R., Chlamtac, I.
          A Generalized Simulator for Computer Networks
          Simulation, Oktober 1982, 123 - 132

/HAW83/   Hayes-Roth, F., Waterman, D.A.
          Building Expert Systems
          Addison-Wesley, London etc. 1983

/IEE83/   Open Systems Interconnection (OSI) - New International
          Standards Architecture And Protocols For Distributed
          Information Systems (Special Issue)
          Proc. IEEE, Vol. 71, Nr. 12, Dezember 1983

/JEN82/   Jensen, W.
          Local Area Networks: Technical Issues - User Services -
          Social Impact
          in: Ravasio et al. (ed.): Local Computer Networks
          Proc. IFIP TC 6 Int. Symp., North Holland, 1982

/KMA82/   Kfoury, A.J., Moll, R.N., Arbib, M.A.
          A Programming Approach to Computability
          Springer-Verlag, Berlin etc. 1982

/KRD84/   Krumm, H., Drobnik, O.
          Transformation of Constructive Specifications of Services
          and Protocols into the Logical Languate of CIL
          IFIP WG6.1 Workshop 'Protocol Specification, Testing,
          and Verification', Skytop 1984

/MÜD84/   Mühlhäuser, M., Drobnik, O.
          The Use of Distributed Simulation for the Development of
          Distributed Software
          Interner Bericht Nr. 1/85,
          Fakultät für Informatik, Universität Karlsruhe

/PEA80/   Peacock, J.K.
          Distributed Simulation Using a Network of Processors
          Dept. of CS and CCNG,
          University of Waterloo, Ont., Canada
          CCNG T-Report T-87, January 1980

/PWM79/   Peacock, J.K., Wong, J.W., Manning, E.G.
          Distributed Simulation Using a Network of Processors
          Computer Networks 3, 1979, 44 - 56

/REI81/   M. Reiser
          Admission Delays on Virtual Routes with Window Flow
          Control
          Proc. of the International Conference on Performance of
          Data Communication Systems and their Applications,
          Paris, 14.-16.9.1981, pp. 67-76

/RIF80/   Riddle, W.E., Fairley, R.E.
          Software Development Tools
          Springer-Verlag, Berlin etc. 1980

/SAM77/   Sauer, C.H., McNair, E.A.
          Application of RESQ: Communication System Evaluation
          IBM Research Report   RC7143, Yorktown Heights, N.Y. 1977

/SGO83/   Sventek, J., Greiman, W., O'Dell, M., Jansen, A.
          Token Ring Local Area Networks - A Comparison of
          Experimental and Theoretical Performance
          DRAFT CS and Math Dept. , Lawrence Berkeley Lab., 1983

/SHE83/   Sheil, B.
          Power Tools for Programmers
          Datamation, Feb. 1983, pp. 131 - 144

/SHO83/   Shooman, M.L.
          Software Engineering - Design, Reliability and
          Management
          McGraw-Hill, New York etc. 1983

/WOM83/   Wolfinger, B., Mühläuser, M.
          Construcion of a Validated Simulator for Performance
          Prediction of DECnet-based Computer Networks
          Performance Evaluation Review, August 1983, 138 - 150

/WÜR82/   Würth, G.
          Entwicklung und Implementierung eines interaktiven
          graphischen Systems zur Darstellung dynamischer Abläufe
          in Rechnernetzen
          Diplomarbeit, Fakultät für Informatik,
          Universität Karlsruhe 1982

/ZAH83/   Zahorjan, J.
          The Influence of Workload Representation on Response Times
          in Queueing Models of Computer Systems
          Performance Evaluation Review, August 1983, 70 - 81

EINFLUSS DER VLSI AUF KOMMUNIKATIONSSYSTEME

H. Stegmeier
Zentrallaboratorium des Unternehmensbereichs
Nachrichten- und Sicherungstechnik der Siemens AG

München, BRD

1. <u>Einführung</u>:

Die neue Kommunikationstechnik bringt, wie durch die Konzeption der
Deutschen Bundespost von 1984 festgelegt:

1985 die Digitalisierung des Fernsprechnetzes, d.h. die Digitali-
     sierung der analogen Fernsprechsignale am Teilnehmer-Eingang
     der Vermittlungsstelle und die Rückwandlung in analoge Fern-
     sprechsignale am Ausgang der Vermittlungsstelle. Damit erfolgt
     die Vermittlung und die Übertragung von Vermittlungs- zu Ver-
     mittlungsstelle digital /¯1_7 (Schritt 1).

1988 den Übergang auf das digitale dienstintegrierte Fernmeldenetz
     (ISDN). Die Digitalisierung erfolgt bereits beim Teilnehmer,
     d. h. die Übertragung vom und zum Teilnehmer erfolgt digital.
     Damit ist es möglich geworden, beim Teilnehmer eine Universal-
     steckdose zum freizügigen Geräteanschluß mit zwei Nutzkanälen
     mit je einer Bitrate von 64 Kbit/s und einem zusätzlichen Steu-
     erkanal mit 16 Kbit/s, einzurichten (Schritt 2).

1990 Integration von Schmal- und Breitbanddiensten der Individual-
     kommunikation im Breitband-ISDN.
     Hier werden zusätzlich Bewegtbildkommunikation und die Übertra-
     gung großer Datenmengen mit Datenraten von 140 Mbit/s angeboten
     (Schritt 3).

1992 Übergang zum Breitbanduniversalnetz (IBFN). Es bietet zusätzlich
     die Verteilkommunikation, HDTV und Mehrprogramme (Schritt 4).

Für die Konzeption und die Machbarkeit dieser Systeme und der dazugehörigen Teilnehmer-Endgeräte ist die Hochintegration (VLSI) von
entscheidender Bedeutung. Dabei nimmt sie massiven Einfluß auf die
Wirtschaftlichkeit und Machbarkeit, auf die System-Architektur und

auf mögliche neue Leistungsmerkmale. Dies wird im Folgenden näher
ausgeführt.

Zum Schluß werden daraus wiederum Konsequenzen für die Technologie
der VLSI gezogen.

2. Einfluß auf die Wirtschaftlichkeit und Machbarkeit von Kommunikationssystemen

Bild 1 zeigt wie die Technik es möglich macht in der Kommunikationstechnik mit zunehmender Zeit immer mehr unterschiedliche Kommunikationsdienste den Benutzern anzubieten. Dies hatte bis heute zur Folge, daß für jeden dieser Dienste eigene Geräte und Systeme entwickelt
und gefertigt werden. Dies hat ein explosionsartiges Ansteigen der
Entwicklungskosten zur Folge.

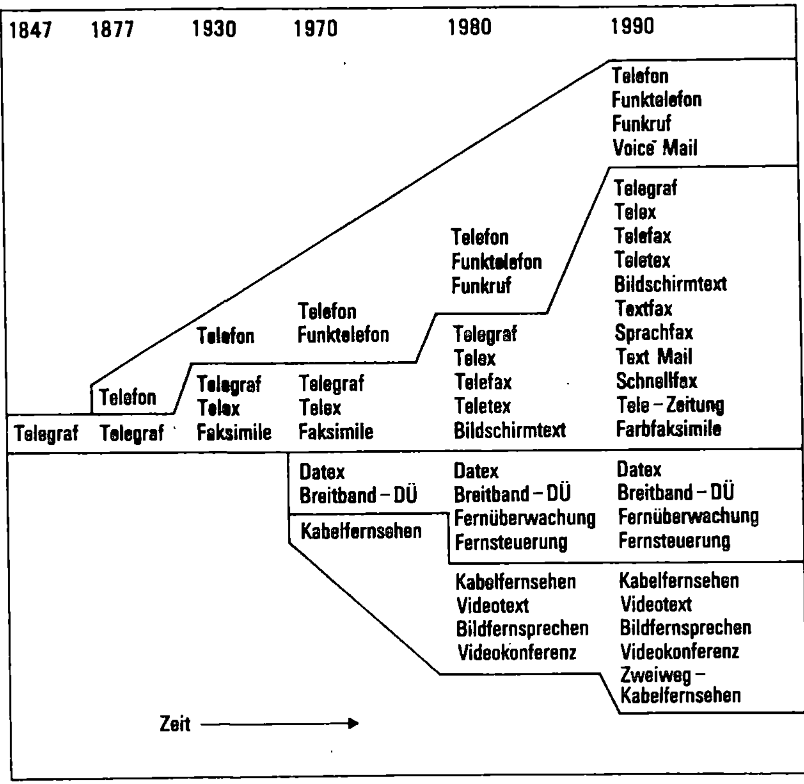

Bild 1: Die Entwicklung der Kommunikationsdienste

Durch die Konzeption der Deutschen Bundespost wird die Systemviel-
falt, ermöglicht durch die Fortschritte, insbesondere der Mikroelektronik, aber auch der Software-Technologie, in den genannten
Schritten auf ein einziges System zurückgeführt.

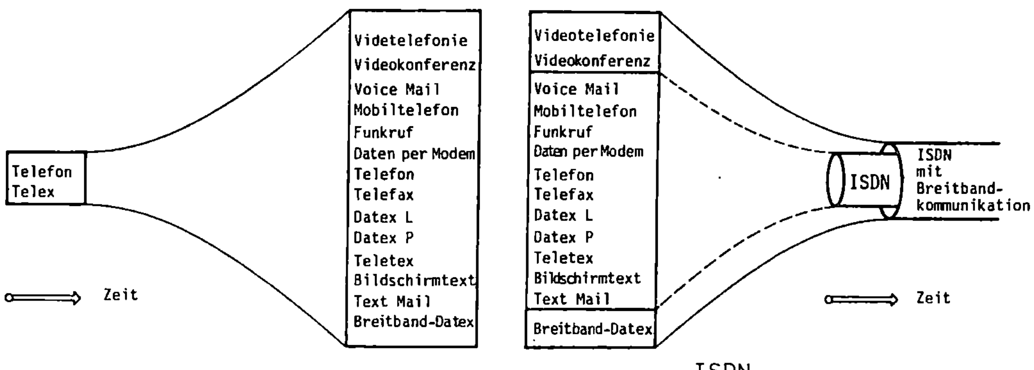

Bild 2: Telekommunikations-Dienste

Schritt 1:
Er wird ermöglicht durch die Digitalisierung der Sprache in der Vermittlungsstelle. Die Schlüsselfunktionen dafür sind das Codieren und
Filtern der analogen Sprachsignale mit einer Bandbreite von 4 KHz in
ein Digitalsignal mit 64 Kbit/s.

Der Codec stellt einen hochpräzisen Analog/Digitalwandler mit einer
Komplexität von ca. 4000 Transistorfunktionen, das Filter ein hoch-
wertiges Filter mit einer Komplexität von ca. 2000 Transistorfunktionen dar. Da beide Funktionen pro Teilnehmeranschluß zweimal zu
realisieren sind, ist der Aufwand (Kosten, Volumen, Verlustleistung,
Zuverlässigkeit) für die Wirtschaftlichkeit und die Machbarkeit entscheidend. Die Schlüsseltechnologie dafür war die Hochintegration in
NMOS-Technologie.

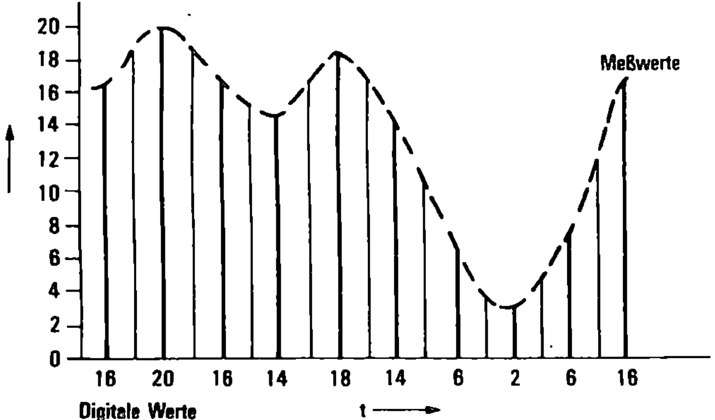

Wertetabelle

digital	binär	digital	binär
16	10000	14	01110
18	10010	10	01010
20	10100	6	00110
18	10010	3	00010
16	10000	2	00001
15	01111	4	00011
14	01110	6	00110
16	10000	10	01010
18	10010	16	10000
16	10000		

Zeitverlauf der binären elektrischen Signale
z.B. für die Folge der digitalen Werte 16, 14, 10, 6

Elektrische Spannung

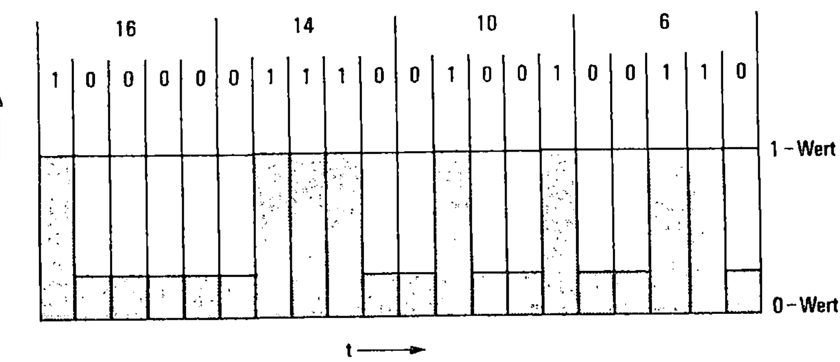

Bild 3: Kodierung eines Fernsprechsignals

Schritt 2:
Hier erfolgt die Digitalisierung beim Teilnehmer. Dies hat 3 Konsequenzen:

1. Die Funktionen Codierung und Filtern wandern in das Endgerät. Damit wird die Verlustleistung, insbesondere wenn man die Fernspeisung der Telefonfunktion vom Amt realisieren will, entscheidend.
2. Die Universalsteckdose erfordert hohe technische Vorleistungen an Komplexität (ca. 80000 Transistorfunktionen).
3. Die Übertragung von 144 Kbit/s zwischen Teilnehmer und Vermittlungsstelle in beiden Richtungen auf einem·Kupferardernpaar erfordert neben hoher Komplexität (ca. 60000 Transistorfunktionen), eine hohe Präzision (AD-Wadler mit einer Auflösung von 12 bit und einer Nichtlinearität, die kleiner als ein niederwertigstes Bit ist) und eine hohe Verarbeitungsleistung (Echokompensation mit einer Abtastrate von 15 MHz). Die Schlüsseltechnologie dafür ist CMOS mit Strukturgrößen von 3 µ und kleiner.

Schritt 3:
Zur Übertragung von Bewegtbildern ist, wenn man keine Redundanzreduktionen vornimmt, eine Datenrate von ca. 140 Mbit/s erforderlich, d. h. alle vorgenannten Funktionen müssen mit dieser Datenrate realisiert werden. Lediglich die Übertragung zwischen Teilnehmern und Vermittlungsstelle wird durch die Glasfasertechnik technisch erleichtert. Diese stellt daher eine weitere Schlüsseltechnologie dar. Die Breitbandkommunikation beinhaltet deshalb 3 große technologische Herausforderungen.

1. Die Realisierung der erforderlichen Datenrate von 140 Mbit/s, d.h. dieser Verarbeitungsleistung. Dies erfordert eine Technologie welche ein Master-Slave Flip/Flop mit einer Toggle Rate von ca. 700 MHz ermöglicht (Machbarkeit).
2. Die Kosten dafür müssen so niedrig sein, daß für die Breitbandkommunikation die doppelte bis dreifache Telefongebühr möglich wird. Dies bedeutet, daß diese hohen Datenraten in hochintegrierten Bausteinen verarbeitet werden müssen. Neben der Anschlußtechnik und den Endgeräten, stellt hier auch das Koppelfeld der Vermittlung eine Herausforderung dar. (Wirtschaftlichkeit)
3. Damit obige Kosten erreichbar sind, muß insbesondere die Übertragungstechnik für das Fernnetz, breitbandige Kanäle entsprechend billig zur Verfügung stellen. Dies erfordert Übertragungssysteme

mit 140 Mbit/s, 565 Mbit/s und höher. (Machbarkeit und Wirtschaftlichkeit.)

Zur Überwindung dieser 3 Herausforderungen sind erforderlich:

1. CMOS-Technologie mit Strukturgrößen von 1µ und kleiner (Punkt 1
   und 2).
2. Silizium bipolar oder GaAs-Technologie (Punkt 3).
3. Lichtwellenleitertechnologie (Punkt 2 und 3).

Der heutige technologische Stand und die Prognosen führender Techno-
logiehäuser zeigen, daß in dem zur Verfügung stehenden Zeitraum diese Herausforderungen gelöst werden.

3. Einfluß auf die System-Architektur

Die Herstellung von leistungsfähigen und preisgünstigen Mikroprozessoren und dazugehörigen Interface- und Speicher-Bausteinen hat es
ermöglicht, Verarbeitungsleistung aus dem zentralen Rechner eines
Kommunikationssystems in pheriphere Steuerungen, die Anschlußtechnik
und die Endgeräte zu dezentralisieren.
Ein weiterer Schritt, ermöglicht durch die Hochintegration, sind
Local Area Networks (LAN), Bild 4.

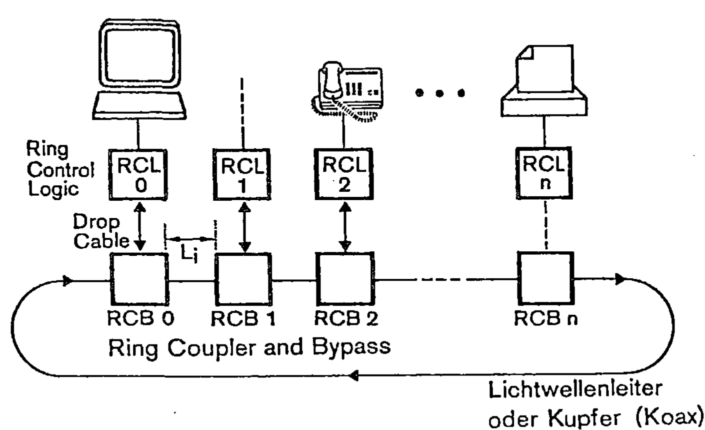

Bild 4: Lokales Optisches Netz (LAN)

Hier ist die Steuerungsfunktion des Systems vollständig in die "Ring-
Control Logic" des Endgeräte-Anschlusses dezentralisiert. Jedes Endgerät kann mit jedem Endgerät ohne Zuhilfenahme einer zentralen, d.h.
für alle Endgeräte gemeinsamen, Steuerung verkehren. Die Steuerungsfunktion wird durch Prozeduren abgewickelt. Die bekanntesten sind
CSMA/CD von Ethernet und Token Passing. Erforderlich ist dafür eine
Ring Control Logic je Endgerät. Sie besteht im wesentlichen aus einem Microprozessor, Speicher, Ein-/Ausgabe zum Gerät und zusätzlicher
Logik mit einer Komplexität von ca. 15000 Gatterfunktionen.

Weiterhin ist eine Ankopplung (RCB) an das Bussystem erforderlich.
Deren Komplexität ist zwar gering (ca. 1000 Gatterfunktionen), jedoch
sind die Geschwindigkeitsanforderungen bei höheren Datenraten auf dem
Bussystem (10 - 100 Mbit/s) recht hoch. Wirtschaftlich wird ein solches System deshalb nur durch den Einsatz von Hochintegration und
schnellen Technologien.

Durch den Einsatz von Hochintegration, derzeit insbesondere in Form
von Microcomputer und Speicher, ist es möglich geworden die Endgeräte mit zusätzlichen Leistungsmerkmalen auszustatten. Beispiele
dafür sind:

- beim Telefon:
    Kurzwahl
    automatischer Rückruf
    Wahlwiederholung
    Anrufumleitung
    Konferenzschaltung
    Verwendung von Displays

- bei Bildschirmgeräten:
    Bedienerführung vom Bildschirm aus.

4. Konsequenzen für die Technologie und die Behandlung der Bausteine

Mit steigendem Integrationsrad verlieren Bausteine mehr und mehr den
Charakter und die Bedeutung von Bauteilen und nehmen mehr und mehr
den Charakter und die Bedeutung von Geräten und Systemteilen an. Wenn
ein Gerät oder ein Systemteil eine Komplexität von beispielsweise
10000 Gatterfunktionen hat und man realisiert diese in einem einzigen

Baustein, dann ist dieser ein Gerät oder ein Systemteil und muß als
solches behandelt werden bezüglich

- know-how,
- technischem Funktionsinhalt,
- Verfügbarkeit,
- Kosten,
- Wertschöpfung.

Dies führt folgerichtig weg von der Anwendung von nur Standardbau-
steinen und hin zur Anwendung von kundenspezifischen Lösungen in Form
von halb-kundenspezifischen Bausteinen $\lceil 2 \rfloor$. Dieser Trend ist derzeit weltweit feststellbar.
Die wichtigste Technologie, für die in den vorhergehenden Kapiteln
genannten Anwendungen, ist derzeit die CMOS-Technologie. Deshalb
wird im Folgenden nur darauf eingegangen.

In neuerer Zeit können auch in dieser Technologie kundenspezifische
Funktionen durch halb-kundenspezifische (Semi-Custom-) Bausteine
realisiert werden. In der Datenverarbeitung ist dies in Form der bi-
polaren "Gate Arrays" schon seit längerem üblich.

Entscheidend für das Eindringen dieser halbkundenspezifischen Bausteine in die Telekommunikation sind:

- die CMOS-Technologie, die sich dafür besonders gut eignet;
- die Verfügbarkeit rechnerunterstützter Verfahren zum Entwerfen,
    Prüfen und Fertigen der Bausteine (CAD, CAM);
- die Verfügbarkeit vorgefertigter und/oder vorentwickelter Struk-
    turen (Gate Arrays und Zellenbibliotheken).

Durch diese vorgefertigten bzw. vorentwickelten Strukturen wird ihr
Entwicklungsaufwand auf viele Anwendungen verteilt. Daher rührt der
Name halb-kundenspezifische (Semi-Custom-)Bausteine. Dies reduziert
den speziellen Aufwand für den kundenspezifischen Baustein und auch
Entwicklungszeit und Risiko. Durch das CAD-System (Computer Aided
Design) zum Design der Bausteine selbst, werden Aufwand, Zeit und
Risiko weiter reduziert.

Im folgenden sind einige Leistungsdaten für halbkundenspezifische
Bausteine, realisiert mit Zellenbibliotheken, im Vergleich mit ande-

ren Technologien angeführt. Da die Zahlenwerte nicht nur von der
Komplexität der Bausteine, ausgedrückt in Gatterfunktionen, sondern
z. B. auch von der Schwierigkeit der Schaltungsstruktur oder der sequentiellen Tiefe der Schaltung abhängen, können diese nur grobe
Richtwerte sein.

In Tabelle 1 sind die Entwicklungszeiten, ab Spezifikation bis fer-
tiggestelltem und geprüftem Baustein, als Funktion der Komplexität
angegeben. Die Entwicklungszeit ist dabei unterteilt in Design, Layout, Fertigung und Prüfung. Weiterhin sind die relativen Entwicklungskosten angegeben; sie enthalten Personal- und·Rechnerkosten. Wie daraus hervorgeht, steigen sowohl die Entwicklungszeit als auch die
Entwicklungskosten weniger als proportional mit der Komplexität an.

	Entwicklungszeiten (in Wochen) für			
	1000 Gatter-funktionen	2000 Gatter-funktionen	3000 Gatter-funktionen	6000 Gatter-funktionen
Logikdesign	12	15	18	22
Layout und Fertigung	13	14	16	19
Prüfung	2	3	4	6
Insgesamt	27	32	38	47
	Relative Entwicklungskosten			
	1	1,36	1,7	3

Tabelle 1: Entwicklungszeiten und relative Entwicklungskosten für
Bausteine, basierend auf Zellenbibliotheken in CMOS-
Technologie

Ein Vergleich mit voll-kundenspezifischen Bausteinen zeigt, daß die
Entwicklungsproduktivität (ausgedrückt in entwickelten Gatterfunktionen je Mann und Jahr) bei mittlerer Komplexität (etwa 2000 Gatterfunktionen) bei Bausteinen auf der Basis von Zellenbibliotheken etwa um den Faktor 10 höher und die Entwicklungszeit um den Faktor 3
kürzer ist.

Nachteile der Bausteine auf der Basis von Zellenbiblotheken sind,
daß die Fläche um etwa 50 bis 100 % größer ist, was höhere Herstell-
kosten bedeutet. Weiterhin kann bei halb-kundenspezifischen Bausteinen die Technologie bezüglich Verarbeitungsgeschwindigkeit (Taktfre-
quenz, Gatterlaufzeit) nicht voll ausgenutzt werden.

Ein weiterer Vergleich hat ergeben, daß zwischen der Realisierung
einer Funktion mit TTL-Bausteinen und einer Realisierung auf der Basis von Zellenbibliotheken bezüglich Entwicklungskosten und Zeit
kein wesentlicher Unterschied besteht, Bausteine auf der Basis von
Zellenbibliotheken aber große Vorteile bezüglich Kosten, Leiterplat-
tenfläche (d. h. Volumen), Verlustleistung und Zuverlässigkeit haben.

<u>Schrifttum:</u>

/ 1_7   Bocker, P.: Konzept und Grundmerkmale des diensteintegrieren-
        den digitalen Nachrichtennetzes ISDN.
        telcom report 6 (1983) S. 64 bis 169

/ 2_7   Stegmeier, H.: Mikroelektronik - Chancen für die Telekommuni-
        kation.
        telcom report 7 (1984) Heft 2, S. 55 bis 61

Perspektiven des Einsatzes und der Entwicklung
Integrierter Bürosysteme

Prof. Dr.-Ing. habil. H.-J. Bullinger, Dipl.-Math. K.-P. Fähnrich

Fraunhofer-Institut für Arbeitswirtschaft und Organisation (IAO),
Stuttgart

Zusammenfassung:

Integrierte Bürosysteme bieten das Nutzungspotential, einen Anwendungs- und Rationalisierungsschub im Bürobereich unserer Unternehmen auszulösen. Über reine kostenorientierte Rationalisierungsstrategien hinaus sollen diese Systeme unsere Unternehmen leistungsfähiger und innovativer machen. Entsprechend sollte ihr Einsatz verstärkt auch für Fachspezialisten und mittleres Management geplant werden. Bei der Weiterentwicklung der entsprechenden Techniksysteme kann konstatiert werden, daß der Bereich der Bearbeitung von Büroinformation sowie der Bereich der Kommunikation soweit fortgeschritten ist, daß der verstärkte Praxiseinsatz möglich ist. Ein Engpaß ist bei der elektronischen Ablage zu sehen. Integrierte Bürosysteme benötigen besonders sorgfältig gestaltete Benutzerschnittstellen. Hier wurden in den letzten Jahren etliche Fortschritte erzielt. Der Trend geht zu Benutzerschnittstellen, die eine eigene Schicht in den entsprechenden Softwarearchitekturen bilden. Wesentlich beeinflußt werden können zukünftige Bürosysteme von Softwarearchitekturen aus dem Bereich der künstlichen Intelligenzforschung. Der breite Einsatz entsprechender Systeme wird jedoch erst für die 90er Jahre prognostiziert.

Abstract:

Integrated office systems offer the potential for an application and rationalization shift in the administrative field. Apart from consideration of cost efficiency, these systems should help to increase general efficiency and innovativeness. Accordingly, their introduction should also be reinforced for professional workers and first line management. Considering the further development of the relevant technical systems, it can be observed that the fields of office information manipulation and of office communication have developed to a degree which allows reinforced practical application. A bottleneck is at present electronic filing. Integrated office systems are in need of very carefully designed interfaces. Recently great progress was made in this field. The trend here is towards interfaces, which are defined as a separate layer in the relevant software architecture. It is to be expected that the future office systems will be influenced by the software architecture from the field of Artificial Intelligence. A broader application of such systems can be predicted no sooner than the nineties.

1: Einleitung

Richtig verstanden, müßte es mit einem Integrierten Bürosystem möglich sein, alle im Bürobereich anfallenden Aufgaben abzubilden und zu bearbeiten. Es muß sich dabei nicht notwendigerweise um ein automatisiertes System handeln - etwa das papierlose Büro, von dem wir alle wissen, daß es für die meisten Betriebe noch lange Zeit Utopie bleiben wird. Integrierte Bürosysteme sehen deshalb Vernetzungen von Technikbausteinen einerseits, aber auch von Arbeitsinhalten der Mitarbeiter andererseits vor. Beides muß zusammen geplant und durchdacht werden, wenn man die mit den Integrierten Bürosystemen erhofften Vorteile erreichen will: Wirtschaftlichere und menschengerechtere Büros. Da es das Büro nicht gibt, wird es auch die Lösung nicht zu kaufen geben. Aus den vorhandenen Bausteinen das beste System zu bauen, setzt einen Plan voraus. Von Informationsmanagement ist deshalb in immer mehr Unternehmen die Rede. Zunächst bedeutet dies nicht mehr und nicht weniger als eine Strategie für die Rationalisierung im Bürobereich zu entwickeln, die durchaus nicht nur auf Kostensenkung allein, sondern auch auf mögliche Nutzensteigerungen hin gerichtet sein sollte.

2: Perspektiven für die Durchdringung der Unternehmen mit Integrierten Bürosystemen

Integrierte Bürosysteme bieten das Anwendungspotential, um den nächsten Rationalisierungsschub im Bürobereich unserer Unternehmen auszulösen. Sie sollen jedoch nicht nur vornehmlich dafür genutzt werden, "Stagnation des Unternehmens zu organisieren", sondern sollten eine Initialzündung für unsere Unternehmen zu weiterem qualitativem und quantitativem Wachstum darstellen.

2.1: Integrierte Bürosysteme - eine Initialzündung

Die momentane Situation ist für viele Unternehmen geprägt von Begriffen wie Kostenreduktion, Erhöhung der Rentabilität durch Mitarbeitereinsparungen und reaktives Marktverhalten.

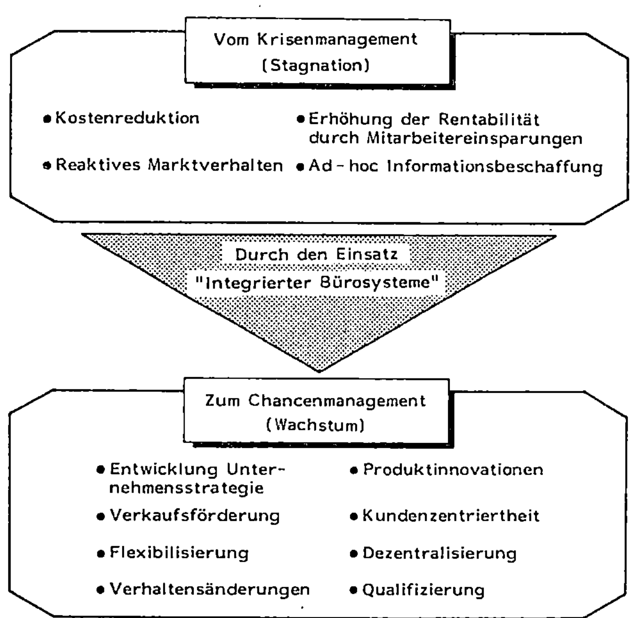

Bild 1: Integrierte Bürosysteme - eine Initialzündung für
unsere Unternehmen

Integrierte Bürosysteme sollen unternehmensinterne und -externe Kommunikation mit dem Ziel der Entwicklung marktangepaßter Unternehmensstrategien, der Produktinnovation, der Verkaufsförderung und der Kundenzentriertheit fördern. Sie sollen dazu beitragen, unsere Unternehmen im Bürobereich zu flexibilisieren und, wo sinnvoll und notwendig, weiter zu dezentralisieren /BULLINGER 84/.

In Kalifornien erleben wir momentan einen erstaunlichen Boom. Hier bilden sich neue Wirtschaftsstrukturen heraus, die - zumindestens in diesem lokalen Bereich - viele Prognosen vom Ende des Wachstums als überdenkenswert erscheinen lassen. Getragen wird dieser Boom von der Herstellung, der Vermarktung und dem Einsatz neuer Informations- und Kommunikationstechnologien vornehmlioch von kleinen und mittelständischen Unternehmen /FERGUSSON 84, NAISBITT 82/

2.2: Rationalisierungsstrategien

Rationalisierungsstrategien .sollten nicht nur pauschal von den momentanen Personal- bzw. Personalnebenkosten ausgehen. Sie sollten sich vielmehr auch an Größen wie der Kostenverantwortlichkeit und dem langfristigen Kostentrend orientieren.

Tätigkeiten	Personal-kosten	Kosten-verantwortlichkeit	Kostentrend
Führungs-tätigkeiten	O O	O O O	⬆
Fachtätigkeiten	O O O	O O O	⬆
Sachbearbeitung	O O O	O O	⬆
Unterstützungs-tätigkeiten	O	O	⬇

Bild 2: Kostentrends im Bürobereich (vgl. dazu auch HIRSCH 84)

Unterstützungstätigkeiten, die bisher einen Schwerpunkt der Rationalisierungsansätze bildeten, weisen bezüglich dieser drei Kenngrößen eher schwache Effekte bei der Rationalisierung aus. Der Sachbearbeiter ist zumeist der Initiator von Arbeitsabläufen, die auch Unterstützungstätigkeiten beinhalten. Personalkosten liegen in diesem Bereich relativ hoch. Der Kostentrend ist steigend. Fachtätigkeiten bzw. gehobene Sachbearbeitungstätigkeiten sind in den meisten Unternehmen für einen relativ hohen Anteil der Personalkosten verantwortlich. In diesem Bereich werden außerdem bei der Produktinnovation wesentliche Vorentscheidungen über spätere Kosten getroffen. Zudem ist der Kostentrend in diesem Bereich stark steigend. Ähnliches gilt in natürlicher Weise für Führungstätigkeiten. Das folgende Bild stellt aus diesen Überlegungen abgeleitet, die entsprechenden Rationalisierungsstrategien für die einzelnen Bereiche vor.

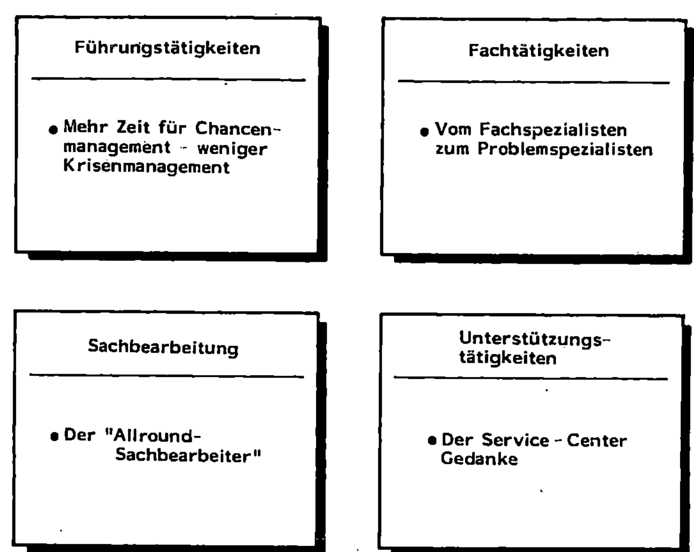

Bild 3: Rationalisierungsstrategien für verschiedene Tätigkeitsberei-
che /vgl. auch BULLINGER 84/

Im Bereich der Führungstätigkeiten ist ein verbessertes Informations-
managment im Unternehmen generell anzustreben. Führungskräfte werden
am sinnvollsten durch massive technische Unterstützung der direkten
Service-Geber wie Sekretariate, Assistenzfunktionen etc. unterstützt.
Das Schwergewicht direkter technischer Unterstützung für Führungstätigkeiten liegt im Bereich der Kommunikationsfunktionen /vergl. HIRSCH
84/.

Bei Fachtätigkeiten wird eine Weiterentwicklung vom Fachspezialisten
zum Problemspezialisten gefordert. Dies beinhaltet verbesserte lokale
Informationsverarbeitung und verbesserte Kommunikation durch kostenin-
tensive Unterstützungstechnologien. Weiterhin verbessertes Projektma-
nagement - vor allen Dingen auch horizontal - sowie den Einsatz hoch-
wertiger, vernetzter und multifunktionaler Arbeitsplatzcomputer.

Bei der Sachbearbeitung ist ein starker Trend zum Allround-Sachbear-
beiter festzustellen. Dieser Trend wird unterstützt durch Ansätze wie:
" Weg von der Sparten- und hin zur Kunden- oder Produktorganisation"
/NATZKE 84/. Der Kunde hat einen Ansprechpartner. Die anlaßgesteuerte

Sachbearbeitung wird propagiert. Das Fernziel lautet: Einmal in die Hand nehmen und sofort abschließen. Dazu sind nachgeordnete Unterstützungstätigkeiten zu automatisieren bzw. in das Tätigkeitsbild des Sachbearbeiters zu reintegrieren. Es wird in diesem Zusammenhang auch von "Mischarbeit" gesprochen /vergl. MAST 84, BULLINGER, FÄHNRICH 83/. Von den technischen Unterstützungssystemen sind moderat funktionsintegrierte Unterstützungssysteme unter Anbindung an klassische Datenbankkonzepte zu empfehlen.

Im Bereich der Unterstützungstätigkeiten setzt sich der Service-Center-Gedanke immer mehr durch. Die Zielsetzung heißt hier: bereichsnahe Zusammenfassung von Servicefunktionen zu einer Servicepalette und massive technische Unterstützung durch Integrierte Bürosysteme in den Bereichen Schriftguterstellung, Ablage, Kommunikation und Verwaltung lokaler Datenbanken. Dabei werden repetitive Aufgabenelemente zunehmend automatisiert. Für die verbleibenden höherwertigen Tätigkeitselemente sind Strategien zur Höherqualifizierung - auch teilweise zu sachbearbeitenden Funktionen - notwendig /KIESMÜLLER et al. 83/.

Die Schwerpunkte der Rationalisierungsbemühungen der Unternehmen liegen momentan noch bei den unterstützenden und sachbearbeitenden Funktionen; Planungen laufen für Fachspezialisten und Management.

2.3.: Organisation vor Technik - zumindest für den Anwender

Weitgehende Automatisierung wie in einigen Bereichen der Produktion ("Vollautomatische Fabrik - Fabrik 2000") ist auf absehbare Zeit im Bürobereich nicht realisierbar. Dies liegt wohl vorwiegend daran, daß informationsverarbeitende Prozesse letztlich wesentlich komplexer sind als Produktionsprozesse, daß im Zuge der Automatisierung der Produktion mehr und mehr informationmsverarbeitende Tätigkeiten in technische und administrative Büros ausgelagert wurden, und daß wir im Bürobereich auf eine wesentlich kürzere Rationalisierungshistorie zurückblicken als in der Produktion. Die Konsequenz daraus kann eigentlich nur lauten, daß wir Bürosysteme primär als funktionale Systeme von arbeitenden Menschen auffassen sollten.

Nun hat sich in unserer Gesellschaft in den letzten Jahren ein gewisser Wandel in der Einstellung zur Innovation und zur Technik vollzogen. Es herrscht laut entsprechenden Untersuchungen eine überwiegend positive Einstellung der Technik gegenüber vor, "stark gestiegen ist jedoch die Besorgnis über die Folgen der Technik ... . Daher werden technische Fortschritte generell kritischer beurteilt.". sowie: "Man steht heute heute im Betrieb Menschen gegenüber, die selbstbewußter und kritischer sind und eine bessere Schulbildung haben. Sie wissen ihre Forderungen durchaus zu formulieren ... ." /MAST 84/. In anderen Gesellschaften (z.B. den USA und Japan) herrschen andere Einstellungen zur Technik vor. Laut unseren intensiven Erfahrungen mit amerikanischen Unternehmen ist dort die Bereitschaft zur Innovation per se höher als bei unseren Unternehmen. Stärker als der Rationalisierungsgesichtspunkt steht hier bei der Einführung neuer, hochkomplexer Techniksysteme der Gesichtspunkt der Innovation, der Transformation des Unternehmens, des "Chancenmanagements" im Fordergrund (vergl. auch 2.4).

Die lange und anerkannte Tradition unserer Unternehmen im organisatorischen Bereich führt zur Globalaussage: Organisation vor Technik, die (vergl. 2.4.) zumindest aus Unternehmersicht erweitert werden sollte zu: Unternehmensstrategie vor. Organisation und Technik. Organisation vor Technik bedeutet vor allen Dingen das Primat der Arbeitsaufgabe, die ein System zu erfüllen hat, vor der einzusetzenden Unterstützungstechnologie. Es bedeutet weiterhin die langfristige strategische Planung ("Informationsmanagement") in diesem Bereich /BULLINGER, FÄHNRICH 84/, das Beherrschen entsprechender Einführungsprozesse /BULLINGER, FÄHNRICH 83/ einschließlich der notwendigen Instrumentarien im Bereich der Analyse sowie der Arbeitsgestaltung und letzlich ein klares Qualifikationskonzept für die Mitarbeiter/MAST 84/ in allen Qualifikationsstufen. Stark in der Diskussion sind momentan Fragen der Wirtschaftlichkeit /REICHWALD 84; FRAHM 84/. Die entscheidende Frage lautet hier: "rechnen oder entscheiden?".

Wirtschaftlichkeitsrechnungen dienen der Planungsunterstützung – sie sollen und können Managemententscheidungen und Konsensbildungsprozesse nicht ersetzen.

Ein Resümee: Unsere Unternehmen werden sich vor immer weniger jedoch immer höher qualifizierten Mitarbeitern immer stärker abhängig machen.

2.4: Ein Fazit für das Informationsmanagment der Zukunft

Integrierte Bürosysteme sind nicht nur im klassischen Sinne eine Ra-
tionalisierungstechnologie, sie fördern und unterstützen vielmehr den
breiten Trend hin "zu einem anderen Unternehmen" /FERGUSSON 84/. Dieser Trend kann im Wesentlichen charakterisiert werden durch:" Vom Produktionsunternehmen zum Dienstleistungsunternehmen". Dies in zweifa-
cher Hinsicht – zum einen etablieren sich neue Dienstleistungsunternehmen bzw. informationsverarbeitende Unternehmen, zum anderen verändern sich Produktionsunternehmen immer mehr in Richtung zu informa-
tionsverarbeitungs-intensiver Kleinserien- und Einzelserienfertigung.
Der klassische Hersteller wird bis zum Ende dieses Jahrhunderts nicht
mehr nur von ihm vorgeplante und vorproduzierte Produkte anbieten,
sondern viel stärker als bisher seine Fähigkeit verkaufen, selbst im
Konsumgüterbereich wunschgerechte individuelle Problemlösungen zu pro-
duzieren /FERGUSSONS 84/. Diese Entwicklung wird auch durch die Entwicklung der Produktionstechnik gefördert. Maßgeschneiderte und auf
immer kleinere Marktsegmente ausgelegte Produkte werden durch die Integration von computergestützten Informations- und Kommunikationssystemen und neuester Produktionstechnologie (Roboter, flexible Ferti-
gungssysteme, automatisierte Materialflußsysteme etc.) möglich.

War gerade das schnelle und flexible Reagieren auf Marktbedürfnisse
und das entsprechende Nachführen der innerbetrieblichen Organisations-
struktur bisher ein wesentliches Gütezeichen von mittelständischen Unternehmen, so zeichnet sich dies zumindest langfristig auch für große
Unternehmen durch den Einsatz von Integrierten Büro- und Produktions-
systemen ab. Damit könnten Wettbewerbsvorteile kleinerer und mittstän-
discher Unternehmen in Zukunft nivelliert werden.

Alle diese Entwicklungen deuten darauf hin, daß es nicht mehr alleini-
ge Ziel im Bereich des Einsatzes Integrierter Bürosysteme sein sollte,
die eher auf operative Tätigkeiten hin ausgerichtete Wirtschaftlichkeit von Bürosystemen zu verbessern:

"The benefit of information systems used to fall into two categories –
efficiency, ... and effectiveness ... . Now we´re beginning to see a
third: transformation and innovation." /HAMMER 84/.

3: Trends der technologischen Entwicklung - aus Anwendersicht gesehen

Technologische Entwicklung · im Bereich Integrierter Bürosysteme wird bei den Basissystemen auf drei Kerngebieten schwerpunktmäßig vorangetrieben - der Bearbeitung, der Ablage und der Kommunikation /CEC 84/. Die Entwicklung in den Basisbereichen ausnutzend, werden integrierte Anwendungssysteme realisiert. Besondere Beachtung wird in den letzten Jahren den Benutzerschnittstellen geschenkt. Am Horizont tauchen für die Breitenanwendung "Systeme der fünften Generation" auf.

3.1: Dezentral bearbeiten

Hard- und softwareseitig finden sich verschiedenste Realisierungen für Integrierte Bürosysteme , z.B. auf der Groß-EDV oder aber im Verbund mit ihr. Dabei realisieren moderat intelligente Terminals und Personalcomputer /LÖFFLATH 84, DÜMMLER 84/ dezentrale Leistung am Arbeitsplatz. Die Anwendungsintegration erfolgt auf Datenbanklevel. Innovativ sind Integrierte Bürosysteme , auf der Basis von Arbeitsplatzcomputern /HIRSCH 84, NATZKE 84/. Es handelt sich um Hochleistungsprozessoren mit großen lokalen Speichern, evtl. Druckern und einen LAN-Anschluß.

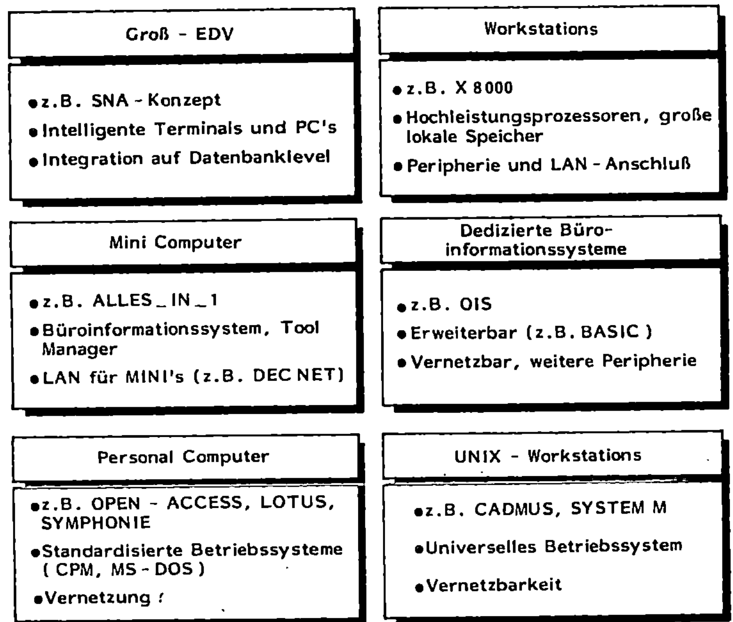

Bild 4: Dezentral bearbeiten - Integrierte Bürosysteme

In der Praxis bestehen noch Probleme beim Anwender bei der Übernahme
bzw. Einbindung bewährter Anwendungen bzw. bei einer Ankopplung an die
von Datenbanken geprägte Groß-EDV-Welt.

Alternativen zu diesen technologisch sehr anspruchsvollen Lösungen
bieten vernetzte Minicomputersysteme /WINTER 84/. Auf diesen Minicom-
putern sind verteilte Büroinformationssysteme installiert - auch bei
diesen Systemen stellt sich die Frage nach der Integration mit der
Groß-EDV. Gegenüber Workstationkonzepten bieten sie den Vorteil, daß
Systemerweiterungen flexibel durchgeführt werden können. Eng verwandt
mit den vorher geschilderten Minicomputerlösungen sind dedizierte
Büroinformationssysteme /BENKE 84, BÜSCHER 84/. Sie haben Wurzeln im
Bereich der komfortablen Mehrplatztextsysteme.

Im unteren Preisbereich angesiedelt sind Integrierte Bürosysteme auf
der Basis von Personalcomputern /FÄHNRICH 84/. Hier wird von Beginn an
auf standardisierte Betriebssysteme gesetzt. Es wird für Integrierte
Anwendungspakete wie Open Access, Lotus, Symphony, Framework etc. von
einem Markt von mehreren Millionen möglichen Installationen ausgegangen. Es stellt sich hierbei die Frage nach einer ausreichend mächtigen
Vernetzbarkeit dieser Systeme, um ganze Arbeitsprozesse unterstützen
zu können.

Eine besondere Spielart von Büroinformationssystemen auf Personalcomputern sind UNIX-Workstations. Sie kommen in ihrer Leistungsfähigkeit
den vorher erwähnten Workstations nahe. Es stellt sich jedoch die Frage, wie schnell integrierte Software für den Bürobereich auf diesen
Systemen vorhanden sein wird.

3.2: Intern und extern multimodal kommunizieren

ISDN wird die Integration der wesentlichen elektronischen Dienste der
Bundespost leisten. Diese Entwicklung wird mit hohem Aufwand vorange-
trieben. Sie soll die landesweite Infrastruktur schaffen, auf der u.a.
Integrierte Bürosysteme aufsetzen können. Die Standardisierung in die-
sem Bereich kann nicht zuletzt die Verfügbarkeit von preisgünstiger
Basissoftware fördern. Diskutiert wird die Frage, wieweit ISDN stan-
dardisierend in den Inhousebereich über digitale Nebenstellenanlagen,
ISDN-Endgeräte oder das Konstrukt der K-Maschine hineinwirken kann.
Einige kritische Stimmen befürchten jedoch, daß durch die Anwendungs-

integration in großen Rechnernetzen ISDN aufgrund der Unsicherheit bezüglich der Verfügbarkeit adäquater Endgeräte sehr spät kommt.

Von vielen Experten werden bereits heute digitale Nebenstellenanlagen als das "Arbeitspferd der Zukunft" im Bürobereich angesehen. Den pro Einzelkanal schmalbandigen Nebenstellenanlagen werden dabei durchaus gute Chancen eingeräumt, sich gegenüber breitbandigen LAN-Konzepten als generelle Infrastuktur in den Bürobereichen unserer Unternehmen zu etablieren. LANs bieten sich in jedem Fall für Subnetze oder Hochleistungsnetze in einzelnen Funktionsbereichen oder aber auch speziell im Produktionsbereich oder vorgeschalteten technischen Bürobereichen an.

3.3: Multimodale Dokumente vorgangsbezogen ablegen und wieder finden

Im Vergleich zu den Endgeräten und Netzwerken hat bisher der Bereich der Ablage (Filing & Retrieval) relativ wenig Aufmerksamkeit gefunden /FÄHNRICH 84/. Dieser Bereich stellt jedoch das unverzichtbare dritte Bein dar, auf dem Integrierte Bürosysteme stehen werden. Im Vergleich zu den Möglichkeiten der multimodalen Verarbeitung und Kommunikation ist er in Theorie und Realisierung noch nicht soweit fortgeschritten.

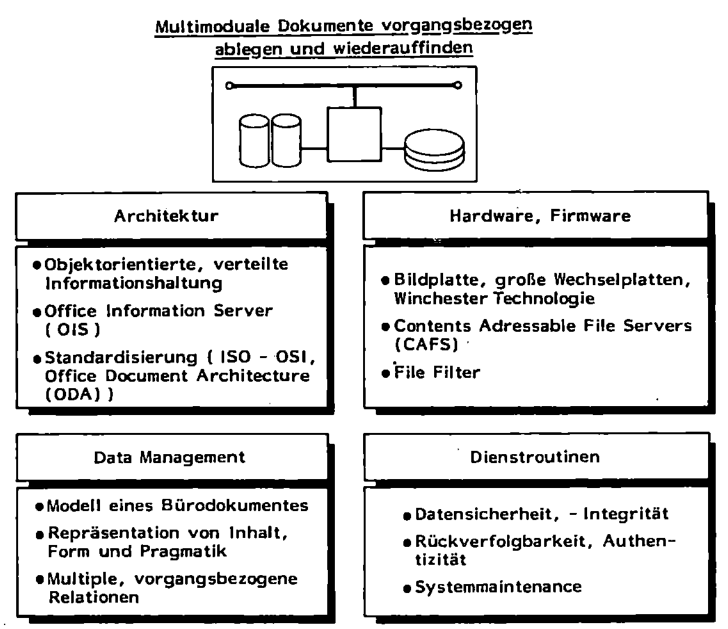

Bild 5: Filing & Retrieval - Trends

Hier sind wesentliche theoretische Arbeiten im Bereich der verteilten Datenhaltung, der intelligenten Ablage, der Zugriffsmethoden sowie der effizienten Speicherung von multimodalen Dokumenten zu leisten. Das Ziel ist: Die objektorientierte, vorgangsbezogene, verteilte Haltung multimodaler Dokumente in einem Netzwerk. Dabei sind auf Bereichsebene Volumen von ca. 10.000 bis 50.000 multimodaler Dokumente und auf Unternehmensebene 1 Mio. bis 10 Mio. multimodaler Dokumente pro System zu archivieren. Spezialprozessoren werden vorgangsbezogenes Suchen, Suchen mit variablen Suchbegriffen effizienter machen. Wesentliche Aufmerksamkeit ist der Datensicherheit und der Integrität, der Rückverfolgbarkeit von Transaktionen und generell der Autenzität zu schenken. Elektronische Ablagesysteme werden nur akzeptiert werden, wenn diese Punkte zur vollsten Zufriedenheit für die praktische Anwendung gelöst sind, da sich ein gesamtes Unternehmen von der elektronischen Archivierung seiner Informationen bzw. Daten abhängig macht. Grundlegende Arbeit ist auch im Bereich des Datenmanagements zu leisten. Hier muß ein weit gefaßtes Modell eines Bürodokumentes mit gleichzeitiger Repräsentation von Inhalt, Layout und Semantik bzw. Pragmatik des Dokumentes geleistet werden. Multiple, vorgangsbezogene Relationen eines Dokumentes bzw. von Teilen von Dokumenten zu anderen sind zu realisieren. Wesentliche Vorarbeiten sind dabei momentan im Bereich der Normierung geleistet, es handelt sich dabei um die sog. Office Document Architecture (ODA) /KRÖNERT 84/.

Auf diese Office Document Architecture bauen des weiteren Kommunikationsprotokolle für multimodale Dokumente entsprechend dem ISO/OSI-Modell auf. Die dritte Arbeitstagung des Fraunhofer-Instituts für Arbeitswirtschaft und Organisation (IAO), Stuttgart mit dem Titel "Integrierte Bürosysteme - Zukunftssichere Strategien und erfolgreiche Anwendungen" hatte als eines der wesentlichen Ergebnisse, daß sowohl hersteller- als auch anwenderseitig der Bereich des Filing & Retrieval als am wenigsten weit entwickelt bei den Basisbausteinen für integrierte Bürosysteme angesehen wird.

3.4: Die Architektur - Organisationsgerechte Dezentralisierung

Unbeschadet der Entscheidung, ob PBX-Konzepte oder LAN-Konzepte bevorzugt werden, zeichnet sich eine generelle Architektur für den Bürobereich ab. Dabei sind vorwiegend Funktionen der Bearbeitung und Kommunikation am Arbeitsplatz lokalisiert. In einzelnen Fällen stehen lokale Speicher zur Verfügung. Auf Funktionsbereichsebene sind größere Ablageeinheiten, Druckereinheiten, Einleseeinheiten und Bereichsrechner realisiert. Auf Gesamtunternehmensebene sind zentralisierte Kommunikationsverteilungsfunktionen, Großrechner und Archivierungsrechner installiert. Eine der wesentlichen Fragen im Rahmen dieses Konzeptes lautet, inwieweit die bestehenden Großrechnerinstallationen mit den dezentral arbeitenden Komponenten verknüpft werden können. Diese Frage ist für viele Unternehmen von immenser Wichtigkeit, da z.B im Banken- und Versicherungsbereich bereits heute ca. 40% bis 60% der unternehmensrelevanten Daten auf Groß-EDV-Systemen gehalten werden /BENKE 84/.

3.5: Integrierte Anwendungssysteme

Das Ziel des Einsatzes integrierter Bürosysteme ist die arbeitsprozessorientierte Funktionsintegration. Dabei wird die Integration bei Bürosystemen durch standardisierte Software bei PC's z.B. auf der Basis standardisierter Betriebssysteme vorangetrieben. Von Bedeutung sind weiterhin standardisierte Dienst-Routinen im Bereich File Management, Database Management etc.. Auch hier setzen sich am Softwaremarkt, insbesondere für PCs, defacto Standards in Form von erfolgreichen Paketen durch. Im Bereich der Kommunikation wird sich standardisierte Basissoftware in Form von Basisdienstroutinen für ISDN-Endgeräte sicherlich etablieren. Neben diesen Fortschritten hin zu standardisierter Software im Bereich der Betriebssoftware setzt sich mehr Kompatiblität und Einheitlichkeit auch im Bereich der Anwendungen durch. Es gibt gerade für PC-Systeme Standardsoftware wie z.B. Spread Sheet, Business Graphic oder Projektverfolgungssysteme. Integrierte Softwaresysteme wie Alles_In_1, Lotus, Synphony, Framework etc. stellen die höchste momentane Integrationsform bei Bürosystemen dar.

Oberhalb dieses reinen Softwarelevels stellen sich für den Organisator Fragen der arbeitsprozessorientierten Funktionsintegration. Hier können in Zukunft formalisierte Bürobeschreibungssprachen ein organisatorisches Hilfsmittel bereitstellen, um den Einsatz von Bürosystemen zu planen. Entsprechend Entwicklungen im Produktionsbereich werden sich, aufbauend auf dezentraler Intelligenz, in den neunziger Jahren Transaktionsmonitore und Scheduler etablieren. Diese Systeme erlauben zur Laufzeit eines Arbeitsprozesses die Redefinition von Betriebsmitteln, die Redefinition der Zuordnung zwischen Bearbeitern und Aufträgen sowie generell eine arbeitsprozessorientierte Auslastungsübersicht über das Gesamtsystem. Von besonderer Bedeutung sind in diesem Zusammenhang natürlich Fragenstellungen des Datenschutzes und der politischen Durchsetzbarkeit dieser Systeme im Betrieb. Für die späten neunziger Jahre wird damit gerechnet, daß dem Organisator rechnergestützte Bürosimulatoren zur Simulation mittelkomplexer Arbeitsabläufe im Bürobereich zur Verfügung stehen. Mit diesen Bürosimulatoren wäre der Organisator wesentlich besser als heute in der Lage, Terminplanungen, Kapazitätsplanungen sowie technische Spezifikationen von Bürosystemen simulierend durchzuführen /vergl. zu diesem Themenkomplex auch: Wißkirchen et al., 83/

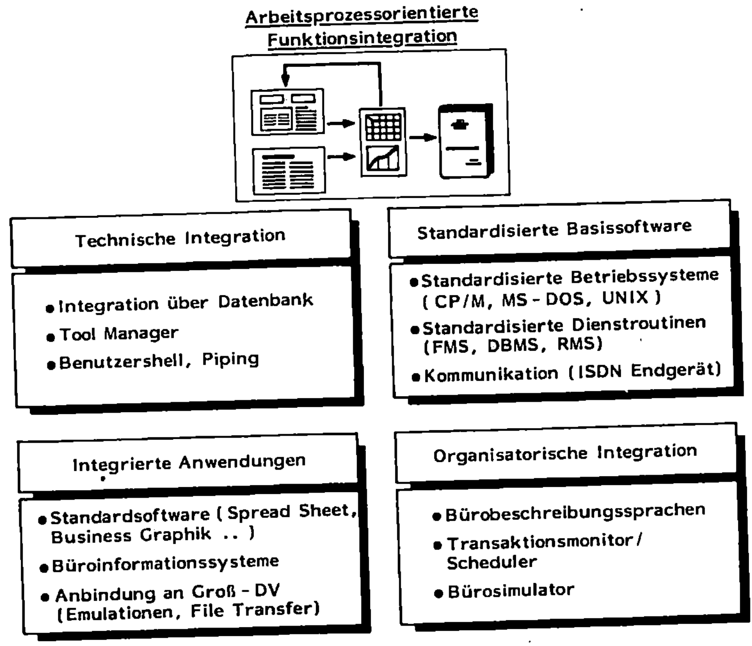

Bild 6: Trends bei der Integration

Am Fraunhofer-Institut für Arbeitswirtschaft und Organisation (IAO), Stuttgart wurden etliche am Markt verfügbaren Integrierten Bürosysteme untersucht. Dabei mußte festgestellt werden, daß im Sinne einer strengen Definition noch keines der untersuchten Systeme als idealtypisches Integriertes Bürosystem bezeichnet werden kann. Die untersuchten Systeme boten stark verschiedene Leistungsprofile. Hatten einige Systeme Stärken bei der hervorragend realisierten Benutzerschnittstelle bzw. der multimodalen Dokumentbearbeitung, andere Stärken im Bereich der Anwendungsintegration auf Datenbanklevel sowie der Definition prozeduraler Abläufe an einem Arbeitsplatz, andere wiederum bei der Abbildung von ganzen Arbeitsprozessen in verteilten Systemen, so wurde jedoch kein System gefunden, bei dem alle diese wichtigen Funktionen idealtypisch in einem System einheitlich gelöst sind /FÄHNRICH 84/.

3.6: Integrierte multimodale Benutzerschnittstellen

Die Grundidee lautet hier: Verbreiteter Informationsfluß zwischen Benutzer und System – sowohl expliziter als auch impliziter Informationsfluß.

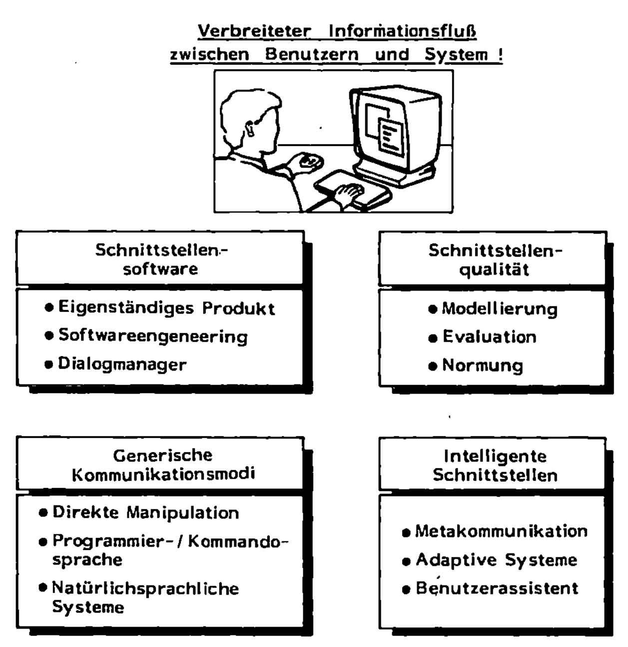

Bild 7: Trends – Integrierte multimodale Benutzerschnittstellen

Schnittstellensoftware hat sich zu einem eigenständigen Produkt entwickelt (Masken- und Menügeneratoren, generische Systeme für die Erstellung von Grafik-Schnittstellen). Dabei werden im Augenblick Schichtenarchitekturen für Benutzerschnittstellensoftware diskutiert. Zusammengefaßt werden diese Bestrebungen in dem Begriff des Dialogmanagers. Der Dialogmanager ist eine eigenständige Softwareinstanz, die Interaktionen mit dem Benutzer abwickelt.

Im Bereich der Dialoggestaltung kristallisieren sich immer 3 sog. generische Kommunikationsmodi zwischen Systembenutzern und System heraus /FÄHNRICH, ZIEGLER 84/. Dies sind:

o Direkte Manipulation

o Programmier- bzw. Kommandosprachen

o Natürlichsprachliche Systeme.

Dabei wird momentan davon ausgegangen, das sich für ungeübte Benutzer die direkte Manipulation besonders gut eignet. Diese Systeme bieten dem Benutzer eine reiche grafische Umgebung und hohes Feedback, was wiederum die Erlernbarkeit und Selbsterklärungsfähigkeit der Systeme stark begünstigt. Kommandosprachen eignen sich stärker für geübte Benutzer. Hier ist momentan ein Trend zu programmierähnlichen Kommandosprachen festzustellen. Dabei wird davon ausgegangen, daß gewisse Anwendungsprogrammierungen (z.B. Tabellenkalkulationen, Formularhandling) funktionsbereichsnah von Büroorganisatoren oder sogar vom Endbenutzer selbst durchgeführt werden können.

Am futuristischsten sind natürlichsprachliche Systeme. Die Forschung auf diesem Gebiet ist nun ca. 20 Jahre alt /FAUSER, RATHKE 82/. Erste Prototypen, z.B. für den Zugang zu Datenbanksystemen, werden kommerziell vertrieben. Wirklich robuste natürlichsprachliche Systeme lassen jedoch momentan noch auf sich warten.

Forschungsschwerpunkte sind im Augenblick "intelligente" Mensch-Maschine-Schnittstellen. Dabei werden Techniken aus der Künstlichen-Intelligenz-Forschung verwendet. Die Grundidee ist die Benutzerschnittstelle als Expertensystem (vergl. 3.7.) aufzufassen. Damit wären mehr implizite Kommunikation und adaptive Systeme (Systeme die während der Laufzeit ihr Verhalten ändern bzw. anpassen) möglich. Zusammenfassend diskutiert wird dieser Bereich unter dem Begriff eines Benutzerassistenten (User Assistent), der das "intelligente Pendant" zum Dialogmanager darstellt /BULLINGER, FÄHNRICH 84/.

3.7: Intelligente, integrierte Bürosysteme

Wesentlich für zukünftige Integrierte Bürosysteme kann eine Softwarearchitektur werden, die Softwarekomponenten als sog. "Expertensysteme" realisiert.

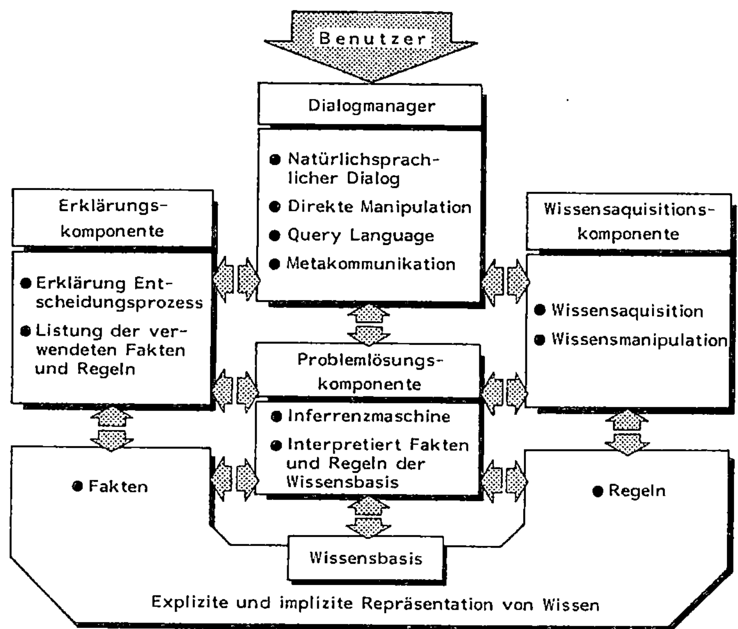

Bild 8: Der prinzipielle Aufbau von Expertensystemen

Zentral in diesem Konzept ist eine Wissensbasis, in der neben Fakten auch Schlußregeln abgelegt sind. Damit ist, neben der expliziten, auch eine implizite Repräsentation von Information gegeben. Die Ablaufsteuerung bei der Interpretation der Wissensbasis nach einer Anfrage übernimmt die sog. Problemlösungskomponente.

Da es sich in Wissensbasen sehr oft um komplexes, teilweise unscharfes Wissen handeln kann, steht dem Benutzer eine Erklärungskomponente zur Seite, die die verwendeten Fakten und Schlüsse transparent macht. Der große Vorteil von Wissensbasen ist, das hier Fakten und Schlußregeln editierbar werden – ein weiterer Schritt zur Dezentralisierung der Erstellung und Wartung der Applikationen. Dies wird durch die Wissensaquisitionskomponente geleistet (einem Wissenseditor), durch den Eingabe, Erweiterung und Konsistenzprüfung möglich werden. Entsprechend den in 3.6 vorgestellten Grundsätzen zentralisiert ein Dialogmanager die Kommunikation des Benutzers mit dem System. Die breite Realisierung dieser Architekturen /CEC 84/ bleibt jedoch den 90er Jahren vorbehalten.

4 : Literatur

Die folgenden Arbeiten sind dem Tagungsband der 3. IAO-Arbeitstagung vom 27./ 28. November 1984 in Stuttgart entnommen. Der Literaturnachweis lautet:

H.-J. Warnecke; H.-J. Bullinger (Hrsg): IPA - IAO - Forschung und Praxis, Band T2; Titel: "Integrierte Bürosysteme - zukunftssichere Strategien und erfolgreiche Anwendungen"; Springer Verlag, 1984.

BENKE, H.:
Einführungsstrategien für Integrierte Bürosysteme - die Diskrepanz zwischen Planung und Realisierung;

BULLINGER, H.-J.:
Die Durchdringung der Unternehmen mit integrierten Bürosystemen - die Phasen der Bürorationalisierung;

BÜSCHER, I.:
Von der Textverarbeitung über Büroinformationssysteme zu integrierten Bürosystemen - 12 Jahre Planungs- und Einsatzerfahrung;

DÜMMLER, A.:
Integrierte Informationsverarbeitung in einem Fertigungsunternehmen;

FÄHNRICH, K.-P.:
Die technologiische Basis für das Informationsmanagement der Zukunft;

FRAHM, T:
Integrierte Textverarbeitung - die Basis für effiziente Büroautomation;

HIRSCH, E.:
Einsatz und Nutzen von Arbeitsplatzrechnern zur Integration der Vorgangsbearbeitung im Büro;

LÖFFLATH, B.:
Bürokommunikation - fester Bestandteil der Informationsverarbeitung;

MAST, C:
Zwischen Kopf und Kabel - Mensch und Büro;

NATZKE, R.:
Unterstützung von Fachspezialisten - von der Sparten- zur Kundenorganisation;

WINTER, H.:
Integration von Anwendungssoftware auf unterschiedlichen Rechnerklassen;

außerdem:

BULLINGER, H.-J.; FÄHNRICH, K.-P.:
Informationsmanagement -Von isolierten Einsparungen zu Wettbewerbsent-
scheidenden Produktivitätssteigerungen. Konferenzunterlagen der 7.
Europäischen Kongreßmesse für Technische Kommunikation. Berlin: ONLINE
1984.

BULLINGER, H.-J.; FÄHNRICH, K.-P.:
Informationsmanagement - Von isolierten Einsparungen zu Wettbewerbs-
entscheidenden Produktivitätssteigerungen. Konferenzunterlagen der 7.
Europäischen Kongreßmesse für Technische Kommunikation. Berlin: ONLINE
1984.

BULLINGER, H.-J.; FÄHNRICH, K.-P.:
Synbiotic Man-Computer Interfaces and the User Assistent Conmcept. In:
Human-Computer Interaction: Proc. 1st U.S.A. - Japan Conference on
Human Computer Interaction.
Honululu, Hawai: August 18 - 20, 1984.
Ed. by G. Salvendy. Amsterdam u.a.: Elsevier, 1984, S. 17-26.

COMISSION OF THE EUROPEAN COMMUNITIES (CEC):
Draft Council adopting the 1984 work programme for the European stra
tegic programme for research and development in ionformation technologies
Official Journal of the European Communities, C47, Vol.27, Feb. 1984

FÄHNRICH, K.-P.; ZIEGLER, J.:
Workstations Using Direct Manipulation as Interaction Mode - Aspects
of Design, Application and Evaluation; in: Shackel (Hrsg.): Procedings
of first IFIP Conference on "Human Computer Interaction - INTERACT
´84"; London 1984

FERGUSSON, M.;
Informationen zum sozialen Wandel, Leading Edge - präsentiert von Ra-
dar System Trend Service, Juni/Juli 1984)

KIESMÜLLER; KOHL; WITTSTOCK; BOLLINGER; WELTZ:
Arbeitsstrukturierung in der Textverarbeitung und den benachbarten
Verwaltungsbereichen eines Industriebetriebes (ASTEX), Schlußbericht
zur Vorphase - Bundesministerium für Forschung und Technologie - Huma-
nisierung des Arbeitslebens, 1983.

KRÖNERT, G.:
Textverarbeitung auf Basis der standardisierten Dokumentenarchitektur;
in: Unterlagen zum Tutorium über "Offene Multifunktionale Büroarbeits-
plätze und Bildschirmtext", TU Berlin, 1984

NAISBITT, J.:
Megatrends, 10 Perspektiven, die unsere Welt verändern werden, Hestia,
2. Auflage, 1984

REICHWALD, R.:
Bürokommunikation im Teletex-Dienst; Organisationsmodelle und Wirtschaftlichkeit; in Bürotechnik, 29. Jahrgang, 1984, Nr.11, Seite
430-440.

WISSKIRCHEN, KREIFELTS, KRÜCKEBERG, RICHTER, WURCH:
Informationstechnik und Bürosysteme
Teubner-Verlag, Stuttgart, 1983